Microbiome and Metabolome in Diagnosis, Therapy, and other Strategic Applications

Microbiome and Metabolome in Diagnosis, Therapy, and other Strategic Applications

Joel Faintuch

Senior Professor, Department of Gastroenterology, Sao Paulo University Medical School, Sao Paulo, Brazil

Salomao Faintuch

Clinical Director of Vascular and Interventional Radiology; Co-Director, Advanced Vascular Care Center; Assistant Professor of Radiology, Beth Israel Deaconess Medical Center, Harvard Medical School, Boston, MA, United States

ACADEMIC PRESS

An imprint of Elsevier

Academic Press is an imprint of Elsevier
125 London Wall, London EC2Y 5AS, United Kingdom
525 B Street, Suite 1650, San Diego, CA 92101, United States
50 Hampshire Street, 5th Floor, Cambridge, MA 02139, United States
The Boulevard, Langford Lane, Kidlington, Oxford OX5 1GB, United Kingdom

Notices
Knowledge and best practice in this field are constantly changing. As new research and experience broaden our understanding,
changes in research methods, professional practices, or medical treatment may become necessary.

Practitioners and researchers must always rely on their own experience and knowledge in evaluating and using any information,
methods, compounds, or experiments described herein. In using such information or methods they should be mindful of their
own safety and the safety of others, including parties for whom they have a professional responsibility.

To the fullest extent of the law, neither the Publisher nor the authors, contributors, or editors, assume any liability for any injury
and/or damage to persons or property as a matter of products liability, negligence or otherwise, or from any use or operation of
any methods, products, instructions, or ideas contained in the material herein.

Library of Congress Cataloging-in-Publication Data
A catalog record for this book is available from the Library of Congress

British Library Cataloguing-in-Publication Data
A catalogue record for this book is available from the British Library

ISBN: 978-0-12-815249-2

For information on all Academic Press publications visit our website at
https://www.elsevier.com/books-and-journals

Working together
to grow libraries in
developing countries

www.elsevier.com • www.bookaid.org

Publisher: John Fedor
Senior Acquisition Editor: Stacy Masucci
Editorial Project Manager: Carlos Rodriguez
Production Project Manager: Sreejith Viswanathan
Designer: Vicky Pearson

Typeset by TNQ Technologies

Contents

Block I
Tools, Toolstations and Models

1. Germ-Free Animals as a Tool to Study Indigenous Microbiota

E. Neumann, L.Q. Vieira and J.R. Nicoli

2. Germ-Free Mouse Technology in Cardiovascular Research

Alexandra Grill and Christoph Reinhardt

3. The Gut Microbiome Beyond the Bacteriome—The Neglected Role of Virome and Mycobiome in Health and Disease

Charikleia Stefanaki

4. Techniques for Phenotyping the Gut Microbiota Metabolome

Elisa Zubeldia-Varela, Beata Anna Raczkowska, Manuel Ferrer, Marina Perez-Gordo and David Rojo

5. Metabolomics Tools and Information Retrieval in Microbiome Hacking

Abdellah Tebani and Soumeya Bekri

6. Laboratory Simulators of the Colon Microbiome

M.Carmen. Martínez-Cuesta, Carmen. Peláez and Teresa Requena

7. Potential of Metabolomics to Breath Tests

F. Priego-Capote and Maria Dolores Luque de Castro

Block II
Background Information

8. Metabolome and Microbiome From Infancy to Elderly

Ramon V. Cortez, Luana N. Moreira and Carla R. Taddei

9. The Oral Microbiome

Marcelle M. Nascimento

10. The Gastric Microbiome in Benign and Malignant Diseases

Thais Fernanda Bartelli, Luiz Gonzaga Vaz Coelho and Emmanuel Dias-Neto

11. The Human Vaginal Microbiome

Iara M. Linhares, Evelyn Minis, Renata Robial and Steven S. Witkin

Block IV
Diagnostic and Therapeutical Applications

22. The Microbiome and Metabolome in Metabolic Syndrome

Rigoberto Pallares-Méndez and Carla Fernández-Reynoso

23. The Gut Microbiota as a Therapeutic Approach for Obesity

Trevor O. Kirby, Emily K. Hendrix and Javier Ochoa-Repáraz

24. The Gut Microbiome After Bariatric Surgery

Camila Solar, Alex Escalona and Daniel Garrido

25. Gut Dysbiosis in Arterial Hypertension: A Candidate Therapeutic Target for Blood Pressure Management

*José Luiz de Brito Alves, Evandro Leite de Souza,
Josiane de Campos Cruz, Camille de Moura
Balarini, Marciane Magnani, Hubert Vidal and
Valdir de Andrade Braga*

26. The Emerging Role of Microbiome–Gut–Brain Axis in Functional Gastrointestinal Disorders

*Karolina Skonieczna-Żydecka, Igor Loniewski,
Anastasios Koulaouzidis and Wojciech Marlicz*

27. The Microbiome and Metabolome in Nonalcoholic Fatty Liver Disease

*Silvia M. Ferolla, Cláudia A. Couto, Maria de
Lourdes A. Ferrari, Luciana Costa Faria, Murilo
Pereira and Teresa C.A. Ferrari*

37. Dysbiosis in Benign and Malignant Diseases of the Exocrine Pancreas

Robert Memba, Sinead N. Duggan, Rosa Jorba and Kevin C. Conlon

38. Importance of the Microbiome and the Metabolome in Cancer

Liliane Martins dos Santos, Ana Clara Matoso Montuori de Andrade and Mateus Eustáquio Moura Lopes

39. The Microbiome in Graft Versus Host Disease

Mathilde Payen and Clotilde Rousseau

40. Impact of the Gut Microbiome on Behavior and Emotions

Ingrid Rivera-Iñiguez, Sonia Roman, Claudia Ojeda-Granados and Panduro Arturo

Block V
Applications for Foods, Drugs, and Xenobiotics

41. The Gut Microbiome in Vegetarians

Ana Carolina F. Moraes, Bianca de Almeida-Pittito and Sandra Roberta G. Ferreira

42. Metformin: A Candidate Drug to Control the Epidemic of Diabetes and Obesity by Way of Gut Microbiome Modification

Kunal Maniar, Vandana Singh, Deepak Kumar, Amal Moideen, Rajasri Bhattacharyya and Dibyajyoti Banerjee

43. Deleterious Impact of Smog on the Intestinal Bacteria

L.R. Pace, C.M. Wells, R. Awais, P. Shrestha, R.D. Parker and T.Y. Wong

Block VI
Challenges and Promises for the Future

44. New-Generation Probiotics: Perspectives and Applications

Dinesh Kumar Dahiya, Renuka, Arun Kumar Dangi, Umesh K. Shandilya, Anil Kumar Puniya and Pratyoosh Shukla

Contributors

Nur Elina Abdul Mutalib, Department of Bioprocess Technology, Faculty of Biotechnology and Biomolecular Sciences, Universiti Putra Malaysia, 43400 UPM Serdang, Selangor Darul Ehsan, Malaysia

Bianca de Almeida-Pittito, Department of Preventive Medicine, Federal University of Sao Paulo, São Paulo, Brazil

Ana Clara Matoso Montuori de Andrade, Department of Biochemistry and Immunology, Federal University of Minas Gerais, Belo Horizonte, Brazil

José Luiz de Brito Alves, Department of Nutrition, Health Sciences Center, Federal University of Paraíba, João Pessoa, Brazil

L. Caetano M. Antunes, National Institute of Science and Technology of Innovation on Diseases of Neglected Populations, Center for Technological Development in Health, Oswaldo Cruz Foundation, Rio de Janeiro, Brazil; National School of Public Health Sergio Arouca, Oswaldo Cruz Foundation, Rio de Janeiro, Brazil

M.J. Arias-Tellez, Department of Nutrition, University of Chile, Santiago, Chile

Panduro Arturo, Department of Molecular Biology in Medicine, Civil Hospital of Guadalajara, Guadalajara, Jalisco, Mexico; Health Sciences Center, University of Guadalajara, Guadalajara, Jalisco, Mexico

R. Awais, Department of Biological Engineering, University of Memphis, Memphis, TN, United States

Camille de Moura Balarini, Biotechnology Center, Federal University of Paraíba, João Pessoa, Brazil; Department of Physiology and Pathology, Health Sciences Center, Federal University of Paraíba, João Pessoa, Brazil

Dibyajyoti Banerjee, Department of Experimental Medicine and Biotechnology, Postgraduate Institute of Medical Education and Research, Chandigarh, India

Thais Fernanda Bartelli, Lab. Medical Genomics, A.C.Camargo Cancer Center, São Paulo, Brazil

Paulo José Basso, Department of Immunology, Institute of Biomedical Sciences, University of São Paulo, São Paulo, Brazil

Soumeya Bekri, Department of Metabolic Biochemistry, Rouen University Hospital, Rouen, France; Normandie Univ, UNIROUEN, CHU Rouen, IRIB, INSERM U1245, Rouen, France

Rajasri Bhattacharyya, Department of Experimental Medicine and Biotechnology, Postgraduate Institute of Medical Education and Research, Chandigarh, India

Aleksandr Birg, Medicine Service, New Mexico VA Health Care System and the Division of Gastroenterology and Hepatology, University of New Mexico, Albuquerque, NM, United States

Natália Alvarenga Borges, Graduate Program in Cardiovascular Sciences, Fluminense Federal University (UFF), Niterói, Brazil; Unidade de Pesquisa Clínica, Niterói, Brazil

Valdir de Andrade Braga, Biotechnology Center, Federal University of Paraíba, João Pessoa, Brazil

Nicholas Buys, Menzies Health Institute Queensland, Griffith University, Gold Coast, QLD, Australia

Niels Olsen Saraiva Câmara, Department of Immunology, Institute of Biomedical Sciences, University of São Paulo, São Paulo, Brazil

Agostinho Carvalho, Life and Health Sciences Research Institute (ICVS), School of Medicine, University of Minho, Braga, Portugal; ICVS/3B's - PT Government Associate Laboratory, Braga/Guimarães, Portugal

Maria Dolores Luque de Castro, Department of Analytical Chemistry, University of Córdoba, Córdoba, Spain; Maimónides Institute of Biomedical Research (IMIBIC), Reina Sofia Hospital, University of Córdoba, Córdoba, Spain; CIBER Fragilidad y Envejecimiento Saludable (CIBERfes), Instituto de Salud Carlos III, Spain

Luiz Gonzaga Vaz Coelho, Alfa Institute of Gastroenterology, Clinics Hospital, Federal University of Minas Gerais, Belo Horizonte, Brazil

Nathaniel Aviv Cohen, IBD Center and Bacteriotherapy Clinic, Department of Gastroenterology and Liver Diseases, Tel Aviv Medical Center, Tel Aviv, Israel; Sackler Faculty of Medicine, Tel Aviv University, Tel Aviv, Israel

Kevin C. Conlon, Professorial Surgical Unit, Department of Surgery Trinity College Dublin, Tallaght Hospital, Dublin, Ireland

Ramon V. Cortez, School of Pharmaceutical Sciences, Dep. of Clinical Analysis and Toxicology, University of São Paulo, SP, Brazil

Cláudia A. Couto, Department of Internal Medicine, Faculty of Medicine, Federal University of Minas Gerais, Belo Horizonte, Brazil

Josiane de Campos Cruz, Biotechnology Center, Federal University of Paraíba, João Pessoa, Brazil

Cristina Cunha, Life and Health Sciences Research Institute (ICVS), School of Medicine, University of Minho, Braga, Portugal; ICVS/3B's - PT Government Associate Laboratory, Braga/Guimarães, Portugal

P. D'Amelio, Department of Medical Sciences, Gerontology and Bone Metabolic Disease Section, University of Torino, Torino, Italy

Dinesh Kumar Dahiya, Advanced Milk Testing Research Laboratory, Post Graduate Institute of Veterinary Education and Research (Rajasthan University of Veterinary and Animal Sciences, Bikaner), Jaipur, India

Arun Kumar Dangi, Enzyme Technology and Protein Bioinformatics Laboratory, Department of Microbiology, Maharshi Dayanand University, Rohtak, India

Emmanuel Dias-Neto, Lab. Medical Genomics, A.C.Camargo Cancer Center, São Paulo, Brazil

Cláudio Duarte-Oliveira, Life and Health Sciences Research Institute (ICVS), School of Medicine, University of Minho, Braga, Portugal; ICVS/3B's - PT Government Associate Laboratory, Braga/Guimarães, Portugal

Sinead N. Duggan, Professorial Surgical Unit, Department of Surgery Trinity College Dublin, Tallaght Hospital, Dublin, Ireland

Alex Escalona, Department of Surgery, Faculty of Medicine, Universidad de Los Andes, Santiago, Chile

Joel Faintuch, Department of Gastroenterology, Sao Paulo University Medical School, Sao Paulo, Brazil

Jacob J. Faintuch, Department of Internal Medicine, Hospital das Clinicas, Sao Paulo, Brazil

Salomao Faintuch, Beth Israel Deaconess Medical Center, Harvard Medical School, Boston, MA, USA

Luciana Costa Faria, Department of Internal Medicine, Faculty of Medicine, Federal University of Minas Gerais, Belo Horizonte, Brazil

Carla Fernández-Reynoso, Centro de Estudios Universitarios Xochicalco, Tijuana, México

Silvia M. Ferolla, Department of Internal Medicine, Faculty of Medicine, Federal University of Minas Gerais, Belo Horizonte, Brazil

Maria de Lourdes A. Ferrari, Department of Internal Medicine, Faculty of Medicine, Federal University of Minas Gerais, Belo Horizonte, Brazil

Teresa C.A. Ferrari, Department of Internal Medicine, Faculty of Medicine, Federal University of Minas Gerais, Belo Horizonte, Brazil

Nicole V. Ferreira, National School of Public Health Sergio Arouca, Oswaldo Cruz Foundation, Rio de Janeiro, Brazil

Sandra Roberta G. Ferreira, Department of Epidemiology, School of Public Health, University of Sao Paulo, São Paulo, Brazil

Rosana B.R. Ferreira, Institute of Microbiology, Federal University of Rio de Janeiro, Rio de Janeiro, Brazil

Manuel Ferrer, Institute of Catalysis, Consejo Superior de Investigaciones Científicas (CSIC), Madrid, Spain

Jarlei Fiamoncini, Department of Food and Experimental Nutrition, School of Pharmaceutical Sciences, University of São Paulo, São Paulo, Brazil

Hooi Ling Foo, Department of Bioprocess Technology, Faculty of Biotechnology and Biomolecular Sciences, Universiti Putra Malaysia, 43400 UPM Serdang, Selangor Darul Ehsan, Malaysia; Institute of Bioscience, Universiti Putra Malaysia, 43400 UPM Serdang, Selangor Darul Ehsan, Malaysia

Damodar Gajurel, Menzies Health Institute Queensland, Griffith University, Gold Coast, QLD, Australia

Daniel Garrido, Department of Chemical and Bioprocess Engineering, School of Engineering, Pontificia Universidad Catolica de Chile, Vicuñ, Santiago, Chile

A. Gil, Department of Biochemistry and Molecular Biology II, School of Pharmacy, University of Granada, Granada, Spain; Institute of Nutrition & Food Technology "Jose Mataix", Biomedical Research Center, University of Granada, Armilla, Spain

Shailendra Giri, Department of Neurology, Henry Ford Health System, Detroit, MI, United States

Thaís Glatthardt, Institute of Microbiology, Federal University of Rio de Janeiro, Rio de Janeiro, Brazil

Samuel M. Gonçalves, Life and Health Sciences Research Institute (ICVS), School of Medicine, University of Minho, Braga, Portugal; ICVS/3B's - PT Government Associate Laboratory, Braga/Guimarães, Portugal

Alexandra Grill, Center for Thrombosis and Hemostasis (CTH), University Medical Center Mainz, Johannes Gutenberg University Mainz, Mainz, Germany; German Center for Cardiovascular Research (DZHK), Partner Site RheinMain, Mainz, Germany

Emily K. Hendrix, Department of Biology, Eastern Washington University, Cheney, WA 99004, United States

Aline Ignacio, Department of Immunology, Institute of Biomedical Sciences, University of São Paulo, São Paulo, Brazil

M. Íñiguez, Infectious Diseases, Microbiota and Metabolism Unit, Infectious Diseases Department, Center for Biomedical Research of La Rioja (CIBIR), Logroño, Spain

Rosa Jorba, Hepatobiliary and Pancreatic Surgery Unit, Department of Surgery, Joan XXIII University Hospital, Tarragona, Spain

N. Kapel, Faculté de pharmacie, Université Paris Descartes, Sorbonne Paris Cité, Paris, France; Laboratoire de Coprologie Fonctionnelle, APHP, Hôpitaux Universitaires Pitié Salpêtrière-Charles Foix, Paris, France

Trevor O. Kirby, Department of Biology, Eastern Washington University, Cheney, WA 99004, United States

Anastasios Koulaouzidis, Endoscopy Unit, The Royal Infirmary of Edinburgh, Edinburgh, United Kingdom

Deepak Kumar, Department of Experimental Medicine and Biotechnology, Postgraduate Institute of Medical Education and Research, Chandigarh, India

Fulvio Lauretani, Department of Medicine and Surgery, University of Parma, Parma, Italy; Dipartimento Medico-Geriatrico-Riabilitativo, Azienda Ospedaliero-Universitaria di Parma, Parma, Italy

Henry C. Lin, Medicine Service, New Mexico VA Health Care System and the Division of Gastroenterology and Hepatology, University of New Mexico, Albuquerque, NM, United States

Iara M. Linhares, Department of Gynecology and Obstetrics, University of Sao Paulo Medical School, Sao Paulo, Brazil; Division of Immunology and Infectious Diseases, Department of Obstetrics and Gynecology, Weill Cornell Medicine, New York, NY, United States

Teck Chwen Loh, Department of Animal Science, Faculty of Agriculture, Universiti Putra Malaysia, 43400 UPM Serdang, Selangor Darul Ehsan, Malaysia; Institute of Tropical Agriculture and Food Security, Universiti Putra Malaysia, 43400 UPM Serdang, Selangor Darul Ehsan, Malaysia

Igor Loniewski, Department of Biochemistry and Human Nutrition, Pomeranian Medical University, Szczecin, Poland; Sanprobi Sp. z o.o. Sp. k., Szczecin, Poland

Mateus Eustáquio Moura Lopes, Department of Biochemistry and Immunology, Federal University of Minas Gerais, Belo Horizonte, Brazil

Denise Mafra, Graduate Program in Cardiovascular Sciences, Fluminense Federal University (UFF), Niterói, Brazil; Graduate Program in Medical Sciences, Fluminense Federal University (UFF), Niterói, Brazil

Marciane Magnani, Department of Food Engineering, Technology Center, Federal University of Paraíba, João Pessoa, Brazil

Nitsan Maharshak, IBD Center and Bacteriotherapy Clinic, Department of Gastroenterology and Liver Diseases, Tel Aviv Medical Center, Tel Aviv, Israel; Sackler Faculty of Medicine, Tel Aviv University, Tel Aviv, Israel

Danish J. Malik, Chemical Engineering Department, Loughborough University, Loughborough LE11 3TU, United Kingdom

Ashutosh Mangalam, Department of Pathology, University of Iowa Carver College of Medicine, Iowa City, IA, United States

Kunal Maniar, Department of Pharmacology, Postgraduate Institute of Medical Education and Research, Chandigarh, India

Wojciech Marlicz, Department of Gastroenterology, Pomeranian Medical University, Szczecin, Poland

M.Carmen. Martínez-Cuesta, Department of Food Biotechnology and Microbiology, Institute of Food Science Research, CIAL (CSIC), Madrid, Spain

Robert Memba, Professorial Surgical Unit, Department of Surgery Trinity College Dublin, Tallaght Hospital, Dublin, Ireland; Hepatobiliary and Pancreatic Surgery Unit, Department of Surgery, Joan XXIII University Hospital, Tarragona, Spain

Tiziana Meschi, Microbiome Research Hub, University of Parma, Italy; Department of Medicine and Surgery, University of Parma, Parma, Italy; Dipartimento Medico-Geriatrico-Riabilitativo, Azienda Ospedaliero-Universitaria di Parma, Parma, Italy

Christian Milani, Microbiome Research Hub, University of Parma, Italy; Laboratory of Probiogenomics, Department of Chemistry, Life Sciences and Environmental Sustainability, University of Parma, Parma, Italy

Evelyn Minis, Division of Immunology and Infectious Diseases, Department of Obstetrics and Gynecology, Weill Cornell Medicine, New York, NY, United States

Amal Moideen, Department of Pharmacology, Postgraduate Institute of Medical Education and Research, Chandigarh, India

Ana Carolina F. Moraes, Department of Epidemiology, School of Public Health, University of Sao Paulo, São Paulo, Brazil

Luana N. Moreira, School of Pharmaceutical Sciences, Dep. of Clinical Analysis and Toxicology, University of São Paulo, SP, Brazil

Carlos G. Moscoso, Division of Gastroenterology, Hepatology and Nutrition, University of Minnesota, Minneapolis, MN, United States

Marcelle M. Nascimento, Associate Professor, Department of Restorative Dental Sciences, Division of Operative Dentistry, College of Dentistry, University of Florida, Gainesville, FL, United States

Andrew Nelson, Northumbria University, Faculty of Health and Life Sciences, Newcastle upon Tyne, United Kingdom

E. Neumann, Department of Microbiology, Institute of Biological Sciences, Federal University of Minas Gerais, Belo Horizonte, Brazil

Krista M. Newman, Division of Gastroenterology, Hepatology and Nutrition, University of Minnesota, Minneapolis, MN, United States

J.R. Nicoli, Department of Microbiology, Institute of Biological Sciences, Federal University of Minas Gerais, Belo Horizonte, Brazil

Antonio Nouvenne, Microbiome Research Hub, University of Parma, Italy; Department of Medicine and Surgery, University of Parma, Parma, Italy; Dipartimento Medico-Geriatrico-Riabilitativo, Azienda Ospedaliero-Universitaria di Parma, Parma, Italy

Javier Ochoa-Repáraz, Department of Biology, Eastern Washington University, Cheney, WA 99004, United States

Claudia Ojeda-Granados, Department of Molecular Biology in Medicine, Civil Hospital of Guadalajara, Guadalajara, Jalisco, Mexico; Health Sciences Center, University of Guadalajara, Guadalajara, Jalisco, Mexico

Gislane Lellis Vilela de Oliveira, Microbiome Study Group, School of Health Sciences Paulo Prata (FACISB), Barretos, Brazil; Microbiology Department, São Paulo State University (UNESP), Institute of Biosciences, Humanities and Exact Sciences (IBILCE), Sao Jose do Rio Preto, Sao Paulo, Brazil

J.A. Oteo, Infectious Diseases, Microbiota and Metabolism Unit, Infectious Diseases Department, Center for Biomedical Research of La Rioja (CIBIR), Logroño, Spain; Infectious Diseases Department, Hospital San Pedro, Logroño, Spain

L.R. Pace, Department of Biological Engineering, University of Memphis, Memphis, TN, United States

Rigoberto Pallares-Méndez, Hospital Universitario "Dr. José Eleuterio Gonzalez" Universidad Autónoma de Nuevo León, Monterrey, México

R.D. Parker, Department of Biological Sciences, University of Memphis, Memphis, TN, United States

Kalpesh G. Patel, Division of Gastroenterology and Hepatology Rutgers, The State University of New Jersey, Newark, NJ, United States

Heidi Pauer, National Institute of Science and Technology of Innovation on Diseases of Neglected Populations, Center for Technological Development in Health, Oswaldo Cruz Foundation, Rio de Janeiro, Brazil

Mathilde Payen, EA4065, Ecosystème intestinal, probiotiques, antibiotiques, Faculté de Pharmacie, Université Paris Descartes. Paris, France

Carmen. Peláez, Department of Food Biotechnology and Microbiology, Institute of Food Science Research, CIAL (CSIC), Madrid, Spain

Murilo Pereira, Post graduation in Functional Clinical Nutrition, VP Institute, Cruzeiro do Sul University, São Paulo, Brazil

Marina Perez-Gordo, Basic Medical Sciences Department, Faculty of Medicine, Universidad CEU San Pablo, Campus Montepríncipe, Madrid, Spain; Institute of Applied and Molecular Medicine (IMMA), Faculty of Medicine, Universidad CEU San Pablo, Campus Montepríncipe, Madrid, Spain

P. Pérez-Matute, Infectious Diseases, Microbiota and Metabolism Unit, Infectious Diseases Department, Center for Biomedical Research of La Rioja (CIBIR), Logroño, Spain

J. Plaza-Diaz, Department of Biochemistry and Molecular Biology II, School of Pharmacy, University of Granada, Granada, Spain; Institute of Nutrition & Food Technology "Jose Mataix", Biomedical Research Center, University of Granada, Armilla, Spain

F. Priego-Capote, Department of Analytical Chemistry, University of Córdoba, Córdoba, Spain; Maimónides Institute of Biomedical Research (IMIBIC), Reina Sofia Hospital, University of Córdoba, Córdoba, Spain; CIBER Fragilidad y Envejecimiento Saludable (CIBERfes), Instituto de Salud Carlos III, Spain

Anil Kumar Puniya, College of Dairy Science & Technology, Guru Angad Dev Veterinary & Animal Sciences University, Ludhiana, India; Dairy Microbiology Division, ICAR-National Dairy Research Institute, Karnal, India

Nikolaos T. Pyrsopoulos, Division of Gastroenterology and Hepatology Rutgers, The State University of New Jersey, Newark, NJ, United States

Beata Anna Raczkowska, Department of Endocrinology, Diabetology and Internal Medicine, Medical University of Bialystok, Bialystok, Poland

Raha Abdul Rahim, Institute of Bioscience, Universiti Putra Malaysia, 43400 UPM Serdang, Selangor Darul Ehsan, Malaysia; Department of Cell and Molecular Biology, Faculty of Biotechnology and Biomolecular Sciences, Universiti Putra Malaysia, 43400 UPM Serdang, Selangor Darul Ehsan, Malaysia

Christoph Reinhardt, Center for Thrombosis and Hemostasis (CTH), University Medical Center Mainz, Johannes Gutenberg University Mainz, Mainz, Germany; German Center for Cardiovascular Research (DZHK), Partner Site RheinMain, Mainz, Germany

Renuka, Department of Veterinary Physiology & Biochemistry, Post Graduate Institute of Veterinary Education and Research, (Rajasthan University of Veterinary and Animal Sciences, Bikaner), Jaipur, India

Teresa Requena, Department of Food Biotechnology and Microbiology, Institute of Food Science Research, CIAL (CSIC), Madrid, Spain

J. Reygner, Faculté de pharmacie, Université Paris Descartes, Sorbonne Paris Cité, Paris, France

Nathaniel L. Ritz, Medicine Service, New Mexico VA Health Care System and the Division of Gastroenterology and Hepatology, University of New Mexico, Albuquerque, NM, United States

Ingrid Rivera-Iñiguez, Department of Molecular Biology in Medicine, Civil Hospital of Guadalajara, Guadalajara, Jalisco, Mexico; Health Sciences Center, University of Guadalajara, Guadalajara, Jalisco, Mexico

Renata Robial, Department of Gynecology and Obstetrics, University of Sao Paulo Medical School, Sao Paulo, Brazil

David Rojo, Centro de Metabolómica y Bioanãlisis (CEMBIO), Facultad de Farmacia, Universidad CEU San Pablo, Campus Montepríncipe, Madrid, Spain

Sonia Roman, Department of Molecular Biology in Medicine, Civil Hospital of Guadalajara, Guadalajara, Jalisco, Mexico; Health Sciences Center, University of Guadalajara, Guadalajara, Jalisco, Mexico

Clotilde Rousseau, EA4065, Ecosystème intestinal, probiotiques, antibiotiques, Faculté de Pharmacie, Université Paris Descartes. Paris, France; Laboratoire de microbiologie, Hôpital Saint-Louis, Paris, France

F.J. Ruiz-Ojeda, Department of Biochemistry and Molecular Biology II, School of Pharmacy, University of Granada, Granada, Spain; Institute of Nutrition & Food Technology "Jose Mataix", Biomedical Research Center, University of Granada, Armilla, Spain

M.J. Saez-Lara, Institute of Nutrition & Food Technology "Jose Mataix", Biomedical Research Center, University of Granada, Armilla, Spain; Department of Biochemistry and Molecular Biology I, School of Sciences, University of Granada, Granada, Spain

Liliane Martins dos Santos, Department of Biochemistry and Immunology, Federal University of Minas Gerais, Belo Horizonte, Brazil

F. Sassi, Department of Medical Sciences, Gerontology and Bone Metabolic Disease Section, University of Torino, Torino, Italy

Umesh K. Shandilya, Animal Biotechnology Division, National Bureau of Animal Genetic Resources, Karnal, India

P. Shrestha, Department of Biological Sciences, University of Memphis, Memphis, TN, United States

Pratyoosh Shukla, Enzyme Technology and Protein Bioinformatics Laboratory, Department of Microbiology, Maharshi Dayanand University, Rohtak, India

Vandana Singh, Department of Experimental Medicine and Biotechnology, Postgraduate Institute of Medical Education and Research, Chandigarh, India

Karolina Skonieczna-Żydecka, Department of Biochemistry and Human Nutrition, Pomeranian Medical University, Szczecin, Poland

Camila Solar, Department of Chemical and Bioprocess Engineering, School of Engineering, Pontificia Universidad Catolica de Chile, Vicuñ, Santiago, Chile

P. Solis-Urra, PROFITH "PROmoting FITness and Health through physical activity" research group, Department of Physical Education and Sport, Faculty of Sport Sciences, University of Granada, Granada, Spain; IRyS Group, School of Physical Education, Physical Activity School, Pontificia Universidad Católica de Valparaíso, Valparaiso, Chile

Evandro Leite de Souza, Department of Nutrition, Health Sciences Center, Federal University of Paraíba, João Pessoa, Brazil

Charikleia Stefanaki, Department of Pediatrics, General Hospital of Nikaia "Agios Panteleimon", Piraeus, Greece; Choremeion Research Laboratory, 1st Department of Pediatrics, Medical School, National and Kapodistrian University of Athens, NKUA, Athens, Greece

Christopher J. Stewart, Newcastle University, Institute of Cellular Medicine, Newcastle upon Tyne, United Kingdom; Baylor College of Medicine, The Alkek Center for Metagenomics and Microbiome Research, Houston TX, United States

Jing Sun, School of Medicine, Griffith University, Gold Coast, QLD, Australia

Carla R. Taddei, School of Pharmaceutical Sciences, Dep. of Clinical Analysis and Toxicology, University of São Paulo, SP, Brazil

Claudio Tana, Department of Medicine and Surgery, University of Parma, Parma, Italy; Dipartimento Medico-Geriatrico-Riabilitativo, Azienda Ospedaliero-Universitaria di Parma, Parma, Italy

Abdellah Tebani, Department of Metabolic Biochemistry, Rouen University Hospital, Rouen, France

Fernanda Fernandes Terra, Department of Immunology, Institute of Biomedical Sciences, University of São Paulo, São Paulo, Brazil

Andrea Ticinesi, Microbiome Research Hub, University of Parma, Italy; Department of Medicine and Surgery, University of Parma, Parma, Italy; Dipartimento Medico-Geriatrico-Riabilitativo, Azienda Ospedaliero-Universitaria di Parma, Parma, Italy

Byron P. Vaughn, Division of Gastroenterology, Hepatology and Nutrition, University of Minnesota, Minneapolis, MN, United States

Marco Ventura, Microbiome Research Hub, University of Parma, Italy; Laboratory of Probiogenomics, Department of Chemistry, Life Sciences and Environmental Sustainability, University of Parma, Parma, Italy

Hubert Vidal, Univ-Lyon, CarMeN (Cardio, Metabolism, Diabetes and Nutrition) Laboratory, Université Claude Bernard Lyon 1, INSA Lyon, Oullins, France

L.Q. Vieira, Department of Biochemistry-Immunology, Institute of Biological Sciences, Federal University of Minas Gerais, Belo Horizonte, Brazil

M.J. Villanueva-Millán, Infectious Diseases, Microbiota and Metabolism Unit, Infectious Diseases Department, Center for Biomedical Research of La Rioja (CIBIR), Logroño, Spain

Ingrid Kazue Mizuno Watanabe, Department of Medicine, Federal University of São Paulo, São Paulo, Brazil

C.M. Wells, Department of Biological Engineering, University of Memphis, Memphis, TN, United States

Steven S. Witkin, Division of Immunology and Infectious Diseases, Department of Obstetrics and Gynecology, Weill Cornell Medicine, New York, NY, United States

T.Y. Wong, Department of Biological Sciences, University of Memphis, Memphis, TN, United States

Chenghong Yin, Beijing Obstetrics and Gynecology Hospital, Capital Medical University, Beijing, China

Elisa Zubeldia-Varela, Centro de Metabolómica y Bioanálisis (CEMBIO), Facultad de Farmacia, Universidad CEU San Pablo, Campus Montepríncipe, Madrid, Spain; Basic Medical Sciences Department, Faculty of Medicine, Universidad CEU San Pablo, Campus Montepríncipe, Madrid, Spain; Institute of Applied and Molecular Medicine (IMMA), Faculty of Medicine, Universidad CEU San Pablo, Campus Montepríncipe, Madrid, Spain

Introduction to the Microbiome and Metabolome

Joel Faintuch[1] and Salomao Faintuch[2]

[1]*Department of Gastroenterology, Sao Paulo University Medical School, Sao Paulo, Brazil;* [2]*Beth Israel Deaconess Medical Center, Harvard Medical School, Boston, MA, USA*

HISTORY

Prokaryotes represent one-half of the planet biomass [1]. They have been isolated from high mountains to the depth of the oceans and from hot deserts to the Antarctic ice sheet. With such an overwhelming presence, it would be surprising if these microscopic organisms did not develop a longstanding relationship with plants and animals. Indeed, prokaryotes encompass two domains, archea and bacteria, both of which are constituents of the human microbiome. Bacteria predominate, with >90% of all identified gene sequences in the gastrointestinal tract, followed by archea, virus/phages, fungi, and protists (unicellular eukaryotes) [2]. Multicellular organisms, such as intestinal worms and other parasites, are usually excluded.

They massively colonize all accessible external and internal surfaces in humans, particularly those rich in fluids and nutritional substrates, such as the digestive, respiratory, and urogenital mucosas, as well as the plasma, placenta, and other "sterile" fluids and tissues, although in proportions several orders of magnitude lower.

The advent of the microscope (Antonie van Leeuwenhoek, 1632–1723) provided the first glimpses of such ecological niches. In the 1683 van Leeuwenhoek already reported on the remarkable differences between his oral (dental) and fecal microbiome. He was also able to appreciate, in patients, the major impact of disease on such findings [3].

The development of the science of microbiology in the 19th century expanded the knowledge. In 1885 Theodor Escherich described bacterium coli commune, now known as *Escherichia coli*, in the stools of healthy infants [4], subsequently identifying as many as 19 different cocci and bacilli in those samples. Still investigators were uncertain whether such microbes were beneficial, detrimental, or bystanders, as information was scarce and conflicting.

Francis I, also known as François Angoulême (1494–1547), was the emblematic renaissance king of France, with remarkable achievements as patron of the arts and literature, along with major military and diplomatic victories. He possibly suffered from inflammatory bowel disease, with bouts of abdominal pain, anorectal fistula, and abscesses. His Turkish ally, Suleiman the Magnificent (1520–66), heard about the illness and was aware that in Western countries, yoghurt was unavailable. Therefore he sent to France his personal physician, with generous shipments of the fermented product, which reportedly much alleviated the gastrointestinal symptoms of the French king.

SCIENTIFIC APPROACHES

The advent of the germ theory of disease, anticipated by Ignaz Semmelweis (1818−65) and consolidated by Louis Pasteur (1822−95), strongly influenced medical opinion toward an antibacterial stance. This did not prevent Elie Metchnikoff (1845−1916), Nobel Prize in Medicine (1908), from staunchly defending the intake of supplements with the "good" intestinal bacteria, or probiotics, in his day represented by *Lactobacillus bulgaricus*−containing yoghurt.

Nonetheless, the 20th century can more properly be described as the era of antibiotics, than of probiotics. Indeed, within less than 100 years, vaccines and notably antibacterial agents changed the profile of world public health. Instead of dying young from infections, people typically succumbed at a mature age from chronic, non-transmissible conditions.

Without antiseptic and aseptic environments and procedures, along with antibiotics and antiviral and anti-fungal agents, modern medicine would be nearly impossible. Hospitals would regress to sanatoria and almshouses, and the health-care profession would not perform much better than at the time of alchemists, herbalists, and barber surgeons. Indeed, current hospitalized patients are often immune suppressed on account of age, malnutrition, drugs, diseases, or trauma and commonly undergo invasive diagnostic and therapeutic procedures, all of which are fraught with the risk of contamination and sepsis.

This does not necessarily mean victory of good (antimicrobials) against evil (germs). For one thing, antibiotics were definitely oversold and improperly used in the past. Every microorganism lives in an ecological niche, in relative harmony with the neighboring environment. If disruptions occur, balance should be restored with little impact on the larger community, not with blanket eradication of commensals and pathobionts alike. The recovery of certain intestinal bacteria can be delayed for up to 4 years, after antibiotic treatment [5]. In the 21th century, the growth of multiresistant pathogens increasingly threatens mankind with a return to the preantibiotic past.

At the same time, the remarkably useful roles of commensals in the microbiome could never be as precisely unveiled and monitored, as with modern metagenomic sequencing and metabolomics. This paves the way for antibiotic-independent manipulation of bacterial communities, with ample potential benefit for infectious, in-flammatory, autoimmune, rheumatological, metabolic, cardiovascular, oncological, and even neuropsychiatric conditions. Indeed, it has been advocated that human beings should be envisaged as a man−microbe symbiotic supraorganism, endowed with a supragenome, which should be considered as a whole, instead of addressing just the host's genome [6].

THE BIRTH OF CURRENT MICROBIOME STUDIES

Culturing methods are a mainstay in microbiology and provide invaluable information about both the genotype and phenotype of individual microorganisms. However, they are not cost-efficient or trustworthy for complex environments involving hundreds or thousands of species. Many microorganisms are still nonculturable, whereas certain fastidious bacteria, including many anaerobes that represent by far the majority of gut commensals, may demand time-consuming and labor-intensive methods to form colonies. During at least half of the 20th century, *E. coli* was regarded as the most abundant bacteria in human feces because of easy and prolific growth in vitro. Now it is known as merely one of the species of proteobacteria, a recognized category which nevertheless does not even come close to the major phyla, Bacteroidetes and Firmicutes, which together forms over 90% of the gut microbiome [7].

Metagenomic bacterial studies were kick-started by the famous Human Genome Project of the 1980's and 1990's. Even though unrelated and strictly targeting human cells, one of the by-products of the human initiative was high-throughput DNA-sequencing equipment, along with statistical methods to deal with the generated big data. Adaptation of these platforms to bacterial requirements did not take much, truly revolutionizing the microbiome field. Major collaborative microbiome protocols are going on in many parts of the world, mapping the gut microbiome in different physiological, pathological, and environmental conditions. Other nongut micro-biomes, such as oral, vaginal, respiratory, and dermatological microbiomes, have not been overlooked either, along with studies targeting the holomicrobiome, including fungi and viruses.

This does not mean that all obstacles have been eliminated. Whereas the host genome is 99.9% identical in humans, as little as 10%–20% homology can be encountered in the intestinal microbiome, when severe cases of dysbiosis are examined, thus posing formidable challenges for the definition of the "normal" gut pattern [8]. Yet, for diagnostic purposes, consensus about normality is not essential, provided a signature or fingerprint can be associated with a given disease.

THE METABOLOME AND OTHER "OMICS"

All the priceless advances notwithstanding, there is still something missing in the metagenomic universe, namely functional repercussions. DNA sequencing provides a rich and informative, however, portrait, depicting the genotype. Function can be anticipated, based on known gene sequences. However, many genes remain unclassified. What is more is that there is no clue as to which genes are being transcribed and translated into proteins. In this essentially two-dimensional snapshot, one is not even sure whether identified bacteria are alive or dead. Shotgun sequencing apparatuses can accept fragmented and denatured DNA samples. In ordinary conditions, culture of the microorganisms should be added, to confirm viability [9].

The current option for filling the gap, at the same time extraordinarily enriching genotype and phenotype analysis, is by means of multiomics designs, aiming at the metabolome, transcriptome, proteome, and lipidome. Thus actively expressed genes, along with proteins, lipids, and other molecules secreted into body fluids and tissues, can be unveiled. Although such approach is not devoid of its own costs and difficulties, it is the most rewarding pathway to truly translational science or precision medicine as now preferred.

The historical origins of the human metabolome are independent of the gut microbiome, and few anticipated that these tools would ever cross paths and work in synergy. Small molecules generated by cell metabolism during health and disease, encompassing those absorbed from the environment, in the form of breathed air, food, food additives, drugs, and pollutants (exposome/xenobiotics), are inherent to the human phenotype. Yet, their links to the host genotype, to external mechanisms, and specifically to the bacterial microbiome were largely ignored in the past. Their identification, often serendipitous, stems from progress in chemistry, notably, organic chemistry, in the 18th and 19th century, long before nucleic acids and modern metagenomics were codified.

Urea for instance, the major downstream product of protein metabolism, was isolated in urine and named by Fourcroy and Vauquelin, in 1799 [10]. CO_2, the gaseous catabolite of energy-providing carbonic chains of carbohydrates, lipids, and proteins, was documented even earlier. Joseph Black (1728–99) extracted inorganic CO_2 from limestone and precipitated it with calcium hydroxide. He did not fail to recognize that this was the same gas produced by both animal respiration and microbial fermentation [11].

Of course, modern metabolomics aim much higher, with hundreds of molecules identified in biological fluids and samples. They obviously count with sensitive analytical instruments, such as ultrahigh-performance liquid and gas chromatography, time-of-flight mass spectrometry, tandem mass spectrometry, nuclear magnetic resonance spectroscopy, capillary electrophoresis, and combinations thereof.

One specialized subarea focuses volatile organic compounds (VOCs), basically in the breath, but also in urinary and fecal samples. Indeed, the food industry already uses hand-held devices or electronic noses for VOC analysis of wines, cheeses, and other products. Adaptation of these tools to clinical breath survey could mean rapid and low-cost bedside diagnosis in certain circumstances. Often these instruments are quite simple and technologically limited, thus not actually identifying the small molecules. They just generate metabolic profiles or fingerprints, which may nevertheless be sufficient for diagnostic purposes [12].

Among the multiple small chemical structures alluded to, many have reliably been linked to specific microorganisms, or at least to the collective gut microbiome, in defined conditions of health or disease. Not less than 10% of circulating metabolites are ascribed to the microbiome [13]. Metabolomics often have the primary aim of measuring well-characterized bacterial metabolites, such as short-chain fatty acids, and microbial siderophores and also biotransformation products from host metabolites and xenobiotics [14]. Many have been traced to certain foods, drugs, xenobiotics, or atmospheric gases. Of course the host cells are a major source of metabolites, both in health and in the presence of illness. The integration of metabolomics with microbial metagenomics has taken in account these pathways, and algorithms for specific diseases are being designed [15–17].

THE ERA OF MULTIOMICS

If paired microbiome and metabolome analysis is not mind-boggling enough for the uninitiated, and indeed not many professionals are conversant with both fields, much more is coming. A recent study investigating prospective weight gain and loss, in both insulin-sensitive and insulin-resistant voluntaries, simultaneously documented genomics, transcriptomics, multiple proteomics assays, metabolomics from peripheral mononuclear cells, plasma and serum, and stool microbiomics.

Sure enough, gut bacterial changes were identified concerning *Oxalobacteraceae formigenes* and *Alistipes* as markers of glucose homeostasis. However, the harvest was substantially enriched by the metabolome, proteome, and transcriptome, as multiple activated immunoinflammatory pathways and biomarkers were unearthed in this deep phenotyping model [18].

Is multiomics too cumbersome and expensive for health-care routine? And perhaps are they redundant as well? Comprehensive exploratory studies will always be a challenge, and the mentioned protocol enlisted a team of 34 [18]. However, as diagnostic fingerprints are cataloged and validated, only disease-relevant "omics" should be mobilized, seeking a single specific marker or, at most, a small list of them. A simple and affordable electronic nose could be sufficient in some circumstances [12]. In others, a gut-on-a-chip device, able to test the interaction between a single patients's microbiome, and an *in vitro* simulated intestinal system can be used [19].

That should dispel the fear of unrewarding and confusing overdiagnosis and the corresponding waste of time and resources. On the contrary, multiomics should represent a new threshold in precision medicine, with more technical tools and ultimately a more favorable cost-benefit ratio than existing alternatives.

REFERENCES

[1] Huang WE, Ferguson A, Singer AC, Lawson K, Thompson IP, Kalin RM, Larkin MJ, Bailey MJ, Whiteley AS. Resolving genetic functions within microbial populations: in situ analyses using rRNA and mRNA stable isotope probing coupled with single-cell Raman-fluorescence in situ hybridization. Appl Environ Microbiol 2009;75:234—41.

[2] Arumugam M, Raes J, Pelletier E, Le Paslier D, Yamada T, Mende DR, Fernandes GR, Tap J, Bruls T, Batto JM, Bertalan M, Borruel N, Casellas F, Fernandez L, Gautier L, Hansen T, Hattori M, Hayashi T, Kleerebezem M, Kurokawa K, Leclerc M, Levenez F, Manichanh C, Nielsen HB, Nielsen T, Pons N, Poulain J, Qin J, Sicheritz-Ponten T, Tims S, Torrents D, Ugarte E, Zoetendal EG, Wang J, Guarner F, Pedersen O, de Vos WM, Brunak S, Doré J, , MetaHIT Consortium, Antolín M, Artiguenave F, Blottiere HM, Almeida M, Brechot C, Cara C, Chervaux C, Cultrone A, Delorme C, Denariaz G, Dervyn R, Foerstner KU, Friss C, van de Guchte M, Guedon E, Haimet F, Huber W, van Hylckama-Vlieg J, Jamet A, Juste C, Kaci G, Knol J, Lakhdari O, Layec S, Le Roux K, Maguin E, Mérieux A, Melo Minardi R, M'rini C, Muller J, Oozeer R, Parkhill J, Renault P, Rescigno M, Sanchez N, Sunagawa S, Torrejon A,Turner K, Vandemeulebrouck G, Varela E, Winogradsky Y, Zeller G, Weissenbach J, Ehrlich SD, Bork P. Enterotypes of the human gut microbiome. Nature 2011;473(7346):174—80.

[3] van Leeuwenhoek A. (An abstract of a letter from Antonie van Leeuwenhoek, Sep. 12, 1683). About animals in the scruff of the teeth. Phil Trans R Soc Lond 1684;14:568—574

[4] Escherich T. Die darmbakterien des neugeborenen und säuglings (The intestinal bacteria of the neonate and breast-fed infant). Fortsch Med 1885;3:515—22.

[5] Jernberg C, Lofmark S, Edlund C, Jansson JK. Long-term ecological impacts of antibiotic administration on the human intestinal microbiota. ISME J 2007;1:56—66.

[6] Doré J, Multon MC, Béhier JM. Participants of Giens XXXII, round table no. 2. The human gut microbiome as source of innovation for health: which physiological and therapeutic outcomes could we expect? Therapie 2017;72:21—38.

[7] Lloyd-Price J, Abu-Ali G, Huttenhower C. The healthy human microbiome. Genome Med 2016;8:51—61.

[8] Turnbaugh PJ, Hamady M, Yatsunenko T, Cantarel BL, Duncan A, Ley RE, Sogin ML, Jones WJ, Roe BA, Affourtit JP, et al. A core gut microbiome in obese and lean twins. Nature 2009;457:480—4.

[9] Patro JN, Ramachandran P, Barnaba T, Mammel MK, Lewis JL, Elkins CA. Culture-independent metagenomic surveillance of commercially available probiotics with high-throughput next-generation sequencing. mSphere 2016;1(2).

[10] Fougere P. Grands pharmaciens. Paris: Buchet/Chastel Ed; 1956.

[11] Lenard P. Great men of science. London: G. Bell and Sons; 1950.

[12] Dryahina K, Smith D, Bortlík M, Machková N, Lukáš M, Spanel P. Pentane and other volatile organic compounds, including carboxylic acids, in the exhaled breath of patients with Crohn's disease and ulcerative colitis. J Breath Res Nov 29, 2017;12(1):016002.

[13] Wikoff W, Anfora A, Liu J, Schultz P, Lesley S, Peters E, Siuzdak G. Metabolomics analysis reveals large effects of gut microflora on mammalian blood metabolites. Proc Natl Acad Sci U S A 2009;106:3698—703.

[14] Shaffer M, Armstrong AJS, Phelan VV, Reisdorph N, Lozupone CA. Microbiome and metabolome data integration provides insight into health and disease. Transl Res 2017;189:51—64.

[15] Lustri BC, Sperandio V, Moreira CG. Bacterial chat: intestinal metabolites and signals in host-microbiota-pathogen interactions. Infect Immun September 25, 2017;85(12). pii:e00476-17.

[16] Santoru ML, Piras C, Murgia A, Palmas V, Camboni T, Liggi S, Ibba I, Lai MA, Orrù S, Loizedda AL, Griffin JL, Usai P, Caboni P, Atzori L, Manzin A. Cross sectional evaluation of the gut-microbiome metabolome axis in an Italian cohort of IBD patients. Sci Rep 2017;7:952.

[17] Dumas ME, Rothwell AR, Hoyles L, Aranias T, Chilloux J, Calderari S, Noll EM, Péan N, Boulangé CL, Blancher C, Barton RH, Gu Q, Fearnside JF, Deshayes C, Hue C, Scott J, Nicholson JK, Gauguier D. Microbial-host Co-metabolites are prodromal markers predicting phenotypic heterogeneity in behavior, obesity, and impaired glucose tolerance. Cell Rep 2017;20:136—48.

[18] Piening BD, Zhou W, Contrepois K, Röst H, Gu Urban GJ, Mishra T, Hanson BM, Bautista EJ, Leopold S, Yeh CY, Spakowicz D, Banerjee I, Chen C, Kukurba K, Perelman D, Craig C, Colbert E, Salins D, Rego S, Lee S, Zhang C, Wheeler J, Sailani MR, Liang L, Abbott C, Gerstein M, Mardinoglu A, Smith U, Rubin DL, Pitteri S, Sodergren E, McLaughlin TL, Weinstock GM, Snyder MP. Integrative personal omics profiles during periods of weight gain and loss. Cell Syst January 16, 2018;6(2):157—170.e8.

[19] Kim HJ, Lee J, Choi JH, Bahinski A, Ingber DE. Co-culture of living microbiome with microengineered human intestinal villi in a gut-on-a-chip microfluidic device. J Vis Exp 2016;114.

Useful Internet Sites

Joel Faintuch[1] and Jacob J. Faintuch[2]

[1]Senior Professor, Department of Gastroenterology, Sao Paulo University Medical School, Sao Paulo, Brazil; [2]Professor, Department of Internal Medicine, Hospital das Clinicas, Sao Paulo, Brazil

TABLE 1 General Microbiome Sources

Address	Microbiome
www.metahit.eu	METAHIT Metagenomics
www.hmpdacc.org/	Human Microbiome Project-Data
www.kegg.jp/kegg	KEGG Encyclop. Genes Genomes
http://www.mg-rast.org/	Metagenomics server MG-RAST
www.theseed.org/servers	The Network-Based SEED API
https://press3.mcs.anl.gov/gensc/	Genomic standarts consortium
www.mybiosoftware.com/	Functional gene ecological analysis
www.micca.org	Microbial Community Analysis
https://omictools.com (copyrighter-tool)	Gene copy number correcting tool
https://img.jgi.doe.gov	IMG/M Integrated Microbial Genomes
https://bioinformatics.ca/	CAMERA Microbial Res. Analysis
https://github.com/jcvi/metarep	METAREP Metagenome Analyzer
www.metavelvet.dna.bio.keio.ac.jp/MSL.html	de novo metagenomic assembler
https://bitbucket.org/biobakery/metaphlan2	MetaPhlAn2 Metagen.Phylog.Anal.
http://g2im.u-clermont1.fr/phylopdb/	PhylOPDb: 16S rRNA database
http://eggnog-mapper.embl.de	Genome-wide funct. orthology annotation
www.insdc.org	Int. Nucleotide sequence Database
http://omim.org	Gene locus information
www.ncbi.nlm.nih.gov/books/NBK1116/	Gene reviews

TABLE 2 Sources for Metabolome and Transcriptome

Address	Metabolome and Omics Tools
www.hmdb.ca	Human Metabolome Data Base
http://hmp2.org	Integrative Human Metabolome Data
www.lipidmaps.org	Lipid Maps Lipidomic Gateway
https://bmcbioinformatics.biomedcentral.com/	Metabolome searcher
https://biocyc.org	BioCyc Database Collection
http://fiehnlab.ucdavis.edu	SetupX Metabolomics Base
https://metacyc.org/	Metabolic Pathway Database
https://pubchem.ncbi.nlm.nih.gov/	Substance, Compound, Bioassay
https://omictools.com	OMIC Tools
www.qiagenbioinformatics.com/products	Ingenuity pathway analysis/ IPA
gmd.mpimp-golm.mpg.de	Golm Metabolome Database
https://omictools.com (transcriptomics)	Bioinformatics for transcriptome
https://github.com/kwanjeeraw/metabox	Metabolomics and omics integration

TABLE 3 Bioinformatics Platforms

Address	Bioinformatics
http://ab.inf.uni-tuebingen.de	MEGAN Algorithm in Bioinformatics
http://comet2.gobics.de	COMET Comparative Anal. Metagenomes
http://www.mg-rast.org/	MG RAST Metagen. Rapid Annotation
https://github.com/biocore(Evident)	Evident software package
http://picrust.github.io	PICRUSt Phylogenetic Investigation
www.bioinformatics.babraham.ac.uk	Babraham bioinformatics
http//hannonlab.cshl.edu/	FASTX Toolkit Hannon laboratory
https://omictools.com (transcriptomics)	Bioinformatics for transcriptome

TABLE 4 Specialized Clinical Sites

Address	Diseases and Other Specialized Sites
http://vmc.vcu.edu/momspi	Multi-Omic Microbiome Study: Pregnancy
www.americangut.org	American Gut Project
http://ibdmdb.org	Inflammatory Bowel Disease Multi-Omic
www.pharmacomicrobiomics.org	The PharmacoMicrobiomics Portal
www.brenda-enzymes.org	Braunschweig Enzyme Database
www.github.com/alifar76	Inflammatory bowel disease prediction
http://www.gastro.org	Am. Gastr.Assoc. Fecal Microbiota Registry
http://www.hpmcd.org	Human Pan-Microbe Communities
www.homd.org	Human Oral Microbiome Database
www.omim.org	Gene catalog and genetic diseases

TABLE 5 Metaproteomic Platforms

Address	Site Resources
https://omictools.com (peptide-intens)	Pipasic peptide intens. weighted proteome
https://github.com/compomics/meta	Metaproteomics data analysis and interpretation
https://unipept.ugent.be OK	Biodiversity of complex metaproteomes
https://mikrobiologie.uni-greifswald.de	Amino acid sequences of a given protein database
www.uniprot.org/proteomes	Uniprot metaproteomic databases
https://ab.inf.uni-tuebingen.de	Metaproteome dataset
www.pfam.xfam.org/	Protein families database
www.ebi.ac.uk/pride/	PRIDE proteomics Database

TABLE 6 Designer Bacterial and Human Genome Tools

Address	Site Resources
https://crispr.mit.edu	Optimized CRISPR sgRNA design
www.e-crisp.org/E-CRISP OK	Design of CRISPR constructs
www.multicrispr.net/basic_input.html	RNA target search in input sequences
https://portals.broadinstitute.or/gpp/public/ analysis-tools/sgrna-design	Picks candidate sgRNA sequences for targets (CRISPRa and CRISPRi)
https://cm.jefferson.edu/off-spotter	Identifier of all genome hits of gRNAs
https://crispr.cos.uni-heidelberg.de/	CRISPR/Cas9 target online predictor
http://chopchop.cbu.uib.no/index.ht	CHOPCHOP genome editing

Block I

Tools, Toolstations and Models

Chapter 1

Germ-Free Animals as a Tool to Study Indigenous Microbiota

E. Neumann[1], L.Q. Vieira[2] and J.R. Nicoli[1]

[1]Department of Microbiology, Institute of Biological Sciences, Federal University of Minas Gerais, Belo Horizonte, Brazil; [2]Department of Biochemistry-Immunology, Institute of Biological Sciences, Federal University of Minas Gerais, Belo Horizonte, Brazil

INTRODUCTION

The use of the germ-free (GF) animal model was the direct results of a statement by Pasteur in 1885 who voiced his interest in raising experimental animals in complete absence of interfering microorganisms, but with "the preconceived idea that life under this condition would be impossible." Shortly thereafter (1895) and contradicting Pasteur, Nuttall and Thierfelder had been able to produce a GF guinea pig, albeit short-lived. However, it took more than 50 years until the first breeding rat colonies were successfully established in the late 1940s simultaneously in United States (Reyniers), Sweden (Gustafsson), and Japan (Miyakawa). This delay was caused largely by a lack of knowledge about nutrition in the absence of indigenous microbiota. Afterward, germ-free research, also called gnotobiology (a word derived from the Greek "gnotos" and "biota" meaning known flora and fauna), has evolved out of its technological era and was rapidly being employed in many diverse areas of biological and medical research. Gnotobiology is an important valuable tool to address specific questions about isolated host—microbe interactions. However, when interpreting results from gnotobiotic (GN) studies, one must keep in mind that the observed effect of a specific microbe or microbial consortium on the host may be affected by the altered mucosal immune system, intestinal milieu, and/or metabolism in GF animals. The results obtained from studies in GN animals therefore require additional verification to support their physiological relevance in normally colonized hosts (conventional: CV).

OBTAINING AND MAINTAINING GERM-FREE ANIMALS

Until today, depending on the purpose of use, innumerous and very different animal species have been obtained under GF conditions, such as invertebrates (mosquitoes, flies, snails), birds (chickens, quail), fishes (trout), and mammals (mice, rats, dogs, pigs, sheep, calf). The production of GF animals depends on the fact that embryos developing in eggs or uterus of healthy female are considered microbiologically sterile and can be aseptically hatched or delivered and then maintained into a sterile environment. GF invertebrates or birds were obtained by disinfection (hypochlorite and antibiotics bath) of eggs. GF mammals are derived from aseptic hysterectomy or embryo transfer using GF foster mothers. Originally, these animals were cesarean-derived and had to be hand-reared. Technology for maintaining GF or GN animals is well established. Flexible film isolators are used that are initially sterilized by an internal mist of 2% peracetic acid and then ventilated with high-efficiency particulate air—filtered air. Sterilized cages, bedding, food, and water can be imported into this environment from large steel drums, which have been passed through a high vacuum autoclave with an appropriately large sterilization chamber.

Purpose of Use

Table 1.1 shows the different models derived from the GF animal used in gnotobiology, their microbial status, and purpose of use. Following, some results obtained with each of these models were described.

Germ-Free Animals

The most straightforward way to investigate the role of the microbiota in the context of specific host functions or complex diseases is the comparison between CV and GF animals in context of health or experimental disease models. If the host function or disease of interest is unchanged between CV and GF animals, it can be assumed that the microbiota is of minor relevance for the function or disease development. Table 1.2 shows the influence of indigenous microbiota on the host biology in health. As it can be seen, indigenous microbiota has an impact on practically all aspects of the host biology (anatomy, physiology, metabolism, immunology, nervous system, and behavior).

In the case of disease, if the respective pathology is aggravated or reduced in the GF state, the microbiota can be assumed to be disease relevant. Currently available data (Table 1.3) indicate that indigenous microbiota almost always has a profound influence on host–parasite relationships during bacterial, viral, and protozoan infections [1]. It is well known that the presence of the indigenous microbiota is essential for the pathogenicity of *Entamoeba histolytica* [2], *Eimeria tenella* [3], and *Giardia duodenalis* [4]. In contrast, the microbiota can reduce the pathological consequences of other infectious diseases, as described for experimental infections with protozoa, fungi, and helminths [5,6] and almost all enteropathogenic bacteria [7].

Many experimental animal models (e.g., inflammatory bowel disease [IBD]-like chronic intestinal inflammation) show reduced or lack of disease under GF conditions, indicating that the intestinal microbiota is involved in the development of most complex diseases. Similarly, pathology due to alcohol consumption and gout is reduced and skin wound healing accelerated in GF mice when compared with their CV counterparts [8–10]. Interestingly, nonobese diabetic (NOD) mice, a model for type 1 diabetes, show unchanged pathology under GF conditions, but disease development can be modulated by association with specific bacteria [11].

Gnotobiotic Animals

After demonstration of an effect of the indigenous microbiota on host phenotypes in health or illness, subsequent monoassociation studies or studies using selected consortia of microorganisms in GN animals allow important mechanistic insights into specific microbe–microbe and microbe–host interactions under highly controlled and standardized

TABLE 1.1 Microbial Status of Different Animal Models, Purpose of Their Use to Study Indigenous Microbiota and Maintenance Conditions

Microbial Status	Purpose of Use	Maintenance Conditions
Germ-free (GF) animals	To be compared with its conventional (CV) counterpart to evaluate the effect of the whole microbiota on some aspects of the host biology	Isolator
GF animals associated with one or more known microorganisms (gnotobiotic: GN)	To evaluate the effect of some known components of the microbiota on one aspect of the host biology	Isolator
GF animals associated with simplified microbial consortia (SMC)	To obtain a simplified and controlled model that presents the same biological characteristics than the CV animal	Isolator
GF animals with transplanted microbiota from the same or other animal species (TM or HAM)	To evaluate if the eubiotic or dysbiotic microbiota transplanted from animal or human donors to GF animals provokes the same beneficial or prejudicial effect in the receptor, or to obtain a humanized animal model (HAM)	Animal house
Conventionalized (GF animal associated with microbiota of its own species) or conventional animals (CVZ or CV)	To be used as control when compared with GF, GN, or SMC animals	Animal house

TABLE 1.2 Differences in Some Biological Characteristics of Germ-free (GF) Animals When Compared With Their Conventional (CV) Counterparts

Biological Aspect	Difference Observed in GF When Compared With CV Animals
Anatomy	Megacecum only in rodents Total mass of intestines decreased Cecal wall thinner Decreased spleen and liver size
Intestinal physiology	Duodenal villi longer Lamina propria of small intestine thinner Slower rate of cell turnover of small intestine epithelium Increased amounts of mucin Increased intestinal transit time Increased enterohepatic cycle of bile salts Higher fecal water content Decreased fatty acids in intestinal contents Increased bilirubin in feces Increased oxidation–reduction potential of intestinal contents Increased excretion of free amino acids, urea, and nitrogen
Nutrition	Require more vitamins (K and B complex) Decreased body fat percentage Reduced energy harvest Increased trypsin and chymotrypsin in feces Increased water intake
Metabolism	Decreased basal metabolism rate Increased blood cholesterol Decreased cardiac output Decreased regional blood flow to skin, liver, lungs, digestive tract
Immunology	Unable to inflame Lower levels of IgA in intestinal contents Lower blood levels of IgM and IgG Rare plasma cells in small intestine Peyer's patch reduced in number and size Mesenteric lymph nodes smaller, less cellular, and without germinal centers Decreased response to T-cell mitogens
Enteric nervous system	Smaller nerve density Smaller ganglia Less abundant connective nerve fibers between ganglia Fewer neurons in myenteric ganglia Decreased intestinal motility
Behavior	Less sociability Decreased anxiety-like behavior Reduce ability for learning and short-term memory

conditions regarding genotype, diet, and microbial colonization. Such animal models have allowed to identify and to characterize key microorganisms responsible for immunological system priming and modulation, colonization resistance function, and nutritional contribution. As examples, monoassociation of GF mice with segmented filamentous bacteria (SFB) greatly increased the number of IgA-producing cells in the lamina propria in the small intestine [12], whereas a *Lactobacillus* sp. or a *Peptostreptococcus* sp., isolated from human feces showing an ex vivo antagonism against *Vibrio cholerae,* were able to drastically eliminate this pathogenic bacterium when monoassociated with GF mice [13]. With regard to IBD, monoassociation studies in IL10−/− mice, a spontaneous colitis model, revealed that different microorganisms, even including some opportunistic pathogens (*Helicobacter hepaticus*) do not trigger intestinal inflammation. However, *Lactobacillus reuteri* and *H. hepaticus* trigger severe colitis when used in a dual-association setup [14]. In

TABLE 1.3 Beneficial, Harmful, or Neutral Influence Due to the Indigenous Microbiota as Demonstrated by Comparison of the Course of Infectious Diseases in Germ-Free and Conventional Animals

Infectious Agent	Beneficial	Harmful	Neutral
Prokaryote	*Clostridium difficile* *Escherichia coli* *Listeria monocytogenes* *Pasteurella haemolytica* *Salmonella typhimurium* *Shigella flexneri* *Vibrio cholerae*	*Treponema hyodysenteriae*	–
Eukaryote	*Aspiculuris tetraptera* *Candida albicans* *Leishmania major* *Strongyloides venezuelensis* *Trypanosoma cruzi*	*Eimeria falciformis* *Eimeria ovinoidalis* *Eimeria tenella* *Entamoeba histolytica* *Giardia duodenalis* *Leishmania mexicana*	*Raillietina cesticillus* *Isospora suis*
Virus	*Vaccinia virus* Hepatitis B virus Dengue Influenza	Poliovirus/reovirus MulV	–

contrast, monoassociation with commensals such as *Enterococcus faecalis* can be sufficient to trigger inflammation in IL10−/− mice [15]. Numerous commensal bacteria (the majority of which is derived from humans) have been used to evaluate microbe–host interactions in GN models. Although models harboring only few bacteria have been very helpful, care must be taken when interpreting data. In simplified bacterial communities, the context of a fully diverse and competitive bacterial ecosystem is lacking. Several studies have therefore started using defined bacterial consortia of higher complexity.

Gnotobiotic Animals With Simplified Microbial Consortia

The use of mice colonized with a consortium of known bacteria that naturally inhabit the murine gut (simplified microbial consortia: SMC) offers a powerful system to investigate mechanisms governing host–microbiota relationships and how members of the GI microbiota interact with one another. The SMC community offers significant advantages to study homeostatic as well as disease-related interactions by taking advantage of a well-defined, limited community of microorganisms. For example, quantification and spatial distribution of individual members, microbial genetic manipulation, genomic-scale analysis, and identification of microorganism-specific host immune responses are all achievable using the SMC model. Although technically challenging and rather expensive, the use of SMC mice has several important advantages compared with the use of GF mice or mice with a complex microbiota (SPF or CV). In contrast to GF mice, SMC animals can be morphologically and functionally similar to normal mice associated with a complex microbiota. Association with specific consortia or even monoassociation with a single commensal (*Bacteroides fragilis,* for example) was found to be sufficient to restore many of the disturbed morphological characteristics, metabolic functions, and immune functions in GF animals. Therefore, results obtained in SMC models can be assumed to be of more physiological relevance than the ones obtained in GF settings. However, the capacity of any specific microbe or consortium to "normalize" specific parameters in GF animals cannot be anticipated but needs to be confirmed. To our knowledge, there is no SMC available that fully normalizes the luminal milieu of the intestine with regard to microbial functions or enzymatic activity [16,17]. An obvious advantage of the use of SMC is the restriction of the complexity of the intestinal microbiota to a more manageable level. The high level of standardization of the intestinal microbiota in these SMC animals reduces the interindividual variability and enables the targeted investigation of functional and mechanistic questions related to microbe–host interactions [18]. Nowadays, the most frequently used microbial combinations to obtain SMC animals are the modified Schaedler microbiota and the simplified human intestinal microbiota (SIHUMI), both consisting of about seven bacterial species.

Germ-Free Animals With Transplanted Eubiotic or Dysbiotic Microbiota From the Same or Other Animal Species

The association of GF mice with a complex microbiota derived from mice or human beings is an attempt to investigate the physiological impact of a given microbiota on a "standardized" host (analogous genotype, environment). The fecal transfer (FT) of a complex microbiota from healthy or diseased organisms into GF animals is a valuable tool to study the relative contribution of the respective microbiota to host dysfunctions or disease phenotypes. Several studies already showed that disease susceptibility or clinical symptoms can be transmitted to GF hosts by FT. Additionally, FT is the best experimental model to clarify if a dysbiotic microbiota is the result or the cause in the context of complex diseases. However, it remains unclear how well different human gut bacterial taxa establish in the mouse intestinal milieu. Table 1.4 shows some examples of studies where FT of a eubiotic or dysbiotic microbiota induced in the recipient GF mice similar phenotypes to those observed in the donor. Food-dependent susceptibility to DSS colitis was, for example, found to be transmissible to GF recipient mice [19]. Alterations in the intestinal microbiota of mice induced by Roux-en-Y gastric bypass (RYGB) surgery were also found to be transferrable to GF recipient mice. Mice receiving microbiota from RYGB-operated animals showed increased weight loss compared with mice associated with the microbiota from sham-operated mice [20]. Furthermore, increased susceptibility to atherosclerosis was found to be transferable by FT in recipient mice [21]. In the field of autoimmune and inflammatory diseases, data on the transferability of disease susceptibility or clinical symptoms on "humanized" mice (HAM) are still scarce. However, there are already interesting data concerning to the impact of the microbiota on metabolic dysfunctions, irritable bowel syndrome (IBS), and colorectal cancer. One study showed that diet has not only a major impact on the microbiota in HAM mice, but that the Western diet-mediated host phenotype, adiposity, can be transferred to GF mice receiving a nonadipogenic diet by FT [22]. Interestingly, increased body and fat mass were also found to be directly transferable by humanization of mice with fecal microbiota from four obese twins compared with their lean twin counterparts, indicating that the intestinal microbiota in obese individuals significantly contributes to the progression of metabolic dysfunctions [23]. CV mice consume less food than their GF counterparts, and so increased food consumption cannot explain the obese phenotype in colonized mice [24]. One study showed that microbiota alterations that are characteristic for IBS can be transmitted and maintained in GF mice associated with fecal microbiota from IBS patients. In addition, mice that were humanized with IBS microbiota showed increased visceral sensitivity compared with mice humanized with microbiota from healthy individuals [25].

GF mice can also be used to investigate functions of complex human-derived microbiota. Human fecal microbiota or culture collections can be stably transplanted into GF mice [22,26]. The microbiota in these models is complex and thus not fully characterized, and therefore these models are not GN. Nevertheless, they allow mining the human microbiome for specific functions and address microbiota-specific effects on the immune system and metabolome and study interindividual differences [27,28]. Moreover, HAM mice make it possible to test whether a complex human disease phenotype can be transmitted by microbiota transplantation [23,29]. Investigation was done to evaluate if an altered microbiota conferred any of the metabolic effects by transferring the gut microbiota from women in the first and third trimester to GF female mice, and it was observed that mice transplanted with microbes from the third trimester had increased body fat and exhibited some impairment of glucose metabolism [30].

Transplantation of human fecal microbiota into GF mice can be viewed as capturing an individual's microbial community at a fixed moment in time. Importantly, the structure and composition of the transplanted human microbiota is well maintained in the mouse. Thus, HAM can be monitored over time and under highly controlled conditions. Therefore, potentially confounding variables can be constrained in ways that are not achievable in human studies to demonstrate whether specific phenotypes are transferred and mediated by the gut microbiota. The gut microbiota has been found to subject pharmaceuticals and other xenobiotics to numerous chemical processes, and these metabolic reactions can alter their pharmacokinetics or bioavailability, resulting in an inappropriate dose or the production of toxic metabolites [31]. Metabolic reactions mediated by the microbiota that are known to significantly impact the biologic activity of drugs and xenobiotics involve reduction, hydrolysis, dihydroxylation, acetylation, deacetylation, proteolysis, deconjugation, and deglycosylation processes [32]. Although more than 30 commonly prescribed drugs have been shown to be metabolically altered by the gut microbiota, an increasing body of literature continues to extend the number of drugs that are subject to bacterial metabolism in the gut and other tissues. In this myriad of activities, the impact of the microbiota may be beneficial or detrimental to the host. However, the composition and metabolism of human microbiota is very different than in CV laboratory animals. Therefore, HAM is a more effective model to simulate metabolic transformation of drugs or xenobiotics ingested by human beings than animals harboring its own microbiota. Furthermore, the implementation of human-derived microbiota enables interventional studies using, for example, different diets, microbes, or pharmacological agents in a controlled environment, which would be highly complicated or impossible in humans. Interestingly, it has been shown

TABLE 1.4 Main Results Obtained With the Use of Fecal Transplantation (FT) From Human or Other Animal Species to Germ-Free (GF) Mice

Donor	Research	Observation
Human	Colorectal cancer	GF mice receiving FT from patients with colorectal cancer exhibited higher proliferating cells in colons than GF mice associated with stool from healthy donors[a]
Human	Irritable bowel syndrome (IBS)	GF mice receiving fecal transplant from patient with IBS exhibited faster intestinal transit, intestinal barrier dysfunction, and anxiety-like behavior[b]
Human	Hypertension	FT from hypertensive human donors to GF mice elevated blood pressure in animals[c]
Human	Additives and xenobiotics	Activity of human intestinal microbiota on mutagens was transferred to GF mice by FT[d]
Human or mice	Obesity	FT in GF mice from genetically obese mice resulted in a significantly greater increase in total body fat than FT with a "lean microbiota." Similar results were obtained with FT in GF mice from obese and lean human [22,23]
Human	Malnutrition	FT from undernourished Malawian infants in young GF mice fed a Malawian diet transmits impaired growth phenotypes[e]
Human	Constipation	GF mice receiving fecal transplant from patient with constipation showed lower evacuation frequency, delayed intestinal transit time, and weaker spontaneous contractions of colonic smooth muscle[f]
Human	Major depressive disorder (MDD)	FT from human with MDD in GF mice resulted in depression-like behavior when compared with FT from healthy donor[g]
Pig	Gut anatomy	Gut microbiota and gut phenotypes were different between three pig breeds, and the fecal transplant partially conveys host characteristics from pig to GF mice[h]
Mice	Probiotics and inflammatory diseases	FT from CV mice treated with a probiotic in GF mice ameliorates DSS-induced colitis[i]

[a]Wong SH, Zhao L, Zhang X, Nakatsu G, Han J, Xu W, Xiao X, Kwong TN, Tsoi H Wu WK, Benhua Z, Chan FK, Sung JJ, Wei H, Yu J. Gavage of fecal samples from patients with colorectal cancer promotes intestinal carcinogenesis in germ-free and conventional mice. Gastroenterology 2017. https://doi.org/10.1053/j.gastro.2017.08.022.

[b]De Palma G, Lynch MD, Lu J, Dang VT, Deng Y, Jury J, Umeh G, Miranda PM, Pigrau PM, Sidani S, Pinta-Sanchez MI, Philip V, McLean PG, Hagelsieb MG, Surette MG, Bergonzelli GE, Verdu EF, Britz-McKibbin P, Neufeld JD, Collins SM, Bercik P. Transplantation of fecal microbiota from patients with irritable bowel syndrome alters gut function and behavior in recipient mice. Sci Transl Med 2017. https://doi.org/10.1126/scitranslmed.aaf6397.

[c]Li J, Zhao F, Wang Y, Chen J, Tao J, Tian G, Wu S, Liu S, Cui Q, Geng B, Zhang W, Weldon R, Auguste K, Yang L, Liu X, Chen L, Yang X, Zhu B, Cai J. Gut microbiota dysbiosis contributes to the development of hypertension. Microbiome 2017;5:14.

[d]Hirayama K, Baranczewski P, Akerlund JE, Midtvedt T, Möller L, Rafter J. Effects of human intestinal flora on mutagenicity of and DNA adduct formation from food and environmental mutagens. Carcinogenesis 2000;21:2105—11.

[e]Blanton LV, Charbonneau MR, Salih T, Barratt MJ, Venkatesh S, Ilkaveya O, Subramanian S, Manary MJ, Trehan I, Jorgensen JM, Fan YM, Henrissat B, Leyn SA, Rodionov DA, Osterman AL, Maleta KM, Newgard CB, Ashorn P, Dewey KG, Gordon JI. Gut bacteria that prevent growth impairments transmitted by microbiota from malnourished children. Science 2016;19:351.

[f]Ge X, Zhao W, Ding C, Tian H, Xu L, Wang H, Ni L, Jiang J, Gong J, Zhu W, Zhu M, Li N. Potential role of fecal microbiota from patients with slow transit constipation in the regulation of gastrointestinal motility. Sci Rep 2017;7:441.

[g]Zheng P, Zeng B, Zhou C, Liu C, Liu M, Fang Z, Xu X, Zeng L, Chen J, Fan S, Du X, Zhang X, Yang D, Yang Y, Meng H, Li W, Mengiri ND, Licinio J, Wei H, Xie P. Gut microbiome remodeling induces depressive-like behaviors through a pathway mediated by the host's metabolism. Mol Psychiat 2016;21:786—96.

[h]Diao H, Yan HL, Xiao Y, Yu B, Yu J, He J, Zheng P, Zeng BH, Wei H, Mao XB, Chen DW. Intestinal microbiota could transfer host gut characteristics from pig to mice. BMC Microbiol 2016;16:238.

[i]Souza ELS, ElianSDA, Paula LMR, Garcia CC, Vieira AT, Teixeira MM, Arantes RME, Nicoli JR, Martins FS. Escherichia coli strain Nissle 1917 ameliorates experimental ulcerative colitis in a faecal transplantation model. J Med Microbiol 2016;65:201—10.

that the transfer of a given murine or even human microbiota into GF recipients can result in the establishment of a rather stable and heritable intestinal microbiota that is compositionally and functionally similar to the donor's microbiota [22,28]. With respect to the transfer of human microbiota into GF mice, this finding is surprising, because the diets of humans and mice differ enormously and because diet is known to have a major impact on the composition of the intestinal microbiota [22,33]. In addition, one must keep in mind that commensal microorganisms can be highly adapted to their respective host, which may result in altered microbial functionality of a given microbe in a different mammalian host. In this context, Chung et al. [34] showed that the microbiota-mediated maturation of host immunity upon colonization of GF mice is dependent on specific mouse-derived microbes, as human microbiota transplantation did not result in analogous full maturation of the murine immune system.

Conventionalized or Conventional Animals

Today, all well-controlled animal husbandry possesses GN facilities from which specific pathogen—free (SPF) animal nucleus stocks can be derived when necessary. SPF status refers to animals presenting an absence of known pathogens but with an indigenous microbiota with an unknown composition. For this reason, SPF is not GN. On the other hand, SMC animals can be considered as GN and APF (all pathogenic free).

All newborn mammals, as well as adult GF host, are immature for numerous essential aspects of its biology, and indigenous microbiota is responsible for their maturation during colonization or conventionalization. Interestingly, a correct priming of these biological aspects is complete and correct only if performed during a short window period soon after birth. For this reason, after conventionalization of GF animals to obtain SPF colonies, only the second generation can be used for experiments.

As various aspects of host biology (immune system, enteric nervous system, physiology, anatomy) are known to be primed by the indigenous microbiota during the colonization just after birth, GF animals can be considered as eternal newborns with very different characteristics when compared with CV animals. This is a common, but important, concern with the use of GF animals. If the sole influence of microbiota in a primed host must be evaluated, the use of antibiotic-treated model can be considered. However, direct effects of antibiotics on host physiology, such as repression of mitochondrial and ribosomal function [35], cannot be ignored. Nevertheless, GF animals still seem to be the best controlled model systems for fecal microbial transplantation.

CONCLUSION

In conclusion, highly sophisticated animal models for the in-depth investigation of microbe—host interactions are available. Influence and functions of the indigenous microbiota on the host biology can be evaluated without the interference of the microbiota (GF), in the presence of some known components of the microbiota (GN) or in the presence of eubiotic or dysbiotic microbiota transplanted from the same animal species or from other animal species (FT—HAM). The use of these models showed that indigenous microbiota is fundamental to maturate immunological system, digestive physiology, and enteric nervous system during the colonization of the newborn and to contribute to colonization resistance, immunomodulation, and nutrition of the adult host. All these functions are fundamental for host health, but unfortunately, under dysbiotic conditions the indigenous microbiota is also involved in numerous disorders such as IBDs, allergy, depression, cancer, hypertension, obesity, undernutrition, and constipation. For these reasons, the GF animal and its derivations are important tools to evaluate the relationships between the host and its microbiota, and the knowledge obtained with its use can be applied to develop intervention methods on microbial ecosystem inhabiting host superficies and mucosa to improve the microbiota—host relationship (probiotics, prebiotics, antibiotics, fecal transplantation). However, knowledge about the advantages and caveats of each experimental model is of pivotal importance for the generation of reproducible results and the interpretation of study outcomes. Studies need to be expanded, and the use of genetically modified GF mice may facilitate to delineate the molecular mechanisms by which specific microbes or consortia affect host metabolism.

REFERENCES

[1] Stecher B, Hardt WD. The role of microbiota in infectious disease. Trends Microbiol 2008;16:107—14.
[2] Phillips BP, Wolfe PA. The use of germ-free Guinea pigs in studies on the microbial interrelationships in amoebiasis. Ann NY Acad Sci 1959;78:308—14.
[3] Gouet P, Yvore P, Naciri M, et al. Influence of digestive microflora on parasite development and the pathogenic effect of Eimeria ovinoidalis in the axenic, gnotoxenic and conventional lamb. Res Vet Sci 1984;36:21—3.

[4] Torres MF, Uetanabaro AP, Costa AF, et al. Influence of bacteria from the duodenal microbiota of patients with symptomatic giardiasis on the pathogenicity of Giardia duodenalisin gnotoxenic mice. J Med Microbiol 2000;49:209−15.

[5] Silva ME, Evangelista EA, Nicoli JR, et al. American trypanosomiasis (Chagas' disease) in conventional and germ-free rats and mice. Rev Inst Med Trop Sao Paulo 1987;29:284−8.

[6] Martins WA, Melo AL, Nicoli JR, et al. A method of decontaminating strongyloides venezuelensislarvae for the study of strongyloidiasis in germ-free and conventional mice. J Med Microbiol 2000;49:387−90.

[7] Wilson KH. Ecological concepts in the control of pathogens. In: Roth JA, editor. Virulence mechanisms of bacterial pathogens. Washington (DC): American Society for Microbiology; 1995. p. 245−56.

[8] Canesso MCC, Queiroz NL, Marcantonio C, Lauar J, Almeida D, Gamba C, Cassali G, Pedroso SH, Moreira C, Martins FS, Nicoli JR, Teixeira MM, Godard ALB, Vieira AT. Comparing the effects of acute alcohol consumption in germ-free and conventional mice: the role of the gut microbiota. BMC Microbiol 2014a;14:240.

[9] Canesso MCC, Vieira AT, Castro TBR, Schirmer BGA, Almeida CP, Cisalpino D, Martins FS, Rachid MA, Nicoli JR, Teixeira MM, Barcelos LS. Skin wound healing is accelerated and scarless in the absence of commensal microbiota. J Immunol 2014b;193:5171−80.

[10] Vieira AT, Macia M, Martins FS, Canesso MC, Amaral FA, Garcia CC, de Leon E, Shim D, Maslowski K, Nicoli JR, Vieira LQ, Harper F, Teixeira MM, Mackay C. A role for the gut microbiota and the metabolite sensing receptor GPR43 in a murine model of gout. Arthritis Rheumatol 2015;67:1646−56.

[11] King C, Sarvetnick N. The incidence of type-1 diabetes in NOD mice is modulated by restricted flora not germfree conditions. PLoS One 2011;6:e17049.

[12] Umesaki Y, Setoyama H. Structure of the intestinal flora responsible for development of the gut immune system in a rodent model. Microbes Infect 2000;2:1343−51.

[13] Silva SH, Vieira EC, Nicoli JR. Antagonism against *Vibrio cholerae* through the production of diffusible substance by bacterial components from the human fecal microbiota. J Med Microbiol 2001;50:161−4.

[14] Whary MT, Taylor NS, Feng Y, Ge Z, Muthupalani S, Versalovic J, Fox JG. *Lactobacillus reuteri* promotes Helicobacter hepaticus-associated typhlocolitis in gnotobiotic B6.129P2-IL-10(tm1Cgn) (IL-10(-/-)) mice. Immunology 2011;133:165−78.

[15] Steck N, Hoffmann M, Sava IG, Kim SC, Hahne H, Tonkonogy SL, Mair K, Krueger D, Pruteanu M, Shanahan F, Vogelmann R, Schemann M, Kuster B, Sartor RB, Haller D. *Enterococcus faecalis* metalloprotease compromises epithelial barrier and contributes to intestinal inflammation. Gastroenterology 2011;141:959−71.

[16] Norin E, Midtvedt T. Intestinal microflora functions in laboratory mice claimed to harbor a "normal" intestinal microflora. Is the SPF concept running out of date? Anaerobe 2010;16:311−3.

[17] Becker N, Kunath J, Loh G, Blaut M. Human intestinal microbiota: characterization of a simplified and stable gnotobiotic rat model. Gut Microbes 2011;2:25−33.

[18] Eun CS, Mishima Y, Wohlgemuth S, et al. Induction of bacterial antigen-specific colitis by a simplified human microbiota consortium in gnotobiotic interleukin-10-/- mice. Infect Immun 2014;82:2239−46.

[19] Fuhrer A, Sprenger N, Kurakevich E, Borsig L, Chassard C, Hennet T. Milk sialyllactose influences colitis in mice through selective intestinal bacterial colonization. J Exp Med 2010;207:2843−54.

[20] Liou AP, Paziuk M, Luevano Jr JM, Machineni S, Turnbaugh PJ, Kaplan LM. Conserved shifts in the gut microbiota due to gastric bypass reduce host weight and adiposity. Sci Transl Med 2013. https://doi.org/10.1126/scitranslmed.3005687.

[21] Gregory JC, Buffa JA, Org E, Wang Z, Levison BS, Zhu W, Wagner MA, Bennett BJ, Li L, DiDonato JA, Lusis AJ, Hazen SL. Transmission of atherosclerosis susceptibility with gut microbial transplantation. J Biol Chem 2015;290:5647−60.

[22] Turnbaugh PJ, Ridaura VK, Faith JJ, Rey FE, Knight R, Gordon JI. The effect of diet on the human gut microbiome: a metagenomic analysis in humanized gnotobiotic mice. Sci Transl Med 2009. https://doi.org/10.1126/scitranslmed.3000322.

[23] Ridaura VK, Faith JJ, Rey FE, Cheng J, Duncan AE, Kau AL, Griffin NW, Lombard V, Henrissat B, Bain JR, Muehlbauer MJ, Ilkayeva O, Semenkovich CF, Funai K, Hayashi DK, Lyle BJ, Martini MC, Ursell LK, Clemente JC, Van Treuren W, Walters WA, Knight R, Newgard CB, Heath AC, Gordon JI. Gut microbiota from twins discordant for obesity modulate metabolism in mice. Science 2013. https://doi.org/10.1126/science.1241214.

[24] Gazzinelli RL, Silva ME, Moraes-Santos T, Nicoli JR, Vieira EC. Effect of high sucrose diets on carcass composition in conventional and germfree mice. Arch Latinoam Nutr 1991;XLI:539−45.

[25] Crouzet L, Gaultier E, Del'Homme C, Cartier C, Delmas E, Dapoigny M, Fioramonti J, Bernalier-Donadille A. The hypersensitivity to colonic distension of IBS patients can be transferred to rats through their fecal microbiota. Neuro Gastroenterol Motil 2013;25:e272−82.

[26] Wos-Oxley M, Bleich A, Oxley AP, Kahl S, Janus L, Smoczek A, Nahrstedt H, Pils M, Taudien S, Platzer M, Hedrich HJ, Medina E, Pieper D. Comparative evaluation of establishing a human gut microbial community within rodent models. Gut Microbes 2012;3:234−49.

[27] Ahern PP, Faith JJ, Gordon JI. Mining the human gut microbiota for effector strains that shape the immune system. Immunity 2014;40:815−23.

[28] Marcobal A, Kashyap PC, Nelson TA, Aronov PA, Donia MS, Spormann A, Fischbach MA, Sonnenburg JL. A metabolomic view of how the human gut microbiota impacts the host metabolome using humanized and gnotobiotic mice. ISME J 2013;7:1933−43.

[29] Subramanian S, Huq S, Yatsunenko T, Haque R, Mahfuz M, Alam MA, Benezra A, DeStefano J, Meier MF, Muegge BD, Barratt MJ, Van Arendonk LG, Zhang Q, Province MA, Petri Jr WA, Ahmed T, Gordon JI. Persistent gut microbiota immaturity in malnourished Bangladeshi children. Nature 2014;510:417−21.

[30] Koren O, Goodrich JK, Cullender TC, et al. Host remodeling of the gut microbiome and metabolic changes during pregnancy. Cell 2012;150:470−80.

[31] Sousa T, Paterson R, Moore V, Carlsson A, Abrahamsson B, Basit AW. The gastrointestinal microbiota as a site for the biotransformation of drugs. Int J Pharm 2008;363:1−25.

[32] Spanogiannopoulos P, Bess EN, Carmody RN, Turnbaugh PJ. The microbial pharmacists within us: a metagenomic view of xenobiotic metabolism. Nat Rev Microbiol 2016;14:273−87.

[33] Kashyap PC, Marcobal A, Ursell LK, Larauche M, Duboc H, Earle KA, Sonnenburg ED, Ferreyra JA, Higginbottom SK, Million M, Tache Y, Pasricha PJ, Knight R, Farrugia G, Sonnenburg JL. Complex interactions among diet, gastrointestinal transit, and gut microbiota in humanized mice. Gastroenterology 2013;144:967−77.

[34] Chung H, Pamp SJ, Hill JA, Surana NK, Edelman SM, Troy EB, Reading NC, Villablanca EJ, Wang S, Mora JR, Umesaki Y, Mathis D, Benoist C, Relman DA, Kasper DL. Gut immune maturation depends on colonization with a host-specific microbiota. Cell 2012;149:1578−93.

[35] Morung A, Dzutsev A, Dong X, Greer RL, Sexton DJ, Ravel J, Schuster M, Hsiao W, Matzinger P, Shulzhenko N. Uncovering effects of antibiotics on the host and microbiota using transkingdom gene methods. Gut 2015;64:1732−43.

Chapter 2

Germ-Free Mouse Technology in Cardiovascular Research

Alexandra Grill[1,2] and Christoph Reinhardt[1,2]

[1]*Center for Thrombosis and Hemostasis (CTH), University Medical Center Mainz, Johannes Gutenberg University Mainz, Mainz, Germany;* [2]*German Center for Cardiovascular Research (DZHK), Partner Site RheinMain, Mainz, Germany*

THE VARIOUS WAYS TO STUDY HOST—MICROBIAL INTERACTIONS

A Short History of Germ-Free Research

Arising from the knowledge of microorganisms inhabiting the gut, Louis Pasteur in 1885 was the first to think of feeding an animal from birth, with chow completely free of any microbiota, and postulated that life would not be possible without them [1]. It took 10 years until Nuttall and Thierfelder raised the first germ-free guinea pigs [2]. In the following decades, several kinds of animals were successfully transferred to germ-free conditions [3]. A major breakthrough in making isolator technology more robust was the development of steel isolators by Reyniers.

Within the 1950s and 1960s, first classical laboratory animals such as rats and mice were raised by hand, and subsequent generations were bred within isolators [4,5]. In the late 1960s, Trexler overcame the disadvantages of steel isolators, such as heavy weight, great cost, and limited sight, by inventing the first plastic isolator. The use of plastic instead of steel not only reduced the cost, but more importantly, the improved visibility allowed more complex procedures to be performed and quickly promoted germ-free technology in various facilities [1].

Production of germ-free animals by caesarean section and closed hysterotomy has been the method of choice for several decades. By now, it is clear that some protozoa, viruses, and bacteria can penetrate through the placental barrier and reach the fetus [6—10]. Most of these contaminants can be overcome by the use of embryo transfer into pseudopregnant germ-free recipients. Application of embryo transfer presents a promising technique, to reduce the risks of intrauterine contamination with microorganisms, and increases survival rates [11].

Some Thoughts and Concerns on the Present Status of Germ-Free Research

These days, the focus in germ-free animal research, lies on the understanding of the interaction between bacteria and host. Even though the clinical relevance of germ-free animals has been questioned, because humans would never be free of bacteria, they provide a powerful tool to answer the question of microbiota influence on host development and function. Hence, these animals should not be seen as a model to mimic human conditions, but as a white page of paper, which enables studies in the absence of microbiota, and on the other hand the addition of any microorganism or distinct microbial product of interest for subsequent investigation. For such colonization experiments, only one bacterial subspecies or a whole cocktail of bacteria can be used.

However, because the generation and maintenance of a germ-free colony is a cost and work-intensive matter, a prevalent alternative approach is the treatment of specific pathogen—free (SPF) mice, with broad-spectrum antibiotics [12]. These antibiotic-treated animals present many phenotypic properties of germ-free animals, such as enlarged cecum, with the advantage that treatment of adult mice does not alter the microbiota-dependent host development during infancy [13]. Nevertheless, the treatment does not remove all bacteria from the gut, and it is impossible to

Microbiome and Metabolome in Diagnosis, Therapy, and other Strategic Applications. https://doi.org/10.1016/B978-0-12-815249-2.00002-6

predict the degree of bacterial elimination or to calculate which bacterial species might overgrow due to lack of competition [13,14].

Further, germ-free mice are seen completely free of any contamination, including viruses, fungi, and bacteriophages. All these microorganisms have an impact on microbiota homeostasis and are unimpaired by antibacterial treatment [15,16]. In the end, the nontherapeutic use of antibiotics to create these models is a relevant point, as there is growing attention toward antibiotic resistance of bacteria and spillage of antibiotics into the environment, which one should not ignore when working with such models [17].

THE COMMENSAL IMPACT ON THE MATURATION AND VIGILANCE OF IMMUNITY

Recent findings in cardiovascular research and new insights into the interaction of the microbiota with the host draw attention to the interplay between microbiota, the immune system, and cardiovascular diseases (CVD). Pattern-recognition receptor (PRR) signaling senses gut microbiota-derived signatures not only in the intestine [18] but also in remote tissues [19]. It has become increasingly clear that cardiovascular disease development is closely linked to changes in host immunity, and therefore it is essential to delineate how the microbiota shapes host immune functions in order to understand its role in cardiovascular disease.

Adaption of Lymphoid Tissues

The immune status of germ-free mice shows the importance of the commensal microbiota for the development of the mammalian immune system. Regarding the gut, impaired formation of lymphoid tissue and the lack of isolated lymphoid follicles were first essential findings in germ-free mice [20,21]. Peyer's patches are smaller, and the number of immunoglobulin A (IgA)-producing plasma cells is reduced [22]. This is conclusive, as Peyer's patches are the predominant place where interstitial dendritic cells, which catch penetrating bacteria, interact with T and B cells, to induce T-cell-dependant IgA class switch recombination [23].

IgA$^+$ B cells recirculate to the lamina propria and secrete IgA, which opsonizes bacteria to prevent translocation through the epithelium. It should be noted that bacteria-laden dendritic cells do not penetrate to secondary lymphatic organs but induce an immune mechanism that by anatomical exception is limited to the mucosal tissue [23].

Keeping Commensals at Bay

To maintain a beneficial and homeostatic relationship, the mammalian immune system had to take control over the microbial interactions with the host tissues, as well as the microbial composition. Therefore, microbial interactions are controlled by stratification and compartmentalization [24]. Goblet cells play an important role in minimizing bacteria–epithelial contact (stratification), as they produce mucin glycoproteins that form a thick mucus layer [25,26], mostly resistant to penetration. In addition, luminal bacteria are bound by IgA from plasma cells to prevent translocation through the epithelial lining of the gut [27].

Compartmentalization is achieved by lamina propria macrophages that typically phagocyte the residual penetrating bacteria, or alternatively, these bacteria can be taken up by dendritic cells [23,28]. This protects the host from bacterial exposure to the systemic part of the immune system (compartmentalization) [24]. Alterations in microbiota composition are mainly caused by alpha-defensins. These small antibacterial proteins are secreted by Paneth cells at the crypt base and reach the lumen where they bind to microbes. Defensin secretion kills certain bacteria subspecies and thus shapes microbiota composition [29].

Alterations of Immune Cell Function

In addition to lymphatic organs, the microbiota shapes the development of immune cells and their functions. By the presence of gut microbiota, myelopoiesis and neutrophil aging are fueled through PRR signaling. Antigen presenting cells of germ-free mice exhibit a compromised chemotaxis, phagocytosis, and antimicrobial activity [19,30–32].

The maturation of the adaptive immune system is also influenced by the microbiota. The reduced number of B cells in germ-free mice arises from an affected spleen function. The spleen of germ-free mice contains lower numbers of germinal centers, where B cells undergo differentiation and affinity maturation [21,33]. This results in a critically low IgA level and shows the importance of the microbiota for immune development.

Newer findings demonstrate that defined commensal species influence the distribution of different T-cell subsets. While some bacteria induce proinflammatory cell types, like *Bacillus fragilis*, others are capable to enhance antiinflammatory immune cell populations, like *Clostridia* strains. Furthermore, the overall T cell proliferation is balanced by the microbiota and their effector molecules [24]. Recent findings suggest an important role of intestinal T helper 17 (Th17) cells, induced by commensal microbiota, in systemic autoimmune diseases, like arthritis and experimental autoimmune encephalomyelitis.

In germ-free mouse models of these diseases, both were inducible by monocolonization with *segmented filamentous bacteria* [34,35], typical of rodents, and with immune modulating functions. There is increasing evidence that microbiota-dependent effects on Th17 cell function are linked to vascular dysfunction and hypertension.

Innate lymphoid cells (ILC) are the most recently described part of immunity, interacting with the microbiome. These immune cells lack rearranged antigen-specific receptors and are believed to represent an innate counterpart for T helper and cytotoxic T cells [36]. Although not much is known about ILC1, it was found that germ-free mice show an impaired natural killer (NK) cell function, with reduced cytotoxicity and interferon gamma (IFN γ) production [37].

ILC2 indirectly influence microbiota, by controlling T- and B-cell responses. Effects of the microbiome on ILC2 could be mediated via Toll-like receptors (TLR) such as TLR1, TLR4, or TLR6, expressed on ILC2, but the exact mechanism remains to be elucidated [36].

Just like ILC2, ILC3 does not affect the microbiota directly. Instead, ILC3 indirectly shapes microbiota composition, by guiding adaptive immune response and controlling epithelial barrier integrity and epithelial colonization of skin, oropharynx and nasopharynx, mainly via IL-22 [38]. Depending on microbiota, lamina propria ILC3 clusters form so-called cryptopatches, which evolve into isolated lymphoid follicles [39]. This demonstrates the role of ILC3 in proper lymphoid organ development. These isolated lymphoid follicles contain B cells and contribute to the T-cell-independent IgA production [40].

GNOTOBIOTIC COLONIZATION AND METAORGANISMAL PATHWAYS

As it is well known that some intestinal bacteria can modulate the immune status of the host [41,42], defined associated gnotobiotic mouse models are in particular advantageous to study the interplay between the microbiota, vascular function, and cardiovascular disease development. Gnotobiotic isolator conditions with minimal microbial consortia bear the advantage to provide causal evidence for the implication of specific microbial community members with the development of certain host phenotypes (Table 2.1).

DEFINED FLORA AND IMMUNE STIMULATION

Defined microbial flora, such as the altered Schaedler flora, often utilized for selected colonization of germ-free mice, are instrumental to investigate the functional relationship between specific colonizers and host responses [43]. Altered Schaedler flora was suggested to represent a normal murine gastrointestinal microbiota that can be stably maintained for generations, under isolator conditions [44–46]. Each microorganism of the altered Schaedler flora can be individually cultured, which allows for the production of specific bacterial antigens, or genetic manipulation of the organism [47]. Furthermore, each microorganism of this flora can be analyzed by organism-specific polymerase chain reactions [48].

Meanwhile, collections of cultivable bacterial strains isolated from the mouse intestine, which comprehensively resemble the composition of the mouse microbiome, were constructed and these strains are commercially available [49].

REGULATORY T CELLS

Colonization experiments with altered Schaedler flora demonstrated that this defined flora is sufficient for the activation and de novo generation of regulatory T cells (Tregs) [50]. In this regard, it has been shown that *Bacteroides fragilis*, and more specifically its immunomodulatory polysaccharide A, directs the development of Foxp3$^+$ regulatory T cells, to increase their suppressive capacity, which in mouse models are protective against acute intestinal inflammation [51–53]. Furthermore, nutrition-derived metabolites of the gut microbiota also influence the generation of Tregs. The short-chain fatty acids butyrate and propionate, which are produced by commensal microbiota during starch fermentation, support extrathymic formation of Tregs [54].

TABLE 2.1 Examples for the Use of Minimal Microbial Flora and Gnotobiotic Mouse Models and Their Effects on Host Phenotype

Flora/Monoassociation	Effects on Host Phenotype	References
Altered Schaedler flora (Lactobacillus acidophilus, Lactobacillus murinus, Bacteroides distasonis, Mucispirillum schaedleri, ASF519, ASF356, ASF492, ASF502)	Activation and de novo generation of regulatory T cells Maturation of lymphoid follicles into large B cell clusters	[39,50]
Bacteroides fragilis	Development of Foxp3$^+$ regulatory T cells	[51]
Segmented filamentous bacterium (Candidatus arthromitus)	Induction of Th17 cells	[57]
Bifidobacterium breve	Induction of the antimicrobial peptide REGIII-γ	[64]
Bacteroides thetaiotaomicron	Promotes small intestinal villus vascularization	[66]

T HELPER CELLS TH1 AND TH17

As Th17 cell development depends on the gut microbiota, the gut microbial induction of the Th17 cell response could be relevant for the development of vascular dysfunction. Interestingly, failure of Treg activation led to Th17 and Th1 cell responses [50]. The differentiation of Th1 cells into proinflammatory IL17-producing T helper cells (Th17) in the lamina propria of the small intestine requires specific commensal microbiota, and the depletion of gut microbiota by antibiotics prevents Th17 development [55]. In rodents, *segmented filamentous bacteria* (*Candidatus arthromitus*) [56] are effective inducers of Th17 cells [57], which are potent mediators of autoimmunity [58].

Furthermore, monoassociation with *segmented filamentous bacterium* has been described to increase the mitotic activity and the ratio of the number of columnar cells to those of goblet cells [59]. The decision of antigen-stimulated T cells to differentiate into Th17 or Treg cells depends on the cytokine-regulated balance of the transcription factors RORγT and Foxp3 [60]. IL6, IL21, and IL23 relieve Foxp3-mediated inhibition of RORγT, supporting differentiation into Th17 cells.

Commensal-induced IL1β production was uncovered as a critical step for Th17 differentiation in the small intestine [61]. Another central hub that controls the homeostasis of type 3 innate lymphoid cells and Th17 cells is the aryl hydrocarbon receptor pathway activity [62]. In humans, *Bifidobacterium adolescentis* has recently been identified as an inducer of Th17 cells, with a mechanism that is distinct from that triggered by *segmented filamentous bacterium* [63].

REGIII-γ PEPTIDE

Monocolonization experiments also revealed that the antimicrobial peptide REGIII-γ is efficiently induced in the ileum and the colon by *Bifidobacterium breve*, but not by the *Escherichia coli* strain JM83 [64]. While germ-free mice had low REGIII-γ levels, altered Schaedler flora did not induce the expression of REGIII-γ as effectively as a complex SPF community.

INTESTINAL ANGIOGENESIS

In addition to immune cell homeostasis, colonizing gut microbes support the development of intricate capillary networks in small intestinal villus structures [65]. In this context, monocolonization of 6-week-old germ-free NMRI mice with the gram-negative symbiont *Bacteroides thetaiotaomicron*, over a time period of 10 days, has demonstrated a twofold increase in the density of villus capillaries [66], suggesting that gnotobiotic monocolonization experiments are a powerful tool to study mucosal angiogenesis and vascular remodeling processes.

IMPACT OF THE COMMENSAL MICROBIOTA ON VASCULAR FUNCTION

Vascular Remodeling

The vasculature of the small intestine is pivotal for nutrient absorption and intestinal immune function. Specialized lymphatic vessels, the lacteals, form during late embryogenesis and in the early postnatal period [67]. They are surrounded by capillary networks and are continuously regenerating [68]. Adult germ-free mice have an arrested capillary network formation. Experiments with germ-free mice have shown that colonization with a gut microbiota evokes mucosal angiogenesis and adaptive vascular remodeling of intricate capillary networks in the villus structures of the distal small intestine [66,69].

It has been suggested that microbial regulation of small intestinal angiogenesis depends on Paneth cells [66]. In addition, protease-activated receptor-1 (PAR1), which is primarily activated by coagulation factor signaling, was identified as a determinant of mucosal vascularization in the small intestine [69]. So far, the linkage of microbiota-derived patterns to intestinal vascular remodeling remains elusive.

Brain Capillary Network

Supporting a role for microbiota-dependent remote signaling in vascular development of the brain microvasculature, it has been revealed with germ-free mouse models that the formation of cerebral cavernous malformations, a cause of stroke and seizure, critically depends on microbiota-triggered TLR4 signaling in the endothelium [70]. Interestingly, the susceptibility to develop cerebral cavernous malformations was associated with specific gram-negative bacteria in the gut microbiome.

Vascular Dysfunction and Hypertension

A dysbiotic gut microbiota with decreased microbial richness, diversity, evenness, and an increased Firmicutes/Bacteroidetes ratio, was found in hypertensive Wistar—Kyoto rats and in spontaneously hypertensive rats [71]. Furthermore, antibiotic treatment reversed Western-diet—induced arterial stiffness and endothelial dysfunction in C57BL/6J mice [72]. In line, a reduced microbial richness and diversity was also found in hypertensive patients [71,73].

Functional studies in germ-free mice treated with angiotensin-II (AngII) gave some indication on the influence of microbes on vascular function [74]. In contrast to SPF mice, when germ-free mice were infused with AngII, blood pressure did not increase, and the formation of ROS in the vessel wall and whole blood was not augmented. Likewise, Ang-II treated germ-free mice, compared with Ang-II treated conventionally raised SPF mice, are protected from Ang-II induced vascular dysfunction, as well as vascular inflammation.

As mentioned above, the development of Th17 cells depends on the gut microbiota [75], and functional studies implied this inflammatory cell type in vascular dysfunction. Importantly, vascular RORγT transcript levels, the key transcription factor for IL-17 synthesis, were also decreased in Ang-II infused germ-free mice, compared with their conventionally raised counterparts [74]. In line with various other autoimmune diseases, a dysregulation of the microbiota-influenced balance between Tregs and Th17 cells was associated with hypertension [76].

Latest findings demonstrate the influence of high-salt diet on the gut microbiome of conventionally raised mice. The higher salt concentration in chow primarily depleted *Lactobacillus* spp. According to that, treatment with *Lactobacillus murinus* protected the mice from salt-induced hypertension. This was mainly due to reduced numbers of Th17 cells, which increase under high-salt conditions and contribute to the pathogenesis of hypertension [77].

THE EMERGING LINK BETWEEN MICROBIOTA AND CARDIOVASCULAR DISEASE

It is well established that germ-free mice are protected from diet-induced obesity [78,79]. High-fat diet does not only affect the composition of the gut microbiota [80,81] but also critically determines intestinal barrier function [82—85] and gut microbial production of bioactive metabolites, influencing cardiovascular physiology [86]. High-fat diet intake results in increased plasma levels of gut microbial products [83,87]. In humans and in mouse models, the gut microbiota was proposed as a source of peptidoglycan [30] and lipopolysaccharides (LPS) [85], leaking into the portal circulation.

In case of liver injury, the central firewall function, limiting the systemic spreading of commensal microbes, is lost [88]. For instance, in patients with liver cirrhosis, serum LPS correlated significantly with coagulation factor VIII and von Willebrand factor (VWF) levels [89]. However, the hepatic endothelium may also sense microbial patterns, taken up via the portal circulation under stead-state conditions, thus increasing plasmatic VWF levels [90]. Hence, microbiota-derived remote signaling effects on the vascular endothelium may promote the development of cardiovascular disease and arterial thrombosis.

Influence of Innate Immune Signaling on Atherosclerosis

In a rabbit model, intravenously injected LPS have been demonstrated to accelerate atherogenesis [91]. In line, atherogenesis on a low-fat diet was dependent on TLR4 signaling in myeloid cells, as bone marrow transplantation of *Tlr4*-deficient bone marrow into atherosclerosis-prone low-density lipoprotein receptor (*Ldlr*)-deficient mice resulted in reduced atherosclerotic lesion size [92]. In support of the involvement of innate immune receptor signaling in early atherosclerosis, mice on a Western diet, which were deficient in both apolipoprotein E (*Apoe*) and the Toll-like receptor adapter protein MyD88, showed reduced atherosclerosis and decreased macrophage recruitment to the arterial wall [93].

Antiatherogenic Bacteria

Apoe-deficient mice on a Western diet, which had a probiotic intervention with the gut bacterium *Akkermansia muciniphila*, showed a reduced atherosclerotic lesion size in the aortic root and reduced macrophage infiltration into aortic atherosclerotic lesions [94]. In this study, immunofluorescence analysis of aortic sections revealed increased staining for MCP-1 and ICAM-1 in the lesion endothelium, but the serum levels of total cholesterol, total triglyceride, low-density lipoprotein (LDL), and high-density lipoprotein (HDL) were unchanged. Furthermore, it was demonstrated that intervention *A. muciniphila* decreased intestinal permeability and reduced serum LPS levels. In contrast, LPS infusion reversed the protective effects of *A. muciniphila*.

TLR2 Signaling, Myeloid Cells, and Atherosclerosis

In addition to TLR4, which senses LPS, a complete deficiency of TLR2, which mainly responds to lipoteichoic acid and bacterial lipopeptides in atherosclerosis-prone *Ldlr*-deficient mice, resulted in reduced atherosclerosis [95]. In contrast to the transfer of *Tlr4*-deficient bone marrow into *Ldlr*-deficient mice [92], the transplantation of *Tlr2*-deficient bone marrow into *Ldlr*-deficient recipient mice had no influence on atherosclerosis development [95]. Intraperitoneal administration of the synthetic TLR2/TLR1 agonist Pam3CSK4 increased atherosclerosis in the Ldlr-deficient mouse model, an effect that was not found in *Ldlr*-deficient mice with global *Tlr2*-deficiency or *Tlr2*-deficiency in the myeloid cells, indicating that TLR2 signaling in the vascular endothelium could be a decisive factor for increased atherogenesis.

Influence of the Microbiota Versus Germ-free or Antibiotic Treatment

Atherosclerosis-prone germ-free *Apoe*-deficient mice on a normal chow diet at 20 weeks of age were reported to have increased total plasma cholesterol levels and decreased triglycerides compared with conventionally raised *Apoe*-deficient mice [96]. Nevertheless, in this study, plaque size and plaque lipid area, as well as intraplaque macrophages, were found decreased in the germ-free *Apoe*-deficient mouse model, compared with conventionally raised *Apoe*-deficient controls.

In line, treatment of Western diet–fed *Apoe*-deficient mice with ampicillin reduced aortic plaque burden but did not affect the plaque lipid area [97]. In contrast to *Apoe*-deficient mice fed with chow diet, ampicillin treatment of Western diet–fed *Apoe*-deficient mice reduced total plasma cholesterol, total LDL, and total very low–density lipoprotein (VLDL) levels. In contrast, a previous study has reported that microbiota protected *Apoe*-deficient mice fed an irradiated low-cholesterol diet against atherosclerotic lesion development [98].

Microbiota-Derived Metabolite Trimethylamine-N-Oxide

In addition to the microbiota-dependent activation of innate immune pathways, the microbiota-dependent metaorganismal trimethylamine-N-oxide (TMAO) pathway has been associated with atherogenesis [99,100]. Bacteria of the gut microbiota express choline TMA-lyases (cutC) and carnitine oxygenases (cntA), enzymes that convert choline, phosphatidylcholine, and carnitine into trimethylamine (TMA), which is further metabolized to TMAO by flavin-containing monooxygenases (i.e., FMO3) in the liver [86]. In human feces, cutC are predominantly encoded by *Clostridium IXa* strains and *Eubacterium* sp. strain AB3007, whereas cntA was associated with γ-proteobacteria references [101].

Hyperlipidemic *Apoe*-deficient mice kept for more than 3 weeks on a choline-supplemented diet (1% w/w), showed increased atherosclerotic plaque size and macrophage foam cells, compared with *Apoe*-deficient mice kept on a control diet (0.08% w/w choline) [86]. The decimation of the intestinal microbiota by antibiotic treatment reduced macrophage foam cell formation. Furthermore, TMA formation by bacterial TMA-lyases could be inhibited by 3,3-dimethyl-1-butanol treatment, reducing the choline-diet promoted development of atherosclerotic lesions and foam cells in the *Apoe*-deficient mouse model [102].

Trimethylamine-N-Oxide and Human Disease

In patients, L-carnitine was associated with cardiovascular disease risk [99], and a number of clinical studies found an association of plasma TMAO levels with atherosclerotic burden and mortality risk in coronary artery disease [103,104]. However, the association-based evidence on the involvement of TMAO in cardiovascular disease remains controversial, as other studies did not find this association between plasma TMAO levels and coronary heart disease or carotid athero-sclerosis [105–107]. Of note, also contradicting studies with *Apoe*-deficient mice on L-carnitine-rich diet and elevated plasma TMAO levels, reporting reduced atherosclerotic lesion size in the aortic root and the thoracic aorta, exist [108].

Implications of the Microbiota in Thrombosis

Recent work on the metaorganismal TMAO pathway has uncovered a correlation of microbiota-derived plasma TMAO levels and thrombotic events, i.e., myocardial infarction and stroke [109]. This study also identified that TMAO augments platelet aggregation, independent of whether thrombin or ADP was used as a stimulus. In additional analyses, TMAO was shown to increase platelet adhesion to collagen coatings under flow conditions, and by intravital imaging of ferric chloride–induced thrombus formation at the carotid artery, it was demonstrated that mice intraperitoneally injected with TMAO had reduced times to cessation of blood flow.

By feeding experiments with choline-rich diet and broad-spectrum antibiotic treatment, the functional role of micro-biota in enhancing maximal platelet aggregation triggered by ADP and in reducing the time to cessation of blood flow in the carotid ferric chloride injury model was demonstrated. The microbiota-dependent effect on arterial thrombus formation was further corroborated by comparing germ-free, conventionally raised and conventionalized mice on a choline-rich diet, where germ-free mice had prolonged times to cessation of blood flow.

Interestingly, these authors found a negative correlation of plasma TMAO levels and the time to cessation of blood flow in the ferric chloride injury model of the mouse carotid artery. In this study, the choline-diet characteristic taxon *Allo-baculum* was positively associated with occlusion time, whereas the taxon *Candidatus arthromitus* showed a negative correlation, for choline-rich diet and occlusion time, in the ferric chloride thrombosis model. In addition, based on cecal microbiota transplantation, from the C57BL/6 high-TMA producing mouse strain into the NZW/LacJ low-TMA producing strain, this study proposed that the reduced occlusion times in the ferric chloride thrombosis model is a transferrable trait. Importantly, these findings hold promise of the development of selective drugs to target the biosynthesis pathways of prothrombotic community members in the human gut microbiota.

Endothelial Platelet Deposition

As microbiota-derived patterns were identified to leak into the portal circulation [83,87], hepatic and also extrahepatic endothelial function, they can be influenced by the colonization status of the host [90]. On a chow diet, germ-free C57BL/6 mice showed reduced platelet deposition on the ligation-injured carotid artery, compared with their conventionally raised counterparts. Incubation of germ-free platelets in the plasma of conventionally raised mice rescued defective platelet deposition to laminin coatings in a static platelet deposition model dependent of plasmatic VWF levels.

In the carotid artery ligation model, platelet deposition on the vascular injury site was dependent on *Tlr2* but was not influenced by *Tlr2* deficiency of platelets. The defect in arterial thrombus growth was due to impaired hepatic VWF synthesis and platelet VWF-integrin ligation in GF and *Tlr2*-deficient mice.

Vascular Events in Inflammatory Bowel Disease

Thrombosis is a life-threatening complication of inflammatory bowel diseases (IBD), and the risk for thromboembolism is threefold to sixfold increased compared with the general population [110,111]. Hematological changes and enhanced thrombus formation were also found, in mouse models of acute intestinal inflammation, where tissue factor was identified as a key mediator, of leukocyte and platelet recruitment to postcapillary venules of the inflamed colon [112]. Probiotic preparations, such as VSL#3, including different strains of *Bifidobacterium*, *Lactobacillus*, and *Streptococcus*, were demonstrated to reduce colitis symptoms in mouse and rat models [113–115].

A recent study comparing male germ-free Swiss Webster mice, with conventionally raised controls, demonstrated that the germ-free group showed an accelerated thrombosis response in cremaster arterioles after 6 days of 3% dextran sodium sulfate (DSS) treatment in a model of acute intestinal inflammation [115]. The administration of VSL#3 to germ-free mice showed protective effects, and the pretreatment of conventionally raised mice by VSL#3 reversed the reduced life span of

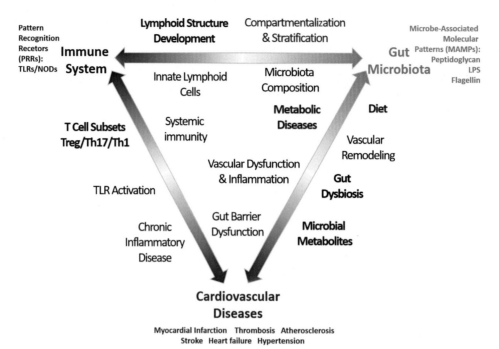

FIGURE 2.1 Illustration of the various ways of interaction between the microbiota, the immune system, and the development of cardiovascular diseases. While the immune system controls microbiota by compartmentalization, stratification, and composition, the microbiota shapes the development of lymphoid structures, innate lymphoid cells, systemic immunity, and T-cell subsets. Toll-like receptor (TLR) activation and chronic inflammatory diseases link the immune system to the development of cardiovascular diseases (CVDs). Diet, gut dysbiosis, metabolic diseases, and microbial metabolites represent the connection between microbiota and CVDs. In addition, vascular remodeling, vascular dysfunction, and inflammation, as well as gut barrier dysfunction, are CVD-predisposing microbiota-dependent factors. *LPS*, lipopolysaccharides.

platelets, the appearance of more immature platelets, and accelerated thrombin-induced platelet aggregation. The TLR adapter protein MyD88 was demonstrated to be involved in the protective effect of VSL#3. Thus, in IBD, improved control of the enteric microbiota may prevent hypercoagulability and thrombosis.

PERSPECTIVE

The wealth of taxonomic microbiome analyses are an invaluable source for hypothesis generation, but since they are association based, it should be kept in mind that they are of limited value to draw conclusions on microbiota-influenced metabolic and immune pathways, which may contribute to disease development. In contrast, germ-free animal models and colonization experiments with selected minimal microbial consortia under isolator conditions, combined with standardized models of vascular function and cardiovascular disease, have the capacity to prove causality.

Future gnotobiotic animal studies on cardiovascular disease should consider the influence of host nutrition and give information on proteomics and metabolomics profiles. To generate a translational workflow, it is of advantage to combine gnotobiotic mouse studies with the isolation and culture of specific bacteria from patient stool. Considering the achieved knowledge on microbiota—host interactions, the selective targeting of metaorganismal pathways that will not interfere with host metabolism is an attractive option for future interventions to prevent cardiometabolic diseases (Fig. 2.1).

LIST OF ACRONYMS AND ABBREVIATIONS

ADP Adenosine triphosphate
AngII Angiotensin II
ApoE Apolipoprotein E protein
Apoe Apolipoprotein E gene
cntA Carnitine oxygenase
cutC choline TMA-lyase
CVD Cardiovascular disease
DSS Dextran sodium sulfate

FMO3 Flavin-containing monooxygenase 3
Foxp3 Forkhead box P3
GF Germ-free
HDL High-density lipoprotein
IBD Inflammatory bowel diseases
ICAM Intracellular adhesion molecule
IFN γ Interferon gamma
IgA Immunoglobulin A
IL Interleukin
ILC Innate lymphoid cells
LDL Low-density lipoprotein
Ldlr Low-density lipoprotein receptor gene
LPS Lipopolysaccharide
MCP Methyl-accepting chemotaxis protein
MyD88 Myeloid differentiation primary response 88
NK Natural killer
PAR Protease-activated receptor
PRR Pattern-recognition receptor
REGIII-γ Regenerating islet-derived protein 3 gamma
RORγT RAR-related orphan receptor gamma T
ROS Reactive oxygen species
SPF Specific pathogen free
Th T helper cell
TLR Toll-like receptor protein
Tlr Toll-like receptor gene
TMA Trimethylamine
TMAO Trimethylamine-N-oxide
Treg Regulatory T cell
VLDL Very low−density lipoprotein
VWF von Willebrand factor

REFERENCES

[1] Kirk RG. Life in a germ-free world": isolating life from the laboratory animal to the bubble boy. Bull Hist Med 2012;86(2):237−75.

[2] Nuttall GHF TH. Thierisches Leben ohne Bakterien im Verdauungskanal. Hoppe Seylers Z Physiol Chem 1897;22(II. Mitteilung):62−73.

[3] Wostmann BS. The germfree animal in nutritional studies. Annu Rev Nutr 1981;1:257−79.

[4] Reyniers JA, Trexler PC, Ervin RF. Rearing germ-free albino rats. Lobund Rep 1946;(1):1−84.

[5] Gustafsson BE. Lightweight stainless steel systems for rearing germfree animals. Ann NY Acad Sci 1959;78:17−28.

[6] Katami K, Taguchi F, Nakayama M, Goto N, Fujiwara K. Vertical transmission of mouse hepatitis virus infection in mice. Jpn J Exp Med 1978;48(6):481−90.

[7] Schwanzer V, Deerberg F, Frost J, Liess B, Schwanzerova I, Pittermann W. Intrauterine infection of mice with ectromelia virus. Z Versuchstierkd 1975;17(2):110−20.

[8] McCance DJ, Mims CA. Transplacental transmission of polyoma virus in mice. Infect Immun 1977;18(1):196−202.

[9] Mims CA. Effect on the fetus of maternal infection with lymphocytic choriomeningitis (LCM) virus. J Infect Dis 1969;120(5):582−97.

[10] Hill AC, Stalley GP. Mycoplasma pulmonis infection with regard to embryo freezing and hysterectomy derivation. Lab Anim Sci 1991;41(6):563−6.

[11] Inzunza J, Midtvedt T, Fartoo M, et al. Germfree status of mice obtained by embryo transfer in an isolator environment. Lab Anim 2005;39(4):421−7.

[12] Naik S, Bouladoux N, Wilhelm C, et al. Compartmentalized control of skin immunity by resident commensals. Science 2012;337(6098):1115−9.

[13] Lundberg R, Toft MF, August B, Hansen AK, Hansen CH. Antibiotic-treated versus germ-free rodents for microbiota transplantation studies. Gut Microbes 2016;7(1):68−74.

[14] Hansen AK. Antibiotic treatment of nude rats and its impact on the aerobic bacterial flora. Lab Anim 1995;29(1):37−44.

[15] Virgin HW. The virome in mammalian physiology and disease. Cell 2014;157(1):142−50.

[16] Underhill DM, Iliev ID. The mycobiota: interactions between commensal fungi and the host immune system. Nat Rev Immunol 2014;14(6):405−16.

[17] Modi SR, Lee HH, Spina CS, Collins JJ. Antibiotic treatment expands the resistance reservoir and ecological network of the phage metagenome. Nature 2013;499(7457):219−22.

[18] Hormann N, Brandao I, Jackel S, et al. Gut microbial colonization orchestrates TLR2 expression, signaling and epithelial proliferation in the small intestinal mucosa. PLoS One 2014;9(11):e113080.

[19] Balmer ML, Schurch CM, Saito Y, et al. Microbiota-derived compounds drive steady-state granulopoiesis via MyD88/TICAM signaling. J Immunol 2014;193(10):5273–83.

[20] Abrams GD, Bauer H, Sprinz H. Influence of the normal flora on mucosal morphology and cellular renewal in the ileum. A comparison of germ-free and conventional mice. Lab Investig 1963;12:355–64.

[21] Bauer H, Horowitz RE, Levenson SM, Popper H. The response of the lymphatic tissue to the microbial flora. Studies on germfree mice. Am J Pathol 1963;42:471–83.

[22] Crabbe PA, Bazin H, Eyssen H, Heremans JF. The normal microbial flora as a major stimulus for proliferation of plasma cells synthesizing IgA in the gut. The germ-free intestinal tract. Int Arch Allergy Appl Immunol 1968;34(4):362–75.

[23] Macpherson AJ, Uhr T. Induction of protective IgA by intestinal dendritic cells carrying commensal bacteria. Science 2004;303(5664):1662–5.

[24] Hooper LV, Littman DR, Macpherson AJ. Interactions between the microbiota and the immune system. Science 2012;336(6086):1268–73.

[25] Johansson ME, Phillipson M, Petersson J, Velcich A, Holm L, Hansson GC. The inner of the two Muc2 mucin-dependent mucus layers in colon is devoid of bacteria. Proc Natl Acad Sci USA 2008;105(39):15064–9.

[26] Johansson ME, Larsson JM, Hansson GC. The two mucus layers of colon are organized by the MUC2 mucin, whereas the outer layer is a legislator of host-microbial interactions. Proc Natl Acad Sci USA 2011;108(Suppl. 1):4659–65.

[27] Macpherson AJ, Gatto D, Sainsbury E, Harriman GR, Hengartner H, Zinkernagel RM. A primitive T cell-independent mechanism of intestinal mucosal IgA responses to commensal bacteria. Science 2000;288(5474):2222–6.

[28] Kelsall B. Recent progress in understanding the phenotype and function of intestinal dendritic cells and macrophages. Mucosal Immunol 2008;1(6):460–9.

[29] Nakamura K, Sakuragi N, Takakuwa A, Ayabe T. Paneth cell alpha-defensins and enteric microbiota in health and disease. Biosci Microbiota Food Health 2016;35(2):57–67.

[30] Clarke TB, Davis KM, Lysenko ES, Zhou AY, Yu Y, Weiser JN. Recognition of peptidoglycan from the microbiota by Nod1 enhances systemic innate immunity. Nat Med 2010;16(2):228–31.

[31] Khosravi A, Yanez A, Price JG, et al. Gut microbiota promote hematopoiesis to control bacterial infection. Cell Host Microbe 2014;15(3):374–81.

[32] Zhang D, Chen G, Manwani D, et al. Neutrophil ageing is regulated by the microbiome. Nature 2015;525(7570):528–32.

[33] Mazmanian SK, Liu CH, Tzianabos AO, Kasper DL. An immunomodulatory molecule of symbiotic bacteria directs maturation of the host immune system. Cell 2005;122(1):107–18.

[34] Lee YK, Menezes JS, Umesaki Y, Mazmanian SK. Proinflammatory T-cell responses to gut microbiota promote experimental autoimmune encephalomyelitis. Proc Natl Acad Sci USA 2011;108(Suppl. 1):4615–22.

[35] Wu HJ, Ivanov II, Darce J, et al. Gut-residing segmented filamentous bacteria drive autoimmune arthritis via T helper 17 cells. Immunity 2010;32(6):815–27.

[36] Britanova L, Diefenbach A. Interplay of innate lymphoid cells and the microbiota. Immunol Rev 2017;279(1):36–51.

[37] Ganal SC, Sanos SL, Kallfass C, et al. Priming of natural killer cells by nonmucosal mononuclear phagocytes requires instructive signals from commensal microbiota. Immunity 2012;37(1):171–86.

[38] Wolk K, Kunz S, Witte E, Friedrich M, Asadullah K, Sabat R. IL-22 increases the innate immunity of tissues. Immunity 2004;21(2):241–54.

[39] Bouskra D, Brezillon C, Berard M, et al. Lymphoid tissue genesis induced by commensals through NOD1 regulates intestinal homeostasis. Nature 2008;456(7221):507–10.

[40] Kruglov AA, Grivennikov SI, Kuprash DV, et al. Nonredundant function of soluble LTalpha3 produced by innate lymphoid cells in intestinal homeostasis. Science 2013;342(6163):1243–6.

[41] Tabas I, Lichtman AH. Monocyte-Macrophages and T Cells in atherosclerosis. Immunity 2017;47(4):621–34.

[42] Wells CL, Balish E. Modulation of the rats' immune status by monoassociation with anaerobic bacteria. Can J Microbiol 1980;26(10):1192–8.

[43] Wymore Brand M, Wannemuehler MJ, Phillips GJ, et al. The altered schaedler flora: continued applications of a defined murine microbial community. ILAR J 2015;56(2):169–78.

[44] Dewhirst FE, Chien CC, Paster BJ, et al. Phylogeny of the defined murine microbiota: altered Schaedler flora. Appl Environ Microbiol 1999;65(8):3287–92.

[45] Sarma-Rupavtarm RB, Ge Z, Schauer DB, Fox JG, Polz MF. Spatial distribution and stability of the eight microbial species of the altered schaedler flora in the mouse gastrointestinal tract. Appl Environ Microbiol 2004;70(5):2791–800.

[46] Stehr M, Greweling MC, Tischer S, et al. Charles River altered Schaedler flora (CRASF) remained stable for four years in a mouse colony housed in individually ventilated cages. Lab Anim 2009;43(4):362–70.

[47] Jergens AE, Wilson-Welder JH, Dorn A, et al. Helicobacter bilis triggers persistent immune reactivity to antigens derived from the commensal bacteria in gnotobiotic C3H/HeN mice. Gut 2007;56(7):934–40.

[48] Deloris Alexander A, Orcutt RP, Henry JC, Baker Jr J, Bissahoyo AC, Threadgill DW. Quantitative PCR assays for mouse enteric flora reveal strain-dependent differences in composition that are influenced by the microenvironment. Mamm Genome 2006;17(11):1093–104.

[49] Lagkouvardos I, Pukall R, Abt B, et al. The Mouse Intestinal Bacterial Collection (miBC) provides host-specific insight into cultured diversity and functional potential of the gut microbiota. Nat Microbiol 2016;1(10):16131.

[50] Geuking MB, Cahenzli J, Lawson MA, et al. Intestinal bacterial colonization induces mutualistic regulatory T cell responses. Immunity 2011;34(5):794−806.

[51] Round JL, Mazmanian SK. Inducible Foxp3+ regulatory T-cell development by a commensal bacterium of the intestinal microbiota. Proc Natl Acad Sci USA 2010;107(27):12204−9.

[52] Chiu CC, Ching YH, Wang YC, et al. Monocolonization of germ-free mice with Bacteroides fragilis protects against dextran sulfate sodium-induced acute colitis. BioMed Res Int 2014;2014:675786.

[53] Chang YC, Ching YH, Chiu CC, et al. TLR2 and interleukin-10 are involved in Bacteroides fragilis-mediated prevention of DSS-induced colitis in gnotobiotic mice. PLoS One 2017;12(7):e0180025.

[54] Arpaia N, Campbell C, Fan X, et al. Metabolites produced by commensal bacteria promote peripheral regulatory T-cell generation. Nature 2013;504(7480):451−5.

[55] Ivanov II, Frutos Rde L, Manel N, et al. Specific microbiota direct the differentiation of IL-17-producing T-helper cells in the mucosa of the small intestine. Cell Host Microbe 2008;4(4):337−49.

[56] Klaasen HL, Koopman JP, Van den Brink ME, Bakker MH, Poelma FG, Beynen AC. Intestinal, segmented, filamentous bacteria in a wide range of vertebrate species. Lab Anim 1993;27(2):141−50.

[57] Goto Y, Panea C, Nakato G, et al. Segmented filamentous bacteria antigens presented by intestinal dendritic cells drive mucosal Th17 cell differentiation. Immunity 2014;40(4):594−607.

[58] Bettelli E, Oukka M, Kuchroo VK. T(H)-17 cells in the circle of immunity and autoimmunity. Nat Immunol 2007;8(4):345−50.

[59] Umesaki Y, Okada Y, Matsumoto S, Imaoka A, Setoyama H. Segmented filamentous bacteria are indigenous intestinal bacteria that activate intraepithelial lymphocytes and induce MHC class II molecules and fucosyl asialo GM1 glycolipids on the small intestinal epithelial cells in the ex-germ-free mouse. Microbiol Immunol 1995;39(8):555−62.

[60] Zhou L, Lopes JE, Chong MM, et al. TGF-beta-induced Foxp3 inhibits T(H)17 cell differentiation by antagonizing RORgammat function. Nature 2008;453(7192):236−40.

[61] Shaw MH, Kamada N, Kim YG, Nunez G. Microbiota-induced IL-1beta, but not IL-6, is critical for the development of steady-state TH17 cells in the intestine. J Exp Med 2012;209(2):251−8.

[62] Schiering C, Wincent E, Metidji A, et al. Feedback control of AHR signalling regulates intestinal immunity. Nature 2017;542(7640):242−5.

[63] Tan TG, Sefik E, Geva-Zatorsky N, et al. Identifying species of symbiont bacteria from the human gut that, alone, can induce intestinal Th17 cells in mice. Proc Natl Acad Sci USA 2016;113(50):E8141−50.

[64] Natividad JM, Hayes CL, Motta JP, et al. Differential induction of antimicrobial REGIII by the intestinal microbiota and Bifidobacterium breve NCC2950. Appl Environ Microbiol 2013;79(24):7745−54.

[65] Khandagale A, Reinhardt C. Gut microbiota - architects of small intestinal capillaries. Front Biosci (Landmark Ed) 2018;23:752−66.

[66] Stappenbeck TS, Hooper LV, Gordon JI. Developmental regulation of intestinal angiogenesis by indigenous microbes via Paneth cells. Proc Natl Acad Sci USA 2002;99(24):15451−5.

[67] Kim KE, Sung HK, Koh GY. Lymphatic development in mouse small intestine. Dev Dyn 2007;236(7):2020−5.

[68] Bernier-Latmani J, Cisarovsky C, Demir CS, et al. DLL4 promotes continuous adult intestinal lacteal regeneration and dietary fat transport. J Clin Investig 2015;125(12):4572−86.

[69] Reinhardt C, Bergentall M, Greiner TU, et al. Tissue factor and PAR1 promote microbiota-induced intestinal vascular remodelling. Nature 2012;483(7391):627−31.

[70] Tang AT, Choi JP, Kotzin JJ, et al. Endothelial TLR4 and the microbiome drive cerebral cavernous malformations. Nature 2017;545(7654):305−10.

[71] Yang T, Santisteban MM, Rodriguez V, et al. Gut dysbiosis is linked to hypertension. Hypertension 2015;65(6):1331−40.

[72] Battson ML, Lee DM, Jarrell DK, et al. Suppression of gut dysbiosis reverses western diet-induced vascular dysfunction. Am J Physiol Endocrinol Metab 2018;314(5):E468−77.

[73] Adnan S, Nelson JW, Ajami NJ, et al. Alterations in the gut microbiota can elicit hypertension in rats. Physiol Genom 2017;49(2):96−104.

[74] Karbach SH, Schonfelder T, Brandao I, et al. Gut microbiota promote angiotensin II-induced arterial hypertension and vascular dysfunction. J Am Heart Assoc 2016;5(9).

[75] Ivanov II, Atarashi K, Manel N, et al. Induction of intestinal Th17 cells by segmented filamentous bacteria. Cell 2009;139(3):485−98.

[76] Kim S, Rodriguez V, Santisteban M, et al. 6b.07: hypertensive patients exhibit gut microbial dysbiosis and an increase in Th17 cells. J Hypertens 2015;33(Suppl. 1):e77−8.

[77] Wilck N, Matus MG, Kearney SM, et al. Salt-responsive gut commensal modulates TH17 axis and disease. Nature 2017;551(7682):585−9.

[78] Drouin-Chartier JP, Cote JA, Labonte ME, et al. Comprehensive review of the impact of dairy foods and dairy fat on cardiometabolic risk. Adv Nutr 2016;7(6):1041−51.

[79] Backhed F, Manchester JK, Semenkovich CF, Gordon JI. Mechanisms underlying the resistance to diet-induced obesity in germ-free mice. Proc Natl Acad Sci USA 2007;104(3):979−84.

[80] Kabeerdoss J, Devi RS, Mary RR, Ramakrishna BS. Faecal microbiota composition in vegetarians: comparison with omnivores in a cohort of young women in southern India. Br J Nutr 2012;108(6):953−7.

[81] Amato KR, Yeoman CJ, Cerda G, et al. Variable responses of human and non-human primate gut microbiomes to a Western diet. Microbiome 2015;3:53.

[82] Zhang C, Zhang M, Wang S, et al. Interactions between gut microbiota, host genetics and diet relevant to development of metabolic syndromes in mice. ISME J 2010;4(2):232−41.

[83] Cani PD, Amar J, Iglesias MA, et al. Metabolic endotoxemia initiates obesity and insulin resistance. Diabetes 2007;56(7):1761−72.

[84] Cani PD, Neyrinck AM, Fava F, et al. Selective increases of bifidobacteria in gut microflora improve high-fat-diet-induced diabetes in mice through a mechanism associated with endotoxaemia. Diabetologia 2007;50(11):2374−83.

[85] Amar J, Burcelin R, Ruidavets JB, et al. Energy intake is associated with endotoxemia in apparently healthy men. Am J Clin Nutr 2008;87(5):1219−23.

[86] Wang Z, Klipfell E, Bennett BJ, et al. Gut flora metabolism of phosphatidylcholine promotes cardiovascular disease. Nature 2011;472(7341):57−63.

[87] Hapfelmeier S, Lawson MA, Slack E, et al. Reversible microbial colonization of germ-free mice reveals the dynamics of IgA immune responses. Science 2010;328(5986):1705−9.

[88] Balmer ML, Slack E, de Gottardi A, et al. The liver may act as a firewall mediating mutualism between the host and its gut commensal microbiota. Sci Transl Med 2014;6(237):237ra66.

[89] Carnevale R, Raparelli V, Nocella C, et al. Gut-derived endotoxin stimulates factor VIII secretion from endothelial cells. Implications for hypercoagulability in cirrhosis. J Hepatol 2017;67(5):950−6.

[90] Jäckel S, Kiouptsi K, Lillich M, et al. Gut microbiota regulate hepatic von Willebrand factor synthesis and arterial thrombus formation via Toll-like receptor-2. Blood 2017;130(4):542−53.

[91] Lehr HA, Sagban TA, Ihling C, et al. Immunopathogenesis of atherosclerosis: endotoxin accelerates atherosclerosis in rabbits on hypercholesterolemic diet. Circulation 2001;104(8):914−20.

[92] Coenen KR, Gruen ML, Lee-Young RS, Puglisi MJ, Wasserman DH, Hasty AH. Impact of macrophage toll-like receptor 4 deficiency on macrophage infiltration into adipose tissue and the artery wall in mice. Diabetologia 2009;52(2):318−28.

[93] Bjorkbacka H, Kunjathoor VV, Moore KJ, et al. Reduced atherosclerosis in MyD88-null mice links elevated serum cholesterol levels to activation of innate immunity signaling pathways. Nat Med 2004;10(4):416−21.

[94] Li J, Lin S, Vanhoutte PM, Woo CW, Xu A. Akkermansia muciniphila protects against atherosclerosis by preventing metabolic endotoxemia-induced inflammation in Apoe-/- mice. Circulation 2016;133(24):2434−46.

[95] Mullick AE, Tobias PS, Curtiss LK. Modulation of atherosclerosis in mice by Toll-like receptor 2. J Clin Investig 2005;115(11):3149−56.

[96] Kasahara K, Tanoue T, Yamashita T, et al. Commensal bacteria at the crossroad between cholesterol homeostasis and chronic inflammation in atherosclerosis. J Lipid Res 2017;58(3):519−28.

[97] Rune I, Rolin B, Larsen C, et al. Modulating the gut microbiota improves glucose tolerance, lipoprotein profile and atherosclerotic plaque development in ApoE-deficient mice. PLoS One 2016;11(1):e0146439.

[98] Stepankova R, Tonar Z, Bartova J, et al. Absence of microbiota (germ-free conditions) accelerates the atherosclerosis in ApoE-deficient mice fed standard low cholesterol diet. J Atheroscler Thromb 2010;17(8):796−804.

[99] Koeth RA, Wang Z, Levison BS, et al. Intestinal microbiota metabolism of L-carnitine, a nutrient in red meat, promotes atherosclerosis. Nat Med 2013;19(5):576−85.

[100] Tang WH, Wang Z, Levison BS, et al. Intestinal microbial metabolism of phosphatidylcholine and cardiovascular risk. N Engl J Med 2013;368(17):1575−84.

[101] Rath S, Heidrich B, Pieper DH, Vital M. Uncovering the trimethylamine-producing bacteria of the human gut microbiota. Microbiome 2017;5(1):54.

[102] Wang Z, Roberts AB, Buffa JA, et al. Non-lethal inhibition of gut microbial trimethylamine production for the treatment of atherosclerosis. Cell 2015;163(7):1585−95.

[103] Senthong V, Li XS, Hudec T, et al. Plasma trimethylamine N-Oxide, a gut microbe-generated phosphatidylcholine metabolite, is associated with atherosclerotic burden. J Am Coll Cardiol 2016;67(22):2620−8.

[104] Senthong V, Wang Z, Fan Y, Wu Y, Hazen SL, Tang WH. Trimethylamine N-Oxide and mortality risk in patients with peripheral artery disease. J Am Heart Assoc 2016;5(10).

[105] Mueller DM, Allenspach M, Othman A, et al. Plasma levels of trimethylamine-N-oxide are confounded by impaired kidney function and poor metabolic control. Atherosclerosis 2015;243(2):638−44.

[106] Skagen K, Troseid M, Ueland T, et al. The Carnitine-butyrobetaine-trimethylamine-N-oxide pathway and its association with cardiovascular mortality in patients with carotid atherosclerosis. Atherosclerosis 2016;247:64−9.

[107] Meyer KA, Benton TZ, Bennett BJ, et al. Microbiota-dependent metabolite trimethylamine N-Oxide and coronary artery calcium in the coronary artery risk development in young adults study (CARDIA). J Am Heart Assoc 2016;5(10).

[108] Collins HL, Drazul-Schrader D, Sulpizio AC, et al. L-Carnitine intake and high trimethylamine N-oxide plasma levels correlate with low aortic lesions in ApoE(-/-) transgenic mice expressing CETP. Atherosclerosis 2016;244:29−37.

[109] Zhu W, Gregory JC, Org E, et al. Gut microbial metabolite TMAO enhances platelet hyperreactivity and thrombosis risk. Cell 2016;165(1):111−24.

[110] Koutroumpakis EI, Tsiolakidou G, Koutroubakis IE. Risk of venous thromboembolism in patients with inflammatory bowel disease. Semin Thromb Hemost 2013;39(5):461−8.

[111] Danese S, Semeraro S, Papa A, et al. Extraintestinal manifestations in inflammatory bowel disease. World J Gastroenterol 2005;11(46):7227—36.

[112] Anthoni C, Russell J, Wood KC, et al. Tissue factor: a mediator of inflammatory cell recruitment, tissue injury, and thrombus formation in experimental colitis. J Exp Med 2007;204(7):1595—601.

[113] Uronis JM, Arthur JC, Keku T, et al. Gut microbial diversity is reduced by the probiotic VSL#3 and correlates with decreased TNBS-induced colitis. Inflamm Bowel Dis 2011;17(1):289—97.

[114] Mennigen R, Nolte K, Rijcken E, et al. Probiotic mixture VSL#3 protects the epithelial barrier by maintaining tight junction protein expression and preventing apoptosis in a murine model of colitis. Am J Physiol Gastrointest Liver Physiol 2009;296(5):G1140—9.

[115] Souza DG, Senchenkova EY, Russell J, Granger DN. MyD88 mediates the protective effects of probiotics against the arteriolar thrombosis and leukocyte recruitment associated with experimental colitis. Inflamm Bowel Dis 2015;21(4):888—900.

Chapter 3

The Gut Microbiome Beyond the Bacteriome—The Neglected Role of Virome and Mycobiome in Health and Disease

Charikleia Stefanaki[1,2]

[1]Department of Pediatrics, General Hospital of Nikaia "Agios Panteleimon", Piraeus, Greece; [2]Choremeion Research Laboratory, 1st Department of Pediatrics, Medical School, National and Kapodistrian University of Athens, NKUA, Athens, Greece

GENERAL FEATURES OF THE GUT MICROBIOME

The complex assemblage of genomes of the microorganisms, which populate the human gastrointestinal tract, is called gut microbiome. Studies that aim to define the composition and function of uncultured microbial communities are often referred to as "metagenomic," although this refers more specifically to sequencing-based assays.

First, for bacteria, and fungi, community DNA is extracted from a sample, typically uncultured, containing multiple microbial members. The bacterial taxa present in the community are most frequently defined by amplifying the 16S rRNA gene and sequencing it. Highly similar sequences are grouped into operational taxonomic units (OTUs), which can be compared to 16S databases. The community can be described in terms of which OTUs are present, their relative abundance, and/or their phylogenetic relationships. An alternate method of identifying community taxa is to directly metagenomically sequence community DNA and compare it to reference genomes or gene catalogs [1].

The composition of the gut microbiome is not the same during the several stages of life, or even during the hours of the same day. We are born consisting, only, of our own eukaryotic human cells, but over the first several years of life, our skin surface, oral cavity, and gut are colonized by a tremendous diversity of bacteria, archaea, fungi, and viruses. The gut microbiome of a neonate seems to differ from that of a toddler, of a child, and the list goes on [2]. Some believe in a core microbiome for all humans, as similar taxa, phyla, and genera tend to be found in the microbiota of the gastrointestinal tract of humans, regardless of their conditions under testing [3], even though this concept is still debated.

Both healthy and diseased populations seem to carry the same core microbiome, but the proportions of its components are not the same, either for bacteriome, virome, and mycobiome. What is currently known about the human gut microbiome derives of studies that evaluated mainly adults with a Westernized lifestyle, and their results and conclusions are mainly based on their bacteriome. Even if the bacteriome entails one important pillar of the composition of microbiome, the majority of the microbial genome is that of the virome. The mycobiome seems to represent a very modest part [4,5].

Within the collective gut microbiome, bacteria are the most relevant for fermentation of indigestible components of the diet, providing absorbable metabolites and producing essential vitamins, while removing toxic compounds (xenobiotics). Gut flora, also, outcompetes pathogens, simultaneously stimulating and regulating the response of the immune system. Current medical literature has demonstrated that the gut microbiome is responsible for the integrity of the gut lumen barrier [6].

Microbiome and Metabolome in Diagnosis, Therapy, and other Strategic Applications. https://doi.org/10.1016/B978-0-12-815249-2.00003-8

CORE GUT VIROME AND ITS PHYSIOLOGIC IMPORTANCE

Characterizing the human virome requires a different approach because, unlike cellular life, no gene or genomic region is homologous across all viruses. The current approach for studying these communities is to isolate virus-like particles (VLPs), using size fractionation, followed by sequencing those using metagenomics [7–9]. Alternatively, viruses can be characterized using DNA microarrays. All these methods have rendered the characterization of the gut viruses feasible, but still the studies that concern the human gut virome are not numerous.

The human gut virome contains various commensal and pathogenic viruses, evoking several immune responses from the host. Bakhshinejad and Ghiasvand also coined the term "gut phageome," as the majority of gut viruses belong to the group of phages [10]. Constant viral modulation of immunity is not only associated with several low-grade, inflammatory diseases, but also confers unexpected benefits to the host. These outcomes of viral infections are often dependent on host genotype. Moreover, it is becoming clear that the virome is part of a dynamic network of microorganisms that inhabit the body. Therefore, viruses can be viewed as a component of the microbiome, and interactions with commensal bacteria and other microbial agents influence their behavior [11].

Inflammatory Responses

Virus-derived DNA and RNA trigger the production of cytokines, via the innate immune response of T-cytotoxic lymphocytes, and natural killer cells, which are counterbalanced by the production of immunosuppressive cytokines from T-regulatory lymphocytes. This function is also known as immunomodulation and is simplistically divided in immunopotentiation and immunosuppression.

Immunosuppression signifies a state of decreased immunity, while immunopotentiation represents the shifting to increased immune response. Immunomodulation can be observed locally and/or systemically and may change over time or alter the immunogenicity of antigens, being differentiated between anatomical compartments [12,13].

The human gut virome includes viruses that infect eukaryotic and prokaryotic cells. It is established after birth and is dominated by viruses that infect bacteria (a.k.a. phages). The virome establishes a mutualistic, commensal relationship with eukaryotes/prokaryotes, contributing to the homeostasis of the gut by influencing microbial ecology and host immunity as described above. Composition of the virome is influenced by numerous factors that affect viruses directly, such as an infection, or changed host–cell populations, e.g., antibiotic consumption, or diet [14]. Members of the virome may contribute to the pathogenesis of certain diseases via microbial host lysis leading to dysbiosis or infection of epithelial cells, as well as translocation via the compromised damaged mucosal barrier to enter the underlying tissues, and immune cells, forcing the activation of the immune system [12,15].

Actions of the Gut Virome on the Host Genome

The virome exists within the microbiome network and interacts with the bacterial microbiome and other organisms that inhabit the host such as fungi (mycobiome), archaea, protozoans, and helminths. The effect of the virome on the host genome represents a subset of the host genome–microbiome interaction. A simple relationship between the virome and host genome is the virus-plus-susceptibility-gene interaction, where a phenotypic outcome, such as a disease pathology or symptom, is evoked by the combination of a viral infection and host gene variant [16]. In addition, to this synthetic interaction, another type of virus-plus-susceptibility-gene interaction is phenotypic complementation, in which the viral infection masks the effect of a host gene variant [17].

Although examples are lacking, a virus could induce benefits in a manner dependent on a host gene variant or negate the beneficial effect of a gene variant. In situations where the outcome is a complex disease or trait, the virome–genome interaction involves multiple viruses and host genetic variants. Some genetic variants exist in noncoding region and might influence gene expression in a manner dependent on viral infection. These interactions are influenced by other members of the microbiome, which regulate the activities of both the host and the virus [14,18,19].

CORE GUT MYCOBIOME AND ITS PHYSIOLOGIC ROLE

The limitations of culture-dependent methods for mycobiome studies have led to the introduction, over the past 20 years, of culture-independent approaches. Methods for classifying fungi that do not rely on microbial culture include restriction fragment length polymorphism (RFLP) analysis, oligonucleotide fingerprinting of rRNA genes (OFRG), denaturing gradient gel electrophoresis (DGGE), and in situ hybridization. These techniques are useful for comparing fungal diversities between different groups, but they lack the specificity necessary to identify the different fungal species in a large-scale study. Direct sequencing of fungal genes has proven to be the most efficient method for classifying the mycobiome [20].

Gut mycobiome represents the minority of the core gut microbiome. The gut mycobiome appears less stable than the bacterial microbiome and depends heavily on environmental factors [21]. The diet is a supplier of both a means of entry for microorganisms to the gut and is a major source of nutrients for commensal microbes, providing noticeable effects on gut microbial composition. David et al. found broad alterations in the composition of gut microbiome, depending on the type of diet their volunteers consumed [22,23].

Lifestyle and the Mycobiome

In their study, fungal composition seemed to be driven by food colonization. Additionally, another study has demonstrated that mycobiome also relies on the environment in which its host resides [24]. Also, antibacterials can work to promote or inhibit pathogenic fungal growth [25]. Fungal communities demonstrated high intra- and intervolunteer dissimilarity. Nash et al. found great variability in the gut mycobiome among healthy adults, while three fungal species dominated their samples: *Saccharomyces, Malassezia,* and *Candida* [26].

Currently, the methods promoted by the microbiome consortia are suboptimal for detection of the most important fungal pathogens and ecologically important degraders [27]. Understanding what constitutes a "normal" or "healthy" gut mycobiome could assist in future research efforts, to determine contributions of commensal fungi to the health of the host or the exacerbation of disease [26].

In humans, the use of broad-spectrum antibiotics that affect anaerobic bacteria are associated with increased yeast flora in the gut, compared with antibiotics with poor anaerobic activity [28]. Additionally, detection of *Aspergillus* spp. in human sputum seems to be correlated with detection of the fungus in human feces, a finding reported years ago [29], even if studies are conflicting. In a recent study, gut microbiome had the potential to influence the anti-*Aspergillus* host response, making the gut microbiome an attractive target of study with the eventual goal of modulating host defense during invasive aspergillosis [30–32].

Systemic Impact

The gut mycobiome may influence immune responses in other parts of the body [33,34]. Bacterial microbiota and the intestinal epithelium influence the ability of fungi to colonize the gut. Commensal fungi produce metabolites and products that influence immunity and inflammation at local and distal sites. Immune cells in the gut respond to commensal fungi to influence inflammation and immune homeostasis [34].

Fungal Virulence

The mycobiome and microbiome may be perceived as being in a state of homeostasis. One of the ways the bacteria in the microbiome to keep the mycobiome in check is the production of extracellular substances that inhibit the growth or yeast-to-hyphae transition of pathobionts, such as *Candida albicans*. Clinically, the use of these substances to prevent pathogenicity may potentially offer an alternative to antifungal drugs, as these skew the mycobiota and select for drug-resistant strains.

The commensal microbiota is known to produce a wide range of metabolites, which have inhibitory effects to pathogens, such as short-chain fatty acids (SCFAs), medium-chain fatty acids (MCFAs), secondary bile acids, bacteriocins, and antimicrobial peptides [35,36].

Yeast-to-Hyphae Transition

Acetic, propionic, and butyric acid have been demonstrated to induce transcriptional changes in *C. albicans* [37], and butyric acid was found to inhibit yeast–hyphal (Y–H) transition [38]. The microbiota also produces gases from fermentation, such as hydrogen, methane, and hydrogen sulphide, which can be metabolically active [39].

The members of the gut mycobiome may produce small molecules such as farnesol and fusel alcohols, which could regulate on their own Y–H transition. Similarly, capric acid from *Saccharomyces boulardii*, a probiotic yeast, could influence Y–H transition, adhesion, and biofilm formation [38]. In turn, ethanol from *Saccharomyces cerevisiae* has been shown to stimulate the growth of bacterial pathobiont *Acinetobacter* in vitro [40], and similar yeasts are found in the gut mycobiome.

ROLES IN DISEASE STATES

In a patient suffering from *Clostridium difficile* gut infection, managed in Zurich by fecal transplantation, virome and mycobiome shifts were followed [41−44]. Broecker and his team have confirmed that enteric phages normally influence the complexity of the host community [44], and they assumed that in a healthy microbiome/virome, phages are less abundant than during inflammatory gut disorders, existing predominantly as integrated prophages.

In the gut of the "Zurich patient," the virome did not exhibit heterogeneity and variability [45]. They advocate that the gut virome may be involved in the composition of gut microbiome [41,46]. Others found that there is a specific virome profile for inflammatory bowel disease (IBD) [46], insinuating a protective role for the virome in gut health [47].

Virus Latent State

Many viruses establish latency, a state in which the viral genome persists within a cell, without producing infectious viral particles in the gut. During latency, the virus is less visible due to decreased metabolic activity and less antigens available for detection by receptors of the innate and adaptive immune system. In situations where the outcome is a complex disease or trait, the virome−genome interaction involves multiple viruses and host genetic variants. Some genetic variants exist in noncoding regions and may influence gene expression in a manner dependent on viral infection [48].

Nevertheless, animal models support a role for molecular mimicry and bystander activation, as mechanisms by which autoimmunity arises downstream of immunomodulation, by Epstein−Barr or other viruses. Viruses, such as murine norovirus, can replace many of the benefits provided by commensal bacteria in the intestine [13,49].

Commensal Fungi and Progression of Infectious Diseases

Alterations in the mycobiome are frequently reported to be associated with disease progression, but it remains to be elucidated whether this is a cause or epiphenomenon. Studies about HIV and oral mycobiome have revealed that the core bacteriome was similar between infected and uninfected groups, whereas the mycobiome differed significantly, suggesting differences in innate immune and adaptive cellular immunity [50]. In contrast vaginal fungal composition, within the context of medical history and lifestyle features, was inconclusive [51,52].

Although vaginal pH and discharge status correlated with the bacterial community composition, these and other features did not correlate with the fungal composition [52]. In the author's opinion, a major setback of that study was the lack of examination of the gut mycobiome.

Fungal Signature in Obesity and Inflammatory Bowel Disease

In obesity two major phyla, Ascomycota and Basidiomycota, were not significantly different between obese and nonobese subjects. However, the minor phylum Zygomycota was significantly underrepresented in fecal samples of obese subjects. Family biodiversity was significantly lower, and a tendency toward increased biodiversity at other levels was also found in nonobese individuals [53].

A few studies anticipate an increase in mycobiome diversity in IBD [54]. A causal relationship could be established if antimicrobial treatment targeting certain fungal groups were to lead to either exacerbation or cure. On the other hand, if treatment of the disease were to concurrently lead to modulation of the mycobiome, then it would seem more likely that the mycobiome is being affected by the disease status [55].

Virome−Mycobiome Interactions

Recent findings indicate cooperation between the host and resident viruses, in controlling both specific members of the microbiota, and their precise geographic areas of colonization within the host [56]. Upon attachment and entry into host cells, there is a turnover in available receptors, along with alteration in cell membrane architecture, due to viral entry perturbation. In addition, virus entry typically results in triggering of cell signaling, which results not only in alterations in host cell metabolism, - which could affect invasive fungal elements when present-but, also, in secretion of a variety of cytokines and chemokines, which may also have profound effects on the surrounding members of microbiome [57].

Prophage−Bacteria−Host Synergy

The phage-mediated genetic contribution shapes symbiosis for the triad phage−bacterium−human [10]. While gut bacteria and their prophages may coexist in a passive state under normal physiological conditions, the induction of prophages can occur in response to various environmental stresses, such as antibiotics, or when the survival of the host bacterium is endangered [58].

FUTURE ASPECTS OF THE MANIPULATION OF GUT VIROME AND GUT MYCOBIOME

Administration of specific phages to chickens, calves, pigs, and cattle, or spraying on food contact surfaces in food processing facilities, and on ready-to-eat foods, has been considered as a potential strategy to prevent contamination by pathogenic bacteria. Phages also seem to mediate inhibition of the production of reactive oxygen species (ROS), by immune cells exposed to endotoxins, indicating potential protective roles against oxidative stress [59].

Mapping the human gut virome seems to have the potential of providing markers for disease and health [60]. Phages seem to represent an encouraging substitute for antibiotics, in the era of antibiotic stewardship [44,61,62].

Also mycobiome mapping seems to be auspicious, since members of the fungal gut microflora may be used to detect disorders and act as disease markers [63].

REFERENCES

[1] Morgan XC, Huttenhower C. Chapter 12: human microbiome analysis. PLoS Comput Biol 2012;8(12):e1002808.

[2] Lim ES, Zhou Y, Zhao G, et al. Early life dynamics of the human gut virome and bacterial microbiome in infants. Nat Med 2015;21(10):1228−34.

[3] Rodriguez JM, Murphy K, Stanton C, et al. The composition of the gut microbiota throughout life, with an emphasis on early life. Microb Ecol Health Dis 2015;26:26050.

[4] Stefanaki C, Peppa M, Mastorakos G, Chrousos GP. Examining the gut bacteriome, virome, and mycobiome in glucose metabolism disorders: are we on the right track? Metab Clin Exp 2017;73:52−66.

[5] Heintz-Buschart A, Wilmes P. Human gut microbiome: function matters. Trends Microbiol Jul. 2018;26(7):563−74.

[6] Kelly JR, Kennedy PJ, Cryan JF, Dinan TG, Clarke G, Hyland NP. Breaking down the barriers: the gut microbiome, intestinal permeability and stress-related psychiatric disorders. Front Cell Neurosci 2015;9:392.

[7] Caporaso JG, Knight R, Kelley ST. Host-associated and free-living phage communities differ profoundly in phylogenetic composition. PLoS One 2011;6(2):e16900.

[8] Jack PJ, Amos-Ritchie RN, Reverter A, et al. Microarray-based detection of viruses causing vesicular or vesicular-like lesions in livestock animals. Vet Microbiol 2009;133(1−2):145−53.

[9] Handley SA. The virome: a missing component of biological interaction networks in health and disease. Genome Med 2016;8(1):32.

[10] Bakhshinejad B, Ghiasvand S. Bacteriophages in the human gut: our fellow travelers throughout life and potential biomarkers of heath or disease. Virus Res 2017;240:47−55.

[11] Ogilvie LA, Jones BV. The human gut virome: a multifaceted majority. Front Microbiol 2015;6:918.

[12] Freer G, Maggi F, Pistello M. Virome and inflammasomes, a finely tuned balance with important consequences for the host health. Curr Med Chem 2017.

[13] Cadwell K. The virome in host health and disease. Immunity 2015;42(5):805−13.

[14] Carding SR, Davis N, Hoyles L. Review article: the human intestinal virome in health and disease. Aliment Pharmacol Ther 2017;46(9):800−15.

[15] Karst SM. Viral safeguard: the enteric virome protects against gut inflammation. Immunity 2016;44(4):715−8.

[16] Grasis JA. The intra-dependence of viruses and the holobiont. Front Immunol 2017;8:1501.

[17] Zou S, Caler L, Colombini-Hatch S, Glynn S, Srinivas P. Research on the human virome: where are we and what is next. Microbiome 2016;4(1):32.

[18] Scarpellini E, Ianiro G, Attili F, Bassanelli C, De Santis A, Gasbarrini A. The human gut microbiota and virome: potential therapeutic implications. Dig Liver Dis 2015;47(12):1007−12.

[19] Virgin HW. The virome in mammalian physiology and disease. Cell 2014;157(1):142−50.

[20] Cui L, Morris A, Ghedin E. The human mycobiome in health and disease. Genome Med 2013;5(7):63.

[21] Hallen-Adams HE, Suhr MJ. Fungi in the healthy human gastrointestinal tract. Virulence 2017;8(3):352−8.

[22] David LA, Maurice CF, Carmody RN, et al. Diet rapidly and reproducibly alters the human gut microbiome. Nature 2014;505(7484):559−63.

[23] Suhr MJ, Hallen-Adams HE. The human gut mycobiome: pitfalls and potentials—a mycologist's perspective. Mycologia 2015;107(6):1057−73.

[24] Xing F, Ding N, Liu X, et al. Variation in fungal microbiome (mycobiome) and aflatoxins during simulated storage of in-shell peanuts and peanut kernels. Sci Rep 2016;6:25930.

[25] Azevedo MM, Teixeira-Santos R, Silva AP, et al. The effect of antibacterial and non-antibacterial compounds alone or associated with antifugals upon fungi. Front Microbiol 2015;6:669.

[26] Nash AK, Auchtung TA, Wong MC, et al. The gut mycobiome of the Human Microbiome Project healthy cohort. Microbiome 2017;5(1):153.

[27] Tedersoo L, Lindahl B. Fungal identification biases in microbiome projects. Environ Microbiol Rep 2016.

[28] Samonis G, Gikas A, Anaissie EJ, et al. Prospective evaluation of effects of broad-spectrum antibiotics on gastrointestinal yeast colonization of humans. Antimicrob Agents Chemother 1993;37(1):51−3.

[29] Amitani R, Murayama T, Nawada R, et al. Aspergillus culture filtrates and sputum sols from patients with pulmonary aspergillosis cause damage to human respiratory ciliated epithelium in vitro. Eur Respir J 1995;8(10):1681−7.

[30] Kolwijck E, van de Veerdonk FL. The potential impact of the pulmonary microbiome on immunopathogenesis of Aspergillus-related lung disease. Eur J Immunol 2014;44(11):3156−65.

[31] Nguyen LD, Viscogliosi E, Delhaes L. The lung mycobiome: an emerging field of the human respiratory microbiome. Front Microbiol 2015;6:89.

[32] Goncalves SM, Lagrou K, Duarte-Oliveira C, Maertens JA, Cunha C, Carvalho A. The microbiome-metabolome crosstalk in the pathogenesis of respiratory fungal diseases. Virulence 2017;8(6):673−84.

[33] Sam QH, Chang MW, Chai LY. The fungal mycobiome and its interaction with gut bacteria in the host. Int J Mol Sci 2017;18(2).

[34] Paterson MJ, Oh S, Underhill DM. Host-microbe interactions: commensal fungi in the gut. Curr Opin Microbiol 2017;40:131−7.

[35] Donia MS, Fischbach MA. Human Microbiota. Small molecules from the human microbiota. Science 2015;349(6246):1254766.

[36] Lee WJ, Hase K. Gut microbiota-generated metabolites in animal health and disease. Nat Chem Biol 2014;10(6):416−24.

[37] Cottier F, Tan AS, Chen J, et al. The transcriptional stress response of Candida albicans to weak organic acids. G3 2015;5(4):497−505.

[38] Shareck J, Belhumeur P. Modulation of morphogenesis in Candida albicans by various small molecules. Eukaryot Cell 2011;10(8):1004−12.

[39] Pimentel M, Mathur R, Chang C. Gas and the microbiome. Curr Gastroenterol Rep 2013;15(12):356.

[40] Smith MG, Des Etages SG, Snyder M. Microbial synergy via an ethanol-triggered pathway. Mol Cell Biol 2004;24(9):3874−84.

[41] Broecker F, Klumpp J, Moelling K. Long-term microbiota and virome in a Zurich patient after fecal transplantation against *Clostridium difficile* infection. Ann NY Acad Sci 2016;1372(1):29−41.

[42] Broecker F, Kube M, Klumpp J, et al. Analysis of the intestinal microbiome of a recovered *Clostridium difficile* patient after fecal transplantation. Digestion 2013;88(4):243−51.

[43] Moelling K, Broecker F. Fecal microbiota transplantation to fight *Clostridium difficile* infections and other intestinal diseases. Bacteriophage 2016;6(4):e1251380.

[44] Dalmasso M, Hill C, Ross RP. Exploiting gut bacteriophages for human health. Trends Microbiol 2014;22(7):399−405.

[45] Broecker F, Russo G, Klumpp J, Moelling K. Stable core virome despite variable microbiome after fecal transfer. Gut Microbes 2017;8(3):214−20.

[46] Norman JM, Handley SA, Baldridge MT, et al. Disease-specific alterations in the enteric virome in inflammatory bowel disease. Cell 2015;160(3):447−60.

[47] Yang JY, Kim MS, Kim E, et al. Enteric viruses ameliorate gut inflammation via toll-like receptor 3 and toll-like receptor 7-mediated interferon-beta production. Immunity 2016;44(4):889−900.

[48] Foxman EF, Iwasaki A. Genome-virome interactions: examining the role of common viral infections in complex disease. Nat Rev Microbiol 2011;9(4):254−64.

[49] Cadwell K, Patel KK, Maloney NS, et al. Virus-plus-susceptibility gene interaction determines Crohn's disease gene Atg16L1 phenotypes in intestine. Cell 2010;141(7):1135−45.

[50] Mukherjee PK, Chandra J, Retuerto M, et al. Oral mycobiome analysis of HIV-infected patients: identification of Pichia as an antagonist of opportunistic fungi. PLoS Pathog 2014;10(3):e1003996.

[51] Ma B, Forney LJ, Ravel J. Vaginal microbiome: rethinking health and disease. Annu Rev Microbiol 2012;66:371−89.

[52] Noyes N, Cho KC, Ravel J, Forney LJ, Abdo Z. Associations between sexual habits, menstrual hygiene practices, demographics and the vaginal microbiome as revealed by Bayesian network analysis. PLoS One 2018;13(1):e0191625.

[53] Mar Rodriguez M, Perez D, Javier Chaves F, et al. Obesity changes the human gut mycobiome. Sci Rep 2015;5:14600.

[54] Ott SJ, Kuhbacher T, Musfeldt M, et al. Fungi and inflammatory bowel diseases: alterations of composition and diversity. Scand J Gastroenterol 2008;43(7):831−41.

[55] Seed PC. The human mycobiome. Cold Spring Harbor Perspect Med 2014;5(5):a019810.

[56] Plotkin BJ, Sigar IM, Tiwari V, Halkyard S. Herpes simplex virus (HSV) modulation of *Staphylococcus aureus* and Candida albicans initiation of HeLa 299 cell-associated biofilm. Curr Microbiol 2016;72(5):529−37.

[57] Plotkin BJ. Contribution of host signaling and virome to the mycobiome. Fungal Genom Biol 2016;6(1):137.

[58] Allen HK, Looft T, Bayles DO, et al. Antibiotics in feed induce prophages in swine fecal microbiomes. mBio 2011;2(6).

[59] Miedzybrodzki R, Switala-Jelen K, Fortuna W, et al. Bacteriophage preparation inhibition of reactive oxygen species generation by endotoxin-stimulated polymorphonuclear leukocytes. Virus Res 2008;131(2):233−42.

[60] Xu GJ, Kula T, Xu Q, et al. Viral immunology. Comprehensive serological profiling of human populations using a synthetic human virome. Science 2015;348(6239):aaa0698.

[61] Kang DW, Adams JB, Gregory AC, et al. Microbiota Transfer Therapy alters gut ecosystem and improves gastrointestinal and autism symptoms: an open-label study. Microbiome 2017;5(1):10.

[62] De Vlaminck I, Khush KK, Strehl C, et al. Temporal response of the human virome to immunosuppression and antiviral therapy. Cell 2013;155(5):1178−87.

[63] Suhr MJ, Banjara N, Hallen-Adams HE. Sequence-based methods for detecting and evaluating the human gut mycobiome. Lett Appl Microbiol 2016;62(3):209−15.

Chapter 4

Techniques for Phenotyping the Gut Microbiota Metabolome

Elisa Zubeldia-Varela[1,2,3], Beata Anna Raczkowska[4], Manuel Ferrer[5], Marina Perez-Gordo[2,3] and David Rojo[1]

[1]Centro de Metabolómica y Bioanálisis (CEMBIO), Facultad de Farmacia, Universidad CEU San Pablo, Campus Montepríncipe, Madrid, Spain; [2]Basic Medical Sciences Department, Faculty of Medicine, Universidad CEU San Pablo, Campus Montepríncipe, Madrid, Spain; [3]Institute of Applied and Molecular Medicine (IMMA), Faculty of Medicine, Universidad CEU San Pablo, Campus Montepríncipe, Madrid, Spain; [4]Department of Endocrinology, Diabetology and Internal Medicine, Medical University of Bialystok, Bialystok, Poland; [5]Institute of Catalysis, Consejo Superior de Investigaciones Científicas (CSIC), Madrid, Spain

METABOLOMICS IN THE CONTEXT OF SYSTEMS BIOLOGY

We are in the "microbiome revolution" [1], which has been possible thanks to the new high-throughput analytical approaches, which measure the expressed genes, proteins, and metabolites [2].

Nowadays, the relationship between the gut microbiota and our health status is clearly established, distinguishing between functional and nonfunctional situations [3]. Élie Metchnikoff, awarded in 1908 with the Nobel Prize for Physiology or Medicine, was one of the first scientists to call the attention on resident microorganisms and their different effects on the human body, particularly the ones inhabiting our large intestine [4]. In 1920, Carl Arthur Scheunert linked the intestinal "flora" to bone inflammation in horses [5], using for the first time the term "dysbiosis" as a synonym of a nonhealthy microbiota. In a more recent context, Helmut Haenel defined "dysbiosis" as a change or imbalance in the gut microorganisms, which could be contrasted to the positive state called "eubiosis" [6]. He even proposed a criterion for detecting these abnormalities, based on the quantification of some typical bacterial species in feces [7].

All of these represent some of the turning points in the history of the gut microbiota research [8], in which the results were based on the taxonomic composition. In this sense, it is important to point out that the current state of art stresses the need to also pay attention to the proteins and the metabolites, the truly active agents of any biological system [9–11].

WORKFLOW IN METABOLOMICS: TARGET ANALYSIS VERSUS FINGERPRINTING

The aim of metabolomics is to identify and quantify the variation on the concentration of the metabolites present in a biological system [12,13]. Since the metabolites are the final products of any biochemical reaction, they are considered to be the best markers of the microbial activity [14]. Metabolomics has two different approaches [9]. First, untargeted analysis, also known as global profiling or fingerprinting, has emerged as a new field in analytical chemistry research. It is used when the goal is to obtain a global overview of the metabolic profile, looking for differences between cases and controls.

It requires a strong experimental design to minimize the interindividual variability. On the other hand, targeted metabolomics focuses on the quantification of a small set of molecules, trying to optimize the analytical procedures in order to maximize the signal-to-noise ratio, following the classic analytical chemistry view.

In diagnosis and therapy, both methodologies are complementary. If fingerprinting is ideal for generating new hypotheses and knowledge, for example, new biomarkers for any disease, target analysis is useful for validation, confirming the potential benefits of any treatment under study. Herein, we will focus on the workflow of metabolomic fingerprinting, including sampling, laboratory procedures, and available analytical techniques, as well as on some applications for gut microbiota's diagnosis and therapeutic implications.

SAMPLING, COLLECTION, STORAGE, AND TRANSPORTATION

Preanalytical considerations are critical in metabolomics research. Group selection and sampling must be carefully designed, considering that the number of individuals should minimize intragroup differences. Sample type, collection, storage, and transport must be well planned to avoid unwanted biases that might reduce the strength of the results.

For the study of gut microbiota, feces are the biological sample of election because of its uncomplicated and noninvasive collection [15,16]. Although feces do not represent the total composition of the microbiota, they are considered representative of the colon segment, where the concentration of bacteria reaches about 10^{11} cells [17]. Other samples such as biopsies can be used for higher segments [18,19], but the number and importance of the disadvantages they have is considerably higher, being the most important its invasive collection technique [20]. In blood and urine, intestinal metabolites can be indirectly measured [21,22], but its concentration is lower than in fecal samples.

Different problems need to be addressed, when considering feces sample handling. Variations in storage temperature, the correct transmission of the guidelines of collection to clinicians and patients, or the maintenance of sterile conditions might produce a significant variability in the results of metabolomic studies of gut microbiota. Few articles have developed the optimization of this procedure [23], and the adoption of a standardized protocol is needed.

Transportation, storage temperature, and storage duration are critical issues that influence stability. For this reason, during transportation the samples should be kept at 4°C with a cold pack or ice. Usually, feces cannot be analyzed immediately after collection, especially in large-scale studies, and samples must be stored at −80°C, as freezing prevents possible changes in microbial communities. Moreover, to reduce the variability induced by the analyst, the storage should end at the same time for all samples.

Human stool samples can be collected at the patient's home [24], with a feces collection kit to facilitate the process and prevent contamination [25,26]. This kit commonly includes collection paper, a stool collection container, gloves, a cold pack, and an easy-to-read guideline that patients can read at home. However, the instructions must be carefully explained beforehand by the clinicians to the patients. In other cases, feces are collected in the hospital during hospitalization or before an endoscopic examination or a surgical intervention [27].

In the case of an animal model, there are two main options: (1) picking up the gut content directly from the large intestine after euthanasia [15,28] or (2) collecting the stool of their cages with a previous abdominal massage to facilitate defecation [29].

Immediately after collection, samples must be stored in the proper conditions. The most common procedure is freezing them at −20°C [30,31] or −80°C [32]. Some authors reported flash freezing of samples with liquid nitrogen before storage [15,32]. However, if collection is made at home, samples should be maintained in the freezer (−20°C approximately) and delivered before 3 days to the center, where they will be stored at −80°C and/or processed. Some studies investigating the effect of storage conditions have shown that there is no need to use a stabilization agent [25]. Indeed, freezing prevents possible changes in microbial communities.

It has been shown that frozen feces can undergo up to four freeze—defrost cycles, without suffering significant changes in the composition of the microbiota [20]. Despite this, to prevent defrosting taking place abruptly during the transportation of the sample, it is recommended to use the cold pack included in the collection kit to keep the temperature approximately at 4°C.

FECAL COMPOSITION, METABOLITE ENRICHMENT, AND COMPLETE METABOLITE EXTRACTION

Feces are a heterogeneous and complex matrix that contains other substances in addition to the metabolic products of microorganisms, such as eukaryotic cells and food debris. On one hand, several authors have reported the advantages of separating bacteria from feces, before the metabolites extraction [33—38]. Fig 4.1 summarizes the complete recommended procedure.

It is interesting to note that the use of bacteria enrichment methods, such as differential centrifugation, reduce contamination avoiding variability [39,40]. They also improve the richness of microbes being detected, while minimizing the contamination with nonmicrobial components [41]. On the other hand, it is also possible to extract the metabolites directly from the total feces [42]. However, it should be mentioned that feces consistency (measured by the Bristol Stool

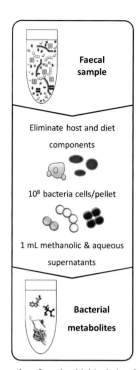

Faecal debris removal

- 0.4 g faeces + 1.2 mL PBS (1:3 w/v)
- 1 min vortex (re-suspension)
- 1000 *g*, 4 °C, 1 min

Prokaryotic cells isolation

- Previous supernatant
- Bacterial pellets: 13000 *g*, 4 °C, 5 min

Metabolites double solvent extraction

I) 1.2 mL of cold MeOH (-80 °C), vortex 10 s

- Sonicate 30 s, 15 W on ice bath (-20 °C)
- Repeat 3 times, 5 min on ice bath in between
- Debris pelleting: 12000 *g*, 4 °C, 10 min

II) 1.2 mL of cold H_2O (4 °C), vortex 10 s

- Sonicate 20 s, 15 W on ice bath (-20 °C)
- Repeat 3 times, 2 min on ice bath in between
- Debris pelleting: 12000 *g*, 4 °C, 10 min

Faecal sample

Eliminate host and diet components

10^8 bacteria cells/pellet

1 mL methanolic & aqueous supernatants

Bacterial metabolites

FIGURE 4.1   Robust procedure for microbial isolation from fecal samples and subsequent metabolites extraction. It encompasses several centrifugation steps, in order to eliminate fecal debris and eukaryotic cells. Finally, the metabolite extraction is done using two solvents, in order to enlarge the metabolite coverage.

Scale) strongly interferes in the taxonomy of the microbiota [43]. When referring to metabolome, with the same feces consistency, the important point is the number of cells. The lyophilization procedure allows to standardize fecal water content [44,45], but it is a laborious procedure, and it can lead to the rupture of prokaryotic cells [9].

Finally, the metabolite extraction is the last step before the laboratory analysis. The selection of the extraction solvent strongly biases the results of the study, due to its huge impact on the loss, underestimation or overestimation of the metabolites' recovery. Homogenization of the sample is crucial before adding any extraction solvent [46]. Furthermore, in a fingerprinting study, it is highly recommended a double extraction, with polar and nonpolar solvents, to enlarge the metabolite coverage. This two-step extraction method, as well as the differential centrifugation mentioned above, has been used in several studies with different pairs of solvents, such as methanol and water [33], trichloromethane and ethanol/methanol [47], or acetonitrile and water [48].

CRITICAL ANALYSIS OF AVAILABLE METABOLOMIC TECHNIQUES

The metabolome hugely varies in chemical properties; therefore, there is not one analytical technique that allows the identification of the entire set [49]. Focusing on gut microbiota fingerprinting, the main analytical platforms are mass spectrometry (MS) and nuclear magnetic resonance (NMR). MS is a powerful analytical technique based on the ionization, and subsequent separation of the molecules, according to their *m/z* ratio. This technique can be applied for both qualitative and quantitative analysis, usually coupled to a separation technique such as gas chromatography (GC), liquid chromatography (LC), or capillary electrophoresis (CE).

Around 70% of the reviewed literature used just one analytical platform, although the ideal would be to use more, to enlarge the metabolite coverage [9], particularly combining NMR and MS [50]. Each technique is devoted to a certain part of the metabolome (Fig 4.2A). For example, GC-MS is ideal for thermally stable and volatile compounds, such as short-chain fatty acids, esters, ketones, alcohols, aldehydes, or amino acids [9], although it usually requires derivatization of the samples. LC-MS is mainly devoted to bile acids and lipids, with minimum sample treatment [34,48,51].

NMR is used to detect short-chain fatty acids, small lipids, and amino acids, requiring the addition of a deuterated buffer to samples, being the most commonly used phosphate buffered saline [52—54]. Finally, CE-MS is ideal for the detection and quantification of small ionizable molecules [55], such as amines, amino acids, and carboxylic acids. Nevertheless, it should be pointed out that sample clusters, which can be identified by different analytical techniques, are

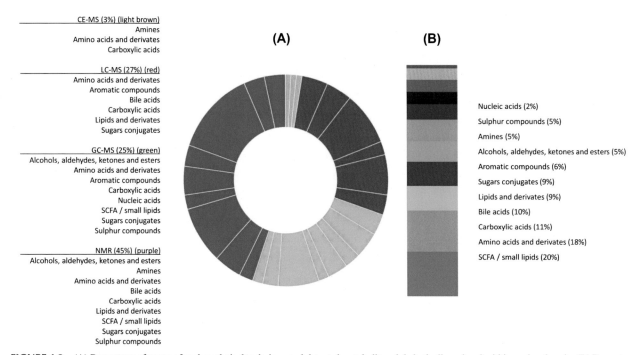

FIGURE 4.2 (A) Percentage of usage of each analytical technique and detected metabolites alphabetically ordered within each color pie. (B) Percentage of each metabolite type identified in gut microbiota. *(The figure is based on the dataset previously published in the review by Rojo D, Méndez-García C, Raczkowska BA, Bargiela R, Moya A, Ferrer M, et al. Exploring the human microbiome from multiple perspectives: factors altering its composition and function. FEMS Microbiol Rev 2017;41(4):453–78. For the current figure, only the articles that fingerprint the gut microbiota, with one analytical technique, have been considered.)*

usually congruent and can be considered as a validation of the results. Table 4.1 shows the comparative description of the main analytical techniques, describing the basis of their mechanism of separation and their pros and cons.

It is important to note that NMR spectroscopy is a powerful nondestructive technique, which requires little sample preparation [53], and it has been the most used analytical tool in the fecal analysis [42] (Table 4.1). Per biochemical category (Fig 4.2B), in gut microbiota the most detected groups of metabolites are the short-chain fatty acids, followed by amino acids and derivatives, carboxylic acids, and bile acids.

DATA PROCESSING AND INTERPRETATION

In the analytical chemistry laboratory, a crucial point is the signal stability that must be assessed in order to ensure that no spurious variability is biasing the result. For this reason, it is highly recommended to use quality controls (QCs) [56] and, in the case that it would be required, any of the available normalization strategies [57]. If possible, QCs should be made by pooling aliquots of all the samples, and they must be injected several times, at the beginning of the analytical run, to guarantee the stability of the system, and during the sequence at periodic intervals, for a proper monitoring [33,58,59]. Nevertheless, other alternatives have been proposed as surrogate QCs, such as the use of external samples [60].

Subsequently to the data acquisition, the deconvolution of the data converts the analytical signal into metabolic features, which are characterized by a retention time, a chemical formula, and a peak area. This raw set of features needs to be aligned, in order to create an operative data matrix, in which the columns are the samples, the rows are the metabolites, and each cell contains the peak area, which corresponds to the metabolite concentration in arbitrary units.

The quality of this data set is crucial and, besides the manual inspection of the integration of each peak, it is important to get rid of deconvolution artifacts, rejecting the features that are not reproducibly detected across the full data set [60]. Prior to the differential analysis, the data clustering in a unsupervised multivariate model reveals the biological grouping of the samples. For this purpose, the principal components analysis (PCA) is a powerful tool, in order to mathematize a biological system. Each principal component corresponds to a biological characteristic under study and, in a two-dimensional plot, they graphically represent the maximal dispersion of the samples. Therefore, the distance between groups can be directly correlated with their biochemical similarity or difference.

TABLE 4.1 Main Analytical Platforms Used in Gut Microbiota Metabolomics Fingerprinting

Analytical Technique	Basis	Advantages	Disadvantages	References
NMR	The determination of the molecular structure by the absorption of electromagnetic radiation from the atomic nucleus (with nonzero magnetic momentum) subjected to radiofrequency pulses placed in a magnetic field.	Highly reproducible spectra. Absolute quantification. Fast speed acquisition rates.	Low sensitivity. Overlay of signals. Strict control of pH.	[42,53,76,90]
LC-MS	The separation relies on the different affinity of the analytes to the mobile phase and the stationary phase. The chemical nature of the stationary phase determines the polarity range of the analytes that are more retained and subsequently the elution order.	High sensitivity. Broad spectrum of polarities. Easy sample pretreatment.	High number of signals. Identification requires time and expertise.	[26,40,91,92]
GC-MS	The metabolites are separated due to their differences in boiling point and their affinity to the stationary phase.	Peak resolution and efficiency. Reproducibility in the fragmentation profile. Powerful spectral libraries (e.g., FiehnLib, NIST).	Pretreatment (derivatization). Limited number of metabolites.	[93–95]
CE-MS	The separation is based on the electroosmotic flow (due to the surface charge of the wall of the capillary) and the electrophoretic mobility (due to the charge to volume ratio in the compound).	Low consumption of sample. Capacity to deal with complex samples.	Reproducibility. Limited to ionic and polar compounds.	[34,55]

The table summarizes the basis of the analyte separation, the main advantages and disadvantages of each acquisition technique and some key references in the art.

A principal component is calculated based on the trend of the individual variables (the metabolites), and the differential analysis reveals their contribution to the model, quantified by some mathematical parameter like the p-value or the variable importance in projection. For this purpose, both univariate and multivariate statistics can be used [61,62].

DATA VALIDATION

Metabolomics data treatment is challenging [9,63], and the validation of results is important to reinforce the conclusions. It should be pointed out that the term "validation" encompasses four aspects: (1) analytical parameters (e.g., precision, accuracy, linearity, limit of quantification, or limit of detection), (2) mathematical model suitability (e.g., permutation test or cross-validation); (3) confirmation of the identity/biological role of the metabolites (e.g., MS/MS analysis or multiomics experiment); and (4) biomarkers discovery (confirmed identification, same finding in a second cohort, and suitable sensitivity and specificity).

Furthermore, in order to harmonize the identification confidence level, it would be recommended to follow internationally standardized criteria [64]: identified metabolites (level 1), putatively annotated compounds (level 2), putatively characterized compound classes (level 3), and unknown compounds (level 4). Different strategies for metabolite identification exist, from structural elucidation by NMR or MS/MS, to the usage of exact mass databases or fragmentation libraries. A comprehensive summary of all of them can be found in the literature [65,66].

DIAGNOSIS AND THERAPEUTIC IMPLICATIONS OF GUT MICROBIOTA METABOLITES

It has been reported that at least 105 different diseases and disorders studied by 16S rDNA sequencing and 37 studied by metabolomics have an impact on the microbiota [9]. To date, the most studied by metabolomics are colorectal cancer

[24,67], Crohn's disease [35,68], irritable bowel syndrome [69], and *Clostridium difficile* infection [33,48]. In addition, the relation between the microbiota and other diseases, such as in the pathophysiology of HIV [36,70,71] and in allergic diseases [72,73], is emerging and growing in importance.

Focusing on allergy, potential groups of biomarkers that might help to diagnose individuals with asthma have been described [74]. Furthermore, the identification of food allergy biomarkers will allow future intervention studies with probiotics, which have shown the ability to restore intestinal microbial balance and modulate activation of immune cells [75]. Nevertheless, it has been reported that, beside diseases, the microbiota could be affected by the usage of antibiotics, the delivery mode at birth, the geographical origin, and the supplementation with probiotics, among others [9].

Metabolites are the closest level to the phenotype, and they are the best markers of the microbial activity [14]. To date, at least 33 bacterial groups and 52 metabolites have been reported, as being highly affected by any kind of perturbation, including diseases (e.g., irritable bowel syndrome, Crohn's disease, colorectal cancer) and environmental factors. Metabolomics can be useful for finding biomarkers of pathological states [76]. However, in gut microbiota, it should be pointed out that, because functional redundancy, the same metabolites can be linked with different bacterial groups depending on the perturbation and that the same species can express different metabolites depending on the circumstances. An ideal biomarker should be reproducible, robust, specific, sensitive, and easy to measure [77], and it always requires to be validated before been accepted.

Fecal microbiota transplant (FMT) constitutes an important therapeutic option for *C. difficile* infection [78,79]. Also prebiotics [80–82], or for instance high-fiber diets during pregnancy and lactation, can improve immunity and metabolism and increase the proportions of *Lactobacillus* and *Clostridium leptum*, which seem to prevent allergy in infants, as described in experimental models [83].

Also probiotics should be mentioned [84,85]; however, the health benefits attributed to one probiotic strain may not be necessarily applicable to another strain, even within one given species [86]. It has been shown that time of administration is also crucial, being more beneficial when the immune system is maturing, i.e., before birth and several months after [82]. In the case of food allergy, coadministration of bacterial adjuvants with oral immunotherapy has been suggested. In fact, probiotic therapy with *Lactobacillus rhamnosus* has demonstrated efficacy when coadministered with peanut oral immunotherapy [87]. Finally, synbiotics, such as extensively hydrolyzed casein formula plus *L. rhamnosus* GG promote tolerance through changes in the infant's gut microbiome [88,89].

ACKNOWLEDGMENTS

This work was supported by the Spanish Ministry of Economy, Industry and Competitiveness (CTQ2014-55279-R) to which D.R. acknowledges funding and by ISCIII (project number PI17/01087) cofounded by FEDER for the thematic network and cooperative research centers ARADyAL (RD16/0006/0015) to which M.P.-G. acknowledges the Spanish Society of Allergy and Clinical Immunology (SEAIC) for financial support. M.F. acknowledges the Instituto de Salud Carlos III and the Fundación Agencia Española contra el Cáncer (grant number AC17/00022 within the ERA NET TRANSCAN-2 program). All the authors acknowledge Tomás Barker for his revision and professor Coral Barbas for her intellectual inputs and critical revision of the manuscript.

REFERENCES

[1] Blaser MJ. The microbiome revolution. J Clin Invest 2014;124(10):4162–5.

[2] Mendez-Garcia C, Barbas C, Ferrer M, Rojo D. Complementary methodologies to investigate human gut microbiota in host health, working towards integrative systems biology. J Bacteriol 2018;200:e00376–417.

[3] Moya A, Ferrer M. Functional redundancy-induced stability of gut microbiota subjected to disturbance. Trends Microbiol 2016;24(5):402–13.

[4] Metchnikoff E. Etudes sur la flore intestinale: putrefaction intestinale. Ann Int Pasteur 1908;22:930–55.

[5] Scheunert A. Über Knochenweiche bei Pferden und "Dysbiose" der Darmflora. Z InfektKr Haustiere 1920;21:105.

[6] Haenel H. Some rules in the ecology of the intestinal microflora on man. J Appl Bacteriol 1961;24(3):242–51.

[7] Haenel H. Human normal and abnormal gastrointestinal flora. Am J Clin Nutr 1970;23(11):1433–9.

[8] Hooks K, O'Malley M. Dysbiosis and its discontents. Am Soc Microbiol 2017;8(5).

[9] Rojo D, Méndez-García C, Raczkowska BA, Bargiela R, Moya A, Ferrer M, et al. Exploring the human microbiome from multiple perspectives: factors altering its composition and function. FEMS Microbiol Rev 2017;41(4):453–78.

[10] Holmes E, Li JV, Athanasiou T, Ashrafian H, Nicholson JK. Understanding the role of gut microbiome-host metabolic signal disruption in health and disease. Trends Microbiol 2011;19(7):349–59.

[11] Goodacre R. Metabolomics of a superorganism. J Nutr 2007;137:259S–66S.

[12] Nicholson J, Lindon J, Holmes E. "Metabonomics": understanding the metabolic responses of living systems to pathophysiological stimuli via multivariate statistical analysis of biological NMR spectroscopic data. Xenobiotica 1999;29(11):1181–9.

[13] Fiehn O. Metabolomics - the link between genotypes and phenotypes. Plant Mol Biol 2002;48:155–71.

[14] Priori R, Scrivo R, Brandt J, Valerio M, Casadei L, Valesini G, et al. Metabolomics in rheumatic diseases: the potential of an emerging methodology for improved patient diagnosis, prognosis, and treatment efficacy. Autoimmun Rev 2013;12(10):1022−30.

[15] Theriot C, Bowman A, Young V. Antibiotic-induced alterations of the gut microbiota alter secondary bile acid production and allow for *Clostridium difficile* spore germination and outgrowth in the large intestine. MSphere 2016;1(1). e00045−15.

[16] Quifer-Rada P, Choy Y, Calvert C, Waterhouse A, Lamuela-Raventos R. Use of metabolomics and lipidomics to evaluate the hypocholestreolemic effect of Proanthocyanidins from grape seed in a pig model. Mol Nutr Food Res 2016;60(10).

[17] Sartor R. Microbial influences in inflammatory bowel diseases. Gastroenterology 2008;134(2):577−94.

[18] Shi X, Wei X, Yin X, Wang Y, Zhang M, Zhao C, et al. Hepatic and fecal metabolomic analysis of the effects of *Lactobacillus rhamnosus* GG on alcoholic fatty liver disease in mice. J Proteome Res 2015;14(2):1174−82.

[19] Zhang L, Nichols RG, Correll J, Murray IA, Tanaka N, Smith PB, et al. Persistent organic pollutants modify gut microbiota−host metabolic homeostasis in mice through aryl hydrocarbon receptor activation. Environ Health Perspect 2015;123(7):679−88.

[20] Alarcón Cavero T, D'Auria G, Delgado Palacio S, Del Campo Moreno R, Ferrer M. Microbiota. In: Procedimientos en Microbiología Clínica; 2016.

[21] Peng S, Zhang J, Liu L, Zhang X, Huang Q, Alamdar A, et al. Newborn meconium and urinary metabolome response to maternal gestational diabetes mellitus: a preliminary case-control study. J Proteome Res 2015;14(4):1799−809.

[22] Dior M, Delagrèverie H, Duboc H, Jouet P, Coffin B, Brot L, et al. Interplay between bile acid metabolism and microbiota in IBS. Neuro Gastroenterol Motil 2016;28(9):1330−40.

[23] Gratton J, Phetcharaburanin J, Mullish BH, Williams HRT, Thursz M, Nicholson JK, et al. Optimized sample handling strategy for metabolic profiling of human feces. Anal Chem 2016;88(9):4661−8.

[24] Sinha R, Ahn J, Sampson JN, Shi J, Yu G, Xiong X, et al. Fecal microbiota, fecal metabolome, and colorectal cancer interrelations. PLoS One 2016;11(3):e0152126.

[25] Al KF, Bisanz JE, Gloor GB, Reid G, Burton JP. Evaluation of sampling and storage procedures on preserving the community structure of stool microbiota: a simple at-home toilet-paper collection method. J Microbiol Methods 2018;144:117−21.

[26] Wan J-Y, Wang C-Z, Liu Z, Zhang Q-H, Musch MW, Bissonnette M, et al. Determination of American ginseng saponins and their metabolites in human plasma, urine and feces samples by liquid chromatography coupled with quadrupole time-of-flight mass spectrometry. J Chromatogr B 2016;1015−1016:62−73.

[27] Monleón D, Morales JM, Barrasa A, López JA, Vázquez C, Celda B. Metabolite profiling of fecal water extracts from human colorectal cancer. NMR Biomed 2009;22(3):342−8.

[28] Fröhlich EE, Farzi A, Mayerhofer R, Reichmann F, Jačan A, Wagner B, et al. Cognitive impairment by antibiotic-induced gut dysbiosis: analysis of gut microbiota-brain communication. Brain Behav Immun 2016;56:140−55.

[29] Etxeberria U, Arias N, Boqué N, Romo-Hualde A, Macarulla MT, Portillo MP, et al. Metabolic faecal fingerprinting of trans-resveratrol and quercetin following a high-fat sucrose dietary model using liquid chromatography coupled to high-resolution mass spectrometry. Food Funct 2015;6(8):2758−67.

[30] García-Villalba R, Espín JC, Tomás-Barberán FA. Chromatographic and spectroscopic characterization of urolithins for their determination in biological samples after the intake of foods containing ellagitannins and ellagic acid. J Chromatogr A 2016;1428:162−75.

[31] Su X, Wang N, Chen D, Li Y, Lu Y, Huan T, et al. Dansylation isotope labeling liquid chromatography mass spectrometry for parallel profiling of human urinary and fecal submetabolomes. Anal Chim Acta 2016;903:100−9.

[32] Lin H, An Y, Hao F, Wang Y, Tang H. Correlations of fecal metabonomic and microbiomic changes induced by high-fat diet in the pre-obesity state. Nat Sci Rep 2016;6:1−14.

[33] Rojo D, Gosalbes MJ, Ferrari R, Pérez-Cobas AE, Hernández E, Oltra R, et al. *Clostridium difficile* heterogeneously impacts intestinal community architecture but drives stable metabolome responses. ISME J 2015;9(10):2206−20.

[34] Rojo D, Hevia A, Bargiela R, López P, Cuervo A, González S, et al. Ranking the impact of human health disorders on gut metabolism: systemic lupus erythematosus and obesity as study cases. Sci Rep 2015;5:1−9.

[35] Bjerrum JT, Wang Y, Hao F, Coskun M, Ludwig C, Günther U, et al. Metabonomics of human fecal extracts characterize ulcerative colitis, Crohn's disease and healthy individuals. Metabolomics 2015;11(1):122−33.

[36] Serrano-Villar S, Rojo D, Martínez-Martínez M, Deusch S, Vázquez-Castellanos JF, Bargiela R, et al. Gut bacteria metabolism impacts immune recovery in HIV-infected individuals. EBioMedicine 2016;8:203−16.

[37] Serrano-Villar S, Rojo D, Martinez-Martinez M, Deusch S, Vazquez-Castellanos JF, Sainz T, et al. HIV infection results in metabolic alterations in the gut microbiota different from those induced by other diseases. Nat Sci Rep 2016;6:1−10.

[38] Pérez-Cobas AE, Artacho A, Knecht H, Ferrús ML, Friedrichs A, Ott SJ, et al. Differential effects of antibiotic therapy on the structure and function of human gut microbiota. PLoS One 2013;8(11):e80201.

[39] Underwood MA, Gaerlan S, De Leoz MLA, Dimapasoc L, Kalanetra KM, Lemay DG, et al. Human milk oligosaccharides in premature infants: absorption, excretion, and influence on the intestinal microbiota. Pediatr Res 2015;78(6):670−7.

[40] Xu W, Chen D, Wang N, Zhang T, Zhou R, Huan T, et al. Development of high-performance chemical isotope labeling LC−MS for profiling the human fecal metabolome. Anal Chem 2015;87(2):829−36.

[41] Tanca A, Palomba A, Pisanu S, Addis MF, Uzzau S. Enrichment or depletion? The impact of stool pretreatment on metaproteomic characterization of the human gut microbiota. Proteomics 2015;15(20):3474−85.

[42] Deda O, Gika HG, Wilson ID, Theodoridis GA. An overview of fecal sample preparation for global metabolic profiling. J Pharm Biomed Anal 2015;113:137−50.

[43] Vandeputte D, Falony G, Vieira-Silva S, Tito RY, Joossens M, Raes J. Stool consistency is strongly associated with gut microbiota richness and composition, enterotypes and bacterial growth rates. Gut 2016;65(1):57−62.

[44] Zheng H, Lorenzen JK, Astrup A, Larsen LH, Yde CC, Clausen MR, et al. Metabolic effects of a 24-week energy-restricted intervention combined with low or high dairy intake in overweightwomen: an NMR-based metabolomics investigation. Nutrients 2016;8(3).

[45] Abdulkadir B, Nelson A, Skeath T, Marrs ECL, Perry JD, Cummings SP, et al. Routine use of probiotics in preterm infants: longitudinal impact on the microbiome and metabolome. Neonatology 2016;109:239−47.

[46] Moosmang S, Pitscheider M, Sturm S, Seger C, Tilg H, Halabalaki M, et al. Metabolomic analysis-addressing NMR and LC-MS related problems in human feces sample preparation. Clin Chim Acta 2017, in press.

[47] De Leoz MLA, Kalanetra KM, Bokulich NA, Strum JS, Underwood MA, German JB, et al. Human milk glycomics and gut microbial genomics in infant feces show a correlation between human milk oligosaccharides and gut microbiota: a proof-of-concept study. J Proteome Res 2015;14(1):491−502.

[48] Weingarden AR, Chen C, Bobr A, Yao D, Lu Y, Nelson VM, et al. Microbiota transplantation restores normal fecal bile acid composition in recurrent *Clostridium difficile* infection. AJP Gastrointest Liver Physiol 2014;306(4):G310−9.

[49] Bingol K, Bruschweiler-Li L, Yu C, Somogyi A, Zhang F, Brüschweiler R. Metabolomics beyond spectroscopic databases: a combined MS/NMR strategy for the rapid identification of new metabolites in complex mixtures. Anal Chem 2015;87(7):3864−70.

[50] Bingol K, Brüschweiler R. Two elephants in the room: new hybrid nuclear magnetic resonance and mass spectrometry approaches for metabolomics. Curr Opin Clin Nutr Metab Care 2015;18(5):471−7.

[51] Jiménez-Girón A, Ibáñez C, Cifuentes A, Simó C, Muñoz-González I, Martín-Álvarez PJ, et al. Faecal metabolomic fingerprint after moderate consumption of red wine by healthy subjects. J Proteome Res 2015;14(2):897−905.

[52] Martin F-PJ, Lichti P, Bosco N, Brahmbhatt V, Oliveira M, Haller D, et al. Metabolic phenotyping of an adoptive transfer mouse model of experimental colitis and impact of dietary fish oil intake. J Proteome Res 2015;14(4):1911−9.

[53] Chai Y, Wang J, Wang T, Yang Y, Su J, Shi F, et al. Application of 1H NMR spectroscopy-based metabonomics to feces of cervical cancer patients with radiation-induced acute intestinal symptoms. Radiother Oncol 2015;117:294−301.

[54] Yang Y, Wang L, Wang S, Huang R, Zheng L, Liang S, et al. An integrated metabonomic approach to studying metabolic profiles in rat models with insulin resistance induced by high fructose. Mol Biosyst 2014;10(7):1803−11.

[55] Ferrer M, Raczkowska BA, Martínez-Martínez M, Barbas C, Rojo D. Phenotyping of gut microbiota: focus on capillary electrophoresis. Electrophoresis 2017;38(18):2275−86.

[56] Wehrens R, Hageman JA, van Eeuwijk F, Kooke R, Flood PJ, Wijnker E, et al. Improved batch correction in untargeted MS-based metabolomics. Metabolomics 2016;12(5):88.

[57] Gagnebin Y, Tonoli D, Lescuyer P, Ponte B, de Seigneux S, Martin P, et al. Metabolomic analysis of urine samples by UHPLC-QTOF-MS: impact of normalization strategies. Anal Chim Acta 2017;955:27−35.

[58] Cheema AK, Maier I, Dowdy T, Wang Y, Singh R, Ruegger PM, et al. Chemopreventive metabolites are correlated with a change in intestinal microbiota measured in A-T mice and decreased carcinogenesis. PLoS One 2016;11(4):e0151190.

[59] Preidis GA, Ajami NJ, Wong MC, Bessard BC, Conner ME, Petrosino JF. Microbial-derived metabolites reflect an altered intestinal microbiota during catch-up growth in undernourished neonatal mice. J Nutr 2016;146(5):940−8.

[60] Godzien J, Alonso-Herranz V, Barbas C, Armitage EG. Controlling the quality of metabolomics data: new strategies to get the best out of the QC sample. Metabolomics 2015;11(3):518−28.

[61] Steuer R, Morgenthal K, Weckwerth W, Selbig J. A gentle guide to the analysis of metabolomic data. Methods Mol Biol 2007;358:105−26.

[62] Korman A, Oh A, Raskind A, Banks D. Statistical methods in metabolomics. Methods Mol Biol 2012;856:381−413.

[63] Smirnov KS, Maier TV, Walker A, Heinzmann SS, Forcisi S, Martinez I, et al. Challenges of metabolomics in human gut microbiota research. Int J Med Microbiol 2016;306(5):266−79.

[64] Salek RM, Steinbeck C, Viant MR, Goodacre R, Dunn WB. The role of reporting standards for metabolite annotation and identification in metabolomic studies. Gigascience December 16, 2013;2(1):13.

[65] Vinaixa M, Schymanski EL, Neumann S, Navarro M, Salek RM, Yanes O. Mass spectral databases for LC/MS- and GC/MS-based metabolomics: state of the field and future prospects. Trends Anal Chem 2016;78:23−35.

[66] Gil de la Fuente A, Grace Armitage E, Otero A, Barbas C, Godzien J. Differentiating signals to make biological sense. A guide through databases for MS-based non-targeted metabolomics. Electrophoresis 2017;38(18):2242−56.

[67] Goedert JJ, Sampson JN, Moore SC, Xiao Q, Xiong X, Hayes RB, et al. Fecal metabolomics: assay performance and association with colorectal cancer. Carcinogenesis 2014;35(9):2089−96.

[68] Jansson J, Willing B, Lucio M, Fekete A, Dicksved J, Halfvarson J, et al. Metabolomics reveals metabolic biomarkers of Crohn's disease. PLoS One 2009;4(7).

[69] Duboc H, Rainteau D, Rajca S, Humbert L, Farabos D, Maubert M, et al. Increase in fecal primary bile acids and dysbiosis in patients with diarrhea-predominant irritable bowel syndrome. Neuro Gastroenterol Motil 2012;24. 513−e247.

[70] Serrano-Villar S, Moreno S, Ferrer M. The functional consequences of the microbiome in HIV: insights from metabolomic studies. Curr Opin HIV AIDS 2017;(1).

[71] Koay W, Siems L, Persaud D. The microbiome and HIV persistence: implications for viral remission and cure. Curr Opin HIV AIDS 2018;13.

[72] Pité H, Morais-Almeida M, Rocha S. Metabolomics in asthma: where do we stand? Curr Opin Pulm Med 2018;24(1):94−103.

[73] Matsumoto M, Ebata T, Hirooka J, Hosoya R, Inoue N, Itami S, et al. Antipruritic effects of the probiotic strain LKM512 in adults with atopic dermatitis. Ann Allergy Asthma Immunol 2014;113(2):209−16.

[74] Checkley W, Deza M, Klawitter J, Romero K, Pollard S, Wise R, et al. Identifying biomarkers for asthma diagnosis using targeted metabolomics approaches. Respir Med 2016;121:59−66.

[75] Castellazzi A, Valsecchi C, Caimmi S, Licari A, Marseglia A, Leoni M, et al. Probiotics and food allergy. Ital J Pediatr 2013;39(1):47.

[76] Smolinska A, Blanchet L, Buydens LMC, Wijmenga SS. NMR and pattern recognition methods in metabolomics: from data acquisition to biomarker discovery: a review. Anal Chim Acta 2012;750:82−97.

[77] Bonassi S, Fenech M, Lando C, Lin Y, Ceppi M, Chang WP, et al. Human micronucleus project: international database comparison for results with the cytokinesis-block micronucleus assay in human lymphocytes: effect of laboratory protocol, scoring criteria, and host factors on the frequency of micronuclei. Environ Mol Mutagen 2001;37(1):31−45.

[78] Cammarota G, Ianiro G, Tilg H, et al. European consensus conference on faecal microbiota transplantation in clinical practice. Gut 2017;66(4):569−80.

[79] De Palma G, Lynch M, Lu J, Dang V, Jury J, Umeh G, et al. Transplantation of fecal microbiota from patients with irritable bowel syndrome alters gut function and behavior in recipient mice. Sci Transl Med 2017;9(379).

[80] Windey K, François I, Broekaert W, De Preter V, Delcour JA, Louat T, et al. High dose of prebiotics reduces fecal water cytotoxicity in healthy subjects. Mol Nutr Food Res 2014;58(11):2206−18.

[81] Martin FJ, Sprenger N, Montoliu I, Rezzi S, Kochhar S, Nicholson JK. Metabonomics research articles. J Proteome Res 2010:5284−95.

[82] West CE, Jenmalm MC, Prescott SL. The gut microbiota and its role in the development of allergic disease: a wider perspective. Clin Exp Allergy 2015;45(1):43−53.

[83] Bouchaud G, Castan L, Chesné J, Braza F, Aubert P, Neunlist M, et al. Maternal exposure to GOS/inulin mixture prevents food allergies and promotes tolerance in offspring in mice. Allergy 2016;71(1):68−76.

[84] Spinler J, Auchtung J, Brown A, Boonma P, Oezguen N, Ross C, et al. Next-generation probiotics targeting *Clostridium difficile* through precursor-directed antimicrobial biosynthesis. Infect Immun 2017;85(10).

[85] Pinto-Sánchez M, Gastroenterology E, Temprano M, Sugai E, González A, Moreno M, et al. Bifidobacterium infantis NLS super strain reduces the expression of α-defensin-5, a marker of innate immunity, in the mucosa of active celiac disease patients. J Clin Gastroenterol 2017;51(9):814−7.

[86] Williams N. Probiotics. Am J Heal Pharm 2010;67(6):449−58.

[87] Tang M, Ponsonby A, Orsini F, Tey D, Robinson M, Su E, et al. Administration of a probiotic with peanut oral immunotherapy: a randomized trial. J Allergy Clin Immunol 2015;135(3):737−44.

[88] Canani R, Sangwan N, Stefka A, Nocerino R, Paparo L, Aitoro R, et al. Lactobacillus rhamnosus GG-supplemented formula expands butyrate-producing bacterial strains in food allergic infants. ISME J 2016;10(3):742−50.

[89] Manes N, Shulzhenko N, Nuccio A, Azeem S, Morgun A, Nita-Lazar A. Multi-omics comparative analysis reveals multiple layers of host signaling pathway regulation by the gut microbiota. mSystems 2017;2(5):e00107−17.

[90] Markley JL, Brüschweiler R, Edison AS, Eghbalnia HR, Powers R, Raftery D, et al. The future of NMR-based metabolomics. Curr Opin Biotechnol 2017;43:34−40.

[91] Cao H, Huang H, Xu W, Chen D, Yu J, Li J, et al. Fecal metabolome profiling of liver cirrhosis and hepatocellular carcinoma patients by ultra performance liquid chromatography-mass spectrometry. Anal Chim Acta 2011;691:68−75.

[92] Roager HM, Sulek K, Skov K, Frandsen HL, Smedsgaard J, Wilcks A, et al. Lactobacillus acidophilus NCFM affects vitamin E acetate metabolism and intestinal bile acid signature in monocolonized mice. Gut Microb 2014;5(3):296−303.

[93] Ponnusamy K, Choi JN, Kim J, Lee SY, Lee CH. Microbial community and metabolomic comparison of irritable bowel syndrome faeces. J Med Microbiol 2011;60(6):817−27.

[94] Ordiz MI, May TD, Mihindukulasuriya K, Martin J, Crowley J, Tarr PI, et al. The effect of dietary resistant starch type 2 on the microbiota and markers of gut inflammation in rural Malawi children. Microbiome 2015;3:37.

[95] Amer B, Nebel C, Bertram HC, Mortensen G, Dalsgaard TK. Direct derivatization vs aqueous extraction methods of fecal free fatty acids for GC-MS analysis. Lipids 2015;50:681−9.

Chapter 5

Metabolomics Tools and Information Retrieval in Microbiome Hacking

Abdellah Tebani[1] and Soumeya Bekri[1,2]

[1]*Department of Metabolic Biochemistry, Rouen University Hospital, Rouen, France;* [2]*Normandie Univ, UNIROUEN, CHU Rouen, IRIB, INSERM U1245, Rouen, France*

INTRODUCTION: WHAT IS METABOLOMICS?

The microbiome started to get rid of static describing of culturable species and move to the culture-independent characterization using next-generation sequencing (NGS) and metagenomic techniques [1]. However, this approach has its limits to describe members of a microbial community, relying only on their genome sequence. Moreover, study comparison and meta-analyses are challenging, NGS results are method-dependent, and several steps may impact the final result as DNA extraction and library preparation and contamination issues [2]. Given these limits, using new high-throughput omics technologies triggered this huge move to comprehensive biological information recovery. Moreover, the development of computational frameworks allowed multiscale systems modeling and immersive visualization [3,4]. Omics strategies mainly allow a comprehensive and global analysis of biomolecules (genes, proteins, metabolites, etc.) of a given biological system (i.e., tissue, cell, fluid, or organism).

Metabolites are small organic molecules produced by enzymatic reactions. They are the last layer in the biological information flow. So, it is the closest to the phenotype. Olivier et al. defined, in 1998, the set of metabolites synthesized by an organism and coined it the term metabolome [5]. Metabolome refers to all metabolites present in a biological system [6]. Other terms have been used, including metabolic fingerprinting, metabotyping, or metabolic phenotyping. Metabolomics is an "omic" technology that is based on molecular characterizations and biochemistry of the metabolome and its variations in response to genetic, environmental, drug, or dietary stimuli [7].

Metabolomics has been applied to many diseases, including microbiome studies, with promising perspectives in screening, diagnosis, prognosis, patient stratification, and treatment follow-up [8,9]. The functional characterization approach moves from the organisms forming up the microbiome (community composition) to the genes, proteins, and metabolites making up the microbiome (community function) [10]. The communications inside microorganism communities and between this set of organisms and the host take place predominantly via metabolite pathways. Thus, cracking these communication pathways aims at uncovering the interactive function of biology, using biological information and its contextualization (Fig. 5.1). Systems biology can be defined as a global and systemic analysis of complex system interconnections and their functional interrelationships [11].

ANALYTICAL STRATEGIES: FROM BIOLOGICAL SAMPLES TO SIGNALS

A metabolic fingerprint is made of a set of changed metabolites, rather than just a single metabolite, given the network theory underpinning biological systems [8]. Analytical technologies need to be reliable and robust, for high-throughput analyses to track as many metabolites as possible. Furthermore, metabolites exhibit a qualitative and quantitative heterogeneity. Multiple sample preparation procedures and analytical strategies are needed to recover as much as possible of the metabolome [12]. The sequential steps of a metabolomics workflow are depicted in Fig. 5.2.

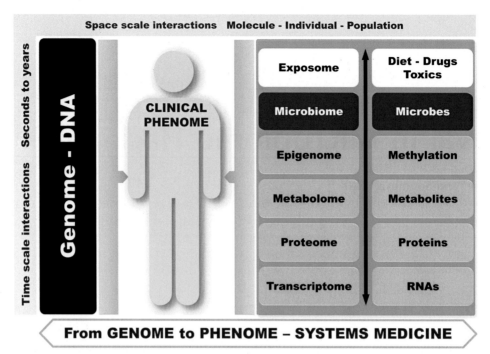

FIGURE 5.1 Systems medicine overview. Three main drivers shape the phenotype; the molecular phenome, the exposome, and the clinical phenome. Thanks to the ever-evolving high-throughput technologies and digital sensors, these different informational layers could be interrogated through biological and clinical metrics to define a highly resolved multidimensional spatiotemporal space. This will pave the way to characterize disease network dynamics from the molecular to the population level.

Multivariate data analyses are widely used to extract information from large metabolomics data sets [13]. The metabolomics workflow generally includes biological question, experimental design, sample preparation, data acquisition, data preprocessing, data cleaning and analysis, network and pathway analysis, and finally biological interpretation and medical decision (Fig.5.2).

Experimental Design Considerations: Doing It Right From the Beginning

Depending of the biological problem, the investigator has to define the metabolomics approach (targeted vs. untargeted), biological samples (biofluids, tissues, cells, and/or intact organisms), sample size, pooling, experimental conditions (i.e., observational studies, exploratory studies, time series), sampling conditions (frequency of sample collection, quenching methods to stop enzymatic activity, storage), analytical platforms, and standardized sample preparation protocols. Sample randomization and sample analysis order can be used to reduce confounders. Some advanced statistical experimental design strategies can be used to manage these issues [14].

Biological Samples

Sample collection (time and type), storage, and handling have deep impacts on the recovered metabolic data. The investigator has to standardize these aspects to avoid artifactual biomarker discovery and misleading metabolic in-terpretations. Samples can be divided into two classes: (1) metabolically active samples (intracellular metabolome) and (2) metabolically inactive samples (extracellular metabolome) [15]. For microbiome metabolomics, different samples can be considered: natural cavities, skin, hair follicles, external ear, mucous membranes, gastrointestinal tract, respiratory tract, urogenital tract, vagina, and placenta.

Sample Preparation

In untargeted metabolomics, minimal or no sample preparation is recommended to avoid metabolite loss. Tissues are homogenized using a mortar and pestle, or ball grinding with silica particles or stainless steel. An extraction solvent is often used in the homogenization process, which leads to cell lysis and metabolite extraction. Monophasic (water/methanol,

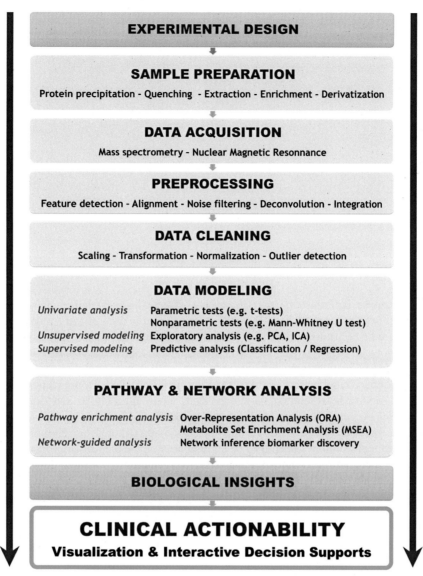

FIGURE 5.2 General metabolomics workflow. Metabolomics is divided into two main strategies. A targeted metabolomics is a quantitative analysis or a semiquantitative analysis of a set of metabolites that might be linked to common chemical classes or a selected metabolic pathway. An untargeted metabolomics approach is primarily based on the qualitative or semiquantitative analysis of the largest possible number of metabolites from diverse chemical and biological classes contained in a biological sample. The generated data undergo data analysis step (univariate and multivariate) and functional analysis to get actionable biological insight.

water/acetonitrile) or biphasic solvent (water and methanol often along with a nonpolar solvent such as chloroform, dichloromethane, or methyl tert-butyl ether) can be used in extraction systems, depending on the considered analysis [16]. The solvent systems choice depends on whether polar or nonpolar molecules are to be analyzed. For cell metabolomics, a quenching step is required before extraction to reduce metabolite modifications. Extraction from cells must be performed quickly to avoid enzymatic reactions and to enhance reproducibility [17].

For microbiome metabolomics analysis, different solvent systems have been described and presented in Table 5.1. Procedures for preparing microbial samples vary, depending on the chosen metabolomics strategy [18]. To analyze metabolites secreted outside the cells, this may include few steps, e.g., filtration, centrifugation, or extraction, while to study the intracellular metabolites, we have to set complex and time-consuming procedures. A huge diversity characterizes microbial samples. Bacterial strains differ regarding size, physiology, cell wall structure, growth conditions, and nutrient requirement. Thus, every analytical experiment concerning bacteria requires individually planned research methodology.

TABLE 5.1 Metabolites Extraction Procedures of Particular Interest in Microbiome Metabolomics Studies [18]

Extraction Solvent	Ratio
Methanol−Chloroform−Water (cold)	10:3:1, v/v
	3:1:1, v/v
	2:1, v/v
	2:2:1, v/v
Ethanol−Water (boiled)	75:25, v/v
Methanol−Water (cold)	80:20, v/v
	50:50, v/v
Acetonitrile−Methanol−Water	2:2:1, v/v
Methanol	/

Extraction from cells can be performed by two approaches, depending on whether quenching and extraction are simultaneous or sequential. In simultaneous cases, the quenching solution is also an extracting solvent. However, in the sequential approach, cells are first separated from their surrounding medium, then metabolites are isolated from cells using an appropriate extracting solvent. In the sequential mode, quenching is followed by sample dividing in two fractions, usually by cold centrifugation or fast filtration. This step reduces the cell dilution effect in the supernatant and separately quantifies extracellular and intracellular metabolites. An extraction step is usually performed after cell isolation using a hot or cold extracting solvent, acid or base solution. In simultaneous mode, intracellular and extracellular metabolites are analyzed together. Nevertheless, the endometabolome and the exometabolome cannot be distinguished.

Extraction pH, storage time, and temperature are important issues that should be rigorously handled [19]. Sample preparation may be costly in terms of time, which may be quite limiting in the clinical environment. Thus, collecting data without sample preparation, along with real-time data analysis, is very appealing for rapid clinical decision-making. This would make metabolomics highly clinically actionable (i.e., surgery, pathology, and microbiology) [20]. For example, the intelligent knife (iKnife) [21] and the MasSpec Pen [22] are huge advances for in vivo sampling and real-time metabolomics. The process vaporizes tissue, and the resultant smoke is guided into a mass spectrometer to provide real-time clinical decision-making in the operating room.

Analytical Platforms

Compared with other omics analyses, metabolome analysis exhibits different challenges. Subunit order is what matters for identification and functional analysis of DNA (genomics), RNAs (transcriptomics), and proteins (proteomics). Thus, analytical techniques based on sequencing essentially rely on incremental detection of these subunits. For metabolomics the strategy is simultaneous, individual, selective, and sensitive differentiation of metabolites across a heterogenous chemical space. In an early paper in the field, Pauling et al. described a method using gas chromatographic separation with flame ionization detection to analyze the breath [23]. Huge analytical developments have been made since then.

The dominant technologies in metabolomics are nuclear magnetic resonance/NMR spectroscopy and mass spectrometry/MS, sometimes in combination with a gas-phase or liquid-phase separation method [24]. These technologies extract global, comprehensive, and unbiased chemical information from complex biological mixtures. To translate this raw information, the extracted high-dimensional spectral data are typically analyzed using chemometric methods to unveil relevant metabolic signals that can be used for either sample classification or biomarker discovery.

Nuclear Magnetic Resonance Spectroscopy: Capturing the Dance of Atoms

NMR spectroscopy offers atom-centered information, which is vital for molecular structures elucidation [25]. NMR is based on the absorption and reemission of energy by the atom nuclei, due to variations in an external magnetic field. Hydrogen is the most commonly used (^{1}H-NMR) because of its naturally high abundance in biological samples. Carbon-13

(^{13}C-NMR) and phosphorus-31 (^{31}P-NMR) can also be used to provide additional information on specific metabolites. The NMR spectral patterns are used for metabolite identification, and peak areas allow quantification.

Frequency axes depicts the chemical shift expressed in parts per million (ppm). The chemical shift is calculated as the difference between the metabolite resonance frequency and that of a reference substance [26]. One-dimensional NMR (1D-NMR) spectra are based on a single-frequency axis, where the peaks of each molecule occur within the resonant frequencies of that axis. This method is the most used in high-throughput metabolomics. Two-dimensional NMR (2D-NMR) is based on two-frequency axes and can be used as 1D-NMR complement. Of note, nuclei with low natural abundance such as ^{2}H (deuteron), ^{13}C, and ^{15}N may serve as excellent metabolic tracers [27].

NMR spectroscopy offers many advantages, because it allows rigorous quantification of highly abundant metabolites present in biological fluids, cell extracts, and tissues, with minimal or no sample preparation [28]. NMR is also a rapid, nondestructive, highly reproducible, and a robust technique. It also allows isomers identification and is the gold standard for determining structures of unknown compounds. Using stable isotope labels, NMR spectroscopy can be used for dynamic assessment of compartmentalization of metabolic pathways, such as metabolite transformations and drug metabolism. Finally, intact tissue NMR imaging and spectroscopy are very appealing for in vivo metabolic investigations [29]. Compared to MS-based methods, NMR main drawback is its low sensitivity and resolution.

Mass Spectrometry: Weighing the Moving Ions

Mass spectrometry is an analytical method, which recovers chemical information from the gas-phase ions produced from a sample. The ions generate different peak patterns that define the fingerprint of the original molecule in the form of a mass-to-charge ratio (m/z) and a relative intensity of the measured chemical features (e.g., metabolites). The sample is introduced into the mass spectrometer via a sample inlet. An ion source produces gas-phase ions, a mass analyzer separates the ions according to their m/z, and a detector generates an electric current from the incident ions. The current is proportional to ion abundances.

A sample can be directly injected into a mass spectrometer depicted as direct infusion mass spectrometry (DI−MS) [30]. However, ion suppression effect is a major drawback that leads to miss metabolite information such as isomers. Mass analyzers can be used alone or in combination with the same type of mass analyzer or with different ones (hybrid instruments). Such combinations define the analytical mode of tandem mass spectrometry (MS/MS). In MS/MS, the ions that arrive at the first mass analyzer (precursor ions) are selected and then fragmented in a collision cell. The fragmented ions are, then, separated according to their m/z in a second mass analyzer and finally detected.

Data-dependent analysis (DDA) modes include single reaction monitoring (SRM) or multiple reaction monitoring (MRM), which is a parallel detection of all SRM transitions in a single analysis. For some mass analyzers such as quadrupole ion traps, several steps of MS/MS can be performed. For example, the fragmented ions can be further fragmented and detected. These experiments are called multiple-stage mass spectrometry (MS^n, n refers to the number of MS steps). MS/MS and MS^n improve structural identification, by combining information from both molecular and fragmentation patterns of precursor ions.

Choosing a mass analyzer is mainly dependent on the type of metabolomics approach to undertake, targeted or untargeted. Single quadrupole, triple quadrupole, quadrupole ion trap, and Orbitrap (OT) are suitable for targeted metabolomics because of their sensitivity and duty cycle characteristics. In comparison, dynamic range, mass accuracy, and resolution power are important for a mass analyzer to be used in untargeted metabolomics. Time of flight (TOF), quadrupole time of flight (QTOF), Fourier transform ion cyclotron resonance (FTICR), and OT are the most used mass analyzers for this purpose. A QTOF mass analyzer is a hybrid instrument that can generate high-resolution MS/MS spectra.

FTICR is an ultrahigh-resolution mass analyzer that uses cyclotron frequency in a fixed magnetic field to measure m/z ions at the cost of relatively slow acquisition rates (typically 1 Hz). OT is also an FTMS instrument, which is based on harmonic ion oscillations in an electrostatic field. A wide range of instrumental and technical variants are currently available for MS spectrometry. These variants are mainly characterized by different ionization and mass selection methods.

Hyphenated Mass Spectrometry Techniques

Because of the matrix effects, MS is generally combined to a separation step in metabolomics. This step allows sequential MS analysis of the different molecules, by reducing the complexity of a biological sample prior to analysis. Different separation methods are used with MS, such as liquid chromatography (LC−MS), gas chromatography (GC−MS), and capillary electrophoresis (CE−MS). Metabolites with different physicochemical properties spend different times (retention time, t_R) in the separation dimension. These separation methods widen MS dynamic and sensitivity ranges and provide orthogonal molecular information.

LC–MS is widely used in metabolomics because of its analytical versatility, covering separation performance of different molecular classes, spanning from very polar to very lipophilic compounds. This is possible through the variety of chromatographic columns and their stationary phases [31]. When polar stationary phase columns are used, the method is referred to as normal-phase liquid chromatography (NPLC); when nonpolar stationary phase columns are used, the method is called reversed-phase liquid chromatography (RPLC).

For lipophilic metabolites, nonpolar C18 and C8 columns are mostly used [32] while hydrophilic interaction liquid chromatography (HILIC) is recommended for polar compounds. Multiple-column strategies could also be used for higher metabolome coverage [33]. Recently, RPLC and HILIC columns with a smaller internal diameter (e.g., 1 mm) and shorter length have been used in metabolomics. Thus, instruments that can operate at very high pressure, ultrahigh-performance liquid chromatography (UHPLC) coupled to mass spectrometry, are employed. UHPLC methods allow increasing resolution and sensitivity, lowering ion suppression.

GC–MS is often used for analysis of volatile or low vapor pressure compounds such as a set of lipids, long-chain alcohols, sugar alcohols, and organic acids. Selective separation, robustness, and reproducibility vary across different systems and vendors. This allowed building comprehensive GC–MS mass spectral libraries. Derivatization is still the main limitation of GC–MS. In metabolomics, derivatization usually uses oximation and a silylation/chloroformate reagent. This step is time-consuming and error-prone and hampers the throughput.

CE–MS is another technique that offers an orthogonal separation dimension. CE-specific characteristics such as high efficiency and resolution, high throughput, and, importantly, the ability to assess the most polar compounds without any derivatization are very appealing in metabolomics [34]. Capillary zone electrophoresis (CZE) is the simplest and most commonly used CE mode because of its principle of separation and its broad application to the analysis of diverse samples, spanning small to large biomolecules. In CZE, analytes are separated according to their intrinsic differential electrophoretic mobility in a capillary filled with separation buffer under the influence of an electric field. Mobilities depend on the ion m/z and the viscosity of the medium [34]. The main drawback of CZE is that neutral molecules are not separated. Thus, other CE modes have been developed, among other capillary isotachophoresis, capillary isoelectric focusing based on pH gradient, and affinity capillary electrophoresis. Given its simplicity, CZE is the preferred CE mode in metabolomics.

While MS analyses alone are very rapid, LC and GC separations lengthen the analysis time to minutes but are necessary to reduce both ionization and detector suppression. Fortunately, new technologies should allow faster and yet comprehensive metabolomics analyses. Ion mobility spectrometry (IMS), a gas-phase separation, is getting interest in metabolomics [35,36]. IMS apprehends spatial molecular conformation information through the determination of the ion collision cross section (CCS) [37–39], a predictable chemical descriptor.

IMS typically performs on a millisecond timescale, which can be perfectly nested between chromatography (seconds) and high-resolution MS detection (microseconds). Hence, coupling IMS with high-resolution mass spectrometry and chromatography (LC–IMS–MS) provides additional analytic selectivity, without significantly hampering the speed of MS-based measurements. As a result, the MS dimension gives access to accurate mass information, while the IMS dimension provides molecular, structural, and conformational information. One important feature of IMS is its ability to separate isomers [40].

The predictability of the CCS and peak width for one isomer mainly depends on ion diffusion [41,42]. Both IMS–MS and LC–IMS–MS analyses have been demonstrated higher sensitivity and throughput than LC–MS studies alone. Of note, the multidimensional coupling of different separation techniques requires that the resolution obtained from each anterior separation must be largely conserved, as the analytes pass to the next dimensions. This conservation is particularly difficult when all analytes travel along the same path during the analysis. Thus, the solution is to incrementally increase the sampling frequency of each subsequent time dimension so that multiple measurements are obtained within a fixed time interval. In this way, the arrival time in each anterior dimension can be grouped based on the integrated signal of subsequent dimensions.

Exploring a multivectorial space containing retention time, CCS and accurate mass obtained by the combination of multiple separation methods with MS affords valuable measurement integration that leads to biomarker discovery enhancement through better molecular characterization [39,43].

Toward Real-Time MS-Based Metabolomics

Recent introduction of ambient ionization sources has enhanced global metabolic profiling throughput. These techniques allow direct sampling of complex matrices under ambient conditions, including atmospheric solids analysis probe (ASAP), desorption electrospray ionization (DESI), and rapid evaporative ionization MS methods (REIMS). These techniques can provide real-time and interpretable MS data on biofluids and tissues in vivo and ex vivo. These methods are reshaping real-time metabolome analysis in various fields [20,44]. For example, in surgery, it is challenging to visually distinguish between healthy and diseased tissues. It often requires biopsies and immunostaining time-consuming procedures, which are performed by trained histopathologists.

Eliminating this need for external tissue histotyping is very appealing in such context. Thus, metabolomics-based tools such as the (iKnife) [21] and the MasSpec Pen [22] could open the way to amazing real-time precision surgery. These techniques also revealed promising potential to gain deeper molecular insight into microbial colonies interactions [45]. Table 5.2 presents a comparison between different analytical strategies used in metabolomics, along with inherent limits and advantages.

QUALITY CONTROL MANAGEMENT: CLEANING DATA TO GET CONSISTENT SIGNALS

Validating an analytical strategy is crucial in quality control/QC strategies. Validation is defined as "establishing documented evidence that provides a high degree of assurance that a specific process will generate a product, meeting its predetermined specifications and quality attributes" [46]. Different guidelines tightly regulate bioanalytical method validation and can be used in targeted metabolomics [46]. These guidelines define different criteria that may include linearity accuracy, precision, specificity, limit of quantitation, and limit of detection. However, in untargeted metabolomics, there are no regulatory guidelines. So, the metabolomics community is still striving to set QC strategies to enhance data integrity [47].

The QC rationale is intrinsically purpose-driven. So, the untargeted metabolomics purpose is to find significant metabolic signals through differential statistical analysis retrieved from as many as possible chemical features. Thus, this purpose drives the validation parameters. On the one hand, comparisons are considered valid, provided all the samples are assessed under the same conditions, using a rigorously precise analytical method. On the other hand, the signal variation should be related to the abundance of the chemical features and should not be dependent on other analytical and instrumental variations. Different filters are proposed to clean and assess the initial data using QC samples. In metabolomics, it is often recommended to analyze QC samples at regular intervals, across an entire batch, in order to monitor the analytical procedure [14].

The QC samples are prepared by pooling the study samples to represent all included samples. Thus, it is possible to assess each feature in the data, regarding its presence in the QCs. The detection percentage evaluates how consistently a feature is detected across all samples. Features that are not present in a defined cutoff number of QC (50% cutoff is often applied [14]) can be filtered out from the data. Instrument response drifts can also be monitored using QC throughout the data acquisition. This drift may lead to intra- and interbatch variation. Many methods are available to remove intensity drifts using feature intensities of the QC and the experimental run order [48].

The repeatability can also be monitored using QC. Each feature in the QCs should show low relative standard deviation (RSD) across all the QC samples to be retained in the final dataset. RSD for each feature is calculated by dividing the sample standard deviation by the sample mean. Features with high RSD values should be cleaned from the data before analysis. Commonly used thresholds are 20%–30% [14]. Finally, a series of QC samples with varying dilutions can be prepared and analyzed within the experimental batch, so it could be used to evaluate features quality. The dilution factors can be regressed against the corresponding intensities of each feature in the data. Features intensity must be correlated to the matrix concentration of the diluted QC samples to be retained for further analysis. Thresholds ranging from 50% to 70% are often suggested [49]. However, inspecting the distribution of the R^2 values may help in setting the threshold.

BIOLOGICAL INFORMATION RECOVERY: FROM DATA TO INFORMATION

Addressing biology as an informational science requires extracting information from high-dimensional data, storing, mining, and visualizing it. Bioinformatics is the field that is born to tackle these challenges [50]. The first challenge is dimensionality reduction by selecting informative signals from the noisy chemical data. To achieve this goal, chemometrics, which is the science that uses data-driven means to extracting useful information from chemical systems, is widely used [51]. Both descriptive and predictive problems could be addressed using chemical data.

Preprocessing: Turning Signals Into Numbers

Targeted metabolomics aims to process clean and ready-to-use data sets retrieved from a subset of the metabolome. It contains predefined and biochemically annotated metabolites. Since only a set of selected metabolites are analyzed, no analytical artifacts are carried in the downstream analysis. Different automated processes have been developed [52] along

TABLE 5.2 Metabolomics Technologies and Related Characterization Descriptors

Platform	Analytical Descriptor	Technique	Principle	Advantages	Limits
Nuclear magnetic resonance	Chemical shift; Chemical shift × chemical shift	One dimension; Two dimensions	Uses interaction of spin active nuclei (^1H, ^{13}C, ^{31}P) with electromagnetic fields, yielding structural, chemical, and molecular environment information	Nondestructive; Highly reproducible; Exact quantification possible; Minimal sample preparation; Molecular dynamic and compartmental information using diffusional methods; Relatively high throughput; Availability of databases for identification	High instrumentation cost; Overlap of metabolites; Low sensitivity
Mass spectrometry	m/z; Isotopic modeling	Direct injection (DI–MS)	Uses a nanospray source directly coupled to MS detector. It does not require chromatographic separation. Isotopic modeling compares the abundances of specific isotopes in the molecular isotopic pattern. It provides rapid indication of amount and identity of heteroatoms	Very high throughput; High sensitivity; No cross-sample contamination; No column carryover; Low-cost analysis; Automated analysis; Low sample volume requirement; Allows MS imaging	Samples not recoverable (destructive); No retention time information, which gives limited specificity; Inability to separate isomers; Subjected to significant ion suppression phenomenon; High ionization discrimination (ESI)
	m/z; Fragmentation pattern	MS/MS	Tandem MS using ion activation to provide characteristic fragment species. It provides metabolite structural information to discriminate multiple isomers for a given elemental formula	Automated analysis; High throughput; High specificity; High sensitivity	Samples not recoverable (destructive); Matrix effect assessment; Need of stable isotope-labeled compounds as ideal internal standards
	Retention time; m/z	Liquid chromatography (LC–MS); Gas chromatography (GC–MS); Capillary electrophoresis (CE–MS)	Uses chromatographic columns that enable liquid-phase chromatographic separation of molecules followed by MS detection (suitable for polar to hydrophobic compounds)	Minimal sample preparation (protein precipitation or dilution of biological sample); High-throughput capability; UHPLC can be coupled to any type of MS; Flexibility in column chemistry widening the range of detectable compounds; High sensitivity	Samples not recoverable (destructive); Very polar molecules need specific chromatographic conditions; Retention times are highly dependent on exact chromatographic conditions; Batch analysis; Lack of large metabolite databases; High ionization discrimination (ESI)
			Uses chromatographic columns that enable gas-phase chromatographic separation of molecules followed by MS detection (suited for apolar and volatile compounds)	Structure information obtained through in-source fragmentation; Availability of universal databases for identification; High sensitivity; Reproducible	Samples not recoverable (destructive); Requires more extensive sample preparation; Only volatile compounds are detected; Polar compounds need derivatization; Low ionization discrimination
			Uses electrokinetic separation of polar molecules paired with a mass spectrometry detector	Excellent for polar analysis in aqueous samples; Measures inorganic and organic anions; Low running costs	Samples not recoverable (destructive); Relatively low-throughput profiling
	Drift time (CCS); m/z	Ion mobility spectrometry (IMS–MS)	Uses a uniform or periodic electric field and a buffer gas to separate ions based on charge, size and shape paired with mass spectrometry	Very robust and reproducible (ability to determine collision cross section, which is a robust chemical descriptor); High peak capacity; High selectivity; Separation of isomeric and isobaric compounds; Very high throughput	Samples not recoverable (destructive); CCS and mass are highly correlated parameters, which limits the orthogonality of the method

with commercial solutions. However, in untargeted metabolomics, data analysis is time-consuming. This approach attempts a comprehensive analysis of all detectable metabolites in a given sample, including unknowns. It requires a holistic analysis of high-dimensional raw data sets. It is then mandatory to convert the data into computationally manageable formats without compromising the carried chemical information.

Due to analytical/instrument noise and sample variation, MS and NMR spectra often exhibit differences in position, peak shape, and width. The goal of preprocessing is to correct these variations for better quantification of metabolites and intersample comparability. Data preprocessing may include some or all of the following steps: peak detection, baseline correction, noise filtering, peak alignment, and spectral deconvolution [53]. A typical output of the preprocessing step is a data matrix, which contains the detected features and the corresponding intensity (abundance) in each sample.

Normalization, Transformation, and Scaling

Metabolomics data have several common characteristics with other omics, such as asymmetric data distribution and analytical, instrumental, and biological noise. Data normalization aims to remove these experimental biases related to the features abundance between samples, along with conserving biological variations [54]. Indeed, the metabolite heterogeneity and interindividual variations lead to differences in extraction and MS ionization patterns, making it difficult to distinguish biological variations from that of analytical biases. Strategies for normalization of metabolomics data are statistically or chemically based. Statistical approaches use mathematical models [55]. Chemical approaches could be based on one or more reference compounds, internal standards, exogenous or endogenous metabolites to normalize the entire chromatogram (single compound), or certain regions of the chromatogram, by normalizing each zone according to a standard that is eluted in that region.

Other strategies use the characteristics of the studied matrix, such as volume (e.g., 24-h urine), osmolality, or dry mass of the samples. Protein or creatinine levels can also be used [56]. A state-of-the-art normalization technique was reported recently [55]. Inference in statistics is often related to a specific type of probability distribution of the data. Hence, the statistical tools to use depend on this data distribution. In metabolomics, MS and NMR data are noisy and the features distributions can be skewed. Thus, transformations before statistical analysis aim to correct skewness and heteroscedasticity. Log or power transformations can be used.

Data scaling aims to remove the offset from the data to focus on the biological variation. Autoscaling (unit variance scaling) and pareto scaling are most popular. The latter is most relevant to metabolomics. Each observation in the mean-centered feature is divided by the square root of the standard deviation in pareto scaling [57].

Data Analysis

Univariate Data Analysis: One Variable at a Time Is Not Enough

The main limitation of univariate methods is that they consider only one variable at a time. This may not fit the metabolomics data high dimensionality. Provided that the normality assumption is verified, parametric tests such as Student's t test and ANOVA are commonly applied to assess the differences between two or more groups, respectively. If normality is not assumed, nonparametric tests such as Mann–Whitney U test or Kruskal–Wallis one-way ANOVA may be used. Another important issue is multiple testing problems due to applying multiple univariate tests in parallel. Since many features are simultaneously analyzed in metabolomics, the probability of accidental statistically significant difference (i.e., true positive) is high. Correction methods can be used to fix this multiple testing problem. In the Bonferroni correction, the significance level for a hypothesis is divided by the number of hypotheses being tested. Hence, the Bonferroni correction is considered a stringent correction method. Less conservative methods are available and are based on lowering the false-discovery rate (FDR). FDR-based methods minimize the expected proportion of false positives among the total number of positives. Confounding factors such as sex, age, or diet may lead to spurious results and should be properly addressed. The main drawback of univariate methods is their lack of feature correlations and interactions. Thus, advanced multivariate tools are more suitable for deeper inference.

Multivariate Data Analysis: Consider It as a Whole

Advanced multivariate tools are more suitable for deeper inference. In multivariate methods, representative samples are presented as dots in the space of the initial variables. The samples can then be projected into a lower dimensionality space, based on components or latent variables, such as a line, a plane, or a hyperplane, which can be seen as the shadow of the initial data set viewed from its best informative perspective. The sample coordinates of the newly defined latent variables are the scores and the loadings of the variance directions to which they are projected. The loadings vector for each latent variable contains the weights of all initial variables (metabolites), for that latent variable. Chemometrics methods are mainly divided into unsupervised and supervised methods. In unsupervised methods, no assumptions are made about the samples and the aim is mainly exploratory. Unsupervised methods attempt to uncover clustering trends in the data that underlie relationships between the samples. Using visualization tools, these methods also highlight the variables that are driving these relationships.

In metabolomics data, metabolic similarity explains the observed clustering. Principal component analysis (PCA) is a widely used pattern recognition method. PCA allows two- or three-dimensional data visualization. Because it contains no assumptions on the data, it is used as an initial visualization and exploratory tool, to detect grouping trends and outliers. Independent component analysis (ICA) is another unsupervised method, which is a blind source separation method, separating multivariate signals into additive subcomponents [58]. Its interpretation is quite similar to PCA, but instead of orthogonal components, it calculates non-Gaussian and mutually independent components. ICA has been successfully used in metabolomics [59]. Other unsupervised methods, such as clustering, aim to identify naturally occurring clusters in the data set, by using similarity defined by distance and linkage metrics. A dendrogram or a heat map can also be created to visualize the similarities between samples. Correlation matrix, k-means clustering, hierarchical cluster analysis, and self-organizing maps are commonly used.

In supervised methods, samples are labeled and each sample is associated with a specific outcome value, and the aim is mainly to explain and to predict that outcome. The task is called classification when the variables are discrete (e.g., control group vs. diseased group). It is called regression when the variables are continuous (e.g., metabolite concentration).

In metabolomics biomarker discovery, it is important to find the simplest combination of metabolites that can produce a suitably effective predictive outcome. The main challenges are therefore predictor selection and the evaluation of the fitness and predictive power of the built model. Different predictor selection techniques have been described. Some of these suggested strategies are based on univariate or multivariate statistical proprieties of variables used as filters (loading weights, variable importance on projection scores, or regression coefficients), while others are based on optimization algorithms [53]. Recently, another elegant method has been reported, which essentially combines estimation of Mahalanobis distances, with principal component analysis and variable selection using a penalty metric [60].

Finally, goodness-of-fit metrics assess the model predictive power. Commonly used metrics may include root mean square error (RMSE) for regression problems and sensitivity, specificity, and the area under the receiver-operating characteristic (ROC) curve, for classification models. Various resampling methods are also used to efficiently use the available data set such as cross-validation, bootstrapping, and jackknifing [61]. Regarding the supervised methods, the most used techniques include linear discriminant analysis (LDA) and partial least squares (PLS) methods, such as PLS-discriminant analysis (PLS-DA) and orthogonal-PLS-DA (OPLS-DA), IC-DA, as well as support vector machines and random forest.

Furthermore, new methods based on topology data analysis seem promising for data analysis because of their intrinsic flexibility, exploratory, and predictive abilities. Recently, a new method, called statistical health monitoring (SHM), has been adapted from industrial statistical process control. An individual metabolic profile is compared to a healthy one in a multivariate fashion. Abnormal metabolite patterns are thus detected in a more intelligible interpretative way [62]. Multivariate and univariate data analysis pipelines are not mutually exclusive and are often used together for better information recovery.

Metabolite Annotation and Characterization

The introduction of high-resolution mass spectrometers and accurate mass measurements facilitate access to the chemical formula of the detected peaks and thus have considerably accelerated the identification of the discriminant metabolites. The combined use of quadrupole ion traps for sequential fragmentation experiments provides additional structural information. However, MS using soft ionization techniques such as electrospray methods exhibits high variability in fragmentation profiles generated on different devices, thus limiting the construction of universal spectral libraries such as those obtained by electron ionization or NMR.

This issue could be addressed using standardized ionization conditions, such as electron-based ionization techniques that are highly reproducible across MS systems. The acquisition of fragmentation spectra at this stage enables us to discriminate the responses obtained previously based on the produced ions or neutral losses and characteristic of chemical groups. Identification standards have been defined within the framework of the Metabolomics Standards Initiative, in regard to the available information on the metabolite to be characterized [63].

Computational tools have been developed for metabolite annotation such as CAMERA, ProbMetab, AStream, and MetAssign [64]. These methods mainly use *m/z*, retention time, adduct patterns, isotope patterns, and correlation methods for metabolite annotation. However, in MS the detected *m/z* ion and MS database matching is not enough for unambiguous characterization. Thus, complementary orthogonal information is required for reliable assignment of chemical identity, such as retention time matching and molecular dissociation patterns, compared to authentic standards [63]. For more reliable characterization, a solution may lie in a multidimensional framework based on orthogonal information integration, which may include accurate mass *m/z*, chromatographic retention time, MS/MS spectra patterns, CCS, chiral form, and peak intensity.

Hybrid strategies merging chemical and biological information, such as network analysis methods, may enhance metabolite identification through different metrics integration, including data-driven network topology, chemical features correlation, omics data, and biological databases. A list of databases is presented in Table 5.3.

Functional Analysis: Connecting the Dots

Any biological network can be pictured as a collection of linked nodes. The nodes may be genes, proteins, metabolites, diseases, microbes, or even individuals. The links or edges represent the interactions between the nodes: metabolic reactions, protein—protein interactions, gene—protein interactions, host—microbe interactions, or between individuals' interactions. The node distribution ranges from highly clustered to random. However, biological networks are not random. They are collections of nodes and links evolving as clusters. Such network topology is referred to as scale-free, which means that these networks contain few highly connected nodes called hubs. Moreover, the core idea of the biological network theory is the modularity.

Three building modules can be defined: topological, functional, and disease modules [65]. A topological module represents a local subset of nodes and links in the network. A functional module is a collection of nodes with similar or correlated function in the same network zone. Finally, a disease module represents a set of network components that contributes together to a cellular function, whose disturbance leads to a disease phenotype. Of note, these three modules may correlate and overlap. To fully embrace this new network perspective, computational biology is taking over more and more space in modern biology. It can be divided into two main building fields: knowledge discovery (or data mining) and simulation. The former generates hypotheses by extracting hidden patterns from high-dimensional experimental data. However, the latter tests hypotheses with *in silico* experiments, yielding predictions to be confirmed by *in vitro* and *in vivo* studies [66].

Pathway mapping strategies rely on the generated data for biological inference [67]. Biological databases are important for mapping different metabolic pathways (Table 5.3). Conceptually, pathway analysis or metabolite set enrichment analysis (MSEA) is inspired by the gene set enrichment analysis approach, previously developed for pathway analysis of gene expression data. There are three distinct methods for performing MSEA: overrepresentation analysis (ORA), quantitative enrichment analysis (QEA), and single sample profiling (SSP) [68].

A series of biochemical reactions interconnects metabolites to build networks. Given the relationships of the measured metabolites, network analysis uses the high degree of correlation in metabolomics data to build metabolic networks in which each metabolite represents a node. However, unlike pathway analysis, the links between nodes denote the level of mathematical correlation between each metabolite pair and are called edge [69]. These data-driven methods have been used for metabolic network reconstruction from metabolomics data [69].

As seen above, biological inference often needs prior identification of metabolites. Since this step is very challenging, a novel approach, named Mummichog, has been introduced by Li et al. to reverse this conventional workflow [70]. Without a priori identification of metabolites, this method predicts biological activity directly from MS-based untargeted metabolomics data. The idea behind this strategy is combining network analysis and metabolite prediction under the same computational framework, which significantly reduces the metabolomics workflow time.

Pirhaji et al. described a new network-based approach, using a prize-winning Steiner forest algorithm, for integrative analysis of untargeted metabolomics (PIUMet). This method infers molecular pathways via integrative analysis of metabolites without prior identification. Furthermore, PIUMet enabled elucidating putative identities of altered metabolites and inferring experimentally undetected, disease-associated metabolites and dysregulated proteins [71]. Compared to Mummichog, PIUMet also allows system-level inference by integrating other omics data.

From a metabolic network stand point, flux is defined as the rate (i.e., quantity per unit time) at which metabolites are converted or transported between different compartments [72]. Thus, metabolic fluxes or the fluxome represents a unique and functional readout of the phenotype [72]. To interrogate these fluxome network modeling can be achieved using constraints of mass and charge conservation, along with stoichiometric and thermodynamic limitations [73,74]. Based on

TABLE 5.3 Biological Databases and Functional Analysis Tools*

Tools	Websites
Biological Databases	
KEGG (Kyoto Encyclopedia of Genes and Genomes)	http://www.genome.jp/kegg
HumanCyc (Encyclopedia of Human Metabolic Pathways)	http://humancyc.org
MetaCyc (Encyclopedia of Metabolic Pathways)	https://metacyc.org
Reactome	http://www.reactome.org
Virtual Metabolic Human Database	https://vmh.uni.lu
Wikipathways	http://www.wikipathways.org
Pathway and Networks Analysis and Visualization	
BioCyc—Omics Viewer	http://biocyc.org
iPath	http://pathways.embl.de
MetScape	http://metscape.ncibi.org
Paintomics	http://www.paintomics.org
Pathos	http://motif.gla.ac.uk/Pathos
Pathvisio	http://www.pathvisio.org
VANTED	http://vanted.ipk-gatersleben.de
IMPaLA	http://impala.molgen.mpg.de
MBROLE 2.0	http://csbg.cnb.csic.es/mbrole2
MPEA	http://ekhidna.biocenter.helsinki.fi/poxo/mpea
Mummichog	http://clinicalmetabolomics.org/init/default/software
PIUMet	http://fraenkel-nsf.csbi.mit.edu/PIUMet
InCroMAP	http://www.ra.cs.uni-tuebingen.de/software/InCroMAP/
MetExplore	http://metexplore.toulouse.inra.fr
Omics Integration Tools	
Mixomics	http://mixomics.org
3Omics	http://3omics.cmdm.tw
Linkedomics	http://www.linkedomics.org
xMWAS	https://kuppal.shinyapps.io/xmwas
Omics Discovery Index	https://www.omicsdi.org
Multifunctional Tools	
MetaboAnlayst	http://www.metaboanalyst.com
XCMS online	https://xcmsonline.scripps.edu
MASSyPup	http://www.bioprocess.org/massypup
Workflow4Metabolomics	http://workflow4metabolomics.org
MetaBox	https://github.com/kwanjeeraw/metabox
MAVEN	http://genomics-ubs.princeton.edu/mzroll/index.php
MZmine 2	http://mzmine.github.io/

*A nonexhaustive list.

the stoichiometry of the reactants and products of biochemical reactions, flux balance analysis can estimate metabolic fluxes, without knowledge about the kinetics of the participating enzymes [75].

Recently, Cortassa et al. suggested a new approach distinct from flux balance analysis or metabolic flux analysis that takes into account kinetic mechanisms and regulatory interactions [76]. Advances in atom-tracking technologies and related informatics are valuable for metabolomics-based investigations. Among these technologies, stable isotope resolved metabolomics (SIRM) is a method that allows tracking individual atoms, through compartmentalized metabolic networks, which allowed highly resolved metabolome investigations [27].

New tools, such as causal networks based on integrative omics, seem very promising in interrogating metabolomics data. Bayesian networks (BN) inference has emerged as a creative method for integration and analysis of multi-omics data [77]. BN are able to incorporate prior knowledge if so desired. However, they are not dependent on it. They can also provide a way to learn regulatory relationships directly from the data. The genotype, gene expression, metabolomics data, as well as literature-based knowledge can be integrated as input into causal network models, which leads to an accuracy improvement in the reconstructed networks [78].

These methods are intrinsically unbiased and data-driven. BN algorithms can capture linear, nonlinear, stochastic, and combinatorial relationships between variables and are powerful for handling noisy data given their probabilistic nature [79]. BN track covariation patterns to unveil biological complex relationships between biological components and the studied outcome. They are able to identify mechanisms and structure, by reflecting connections and their influences to understand biological information flow. In contrast, the widely used differential expression approaches track mainly statistically significant component concentrations (gene, transcript, protein, or metabolite) between the studied groups. Moreover, BN algorithms may avoid the multiple testing issue by using global network topology optimization procedures to make statistical inferences considering the whole data [80].

Deep neural network or deep learning (DL) is a new class of machine learning methods that have been successfully applied to various omics data, in particular genomics. Recently, Alkawaa et al. showed that DL methods perform better on metabolomics data than some widely used machine learning methods [81]. Of note, compared to genomics data, metabolomics data exhibits higher noise levels and multicolinearity. These issues have to be addressed when extending genomics data analysis tools for metabolomics research. Table 5.2 presents different functional analysis tools for both pathway analysis and visualization.

HACKING THE HOST–MICROBIOME METABOLIC INTERACTIONS THROUGH METABOLOMICS AND BEYOND

Microbiome-related metabolomics has been used to explore a range of clinical conditions. It is generally useful to characterize fecal/gut metabolome and microbiome, in combination with the host's systemic metabolome such as blood, urine, or saliva [82]. Clayton et al. identified individual patterns of drug susceptibility based on gut-associated metabolite patterns such as urine p-cresol [83]. Janson et al. correlated the fecal metabolome with blood and urine metabolome to better understand peripheral markers of gut microbial metabolism [84]. Mente et al. evaluated the effect of diet inputs on the microbiome/metabolome and their association to disease endpoints such as choline and trimethylamine oxide [85]. Yano et al. studied the patterns of microbiome-derived metabolites that reach systemic circulation [86]. Chumpitazi et al. developed novel small molecule biomarkers for clinical diet prediction in irritable bowel syndrome [87].

An amazing example of microbiome analysis application is the evolving performance of predictive modeling of postprandial glucose response in patients with impaired glucose control [88]. Recent advances in MS imaging (MSI) and three-dimensional mapping of MS data provide a high-resolution image of the very complex chemical landscape of microbe–host interactions, which may pave the way for modulating these chemicals to prevent or treat infectious diseases [89,90]. Bouslimani et al. published a molecular cartography of the human skin surface using MSI [91]. Recently, Grag et al. developed a workflow to enable the cartography of metabolomic and microbiome data onto a three-dimensional organ reconstruction, built of radiological images. This elegantly enabled direct visualization of the microbial and chemical makeup of a human lung from a cystic fibrosis patient. They also detected host-derived molecules, microbial metabolites, medications, and region-specific metabolism of medications and placed it in the context of microbial distributions in the lung [92].

Individual microbial metabolism can be predicted using in silico approaches [93], and combined metabolism from entire microbiota communities can also be integrated into mathematical models that can be interrogated according to a specific condition [1,94]. The main limit of metagenomics is their inability to identify functionally associated microbial traits through direct functional inference. Metabolomics and metaproteomics may unveil functions that are blind to sequence-based strategies [95]. Thus, coupling metabolomics with other omics has great potential to shift current microbiome studies from static to more functional and dynamic interrogative investigations [96].

Similar to the abovementioned description regarding prior knowledge input, multi-omics integrative methods could be either statistics-based or knowledge-driven. In the former case, univariate or multivariate statistics are used to track correlations between biological variables, in the different omics layers [97,98]. The latter interrogates potential mechanistic relationships by projecting the significant biological variables into an existing knowledge base [99]. Recent trends in computational developments aim going a step further by merging the statistical and functional potential of the two methods to improve biological insights [96]. A promising way to deal with massive data would be to address these issues by mining omics and clinical data in real-time fashion for better clinical actionability [100].

CONCLUSION

Metabolomics is obviously taking its deserved place among omics. Translating metabolomic data into actionable knowledge is still the goal. To get there, attention should be paid to computational tools for multidimensional data processing. There is an urgent need for more databases with validated and curated MRM transitions for targeted metabolites. Furthermore, for untargeted metabolomics, larger libraries and curated MS/MS spectra for metabolite identification are needed. Hybrid strategies including pathway and network analysis methods could enhance metabolite characterization, through integration of different metrics, including data-driven network topology, chemical features correlation, omics data, and biological databases. Such multidimensional approaches may improve the chemical characterization by combining both extended chemical information and biological context. With all the high-dimensional data management issues, like other omics, metabolomics clinical implementation should be tackled using big data handling strategies for efficient storage, integration, visualization, and sharing of metabolomics data. NMR and chromatography-based platforms are still the well-established technologies for metabolomics studies. LC−MS and GC−MS are the most adopted analytical platforms in clinical metabolomics. Still, for a more comprehensive metabolome coverage, implementation of multiplatform approaches is necessary. To reach next-generation metabolomics, further advances are urgently needed in analytical strategies for reliable identification and absolute quantification. Finally, standardization regarding sample handling and analytical procedures is a big issue for larger clinical studies and wide adoption of metabolomics, particularly, in clinical environments. The goal is to be able to predict, for example, the interaction of drugs with the host and the microbiota, treated as a single, computationally integrated, genomic and metabolomic entity. Clinically actionable visual analytics with user-friendly tools are mandatory for wide adoption in clinical settings and more effective impact on healthcare.

REFERENCES

[1] Claesson MJ, Clooney AG, O'Toole PW. A clinician's guide to microbiome analysis. Nat Rev Gastroenterol Hepatol 2017;14(10):585−95.

[2] Jones MB, Highlander SK, Anderson EL, et al. Library preparation methodology can influence genomic and functional predictions in human microbiome research. Proc Natl Acad Sci U S A 2015;112(45):14024−9.

[3] Tenenbaum JD, Avillach P, Benham-Hutchins M, et al. An informatics research agenda to support precision medicine: seven key areas. J Am Med Inf Assoc 2016;23(4):791−5.

[4] Ritchie MD, Holzinger ER, Li R, Pendergrass SA, Kim D. Methods of integrating data to uncover genotype-phenotype interactions. Nat Rev Genet 2015;16(2):85−97.

[5] Oliver SG, Winson MK, Kell DB, Baganz F. Systematic functional analysis of the yeast genome. Trends Biotechnol 1998;16(9):373−8.

[6] Nicholson JK, Lindon JC, Holmes E. 'Metabonomics': understanding the metabolic responses of living systems to pathophysiological stimuli via multivariate statistical analysis of biological NMR spectroscopic data. Xenobiotica 1999;29(11):1181−9.

[7] Fiehn O. Metabolomics−the link between genotypes and phenotypes. Plant Molecular Biology 2002;48(1−2):155−71.

[8] Bekri S. The role of metabolomics in precision medicine. Exp Rev Precision Med Drug Dev 2016;1(6):517−32 [just-accepted].

[9] Tebani A, Abily-Donval L, Afonso C, Marret S, Bekri S. Clinical metabolomics: the new metabolic window for inborn errors of metabolism investigations in the post-genomic era. Int J Mol Sci 2016;17(7).

[10] Coyte KZ, Schluter J, Foster KR. The ecology of the microbiome: networks, competition, and stability. Science 2015;350(6261):663−6.

[11] Kitano H. Systems biology: a brief overview. Science 2002;295(5560):1662−4.

[12] Tebani A, Afonso C, Bekri S. Advances in metabolome information retrieval: turning chemistry into biology. Part I: analytical chemistry of the metabolome. J Inherit Metab Dis 2017;41(3):379−91.

[13] Tebani A, Afonso C, Bekri S. Advances in metabolome information retrieval: turning chemistry into biology. Part II: biological information recovery. J Inherit Metab Dis 2017;41(3):393−406.

[14] Dunn WB, Wilson ID, Nicholls AW, Broadhurst D. The importance of experimental design and QC samples in large-scale and MS-driven untargeted metabolomic studies of humans. Bioanalysis 2012;4(18):2249−64.

[15] Chetwynd AJ, Dunn WB, Rodriguez-Blanco G. Collection and preparation of clinical samples for metabolomics. In: Sussulini A, editor. Metabolomics: from fundamentals to clinical applications. Cham: Springer International Publishing; 2017. p. 19−44.

[16] Sitnikov DG, Monnin CS, Vuckovic D. Systematic assessment of seven solvent and solid-phase extraction methods for metabolomics analysis of human plasma by LC-MS. Sci Rep 2016;6:38885.

[17] Ser Z, Liu X, Tang NN, Locasale JW. Extraction parameters for metabolomics from cultured cells. Anal Biochem 2015;475:22−8.

[18] Patejko M, Jacyna J, Markuszewski MJ. Sample preparation procedures utilized in microbial metabolomics: an overview. J Chromatogr B 2017;1043(Supplement C):150−7.

[19] Vuckovic D. Current trends and challenges in sample preparation for global metabolomics using liquid chromatography-mass spectrometry. Anal Bioanal Chem 2012;403(6):1523−48.

[20] Dunham SJB, Ellis JF, Li B, Sweedler JV. Mass spectrometry imaging of complex microbial communities. Acc Chem Res 2017;50(1):96−104.

[21] Balog J, Sasi-Szabo L, Kinross J, et al. Intraoperative tissue identification using rapid evaporative ionization mass spectrometry. Sci Transl Med 2013;5(194):11.

[22] Zhang J, Rector J, Lin JQ, et al. Nondestructive tissue analysis for ex vivo and in vivo cancer diagnosis using a handheld mass spectrometry system. Sci Transl Med 2017;9(406).

[23] Pauling L, Robinson AB, Teranishi R, Cary P. Quantitative analysis of urine vapor and breath by gas-liquid partition chromatography. Proc Natl Acad Sci U S A 1971;68(10):2374−6.

[24] Alonso A, Marsal S, Julia A. Analytical methods in untargeted metabolomics: state of the art in 2015. Front Bioeng Biotechnol 2015;3:23.

[25] Emwas A-H, Salek R, Griffin J, Merzaban J. NMR-based metabolomics in human disease diagnosis: applications, limitations, and recommendations. Metabolomics 2013;9(5):1048−72.

[26] Nagana Gowda GA, Raftery D. Recent advances in NMR-based metabolomics. Anal Chem 2017;89(1):490−510.

[27] Fan TW, Lane AN, Higashi RM. Stable isotope resolved metabolomics studies in ex vivo tissue slices. Bio-protocol 2016;6(3).

[28] Fan TWM, Lane AN. Applications of NMR spectroscopy to systems biochemistry. Prog Nucl Magn Reson Spectrosc 2016;92−93:18−53.

[29] Verma A, Kumar I, Verma N, Aggarwal P, Ojha R. Magnetic resonance spectroscopy — revisiting the biochemical and molecular milieu of brain tumors. BBA Clin 2016;5:170−8.

[30] González-Domínguez R, Sayago A. Fernández-Recamales Á. Direct infusion mass spectrometry for metabolomic phenotyping of diseases. Bioanalysis 2017;9(1):131−48.

[31] Kuehnbaum NL, Britz-McKibbin P. New advances in separation science for metabolomics: resolving chemical diversity in a post-genomic era. Chem Rev 2013;113(4):2437−68.

[32] Forcisi S, Moritz F, Kanawati B, Tziotis D, Lehmann R, Schmitt-Kopplin P. Liquid chromatography−mass spectrometry in metabolomics research: mass analyzers in ultra high pressure liquid chromatography coupling. J Chromatogr A 2013;1292:51−65.

[33] Haggarty J, Burgess KEV. Recent advances in liquid and gas chromatography methodology for extending coverage of the metabolome. Curr Opin Biotechnol 2017;43:77−85.

[34] García A, Godzien J, López-Gonzálvez Á, Barbas C. Capillary electrophoresis mass spectrometry as a tool for untargeted metabolomics. Bioanalysis 2016;9(1):99−130.

[35] Paglia G, Angel P, Williams JP, et al. Ion mobility-derived collision cross section as an additional measure for lipid fingerprinting and identification. Anal Chem 2015;87(2):1137−44.

[36] Smolinska A, Hauschild AC, Fijten RR, Dallinga JW, Baumbach J, van Schooten FJ. Current breathomics−a review on data pre-processing techniques and machine learning in metabolomics breath analysis. J Breath Res 2014;8(2):027105.

[37] Fenn L, McLean J. Biomolecular structural separations by ion mobility−mass spectrometry. Anal Bioanal Chem 2008;391(3):905−9.

[38] Tebani A, Schmitz-Afonso I, Rutledge DN, Gonzalez BJ, Bekri S, Afonso C. Optimization of a liquid chromatography ion mobility-mass spectrometry method for untargeted metabolomics using experimental design and multivariate data analysis. Anal Chim Acta 2016;913:55−62.

[39] Tebani A, Schmitz-Afonso I, Abily-Donval L, et al. Urinary metabolic phenotyping of mucopolysaccharidosis type I combining untargeted and targeted strategies with data modeling. Clin Chim Acta 2017;475:7−14.

[40] Domalain V, Tognetti V, Hubert-Roux M, et al. Role of cationization and multimers formation for diastereomers differentiation by ion mobility-mass spectrometry. J Am Soc Mass Spectrom 2013;24(9):1437−45.

[41] Zhou Z, Shen X, Tu J, Zhu Z-J. Large-scale prediction of collision cross-section values for metabolites in ion mobility-mass spectrometry. Anal Chem 2016;88(22):11084−91.

[42] Zhou Z, Tu J, Xiong X, Shen X, Zhu ZJ. LipidCCS: prediction of collision cross-section values for lipids with high precision to support ion mobility-mass spectrometry-based lipidomics. Anal Chem 2017;89(17):9559−66.

[43] Sherrod SD, McLean JA. Systems-wide high-dimensional data acquisition and informatics using structural mass spectrometry strategies. Clin Chem 2015;62(1):77−83.

[44] Arentz G, Mittal P, Zhang C, et al. Chapter two − applications of mass spectrometry imaging to cancer. In: Richard RD, Liam AM, editors. Advances in cancer research. Academic Press; 2017. p. 27−66.

[45] Li H, Balan P, Vertes A. Molecular imaging of growth, metabolism, and antibiotic inhibition in bacterial colonies by laser ablation electrospray ionization mass spectrometry. Angew Chem Int Ed Engl 2016;55(48):15035−9.

[46] Kadian N, Raju KSR, Rashid M, Malik MY, Taneja I, Wahajuddin M. Comparative assessment of bioanalytical method validation guidelines for pharmaceutical industry. J Pharmaceut Biomed Anal 2016;126:83−97.

[47] Dunn WB, Broadhurst DI, Edison A, et al. Quality assurance and quality control processes: summary of a metabolomics community questionnaire. Metabolomics 2017;13(5):50.

[48] Shen X, Gong X, Cai Y, et al. Normalization and integration of large-scale metabolomics data using support vector regression. Metabolomics 2016;12(5):89.

[49] Lewis MR, Pearce JTM, Spagou K, et al. Development and application of UPLC-ToF MS for precision large scale urinary metabolic phenotyping. Anal Chem 2016;88(18):9004−13.

[50] Hogeweg P. The roots of bioinformatics in theoretical biology. PLoS Comput Biol 2011;7(3):e1002021.

[51] Brereton RG. A short history of chemometrics: a personal view. J Chemometr 2014;28(10):749−60.

[52] Cai Y, Weng K, Guo Y, Peng J, Zhu Z-J. An integrated targeted metabolomic platform for high-throughput metabolite profiling and automated data processing. Metabolomics 2015;11(6):1575−86.

[53] Yi L, Dong N, Yun Y, et al. Chemometric methods in data processing of mass spectrometry-based metabolomics: a review. Anal Chim Acta 2016;914:17−34.

[54] Tebani A, Afonso C, Marret S, Bekri S. Omics-based strategies in precision medicine: toward a paradigm shift in inborn errors of metabolism investigations. Int J Mol Sci 2016;17(9).

[55] Li B, Tang J, Yang Q, et al. Performance evaluation and online realization of data-driven normalization methods used in LC/MS based untargeted metabolomics analysis. Sci Rep 2016;6:38881.

[56] Wu Y, Li L. Sample normalization methods in quantitative metabolomics. J Chromatogr A 2016;1430:80−95.

[57] van den Berg RA, Hoefsloot HC, Westerhuis JA, Smilde AK, van der Werf MJ. Centering, scaling, and transformations: improving the biological information content of metabolomics data. BMC Genomics 2006;7:142.

[58] Bouveresse DJ-R, Rutledge D. Independent components analysis: theory and applications. Resolving spectral mixtures: with applications from ultrafast time-resolved spectroscopy to super-resolution imaging, 30; 2016. p. 7225.

[59] Liu Y, Smirnov K, Lucio M, Gougeon RD, Alexandre H, Schmitt-Kopplin P. MetICA: independent component analysis for high-resolution mass-spectrometry based non-targeted metabolomics. BMC Bioinformatics 2016;17(1):1−14.

[60] Engel J, Blanchet L, Engelke U, Wevers R, Buydens L. Sparse statistical health monitoring: a novel variable selection approach to diagnosis and follow-up of individual patients. Chemometr Intell Lab Syst 2017;164:83−93.

[61] Westad F, Marini F. Validation of chemometric models − a tutorial. Anal Chim Acta 2015;893:14−24.

[62] Engel J, Blanchet L, Engelke UF, Wevers RA, Buydens LM. Towards the disease biomarker in an individual patient using statistical health monitoring. PLoS One 2014;9(4):e92452.

[63] Sumner LW, Amberg A, Barrett D, et al. Proposed minimum reporting standards for chemical analysis. Metabolomics 2007;3(3):211−21.

[64] Misra BB, Fahrmann JF, Grapov D. Review of emerging metabolomic tools and resources: 2015−2016. Electrophoresis 2017;38(18):2257−74.

[65] Barabasi A-L, Gulbahce N, Loscalzo J. Network medicine: a network-based approach to human disease. Nat Rev Genet 2011;12(1):56−68.

[66] Kitano H. Computational systems biology. Nature 2002;420(6912):206−10.

[67] Cazzaniga P, Damiani C, Besozzi D, et al. Computational strategies for a system-level understanding of metabolism. Metabolites 2014;4(4):1034−87.

[68] Xia J, Sinelnikov IV, Han B, Wishart DS. MetaboAnalyst 3.0−making metabolomics more meaningful. Nucleic Acids Res 2015;43(W1):W251−7.

[69] Krumsiek J, Suhre K, Illig T, Adamski J, Theis FJ. Gaussian graphical modeling reconstructs pathway reactions from high-throughput metabolomics data. BMC Syst Biol 2011;5:21.

[70] Li S, Park Y, Duraisingham S, et al. Predicting network activity from high throughput metabolomics. PLoS Comput Biol 2013;9:e1003123.

[71] Pirhaji L, Milani P, Leidl M, et al. Revealing disease-associated pathways by network integration of untargeted metabolomics. Nat Methods 2016;13(9):770−6.

[72] Aon MA, Cortassa S. Systems biology of the fluxome. Processes 2015;3(3):607−18.

[73] Winter G, Kromer JO. Fluxomics − connecting 'omics analysis and phenotypes. Environ Microbiol 2013;15(7):1901−16.

[74] Aurich MK, Thiele I. Computational modeling of human metabolism and its application to systems biomedicine. Meth Mol Biol 2016;1386:253−81.

[75] Cascante M, Marin S. Metabolomics and fluxomics approaches. Essays Biochem 2008;45:67−82.

[76] Cortassa S, Caceres V, Bell LN, O'Rourke B, Paolocci N, Aon MA. From metabolomics to fluxomics: a computational procedure to translate metabolite profiles into metabolic fluxes. Biophys J 2015;108(1):163−72.

[77] Argmann Carmen A, Houten Sander M, Zhu J, Schadt Eric E. A next generation multiscale view of inborn errors of metabolism. Cell Metabol 2015;23(1):13−26.

[78] Zhu J, Sova P, Xu Q, et al. Stitching together multiple data dimensions reveals interacting metabolomic and transcriptomic networks that modulate cell regulation. PLoS Biology 2012;10(4):e1001301.

[79] Yan J, Risacher SL, Shen L, Saykin AJ. Network approaches to systems biology analysis of complex disease: integrative methods for multi-omics data. Briefings Bioinf 2017. https://doi.org/10.1093/bib/bbx066.

[80] Glaab E. Using prior knowledge from cellular pathways and molecular networks for diagnostic specimen classification. Briefings Bioinf 2016;17(3):440−52.

[81] Alkawaa FM, Chaudhary K, Garmire LX. Deep learning accurately predicts estrogen receptor status in breast cancer metabolomics data. J Proteome Res 2017;17(1):337−47.

[82] Beger RD, Dunn W, Schmidt MA, et al. Metabolomics enables precision medicine: "a white paper, community perspective". Metabolomics 2016;12(9):1–15.

[83] Clayton TA, Baker D, Lindon JC, Everett JR, Nicholson JK. Pharmacometabonomic identification of a significant host-microbiome metabolic interaction affecting human drug metabolism. Proc Natl Acad Sci U S A 2009;106(34):14728–33.

[84] Jansson J, Willing B, Lucio M, et al. Metabolomics reveals metabolic biomarkers of Crohn's disease. PLoS One 2009;4(7):e6386.

[85] Mente A, Chalcraft K, Ak H, et al. The relationship between trimethylamine-N-oxide and prevalent cardiovascular disease in a multiethnic population living in Canada. Can J Cardiol 2015;31(9):1189–94.

[86] Yano JM, Yu K, Donaldson GP, et al. Indigenous bacteria from the gut microbiota regulate host serotonin biosynthesis. Cell 2015;161(2):264–76.

[87] Chumpitazi BP, Hollister EB, Cope JL, Luna RA, Versalovic J, Shulman RJ. 164 gut microbiome biomarkers are associated with clinical response to a low FODMAP diet in children with irritable Bowel syndrome. Gastroenterology 2015;148(4):S-44.

[88] Zeevi D, Korem T, Zmora N, et al. Personalized nutrition by prediction of glycemic responses. Cell 2015;163(5):1079–94.

[89] Marcobal A, Kashyap PC, Nelson TA, et al. A metabolomic view of how the human gut microbiota impacts the host metabolome using humanized and gnotobiotic mice. ISME J 2013;7(10):1933–43.

[90] Rath CM, Alexandrov T, Higginbottom SK, et al. Molecular analysis of model gut microbiotas by imaging mass spectrometry and nanodesorption electrospray ionization reveals dietary metabolite transformations. Anal Chem 2012;84(21):9259–67.

[91] Bouslimani A, Porto C, Rath CM, et al. Molecular cartography of the human skin surface in 3D. Proc Natl Acad Sci U S A 2015;112(17):E2120–9.

[92] Garg N, Wang M, Hyde E, et al. Three-dimensional microbiome and metabolome cartography of a diseased human lung. Cell Host Microbe 2017;22(5):705–16.e4.

[93] Bauer E, Laczny CC, Magnusdottir S, Wilmes P, Thiele I. Phenotypic differentiation of gastrointestinal microbes is reflected in their encoded metabolic repertoires. Microbiome 2015;3(1):55.

[94] Heinken A, Thiele I. Systems biology of host–microbe metabolomics. Wiley Interdiscip Rev 2015;7(4):195–219.

[95] Franzosa EA, Morgan XC, Segata N, et al. Relating the metatranscriptome and metagenome of the human gut. Proc Natl Acad Sci U S A 2014;111(22):E2329–38.

[96] Chong J, Xia J. Computational approaches for integrative analysis of the metabolome and microbiome. Metabolites 2017;7(4):62.

[97] Mao SY, Huo WJ, Zhu WY. Microbiome–metabolome analysis reveals unhealthy alterations in the composition and metabolism of ruminal microbiota with increasing dietary grain in a goat model. Environ Microbiol 2016;18(2):525–41.

[98] Garali I, Adanyeguh IM, Ichou F, et al. A strategy for multimodal data integration: application to biomarkers identification in spinocerebellar ataxia. Briefings Bioinf 2017.

[99] Greenblum S, Turnbaugh PJ, Borenstein E. Metagenomic systems biology of the human gut microbiome reveals topological shifts associated with obesity and inflammatory bowel disease. Proc Natl Acad Sci U S A 2012;109(2):594–9.

[100] Shameer K, Badgeley MA, Miotto R, Glicksberg BS, Morgan JW, Dudley JT. Translational bioinformatics in the era of real-time biomedical, health care and wellness data streams. Briefings Bioinf 2016. https://doi.org/10.1093/bib/bbv118.

Chapter 6

Laboratory Simulators of the Colon Microbiome

M.Carmen. Martínez-Cuesta, Carmen. Peláez and Teresa Requena

Department of Food Biotechnology and Microbiology, Institute of Food Science Research, CIAL (CSIC), Madrid, Spain

INTRODUCTION

The evaluation of gut microbiota composition and functionality has used mostly fecal samples, which are considered representative of the distal large intestine. Fecal inoculation has also been used in gnotobiotic humanized mouse models to reproduce the human gut microbiota in vivo [1].

Laboratory simulators of the colon microbiome are considered excellent tools to allow the screening of a wide range of dietary ingredients, pathogens, drugs, and toxic or radioactive compounds, without ethical constraints. In vitro models vary from batch incubations, using anaerobic conditions and dense fecal microbiota, to more complex continuous models, involving one or multiple connected, pH-controlled reactors inoculated with fecal microbiota, and representing different parts of the human colon [2].

Laboratory colonic models are inoculated with fecal microbiota and operated in an anaerobic atmosphere at temperature and pH values mimicking physiological conditions. The simulation in vitro of the representative microbial populations from the human gut encounters the challenges, among others, of reproducing the microbiota complexity and being representative of a general population. The inoculation prepared from a pool of donors has been suggested to integrate a wider variety of bacterial species and to produce a standardized and reproducible microbiota. This should lessen the effect of interindividual variation among donors, for the accurate interpretation of the effects of experimental treatments [3]. In other conditions, such as the exploration of different metabotypes, the reactors should be inoculated with the fecal microbiome from one individual, to characterize bioactive metabolite-producing phenotypes toward specific compounds, for example polyphenols [4].

KEY MICROORGANISMS AND CRUCIAL METABOLITES

Despite interindividual diversity, it is possible to identify a human intestinal core microbiota, representing bacteria that are present among all so far analyzed human populations [5]. The bacterial phyla representative of the human gut microbiota are Bacteroidetes, Firmicutes, Actinobacteria, Proteobacteria, and Verrucomicrobia. Archaeal representatives consist largely of the methanogens *Methanobrevibacter* and *Methanosphaera*. The complexity increases toward the lower taxonomic ranks, from phylum to species level, but analyses of cumulative abundance in metagenomic studies indicated that bacteria related to *Faecalibacterium, Bifidobacterium, Roseburia, Ruminococcus, Subdoligranulum, Prevotella, Ruminococcus, Akkermansia,* and *Oscillospira* represent common core bacteria in the Western adult population [6].

Additionally, the intestinal minimal functional core contains specific pathways involved in resistance to bile, ability to grow in an anaerobic environment, and the metabolic capabilities toward degradation of a wide range of polysaccharides and utilization of available amino acids and lipids.

The major metabolic products of anaerobic fermentation by gut microbial communities that colonize the mammalian gut are short-chain fatty acids (SCFA), mainly acetate, propionate, and butyrate. The release of these microbial metabolites causes decrease of the luminal pH, which reduces the risk of pathogen colonization and favors absorption of some nutrients. SCFA are also absorbed efficiently by the gut mucosa and have important impacts upon human physiology [7].

Butyrate is used preferentially as an energy source by colonocytes and has demonstrated antiinflammatory activity and protective effects against colorectal cancer. Propionate promotes satiety and reduces cholesterol [8]. In addition, butyrate and propionate have also been reported to induce the differentiation of T-regulatory cells, assisting the control of inflammation. Members of the *Clostridium* clusters IV and XIVa are exceptional inducers of T-regulatory cells in the colon and are considered as therapeutic options for intestinal inflammatory diseases and allergies [9].

Intestinal microbiota also produces some biogenic compounds such as free vitamins, bioactive peptides, and neurotransmitters like γ-amino butyric acid (GABA). The intestinal microbiota participates in the detoxification of xenobiotics and contaminants and the biotransformation of drugs or their metabolites [10].

GENERAL MODELS OF THE COLON MICROBIOME

Batch Fermentation Models

Batch models are used for short-term fermentation studies, in which specific strains or fecal microbial communities are maintained under anaerobic conditions. These models range from closed bottles used to grow defined (single or mixed) bacterial strains to controlled reactors inoculated with fecal microbiota suspensions. These batch fermentations are of particular interest, for initial screening assays of microbial metabolism, due to the high interindividual variations in microbiota composition, as well as to test different sources or doses of compounds. The model is also useful to elucidate microbial metabolism pathways, although it is limited by substrate depletion and accumulation of end products (Table 6.1).

Dynamic Fermentation Models

Single or multistage continuous models are relevant for microbial ecological studies, since they allow long-term experiments needed to evaluate the spatial and temporal adaptation of the colonic microbiota to dietary compounds (Table 6.1).

Continuous single-stage fermentation models are often designed to simulate the proximal colon conditions and to reproduce its metabolic activity. The computer-controlled, dynamic in vitro proximal colon model (TIM-2) was developed by the TNO in the Netherlands [11]. This model has unique features, as it includes host functions such as peristaltic mixing and water and metabolite absorption. The TIM-2 model consists of four interconnected glass compartments, with a flexible membrane inside, and water (37 °C) between the glass jacket and the membrane.

Peristaltic movements are achieved, by applying pressure on the water at regular intervals, causing contraction of the flexible membrane, and then peristaltic waves, which mix the luminal content and move it through the system. The TIM-2 contains dialysis membranes (molecular mass cut-off 50 kDa), in the lumen of the reactor, that prevent accumulation of microbial metabolites. The model is equipped with a level sensor to maintain the colon simulator volume at 120 mL and pH control at 5.8.

TABLE 6.1 Advantages and Limitations of In Vitro Laboratory Simulators of the Colon Microbiome

Model	Advantages	Limitations
Batch	Easy to operate. Screening studies and interindividual variability. Matrix/doses and metabolism assessment. Cost-effective.	Short-term studies. Substrate depletion and metabolite accumulation to inhibitory levels.
Dynamic	Control of environmental parameters. Continuous replenishment.	No host functionality, and short-time studies.
Immobilized	High-cell density and long-term studies of a continuous fermentation system, with immobilized fecal microbiota.	No simulation of water and metabolite absorption. Home-designed and expensive.
Multistage	Colon region—specific research. Microbiome stability over a long time frame. Multiparametric control.	No host functionality. Home-designed and expensive.
Host interaction devices	TIM-2, metabolite absorption, and peristalsis.	No host immunoendocrine response. Expensive.
	M-SHIME, intestinal mucus simulation.	No feedback mechanisms or host immunoendocrine response. Expensive.

An in vitro proximal colon model, composed of a single-stage culture, that uses several vessels to allow testing in parallel the effects of different treatments, on the same complex gut microbiota (biological replication), was developed at ETH in Zurich, Switzerland [12]. The model PolyFermS uses immobilized fecal microbiota in a first-stage inoculum reactor, representative of the complex bacterial community, which is present both at planktonic and sessile (biofilm-associated) states in the colon. Thus, the fecal inoculum is immobilized in 1−2 mm diameter gel beads, on a dispersion process in a two-phase system, composed of gellan and xanthan gums. The first-stage inoculum reactor, seeded with immobilized fecal microbiota, is used to continuously inoculate a set of parallel second-stage reactors.

Dynamic multistage-fermentation models are mostly based on the Reading model, first described by Gibson et al. [13]. The model consists of three vessels of increasing size connected in series, replicating the proximal (V1; 0.3 L), transverse (V2; 0.5 L), and distal (V3; 0.8 L) colon regions. The operating conditions include setting of pH for the three vessels at 5.5, 6.2, and 6.8, respectively, and a total retention time of 62.7 h. Each fermenter is magnetically stirred and maintained under an atmosphere of CO_2.

The authors also developed a nutritious medium that has been extensively used in both batch and dynamic fermentation models. The medium consists of protein substrates (casein and peptone), complex carbohydrates (pectin, xylan, arabinogalactan and resistant starches), that are not digested by gastrointestinal enzymes, and a mixture of bile salts, minerals, and vitamins. Vessel V1 is fed with the medium from a reservoir (R1), and can also receive mucin from reservoir R2. The model can operate at different retention times to assess the effect of colonic transit on carbohydrate utilization, metabolite formation, and the ecology of numerically important bacterial populations.

A further development of the Reading model is the Simulator of the Human Intestinal Microbial Ecosystem (SHIME) (Ghent University, Belgium). This system comprises a series of five double-jacketed glass reactors, which simulates the stomach and small intestine (mimicking the ingestion and digestion processes), connected to the three-stage continuous colonic model, through peristaltic pumps [14]. The system operates semicontinuously, with intermittent supplementation of nutritional medium and the removal of microbial suspension. The first two reactors follow a fill-and-draw principle, adding three times a day a defined nutritional medium to the gastric compartment and pancreatic and bile liquid to the small intestine compartment. Upon digestion in the gastric and intestine compartments, the slurry is pumped in the ascending colon vessel, where colon fermentation is initiated. Retention colon time is primarily modulated through a change in compartment volume and, depending on the human target group of interest, may vary from 24 to 72 h.

AUTOMATED SIMULATORS OF THE COLON MICROBIOME: THE SIMGI AND BFBL MODELS

We developed the SIMGI (SIMulator Gastro-Intestinal) model at the Institute of Food Science Research CIAL (Madrid, Spain) as a fully automated gastrointestinal multichamber simulator to dynamically reproduce the physiological processes taking place during digestion in the stomach and small intestine, as well as to simulate the colonic microbiota, responsible for metabolic bioconversions in the large intestine [15].

The gastric compartment consists of two cylindrical transparent and rigid methacrylate plastic modules, covering a reservoir of flexible silicone walls, where the gastric content is mixed by peristaltic movements. The gastric peristalsis is achieved by the pressure of thermostated water that flows in the jacket between the plastic modules and the reservoir and keeps the temperature of the gastric content at 37 °C. The meal received by the stomach compartment is mixed with gastric electrolytes and enzymes, and the decrease of pH is automatically controlled to follow the curve resulting from a linear fit of data, representing experimental in vivo conditions.

The stomach emptying is programmable to allow the modification of the emptying curve shape, depending on liquid, semisolid, or solid foods. The process of digestion of whey proteins in this compartment has recently been demonstrated to represent in vivo data, with a stepped appearance of peptides from serum albumin and α-lactalbumin and the partial passage of undegraded β-lactoglobulin to the upper intestine [16].

The small intestine consists in a double-jacketed glass reactor vessel, continuously stirred (150 rpm), that receives the gastric content, which is then mixed with pancreatic juice and bile salts. The intestinal content is digested during 2 h at 37 °C and kept at neutral pH values. The digestion temperature is regulated by pumping thermostated water, into the space between the glass jackets, and nitrogen is continuously flushed into the medium, to induce anaerobic conditions during digestion. After the small intestinal digestion, the whole content of the vessel is automatically transferred, at a flow rate of 5 mL/min, to the proximal colon compartment.

The stages of the large intestine are simulated in three double-jacketed glass continuous reactors, which are interconnected by pipes and peristaltic valve pumps. The pH in the colonic units, named ascending AC, transverse TC, and descending DC colon, is controlled to 5.6 in the AC, 6.3 in the TC, and 6.8 in the DC compartments. When the digested

FIGURE 6.1 BFBL laboratory simulator of the colon microbiome.

content of the small intestine is transferred to the proximal colon vessel (AC), the transit of colonic content between the AC, TC, and DC compartments is simultaneously initiated at the same flow rate. Reactors are continuously purged with nitrogen, creating a bubbling effect in the colonic content to induce anaerobic conditions for the oxygen-sensitive intestinal microbial communities.

The SIMGI design is aimed to dynamically operate, with the five units simulating the whole gastrointestinal process, and constitute a CIAL Institute Platform for digestion and fermentation services. Based on it, we have proceeded with the development of a specific colon simulator, adapted to laboratory-scale conditions, named Simulator BFBL (after Spanish words for Functional Biology of Lactic acid Bacteria; Fig. 6.1). The BFBL laboratory simulator of the colon microbiome is also a three-stage fermentation model, with the direct feeding of the small intestine with nutritious medium. After digestion, the controlled transit to the AC, DC, and TC vessels of the small intestinal content is activated for reproducing the colon region—specific microbiota.

The system is fully automated and placed in a biological safety cabinet that allows therapeutical protocols involving foodborne pathogens and avoids environmental contamination with microaerophilic opportunists. An additional upgrade in the BFBL simulator is the control of reactor volumes with level sensors that activate transport of the colonic content to guarantee the desired colon retention times. In addition, nitrogen is flushed to remove oxygen from the head space, only during digestion and fluid transit between compartments, and without the bubbling effect. This would avoid a direct nitrogen influence on the bacterial metabolism and fermentation yields, mainly regarding hydrogenotrophic colon organisms such as acetogens, sulfate-reducing bacteria, and methanogens [17].

The BFBL simulator stores the input monitored values from the temperature and pH sensors. The automated representation of pH curves and pumping of acid and base to the compartments are indicative of microbial development and metabolism. This information is crucial as control of the process, as the analyses of metabolites and fermentation products are delayed and not measured online.

The laboratory-scale setting of the BFBL simulator facilitates its flexible-modulating characteristics to operate with individual reactors. Additionally, it can incorporate devices simulating the gut microbiota—host interactions (Fig. 6.1). Assays for evaluating this type of valuable interaction are currently approached by coculturing colon region—specific microbiota suspensions, from the AC, TC, and DC vessels, with epithelial or immune cells. The simulation of bacterial adhesion and intestinal absorption of microbial metabolites is planned with the setting up of microbial/mucosa interfaces in the lumen of the colon reactors and by including dialysis devices between compartments.

PROTOCOLS AND APPLICATIONS

The three-stage colon simulators aligned in series imply that a sequential feeding of growth medium occurs. Thus, the first vessel has a high availability of substrate, representing a rapid bacterial growth rate, and is operated at an acidic pH, similar to events in the proximal colon. In contrast, the final vessel resembles the neutral pH, slow bacterial proliferation rate, and low substrate availability, which is characteristic of more distal regions. The viability of the in vitro gut microbiota is dependent on the continuous replenishment of nutrients and the control of physiological temperature and anaerobic conditions.

The control of these factors allows the establishment of steady-state conditions, with respect to both microbial composition and metabolic activity, whereas the control of defined pH values, downstream nutrient limitations, and retention times in the different vessels allows a region-specific differentiation of the colonic microbiota, in terms of metabolic activity and microbial communities.

The development of a colon region–specific microbial ecosystem occurs after the stabilization period in the simulator, and it is needed before starting any experimental approach [15]. Differences observed between the compartments after stabilization indicate that *Bacteroides* are more representative in the AC and TC compartments, than in the DC reactor, whereas the butyrate-producing groups *Clostridium leptum* and *Ruminococcus* (cluster IV) are less represented in the proximal colon compartment (AC) than in the distal vessels (TC and DC).

Other butyrate-producing bacteria such as *Blautia coccoides–Eubacterium rectale* group (cluster XIVa) remain equivalent in all colonic compartments (8.1–8.6 log copy number/mL at the end of the stabilization period). The carbohydrate-rich conditions of the proximal vessel (AC) favor the predominance of *Bacteroides*, which are characterized by a marked ability to utilize a wide variety of polysaccharides [18]. Interestingly, when the stabilization period was adapted to simulate an obese-associated microbiota, by using an overweight donor, and high-energy content nutritive medium (increased content of high-glycemic index carbohydrates, such as digestible starch and fructose), the microbial community profiles of the three colon compartments clustered together independently of the carbohydrate content of the diets [19]. The study showed that, in average, the distal colon regions were enriched with *Bifidobacterium*, *C. leptum*, *Ruminococcus*, *Akkermansia*, *Faecalibacterium*, and *Roseburia*.

The stabilization is also required to provide a steady-state environment, where the metabolism of the microbial community can be evaluated during long-term experimental dietary interventions. Evaluation of microbial metabolism is usually addressed by measuring SCFA and ammonium contents. Overall, the net SCFA production, up to functional stability, is highest in the AC compartment in correspondence with the carbohydrate-excess conditions, although total content accumulates in the distal compartments due to the lack of absorption steps.

Acetate, propionate, and butyrate are the main metabolites measured in colon simulators and, therefore, are representative of in vivo conditions. Formate, succinate, and lactate can also be detected, although they are only produced in the AC compartment. Succinate and lactate can be further metabolized within the colon and turned into propionate and butyrate, respectively, through cross-feeding by gut bacteria [8]. Ammonium is also evaluated over time and is considered representative of proteolytic microbial metabolism.

Although proteolysis occurs throughout the entire colon model, average values increase along the colon reactors. Depending on the conditions assayed (diet interventions, development of microbial communities representative of dysbiosis associated to intestinal pathologies, food ingredients, etc.), specialized analytical methodologies targeting the formation of specific intermediate and end metabolites are evaluated [4].

THERAPEUTICAL AND OTHER STRATEGICAL APPLICATIONS

The dynamic simulators have been developed with the purpose to establish in vitro a relatively stable microbial ecosystem under physiologically relevant colon conditions. The models mimic physiological parameters, based in temperature, anaerobiosis, pH, and retention times, emulating measurements made in in vivo conditions. Due to this, these validated systems are frequently used to test experimental substrates and could be predictive for in vivo situations [2].

In vitro models are particularly well suited for screening prebiotics or probiotics for special functions in the gut before moving to in vivo investigation. A comparison of TIM-2 and SHIME, to study long-chain arabinoxylans as potential prebiotics, has demonstrated that both models similarly revealed a compound-specific modulation of prebiotics, in terms of SCFA production, and stimulation of specific *Bifidobacterium* species [20]. Since the majority of carbohydrate breakdown occurs in the proximal colon region, dynamic multistage-fermentation models are also valuable, for identifying slowly fermentable prebiotics.

Viability of probiotics can be analyzed during daily feeding of the laboratory simulators and after ending dietary interventions, in order to evaluate competitive and colonization advantages. The incorporating of mucin-covered surfaces in the colon simulators, e.g., M-SHIME, allows differentiation of probiotic adhesion capabilities to the mucosal environment, thereby elucidating their potential role in human health [21].

The suitability of the SIMGI and BFBL laboratory automated simulators of the colon microbiome to reproduce human conditions associated to changes in dietary lifestyles has been evaluated by the microbiological and metabolic changes of an obese-associated colonic microbiota after the supplementation of the nutritive medium with oligosaccharides derived from lactulose. The comparison of SCFA and ammonium formation, during the microbiota stabilization period (high energy diet), and the substitution of easily digestible carbohydrates, by the oligosaccharides derived from

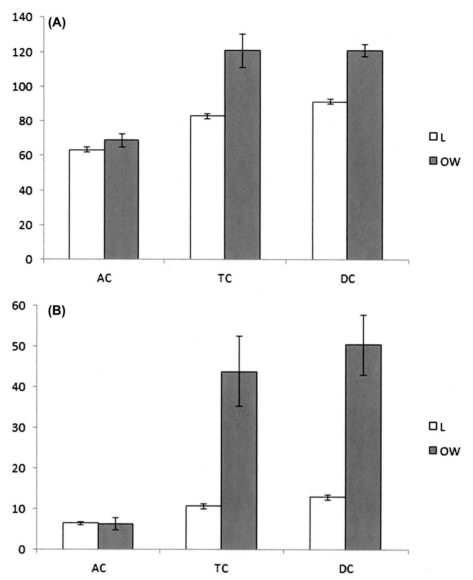

FIGURE 6.2 Total SCFA (A) and ammonium (B) production (mM) by the microbiome from overweight (OW) and lean (L) donors, in the ascending (AC), transverse (TC), and descending colon (DC) reactors of the BFBL laboratory simulator.

lactulose, indicated that the prebiotic allowed the development of fermentative functionality, maintaining the net production of butyrate, with potential beneficial effects on health, and avoiding a full transition to proteolytic metabolism profiles [19].

In addition, we have observed differences in total SCFA and ammonium production, between overweight and lean microbiota donor samples (Fig. 6.2). There are reports describing an intestinal obesogenic microbiome signature, characterized by a higher capacity of energy harvest, which persists after successful dieting and contributes to faster weight regain and metabolic aberrations, commonly referred to as the "yo-yo effect" [22]. The results we have observed in vitro indicate that the higher metabolism of the obesogenic microbiota seems to occur in the transverse and distal colon regions (Fig. 6.2).

CONCLUDING REMARKS

The developing of in vitro fermentation models, able to closely mimic the gut microbial environment, can offer remarkable insights into gut microbiota diversity and key functions, required for healthy intestinal homeostasis. However, the laboratory in vitro simulators are still unsuccessful in the analysis of the mutual communication between the gut bacteria and the intestinal epithelium, or in the simulation of disease conditions of the host. Accordingly, in vitro findings should be fully validated in human studies.

REFERENCES

[1] Martin R, Bermudez-Humaran LG, Langella P. Gnotobiotic rodents: an in vivo model for the study of microbe-microbe interactions. Front Microbiol 2016;7:409.

[2] Aguirre M, Venema K. Challenges in simulating the human gut for understanding the role of the microbiota in obesity. Benef Microbes 2017;8(1):31−53.

[3] O'Donnell MM, Rea MC, O'Sullivan O, Flynn C, Jones B, McQuaid A, Shanahan F, Ross RP. Preparation of a standardised faecal slurry for ex-vivo microbiota studies which reduces inter-individual donor bias. J Microbiol Methods 2016;129:109−16.

[4] Barroso E, Van de Wiele T, Jimenez-Giron A, Munoz-Gonzalez I, Martin-Alvarez PJ, Moreno-Arribas MV, Bartolome B, Pelaez C, Martinez-Cuesta MC, Requena T. *Lactobacillus plantarum* IFPL935 impacts colonic metabolism in a simulator of the human gut microbiota during feeding with red wine polyphenols. Appl Microbiol Biotechnol 2014;98(15):6805−15.

[5] Mancabelli L, Milani C, Lugli GA, Turroni F, Ferrario C, van Sinderen D, Ventura M. Meta-analysis of the human gut microbiome from urbanized and pre-agricultural populations. Environ Microbiol 2017;19(4):1379−90.

[6] Shetty SA, Hugenholtz F, Lahti L, Smidt H, de Vos WM. Intestinal microbiome landscaping: insight in community assemblage and implications for microbial modulation strategies. FEMS Microbiol Rev 2017;41(2):182−99.

[7] Rios-Covian D, Ruas-Madiedo P, Margolles A, Gueimonde M, de Los Reyes-Gavilan CG, Salazar N. Intestinal short chain fatty acids and their link with diet and human health. Front Microbiol 2016;7:185.

[8] Louis P, Flint HJ. Formation of propionate and butyrate by the human colonic microbiota. Environ Microbiol 2017;19(1):29−41.

[9] El Hage R, Hernandez-Sanabria E, Van de Wiele T. Emerging trends in "smart probiotics": functional consideration for the development of novel health and industrial applications. Front Microbiol 2017;8:1889.

[10] Spanogiannopoulos P, Bess EN, Carmody RN, Turnbaugh PJ. The microbial pharmacists within us: a metagenomic view of xenobiotic metabolism. Nat Rev Microbiol 2016;14(5):273−87.

[11] Minekus M, Smeets-Peeters M, Bernalier A, Marol-Bonnin S, Havenaar R, Marteau P, Alric M, Fonty G. Huis in 't Veld JH. A computer-controlled system to simulate conditions of the large intestine with peristaltic mixing, water absorption and absorption of fermentation products. Appl Microbiol Biotechnol 1999;53(1):108−14.

[12] Zihler Berner A, Fuentes S, Dostal A, Payne AN, Vazquez Gutierrez P, Chassard C, Grattepanche F, de Vos WM, Lacroix C. Novel Polyfermentor intestinal model (PolyFermS) for controlled ecological studies: validation and effect of pH. PLoS One 2013;8(10):e77772.

[13] Gibson GR, Cummings JH, Macfarlane GT. Use of a three-stage continuous culture system to study the effect of mucin on dissimilatory sulfate reduction and methanogenesis by mixed populations of human gut bacteria. Appl Environ Microbiol 1988;54(11):2750−5.

[14] Molly K, Vande Woestyne M, Verstraete W. Development of a 5-step multi-chamber reactor as a simulation of the human intestinal microbial ecosystem. Appl Microbiol Biotechnol 1993;39(2):254−8.

[15] Barroso E, Cueva C, Pelaez C, Martinez-Cuesta MC, Requena T. Development of human colonic microbiota in the computer-controlled dynamic SIMulator of the GastroIntestinal tract SIMGI. LWT-Food Sci Technol 2015;61(2):283−9.

[16] Miralles B, Del Barrio R, Cueva C, Recio I, Amigo L. Dynamic gastric digestion of a commercial whey protein concentrate. J Sci Food Agric 2018;98:1873−9.

[17] Feria-Gervasio D, Tottey W, Gaci N, Alric M, Cardot JM, Peyret P, Martin JF, Pujos E, Sebedio JL, Brugere JF. Three-stage continuous culture system with a self-generated anaerobia to study the regionalized metabolism of the human gut microbiota. J Microbiol Methods 2014;96:111−8.

[18] Wexler AG, Goodman AL. An insider's perspective: Bacteroides as a window into the microbiome. Nat Microbiol 2017;2:17026.

[19] Barroso E, Montilla A, Corzo N, Pelaez C, Martinez-Cuesta MC, Requena T. Effect of lactulose-derived oligosaccharides on intestinal microbiota during the shift between media with different energy contents. Food Res Int 2016;89(Pt 1):302−8.

[20] Van den Abbeele P, Venema K, Van de Wiele T, Verstraete W, Possemiers S. Different human gut models reveal the distinct fermentation patterns of arabinoxylan versus inulin. J Agric Food Chem 2013;61(41):9819−27.

[21] Van den Abbeele P, Marzorati M, Derde M, De Weirdt R, Joan V, Possemiers S, Van de Wiele T. Arabinoxylans, inulin and Lactobacillus reuteri 1063 repress the adherent-invasive *Escherichia coli* from mucus in a mucosa-comprising gut model. NPJ Biofilms Microbiomes 2016;2:16016.

[22] Thaiss CA, Itav S, Rothschild D, Meijer MT, Levy M, Moresi C, Dohnalova L, Braverman S, Rozin S, Malitsky S, Dori-Bachash M, Kuperman Y, Biton I, Gertler A, Harmelin A, Shapiro H, Halpern Z, Aharoni A, Segal E, Elinav E. Persistent microbiome alterations modulate the rate of post-dieting weight regain. Nature 2016;540:544−51.

Chapter 7

Potential of Metabolomics to Breath Tests

F. Priego-Capote[1,2,3] and Maria Dolores Luque de Castro[1,2,3]

[1]Department of Analytical Chemistry, University of Córdoba, Córdoba, Spain; [2]Maimónides Institute of Biomedical Research (IMIBIC), Reina Sofia Hospital, University of Córdoba, Córdoba, Spain; [3]CIBER Fragilidad y Envejecimiento Saludable (CIBERfes), Instituto de Salud Carlos III, Spain

INTRODUCTION

Breath profiling analysis started in 1971, when Pauling detected more than 250 volatile organic compounds (VOCs), in breath trapped in a cooled stainless steel tube [1]. The benefits of exhaled breath convert this biofluid in an accessible sample, obtained in a noninvasive manner, easily repeatable, released spontaneously, and in relatively large amounts. The collection of exhaled breath is less expensive than alternative invasive sampling, such as collection of sputum or bronchoalveolar fluid. Exhaled breath composition is of interest in studies dealing with occupational or environmental exposures, and/or clinical applications, based on the diagnostic/prognostic of pathophysiological conditions [2,3].

Breath samples are characterized by the presence of exogenous and endogenous compounds, as well as microbiome-associated metabolites. Exhaled breath contains detectable levels of VOCs, semivolatile organic compounds (SVOCs), water soluble metabolites, lipids, other macromolecules such as proteins and DNA, and microbiota components such as bacteria and viruses. This chemical diversity supports the utility of breath analysis to evaluate environmental exposure of individuals, enzyme activity, disease detection, therapeutic monitoring, and oxidative stress, among others.

Exhaled Breath Versus Exhaled Breath Condensate

There are two main types of samples involving breath. One of them contains essentially VOCs. Around 1800 VOCs has been reported in breath analysis, by using techniques such as ion mobility spectroscopy (IMS), selected ion flow tube mass spectrometry (SIFT−MS), gas chromatography coupled to MS (GC−MS), or electronic noses (e-noses) [4,5]. The second type uses condensed breath, which is obtained by a cooling procedure that allows collecting an aqueous phase. Exhaled breath condensate (EBC) contains mainly nonvolatile compounds and water-soluble volatiles, since most VOCs are lost during sampling [6].

Breath and EBC are two complementary samples able to be used in metabolomics studies. In the first case, the term volatolomics has been used to define the analysis of VOCs, developed either directly or after a preconcentration step. On the other hand, EBC is a sample that can be stored for further analysis or in biobanks, handled for sample preparation according to the final purpose, and fractioned for different determinations.

Technical Hurdles—Dilution and Standardization

Despite all the positive aspects of breath analysis, there are key limitations essentially related with the low concentration of metabolites and the lack of standardization of the analytical steps from sampling to determination. In the case of EBC, the dilution effect is particularly important, as about 99.9% of the sample is water. Dilution also affects breath analysis, since most VOCs are present at low concentrations, and therefore, breath profiles are dominated by the most concentrated compounds. For this reason, the inclusion of preconcentration steps, online for breath analysis and off-line for EBC, is mandatory to obtain wide information on breath composition.

Microbiome and Metabolome in Diagnosis, Therapy, and other Strategic Applications. https://doi.org/10.1016/B978-0-12-815249-2.00007-5

A key factor also related with dilution is pulmonary capacity, which varies between individuals, especially when patients suffering from respiratory difficulties are involved together with healthy controls or with other types of diseases. Several standardization strategies have been proposed, e.g., the use of the exhaled volume, which would seem suitable for independent studies but do not completely correct interindividual differences, due to particles content, or differences in breathing frequency.

The correction by chemical parameters such as ionic strength, conductivity, or concentration of certain metabolites such as urea has not provided the pursued results. Therefore, there is not a standard golden rule to correct dilution effects [2,3,7]. Another aspect that makes standardization difficult is the variety of sampling devices, which sample different parts of the respiratory tract and, therefore, provide samples with diverse physical characteristics and chemical compositions.

Airway Anatomy and Physiology

In the case of exhaled breath, Lawal et al. have recently described three breath portions that can be differently sampled: late respiratory, end-tidal, and mixed respiratory. Mixed expiratory breath sampling encompasses the total exhaled breath, which includes "dead space air"—defined as the air not involved in gaseous exchange included in mouth and potentially nose air—while the other breath portions aim to minimize contamination from this dead space [8].

In the case of EBC, some devices allow collecting separated fractions into different bags, which will be further evaluated.

LABORATORY EQUIPMENT FOR SAMPLING

In EBC sampling, an individual breathes into a sampling device usually equipped with a mouthpiece, a separation valve, and a cooled condensation unit, in which the exhaled breath is collected for analysis. The collection tube is the most critical part of the sampling device, since its surface area and temperature are essential parameters for EBC collection. Generally, the condensate phase is collected by gravity or with a special plunger to increase the sampled volume.

The different devices designed for EBC sampling can be divided into laboratory-made and commercial devices. Examples of laboratory made devices based on a common principle can be found [9−14]. The condensation tube is built with different materials such as glass, polypropylene, polyethylene, PTFE, or aluminum. Cooling of the condensation tube is favored by ice or by a cooling gel. Devices proposed [9−11] were designed to sample in a research facility, while others [12−14] are portable samplers, able to be miniaturized. All these systems are characterized by an open-end configuration, by which the exhaled breath is directed through the cooling surface and exits to avoid recirculation. The estimated collection efficiency of these laboratory-made devices is around 50% [3].

Industrialized Exhaled Breath Condensate Samplers

Three types of commercial EBC samplers have been described in the literature (Fig 7.1). The most widely used is the EcoScreen sampler, currently commercialized as the EcoScreen2 version. This system condensates and collects EBC in disposable bags with polyethylene surface. The sampler allows control of EBC collection, and condensation is fractioned into two separate chambers, depending on the depth of the respiratory tract. Thus, one of the fractions corresponds to the mouth and upper respiratory tract, and the other one comes from the lower respiratory tract. The EcoScreen2 device also contains a spirometer to measure the exhaled volume and time collection. The main limitation of this sampler is its nonportable use.

Another commercial EBC sampler, known as Turbodecs, is a fixed−portable device that contains a disposable collection cell, inserted in a refrigeration unit based on Peltier cooling effect, and allows electronic adjustment of the condensation temperature.

The completely portable RTube EBC sampler, consists of a large T-section furnished with a disposable valve and a saliva separator, a collection tube made of polypropylene and a cooling sleeve that can be refrigerated prior to analysis. A plunger is used to collect the EBC for further analysis. The RTube is the simplest commercial sampler system, but the condensation temperature could change for long collection times.

Exhaled Breath Samplers

There are two basic approaches for sampling of exhaled breath, depending on off-line or online connection between the sampler and the analytical equipment—the latter providing real-time information. Sampler configuration in this case is a critical task that should be properly planned in the light of the pursued aim.

EcoScreen2

FIGURE 7.1 Commercial devices for sampling of exhaled breath condensate.

Turbodecs

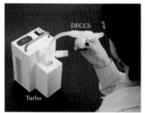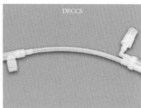

RTube

Physiological parameters such as partial pressure of exhaled CO_2, temperature, flow, or partial pressure of exhaled O_2 can be properly used to control sampling as a function of the respiratory phases. Sampling of systemic compounds (transported by blood perfusion to the lungs for exhalation) is favored by CO_2-controlled sampling [15−18], by restricting the sampling process to the end-tidal phase [19].

Exhaled breath can directly lead to the detection system, which is based on techniques such as PTR−MS, SIFT−MS, IMS, laser spectrometry, photoacoustic spectroscopy, and sensor arrays [4]. In direct analysis of exhaled breath, no preconcentration or storage of breath samples is performed. Direct analysis protocols constitute interesting approaches, for real-time monitoring of certain compounds, but with key limitations such as low analytical sensitivity for detection of low-concentrated compounds and the lack of resolution to profile isomeric components. An alternative to improve sensitivity in direct analysis is the implementation of automated sampling valves, to select representative breath phases, which are subsequently analyzed [17,20].

Off-line sampling for breath analysis solves limitations of direct configurations in terms of sensitivity and selectivity, as it can include sample preparation steps that overcome shortcomings, owing to the inappropriate values of these parameters. Sampling is generally carried out in collection containers, among which polymer bags are the dominating devices, with special emphasis on Tedlar bags. The Bio-VOC and the BCA (Breath Collection Apparatus) samplers are especially suited to collect late expiratory breath, despite operating differently. BCA is an extended tubular structure where air flows downstream, and that flowing proximal to the mouth is collected; while the Bio-VOC system consists of a small reservoir, in which air is continuously displaced as exhalation proceeds [8].

Preconcentration is frequently a mandatory step to detect VOCs in off-line protocols, particularly for mixed expiratory breath samples. They include solid-phase extraction (SPE), with subsequent thermal desorption of the retained compounds before analysis. Disadvantages of sorbent trapping are (1) high sampling volumes (typically 100−5000 mL), which is a key limitation for individuals suffering respiratory deficiencies, and (2) problems created by the high water content.

Microextraction techniques such as solid-phase microextraction (SPME) or extractive needle trap microextraction overcome the mentioned limitations and meet the requirements for controlled and fast sampling and reliable preconcentration of VOCs. For these reasons, microextraction techniques have gained relevance in laboratories for profiling analysis of breath. However, extraction efficiency of these techniques will depend critically on the physicochemical properties of sorbents that in a certain way can provide a fractioned snapshot of the volatilome. Nevertheless, microextraction techniques can suffer from carryover and poor quantification of given compounds, depending on the sample matrix [2,4,8].

Collection Routine

In general, special cautions should be taken in breath and EBC samplings, to avoid cross-contamination from different individuals. Therefore, cleaning of the sampler unit is essential to minimize contamination, and reuse of bags or containers must be avoided. One other key aspect is to reduce the storage time in breath or EBC containers, since some compounds could be adsorbed on the surface material.

According to the differences in samplers for breath and EBC collection, the variability of metabolite profiles found in the two types of samples is justified, by the lack of standardization of the sampling process.

CRUCIAL MOLECULES AND PATHWAYS

Both biological samples (breath and EBC) are complementary, due to the fact that the former essentially contains VOCs and SVOCs, while breath condensation favors the presence of nonvolatile and water-soluble components.

Breath composition has been the subject of comprehensive characterization studies. The existence in breath of 874 VOCs, unequivocally defined by the CAS number, is demonstrated. Spectral library matches and retention indices, identified in exhaled breath approximately 210 volatile compounds, out of 874. These data were confirmed unequivocally by the corresponding standards [4].

Two abundant compounds in breath are isoprene and acetone. The former is not only a by-product of the mevalonate pathway but also produced in the periphery of the human body [21]. Acetone can be formed from acetoacetate by acetoacetate decarboxylase enzyme. This is the final step of the ketone-body pathway, which supplies a secondary energy source [22].

A significant fraction of the compounds detected in exhaled breath is from exogenous sources such as environmental exposition; therefore, they are not detected in all individuals. As an example, it is well-known that compounds such as dienes, alkenes, and alkynes are smoking-related compounds. Also, an important fraction of VOCs may be related to food consumption or medication, and other breath components (such as indole and derivatives coming from tryptophan conversion) are produced by the microbiome [2,4,8].

Concerning EBC composition, it is worth mentioning two critical biological processes: oxidative stress and/or chronic inflammation. Thus, compounds related with both processes can be grouped in two main pathways that lead to the formation of inflammatory markers: (1) production of nitrogen reactive species and (2) formation of polyunsaturated fatty acids (PUFAs) metabolites, with special emphasis on arachidonic acid [3].

Nitrogen Reactive Species

Nitrogen reactive species include inorganic species such as nitrite and nitrate, and organic species such as 3-nitrotyrosine (3-NT) and S-nitrosothiols (RS−NO). All of them are form nitric oxide, NO, formed from arginine. NO is the most extensively studied gas in exhaled breath and also the most powerful marker to evaluate oxidative stress. NO has been found of clinical relevance to aid in the diagnosis of asthma and other lung diseases. Nitrite and nitrate, formed by oxidation of NO, can also react with thiols, which can further form 3-NT.

High levels of nitrate, nitrite, and 3-NT have been often found in EBC from patients with respiratory diseases such as asthma, chronic obstructive pulmonary disease (COPD), or cystic fibrosis (CF) [23,24]. While concentrations of nitrite and nitrate are typically in the low $g\,L^{-1}$ levels and can be easily measured by accessible analytical techniques, the concentrations of 3-NT and RS−NO are significantly lower (ng L^{-1}) and require sensitive analytical techniques.

Arachidonic Acid

PUFAs, with special emphasis on arachidonic acid, are present in glycerophospholipids of the body cell membranes. Arachidonic acid is a precursor of eicosanoids, and it is also considered one of the most important intermediates in the inflammatory response. Fig 7.2 shows the most important pathways by which arachidonic acid is transformed into its metabolites, involving different enzymes leading to different groups of compounds, the most important being isoprostanes,

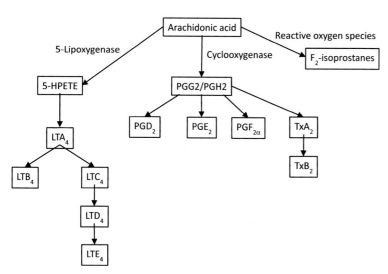

FIGURE 7.2 Main biochemical pathways produced by oxidation of arachidonic acid.

leukotrienes, and prostanoids [3]. Isoprostanes are in vivo formed by free-radical catalyzed peroxidation of fatty acids in a nonenzymatic process. 8-Isoprostane, detected in EBC, is considered one of the most reliable markers of lipid peroxidation and oxidative stress [25].

Leukotrienes are inflammatory mediators synthesized through oxidation of arachidonic acid by 5-lipoxygenase. The presence of leukotrienes in EBC is justified by their release from inflammatory cells of the airways and contract airway smooth muscle. Nevertheless, these metabolites are detected in EBC at really low concentrations; therefore, their determination is a challenge [26].

Prostanoids are also formed from arachidonic acid in a process catalyzed by cyclooxygenases to produce three main groups of compounds: prostaglandins, thromboxanes, and prostacyclins. The role of prostanoids in the inflammatory response is not completely known, because they can act both as pro- and antiinflammatory agents, as a function of the stimulus [27].

Apart from inflammatory and oxidative stress—related markers, other families of metabolites have been reported in EBC. A comprehensive characterization of EBC was carried out by Peralbo-Molina et al., who identified around 50 different compounds in EBC by GC—MS analysis. The identification included aliphatic heterocyclic compounds such as eucalyptol and indole, homomonocyclic aromatic compounds such as cresol and benzyl alcohol, free fatty acids and esters, fatty alcohols and amides, glycerolipids, prenol monoterpenes, such as limonene, linalool, and camphor, and triterpenes such as squalene, among other organic compounds [28].

Complementarily, Fernández-Peralbo et al. developed an optimization study to analyze EBC by LC—MS/MS, which led to the identification of 49 compounds including amino acids (valine, leucine, aspartic and glutamic acid, phenylalanine), fatty acids (octanedioic acid, nonanedioic acid, decanedioic acid) and amides (decanamide and dodecanamide), fatty aldehydes (9,12-tetradecadienal), sphingoid bases (d18:2-sphingosine and d18:3-sphingosine), imidazoles (urocanic acid), hydroxyl acids (lactic acid), aliphatic acyclic acids (urea), among others. The use of both analytical platforms reported a complementary profile of EBC due to the occurrence of chemical families exclusively detected by one of them [29].

APPLICATIONS IN AIRWAY DISEASES

Airway diseases are highly prevalent worldwide. Differentiation among asthma, COPD, and bronchiectasis in the early stage of disease is extremely important, for the adoption of appropriate therapy. However, the high prevalence of these diseases and the common pathophysiological pathways support the fact that some patients with different diseases may present similar symptoms, frequently involving chronic inflammation and/or oxidative stress.

In a recent review article, Davis et al. classified respiratory diseases according to the number of EBC papers published in the last 20 years, which allowed detecting that asthma has been the most studied disease, followed by COPD [30]. The use of breath analysis could be in this context an interesting alternative for disease diagnosis and monitoring, because of the benefits of EBC.

Nitric oxide and metabolites (NOx), as well as pH, have been monitored in breath-derived samples with this purpose [31]. Nevertheless, it should be noted that none of the potential biomarkers detected in breath or EBC has so far been approved for clinical diagnosis. Therefore, further studies are needed to validate promising biomarkers for their implementation in diagnostics.

Untargeted Versus Fingerprinting Metabolomics

The physicochemical characteristics of breath and EBC have influenced the metabolomics approach. Thus, breath analysis has been faced by untargeted approaches, while EBC has been predominantly analyzed by fingerprinting approaches, with direct analysis as the most frequent strategy.

EBC in asthma has been predominantly analyzed by NMR, aiming at metabolomics fingerprinting. Thus, Motta et al. developed a partial least squares discriminant analysis (PLS-DA), with external validation, to discriminate asthmatic individuals from healthy individuals. The best quality parameters were obtained by a targeted model with R^2 and Q^2 values of 0.91 and 0.87, respectively [32].

Sinha et al. also used NMR to classify asthmatic and healthy individuals, with 80% sensitivity and 75% specificity. Unsupervised clustering within the asthma group revealed cluster 1 formed by patients with low long-term exacerbation scores; cluster 3, with patients with low blood eosinophils and high neutrophil levels, and with strong family history of asthma; and cluster 2, with an intermediate position [33].

Bertini et al. used a similar approach, to find discrimination patterns in EBC from COPD patients. Overall accuracy was 86%. The EBC from COPD individuals featured significantly lower levels of acetone, valine, and lysine and higher levels of lactate, acetate, propionate, serine, proline, and tyrosine [34].

Recently, Maniscalco et al. have also proposed a screening approach based on NMR fingerprinting analysis of EBC, to discriminate COPD and asthma patients in a pilot study. An OPLS-DA model allowed differentiating EBC from COPD and asthma patients, with high classification parameters (R^2 and Q^2 values of 0.86 and 0.86, respectively). COPD patients showed higher levels of ethanol and methanol but lower levels of formate and acetone/acetoin ratio. The performance of the model was also blindly tested by using an independent cohort, including patients diagnosed with COPD and asthma (R^2 and Q^2 values of 0.89 and 0.87, respectively) [35].

On the other hand, breath analysis has been essentially carried out by untargeted strategies, such as that adopted by Smolinska et al. [36]. A set of 17 VOCs identified by GC–MS analysis was selected to discriminate preschool asthmatic children from transient wheezing children, with a prediction rate of 80%. These VOCs were related to oxidative stress caused by inflammation in the lungs and consequently by lipid peroxidation. More recently, the same team has proposed a panel of seven VOCs (three aldehydes, one hydrocarbon, one ketone, one aromatic compound, and one unidentified VOC), to predict asthma exacerbations in children, within 14 days after sampling (88% sensitivity and 75% specificity) [37].

Cystic Fibrosis and Bacterial Pneumonia

Montuschi et al. evaluated if NMR-based fingerprinting followed by identification of selected compounds in EBC was able to differentiate stable ($n = 29$) from nonstable ($n = 24$) patients with CF. Acetate, ethanol, 2-propanol, and acetone allowed separating CF from the healthy group [38]. Complementarily, ethanol, acetate, 2-propanol, and methanol were relevant to separate stable versus exacerbated cases in CF. On the other hand, Robroeks et al. compared VOC profiles obtained in breath, from 57 healthy controls and 48 CF cases. A set formed by 22 compounds enabled classification between healthy and CF subjects with 100% specificity, while 14 tentative compounds correctly identified the 2 *P. aeruginosa*-positive individuals, compared with the 25 *P. aeruginosa*-negative individuals [39].

Ventilator-Associated Pneumonia

Van Oort et al. profiled VOCs in breath to diagnose pneumonia in 93 mechanically ventilated intensive care patients. Out of 145 identified VOCs, 12 were significantly altered [40]. In colonized patients, 52 VOCs were significantly different. Partial least square discriminant analysis classified patients with modest accuracy. For determining the colonization status of patients, the model had an area under the ROC curve of 0.69. Therefore, the authors hypothesized that exhaled breath analysis can be used not only to discriminate pneumonia from controls with a relatively good accuracy but also to detect the presence and absence of pathogens in the respiratory tract.

APPLICATIONS IN LUNG CANCER

GC—MS has been the most common analytical platform in metabolomics analysis of exhaled breath and EBC to aid in the diagnostic of lung cancer, even though it requires expert personnel to both handle the samples and to interpret the results. Two criteria have been considered for selecting potential biomarkers: (1) their presence/absence in the patient; (2) their increased/decreased concentration as a consequence of lung cancer (LC).

Breath as Sample for LC Biomarkers Searching

Breath has been the most used sample to look for LC biomarkers—possibly because its sampling requires simple devices for collection. As early as 1985, Gordon et al. found 22 VOCs that showed the largest difference in LC [41]. In further analysis, the authors selected three VOCs (acetone, methyl ethyl ketone, and n-propanol) according to their chromatographic peaks and occurrence in the subjects to accurately classify 93% of the samples. Also 22 VOCs in breath were selected by Phillips et al. to discriminate patients with LC, regardless of stage, among 108 patients with an abnormal chest radiograph [42]. For primary LC, the 22 VOCs panel reported 100% sensitivity and 81.3% specificity.

In a different study, the same authors reported identifying primary LC with a sensitivity of 89.6% and a specificity of 82.9%, using nine VOCs. In primary LC breath test findings were consistent with accelerated catabolism of alkanes and monomethylated alkanes [43]. Poli et al. chose 13 VOCs to compare three groups: non—small cell lung cancer (NSCLC), mild—moderate COPD, and asymptomatic smokers as controls. None of the selected VOCs alone distinguished the NSCLC patients from the other study groups, but overall VOC concentrations were highly discriminant (>70%) [44]. In later studies, the same authors found the levels of all aldehydes increased in NSCLC patients compared with healthy controls, with a discrimination power of >90% [45].

Bajtarevic et al. used 4 VOCs of LC patients to detect LC with a sensitivity of 52% and a specificity of 100% [46]. Ulanowska et al. reported increased concentrations of 11 VOCs in the breath of 137 LC patients, compared with 143 healthy nonsmokers; three VOCs (pentanal, hexanal, and nonane) were identified only in the breath of LC patients [47].

The controversy on the origin of VOCs namely direct volatile emission from the cancerous tissues in lungs into the airways versus collection of these metabolites by the blood and then exchange at the air—blood interface in the lung was solved by Capuano et al. They collected the breath and the air inside both lungs with a modified bronchoscopic probe, then analyzed the two samples by both GC—MS and e-nose. The results obtained by the former analytical platform demonstrated a substantial preservation of the VOCs patterns from inside the lung, to the exhaled breath [48].

Despite the number of publications on LC biomarkers using breath as sample, confidence in their potential general use is scant. A recent publication by Schallschmidt et al. (1) quantified each VOC considered as a potential disease marker on the basis of individual calibration; (2) adopted quality control measures required to maintain reproducibility in breath sampling; (3) analyzed VOCs using SPME—GC—MS; and (4) quantified 24 VOCs. On these bases, the authors concluded that the concentration of aromatic compounds in the breath was increased, as expected, in smokers, while LC patients displayed significantly increased levels of oxygenated VOCs such as aldehydes, 2-butanone, and 1-butanol. The effective selectivity of the breath VOC approach with regard to LC detection was limited [49].

EBC as Sample for LC Biomarkers Searching

The use of EBC to search for LC biomarkers has been very recent and almost so far limited to contributions from the authors' group. A first study, developed by a GC—time-of-flight (TOF) MS platform, was devoted to a cohort of 48 LC patients, 130 risk factor individuals (active smokers and ex-smokers), and 61 healthy controls (nonsmokers and without respiratory diseases), who provided two-fraction EBC (upper and central airways and distant airway). Forty-seven compounds were tentatively identified that contributed to discriminate among the three groups of individuals as Fig 7.3 shows, with a relevant role for lipid metabolites such as monoacylglycerols and squalene, thus opening a door to the use of metabolomics to go inside the study of LC [51].

Marker Interpretation, and Overlapping With Other tumors

Current restrictions are set by the low and variable VOC concentrations, the technical complexity of studies involving breath sampling, the limited capability of current analytical procedures to detect unstable marker candidates, and particularly, the number of VOCs candidates to LC biomarkers, which also increase/decrease their concentration in patients with colon, breast, and prostate cancers [50].

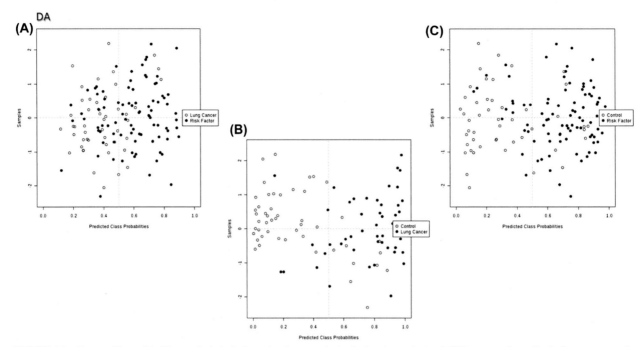

FIGURE 7.3 Support Vector Machine analysis built from the data set, obtained after the analysis of EBC extracts from distal airways, comparing (A) lung cancer patients and risk factor individuals, (B) lung cancer patients and control individuals, and (C) control and risk factor individuals. *Reproduced with permission of IOP Publishing, Peralbo-Molina A, Calderón-Santiago M, Priego-Capote F, Jurado-Gámez B, Luque de Castro MD. Metabolomics analysis of exhaled breath condensate for discrimination between lung cancer patients and risk factor individuals. J Breath Res 2016;10(1):016011—14.*

A subsequent study on the same samples and using the same analytical platform allowed configuration of three panels of five metabolites each, using the Panelomix tool to minimize false negatives (increasing specificity) and false positives (increasing sensitivity). The three panels provided sensitivity and specificity above 77.9% and 67.5%, respectively, with an area under the curve (AUC) above 77.5% [52].

Smoking is a crucial factor in respiratory diseases and lung inflammation. EBC was compared in three groups of individuals: current smokers, former smokers, and never smokers [53]. Twelve significant compounds included monoacylglycerol derivatives, terpenes, and other compounds, the presence of which could be associated to smoking. The highest alteration occurred with indole, p-cresol, and monostearin, the discrimination capability of which is shown in Fig 7.4.

APPLICATIONS IN GASTROINTESTINAL AND SYSTEMIC DISEASES

Gases produced in the intestine diffuse into the systemic circulation and are expired through the lungs. In fact there is a clinical assay, the breath test for small gut bacterial overgrowth, based on this principle. The application of the breath test is gaining interest not only in small intestinal bacterial overgrowth (SIBO) and irritable bowel syndrome—like symptoms, but also in carbohydrate maldigestion and dysfunction or alterations in orocecal transit [54]. Monitoring H_2 and CH_4, which are exclusively produced via microbial fermentation in the gut, is the foundation. Gut microbes readily digest carbohydrates resulting in the production of these gases, which can be further detected in the exhaled breath [55]. The breath test is useful, inexpensive, simple, and safe; however, standardization is lacking regarding indications for testing, test methodology, and interpretation of results [54].

Microbial Fermentation and Inflammatory Bowel Disease

Kurada et al. reviewed the capability of breath analysis in inflammatory bowel disease (IBD). Pentane, ethane, propane, 1-octene, 3-methylhexane, 1-decene, and NO levels were elevated (P-value $< .05$ to $<10^{-7}$), while 1-nonene, 2-nonene, hydrogen sulfide, and methane were decreased. Complementarily, cytokines detected in EBC were higher in IBD cases compared with healthy individuals (P-value $< .008$), while interleukin-1β exhibited an inverse relationship with clinical disease activity. Future trials should be targeted at determining the exact metabolome patterns, linked to diagnosis and phenotype of IBD [56].

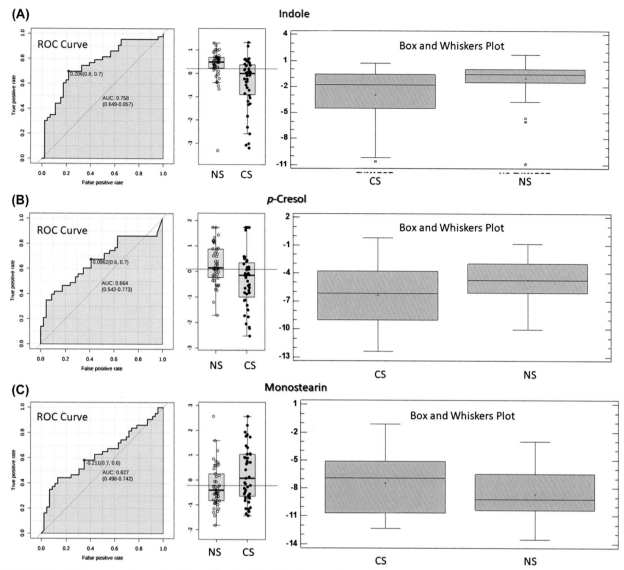

FIGURE 7.4 ROC curves and box and whiskers plots, for (A) indole, (B) *p*-cresol, and (C) monostearin, the three compounds with the highest capability for discrimination in the comparison between current smoker (CS) and never smoker (NS) groups. *Reproduced with permission of Nature, Peralbo-Molina A, Calderón-Santiago M, Jurado-Gámez B, Luque de Castro MD, Priego-Capote F. Exhaled breath condensate to discriminate individuals with different smoking habits by GC–TOF/MS. Sci Rep 2017;7:1421.*

Lipid Peroxidation and Oxidative Stress

Ross and Glen found in healthy volunteers a good positive correlation between breath ethane concentrations and blood hydroperoxide concentrations ($R = 0.60$) and negative correlation with the concentration of vitamin E ($R = -0.65$). These data supported the hypothesis that breath ethane is a marker of systemic lipid peroxidation, a mechanism of oxidative stress, which is elevated in neurodegenerative disease, psychiatric illness, stroke, cardiovascular diseases, and diabetes [57].

Type 2 Diabetes

Exhaled isopropanol could be a biomarker of diabetic individuals after a ketogenic diet. The analysis was carried out by GC–MS, with SPME as sample preparation technique. Isopropanol in the diabetic group was significantly higher than that observed in the healthy group (*P*-value < .001), with promising discriminatory ability (AUC 0.86) and sensitivity and specificity of 75.3% and 85.7%, respectively [58].

SENSORS FOR BREATH ANALYSIS

An ideal nanomaterial-based sensor for breath testing should be sensitive enough at very low concentrations of VOCs, even in the presence of environmental or physiological interferents. Also, it should respond rapidly and selectively to a given VOC. In the absence of the VOC, the sensor should quickly return to its baseline state or be simple and inexpensive enough to be disposable [59].

Electronic Nose

The electronic nose (e-nose) consists of a gas sensor array to provide a fingerprint of exhaled breath (breath print, BP) by detecting VOCs through multiple sensors. The e-nose is able to give quantitative response to a comprehensive VOCs profile, but in this case individual VOCs remain unidentified. Chang et al. have recently designed an e-nose for detection of VOCs in exhaled breath. The sensor system consisted of an array of seven metal oxide gas sensors, a gas flow controlling module, a heating module, a gas adsorption—desorption module, and classifiers for data analysis.

VOCs in exhaled breath potentially discriminated mostly early-stage lung cancer patients before surgery from healthy volunteers. The prognosis after surgery would be predicted by this system [60]. The e-nose has also been able to discriminate asthmatic patients from healthy controls [61] and from COPD patients, likely reflecting the well-known differences in pathogenic mechanisms of asthma and COPD [62].

Quantum Cascade Laser Breath Analyzers

Another interesting approach is the use of quantum cascade lasers (QCLs), which can be arranged in small formats, offering a robust performance. These devices can be made to detect wavelengths over the wide range of 3—20 μm. Important technical parameters for QCLs include high optical power, single frequency operation, good spectral purity, and wide wavelength tunability [63]. Wörle et al. designed an approach based on QCLs for quantitative determination of $^{12}CO_2$ and $^{13}CO_2$ in exhaled breath, which could be implemented in the monitoring of glucose metabolism dysfunction [64].

CONCLUSIONS

The analysis of exhaled medium—either condensate or vapor—may be a feasible approach to support diagnosis of pathologies with special emphasis on respiratory and gastrointestinal diseases. However, it requires further standardization and clinical implementation. The results of the research developed so far, particularly those based on GC—MS platforms, allow concluding that it is not possible to identify single specific breath markers for diseases, which alter the concentrations of a number of compounds and modify the overall chemical composition of exhaled breath. This conclusion should lead to the development of electronic noses, as strategical tools to aid in diagnosis/prognosis of diseases, by typical "breath prints" rather than by a specific component detected in breath.

LIST OF ABBREVIATIONS

3-NT 3-Nitrotyrosine
AUC Area under the curve
BCA Breath collection apparatus
BP Breath print
CAS Chemical Abstract Service
EBC Exhaled breath condensate
CF Cystic fibrosis
COPD Chronic obstructive pulmonary disease
GC—MS Gas chromatography coupled to mass spectrometry
IBD Inflammatory bowel disease
IMS Ion mobility mass spectrometry
LC Lung cancer
LC—MS/MS Liquid chromatography coupled to tandem mass spectrometry
LDA Linear discriminant analysis
NMR Nuclear magnetic resonance

NOx Nitric oxide and metabolites
NSCLC Non−small cell lung cancer
OPLS−DA Orthogonal partial least squares discriminant analysis
PLS−DA Partial least squares discriminant analysis
PTFE Polytetrafluoroethylene
PTR−MS Proton transfer reaction mass spectrometry
PUFAs Polyunsaturated fatty acids
QCL Qquantum cascade laser
ROC Receptor operating characteristics
RS−NO S-nitrosothiols
SIBO Small intestinal bacterial overgrowth
SIFT−MS Selected ion flow tube mass spectrometry
SPME Solid-phase microextraction
SVOCs Semivolatile organic compounds
TOF Time of flight
VOCs Volatile organic compounds

REFERENCES

[1] Teranishi R, Mon TR, Robinson AB, Cary P, Pauling L. Gas chromatography of volatiles from breath and urine. Anal Chem 1972;44:18−20.

[2] Amann A, Miekisch W, Schubert J, Buszewski B, Ligor T, Jezierski T, Pleil J, Risby T. Analysis of exhaled breath for disease detection. Annu Rev Anal Chem 2014;7:455−82.

[3] Kuban P, Foret F. Exhaled breath condensate: determination of non-volatile compounds and their potential for clinical diagnosis and monitoring. A review. Anal Chim Acta 2013;805:1−18.

[4] Amann A, Costello B, Miekisch W, Schubert J, Buszewski B, Pleil J, Ratcliffe N, Risby T. The human volatilome: organic compounds (VOCs) in exhaled breath, skin emanations, urine, feces and saliva. J Breath Res 2014;8:034001.

[5] Costello B, Amann A, Al-Kateb H, Filipiak W, Khalid T, Osborne D, Ratcliffe N. A review of the volatiles from the healthy human body. J Breath Res 2014;8:014001.

[6] Hunt J. Exhaled breath condensate − an overview. Immunol Allergy Clin North Am 2007;27:587−96.

[7] Winters BR, Pleil J, Angrish MM, Stiegel MA, Risby TH, Madden MC. Standardization of the collection of exhaled breath condensate and exhaled breath aerosol using a feedback regulated sampling device. J Breath Res 2017;11:047107.

[8] Lawal O, Ahmed WM, Nijsen TME, Goodacre R, Fowler SJ. Exhaled breath analysis: a review of 'breath-taking'methods for off-line analysis. Metabolomics 2017;13:110.

[9] Rosias PPR, Dompeling E, Hendriks HJE, Heijnens JWCM, Donckerwolcke RAMG, Jöbsis Q. Exhaled breath condensate in children: pearls and pitfalls. Pediatr Allergy Immunol 2004;15:4−19.

[10] Mutlu GM, Garey KW, Robbins RA, Danziger LH, Rubinstein I. Collection and analysis of exhaled breath condensate in humans. Am J Respir Crit Care Med 2001;164:731−7.

[11] Montuschi P. Analysis of exhaled breath condensate in respiratory medicine: methodological aspects and potential applications. Ther Adv Respir Dis 2007;1:5−23.

[12] Kuban P, Kobrin EG, Kaljurand M. Capillary electrophoresis − a new tool for ionic analysis of exhaled breath condensate. J Chromatogr A 2012;1267:239−45.

[13] Ueno T, Kataoka M, Hirano A, Iio K, Tanimoto Y, Kanehiro A, Okada C, Soda R, Takahashi K, Tanimoto M. Inflammatory markers in exhaled breath condensate from patients with asthma. Respirology 2008;13:654−63.

[14] Rosias PPR, Robroeks CM, van de Kant KD, Rijkers GT, Zimmermann LJ, Van Schayck CP, Heynens JW, Jöbsis Q, Dompeling E. Feasibility of a new method to collect exhaled breath condensate in pre-school children. Pediatr Allergy Immunol 2010;21:e235−44.

[15] Cope KA, Watson MT, Foster WM, Sehnert SS, Risby TH. Effects of ventilation on the collection of exhaled breath in humans. J Appl Physiol 2004;96:1371−9.

[16] Birken T, Schubert J, Miekisch W, Noldge-Schomburg G. A novel visually CO_2 controlled alveolar breath sampling technique. Technol Health Care 2006;14:499−506.

[17] Miekisch W, Hengstenberg A, Kischkel S, Beckmann U,Mieth M, Schubert JK. Construction and evaluation of a versatile CO_2 controlled breath collection device. IEEE Sensors J 2010;10:211−5.

[18] Solga SF, Mudalel M, Spacek LA, Lewicki R, Tittel F, Loccioni C, Russo A, Risby TH. Factors influencing breath ammonia determination. J Breath Res 2013;7:037101.

[19] King J, Kupferthaler A, Unterkofler K, Koc H, Teschl S, Teschl G, Miekisch W, Schubert J, Hinterhuber H, Amann A. Isoprene and acetone concentration profiles during exercise on an ergometer. J Breath Res 2009;3:027006.

[20] Kamysek S, Fuchs P, Schwoebel H, Roesner JP, Kischkel S, Wolter K, Loeseken C, Schubert JK, Miekisch W. Drug detection in breath: effects of pulmonary blood flow and cardiac output on propofol exhalation. Anal Bioanal Chem 2011;40:2093−102.

[21] Gelmont D, Stein RA, Mead JF. Isoprene—the main hydrocarbon in human breath. Biochem Biophys Res Commun 1981;99:1456−60.

[22] Anderson JC. Measuring breath acetone for monitoring fat loss: review. Obesity 2015;23:2327—34.

[23] Formanek W, Inci D, Lauener RP, Widhaber JH, Frey U, Hall GL. Elevated nitrite in breath condensates of children with respiratory disease. Eur Respir J 2002;19:487—91.

[24] Horvath I, de Jongste JC, editors. Exhaled biomarkers. Eur Respir Society; 2010. p. 210—5.

[25] Milne GL, Musiek ES, Morrow JD. F2-isoprostanes as markers of oxidative stress in vivo: an overview. Biomarkers 2005;10:S10—23.

[26] Heffler E, Crimi C, Brussino L, Nicola S, Sichili S, Dughera L, Rolla G, Crimi N. Exhaled breath condensate pH and cysteinyl leukotriens in patients with chronic cough secondary to acid gastroesophageal reflux. J Breath Res 2016;22:016002.

[27] Smyth EM, Grosser T, Wang M, Yu Y, FitzGerald GA. Prostanoids in health and disease. J Lipid Res 2009;50:S423—8.

[28] Peralbo-Molina A, Calderon-Santiago M, Priego-Capote F, Jurado-Gámez B, Luque de Castro MD. Development of a method for metabolomic analysis of human exhaled breath condensate by gas chromatography-mass spectrometry in high resolution mode. Anal Chim Acta 2015;887:118—26.

[29] Fernandez-Peralbo MA, Calderon-Santiago M, Priego-Capote F, Luque de Castro MD. Study of exhaled breath condensate simple preparation for metabolomics analysis by LC—MS/MS in high resolution mode. Talanta 2015;144:1360—9.

[30] Davis MD, Montpetit A, Hunt J. Exhaled breath condensate: an overview. Inmunol Allergy Clin North Am 2012;32:363—75.

[31] Ojoo JC, Mulrennan SA, Kastelik JA, Morice AH, Redington AE. Exhaled breath condensate pH and exhaled nitric oxide in allergic asthma and in cystic fibrosis. Thorax 2005;60:22—6.

[32] Motta A, Paris D, D'Amato M, Melck D, Calabrese C, Vitale C, Stanziola AA, Corso G, Sofia M, Maniscalco M. NMR metabolomics analysis of exhaled breath condensate of asthmatic patients at two different temperatures. J Proteome Res 2014;13:6107—20.

[33] Sinha A, Desiraju K, Aggarwal K, Kutum R, Roy S, Lodha R, Kabra SK, Ghosh B, Sethi T, Agrawal A. Exhaled breath condensate metabolome clusters for endotype discovery in asthma. J Transl Med 2017;15:262.

[34] Bertini I, Luchinat C, Miniati M, Monti S, Tenori L. Phenotyping COPD by ^{1}H NMR metabolomics of exhaled breath condensate. Metabolomics 2014;10:302—11.

[35] Maniscalco M, Paris D, Melck DJ, Molino A, Carone M, Ruggeri P, Caramori G, Mottal A. Differential diagnosis between newly diagnosed asthma and chronic obstructive pulmonary disease using exhaled breath condensate metabolomics: a pilot study. Eur Respir J 2018;51:1701824.

[36] Smolinska A, Klaassen EMM, Dallinga JW, vand de Kant KDG, Jobsis Q, Moonen EJC, van Schayck OCP, Dompeling E, van Schooten FJ. Profiling of volatile organic compounds in exhaled breath as a strategy to find early predictive signatures of asthma in children. PLoS One 2014;9:e105447.

[37] van Vliet D, Smolinska A, Jobsis Q, Rosias P, Muris J, Dallinga J, Dompeling E, van Schooten. J Breath Res 2017;11:016016.

[38] Montuschi P, Paris D, Melck D, Lucidi V, Ciabattoni G, Raia V, Calabrese C, Bush A, Barnes PJ, Motta A. NMR spectroscopy metabolomic profili of exhaled breath condensate inpatients with stable and unstable cystic fibrosis. Thorax 2012;67:222.

[39] Robroeks CM, van Berkel JJ, Dallinga JW, Jöbsis Q, Zimmermann LJ, Hendriks HJ, Wouters MF, van der Grinten CP, van de Kant KD, van Schooten FJ, Dompeling E. Metabolomics of volatile organic compounds in cystic fibrosis patients and controls. Pediatr Res 2010;68:75—80.

[40] van Oort PM, de Bruin S, Weda H, Knobel HH, Schultz MJ, Bos LD. Exhaled breath metabolomics for the diagnosis of pneumonia in intubated and mechanically ventilated intensive care unit (ICU) patients. Int J Mol Sci 2017;18:E449.

[41] Gordon SM, Szidon JP, Krotoszynski BK, Gibbons RD, O'Neill HJ. Volatile organic compounds in exhaled air from patients with lung cancer. Clin Chem 1985;31(8):1278—82.

[42] Phillips M, Cataneo RN, Cummin AR, et al. Detection of lung cancer with volatile markers in the breath. Chest 2003;123(6):2115—23.

[43] Phillips M, Gleeson K, Hughes JM, et al. Volatile organic compounds in breath as markers of lung cancer: a cross-sectional study. Lancet 1999;353(9168):1930—3.

[44] Poli D, Carbognani P, Corradi M, et al. Exhaled volatile organic compounds in patients with non-small cell lung cancer: cross sectional and nested short-term follow-up study. Respir Res 2005;6:71.

[45] Poli D, Goldoni M, Corradi M, et al. Determination of aldehydes in exhaled breath of patients with lung cancer by means of on-fiber derivatisation SPME—GC/MS. J Chromatogr B 2010;878(27):2643—51.

[46] Bajtarevic A, Ager C, Pienz M, et al. Noninvasive detection of lung cancer by analysis of exhaled breath. BMC Canc 2009;9:348.

[47] Ulanowska A, Kowalkowski T, Trawińska E, Buszewski B. The application of statistical methods using VOCs to identify patients with lung cancer. J Breath Res 2011;5(4):046008.

[48] Capuano R, Santonico M, Pennazza G, Ghezzi S, Martinelli E, Roscioni C, Lucantoni G, Galluccio G, Paolesse R, Di Natale C, D'Amico A. The lung cancer breath signature:a comparative analysis of exhaled breath and air sampled from inside the lungs. Sci Rep 2015;5:16491. https://doi.org/10.1038/srep16491.

[49] Schallschmidt K, Becker R, Jung C, Bremser W, Walles T, Neudecker J, Leschber G, Frese S, Nehls I. Comparison of volatile organic compounds from lung cancer patients and healthy controls—challenges and limitations of an observational study. J Breath Res 2016;10:046007.

[50] Peng G, Hakim M, Broza YY, et al. Detection of lung, breast, colorectal, and prostate cancers from exhaled breath using a single array of nanosensors. Br J Cancer 2010;103(4):542—51.

[51] Peralbo-Molina A, Calderón-Santiago M, Priego-Capote F, Jurado-Gámez B, Luque de Castro MD. Metabolomics analysis of exhaled breath condensate for discrimination between lung cancer patients and risk factor individuals. J Breath Res 2016;10(1):016011.

[52] Peralbo-Molina A, Calderón-Santiago M, Priego-Capote F, Jurado-Gámez B, Luque de Castro MD. Identification of metabolomics panels by analysis of exhaled breath condensate. J Breath Res 2016;10(2):26002—13.

[53] Peralbo-Molina A, Calderón-Santiago M, Jurado-Gámez B, Luque de Castro MD, Priego-Capote F. Exhaled breath condensate to discriminate individuals with different smoking habits by GC–TOF/MS. Sci Rep 2017;7:1421.

[54] Rezale A, Buresi M, Lembo A, Lin H, McCallum R, Rao S, Schmulson M, Valdovinos M, Zakko S, Pimentel M. Hydrogen and methane-based breath testing in gastrointestinal disorders: the North American consensus. Am J Gastroenterol 2017;112:775–84.

[55] Levitt MD. Volume and composition of human intestinal gas determined by means of an intestinal washout technique. N Engl J Med 1971;284:1394–8.

[56] Kurada S, Alkhouri N, Fiocchi C, Dweik R, Rieder F. Review article: breath analysis in inflammatory bowel diseases. Aliment Pharmacol Ther 2015;41:329–41.

[57] Ross BM, Glen I. Breath ethane concentrations in healthy volunteers correlate with a systemic marker of lipid peroxidation but not with omega-3 fatty acid availability. Metabolites 2014;4:572–9.

[58] Li W, Liu Y, Liu Y, Cheng S, Duan Y. Exhaled isopropanol: new potential biomarker in diabetic breathomics and its metabolic correlations with acetone. RSC Adv 2017;7:17480–8.

[59] Konvalina G, Haick H. Sensors for breath testing: from nanomaterials to comprehensive disease detection. Acc Chem Res 2014;47:66–76.

[60] Chang J, Lee D, Ban S, Oh J, Jung M, Kim S, Park S, Persaud K, Jheon S. Analysis of volatile organic compounds in exhaled breath for lung cancer diagnosis using a sensor system. Sensor Actuator B Chem 2018;255:800–7.

[61] Dragonieri S, Schot R, Mertens BJ, Le Cessie S, Gauw SA, Spanevello A, Resta O, Willard NP, Vink TJ, Rabe KF, Bel EH, Sterk PJ. An electronic nose in the discrimination of patients with asthma and controls. J Allergy Clin Immunol 2007;120:856–62.

[62] Fens N, Zwinderman AH, van der Schee MP, de Nijs SB, Dijkers E, Roldaan AC, Cheung D, Bel EH, Sterk PJ. Exhaled breath profiling enables discrimination of chronic obstructive pulmonary disease and asthma. Am J Respir Crit Care Med 2009;180:1076–82.

[63] Risby TH, Tittel FK. Current status of midinfrared quantum and interband cascade lasers for clinical breath analysis. Opt Eng 2010;49:1–14.

[64] Wörle K, Seichter F, Wilk A, Armacost C, Day T, Godejohann M, Wachter U, Vogt J, Radermacher P, Mizaikoff B. Breath analysis with broadly tunable quantum cascade lasers. Anal Chem 2013;85:2697–702.

Block II

Background Information

Metabolome and Microbiome From Infancy to Elderly

Ramon V. Cortez, Luana N. Moreira and Carla R. Taddei

School of Pharmaceutical Sciences, Dep. of Clinical Analysis and Toxicology, University of São Paulo, SP, Brazil

INTRODUCTION

The formation of the intestinal microbiome is influenced by several external factors, as well as internal factors intrinsic to the host. The external factors include the composition of the maternal microbiome, the type of birth (cesarean section or normal delivery), geographic region, environmental contamination, feeding regimen, and the use of medicines [1,2]. The internal factors are related to physiology, such as gastrointestinal tract anatomy, peristalsis, bile acids, pH of the intestine, and immune responses. Thus, a competition between microorganisms for mucosal receptors and the interactions between microorganisms and the host modulate their composition and function, making the microbiome unique to the host [3].

The microbiome degrades resistant carbohydrates, producing bioactive fatty acids and metabolites, such as conjugated linoleic acid (CLA), short-chain fatty acids (SCFAs) and gamma-amino butyric acid (GABA), which have potential in the treatment of cancer, inflammatory bowel disease (IBD), and obesity [2]. A key feature of SCFAs, especially butyrate, in the physiology of the colon, is the trophic effect on intestinal epithelium.

The Metabolome

The gut metabolome is composed of metabolites produced within the host and microbial community. Thus, metabolomics becomes a practical tool to understand the functioning of the microbial community [4,5]. Metabolomics constitute the identification, quantification, and characterization of small molecule metabolites ($<1500\,Da$) [6–8] in the metabolome. Unlike other "omics" investigations, metabolites and metabolic pathways are relatively preserved across species [5].

While other "omics" offer information regarding only phenotypes and molecular mechanisms, metabolomics plays a key role in connecting host phenotype and microbiome function [6,9]. It also provides a view of dynamic changes affected by external influences, including oxidative stress, inflammatory processes, and energy metabolism [4].

Metabolomics and metabolite analyses are widely used to identify disease biomarkers. Initial microbiome studies endeavored to discover bacterial metabolites that could be linked to diseases, physiological states, drug use, or dietary intake [10–14].

GUT MICROBIOME AND METABOLITES

The intestinal microbiota is responsible for important metabolic functions, such as the production of SCFAs, amino acid synthesis (AAs), bile acid biotransformation, hydrolysis, and fermentation of nondigestible substrates, as well as the production of metabolites involved in intestinal mucosal homeostasis [15,16]. SCFAs are particularly relevant [17]. It is already known that butyrate is the product of bacterial fermentation of nondigestible carbohydrates [18].

Due to the presence of acidic products originating from fermentation processes, the gut luminal pH is approximately 5.5. This slightly acidified environment allows competition between butyrate-producing bacteria and bacteria that use carbohydrates, such as *Bacteroides*, in addition to stimulating the production of butyrate. The decrease of pH impedes the permanence of *Eubacterium* [18], allowing the accumulation of lactic acid. Thus, the interaction between the members of the microbiome induces the transformation of lactate into butyrate, avoiding the excessive acidification of the medium. The

Microbiome and Metabolome in Diagnosis, Therapy, and other Strategic Applications. https://doi.org/10.1016/B978-0-12-815249-2.00008-7

butyric substrates lead to the multiplication of species producing butyrate, allowing a balance between the presence of *Eubacterium* and *Bifidobacterium*.

Metabolic processes of cross-feeding also lead to environmental balance. Bacteria producing butyrate, such as *Roseburia*, *Faecalibacterium prausnitzii*, and *Eubacterium*, are capable of fermenting products of the metabolism of oligosaccharides produced by *Bifidobacteria*, such as lactate. Given the importance of butyrate for metabolism of the colonic epithelium, it is assumed that the bacteria producing butyrate are tolerated by the innate immune system [17]. These metabolic events occur in the gut from the time of birth, which contributes to eubiosis [18–20].

MICROBIOME AND METABOLOME ON INFANCY

The microbiome has a strong relationship with the immune system [3], and its origins can be traced to pregnancy. Bacteria are present in samples of placenta, amniotic fluid, and meconium, in tiny amounts [21]. Despite this intrauterine colonization, it is known that the much heavier microbial burden of childbirth is essential for the establishment of a healthy microbiome. After vaginal delivery, the microbiome is more diverse, predominantly related to the maternal vaginal flora such as *Lactobacillus* and *Prevotella*.

With cesarean section, the microbiome will interact with other sources, including *Staphylococcus* and *Propionibacterium* of the skin, suggesting susceptibility to the hospital environment [22]. The first bacteria that colonize the gastrointestinal tract appear to play an important role in subsequent colonization. The luminal environment is highly oxygenated at the birth, and facultative anaerobic bacteria are predominant at this period. In the days following birth, oxygen is consumed, and the anaerobic bacteria become predominant, increasing the production of SCFA and modulating the composition of the microbiome [23].

Anaerobic bacteria can also modulate the gene expression of host epithelial cells, thus creating a favorable environment for themselves and preventing the growth of others. This may have lifelong health consequences. During the first weeks of life, the microbiome is modified according to the feeding method, namely infant formula or breast milk. Breastfed children have larger amounts of *Lactobacillus* and *Bifidobacterium* in their feces [2]. This correlation between breast milk and *Bifidobacterium* is more evident, since some species have enriched genes, which regulate the processing of human milk oligosaccharides. The importance of this genus also lies in competitive inhibition of other bacteria, adhesion to the intestinal mucosa, and synthesis of compounds that inhibit or destroy pathogenic bacteria. Its effects on immunomodulation favor natural killer cells and promote phagocytosis and IgA secretion.

During weaning, the introduction of solid food such as vegetables, fruits, meats, and grains allows new bacterial genera to reach the infant's gut epithelium, changing the microbial community [2]. An increase in bacterial diversity follows, and the intestinal microbiota will become stable by the end of the second or eventually the third year of life, modulated mostly by diet and environmental factors [23]. These findings are corroborated by similarities between the metabolomic profile of feces from 2-year-old children and of adults, with increased levels of butyrate and acetate.

Preterm infants normally stay in neonatal intensive care units (NICUs) for a period, exposing them to abnormal gut colonization and the heightened risk of antibiotic resistant bacterial clones [24]. The gut metabolomic profile of preterm infants during a stay within the NICU is mostly shifted after discharge, when the contact with a different environment and the introduction of a complex diet will increase the microbial diversity, as expected [25]. The diet will strongly affect early life stool metabolites, not only because of new bacteria genera but also because of the new diet itself [26].

Full-term infants that were exclusively breastfed from obese mothers exhibit high concentrations of fucosylated oligosaccharide and lactic acid, over the first 12 months of life, while in formula-fed infants from obese mothers the metabolome is associated with gut bacteria proteolytic activity [27]. The alluded to oligosaccharides are typical of milk and carbohydrate metabolism, suggesting that in early life, macro- and micronutrients from diet can affect the infant metabolic profile [27].

Late-onset sepsis (LOS) is a form of neonatal sepsis recognized after 72 h of life, which occurs mainly in premature infants [28]. It is frequently caused by bacteria found in the gut microbiota, suggesting translocation of gut bacteria into the bloodstream. However, the gut microbiota of LOS infants is different compared with healthy infants [29,30]. Metabolites such as sucrose, L-glutamate, 18-hydroxycortisol, 18-oxocortisol, and raffinose were decreased in infants with LOS [31].

A positive correlation was found between *Bifidobacterium*- and *Streptococcus*-dominant communities in the controls and an increase of some metabolites, for example, raffinose and L-alpha-acetyl-N-normethadol (associated with prescribed drugs) [31]. Raffinose, an α-galactosyl oligosaccharide, was lower in LOS infants prior to diagnosis, increasing after treatment, whereas it remained consistently higher in the infant control group.

Necrotizing enterocolitis/NEC is another serious concern in newborns [32–34]. The higher levels of specific metabolites in infants who later developed NEC contrast with lower levels in healthy infants, with high relative abundance of

Bifidobacterium. The NEC-associated metabolites were related to C21-steroid hormone biosynthesis, linoleate pathway, and leukotriene/prostaglandin in the arachidonate pathway, known to mediate inflammation.

Bifidobacterium is frequently associated with a healthy status. The reduction of prostaglandin formation in infants with an abundance of *Bifidobacterium* supports the hypothesis that the presence of this genus in early life could be a marker for healthy gut microbiota [32–34].

THE MICROBIOME AND METABOLOME IN HEALTHY ADULTS

Animal investigations highlight the contribution of age [35–37], drug administration [38,39], and the metabolic profiles [40,41] of different diseases [42]. Metabolites from the lumen and intestinal tissues, in mice, are directly related to the microbial degradation of carbohydrates and proteins in the cecum [43]. Reabsorption processes in the colon undergo spatial variation, being influenced by the microbiota, and then promoting feedback for the growth of bacteria along the intestine, triggering signaling pathways to the microbiome and the host [44].

Human studies have focused cancer [40,45], celiac disease [46], nonalcoholic fatty liver disease [47], ulcerative colitis, and Crohn's disease [48,49]. Some foods, such as dairy products, other dietary components, carcinogens, and fruit extracts also contribute to fecal metabolome formations [42].

In IBDs, the imbalance of the gut microbiome is attributed to less abundance of *F. prausnitzii*, an SCFA producer. Increase of amino acids like valine, alanine, isoleucine, leucine, lysine, valine, and phenylalanine, along with depression of butyrate, is similarly observed [42]. Some studies reported the same increase in amino acids, such as lysine, threonine, and valine, and some SCFAs such acetic acid and valeric acid, along with a reduction of butyrate in fecal content of colorectal cancer patients [50].

Colorectal cancer patients can suffer from inflamed intestinal tissue, resulting in difficulties with nutrient absorption [50]. Therefore, these metabolites could be a consequence of inflammation or altered gut microbiota [42].

Microbiome and Metabolome in the Aging Process

In the elderly, interindividual variation is significant. There is a decrease of beneficial bacteria like *Bacteroides*, *F. prausnitzii*, and *Bifidobacterium*, and as a consequence, an increase of members of the phylum Firmicutes, like *Clostridium*. This appears to be related to the general and immunological decline [23].

Aging remains a poorly understood phenomenon [51,52]. The decline in the immune function and cumulative damages due to oxidative stress are usually accepted as the main factors of aging [53]. The progressive decay in physiological functions is characterized by homeostatic imbalance, degenerated capacity for stress responses, and increased risk of aging-related diseases [54].

In mice [55,56], a set of markers confirmed that aging is associated with alterations in nutrient sensing, lipid and amino acid metabolism, and redox homeostasis. This was the outcome of metabolomic nontargeted profiling, which enables analysis of thousands of small molecules, even those with unidentified chemical properties [57,58]. Molecular damage secondary to shifts in metabolite diversity can also be monitored the same way [59].

CONCLUSION

The intestinal symbiont microbiome of the human gastrointestinal tract, in any age group, is influenced by several internal and external factors, with emphasis on diet and environmental variables or interferences. This modulation can affect nutrient acquisition and energy regulation, leading to either a balance in the ecosystem, eubiosis, or an imbalance, dysbiosis, with potential loss of the intestinal barrier, metabolic diseases, inflammatory processes, and infections. While in the past decades we were focused on revealing the composition of the human microbiome, and interfaces concerning hosts and diseases, the future will likely be marked by discovering biochemical products and pathways of host–microbiome interactions.

REFERENCES

[1] Salminen S, Gueimonde M. Gut microbiota in infants between 6 and 24 months of age. In: Hernell O, Schmitz J, editors. Feeding during late infancy and early childhood: impact on health. Nestle Nutr Workshop Ser Pediatr Program, 56; 2005. p. 43–51.

[2] Scholtens PA, Oozeer R, Martin R, Amor KB, Knol J. The early settlers: intestinal microbiology in early life. Annu Rev Food Sci Technol 2012;3:425–47.

[3] Hooper LV, Macpherson AJ. Immune adaptations that maintain homeostasis with the intestinal microbiota. Nat Rev Immunol 2010;10:159−69.

[4] Shaffer M, Armstrong AJS, Phelan VV, Reisdorph N, Lozupone CA. Microbiome and metabolome data integration provides insight into health and disease. Transl Res 2017;189:51−64.

[5] Chong J, Xia J. Computational approaches for integrative analysis of the metabolome and microbiome. Metabolites 2017;7(4):62.

[6] Patti GJ, Yanes O, Siuzdak G. Metabolomics: the apogee of the omics trilogy. Nat Rev Mol Cell Biol 2012;13(4):263−9.

[7] Wishart DS, et al. HMDB: the human metabolome database. Nucleic Acids Res 2007;35:D521−6.

[8] German JB, Hammock BD, Watkins SM. Metabolomics: building on a century of biochemistry to guide human health. Metabolomics 2005;1:3−9.

[9] Fiehn O. Metabolomics—the link between genotypes and phenotypes. In: Functional genomics. Berlin/Heidelberg, Germany: Springer; 2002. p. 155−71.

[10] Smith CA, Want EJ, O'Maille G, Abagyan R, Siuzdak G. XCMS: processing mass spectrometry data for metabolite profiling using nonlinear peak alignment, matching, and identification. Anal Chem 2006;78:779−87.

[11] Zamboni N, Saghatelian A, Patti GJ. Defining the metabolome: size, flux, and regulation. Mol Cell 2015;58:699−706.

[12] Koizumi S, et al. Imaging mass spectrometry revealed the production of lyso-phosphatidylcholine in the injured ischemic rat brain. Neuroscience 2010;168:219−25.

[13] Maurice CF, Haiser HJ, Turnbaugh PJ. Xenobiotics shape the physiology and gene expression of the active human gut microbiome. Cell 2013;152:39−50.

[14] McNulty NP, Yatsunenko T, Hsiao A, et al. The impact of a consortium of fermented milk strains on the gut microbiome of gnotobiotic mice and monozygotic twins. Sci Transl Med 2011;3:106ra106.

[15] Putignani L, Del Chierico F, Vernocchi P, Cicala M, Cucchiara S, Dallapiccola B, et al. Gut microbiota dysbiosis as risk and premorbid factors of IBD and IBS along the childhood-adulthood transition. Inflamm Bowel Dis 2015;22:487−504.

[16] Holmes E, Li JV, Athanasiou T, Ashrafian H, Nicholson JK. Understanding the role of gut microbiome-host metabolic signal disruption in health and disease. Trends Microbiol 2011;19:349−59.

[17] Guilloteau P, Martin L, Eeckhaut V, Ducatelle R, Zabielski R, Van Immerseel F. From the gut to the peripheral tissues: the multiple effects of butyrate. Nutr Res Rev 2010;23(2):366−84.

[18] Flint HJ, Scott KP, Duncan SH, Louis P, Forano E. Microbial degradation of complex carbohydrates in the gut. Gut Microb 2012;3:289−306.

[19] Dicksved J, Halfvarson J, Rosenquist M, Järnerot G, Tysk C, Apajalahti J, et al. Molecular analysis of the gut microbiota of identical twins with Crohn's disease. ISME J 2008;2:716−27.

[20] Jansson J, Willing B, Lucio M, Fekete A, Dicksved J, Halfvarson J, et al. Metabolomics reveals metabolic biomarkers of Crohn's disease. PLoS One 2009;4:e6386.

[21] Funkhouser LJ, Bordenstein SR. Mom knows best: the universality of maternal microbial transmission. PLoS Biol 2013;11(8):e1001631.

[22] Dominguez-Bello MG, Costello EK, Contreras M, Magris M, Hidalgo G, Fierer N, Knight R. Delivery mode shapes the acquisition and structure of the initial microbiota across multiple body habits in newborn. Proc Natl Acad Sci U S A 2010;107:11971−5.

[23] Taddei CR. O microbioma da infância à idade adulta. In: Faintuch J, editor. Microbioma, disbios, probióticos e bacterioterapia. 1st ed. São Paulo, Brazil: Manole.

[24] Moles L, Gómez M, Jiménez E, Fernández L, Bustos G, Chaves F, Cantón R, Rodríguez JM, Del Campo R. Preterm infant gut colonization in the neonatal ICU and complete restoration 2 years later. Clin Microbiol Infect 2015;21(10):936.e1−936.e10.

[25] Stewart CJ, Skeath T, Nelson A, Fernstad SJ, Marrs EC, Perry JD, Cummings SP, Berrington JE, Embleton ND. Preterm gut microbiota and metabolome following discharge from intensive care. Sci Rep 2015;24(5):17141.

[26] Hellmuth C, Uhl O, Kirchberg FF, Grote V, Weber M, Rzehak P, Carlier C, Ferre N, Verduci E, Gruszfeld D, Socha P, Koletzko B, European Childhood Obesity Trial Study Group. Effects of early nutrition on the infant metabolome. Nestle Nutr Inst Workshop Ser 2016;85:89−100.

[27] Abdulkadir B, Nelson A, Skeath T, Marrs EC, Perry JD, Cummings SP, Embleton ND, Berrington JE, Stewart CJ. Routine use of probiotics in preterm infants: longitudinal impact on the microbiome and metabolome. Neonatology 2016;109(4):239−47.

[28] Dong Y, Speer CP. Late-onset neonatal sepsis: recent developments. Arch Dis Child Fetal Neonatal Ed 2015;100(3):F257−63.

[29] Graham 3rd PL, Della-Latta P, Wu F, Zhou J, Saiman L. The gastrointestinal tract serves as the reservoir for Gram-negative pathogens in very low birth weight infants. Pediatr Infect Dis J 2007;26(12):1153−6.

[30] Carl MA, Ndao IM, Springman AC, Manning SD, Johnson JR, Johnston BD, Burnham CA, Weinstock ES, Weinstock GM, Wylie TN, Mitreva M, Abubucker S, Zhou Y, Stevens HJ, Hall-Moore C, Julian S, Shaikh N, Warner BB, Tarr PI. Sepsis from the gut: the enteric habitat of bacteria that cause late-onset neonatal bloodstream infections. Clin Infect Dis 2014;58(9):1211−8.

[31] Stewart CJ, Embleton ND, Marrs ECL, Smith DP, Fofanova T, Nelson A, Skeath T, Perry JD, Petrosino JF, Berrington JE, Cummings SP. Longitudinal development of the gut microbiome and metabolome in preterm neonates with late onset sepsis and healthy controls. Microbiome 2017;12(5):75.

[32] Neu J. Necrotizing enterocolitis. World Rev Nutr Diet 2014;110:253−63.

[33] Dobbler PT, Procianoy RS, Mai V, Silveira RC, Corso AL, Rojas BS, Roesch LFW. Low microbial diversity and abnormal microbial succession is associated with necrotizing enterocolitis in preterm infants. Front Microbiol 2017;15(8):2243.

[34] Mai V, Young CM, Ukhanova M, Wang X, Sun Y, Casella G, Theriaque D, Li N, Sharma R, Hudak M, Neu J. Fecal microbiota in premature infants prior to necrotizing enterocolitis. PLoS One 2011;6(6):e20647.

[35] Arumugam M, Raes J, Pelletier E, Le Paslier D, Yamada T, Mende DR, Fernandes GR, Tap J, et al. Enterotypes of the human gut microbiome. Nature 2011;473:174−80.

[36] Calvani R, Brasili E, Pratico G, Capuani G, Tomassini A, Marini F, Sciubba F, Finamore A, Roselli M, Marzetti E, Miccheli A. Fecal and urinary NMR-based metabolomics unveil an aging signature in mice. Exp Gerontol 2014;49:5−11.

[37] Saric J, Wang Y, Li J, Coen M, Utzinger J, Marchesi JR, Keiser J, Veselkov K, Lindon JC, Nicholson JK, Holmes E. Species variation in the fecal metabolome gives insight into differential gastrointestinal function. J Proteome Res 2008;7:352−60.

[38] Coen M, Goldfain-Blanc F, Rolland-Valognes G, Walther B, Robertson DG, Holmes E, Lindon JC, Nicholson JK. Pharmacometabonomic investigation of dynamic metabolic phenotypes associated with variability in response to galactosamine hepatotoxicity. J Proteome Res 2012;11:2427−40.

[39] Romick-Rosendale LE, Goodpaster AM, Hanwright PJ, Patel NB, Wheeler ET, Chona DL, Kennedy MA. NMR-based metabonomics analysis of mouse urine and fecal extracts following oral treatment with the broad-spectrum antibiotic enrofloxacin (Baytril). Magn Reson Chem 2009;47:S36−46.

[40] Bezabeh T, Somorjai R, Dolenko B, Bryskina N, Levin B, Bernstein CN, Jeyarajah E, Steinhart AH, Rubin DT, Smith ICP. Detecting colorectal cancer by H-1 magnetic resonance spectroscopy of fecal extracts. NMR Biomed 2009;22:593−600.

[41] Trezzi JP, Vlassis N, Hiller K. The role of metabolomics in the study of cancer biomarkers and in the development of diagnostic tools. Adv Exp Med Biol 2015;867:41−57.

[42] Heinzmann SS, Schmitt-Kopplin P. Deep metabotyping of the murine gastrointestinal tract for the visualization of digestion and microbial metabolism. J Proteome Res 2015;14:2267−77.

[43] Delzenne NM, Neyrinck AM, Baeckhed F, Cani PD. Targeting gut microbiota in obesity: effects of prebiotics and probiotics. Nat Rev Endocrinol 2011;7:639−46.

[44] Monleon D, Manuel Morales J, Barrasa A, Antonio Lopez J, Vazquez C, Celda B. Metabolite profiling of fecal water extracts from human colorectal cancer. NMR Biomed 2009;22:342−8.

[45] Di Cagno R, De Angelis M, De Pasquale I, Ndagijimana M, Vernocchi P, Ricciuti P, Gagliardi F, Laghi L, Crecchio C, Guerzoni ME, Gobbetti M, Francavilla R. Duodenal and faecal microbiota of celiac children: molecular, phenotype and metabolome characterization. BMC Microbiol 2009;11.

[46] Michail S, Lin M, Frey MR, Fanter R, Paliy O, Hilbush B, Reo NV. Altered gut microbial energy and metabolism in children with non-alcoholic fatty liver disease. FEMS Microbiol Ecol 2015;91.

[47] Bjerrum JT, Wang Y, Hao F, Coskun M, Ludwig C, Gunther U, Nielsen OH. Metabonomics of human fecal extracts characterize ulcerative colitis, Crohn's disease and healthy individuals. Metabolomics 2015;11:122−33.

[48] Le Gall G, Noor SO, Ridgway K, Scovell L, Jamieson C, Johnson IT, Colquhoun IJ, Kemsley EK, Narbad A. Metabolomics of fecal extracts detects altered metabolic activity of gut microbiota in ulcerative colitis and irritable bowel syndrome. J Proteome Res 2011;10:4208−18.

[49] Smirnov KS, Maier TV, Walker A, Heinzmann SS, Forcisi S, Martinez I, et al. Challenges of metabolomics in human gut microbiota research. Int J Med Microbiol 2016;306(5):266−79.

[50] Weir TL, Manter DK, Sheflin AM, Barnett BA, Heuberger AL, et al. Stool microbiome and metabolome differences between colorectal cancer patients and healthy adults. PLoS One 2013;8(8):e70803.

[51] Gonzalez-Covarrubias V, Beekman M, Uh HW, Dane A, Troost J, Paliukhovich I, et al. Lipidomics of familial longevity. Aging Cell 2013;12(3):426−34.

[52] Beekman M, Blanche H, Perola M, Hervonen A, Bezrukov V, Sikora E, et al. Genome-wide linkage analysis for human longevity: genetics of healthy aging study. Aging Cell 2013;12(2):184−93.

[53] Lee SH, Park S, Kim H, Jung BH. Metabolomic approaches to the normal aging process. Metabolomics 2014;10:1268−92.

[54] Bruunsgaard H, Pedersen M, Pedersen BK. Aging and proinflammatory cytokines. Curr Opin Hematol 2001;8(3):131−6.

[55] Houtkooper RH, Argmann C, Houten SM, Canto C, Jeninga EH, Andreux PA, Thomas C, Doenlen R, Schoonjans K, Auwerx J. The metabolic footprint of aging in mice. Sci Rep 2011;1:134.

[56] Tomas-Loba A, Bernardes de Jesus B, Mato JM, Blasco MA. A metabolic signature predicts biological age in mice. Aging Cell 2013;12:93−101.

[57] Lawton KA, Berger A, Mitchell M, Milgram KE, Evans AM, Guo L, et al. Analysis of the adult human plasma metabolome. Pharmacogenomics 2008;9(4):383−97.

[58] Kristal BS, Shurubor YI. Metabolomics: opening another window into aging. Sci Aging Knowl Environ 2005;2005(26):pe19.

[59] Avanesov AS, Ma S, Pierce KA, Yim SH, Lee BC, Clish CB, Gladyshev VN. Age- and diet-associated metabolome remodeling characterizes the aging process driven by damage accumulation. eLife 2014;3:e02077.

Chapter 9

The Oral Microbiome

Marcelle M. Nascimento

Associate Professor, Department of Restorative Dental Sciences, Division of Operative Dentistry, College of Dentistry, University of Florida, Gainesville, FL, United States

INTRODUCTION

Untreated dental caries is the most common disease, and severe periodontitis is the sixth most common disease affecting humans globally [1] (Figs. 9.1 and 9.2). Of great concern are the serious implications that these oral diseases can have on general health [2]. Increasing evidence shows that oral bacteria may spread and be associated with infections in other parts of the body [3,4]. For example, reports have identified transient bacteremia, which can be caused by poor oral hygiene, periodontitis, dental procedures, and even tooth brushing, as the precursor of some cases of infective endocarditis [3].

Moreover, systemic conditions and their associated treatments may also affect oral health by reducing salivary flow and affecting the ecological balance of the oral microbiome. Not surprisingly, there is growing interest on the composition and metabolic activities of oral microbiome, as well as an evolving trend for dental and medical research to share knowledge on the etiology and pathogenicity of human diseases.

ORAL BIOFILMS

Insights provided from the Human Microbiome Project reveals that ecological balance in biofilms play a significant role in health [5]. As a result, an expanding area of research focuses on therapeutic interventions that modulate microbial ecology to restore homeostasis of human biofilms and thus health [5,6]. It is well accepted today that dental caries and periodontitis are closely related to a dysbiosis of the microbial consortia of oral biofilms driven by environmental changes, such as a sugar-frequent and acidic-pH environment in caries and a protein-rich and alkaline-pH environment in periodontal disease [7–9]. Conventional therapies for caries and periodontitis aim at controlling the development and metabolic activities of supragingival and subgingival oral biofilms (also called dental plaque), respectively. Still, caries and periodontitis remain as major public health problems worldwide.

COMPOSITION OF THE ORAL MICROBIOME

The oral microbiome comprises hundreds of microbial species, including bacteria, viruses, mycoplasmas, archaea, fungi, and protozoa, that coinhabit and functionally interact in oral biofilms to maintain homeostasis or to cause disease [10–13]. Oral microbiology studies have traditionally focused on bacteria and bacterial infections, but the advances of OMICS techniques (e.g., metagenomics, metatranscriptomics, metaproteomics, metabolomics, spectral imaging fluorescence, in situ hybridization) are enabling the study of complex community interactions that includes members of the microbiota from different kingdoms. One of the major challenges facing oral health researchers today is to distinguish which of the potential microbial–microbial and microbial–host interactions are critical for maintenance of dental health.

Following the advancement of gut microbiome research, genome databases specific to oral microorganisms were developed, including the Human Oral Microbiome database (HOMD) [38] and the Core Oral Microbiome (CORE) database [39], which are curated to remove contaminants and improve reliability of the analysis. Also important was the sequencing of genomes of oral bacteria by the Human Microbiome Consortium [40]. This was particularly relevant to

(A)

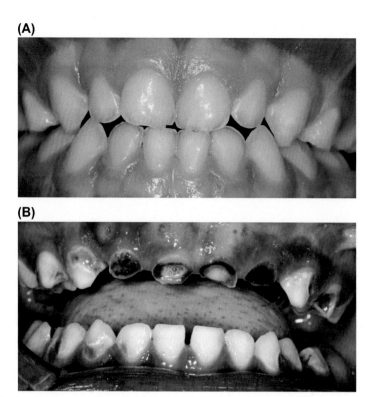

FIGURE 9.1 (A) Child with a healthy oral cavity; (B) Child with early childhood dental caries (ECC). *Photography by Dr. Nascimento.*

(A)

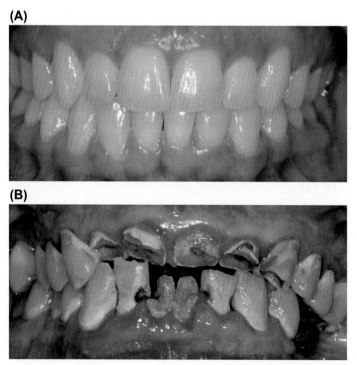

FIGURE 9.2 (A) Adult with a healthy oral cavity; (B) Adult with both dental caries and periodontal diseases. *Photography by Dr. Nascimento.*

reduce the number of unassigned reads in metagenomic analysis. Nevertheless, 40%−50% of the metagenomic reads still remain unassigned or assigned as hypothetical proteins with unknown function; therefore, more work is needed in this area [41−43].

To provide a more realistic picture of the complex interactions contributing to the compositional and functional stability of the oral ecosystem, studies should also take place involving kingdoms other than bacteria, such as viruses, fungi, archaea, and protozoa. It was recently demonstrated that bacterial viruses (bacteriophages) might assist in maintaining a stable, healthy ecosystem compared with the dysbiosis of periodontal diseases [44]. OMICS approaches have been successfully used to assess composition and functionality of complex communities grown in an in vitro biofilm model [45]. OMICS could also be used in search for optimal growth conditions of the so-called "unculturable" organisms, which may grow in the presence of certain helper strains and/or compounds with siderophore activity [46].

Acquisition of the Oral Microbiome

The oral cavity of a newborn baby is typically sterile. Right after birth, the oral cavity becomes a major point of entry of microorganisms into the human body [10]. Oral microorganisms can be acquired from the birth canal, mother's skin, air, food, water, and other fluids, but saliva is the main route of microbial transmission. Salivary transmission occurs primarily from the mother or main caregiver (vertical transmission), but it may also occur from other individuals in close contact with the newborn or child (horizontal transmission). Notably, available evidence suggests that clinical and educational interventions beginning during pregnancy are effective at reducing mother−child levels of cariogenic bacteria (e.g., mutans streptococci) and thus transmission and caries in young children [14].

The Placenta and the Microbiome

The mode of delivery has been associated with specific oral microbiome patterns in 3- to 6-month-old infants [15]. Oral commensals such as *Streptococcus*, *Fusobacterium*, *Neisseria*, *Prevotella*, and *Porphyromonas* have been detected in the human placenta by 16 rRNA gene sequencing analysis, revealing that the oral microbiome may be a potential source of intrauterine infection [16]. The placental microbiome was also shown to be more similar to the mother's oral microbiome compared with the microbiome of any other body site [17].

It has been suggested that oral commensals may reach the amniotic fluid through blood during pregnancy, which can also be aggravated if the mother has periodontal disease [18]. In fact, significant evidence supports an association between adverse outcomes of pregnancy (e.g., preterm birth and preeclampsia) and the virulence properties of periodontal pathogens, such as *Fusobacterium nucleatum*, *Porphyromonas gingivalis*, *Filifactor alocis*, *Campylobacter rectus*, and others [19].

THE ORAL MICROBIOME OF THE INFANT AND CHILD

The diversity of the oral microflora increases during the first months of life. The pioneer oral bacterial species are mainly streptococci, particularly *Streptococcus salivarius*, *Streptococcus mitis*, and *Streptococcus oralis*. These early colonizers are followed by gram-negative anaerobes, including *Prevotella melaninogenica*, *Fusobacterium nucleatum*, and *Veillonella* spp. Bacterial species such as mutans streptococci and *Streptococcus sanguinis* begin to colonize the oral cavity at the time of teeth eruption, when they are able to attach to the nonshedding hard dental tissues and initiate the development and maturation of oral biofilms [17]. From the first months of life into the first years and adulthood, the oral microbiome becomes significantly more diverse and undergoes specific succession mechanisms that are mostly influenced by the environmental factors and immune filtering [20].

The Oral Environment

The oral cavity is moist and warm (35°−37°C), which is suitable to the growth of a broad range of microorganisms. The oral cavity also not only provides abundant and continual influx of nutrients for microbial growth, such as salivary proteins and glycoproteins, but also provides dietary components such as carbohydrates [21,22]. As a result, the mouth harbors over 700 bacterial species and a range of viruses, archaea, fungi, and protozoa [10]. Genome sequencing techniques have identified some of these oral microorganisms, but several others (about one-third of oral bacteria) are yet to be isolated from oral samples and cultivated in the laboratory by using conventional methods. Bacterial cultivation may be challenging because some bacteria have specific requirements for nutrients, while others may be inhibited by substances in the culture media or produced by other bacteria [23].

Oral Microenvironments

The physical and biological properties of each oral habitat can be very distinct from one another, and this results in diverse site-specific microbial communities that have likely adapted to their unique oral environmental niche [24,25]. While some communities colonize the shedding surfaces of the oral mucosa (tongue, cheek, gingiva, and palate), others colonize the hard and nonshedding surfaces of teeth and prosthetic devices [26]. Other factors affecting the communities' composition at the oral sites include differences in nutrient availability and redox potentials, competition among species for binding sites, interspecies antagonism or cooperation, differences in species-specific receptors on different tissues, pH, atmospheric conditions, salinity, and access to saliva [24].

The microbial load on mucosal surfaces is relatively low due to constant desquamation; however, colonization is augmented on the nonshedding surfaces of the oral cavity. More specifically, oral biofilms develop and mature over time in oral sites that are relatively protected from the mechanical actions of the tongue, cheeks, abrasive foods, and tooth brushing [27]. To persist in the oral cavity, oral microorganisms must adhere to a surface or to other organisms. This coadhesion process facilitates nutritional cooperation and food chains, gene transfer and cell–cell signaling, and ultimately the formation of multispecies biofilms [28].

Oral microorganisms gain substantial advantages by growing as a biofilm and by functioning as a microbial community. Biofilms are inherently more tolerant to environmental stresses, host defenses, and antimicrobial agents compared with growth as single microbial cells.

BIOFILM ARCHITECTURE AND COOPERATION

The complex spatial organization or biogeography of supragingival oral biofilms was revealed by spectral imaging fluorescence in situ hybridization (CLASI-FISH) and metagenomic sequence analysis [29]. *Corynebacterium* was shown as the cornerstone of supragingival plaque architecture with long filaments that serve as anchor sites for many other microorganisms [29]. If these findings are confirmed, it could imply that *Corynebacterium*, rather than *Fusobacterium* species, are the bridging bacteria in biofilms.

This study also revealed that individual taxa are localized at the micron scale, in ways suggestive of their functional niche in the biofilm consortium. For example, anaerobic taxa tend to be in the interior, whereas facultative or obligate aerobes tend to be at the periphery of the consortium. Consumers and producers of certain metabolites, such as lactate, tend to be near each other.

Another example of synergistic relationship occurring within oral biofilms is related to oxygen tolerance and metabolism. Despite the fact that the oral cavity is overtly aerobic, obligate anaerobes and facultative anaerobic bacteria comprise the most numerous group of oral bacteria. Survival mechanisms of oral anaerobes include the ability to express a range of enzymes, whose function is to scavenge low levels of oxygen in the environment and the ability to persist in close partnership with oxygen-consuming species [26,30].

PATHOGENESIS OF SUPRAGINGIVAL BIOFILMS

The pH and carbohydrate availability are key environmental factors affecting the physiology, ecology, and pathogenicity of the oral biofilms colonizing the teeth [31]. Many beneficial oral bacteria can tolerate brief conditions of low pH, but their growth is inhibited by prolonged or frequent exposures to acidic conditions. In this context, the buffering activity of saliva plays a major role, in maintaining the intraoral pH at around neutrality, which is optimal for the growth of most members of the oral microbiome [32].

Changes in environmental pH occur following consumption of dietary sugar. Specifically, organic acids produced by the fermentation of dietary carbohydrates by cariogenic bacteria elicit demineralization of tooth enamel. These periods of acid challenge to the tooth are followed by periods of alkalinization, which neutralizes plaque pH and promotes remineralization and enamel surface integrity [31]. Whereas many factors contribute to the alkalinization of oral biofilms (e.g., buffers in saliva, diffusion of acids out of biofilms), alkali generation by oral bacteria plays a major role in pH homeostasis in oral biofilms and inhibits the initiation and progression of dental caries [31,33,34] (Fig. 9.2).

Is It Symbiosis or Dysbiosis?

The most abundant members of the oral microbiota are commensal organisms beneficial for oral health, but pathogens responsible for oral disease also exist. Commensal communities function to maintain the normal development of host tissues and defenses by providing colonization resistance and downregulation of damaging host inflammatory responses. However, the homeostasis or symbiotic relationship between the oral microbiome and the host is highly dynamic, as the composition and metabolic activities of microbial communities fluctuate according to the environmental changes in pH, nutrient availability, oxygen tension and redox environment, shedding effects of oral surfaces, and composition of salivary and crevicular fluids [35].

In addition, both health-associated and disease-associated bacteria display remarkable phenotypic plasticity. Specifically, they can morph rapidly in response to oral environmental changes [35]. Constant changes in the environment can disturb the symbiotic interactions between microbe—microbe and microbe—host and lead to ecological dysbiosis, which is characterized by the outgrowth of pathogens and increased risk of disease development (Fig. 9.3) [26].

Oral Diseases

The factors driving dysbiosis in caries differ from those of periodontal disease. In caries, continuous acid production from the metabolism of dietary carbohydrates results in the emergence of acid-producing and acid-tolerant organisms in supragingival biofilms, a selective process that alters the pH homeostasis of biofilms and shifts the demineralization—remineralization equilibrium toward loss of tooth minerals. Microbial diversity appears to be lower in caries than health, which may reflect the ecological pressure of low environmental pH [13].

Accumulation of subgingival plaque leads to inflammation of the gum tissues, or gingivitis, which may progress to periodontitis. In periodontitis, certain members of the microbial community can destabilize the host immune response, which may result in destruction of periodontal tissues in susceptible individuals. Contrasting with caries, periodontal diseases are associated with an increase in microbial diversity, which could be the result of impaired local immune function, increased availability of nutrients, or a reflection of the diverse environmental niches at the periodontal pocket [36,37].

The Caries Microbiome

The microbiome that naturally colonizes teeth in health is a biofilm community that can counterbalance acid production from dietary intake of carbohydrates to maintain an intact tooth surface, for example, by ammonia production from arginine or urea [31]. Dysbiosis of the biofilm, with change of the bacterial composition, occurs when excessive and frequent intake of carbohydrates exceeds the buffering capacity of the healthy microbiome [47].

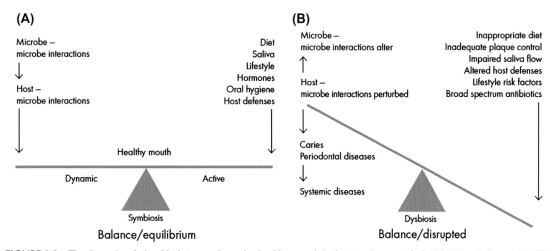

FIGURE 9.3   The dynamic relationship between the oral microbiome and the host environment in health (A) and disease (B) [26].

The renowned Keyes's diagram of dental caries describes three main pillars as responsible for the disease: host features (e.g., immune system, genetic nature that predisposes the enamel structure, salivary composition and buffering effect), environmental components (e.g., dietary sugars, fluoride, oral hygiene habits, as well as personal factors influenced by socioeconomic status and lifestyle), and microbiological features (e.g., acidogenicity of dental plaque, presence of pH-buffering bacteria, levels of pathogenic microorganisms). Numerous oral microbial taxa exhibit association with dental health or caries activity [48–50]. The use of next-generation sequencing (NGS) technologies have revealed the high complexity of the oral microbiome [41,43], metatranscriptome [42,51], metaproteome [52,53], and metabolome [54,55].

Bacterial Signature in Childhood Caries

The microbiome of healthy tooth surfaces differs substantially from that found in caries activity [41,48,49,56]. Changes in the microbial profile with the progressive stages of early childhood caries were observed when supragingival plaque of healthy tooth surfaces was compared with those of enamel and then dentin carious lesions [13]. In particular, plaque communities from dentin carious lesions of caries-active children (CA-PD) showed a very distinctive bacterial profile. Moreover, communities from healthy tooth surfaces of caries-active children (CA-PF) were shown to be more similar to those from enamel carious lesions (CAE-PE and CA-PE) than to those of healthy teeth from caries-free children (CF-PF), suggesting that CA-PF sites appear to be at greater risk of caries development than CF-PF sites (Fig. 9.4). This finding is concordant with previous risk factor studies for childhood caries [57] and highlights the interconnectedness of plaque communities, in which these communities are part of a larger ecosystem, where changes in the structure of one community may eventually affect others.

NEWLY IDENTIFIED BENEFICIAL AND PATHOGENIC MICROBES

One of the most significant findings from OMICS studies was the identification of previously unknown, thus "new" bacterial genera/species, which appeared to be associated with dental health or caries. Such was the case for *Streptococcus* A12 [58] and *Streptococcus dentisani* [59], which were both isolated from supragingival plaque of caries-free individuals. A12 is able to inhibit the growth and intercellular signaling of the caries pathogen *Streptococcus mutans* and to buffer pH

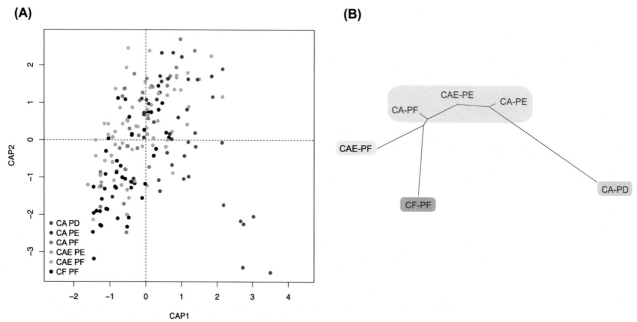

FIGURE 9.4 (A) Ordination analysis: distance-based redundancy analysis (db-RDA) of plaque bacterial communities. (B) Neighbor joining phylogeny based on pairwise PERMANOVA variance components for beta diversity. Branch lengths represent the degree to which bacterial communities are differentiated. Groups shaded in green showed no significant differences in beta diversity. *CF*, caries-free children; *CAE*, caries-active, with enamel carious lesions, children; *CA*, caries-active, with dentin carious lesions, children; *PF*, supragingival plaque samples from caries-free tooth surfaces; *PE*, plaque from active, enamel carious lesion; *PD*, plaque from active, dentin carious lesions [13].

through the arginolytic pathway. Similar to A12, *S. dentisani* presents antimicrobial activity on *S. mutans* and is capable of metabolizing arginine. Current evidence supports the role of arginine metabolism in caries prevention [60−65].

Given their beneficial and health-associated properties, A12 and *S. dentisani* are currently being tested as probiotic strains for caries prevention. Other newly identified species are yet to be characterized, such as *Schlegelella* species, which were detected by high-throughput 16S rRNA gene sequence analysis of carious dentin samples. In some subjects, *Schlegelella* represented a high proportion of the total microbiota detected in the carious samples [66]. *Scardovia wiggsiae* was significantly associated with severe early childhood caries [67].

Lactobacillus species have long been associated with late stages of caries progression [68]. In a metagenome analysis of the bacterial communities found at the different stages of caries development, *Lactobacillus* species were detected only in deep carious dentin [43].

ACTIVE VERSUS INACTIVE BACTERIA

OMICS studies also underpinned a pivotal aspect of the caries process—the presence of bacteria does not necessarily indicate metabolic activity. Specifically, the prevalence of bacteria revealed by metagenome analysis may not correlate well with the patterns of the active microbial community revealed by metatranscriptome analysis of the same oral sample. *Actinomyces*, *Corynebacterium*, and *Neisseria* were the three most abundant taxa in the RNA-based community of a 24-h plaque sample, whereas *Veillonella*, *Streptococcus*, and *Leptotrichia* were the most prevalent in the total DNA-based metagenome of the same sample [51]. This underlines the dynamic nature of microbial activity in biofilms by indicating that some genera were especially active at 24 h of biofilm development, whereas others were less active, albeit being present at higher proportions at that point of sampling.

Gum Microbiome

In periodontitis, an inflammatory response is triggered if biofilm accumulates around the gingival margin beyond levels compatible with oral health. The flow of gingival crevicular fluid (GCF) is increased to deliver components of the host defenses (e.g., immunoglobulins, complement, neutrophils, cytokines) in response to the microbial challenge. This response will inhibit the growth of susceptible species, but a number of subgingival organisms (including *P. gingivalis*) can use specific mechanisms to disturb host defenses, such as by degrading complement, interfering with neutrophil function, and blocking phagocytosis. The increased flow of GCF also provides other host molecules that can serve as nutrients for some of the proteolytic members of the subgingival microbiota.

Early studies linked advanced periodontitis with the presence of the "red complex," which was composed by *P. gingivalis*, *Treponema denticola*, and *Tannerella forsythia*. Other consortia termed the "orange complex," which included *Fusobacterium nucleatum*, *Eubacterium nodatum*, and various *Prevotella* and *Campylobacter* species, often preceded the presence of the "red complex." However, there is currently no great consistency in defining the predominant species implicated in disease, and generally inflammation is associated with diverse polymicrobial communities.

Like in other polymicrobial infections, no single bacterial species has been implicated as a principal pathogen in periodontal disease. The prevalent taxa in periodontal disease fall into eight bacterial phyla: Firmicutes, Proteobacteria, Spirochaetes, Bacteroidetes, Actinobacteria, Synergistetes, Fusobacteria, and TM7 [10]. The genera *Peptostreptococcus*, *Veillonella*, and *Selenomonas* are prevalent in chronic periodontitis, while the genera *Streptococcus*, *Eubacterium*, *Selenomonas*, *Treponema*, *Porphyromonas*, and *Capnocytophaga* were more common in subjects with aggressive periodontitis. It remains uncertain whether phylotypes identified by 16S rDNA as associated with chronic and aggressive periodontitis (*Filifactor alocis*), or chronic and refractory periodontitis (TM7 and Synergistetes), have a meaningful role as periodontal pathogens [24].

Oral Mycobiome

Recent molecular studies revealed a diverse array of fungi as potential oral residents [72]. *Candida albicans* is the most commonly cultivated and studied fungus, and *Candida* species, especially *C. albicans*, have been unequivocally linked to the etiology of oral thrush (candidiasis). Despite considerable research in oral candidiasis, it is not clear why some subjects develop the condition, while others with similar risk factors remain disease-free.

Oral candidiasis can present clinically as a pseudomembranous condition or as erythematous and/or hyperplastic lesions [73]. Most cases of oral candidiasis are associated with *C. albicans*, although some have been linked to *Candida dubliniensis*, *Candida glabrata*, *Candida krusei*, and *Candida tropicalis*, commonly as mixed infections including

C. albicans. Aspergillus, Fusarium, and *Cryptococcus* are other fungi reported to be colonizers of the oral cavity of healthy individuals. While the mechanisms that prevent these pathogenic fungi from causing disease in healthy individuals are unknown, there is evidence that oral bacteria provide a natural defense through the production of antifungal compounds. Components of saliva, such as lactoferrin, also inhibit growth of *Candida* and *Aspergillus* [24].

Recent studies have shown that *C. albicans* may play a role as a secondary agent in the caries process [74]. *C. albicans* and other members of the fungal kingdom have also been recovered from periodontal pockets of patients with chronic and aggressive periodontitis. *C. albicans* can express both cell surface—bound and secreted proteinases capable of degrading major extracellular matrix and basement membrane components such as collagens and fibronectin, as well as promoting strong, chronic, and compact inflammatory cell infiltrates in the periodontal tissue [75,76]. Moreover, these *C. albicans* proteinases can enhance the tissue-destructive host cell proteinase network, in the same way as the proteases expressed by periodontopathic bacteria. However, the role played by *C. albicans* in periodontal diseases is unclear, and further studies are needed to demonstrate the clinical significance of the findings.

Oral Microbiome Associated With Systemic Disease

Severe forms of oral disease may result in systemic repercussions at different body sites, including cardiovascular disease, pneumonia, preterm birth, and diabetes. Chronic inflammatory periodontal diseases, which are among the most prevalent of chronic infections in humans, are thought to constitute a significant inflammatory burden that contributes to cardiovascular disease (e.g., atherosclerosis, myocardial infarction, and stroke). Bacteria that colonize the oropharynx, including *Streptococcus pneumoniae, Haemophilus influenzae,* and *Mycoplasma pneumoniae,* are recognized as a source of community-acquired pneumonia. *Streptococcus gordonii,* a pioneering colonizer of the oral environment and commensal microorganism, is an opportunistic pathogen associated with endocarditis [24].

Of importance, oral diseases share common risk factors with other chronic, systemic diseases/conditions including (1) Western-type diet (risk factor for dental caries, coronary heart disease, stroke, diabetes, cancer, obesity); (2) tobacco smoking/chewing (oral and other cancers, periodontal disease, coronary heart disease, stroke, respiratory diseases, diabetes); (3) alcohol consumption (oral and other cancers, cardiovascular disease, liver cirrhosis, trauma); (4) poor hygiene (periodontal disease and other bacterial and inflammatory conditions); (5) injuries (trauma, including dental trauma); (6) stress (periodontal disease and cardiovascular disease); and (7) socioeconomic status (underlying determinant of other risk factors; information available from: https://www.dentalhealth.ie/dentalhealth/causes/general.html).

CONCLUSIONS

Even though the healthy oral microbiome appears to be more stable than those of other body niches like the gut, there is evidence for substantial degree of within-individual variability in the oral microbiome [69,70].

Genetic diversity and complexities in the adaptive strategies of oral bacteria to fluctuating biofilm conditions diminish the utility of taxa-level correlation of the microbiome with health especially, but also with caries and periodontal disease. The relevant questions may be answered by elucidating what the microbes are doing, rather than focusing primarily on who is performing those actions [9]. Also deserving more attention are the microbial interactions with the host, e.g., adhesion mechanisms between microbes and salivary proteins [71] and the recognition patterns with the oral immune system [66]. Future microbiome and metagenome analysis will certainly contribute to the development of more effective therapeutic and diagnostic techniques and, ultimately, to personalized medicine and personalized dental medicine [77].

REFERENCES

[1] Frencken JE, Sharma P, Stenhouse L, Green D, Laverty D, Dietrich T. Global epidemiology of dental caries and severe periodontitis - a comprehensive review. J Clin Periodontol 2017;44(Suppl. 18):S94—105.

[2] Kim JK, Baker LA, Davarian S, Crimmins E. Oral health problems and mortality. J Dent Sci 2013;8(2).

[3] Cahill TJ, Harrison JL, Jewell P, et al. Antibiotic prophylaxis for infective endocarditis: a systematic review and meta-analysis. Heart 2017;103(12):937—44.

[4] Kriebel K, Hieke C, Muller-Hilke B, Nakata M, Kreikemeyer B. Oral biofilms from symbiotic to pathogenic interactions and associated disease -connection of periodontitis and rheumatic arthritis by peptidylarginine deiminase. Front Microbiol 2018;9:53.

[5] Turnbaugh PJ, Ley RE, Hamady M, Fraser-Liggett CM, Knight R, Gordon JI. The human microbiome project. Nature 2007;449(7164):804—10.

[6] Proctor LM. The National Institutes of health human microbiome project. Semin Fetal Neonatal Med 2016;21(6):368—72.

[7] Marsh PD. Dental plaque as a biofilm and a microbial community - implications for health and disease. BMC Oral Health 2006;6(Suppl. 1):S14.

[8] Nascimento MM, Zaura E, Mira A, Takahashi N, Ten Cate JM. Second era of OMICS in caries research: moving past the phase of disillusionment. J Dent Res 2017. https://doi.org/10.1177/0022034517701902.

[9] Takahashi N. Oral microbiome metabolism: from "who are They?" to "what are they doing? J Dent Res 2015;94(12):1628−37.

[10] Dewhirst FE, Chen T, Izard J, et al. The human oral microbiome. J Bacteriol 2010;192(19):5002−17.

[11] Ghannoum MA, Jurevic RJ, Mukherjee PK, et al. Characterization of the oral fungal microbiome (mycobiome) in healthy individuals. PLoS Pathog 2010;6(1):e1000713.

[12] Pride DT, Salzman J, Haynes M, et al. Evidence of a robust resident bacteriophage population revealed through analysis of the human salivary virome. ISME J 2012;6(5):915−26.

[13] Richards VP, Alvarez AJ, Luce AR, et al. Microbiomes of site-specific dental plaques from children with different caries status. Infect Immun 2017;85(8).

[14] Finlayson TL, Gupta A, Ramos-Gomez FJ. Prenatal maternal factors, Intergenerational transmission of disease, and child oral health outcomes. Dent Clin 2017;61(3):483−518.

[15] Lif Holgerson P, Harnevik L, Hernell O, Tanner AC, Johansson I. Mode of birth delivery affects oral microbiota in infants. J Dent Res 2011;90(10):1183−8.

[16] Fardini Y, Chung P, Dumm R, Joshi N, Han YW. Transmission of diverse oral bacteria to murine placenta: evidence for the oral microbiome as a potential source of intrauterine infection. Infect Immun 2010;78(4):1789−96.

[17] Aagaard K, Ma J, Antony KM, Ganu R, Petrosino J, Versalovic J. The placenta harbors a unique microbiome. Sci Transl Med 2014;6(237):237ra65.

[18] Gomez A, Nelson KE. The oral microbiome of children: development, disease, and implications beyond oral health. Microb Ecol 2017;73(2):492−503.

[19] Cobb CM, Kelly PJ, Williams KB, Babbar S, Angolkar M, Derman RJ. The oral microbiome and adverse pregnancy outcomes. Int J Womens Health 2017;9:551−9.

[20] Lif Holgerson P, Ohman C, Ronnlund A, Johansson I. Maturation of oral microbiota in children with or without dental caries. PLoS One 2015;10(5):e0128534.

[21] Jakubovics NS. Saliva as the sole nutritional source in the development of multispecies communities in dental plaque. Microbiol Spectr 2015;3(3).

[22] Wei GX, van der Hoeven JS, Smalley JW, Mikx FH, Fan MW. Proteolysis and utilization of albumin by enrichment cultures of subgingival microbiota. Oral Microbiol Immunol 1999;14(6):348−51.

[23] Wade W, Thompson H, Rybalka A. Uncultured members of the oral microbiome. J Calif Dent Assoc 2016;44(7):447−56.

[24] Boonanantasarn KG, S R. The oral microbiome. In: Kolenbrander PE, editor. Oral microbial communities: genomic inquiry and interspecies communication. Washington, DC: ASM Press; 2011.

[25] Marsh PD, Head DA, Devine DA. Dental plaque as a biofilm and a microbial community—implications for treatment. J Oral Biosci 2015;57(4):185−91.

[26] Marsh PD. Ecological events in oral health and disease: new opportunities for prevention and disease control? J Calif Dent Assoc 2017;45(10):525−37.

[27] Fejerskov O, Nyvad B, Kidd E. Dental caries: what is it? In: Fejerskov O, Nyvad B, Kidd E, editors. Dental caries: the disease and its clinical management. Wiley-Blackwell; 2015. p. 7−10.

[28] Kolenbrander PE. Multispecies communities: interspecies interactions influence growth on saliva as sole nutritional source. Int J Oral Sci 2011;3(2):49−54.

[29] Mark Welch JL, Rossetti BJ, Rieken CW, Dewhirst FE, Borisy GG. Biogeography of a human oral microbiome at the micron scale. Proc Natl Acad Sci U S A 2016;113(6):E791−800.

[30] Bradshaw DJ, Marsh PD, Allison C, Schilling KM. Effect of oxygen, inoculum composition and flow rate on development of mixed-culture oral biofilms. Microbiology 1996;142(Pt 3):623−9.

[31] Burne RA, Marquis RE. Alkali production by oral bacteria and protection against dental caries. FEMS Microbiol Lett 2000;193(1):1−6.

[32] Marsh PD, Takashashi N, Nyvad B. Biofilms in caries development. In: Fejerskov O, Nyvad B, Kidd E, editors. Dental caries: the disease and its clinical management. 3rd ed. Wiley-Blackwell; 2015. p. 107−31.

[33] Nascimento MM, Gordan VV, Garvan CW, Browngardt CM, Burne RA. Correlations of oral bacterial arginine and urea catabolism with caries experience. Oral Microbiol Immunol 2009;24(2):89−95.

[34] Nascimento MM, Liu Y, Kalra R, et al. Oral arginine metabolism may decrease the risk for dental caries in children. J Dent Res 2013;92(7):604−8.

[35] Burne RA, Zeng L, Ahn SJ, et al. Progress dissecting the oral microbiome in caries and health. Adv Dent Res 2012;24(2):77−80.

[36] Mira A, Simon-Soro A, Curtis MA. Role of microbial communities in the pathogenesis of periodontal diseases and caries. J Clin Periodontol 2017;44(Suppl. 18):S23−38.

[37] Sanz M, Beighton D, Curtis MA, et al. Role of microbial biofilms in the maintenance of oral health and in the development of dental caries and periodontal diseases. Consensus report of group 1 of the Joint EFP/ORCA workshop on the boundaries between caries and periodontal disease. J Clin Periodontol 2017;44(Suppl. 18):S5−11.

[38] Chen T, Yu WH, Izard J, Baranova OV, Lakshmanan A, Dewhirst FE. The Human Oral Microbiome Database: a web accessible resource for investigating oral microbe taxonomic and genomic information. Database 2010;2010:baq013.

[39] Griffen AL, Beall CJ, Firestone ND, et al. CORE: a phylogenetically-curated 16S rDNA database of the core oral microbiome. PLoS One 2011;6(4):e19051.

[40] Gevers D, Knight R, Petrosino JF, et al. The Human Microbiome Project: a community resource for the healthy human microbiome. PLoS Biol 2012;10(8):e1001377.

[41] Belda-Ferre P, Alcaraz LD, Cabrera-Rubio R, et al. The oral metagenome in health and disease. ISME J 2012;6(1):46−56.

[42] Duran-Pinedo AE, Frias-Lopez J. Beyond microbial community composition: functional activities of the oral microbiome in health and disease. Microb Infect 2015;17(7):505—16.

[43] Simon-Soro A, Belda-Ferre P, Cabrera-Rubio R, Alcaraz LD, Mira A. A tissue-dependent hypothesis of dental caries. Caries Res 2013;47(6):591—600.

[44] Wang J, Gao Y, Zhao F. Phage-bacteria interaction network in human oral microbiome. Environ Microbiol 2016;18(7):2143—58.

[45] Edlund A, Yang Y, Yooseph S, et al. Meta-omics uncover temporal regulation of pathways across oral microbiome genera during in vitro sugar metabolism. ISME J 2015;9(12):2605—19.

[46] Vartoukian SR, Adamowska A, Lawlor M, Moazzez R, Dewhirst FE, Wade WG. In vitro cultivation of 'unculturable' oral bacteria, facilitated by community culture and media supplementation with siderophores. PLoS One 2016;11(1):e0146926.

[47] Tanner ACR, Kressirer CA, Rothmiller S, Johansson I, Chalmers NI. The caries microbiome: implications for reversing dysbiosis. Adv Dent Res 2018;29(1):78—85.

[48] Aas JA, Griffen AL, Dardis SR, et al. Bacteria of dental caries in primary and permanent teeth in children and young adults. J Clin Microbiol 2008;46(4):1407—17.

[49] Gross EL, Leys EJ, Gasparovich SR, et al. Bacterial 16S sequence analysis of severe caries in young permanent teeth. J Clin Microbiol 2010;48(11):4121—8.

[50] Tanner AC, Kressirer CA, Faller LL. Understanding caries from the oral microbiome perspective. J Calif Dent Assoc 2016;44(7):437—46.

[51] Benitez-Paez A, Belda-Ferre P, Simon-Soro A, Mira A. Microbiota diversity and gene expression dynamics in human oral biofilms. BMC Genomics 2014;15:311.

[52] Belda-Ferre P, Williamson J, Simon-Soro A, Artacho A, Jensen ON, Mira A. The human oral metaproteome reveals potential biomarkers for caries disease. Proteomics 2015;15(20):3497—507.

[53] Belstrom D, Jersie-Christensen RR, Lyon D, et al. Metaproteomics of saliva identifies human protein markers specific for individuals with periodontitis and dental caries compared to orally healthy controls. PeerJ 2016;4:e2433.

[54] Takahashi N, Washio J, Mayanagi G. Metabolomics of supragingival plaque and oral bacteria. J Dent Res 2010;89(12):1383—8.

[55] Washio J, Ogawa T, Suzuki K, Tsukiboshi Y, Watanabe M, Takahashi N. Amino acid composition and amino acid-metabolic network in supragingival plaque. Biomed Res 2016;37(4):251—7.

[56] Gross EL, Beall CJ, Kutsch SR, Firestone ND, Leys EJ, Griffen AL. Beyond Streptococcus mutans: dental caries onset linked to multiple species by 16S rRNA community analysis. PLoS One 2012;7(10):e47722.

[57] Lee HJ, Kim JB, Jin BH, Paik DI, Bae KH. Risk factors for dental caries in childhood: a five-year survival analysis. Community Dent Oral Epidemiol 2015;43(2):163—71.

[58] Huang X, Palmer SR, Ahn SJ, et al. A highly arginolytic *Streptococcus* species that potently antagonizes Streptococcus mutans. Appl Environ Microbiol 2016;82(7):2187—201.

[59] Camelo-Castillo A, Benitez-Paez A, Belda-Ferre P, Cabrera-Rubio R, Mira A. *Streptococcus dentisani* sp. nov., a novel member of the mitis group. Int J Syst Evol Microbiol 2014;64(Pt 1):60—5.

[60] Burne RA, Parsons DT, Marquis RE. Environmental variables affecting arginine deiminase expression in oral streptococci. Washington, DC. 1991.

[61] Casiano-Colon A, Marquis RE. Role of the arginine deiminase system in protecting oral bacteria and an enzymatic basis for acid tolerance. Appl Environ Microbiol 1988;54(6):1318—24.

[62] Kleinberg I. Prevention and dental caries. J Prev Dent 1978;5(3):9—17.

[63] Kleinberg IA. Mixed-bacteria ecological approach to understanding the role of the oral bacteria in dental caries causation: an alternative to Streptococcus mutans and the specific-plaque hypothesis. Crit Rev Oral Biol Med 2002;13(2):108—25.

[64] Marquis RE. Oxygen metabolism, oxidative stress and acid-base physiology of dental plaque biofilms. J Ind Microbiol 1995;15(3):198—207.

[65] Marquis RE, Bender GR, Murray DR, Wong A. Arginine deiminase system and bacterial adaptation to acid environments. Appl Environ Microbiol 1987;53(1):198—200.

[66] Simon-Soro A, Mira A. Solving the etiology of dental caries. Trends Microbiol 2015;23(2):76—82.

[67] Tanner AC, Mathney JM, Kent RL, et al. Cultivable anaerobic microbiota of severe early childhood caries. J Clin Microbiol 2011;49(4):1464—74.

[68] Becker MR, Paster BJ, Leys EJ, et al. Molecular analysis of bacterial species associated with childhood caries. J Clin Microbiol 2002;40(3):1001—9.

[69] Caporaso JG, Lauber CL, Costello EK, et al. Moving pictures of the human microbiome. Genome Biol 2011;12(5):R50.

[70] Turnbaugh PJ, Hamady M, Yatsunenko T, et al. A core gut microbiome in obese and lean twins. Nature 2009;457(7228):480—4.

[71] Nobbs AH, Jenkinson HF, Jakubovics NS. Stick to your gums: mechanisms of oral microbial adherence. J Dent Res 2011;90(11):1271—8.

[72] Diaz PI, Hong BY, Dupuy AK, Strausbaugh LD. Mining the oral mycobiome: methods, components, and meaning. Virulence 2017;8(3):313—23.

[73] Williams D, Lewis M. Pathogenesis and treatment of oral candidosis. J Oral Microbiol 2011;3.

[74] Pereira D, Seneviratne CJ, Koga-Ito CY, Samaranayake LP. Is the oral fungal pathogen *Candida albicans* a cariogen? Oral Dis 2017;4(4):518—26.

[75] Jarvensivu A, Hietanen J, Rautemaa R, Sorsa T, Richardson M. Candida yeasts in chronic periodontitis tissues and subgingival microbial biofilms in vivo. Oral Dis 2004;10:106—12.

[76] Rodier MH, el Moudni B, Kauffmann-Lacroix C, Daniault G, Jacquemin JL. A *Candida albicans* metallopeptidase degrades constitutive proteins of extracellular matrix. FEMS Microbiol Lett 1999;177:205—10.

[77] Zarco MF, Vess TJ, Ginsburg GS. The oral microbiome in health and disease and the potential impact on personalized dental medicine. Oral Dis 2012;18(2):109—20.

Chapter 10

The Gastric Microbiome in Benign and Malignant Diseases

Thais Fernanda Bartelli[1], Luiz Gonzaga Vaz Coelho[2] and Emmanuel Dias-Neto[1]

[1]Lab. Medical Genomics, A.C.Camargo Cancer Center, São Paulo, Brazil; [2]Alfa Institute of Gastroenterology, Clinics Hospital, Federal University of Minas Gerais, Belo Horizonte, Brazil

THE GASTRIC MICROBIAL COMMUNITY

For many years, the concept of an endogenous stomach microbiota was dismissed by the notion that the stomach is a harsh and inhospitable environment. The low pH, a consequence of hydrochloric acid (HCl) production, the occasional reflux of bile acids, and the presence of nitric oxide (a potent antimicrobial agent), as well as the constant peristalsis of its muscularis layers, led to the belief that the stomach was a sterile organ [1]. In the early 1980s, this belief was challenged by the discovery of several bacteria, including *Campylobacter pyloridis* (later renamed *Helicobacter pylori*), *Streptococcus*, *Lactobacillus*, and *Neisseria*, which were identified by techniques based on microorganism growth [1].

Current estimates point out that whereas the large intestine harbors up to 10^{12} colony-forming units (CFU)/mL, the stomach lodges around $10^3 - 10^4$ CFU/mL [2,3]. Although significantly smaller, this microbiota operates as an important barrier against the colonization by exogenous and pathogenic bacteria that could find its way through the gastrointestinal tract and infect the host. Additionally, recent studies found that the stomach actively releases oxidizing chemicals and may be able to recognize hazardous foreign bodies, thus triggering this response [4,5].

It was only in 2006 that the first molecular study using 16S rDNA by Bik and collaborators [6] would identify at least 128 different bacterial phylotypes on the gastric mucosa of healthy individuals. The five most abundant phyla in both children and adults are Actinobacteria, Bacteroidetes, Firmicutes, Fusobacteria, and Proteobacteria [6–10]. This gastric microbiota is relatively stable and maintained over different populations, possibly because of the high selective pressure exerted by the stomach environment, rather than cultural individual habits [8]. Specifically, the five most common genera, found in healthy and antral gastritis patients without *H. pylori* infection were *Streptococcus* (phylum Firmicutes), *Prevotella* and *Porphyromonas* (phylum Bacteroidetes), and *Neisseria* and *Haemophilus* (phylum Proteobacteria) [8].

H. pylori, a Proteobacteria present in almost 50% of the world population, is the most studied gastric bacteria, with a well-recognized role in gastritis, preneoplastic, and neoplastic lesions. *Lactobacilli*, another common stomach microorganism, is commonly used as probiotic due to its health-promoting properties. Nonetheless, growth-independent methods have identified a vast number of commensals, including *Rothia, Actinobacillus, Caulobacter, Corynebacterium, Gemella, Leptotrichia, Porphyromonas, Capnocytophaga*, TM7, *Flexistipes*, and *Deinococcus* [6].

TECHNICAL LIMITATIONS IN THE GASTRIC MICROBIOME ANALYSIS

A relevant limitation of the traditional molecular microbiome studies may arise by the essentially one-and-only analysis of the bacterial DNA content present in the stomach environment, by 16S rDNA sequencing, which is unable to distinguish between transcriptionally active or inactive, dead or alive bacteria [10]. As a consequence, the sequencing of this bacterial DNA present may not represent the true active community, as it may also contemplate bacterial DNA derived from dead bacteria or migrating organisms derived from the upper aerodigestive tract and/or from the diet.

Microbiome and Metabolome in Diagnosis, Therapy, and other Strategic Applications. https://doi.org/10.1016/B978-0-12-815249-2.00010-5

Additionally, due to the complexity of the still unexplored interactions between bacteria and the gastric environment, the study of gastric juice may lead to inconclusive data, since it can be more representative of the stomach transient bacteria population, rather than the bacterial cells that closely interact with the gastric tissue and mucus layer [1]. This occurs because, while the gastric juice tends to restrict the growth of the majority of the microorganisms, the mucus layer has a protective role allowing a more hospitable environment and the growth of a richer and more diverse tissue-associate microbiome [11]. In the gastrointestinal tract, the mucus barrier is variable among the population and formed by different patterns of host-specific glycosylation products and glycoproteins, which are capable of shaping the identity and the composition of the indigenous microbiota [12].

EUKARYOTES AND VIRUSES IN THE UPPER GASTROINTESTINAL TRACT

Whereas the gut bacteriome is usually the center of the attention, it is important to keep in mind that viruses and eukaryotes such as fungi, helminths, and protozoa are also important members of the human microbial community [13]. A comprehensive analysis using traditional culture methods and molecular study techniques detected a rich eukaryotic community in the human gastrointestinal tract, composed by more than 15 different protozoan and 50 helminthic genera linked to commensalism or parasitism [13,14]. Among them are known human parasites such as *Blastocystis* spp., with high prevalence in healthy individuals, *Cryptosporidium* spp., and *Entamoeba histolytica* [13]. Bacteria–protozoa interactions and host immune system modulation are possible in this context [13].

Fungal communities have also been detected in the upper portion of the gastrointestinal tract as a possible resident, or otherwise transient, members of the human gastric environment in healthy individuals [10,15]. Fungi represent only 0.1% of the microorganisms found in the human microbiota [16–18]; nonetheless, the mycobiome is believed to play a greater role than previously expected in mucosal health, especially when the bacteriome is disturbed [15]. Fungi of the genus *Candida*, known members of the human microbiome and resident members of the intestinal tract, are also frequently identified in the stomach environment [15,19]. Analysis of the gastric fluid of 25 patients with clinically recommended upper endoscopy found *Candida* sp. in all samples, along with the opportunistic fungus *Phialemonium* [10], also involved in infections in immunocompromised hosts.

Viruses also inhabit the stomach. Examples of their importance to the gastric health can be given by phage and the Epstein–Barr virus (EBV). Phages, a group of bacteria-infecting viruses, have an important role in infecting and controlling bacterial populations; however, they also have a role in horizontal gene transfer and antibiotic resistance [20]. EBV is known to play a major role in approximately 10% of stomach cancer cases [21]. The identification of EBV as an important player in gastric carcinogenesis opens the possibility that other viruses may inhabit the stomach and play a role in gastric diseases.

HELICOBACTER PYLORI AND GASTRIC CANCER

H. pylori is thought to be colonizing humans for at least 50,000 years [22]. Chronic infection by *H. pylori* has been established as a major risk factor for gastric cancer (twofold risk), particularly non-cardia and nondiffuse tumor subtypes, with most gastric cancer cases (80%) associated with the chronic presence of this bacterium, making *H. pylori* a definite carcinogen by the World Health Organization, since 1994 [23,24].

The causal or mechanistic links associated to *H. pylori* stomach carcinogenesis are due to putative virulence genes, such as (1) cagPAI (cytotoxin-associated gene pathogenicity island), which encodes proteins that are members of the type IV secretion system, which eventually leads to dephosphorylation of host cell proteins and destabilization of cellular morphology; (2) genes encoding outer-membrane proteins (oipA, *sabA*, *sabB*, *babA*, *babC*, and *hopZ*), responsible mainly for bacterium binding to the gastric mucosal cells or promotion of inflammatory processes; and (3) vacA (vacuolating cytotoxin gene), which encodes a cytotoxic protein capable of inducing vacuolation on the host cells [25]. Other gut bacteria, such as *Escherichia coli*, are also capable of producing virulence factors that directly target the host DNA, such as colibactins, genotoxins capable of inducing DNA double-strand breaks, associated with genomic instability and colon cancer tumorigenesis [26].

Despite the clear role of *H. pylori* in gastric cancer, the occurrence of gastritis, and disease progression to metaplasia and dysplasia, is irrespective of the presence, eradication, or absence of the pathogen [8,27–29]. Only 1%–3% of *H. pylori*-infected patients effectively develop gastric cancer [30].

GENERAL DYSBIOSIS IN GASTRIC CANCER

Other taxa enriched in gastric cancer patients are *Lactococcus*, *Veillonella*, *Fusobacterium*, and *Leptotrichia*. Analysis of microbial metabolic output in those patients identified an increase in lactic acid—producing bacteria, enrichment of short-chain fatty acid production pathways, and enrichment of proinflammatory oral bacterial species [31].

Dysbiosis in intestinal metaplasia and gastric cancer subjects has been characterized, as compared to superficial or atrophic gastritis. Additionally, oral bacteria, such as *Peptostreptococcus stomatis* and *Dialister pneumosintes*, had significantly higher abundance in gastric cancer as compared to the other stages analyzed, and together with *Streptococcus anginosus*, *Parvimonas micra* and *Slackia exigua*, could differentiate gastric cancer patients from superficial gastritis cases [32].

Whether the gastric mucosal dysbiosis is the causative agent or a consequence of altered mucosal physiology/tumor microenvironment is yet to be fully uncovered. Controlled in vivo experiments, through the introduction of specific known bacteria, will be required to elucidate their role [33]. Nonetheless, gastric cancer has been subdivided into four molecular subtypes by the TCGA (The Cancer Genome Atlas) [34], one of them defined according to the presence of EBV. Future studies might well consider the results of gastric microbiome analysis [33].

GASTRIC MICROORGANISMS AND EXTRAGASTRIC DISEASE

Campylobacter concisus, transcriptionally active in the stomach [10], has a significantly higher prevalence in children with Crohn's disease in comparison to healthy controls [35,36]. *H. pylori*, aside from being a known risk factor for stomach carcinogenesis, has a massive influence overall the gastrointestinal tract, including modulation of the resident microbiota and the host immunological response. In the stomach, colonization by *H. pylori* leads to increased pH through mucosal atrophy and changes in nutrient availability, favoring the growth of bacterial species that would otherwise not survive gastric physiology, leading to conditions that alter the stomach homeostasis, consequently favoring carcinogenesis [7,37], and influencing the whole gut. Analysis of the active microbiome, determined by sequencing 16S rRNA transcripts of stomach biopsies, indicated that gastric cancer patients had increased bacterial richness and phylogenetic diversity as compared to functional dyspepsia patients, independent of their *H. pylori* serum-positivity. However, the composition of the gastric microbiota and global microbial metabolic output, in patients testing positive for this bacterium, was different from those tested as negative [31].

EPSTEIN—BARR VIRUS AND GASTRIC CANCER

EBV is present in about 95% of the population in a latent stage [38], yet its oncogenic activity has been associated with the occurrence of several human tumors, comprising about 10% of gastric cancer cases worldwide [39], predominantly nonantrum carcinomas [39—41]. The virus enters the stomach via the oral route, where it is capable of infecting the epithelial mucosal cells, through unknown mechanisms that are different from those used for EBV B-lymphocyte infection.

A series of biological mechanisms operate in EBV-induced tumors, including the interference of EBV with cell-cycle checkpoints as well as cell death pathways and the immortalization and proliferation of EBV-infected cells [14]. EBV-immortalized B-cells are in the basis of B-cell lymphomas, where EBV-infection appears to act as a somatic mutator, due to the activation of DNA and RNA editing enzymes. Some authors have also suggested that gastritis (related or not to *H. pylori* infection) will recruit EBV-infected B-lymphocytes in a process that may increase the epithelial frequency of EBV-infected cells [41].

As EBV-positive tumors have been classified by TCGA as a distinct molecular subtype of gastric cancer (alongside the microsatellite unstable, genomically stable and those with chromosomal instability subtypes), EBV-positive gastric tumors have specific molecular alterations such as mutations in the *PIK3CA* gene, DNA hypermethylation and amplification of *JAK2*, *CD274* and *PDL1* and *PDL2* [34]. These EBV-positive tumors have been associated with the increased expression of programmed cell death 1 ligand (PDL1/2). For this reason, the presence of this virus may indicate good candidates for therapy with immune checkpoint inhibitors [34], with evidence of better outcomes for those patients [42,43].

LIFESTYLE HABITS AND STOMACH DYSBIOSIS

Even though well-defined microbial communities colonize the gastrointestinal tract, its composition is dynamic and can be shaped by several factors, such as lifestyle and dietary habits, use of medications, and eventually predisposition by genetic

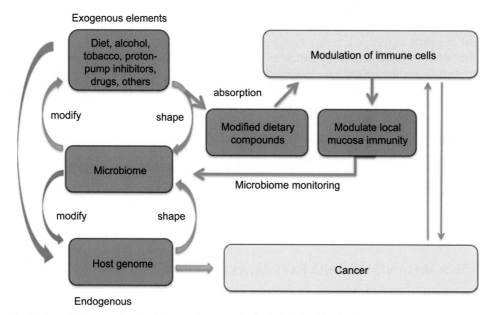

Exogenous elements

Endogenous

FIGURE 10.1 Microbiota and cancer: an integrated view. The central role of the microbiota in the general homeostasis is presented, together with some of its interactions with exogenous and endogenous elements that impact mutation burden and cancer development. Relevant to gastric cancer, the microbiota is shaped by dietary elements such as a salted diet, consumption of red meat, vegetables, as well as alcohol, tobacco, and pH-elevating drugs. The microbiota, including viral infections (such as the EBV), presumably triggers DNA-modifying enzymes, leading to the accumulation of mutations in the cancer genome. On the other hand, these exogenous elements are also able to directly lead to mutations in the host genome and are modified by the microbiota, such as the breakdown of ethanol to acetaldehyde or the synthesis of vitamins and cofactors produced by the microbiota. Modified dietary compounds are important modulators of the immune system that affects the local immune status and impact the development of cancer.

factors. In turn, the microbiota will metabolize diet components and contribute to the shaping of the immune system, with remarkable influence over the microbiota itself and also over the development of diseases such as cancer (Fig 10.1).

Proton pump inhibitors (PPI) are among the top five drugs used by the Western population. Together with antibiotics, PPIs are especially used for the eradication of *H. pylori*, through direct targeting of the bacterial proton pumps or by promoting changes in the pH of the gastric microenvironment, affecting its growth [44].

However, long-term PPI use has been associated with an at least 2.4-fold increase in the risk of gastric cancer development, in a dose- and time-dependent manner, especially risky for daily users for a period ≥3 years [45]. Aside from drastic changes in the stomach pH by PPIs, which influence the gastric microbiome toward inflammation and cancer predisposing bacteria [46,47], colonization by *H. pylori* also compromises the stomach acidity by the active release of urease, hydrolyzing urea, and increasing the pH in the surroundings [48]. How does the pH change caused by these drugs affect the gastric microbiota in medium/long term? *H. pylori* control is likely to make available an ecological niche in the stomach that, together with the modification of gastric pH due to PPIs use, is still an open question.

The gastric mucosa of patients under the use of PPI usually presents a lower microbial diversity, enriched with a few specific bacteria, particularly Streptococcaceae, whose abundance is known to exacerbate or maintain the symptoms associated to dyspepsia [47,49]. In addition, all samples analyzed by Paroni et al. [47] had increased *Capnocytophaga* (Bacteroidetes) and *Actinobacillus* (a Proteobacteria like *H. pylori*). Increased Streptococcaceae has also been described in a cohort of 1827 healthy individuals using PPIs, where additionally, the change in the gastric pH has led to an abundance of oral and upper gastrointestinal tract commensals in their feces [50]. In this sense, the use of PPIs is associated with an increased risk of intestinal infections, including *Clostridium difficile* colitis [51,52].

Aside from the increase in Streptococcaceae, authors have recently observed a decrease in the abundance of *Faecalibacterium* [49,53] in patients using PPIs. This genus is expected to be highly abundant in the healthy gut and has important roles aiding the digestion of dietary fibers, or as a potential probiotic, with anti-inflammatory properties capable of boosting the immune system. Diminished *Faecalibacterium* has been linked to intestinal inflammation-associated diseases (i.e., colitis and Crohn's disease).

The use of diet components or additives may also be associated with dysbiosis and the emergence of epidemics or to the virulence of some bacteria. Some components of the Western diet have been modified by the industry for a series of reasons. An example is the disaccharide trehalose, a sweetening/texturizing ingredient employed to decrease the freezing point of certain processed food, with added properties of reducing sudden glycemic increment after ingestion. In 2018, a group of Baylor College of Medicine showed that the recent epidemics of *C. difficile* could be associated with the rise in the use of the artificial sweetener trehalose, by the food industry. Epidemic *C. difficile* ribotypes are able to metabolize trehalose, which increases *C. difficile* virulent strains [54].

As expected, antibiotics are also modifiers of the stomach microbiome. Although their use has unquestionable benefits to the control of bacterial pathogens, it does not come without some prejudice to the indigenous microbiota. The effect of oral cefoperazone, a third-generation cephalosporin antibiotic, on the gastric bacterial microbiota was observed at least 3 weeks after cessation of its use, causing an especially significant disturbance associated, among other effects, to an increase in the presence of the yeast *Candida albicans* in the stomach, responsible for gastritis during the postantibiotic recovery phase [55].

Immunosuppression linked to elderly patients and the use of PPI also seems to be important risk factors favoring yeast colonization in the stomach. This microbial unbalance and association of *Candida* yeasts and *H. pylori* are believed to be linked to peptic diseases [19] and the development of gastric malignancies [56]. In a study with over 150 patients, at least 17% of them had the fungus *Candida* detected in their gastric mucosa and 11% of those patients were co-colonized with both *H. pylori* and *Candida* [57]. The importance of this coinfection in the rise of gastric diseases is still poorly explored, yet this bacterium−yeast association or yeast infection alone could be causing dyspeptic diseases [19].

H. pylori has been inversely associated with a series of immune-associated diseases, including asthma [58,59], due to direct or indirect influences on gastric and gut microbial community, or the patient immune system [37]. Studies show that the gastric microbiome is influenced by *H. pylori* infection in children but not in adults, as children infected with *H. pylori* tend to have a more diverse bacterial community in the stomach than noninfected ones, with a smaller abundance of Firmicutes and increased richness of non-*Helicobacter* Proteobacteria. On the other hand, *H. pylori*-infected adults have similar bacterial profiles as those noninfected. *H. pylori* in the early life of these children possibly suppress the mucosal inflammation by a T-regulatory response, reducing the symptoms of mucosal damage. Meanwhile, it is still not clear whether this early-life presence of *H. pylori* has an impact on the gastric disease symptoms in adults [58], although in a murine model, the age at which *H. pylori* was acquired influenced, for example, whether they would develop or not precancerous lesions [60].

Aside from having an active role in carcinogenesis, the gut bacteriome may indirectly, through its metabolites, influence the stomach. Conditions that alter the stomach pH knowingly favor the growth of non-*H. pylori*, especially nitrosating bacteria, which convert nitrate, a common food additive, to carcinogenic *N*-compounds [7,61]. In the same manner, a recent meta-analysis reinforced the notion that ethanol consumption leads to an elevated risk of gastric cancer development (odds ratio: 1.39; 95% confidence interval 1.20−1.61) [62], as ethanol metabolization generates the carcinogen acetaldehyde, a process that can be exacerbated by bacteria, making ethanol-induced carcinogenesis also dependent of the host microbiota.

Some natural interventional strategies, such as the use of probiotics, can help in restoring the healthy microbiome. Although this approach is more constantly used aiming the low gastrointestinal tract, the ingestion of probiotic-rich food such as yogurt may improve symptoms linked to gastric dyspepsia [63,64] bringing benefits to the host. Lactic acid bacteria, such as *Lactobacillus*, have shown to be beneficial in preventing or helping eradicate *H. pylori* infection [65,66] or even effectively suppressing dyspeptic symptoms in *H. pylori*-infected patients [64]. Nonetheless, the literature concerning gastric probiotics is still limited and need further investigation.

The importance of endogenous and exogenous factors to shape and to determine the microbiome composition is broadly recognized. However, until recently, the relative contribution of the host genetic versus the environmental factors have not been evaluated. A recent study genotyped 696 healthy individuals who live in similar environments but derived from distinct ancestral origins. The authors found no statistically significant association between microbiome composition and ethnicity. In contrast, genetically unrelated individuals sharing the same household had very similar microbiome compositions. This publication strongly suggests that the human microbiome composition is most importantly determined by environmental factors rather than host genetics [67].

CONCLUSIONS AND FUTURE PERSPECTIVES

A significant challenge that remains is the establishment of a causal link explaining the roles of bacterial populations and their derived metabolites in the development of diseases in humans, including gastric cancer, gastritis, and others. Once this causal link is well established, it raises the possibility of modulating the stomach microbiome, aiming to prevent or intervene in the course of a disease [1]. As the microbiota knowledge advances, including the molecular basis of microbial—host interaction, microbial populations will be translated in a sort of library of chemical/biochemical compounds that may have the potential of revolutionizing a series of human diseases and conditions.

The impact of the gastric microbiota in the future of gastric cancer therapy is expected, as compelling evidence demonstrates its determinant role in tumor response to chemotherapy and immunotherapy. Specifically, there is increasing interest in the synergistic combination of microbiome in the modulation of the host immune system [68]. The literature started to document that the blockage of immune checkpoint inhibitors that target molecules such as cytotoxic T-lymphocyte-associated protein 4 (CTLA4), programmed cell death 1 (PD1) and programmed cell death 1 ligand 1 (PDL1) can benefit from a healthy microbiome status during cancer therapy [33]. In general, higher microbiota diversity is associated with better response to cancer treatment. Recent investigations targeting hundreds of patients, with diverse tumor types (bladder, kidney, lung), clearly show that an antibiotic-disrupted microbiota is correlated with faster tumor relapse and reduced overall survival in patients treated with anti-PD1 immunotherapy.

These observations were reinforced by the finding that germ-free mice receiving fecal transplantation from chemotherapy responders versus nonresponders showed the same response as the donors [67] and the same was observed with regard to melanoma patients, treated by the same drug [68]. Microorganisms that are capable of priming T-cells populations toward better recognition of tumor neoantigens and inducing the release of IL-12 showed to be beneficial for immunotherapy response. However, bacteria pointed out by these two articles were different: *Akkermansia muciniphila* in the first study [67] and *Faecalibacterium* and Clostridiales in the second one [68], a difference that can be explained by diet and lifestyle habits of the different populations enrolled. While the right composition for fecal transplantation is still being vetted, the study of Zitvogel's group is clear in suggesting that by simply avoiding antibiotics during immunotherapy, response rates can jump from 25% to 40% [67]. In this sense the importance of a better understanding of the human microbiome, including the gastric microbiome, to better treat human diseases is becoming a reality.

ACKNOWLEDGMENTS

The authors acknowledge the financial support received from PRONON and Fundação de Amparo à Pesquisa do Estado de São Paulo (FAPESP - 2014/26897-0).

REFERENCES

[1] Nardone G, Compare D. The human gastric microbiota: is it time to rethink the pathogenesis of stomach diseases? United Eur Gastroenterol J June 2015;3(3):255—60.

[2] Lopetuso LR, Scaldaferri F, Franceschi F, Gasbarrini A. The gastrointestinal microbiome - functional interference between stomach and intestine. Best Pract Res Clin Gastroenterol December 2014;28(6):995—1002.

[3] Sender R, Fuchs S, Milo R. Revised estimates for the number of human and bacteria cells in the body. PLoS Biol August 19, 2016;14(8).

[4] Kalantar-Zadeh K, Yao CK, Berean KJ, Ha N, Ou JZ, Ward SA, et al. Intestinal gas capsules: a proof-of-concept demonstration. Gastroenterology January 2016;150(1):37—9.

[5] Ou JZ, Cottrell JJ, Ha N, Pillai N, Yao CK, Berean KJ, et al. Potential of in vivo real-time gastric gas profiling: a pilot evaluation of heat-stress and modulating dietary cinnamon effect in an animal model. Sci Rep September 16, 2016;6:33387.

[6] Bik EM, Eckburg PB, Gill SR, Nelson KE, Purdom EA, Francois F, et al. Molecular analysis of the bacterial microbiota in the human stomach. Proc Natl Acad Sci USA January 17, 2006;103(3):732—7.

[7] Brawner KM, Morrow CD, Smith PD. Gastric microbiome and gastric cancer. Cancer J Sudbury Mass June 2014;20(3):211—6.

[8] Li X-X, Wong GL-H, To K-F, Wong VW-S, Lai LH, Chow DK-L, et al. Bacterial microbiota profiling in gastritis without *Helicobacter pylori* infection or non-steroidal anti-inflammatory drug use. PLoS One November 24, 2009;4(11):e7985.

[9] Llorca L, Pérez-Pérez G, Urruzuno P, Martinez MJ, Iizumi T, Gao Z, et al. Characterization of the gastric microbiota in a pediatric population according to *Helicobacter pylori* status. Pediatr Infect Dis J February 2017;36(2):173—8.

[10] von Rosenvinge EC, Song Y, White JR, Maddox C, Blanchard T, Fricke WF. Immune status, antibiotic medication and pH are associated with changes in the stomach fluid microbiota. ISME J July 2013;7(7):1354—66.

[11] The Gut Microbiome, An issue of gastroenterology clinics of North America, vol. 46-1-1st ed. [Internet]. [cited 2018 Feb 18]. Available from: https://www.elsevier.com/books/the-gut-microbiome-an-issue-of-gastroenterology-clinics-of-north-america/unknown/978-0-323-50978-7.

[12] Yang I, Nell S, Suerbaum S. Survival in hostile territory: the microbiota of the stomach. FEMS Microbiol Rev September 2013;37(5):736—61.

[13] Chabé M, Lokmer A, Ségurel L. Gut protozoa: friends or foes of the human gut microbiota? Trends Parasitol December 2017;33(12):925—34.

[14] Hamad I, Raoult D, Bittar F. Repertory of eukaryotes (eukaryome) in the human gastrointestinal tract: taxonomy and detection methods. Parasite Immunol January 2016;38(1):12—36.

[15] Huffnagle GB, Noverr MC. The emerging world of the fungal microbiome. Trends Microbiol July 2013;21(7):334—41.

[16] Arumugam M, Raes J, Pelletier E, Le Paslier D, Yamada T, Mende DR, et al. Enterotypes of the human gut microbiome. Nature May 12, 2011;473(7346):174—80.

[17] Qin J, Li R, Raes J, Arumugam M, Burgdorf KS, Manichanh C, et al. A human gut microbial gene catalog established by metagenomic sequencing. Nature March 4, 2010;464(7285):59−65.

[18] Underhill DM, Iliev ID. The mycobiota: interactions between commensal fungi and the host immune system. Nat Rev Immunol June 2014;14(6):405−16.

[19] Massarrat S, Saniee P, Siavoshi F, Mokhtari R, Mansour-Ghanaei F, Khalili-Samani S. The effect of *Helicobacter pylori* infection, aging, and consumption of proton pump inhibitor on fungal colonization in the stomach of dyspeptic patients. Front Microbiol 2016;7:801.

[20] Sun CL, Relman DA. Microbiota's "little helpers": bacteriophages and antibiotic-associated responses in the gut microbiome. Genome Biol 2013;14(7):127.

[21] Shibata D, Weiss LM. Epstein-Barr virus-associated gastric adenocarcinoma. Am J Pathol April 1992;140(4):769−74.

[22] Atherton JC, Blaser MJ. Coadaptation of *Helicobacter pylori* and humans: ancient history, modern implications. J Clin Investig September 2009;119(9):2475−87.

[23] Nagini S. Carcinoma of the stomach: a review of epidemiology, pathogenesis, molecular genetics and chemoprevention. World J Gastrointest Oncol July 15, 2012;4(7):156−69.

[24] Eslick G-D. *Helicobacter pylori* infection causes gastric cancer? A review of the epidemiological, meta-analytic, and experimental evidence. World J Gastroenterol May 21, 2006;12(19):2991−9.

[25] Roesler BM, Rabelo-Gonçalves EMA, Zeitune JMR. Virulence factors of *Helicobacter pylori*: a review. Clin Med Insights Gastroenterol 2014;7:9−17.

[26] Gorgoulis VG, Vassiliou L-VF, Karakaidos P, Zacharatos P, Kotsinas A, Liloglou T, et al. Activation of the DNA damage checkpoint and genomic instability in human precancerous lesions. Nature April 14, 2005;434(7035):907−13.

[27] Aviles-Jimenez F, Vazquez-Jimenez F, Medrano-Guzman R, Mantilla A, Torres J. Stomach microbiota composition varies between patients with non-atrophic gastritis and patients with intestinal type of gastric cancer. Sci Rep February 26, 2014;4:4202.

[28] Li TH, Qin Y, Sham PC, Lau KS, Chu K-M, Leung WK. Alterations in gastric microbiota after H. Pylori eradication and in different histological stages of gastric carcinogenesis. Sci Rep March 21, 2017;7:44935.

[29] Wang L, Zhou J, Xin Y, Geng C, Tian Z, Yu X, et al. Bacterial overgrowth and diversification of microbiota in gastric cancer. Eur J Gastroenterol Hepatol March 2016;28(3):261−6.

[30] Wroblewski LE, Peek RM, Wilson KT. *Helicobacter pylori* and gastric cancer: factors that modulate disease risk. Clin Microbiol Rev October 2010;23(4):713−39.

[31] Castaño-Rodríguez N, Goh K-L, Fock KM, Mitchell HM, Kaakoush NO. Dysbiosis of the microbiome in gastric carcinogenesis. Sci Rep November 21, 2017;7(1):15957.

[32] Coker OO, Dai Z, Nie Y, Zhao G, Cao L, Nakatsu G, et al. Mucosal microbiome dysbiosis in gastric carcinogenesis. Gut August 1, 2017;67(6):1024−32.

[33] Shah MA. Gastric cancer: the gastric microbiota - bacterial diversity and implications. Nat Rev Gastroenterol Hepatol December 2017;14(12):692−3.

[34] Cancer Genome Atlas Research Network. Comprehensive molecular characterization of gastric adenocarcinoma. Nature September 11, 2014;513(7517):202−9.

[35] Kaakoush NO, Man SM, Mitchell HM. Functional relationship between Campylobacter concisus and the stomach ecosystem in health and disease. ISME J December 2013;7(12):2245−7.

[36] Man SM, Zhang L, Day AS, Leach ST, Lemberg DA, Mitchell H. Campylobacter concisus and other Campylobacter species in children with newly diagnosed Crohn's disease. Inflamm Bowel Dis June 2010;16(6):1008−16.

[37] Kienesberger S, Cox LM, Livanos A, Zhang X-S, Chung J, Perez-Perez GI, et al. Gastric *Helicobacter pylori* infection affects local and distant microbial populations and host responses. Cell Rep February 16, 2016;14(6):1395−407.

[38] Luzuriaga K, Sullivan JL. Infectious mononucleosis. N Engl J Med May 27, 2010;362(21):1993−2000.

[39] Murphy G, Pfeiffer R, Camargo MC, Rabkin CS. Meta-analysis shows that prevalence of Epstein-Barr virus-positive gastric cancer differs based on sex and anatomic location. Gastroenterology September 2009;137(3):824−33.

[40] Akiba S, Koriyama C, Herrera-Goepfert R, Eizuru Y. Epstein-Barr virus associated gastric carcinoma: epidemiological and clinicopathological features. Cancer Sci February 2008;99(2):195−201.

[41] Iizasa H, Nanbo A, Nishikawa J, Jinushi M, Yoshiyama H. Epstein-Barr Virus (EBV)-associated gastric carcinoma. Viruses December 2012;4(12):3420−39.

[42] Kim JW, Nam KH, Ahn S-H, Park DJ, Kim H-H, Kim SH, et al. Prognostic implications of immunosuppressive protein expression in tumors as well as immune cell infiltration within the tumor microenvironment in gastric cancer. Gastric Cancer Off J Int Gastric Cancer Assoc Jpn Gastric Cancer Assoc 2016;19(1):42−52.

[43] Cho J, Kang M-S, Kim K-M. Epstein-barr virus-associated gastric carcinoma and specific features of the accompanying immune response. J Gastric Cancer March 2016;16(1):1−7.

[44] Minalyan A, Gabrielyan L, Scott D, Jacobs J, Pisegna JR. The gastric and intestinal microbiome: role of proton pump inhibitors. Curr Gastroenterol Rep August 2017;19(8):42.

[45] Cheung KS, Chan EW, Wong AYS, Chen L, Wong ICK, Leung WK. Long-term proton pump inhibitors and risk of gastric cancer development after treatment for *Helicobacter pylori*: a population-based study. Gut 2018;67(1):28−35.

[46] Williams C, McColl KEL. Review article: proton pump inhibitors and bacterial overgrowth. Aliment Pharmacol Ther January 1, 2006;23(1):3−10.

[47] Paroni Sterbini F, Palladini A, Masucci L, Cannistraci CV, Pastorino R, Ianiro G, et al. Effects of proton pump inhibitors on the gastric mucosa-associated microbiota in dyspeptic patients. Appl Environ Microbiol November 15, 2016;82(22):6633—44.

[48] Walker MM, Talley NJ. Review article: bacteria and pathogenesis of disease in the upper gastrointestinal tract—beyond the era of *Helicobacter pylori*. Aliment Pharmacol Ther April 2014;39(8):767—79.

[49] Takagi T, Naito Y, Inoue R, Kashiwagi S, Uchiyama K, Mizushima K, et al. The influence of long-term use of proton pump inhibitors on the gut microbiota: an age-sex-matched case-control study. J Clin Biochem Nutr January 2018;62(1):100—5.

[50] Jackson MA, Goodrich JK, Maxan M-E, Freedberg DE, Abrams JA, Poole AC, et al. Proton pump inhibitors alter the composition of the gut microbiota. Gut May 2016;65(5):749—56.

[51] Janarthanan S, Ditah I, Adler DG, Ehrinpreis MN. Clostridium difficile-associated diarrhea and proton pump inhibitor therapy: a meta-analysis. Am J Gastroenterol July 2012;107(7):1001—10.

[52] Kwok CS, Arthur AK, Anibueze CI, Singh S, Cavallazzi R, Loke YK. Risk of *Clostridium difficile* infection with acid suppressing drugs and antibiotics: meta-analysis. Am J Gastroenterol July 2012;107(7):1011—9.

[53] Tsuda A, Suda W, Morita H, Takanashi K, Takagi A, Koga Y, et al. Influence of proton-pump inhibitors on the luminal microbiota in the gastrointestinal tract. Clin Transl Gastroenterol June 11, 2015;6:e89.

[54] Collins J, Robinson C, Danhof H, Knetsch CW, van Leeuwen HC, Lawley TD, et al. Dietary trehalose enhances virulence of epidemic *Clostridium difficile*. Nature January 18, 2018;553(7688):291—4.

[55] Mason KL, Erb Downward JR, Falkowski NR, Young VB, Kao JY, Huffnagle GB. Interplay between the gastric bacterial microbiota and *Candida albicans* during postantibiotic recolonization and gastritis. Infect Immun January 2012;80(1):150—8.

[56] Wang ZK, Yang YS, Stefka AT, Sun G, Peng LH. Review article: fungal microbiota and digestive diseases. Aliment Pharmacol Ther April 2014;39(8):751—66.

[57] Karczewska E, Wojtas I, Sito E, Trojanowska D, Budak A, Zwolinska-Wcislo M, et al. Assessment of co-existence of *Helicobacter pylori* and Candida fungi in diseases of the upper gastrointestinal tract. J Physiol Pharmacol Off J Pol Physiol Soc December 2009;60(Suppl. 6):33—9.

[58] Brawner KM, Kumar R, Serrano CA, Ptacek T, Lefkowitz E, Morrow CD, et al. *Helicobacter pylori* infection is associated with an altered gastric microbiota in children. Mucosal Immunol September 2017;10(5):1169—77.

[59] Arnold IC, Dehzad N, Reuter S, Martin H, Becher B, Taube C, et al. *Helicobacter pylori* infection prevents allergic asthma in mouse models through the induction of regulatory T cells. J Clin Investig August 2011;121(8):3088—93.

[60] Arnold IC, Lee JY, Amieva MR, Roers A, Flavell RA, Sparwasser T, et al. Tolerance rather than immunity protects from *Helicobacter pylori*-induced gastric preneoplasia. Gastroenterology January 2011;140(1):199—209.

[61] Mowat C, Williams C, Gillen D, Hossack M, Gilmour D, Carswell A, et al. Omeprazole, *Helicobacter pylori* status, and alterations in the intragastric milieu facilitating bacterial N-nitrosation. Gastroenterology August 2000;119(2):339—47.

[62] Ma K, Baloch Z, He T-T, Xia X. Alcohol consumption and gastric cancer risk: a meta-analysis. Med Sci Monit Int Med J Exp Clin Res January 14, 2017;23:238—46.

[63] Ianiro G, Pizzoferrato M, Franceschi F, Tarullo A, Luisi T, Gasbarrini G. Effect of an extra-virgin olive oil enriched with probiotics or antioxidants on functional dyspepsia: a pilot study. Eur Rev Med Pharmacol Sci 2013;17(15):2085—90.

[64] Takagi A, Yanagi H, Ozawa H, Uemura N, Nakajima S, Inoue K, et al. Effects of *Lactobacillus gasseri* OLL2716 on *Helicobacter pylori*-associated dyspepsia: a multicenter randomized double-blind controlled trial. Gastroenterol Res Pract 2016;2016:7490452.

[65] Lu C, Sang J, He H, Wan X, Lin Y, Li L, et al. Probiotic supplementation does not improve eradication rate of *Helicobacter pylori* infection compared to placebo based on standard therapy: a meta-analysis. Sci Rep March 21, 2016;6:23522.

[66] Sakamoto I, Igarashi M, Kimura K, Takagi A, Miwa T, Koga Y. Suppressive effect of *Lactobacillus gasseri* OLL 2716 (LG21) on *Helicobacter pylori* infection in humans. J Antimicrob Chemother May 2001;47(5):709—10.

[67] Routy B, Le Chatelier E, Derosa L, Duong CPM, Alou MT, Daillère R, et al. Gut microbiome influences efficacy of PD-1-based immunotherapy against epithelial tumors. Science January 5, 2018;359(6371):91—7.

[68] Gopalakrishnan V, Spencer CN, Nezi L, Reuben A, Andrews MC, Karpinets TV, et al. Gut microbiome modulates response to anti-PD-1 immunotherapy in melanoma patients. Science January 5, 2018;359(6371):97—103.

Chapter 11

The Human Vaginal Microbiome

Iara M. Linhares[1,2], Evelyn Minis[2], Renata Robial[1] and Steven S. Witkin[2]

[1]Department of Gynecology and Obstetrics, University of Sao Paulo Medical School, Sao Paulo, Brazil; [2]Division of Immunology and Infectious Diseases, Department of Obstetrics and Gynecology, Weill Cornell Medicine, New York, NY, United States

INTRODUCTION

The vagina is continually exposed to microorganisms present in the external environment, as well as by contamination from the rectum, fingers, and sexual partners. Mechanisms have evolved to prevent this exposure from inducing pathological consequences in both nonpregnant and pregnant women. A major factor in preventing foreign microorganisms from establishing a foothold and proliferating in the female genital tract is the presence of an endogenous vaginal microbiome. As diverse mammalian species have unique exposures to exogenous microorganisms, it is perhaps not surprising that the vaginal microbiome is different in women than in nonhuman primates and laboratory animals, such as mice, rats, and rabbits [1].

The distinctive composition of the human vaginal microbiome may have been influenced by specific behaviors, such as engaging in sexual intercourse at any stage of the menstrual cycle, and even during pregnancy, male promiscuity, as well as dietary factors that influenced the composition of vaginal fluid, and thereby selected for growth of specific endogenous bacterial species [1−3]. An additional factor influencing vaginal microbiome composition is the genetic capacity of the individual to effectively respond immunologically to individual microorganisms or their products [4,5].

Elucidation of the composition of the vaginal microbiome has been rapidly advanced by the development and refinement of gene amplification analyses of bacterial DNA, present in samples obtained from the vagina. The majority of bacteria present at this site are difficult, if not impossible to grow in in vitro culture, and so culture-independent protocols have been essential in defining the composition and relative abundance of bacteria, in individual women, and at different stages of life.

In this chapter we will describe the bacterial composition of the human vaginal microbiome, in both nonpregnant and pregnant reproductive age women, and how this knowledge can be used in development of protocols to assess susceptibility to infection and adverse pregnancy outcome, as well as to devise unique prevention and treatment strategies.

NONPREGNANT WOMEN

Numerous articles have detailed the bacterial composition of the human vagina in healthy asymptomatic reproductive age women [6−8]. The findings have been remarkably consistent. In about 75% of woman, the numerically dominant bacterium is one of four species of *Lactobacilli*, either *L. crispatus*, *L. iners*, *L. gasseri*, or *L. jensenii*. Of these, *L. crispatus* or *L. iners* have been identified in the most women. In the remaining 25% of women, a variety of different bacteria may be dominant, most frequently *Gardnerella vaginalis*, *Atopobium vaginae*, or *Streptococcus* species.

The above percentages are typical of white or Asian women. Only about 50% of women who are black or Hispanic have a vaginal microbiota dominated by *Lactobacilli*. In most women only one species of *Lactobacillus* predominates, with the other *Lactobacilli* being either nondetectable or present at very low levels. The reasons for this exclusivity have not been determined but may be due to variations in the composition of vaginal fluid that offer a selective growth advantage to an individual species.

Also, the stability of the vaginal microbiome during various stages of the menstrual cycle and between different cycles appears to be very heterogeneous. In some women the predominant bacterium varies almost constantly, in others the composition changes only during menses or after sexual intercourse, while in a third group the microbiota is almost invariant at all times. Reasons for these differences are not known, but, strikingly, there is no association between the frequency of fluctuation and vaginal health.

While women with or without a *Lactobacillus*-dominated vaginal microbiota are often seemingly equally healthy, many investigations have noted an association between the absence of a *Lactobacillus*-dominated microbiome and increased susceptibility to a variety of infectious disorders. Bacterial vaginosis (BV) is the name given to a disorder, or perhaps more accurately a group of disorders, characterized by the scarcity or absence of vaginal *Lactobacilli* and their replacement by a heterogeneous group of anaerobic and facultative bacteria, most commonly including *G. vaginalis* and *A. vaginae*.

This alteration in the microbiota may in some, but not all women, be associated with the presence of a discharge and foul-smelling odor [9]. The presence of BV has been associated with increased susceptibility to acquire by sexual contact, human immunodeficiency virus (HIV) [10], human papillomavirus [11], *Chlamydia trachomatis* [12], *Neisseria gonorrhoeae* [12], and postsurgical infections [13].

PREGNANT WOMEN

The prevalence of *Lactobacilli* in the vagina increases during pregnancy, as compared to levels in nonpregnant women [14,15]. This is most likely due to increased estrogen production that promotes elevated levels of glycogen in vaginal epithelial cells. The subsequent release of glycogen into the vaginal lumen and its degradation by host alpha amylase results in enhanced levels of compounds preferentially utilized by *Lactobacilli* for its growth. The predominance of endogenous bacteria in the vagina, especially *Lactobacilli*, may successfully compete with exogenous microorganisms for adherence to the vaginal epithelium and, thereby, prevent initiation of an infection.

It remains unclear if the composition of the vaginal microbiome during pregnancy influences the rate of premature delivery. Two large studies have come to opposite conclusions [16,17]. Several studies have reported associations between BV and an elevated rate of adverse pregnancy outcomes, including preterm birth, spontaneous abortion, low birth weight, and postpartum endometritis [18–20]. However, antibiotic treatment of BV-positive pregnant women was shown to have no effect on the rate of premature delivery relative to a control group treated with a placebo [21].

It is likely that differences in composition of the subject groups between studies influenced the outcome. There clearly are variations in the magnitude and direction of immune responses to specific microorganisms or their products between individual women. Lumping together into a single group, all women with a specific vaginal microbiome may very well hide subgroup variations and lead to inaccurate conclusions.

Recent evidence suggests that the vaginal microbiome in first trimester pregnant women differs depending on pregnancy history [22]. Significant variations in the relative proportions of *L. crispatus*, *L. iners*, and *G. vaginalis* were noted between women with a first conception, those with a prior spontaneous or induced pregnancy loss but no deliveries, and women with a prior delivery. This is perhaps not surprising, since a prior conception or delivery results in physical alterations in the genital tract that alter the vaginal environment. In addition, the magnitude of immune responses varies between women with a first or subsequent conception. These changes undoubtedly result in a unique vaginal milieu in each situation and, therefore, would confer a selective growth advantage to different groups of bacteria. Therefore, studies that seek to determine relationships between the vaginal microbiome and pregnancy outcome must be cognizant of their subjects' pregnancy history.

INFLUENCE OF THE VAGINAL MICROBIOME ON MOTHER TO NEWBORN BACTERIAL TRANSMISSION

Vaginal delivery results in transmission of the mother's vaginal microbiota to her babies. In contrast babies born via cesarean section, especially those without a trial of labor, are predominantly colonized with bacteria from the mother's skin [23]. It remains somewhat controversial, whether this difference in initial colonization has any lasting impact on gastrointestinal colonization or in priming neonatal immune responses. A recent comprehensive study has demonstrated that by 6 months of age there were no differences in a babies' microbiome at multiple body sites between those born vaginally and by cesarean section [24].

LACTIC ACID

Lactic acid is the predominant acid in the vagina and responsible for the vaginal pH in most healthy women to be <4.5. The majority of the lactic acid comes from the anaerobic glycolysis of glycogen degradation products by *Lactobacilli*, as well as by other lactic acid—producing bacteria. The vaginal epithelial cells also release a relatively minor amount of lactic acid into the vaginal lumen. *L. crispatus, L. gasseri, and L. jensenii* each produce both the D and L isomeric form of lactic acid. In contrast, *L. iners*, other bacteria, and epithelial cells only produce the L isomer [4,25,26].

Unlike similar acidic compounds, lactic acid possesses unique immunological properties that downregulate proinflammatory immune responses. This has been most elucidated in the cancer literature, where production of lactic acid by tumors has been shown to decrease T lymphocyte activity, inhibit the production of proinflammatory cytokines, and induce macrophages to revert to an antiinflammatory phenotype (summarized in Ref. [27]). It has also been demonstrated that lactic acid production by bacteria present in the gastrointestinal tract reduces local proinflammatory immunity [28].

It is, therefore, reasonable to project that lactic acid in the vagina similarly functions to diminish the magnitude and facilitate the resolution of inflammatory responses at this site that may be induced due to daily exposure to microorganisms and chemicals in the external environment. Especially during pregnancy, inflammation may interfere with the programmed series of events needed for fetal maturation, as well as induce a premature termination of gestation [5,29]. Thus, the ability of lactic acid to inhibit or kill potential microbial pathogens, while concomitantly blocking inflammation, highlights its utility as a key factor to promote successful pregnancy outcomes.

Addition of D lactic acid by vaginal *Lactobacilli* further contributes to well-being. This isomer regulates the rate of lactic acid flow into and out of cells by modulating the activity of extracellular matrix metalloproteinase inducer (EMMPRIN), an essential cofactor for monocarboxylate transporter activity [26]. As its name implies, EMMPRIN also induces production of metalloproteinase enzymes that degrade the cellular matrix and facilitate bacterial migration from the lower to the upper genital tract [26]. Thus, the regulation of EMMPRIN production by D-lactic acid facilitates epithelial cell survival in an acidic environment by preventing an intracellular acid buildup and by downregulating deleterious matrix metalloproteinase production.

MICROBIOME—EPITHELIAL CELL INTERACTIONS

Recent investigations have revealed unique attributes of *L. crispatus* that also contribute to vaginal health. In a study of 154 pregnant women, vaginal samples obtained in their first trimester demonstrated that the stress response in epithelial cells, as determined by measurement of the 70 kDa heat shock protein (hsp70), was lowest when *L. crispatus* was the numerically dominant member of the vaginal microbiome [30]. Hsp70 levels were highest when *Streptococci* or *Bifidobacteria* predominated. Similarly, the level of autophagy in the vaginal epithelial cells was highest when *L. crispatus* predominated and lowest in the presence of *Streptococci* or *Bifidobacteria* dominance.

Autophagy is an essential process that maintains an optimal intracellular environment. Defective or aged mitochondria, aggregated or dysfunctional proteins, as well as bacteria or viruses that have entered the cell cytoplasm, are engulfed by a double membrane structure called an autophagosome. Subsequent fusion of the autophagosome with a lysosome leads to destruction of the engulfed components [31]. Thus, it appears that, at least in pregnancy, *L. crispatus* is the bacterium that is best tolerated by the vaginal epithelial cells and interferes the least with autophagy.

A comparison of epithelial cell products in vaginal secretions, when *L. crispatus* or *L. iners* predominated in the vaginal microbiome, demonstrated that the concentration of multiple compounds that serve protective functions was elevated in the presence of *L. iners* [32]. This demonstrates the dynamic nature of the vaginal epithelium and its ability to respond to potentially adverse conditions.

STRESS

Stress potentiates a loss of *Lactobacilli* in the intestinal microbiome in humans, nonhuman primates, and mice [28,33]. Although not systematically studied in the vaginal microbiome, norepinephrine, the hormone released in response to exposure to stressful stimuli has been shown to bind to receptors on vaginal epithelial cell lines and to potentiate the release of proinflammatory mediators in the concomitant presence of immune system activators [34]. Thus, there is a strong likelihood that a stress response would lead to a diminution in vaginal *Lactobacilli* in some women, resulting in a concomitant increase in the concentration of other bacteria and vaginal inflammation. Variations in the magnitude of a stress response to various stimuli between individual women would be expected to result in differences in the composition of the vaginal microbiota. A more thorough evaluation of this effect, and its potential influence on susceptibility to disease acquisition or pregnancy outcome, is warranted.

CAN VARIATIONS IN THE VAGINAL MICROBIOME BE EASILY DETECTED?

If the composition of the vaginal microbiome is a major contributor to health, in both pregnant and nonpregnant women, can variations that increase susceptibility to disease be identified and corrected? Clearly utilization of gene amplification technology to characterize the vaginal microbiome is not applicable to the overwhelming majority of women, and even tests requiring microscopy or laboratory facilities are beyond the reach of women in disadvantaged cultures. It has been suggested that pregnant women can determine their own vaginal pH, if provided with a pH indicator strip or a specially constructed glove, and those with a pH > 5.0 are assumed to have a low level of lactic acid and, therefore, a deficiency of vaginal Lactobacilli [35].

A problem with this approach is that both D- and L-lactic acid contribute to vaginal pH, and bacteria that have been associated with vaginal dysbiosis, such as *L. iners* and *Streptococcus* species, are potent L-lactic acid producers [26]. Vaginal epithelial cells also release L-lactic acid into the vaginal lumen [25]. Thus, mere vaginal pH determination will provide only a poor assessment of the vaginal microbiome composition. A more recent suggestion focuses on detection of the D-lactic acid isomer in vaginal secretions [5,26,36]. Since D-lactic acid is produced in the vagina, primarily by *L. crispatus*, *L. jensenii*, and *L. gasseri*, then a low or absent D-lactic acid level should be a sensitive indication of whether or not the vaginal microbiome is dominated by one of these *Lactobacillus* species. This testing might be especially beneficial for pregnant women, where the lack of a *Lactobacillus*-dominated microbiota, or the predominance of *L. iners*, has been associated with adverse pregnancy outcomes [37]. Testing this possibility awaits the development of a low-cost device that can be utilized in low resource settings to detect or semiquantitate vaginal D-lactic acid levels.

CAN A WOMAN'S VAGINAL MICROBIOME BE ALTERED?

If women with potentially less than ideal vaginal microbiomes can be readily detected, the next step is manipulation to improve on a long-term basis the composition of their vaginal microbiota. Women with BV are commonly treated with antibiotics in an attempt to selectively kill the non-Lactobacilli and return to a *Lactobacillus*-dominated microbiota. However, numerous studies have shown that most treated women revert again to a BV-dominated microbiome when the antibiotics are discontinued [38,39]. It appears that innate or environmental factors, or most likely a combination of both, are major determinants of the vaginal microbiota composition, and thus it is difficult to achieve a permanent alteration.

A major industry on the use of probiotics to alter the microbiome at different body sites is rapidly expanding. Studies on the short-term efficacy of various probiotic preparations in treating BV, some sponsored by the company producing the preparation, appear with increasing frequency in the medical literature. The problem with oral preparations is the transport of sufficient concentrations of ingested bacteria to the vagina to have a desired effect. Migration of exogenous bacteria from the gastrointestinal tract to the rectum and eventually to the vaginal canal occurs with a variable frequency in individual women, depending on personal hygiene practices as well as differences in immune recognition and intestinal physiology.

A second problem with oral probiotics, as well as with vaginal preparations, is the ability of the exogenous bacterium to successfully compete with endogenous microorganisms, to adhere to the vaginal epithelium for a prolonged period of time, to be tolerated by the vaginal epithelium, and to have a selective advantage to proliferate at that site [36]. Given these limitations, to date there has been no consensus that a particular probiotic preparation has long-term efficacy in altering the composition of the vaginal microbiome and positively influencing vaginal health. A potentially promising possibility is that, in addition to providing the desired bacterium, a prebiotic is concomitantly introduced to improve the likelihood of selective bacterial maintenance and proliferation. Possible prebiotics might include glycogen [5,40] and alpha amylase to provide additional substrates favored by *Lactobacilli* for their replication, along with low levels of progesterone or other compounds that promote glycogen deposition on the surface of vaginal epithelial cells, or exogenous D-lactic acid to inhibit growth of bacteria other than *Lactobacilli* [26,36].

WHAT ABOUT THE HOST?

An important point not usually acknowledged is that host variables, and not only the vaginal bacterial composition, exert a major influence on the consequences of bacterial presence. A particular microbiome may be innocuous to some women but detrimental to the health of other women. A specific combination of bacteria will be a risk factor for pathology only in those women who possess an inadequate immunological, biochemical, or hormonal repertoire to deal with the particular microorganisms. We have proposed that the migration of groups of people to vastly different environments throughout evolution might have resulted in the creation of conditions where specific genetic variants that were favorable to survival and reproductive efficacy in the original environment are now detrimental to dealing with the new altered microbiota [41]. Microbe-rich or microbe-poor environments, due to differences in access to clean water, safe food storage, and hygiene practices, may promote the selection of different gene polymorphisms that best promote survival and fecundity under each condition. The persistence of these genetic differences to the present time perhaps contributes to the observed variations in bacterial predominance in the vaginal microbiota. This might account, at least in part, in the observed distinction in the composition of the vaginal microbiome between white and black women [6,7].

The composition of the vaginal microbiome in first trimester pregnant women is also influenced by their pregnancy history. *Lactobacilli* are more prominent in the vagina of women with their first conception than in those who have had a previous pregnancy [22]. The composition of vaginal secretions, and therefore the relative amount of substrates favoring the growth of selective bacterial species appears to be altered depending on pregnancy history. Clearly, attempts at manipulating the microbiome without a consideration of host factors that may also require modification is likely to be ineffective, at least for a subpopulation of women.

REFERENCES

[1] Witkin SS, Ledger WJ. Complexities of the uniquely human vagina. Sci Transl Med 2012;4:132fs11.

[2] Yildirim S, Yeoman J, Janga SC, Thomas SM, Ho M, Leigh SR, et al. Primate vaginal microbiomes exhibit species specificity without universal Lactobacillus dominance. ISME J 2014;8:2431—44.

[3] Stumpf RM, Wilson BA, Rivera A, Yildirim S, Yeoman CJ, Polk JD, et al. The primate vaginal microbiome: comparative context and implications for human health and disease. Am J Phys Anthropol 2013;152:119—34.

[4] Witkin SS, Linhares IM. Why do lactobacilli dominate the human vaginal microbiota? BJOG 2017;124:606—11.

[5] Witkin SS. The vaginal microbiome, vaginal anti-microbial defence mechanisms and the clinical challenge of reducing infection-related preterm birth. BJOG 2015:213—9.

[6] Ravel J, Gajer P, Abdo Z, Schneider GM, Koenig SSK, McCulle SL, et al. Vaginal microbiome of reproductive-age women. Proc Natl Acad Sci USA 2011;108:4680—7.

[7] Gajer P, Brotman RM, Bai GY, Sakamoto J, Schuette UM, Zhong X, et al. Sci Transl Med 2012;4:132ra52.

[8] Lamont RR, Sobel JD, Akins RA, Hassan SS, Chaiworopongsa T, Kusanovic JP, et al. The vaginal microbiome: new information about genital tract flora using molecular based techniques. BJOG 2011;118:533—49.

[9] Nasioudis D, Linhares IM, Ledger WJ, Witkin SS. Bacterial vagnosis: a critical analysis of current knowledge. BJOG 2017;124:61—9.

[10] Atashili J, Poole C, Ndumbe PM, Adimora AA, Smith JS. Bacterial vaginosis and HIV acquisition: a meta-analysis of published studies. AIDS 2008;22:1493—501.

[11] Gillet E, Meys JF, Verstraelen H, Bosire C, De Sutter P, Temmerman M, et al. Bacterial vaginosis is associated with uterine cervical human papillomavirus infection: a meta-analysis. BMC Infect Dis 2011;11:10.

[12] Gallo MF, Macaluso M, Warner L, Fleenor ME, Hook EW, Brill I, et al. Bacterial vaginosis, gonorrhea, and chlamydial infection among women attending a sexually transmitted disease clinic: a longitudinal analysis of possible causal links. Ann Epidemiol 2012;22:213—20.

[13] McElligott KA, Havnilesky LJ, Myers ER. Preoperative screening strategies for bacterial vaginosis prior to elective hysterectomy: a cost comparison study. Am J Obstet Gynecol 2011;205:500e1—7.

[14] Romero R, Hassan SS, Gajer P, Tarca AL, Fadrosh DW, Nikita L, et al. The composition and stability of the vaginal microbiota of normal pregnant women is different from that of non-pregnant women. Microbiome 2014;2:4.

[15] Mcintyre DA, Chandiramoni M, Lee SY, Kindinger L, Smith A, Angelopoulos N, et al. The vaginal microbiome during pregnancy and the postpartum period in a European population. Sci Rep 2015;5:8988.

[16] DiGiulio DB, et al. Temporal and spatial variation of the human microbiota during pregnancy. Proc Natl Acad Sci USA 2015;112:11060—5.

[17] Romero R, et al. The vaginal microbiota of pregnant women who subsequently have a spontaneous preterm labor and delivery and those with a normal delivery at term. Microbiome 2014;2:18.

[18] Laxmi U, Agrawal S, Raghunandan C, Randhawa VS, Salli A. Association of bacterial vaginosis with adverse fetomaternal outcome in women with spontaneous preterm labor: a prospective cohort study. J Matern Fetal Neonatal Med 2012;25:64—7.

[19] Giakoumelou S, Wheelhouse N, Cushieri K, Entrican G, Howie SE, Howe AW. The role of infection in miscarriage. Hum Reprod Update 2016;22:116—33.

[20] Watts DH, Krohn MA, Hillier SL, Eschenbach DA. Bacterial vaginosis as a risk factor for post-cesarean endometritis. Obstet Gynecol 1990;75:52—8.

[21] Carey JC, Klebanoff MA, Hauth JC, Hillier SL, Thorn EA, Ernest JM, et al. Metronidazole to prevent preterm delivery in pregnant women with asymptomatic bacterial vaginosis. N Eng J Med 2000;342:534—40.

[22] Nasioudis D, Forney LJ, Schneider GM, Gliniewicz K, France M, Boester A, et al. Influence of pregnancy history on the vaginal microbiome of pregnant women in their first trimester. Sci Rep 2017;7:10201.

[23] Dominguez-Bello MG, et al. Delivery mode shapes the acquisition and structure of the initial microbiota across multiple body habitats in newborns. Proc Natl Acad Sci USA 2010;107:11971—5.

[24] Chu DM, Ma J, Prince AL, Antony KM, Seferovic MD, Aagaard KM. Maturation of the infant microbiome community structure and function across multiple body sites and in relation to mode of delivery. Nature Med 2017;23:314−28.

[25] Boskey ER, Cone RA, Whaley KJ, Moench TR. Origins of vaginal acidity: high D/L lactic acid ratio is consistent with bacteria being the primary source. Hum Reprod 2001;16:1809−13.

[26] Witkin SS, Mendes-Soares H, Linhares IM, Jayaram A, Ledger WJ, Forney LJ. Influence of vaginal bacteria and D- and L- lactic acid isomers on vaginal extracellular matrix metalloproteinase inducer: implications for protection against upper genital tract infections. mBio 2013;4. e00460−13.

[27] Witkin SS. Lactic acid alleviates stress: good for female genital tract homeostasis, bad for protection against malignancy. Cell Stress Chaperones 2017. https://doi.org/10.1007/s12192-017-0852-3.

[28] Galle JD, Bailey MT. Impact of stressor exposure on the interplay between commensal microbiota and host inflammation. Gut Microb 2014;5:390−6.

[29] Gravett MG, Witkin SS, Haluska GJ, Edwards JL, Cook MJ, Novy MJ. An experimental model for intraamniotic infection and preterm labor in rhesus monkeys. Am J Obstet Gynecol 1994;171:1660−7.

[30] Nasioudis D, Forney LJ, Schneider GM, Gliniewicz K, France MT, Boester A, et al. The composition of the vaginal microbiome in first trimester pregnant women influences the level of autophagy and stress in vaginal epithelial cells. J Reprod Immunol 2017;123:35−9.

[31] Wang C-W, Klionsky DJ. The molecular mechanism of autophagy. Mol Med 2003;9:65−76.

[32] Leizer J, Nasioudis D, Forney LJ, Schneider GM, Gliniewicz K, Boester A, et al. Properties of epithelial cells and vaginal secretions in pregnant women when *Lactobacillus crispatusor Lactobacillus iners* dominate the vagina microbiome. Reprod Sci 2018;25:854−60.

[33] Jasarevic E, Howerton CL, Howard CE, Bale TL. Alterations in the vaginal microbiome by maternal stress are associated with metabolic reprogramming of the offspring gut and brain. Endocrinol 2015;156:3265−76.

[34] Brosnahan AJ, Vulchanova L, Witta SR, Dai Y, Joones BJ, Brown DR. Norepinephrine potentiates proinflammatory responses of human vaginal epithelial cells. J Neuroimmunol 2013;259:8−16.

[35] Hoyme UB, Saling E. Efficient prematurity prevention is possible by pH-self measurement and immediate therapy of threatening ascending infection. Eur J Obstet Gynecol Reprod Biol 2004;115:148−53.

[36] Tachedjian G, Aldunate M, Bradshaw CS, Cone RA. The role of lactic acid production by probiotic Lactobacillus species in vaginal health. Res Microbiol 2017. https://doi.org/10.1016/j.resmic.2017.04.001.

[37] Maclaim JM, Fernandes AD, DiBella JM, Hammond JA, Reid G, Gloor GB. Comparative meta-RNA-seq of the vaginal microbiota and differential expression by *Lactobacillus iners* in health and dysbiosis. Microbiome 2013;1:12−23.

[38] Ravel J, Brotman RM, Gajer P, Ma B, Nandy M, Fadrosh DW, et al. Daily temporal dynamics of vaginal microbiota before, during and after episodes of bacterial vaginosis. Microbiome 2013;1:29.

[39] Bradshaw CS, Brotman RM. Making inroads into improving treatment of bacterial vaginosis − striving for long-term cure. BMC Infect Dis 2015;15:292.

[40] Mirmonsef P, Hotton AL, Gilbert D, Burgad D, Landay A, Weber KM, et al. Free glycogen in vaginal fluids is associated with Lactobacillus colonization and low vaginal pH. PLoS One 2014;9:e102467.

[41] Jaffe S, Normand N, Jayaram A, Orfanelli T, Doulaveris G, Passos M, et al. Unique variation in genetic selection among Black North American women and its potential influence on pregnancy outcome. Med Hypotheses 2013;81:919−22.

Chapter 12

Bioactive Molecules of the Human Microbiome: Skin, Respiratory Tract, Intestine

Heidi Pauer[1], Thaís Glatthardt[2], Nicole V. Ferreira[3], Rosana B.R. Ferreira[2] and L. Caetano M. Antunes[1,3]

[1]National Institute of Science and Technology of Innovation on Diseases of Neglected Populations, Center for Technological Development in Health, Oswaldo Cruz Foundation, Rio de Janeiro, Brazil; [2]Institute of Microbiology, Federal University of Rio de Janeiro, Rio de Janeiro, Brazil; [3]National School of Public Health Sergio Arouca, Oswaldo Cruz Foundation, Rio de Janeiro, Brazil

NATURAL PRODUCTS AS SOURCES OF BIOLOGICAL ACTIVITY

For the last several decades, scientists have continuously searched for molecules with specific biological activities. Since the discovery of penicillin, the success of medical practice has become at least partly dependent upon the discovery of such active compounds. Although the sources of these compounds are varied, including plants, marine animals, bacteria, and fungi, the vast majority of bioactive molecules has been obtained from microorganisms [1,2]. Given their sources, these molecules have been historically described as "natural products," a loose term used to define relatively small molecules associated with some sort of biological activity.

The importance of natural product discovery for medicine is clearly exemplified by the number of compounds approved by the United States Food and Drug Administration (FDA) that fall within this group. Until 2013, the FDA had approved just over 1400 molecular entities, and almost half of these were natural products [3,4]. Since the discovery of penicillin, over 23,000 natural products have been characterized. Actinomycetes are the most prolific producers, with some strains being able to produce up to 50 of these compounds [1].

At the start of the natural product research era, there was an insatiable search for new antibiotics. In what was called the "golden age of antibiotic discovery," these molecules were relatively easy to find, and simple laboratory experiments using batch cultures of the producers, associated with easy screening methods, allowed the identification of several classes of antibiotics in relatively simple microbiology laboratories [2]. Soon after, the pharmaceutical industry joined these efforts, and there was a fast development in the discovery rate of natural products, as well as the production of semisynthetic derivatives with improved activity.

As these searches began to dry, more sophisticated and expensive methods to screen complex libraries of synthetic compounds became the primary discovery mode, and discovery through natural product research lost strength. Incidentally, the reduced interest in the development of new antibiotics came as drug-resistant strains of the most important human pathogens arose and, in some cases, became widely prevalent [5]. As a consequence, the need for the discovery of new compounds with biological activity is urgent. In this chapter we present accumulating evidence that the communities of microbes inhabiting the human body may hold the key to the discovery of such needed compounds.

THE HUMAN MICROBIOME AS A BLACK BOX FOR NATURAL PRODUCT DISCOVERY

The realization that microbes inhabit our bodies dates back to the 17th century, when Antonie van Leeuwenhoek scrapped the surface of his teeth with a toothpick, and first observed oral microbes under a microscope lens. Although this was the first indication that microbes can peacefully coexist with their hosts, this observation was left unexplored for several decades. For most of the following centuries, the work of medical doctors and microbiologists focused on the microbes that

Microbiome and Metabolome in Diagnosis, Therapy, and other Strategic Applications. https://doi.org/10.1016/B978-0-12-815249-2.00012-9

cause disease in humans. Microbial cultivation and identification methods were developed, allowing the study of these microbes in the laboratory.

Just a few decades ago, in the 1970s, the importance of the microbes that live in and on us was revisited, and studies showed that disruption of these resident microbial communities was detrimental to health [6,7]. However, it was not until the mid-1990s that the relationship between the human host and the microbes it carries started to be studied in molecular detail.

For several years, most of the work in the field of microbiota research was performed in the laboratory of Jeffrey I. Gordon, at Washington State University in St. Louis (USA). In one of their first and most important publications on the subject, Gordon's group studied conventionally raised and germ-free animals, and showed that fucosylation of the intestinal epithelium was altered in the absence of the microbiota [8]. Additionally, colonization of germ-free animals with *Bacteroides thetaiotaomicron*, a prominent member of the human gut microbiota, restored fucosylation, showing that a single component of this complex ecosystem produced significant effects on the host, and that these effects could be studied in the laboratory. This sparked significant interest in the microbiota, and there are now hundreds of laboratories throughout the world dedicated to studying the human microbiome. We now know who these microbes are, where they are, and what they do [9].

Cellular interactions often occur through the production of diffusible molecules that act as signals. This is true for eukaryotic, multicellular organisms, which have been known to use hormones for signaling for over a century, and it is also true for microbes. Bacteria are now known to produce and respond to various signaling molecules [10,11]. Although we have learned a great deal about these phenomena, most of these studies have been performed with microbes grown in pure cultures. In their natural setting, microbes live in complex, multispecies communities, and many other signaling events are likely to occur [12].

Our group has shown that the mammalian gut and its microbiota contain thousands of small molecules, that most of these molecules are currently unknown, and that some of them have readily detectable biological activity [13−15].

BIOACTIVE MOLECULES OF THE SKIN MICROBIOME

Since the skin is directly exposed to the environment, it encounters a large number of microorganisms, many of which will only interact briefly with the host and will be quickly removed. Others, however, may have specialized tools to colonize this niche and overcome its defense barriers to cause disease. In fact, the skin is considered the first barrier to microbial pathogens, in the line of host defense [16]. The microbial composition of the skin microbiome is determined by several environmental and physiological parameters, such as the anatomical site, local humidity, hormone production, and the distribution of sweat and sebaceous glands [17]. *Staphylococcus*, *Propionibacterium*, *Micrococcus*, and *Corynebacterium* are, in general, the most abundant members of the skin microbiome [18].

In an analogous way as is described for the gut microbiome, it is believed that the skin microbiome is essential for protection against microbial pathogens. This protection may happen due to different mechanisms, such as competition for nutrients and colonization sites, immune system activation, or production of antibacterial molecules [19,20].

Staphylococcus

A study performed with *Staphylococcus* strains isolated from the nasal microbiota showed that 84% of the strains tested could produce antimicrobial molecules, with varied spectra of activity. Since *Staphylococcus* species are some of the most abundant microbes in the skin microbiota [17,21], it is possible that these bacteria produce molecules that protect the skin against the colonization by pathogens. *Staphylococcus epidermidis* is one of the most frequently isolated species from the skin, and is considered to be ubiquitous in healthy individuals [21].

Several *S. epidermidis* bacteriocins, proteins with antibacterial activity against closely related bacterial species, have been described, including epidermin, Pep5, epicidin 280, and epilancin K7, with activity against a broad spectrum of Gram-positive bacteria [20,22−24] (Table 12.1). Bacteriocins produced by other *Staphylococcus* species commonly found in the human skin microbiota have also been discovered. *Staphylococcus hominis* has been shown to produce hominicin, a bacteriocin with potent activity against clinically relevant strains of *Staphylococcus aureus*, including MRSA (methicillin-resistant *S. aureus*) and VISA (vancomycin-intermediate *S. aureus*) [25]. The activity of hominicin against these *S. aureus* strains is of great importance, due to the emergence of strains with low susceptibility to vancomycin in recent years [26].

Staphylococcus warneri was also shown to produce molecules with antimicrobial activity, including the bacteriocin nukacin ISK-1 [27] and warnericin RK, which was the first antibacterial peptide with activity against *Legionella* [28,29].

TABLE 12.1 Bioactive Compounds of the Human Microbiome

Compound	Microbiome	Microorganism	Activity	References
Epidermin	Skin	S. epidermidis	Antimicrobial activity, Gram-positives	[20,22]
Pep5	Skin	S. epidermidis	Antimicrobial activity, Gram-positives	[20,22]
Epicidin 280	Skin	S. epidermidis	Antimicrobial activity, Gram-positives	[20,23]
Epilancin K7	Skin	S. epidermidis	Antimicrobial activity, Gram-positives	[20,24]
Hominicin	Skin	S. hominis	Antimicrobial activity, drug-resistant S. aureus	[25]
Nukacin ISK-1	Skin	S. warneri	Antimicrobial activity	[27]
Warnericin RK	Skin	S. warneri	Antimicrobial activity, Legionella	[28,29]
Phenol-soluble modulins	Skin	S. epidermidis	Antimicrobial activity, S. aureus and S. pyogenes	[30,31]
Short-chain fatty acids	Skin	P. acnes	Reduced S. aureus colonization of skin wounds	[37]
Acnecin	Oral	P. acnes	Bacteriostatic activity, Gram-positives and Gram-negative anaerobes	[38]
Serine protease	Respiratory tract	S. aureus	Activation of hemagglutinin from influenza virus	[56,57]
Hydrogen peroxide	Respiratory tract	S. pneumoniae	Antimicrobial activity, S. aureus	[58]
Esp	Nares	S. epidermidis	Antibiofilm activity, inhibits nasal colonization by S. aureus	[63]
Lugdunin	Nares	S. lugdunensis	Antimicrobial activity, drug-resistant S. aureus and Enterococcus	[64]
3,4-dimethylbenzoic acid	Gut	C. citroniae	Antivirulence activity, S. enterica	[14,15]
Dimethyl sulfide	Gut	Intestinal microbiota	Antivirulence activity, S. enterica	[83]
Acetate	Gut	Bifidobacterium	Antimicrobial activity, E. coli O157:H7	[86]
Autoinducer 2	Gut	R. obeum	Inhibits V. cholerae intestinal colonization	[91]
Deoxycholate	Gut	C. scindens C. hylemonae	Stimulates C. difficile spore germination and inhibits growth	[93]
Ruminococcin A	Gut	R. gnavus C. nexile	Antimicrobial activity, Gram-positives	[92]
Heat-stable enterotoxin linaclotide	Gut	E. coli	Treatment of constipation associated with irritable bowel syndrome	[97]
Microcins	Gut	E. coli	Narrow spectra antimicrobials, Gram-negatives	[98]

S. epidermidis isolates from the skin can also secrete phenol-soluble modulins (PSMγ and PSMδ), which have been described to inhibit the growth of skin pathogens, such as *S. aureus* and *Streptococcus pyogenes*, by forming pores on their cytoplasmic membrane [30,31].

The role of *S. epidermidis* in protecting the skin against pathogens has been discussed in atopic dermatitis patients. In such individuals, the presence of *S. epidermidis* can be associated with healing of the skin lesions, whereas the presence of *S. aureus* can be associated with exacerbations [32]. Recently, a study demonstrated that administration of coagulase-negative *Staphylococcus* strains, including *S. epidermidis* and *S. hominis*, suppresses *S. aureus* colonization on patients with atopic dermatitis, and that this activity is associated with the production of antimicrobial peptides [33].

Regarding the antagonism between *S. aureus* and *S. epidermidis*, another hypothesis has been raised considering the *agr* locus. This locus encodes a quorum sensing system that responds to secreted auto-inducing peptides (AIPs), which are in turn sensed by AgrC, a two-component histidine kinase. This system is responsible for the regulation of virulence in *Staphylococcus* spp., and it is known that the AIP secreted by *S. epidermidis* is a potent inhibitor of the *S. aureus agr* system [34]. As *agr* is the major regulatory system of *S. aureus* virulence, AIP secreted by *S. epidermidis* could be considered an antivirulence molecule, and could assist in the search for new therapeutic strategies against this pathogen.

Propionebacterium acnes

Another major member of the resident skin microbiota is *Propionibacterium acnes*. This species can produce bacteriocins and "bacteriocin-like" compounds, and these will affect its interactions with surrounding microbes [35,36]. Additionally, *P. acnes* can produce propionic acid through fermentation, and this short-chain fatty acid (SCFA) is known to have antimicrobial activity [37]. In this study, Shu and colleagues have shown that the fermentation of glycerol present on the skin by *P. acnes* leads to the production of various SCFAs. Interestingly, administration of *P. acnes* fermentation products to laboratory animals resulted in a reduction of MRSA colonization of skin wounds [37].

The authors also described that SCFAs resulting from glycerol fermentation by *P. acnes* lowered the pH of the environment, and resulted in acidification of the intracellular milieu of *S. aureus*. This could be one of the mechanisms by which *P. acnes* fermentation products inhibit *S. aureus* development in vivo.

Regarding bacteriocin activity, an oral isolate of *P. acnes* has been shown to produce a substance called acnecin, with bacteriostatic activity against both Gram-positive and Gram-negative anaerobes [36], while a study involving genomic and transcriptomic analyses showed that a subset of *P. acnes* strains harbor the gene cluster involved in bacteriocin/lanthionine biosynthesis [38]. These results were found by searching for genes involved in the biosynthesis of thiazolyl peptides, antibiotics that act as inhibitors of protein synthesis in Gram-positive bacteria.

P. acnes also produces molecules with antibacterial activity against *S. epidermidis*, probably involved in the competition between these two members of the microbiota to maintain skin balance [39]. However, the bioactive molecules associated with this activity have not yet been characterized. Recently, a novel enzyme produced by *P. acnes* isolated from the skin, with antioxidant activity, was described [40]. Considering that many skin diseases, including psoriasis and atopic dermatitis, are related to oxidative stress [41], and that a lower abundance of *P. acnes* was shown in skin lesions of individuals with psoriasis [42], the presence of the antioxidant enzyme RoxP secreted by *P. acnes* may be important for maintaining the redox homeostasis and the health status of the skin.

Another enzyme produced by *P. acnes* that is known for its benefits to the host is a polyunsaturated fatty acid isomerase, which catalyzes the isomerization of linoleic acid to 10,12-conjugated linoleic acid (CLA) [43]. Some CLAs have been related to carcinogenesis inhibition [44], modulation of immune response [45], modulation of insulin tolerance, and regulation of body fat gain in animals [46].

Corynebacterium

Besides *Propionibacterium*, several species of *Corynebacterium* are also commonly isolated from the skin, and the importance of their secreted molecules for maintaining skin health is of current interest. One work showed that *Corynebacterium jeikeium* encoded a hypothetical protein (AucA) that has 66% similarity with the bacteriocin lacticidin Q, produced by *Lactococcus lactis* [47]. Nevertheless, its bacteriocin activity still needs to be investigated, both in vitro and in vivo. Recently, the effect of *Corynebacterium* on *S. aureus* virulence has been studied, and the results suggest that *Corynebacterium* may be important to prevent infections by this pathogen [48]. When exposed to *Corynebacterium striatum*, *S. aureus* exhibited increased adhesion to epithelial cells and decreased hemolysin activity, reflecting an attenuation of virulence, what the authors considered to be a transition to a commensal state. When tested in vivo, *S. aureus* displayed diminished fitness during coinfection with *C. striatum*, when compared to a monoinfection. Similar to what is seen regarding the antagonism between *S. epidermidis* and *S. aureus*, the reduction of *S. aureus* virulence by *C. striatum* seems to be related to *agr* inhibition, but the effective molecule has yet to be characterized.

BIOACTIVE MOLECULES OF THE RESPIRATORY TRACT MICROBIOME

As is the case with other body sites, the composition of the microbiome of the respiratory tract has a strict connection with one's birth. Some of the first microbial colonizers of the respiratory tract originate from the mother's skin (*Staphylococcus*, *Streptococcus*, and *Dolosigranulum*), as well as the vaginal tract (*Lactobacillus*, *Prevotella*, *Atopobium*, and *Sneathia*) [49]. Additionally, colonization by some other members can be associated with breastfeeding, as is the case of *Moraxella* and *Corynebacterium* [50]. In adults, the respiratory tract microbiome is composed of a diverse set of microorganisms, which includes strictly commensal species, and some potentially pathogenic bacteria, including *S. pneumoniae*, *S. aureus*, *Haemophilus influenza*, and *Moraxella catarrhalis* [51].

Although most of the current knowledge of the respiratory tract microbiome is related to bacterial colonizers, there is also a high prevalence of viruses in the respiratory tract. In fact, some studies have shown a correlation between the bacteria and viruses inhabiting the respiratory tract and the outcome of diseases [52]. For instance, infections with *S. pneumoniae* have increased risk of complications when associated with an infection by the influenza A virus. This was observed when the number of cases of bacterial pneumonia considerably increased during an influenza A epidemic [53].

Since then, several studies describing associations between influenza infection and colonization by bacterial pathogens, such as *S. pneumoniae*, *S. aureus,* and *H. influenzae*, have been reported [52,54,55]. At least in some of these cases, bioactive molecules produced by bacteria are involved in this interaction. For instance, *S. aureus* produces a serine protease capable of promoting the activation of hemagglutinin from the influenza virus, and consequently increase the number of viral particles entering the host and causing the disease [56,57]. Direct inhibition of competing microbes through the production of antimicrobial compounds is also described, and can be exemplified by the production of hydrogen peroxide by *S. pneumoniae*, which has a bactericidal effect on *S. aureus*. Both of these bacteria are common causes of respiratory tract infections, as well as infections at other body sites.

Confirming the ecological role of this antagonism, epidemiological data point to an inverse association between these two species, with *S. pneumoniae* inhibiting *S. aureus* nasal colonization [58]. With this hypothesis in hand, it is important to fathom the possibility that pneumococcal vaccines could suppress the protective effect exerted by *S. pneumoniae* against *S. aureus*. In fact, an increase of *S. aureus* acute otitis media in patients vaccinated with the pneumococcal vaccine has been observed [59]. The protective factor in this case was hydrogen peroxide, a byproduct of the pyruvate oxidase activity, as part of aerobic metabolism of *S. pneumoniae* [58]. Although *S. aureus* produces catalase, an enzyme that degrades hydrogen peroxide, it is not able to survive at the concentrations secreted by *S. pneumoniae*.

S. aureus Carriers

Although *S. aureus* is considered a pathogen, it can also be found in the skin and nasal cavity of asymptomatic individuals, who, in this case, are considered carriers [60]. Roughly one-third of the people are asymptomatic carriers of *S. aureus*, and the carriage state is a risk factor for infections caused by this pathogen [61,62]. Although the reason why the majority of people are not colonized by *S. aureus* is not completely understood, given the role of the microbiota in colonization resistance against enteric pathogens, established since the early seventies [6], microbiota-mediated antagonism could also play a role in pathogen colonization of the respiratory tract.

Unlike *S. aureus*, *S. epidermidis* is a true commensal found ubiquitously in the skin and nasal cavity of healthy individuals. In order to investigate if *S. epidermidis* colonization may lead to a diminished *S. aureus* nasal colonization in hospitalized patients, Iwase and colleagues [63] studied the interaction between these two microbes using coculture experiments. By doing so, the authors demonstrated that *S. epidermidis* is capable of inhibiting biofilm formation by *S. aureus*, as well as destroying preformed biofilms.

Follow-up studies revealed that a 27-kDa serine protease (Esp) secreted by selected *S. epidermidis* isolates was responsible for the inhibition of *S. aureus* biofilm formation and nasal colonization [63]. In another study, a collection of *Staphylococcus* nasal isolates was screened for antimicrobial activity against *S. aureus*. Among the isolates, a *Staphylococcus lugdunensis* strain was able to inhibit *S. aureus* growth by producing an antimicrobial substance under specific nutritional conditions [64]. Transposon mutagenesis of the producer tracked activity to an uncharacterized gene encoding a potential nonribosomal peptide synthase, immersed in an operon containing several genes with antibiotic synthesis functions. The operon, named *lug*, is exclusively found in *S. lugdunensis*.

Using bioactivity-guided fractionation, as well as mass spectrometry and nuclear magnetic resonance, the authors were able to fully characterize the bioactive molecule, termed lugdunin, as a thiazolidine-containing cyclic peptide. Lugdunin acts as an antibiotic that strongly inhibits *S. aureus* and also other pathogens. It has shown potent antimicrobial activity against various Gram-positive bacteria, including MRSA, and vancomycin-resistant *Enterococcus* [64]. As such, lugdunin is one of the few bioactive nonribosomal peptides synthesized by bacteria associated with humans [65].

Because it does not show cytotoxicity against eukaryotic cells and is not able to lyse neutrophils or primary human erythrocytes, lugdunin shows promise as a therapeutic molecule [64]. Another potential therapeutic use would be through the production and administration of a lugdunin-producing probiotic strain. Although *S. lugdunensis* can be a cause of opportunistic infections, a mutant without virulence factors could be engineered to this purpose. Alternatively, *lug* genes could be introduced in commensal species suited for probiotic activity [64].

Corynebacterium

The mucosal moisture of the nose has been a large clinical reservoir for species of *Corynebacterium* and *Staphylococcus*. In line with the studies described above, others have shown that a high presence of *Corynebacterium pseudodiphtheriticum*, at specific microsites in the nasal cavity, is associated with low levels of *S. aureus*, whereas another *Corynebacterium species*, *Corynebacterium accolens*, is positively associated with *S. aureus* in its colonization sites [66]. The authors showed that *S. aureus* strongly enhances growth of *C. accolens*, which, in turn, shows a moderate positive effect on *S. aureus* growth. Conversely, *C. pseudodiphtheriticum* strongly inhibited *S. aureus* growth [66]. *C. pseudodiphtheriticum* is an opportunistic skin pathogen that sharply interferes with the colonization of *S. aureus*, inhibiting its growth in vitro through a yet unidentified mechanism.

Moraxella catarrhalis

As with the relationship between *S. aureus* and *C. accolens* described above, other bacterial species have been described as displaying synergistic interactions. *M. catarrhalis* is a bacterial pathogen commonly found causing respiratory tract infections [67,68]. This species is also clinically relevant due to the fact that most isolates produce a β-lactamase, and are therefore resistant to β-lactam antibiotics [69,70]. As with many Gram-negative bacteria, *M. catarrhalis* releases outer membrane vesicles (OMVs), which are membrane structures that detach from the producing cell, and carry many molecules to neighboring cells.

The material carried in these vesicles varies, but often include molecules with important roles in virulence and microbial signaling [71,72]. Interestingly, Schaar and colleagues have shown that OMVs produced by *M. catarrhalis* are loaded with a β-lactamase [73]. *M. catarrhalis* is commonly found in mixed-species infections, and it is often associated with *H. influenzae* and *S. pneumoniae* [74]. This raised the possibility that OMVs produced by *M. catarrhalis* containing β-lactamase could affect *H. influenzae* and *S. pneumoniae* present at the same infection site. Indeed, Schaar et al. demonstrated that β-lactamase-containing OMVs from *M. catarrhalis* could hydrolyze amoxicillin, as well as protect amoxicillin-susceptible *M. catarrhalis*, *H. influenzae*, and *S. pneumoniae* from antibiotic-induced cell death [73].

Although this study was limited to in vitro observations, the report that OMVs produced by *M. catarrhalis* can be found in vivo supports the notion that OMV-mediated antibiotic protection may occur in an infectious setting [73]. Indeed, a previous, independent study found that mice infected with *S. pneumoniae* and treated with β-lactamase died only if also infected with a strain of *M. catarrhalis* that produced a β-lactamase [75,76].

BIOACTIVE MOLECULES OF THE GUT MICROBIOME

Gut microbes can produce and sense a myriad of bioactive small molecules, and the signaling mechanisms involved are often important for cellular and populational fitness [10,11,77,78]. As a general rule, these small molecules are responsible for delivering biological messages, without the need for cell–cell contact. In this sense, gut microbiota–derived bioactive molecules represent the functional connection between the gut microbiome and the physiology of the human holobiont [79]. An intact gut microbiota protects the host against gastrointestinal pathogens, a phenomenon termed colonization resistance, and the production of protective small molecules may be a contributing factor [80]. Although this hypothesis still needs to be verified, several observations made to date point to various mechanisms being involved in colonization resistance, in contrast with the generalized and simplistic view that the microbiota confers protection solely through competition for nutrients and binding sites [81].

When *Salmonella enterica* serovar Typhimurium (hereafter referred to simply as *S. enterica*) is administered orally to mice, it invades the gut epithelium to cause a systemic, typhoid-like infection, with minimal intestinal colonization. However, when mice are pretreated with streptomycin, with a resulting reduction of 95% of their intestinal microbes, they become susceptible to gastroenteritis, with elevated levels of *S. enterica* colonization and inflammation in the cecum [82]. We have shown that antibiotic treatment results in major disturbance of the metabolic activities of the microbiota, resulting in increased or decreased abundances of several host and bacterial small molecules [13,77].

More recently, we have also shown that small molecules present in human feces can significantly alter the expression of several dozen *S. enterica* genes [14]. Interestingly, many of the differentially regulated genes are involved in processes critical for *S. enterica* virulence, such as motility and chemotaxis genes, whose expression was activated by the fecal extract, and genes required for invasion of nonphagocytic host cells, whose expression was severely inhibited by the fecal extract [14]. Although the human gut microbiome is a complex ecosystem, we were able to culture and characterize an isolate that produced the activity, a strain of *Clostridium citroniae* [14].

Through bioactivity-guided purification, we also found that the active molecules from the human gut metabolome sensed by *S. enterica* are small aromatic compounds [15]. The most active molecule identified in the organic extract, 3,4-dimethylbenzoic acid, acted as a strong inhibitor of both invasion gene expression and host cell invasion in cultured epithelial cells [15]. It is unclear at this time if small aromatic compounds are used by *S. enterica* to regulate its gene expression to maximize its fitness in the host environment, or if these bioactive molecules are produced by the resident microbiota or the host to combat this invading pathogen by disabling its virulence machinery. In either case, it is becoming increasingly clear that small molecules are involved in multiple facets of intercellular interactions in microbiomes. In fact, besides 3,4-dimethylbenzoic acid, several other small molecules produced by microbes have shown equivalent activity. Our group has shown that dimethyl sulfide, a product of dimethyl sulfoxide microbial respiration, can also act as an inhibitor of the expression of *S. enterica* host cell invasion—related genes [83].

Short-Chain Fatty Acids

Dietary fibers resist digestion and absorption in the gastrointestinal tract and can only be fermented by gut commensals, such as certain species from the Bacteroidetes and Firmicutes phyla. This fermentation results in the production of SCFAs, the most abundant products of bacterial fermentation in the gut, reaching concentrations of 50–150 nM in the human colon [84]. Their effects can be either beneficial or inhibitory to the pathogen, depending on their concentration and environmental pH.

High concentrations and low external pH promote the accumulation of SCFAs in the bacterial cytoplasm, where these molecules can exert various toxic effects, such as dissipation of the proton motive force [85]. However, at lower concentrations, SCFAs can be a useful cue for virulence gene regulation, allowing pathogens to determine their position within the gastrointestinal tract [80]. The *Bifidobacterium* metabolite acetate can potently inhibit the growth of enterohemorrhagic *Escherichia coli* O157:H7, and the lethality induced by translocation of its Shiga toxin into the bloodstream [86]. However, exposure of *S. enterica* to acetate can lead to a dramatic increase in its invasive properties, whereas propionate and butyrate exposure leads to decreased invasion of human cells.

SCFAs are currently used in commercial mixtures to control *Salmonella* in poultry, and the addition of prebiotics to the feed to stimulate fermentation reactions in the gut can also drive this process [87]. SCFAs are also known to have a beneficial role as immunomodulatory molecules that contribute to the maintenance of gut homeostasis, and in the regulation of host gene expression, inflammation, differentiation, apoptosis, and cancer inhibition. A decrease in luminal SCFAs is associated with ulcerative colitis and intestinal inflammation in humans, which can be ameliorated with the administration of dietary fiber or SCFAs [88].

Microbiome Vitamins

Other small molecules, which can be either bacterial metabolites or structural components, can diffuse throughout the body, affecting organs either directly, or by hormonal and neuronal signaling [89]. As an example, some gut microorganisms can produce vitamins, such as menaquinone, folate, cobalamin, and riboflavin, which are involved in several biological functions, from blood coagulation to bone metabolism and insulin sensitivity [90].

Quorum Sensing Biomolecules

Bacteria can also produce bioactive molecules to communicate with one another, through hormone-like signals that modulate gene expression, in a process known as quorum sensing. This process can control several important bacterial phenotypes, such as adhesion, biofilm formation, and production of pathogenicity factors (e.g., toxins). For example, the quorum sensing molecule termed autoinducer 2 (AI-2; furanosyl borate diester) is produced by certain gut microbiome components, like *Ruminococcus obeum*, and can interfere with intestinal colonization by *Vibrio cholerae*, repressing the biosynthesis of several *V. cholerae* colonization factors [91].

Bile Acid Metabolism

Bile acids are amphipathic lipids that are important in fat and cholesterol metabolism. Several species of Firmicutes (e.g., *Clostridium scindens* and *Clostridium hylemonae*) are able to transform bile acids in primary and secondary bile acids [92]. It is known that the secondary bile acid deoxycholate, which is less abundant in antibiotic-treated mice, stimulates

Clostridium difficile spore germination, but inhibits its growth [93]. Besides, the secondary bile acids deoxycholate and lithocholic acid have been implicated in hepatotoxicity and colon cancer [94].

Lantibiotics and Bacteriocins

Several natural products are secreted by bacteria, including ribosomally synthesized, posttranslationally modified peptides (RiPP). Many organisms in the microbiota produce diverse RiPPs, and the most commonly isolated are lantibiotics and bacteriocins [95]. Ruminococcin A is a short peptide of <40 amino acids produced by the gut commensals *Ruminococcus gnavus* and *Clostridium nexile*, and is usually active against a narrow spectrum of Gram-positive bacteria [92]. The human pathogen *Enterococcus faecalis* also produces a two-component lantibiotic cytolysin that exerts antibacterial activity against human commensals [96].

Another RiPP produced by *E. coli* is a heat-stable enterotoxin, which is a 14-amino acid peptide that mimics the effect of the host peptide hormones guanylin and uroguanylin, by agonizing guanylate cyclase 2C. In 2012, the US FDA approved a single amino acid variant of the heat-stable enterotoxin linaclotide, for the treatment of constipation associated with irritable bowel syndrome. This peptide can survive the proteolytic milieu of the gut lumen, delivering a potent biological activity, without absorption into host circulation [97].

Microcins

Microcins are narrow-spectrum antibacterials, displaying a wide range of unusual posttranslational modifications. They derive exclusively from enterobacteria, mainly *E. coli*, and have potent antibacterial activity against Gram-negative microbiota [98]. Thiazole/oxazole modified microcins (TOMMs) are similar to microcins, but include a larger family of natural products, generated by both Gram-positive and Gram-negative bacteria, such as clostridiolysin S from *Clostridium sporogenes* [92,99]. Also produced by a subset of enterobacteria, including strains of *E. coli*, as a product of a non-ribosomal peptide synthetase-polyketide synthase (NRPS−PKS), there is colibactin. Exposure of mammalian cells to colibactin induces DNA double-strand breaks, and may influence the risk for colorectal cancer in humans [100,101].

PRACTICAL APPLICATIONS AND CONCLUDING REMARKS

Accumulating evidence over the last several decades has shown that small molecules produced by members of microbiomes have important functions in the cellular interactions that occur in these environments. Additionally, the biological activities described can be harvested for their useful properties. Knowledge obtained from the study of small molecules from microbiomes can be used for our advantage in various ways. First, bioactive molecules themselves can be developed into therapeutics. This is precisely what happened during the discovery of penicillin. A bioactive compound produced by *Penicillium* showed inhibitory properties against bacteria, and this property was explored for human benefit through the use of penicillin and, later on, other antibiotics. The second strategy is to use knowledge generated from the study of the signaling events in which the small molecules participate to design new therapeutics. For instance, by identifying small molecules involved in signaling cascades that are required for virulence, one can then design inhibitors of those signaling pathways, sometimes using the natural molecules themselves as scaffolds. This approach has been exemplified by the many studies describing antivirulence compounds. One such example is the research on quorum sensing inhibition, which shows promise for the development of new drugs, targeting bacterial virulence. The third potential use is through the discovery of biomarkers. Although this does not depend on biological activity, and therefore is not the focus of this chapter, small molecules produce fingerprints of biological samples, and these molecular signatures can be used for diagnostic purposes. For instance, molecular profiles based on metabolites present in biological samples (metabolomic analysis) can be used to determine if an individual has a specific disease, as well as provide information on the type of disease and response to treatment. In summary, bioactive molecules represent the functional lexicon of biological diversity. Future studies will likely focus on expanding the ever-increasing catalogue of interesting molecules, as well as deciphering the intricacies of the biological pathways elicited by them.

REFERENCES

[1] Bérdy J. Thoughts and facts about antibiotics: where we are now and where we are heading. J Antibiot (Tokyo) 2012;65(8):385−95.
[2] Katz L, Baltz RH. Natural product discovery: past, present, and future. J Ind Microbiol Biotechnol 2016;43(2−3):155−76.
[3] Demain AL. Importance of microbial natural products and the need to revitalize their discovery. J Ind Microbiol Biotechnol 2014;41(2):185−201.

[4] Kinch MS, Haynesworth A, Kinch SL, Hoyer D. An overview of FDA-approved new molecular entities: 1827-2013. Drug Discov Today 2014;19(8):1033—9.

[5] Davies J. Where have all the antibiotics gone? Can J Infect Dis Med Microbiol 2006;17(5):287—90.

[6] van der Waaij D, Berghuis-de Vries JM, Lekkerkerk Lekkerkerk v. Colonization resistance of the digestive tract in conventional and antibiotic-treated mice. J Hyg (Lond) 1971;69(3):405—11.

[7] van der Waaij D, Berghuis JM, Lekkerkerk JE. Colonization resistance of the digestive tract of mice during systemic antibiotic treatment. J Hyg (Lond) 1972;70(4):605—10.

[8] Bry L, Falk PG, Midtvedt T, Gordon JI. A model of host-microbial interactions in an open mammalian ecosystem. Science 1996;273(5280):1380—3.

[9] Consortium HMP. Structure, function and diversity of the healthy human microbiome. Nature 2012;486(7402):207—14.

[10] Antunes LC, Ferreira RB. Intercellular communication in bacteria. Crit Rev Microbiol 2009;35(2):69—80.

[11] Antunes LC, Ferreira RB, Buckner MM, Finlay BB. Quorum sensing in bacterial virulence. Microbiology 2010;156(Pt 8):2271—82.

[12] Sekirov I, Russell SL, Antunes LC, Finlay BB. Gut microbiota in health and disease. Physiol Rev 2010;90(3):859—904.

[13] Antunes LC, Han J, Ferreira RB, Lolić P, Borchers CH, Finlay BB. Effect of antibiotic treatment on the intestinal metabolome. Antimicrob Agents Chemother 2011;55(4):1494—503.

[14] Antunes LC, McDonald JA, Schroeter K, Carlucci C, Ferreira RB, Wang M, et al. Antivirulence activity of the human gut metabolome. mBio 2014;5(4). e01183—14.

[15] Peixoto RJM, Alves ES, Wang M, Ferreira RBR, Granato A, Han J, et al. Repression of *Salmonella* host cell invasion by aromatic small molecules from the human fecal metabolome. Appl Environ Microbiol July 28, 2017. https://doi.org/10.1128/AEM.01148-17. pii: AEM.01148—17. [Epub ahead of print].

[16] Rosenthal M, Goldberg D, Aiello A, Larson E, Foxman B. Skin microbiota: microbial community structure and its potential association with health and disease. Infect Genet Evol 2011;11(5):839—48.

[17] Grice EA, Segre JA. The skin microbiome. Nat Rev Microbiol 2011;9(4):244—53.

[18] Gao Z, Perez-Perez GI, Chen Y, Blaser MJ. Quantitation of major human cutaneous bacterial and fungal populations. J Clin Microbiol 2010;48(10):3575—81.

[19] Sanford JA, Gallo RL. Functions of the skin microbiota in health and disease. Semin Immunol 2013;25(5):370—7.

[20] Bastos MC, Ceotto H, Coelho ML, Nascimento JS. Staphylococcal antimicrobial peptides: relevant properties and potential biotechnological applications. Curr Pharm Biotechnol 2009;10(1):38—61.

[21] Byrd AL, Belkaid Y, Segre JA. The human skin microbiome. Nat Rev Microbiol January 15, 2018. https://doi.org/10.1038/nrmicro.2017.157 [Epub ahead of print].

[22] Nascimento JS, Ceotto H, Nascimento SB, Giambiagi-Demarval M, Santos KR, Bastos MC. Bacteriocins as alternative agents for control of multiresistant staphylococcal strains. Lett Appl Microbiol 2006;42(3):215—21.

[23] Heidrich C, Pag U, Josten M, Metzger J, Jack RW, Bierbaum G, et al. Isolation, characterization, and heterologous expression of the novel lantibiotic epicidin 280 and analysis of its biosynthetic gene cluster. Appl Environ Microbiol 1998;64(9):3140—6.

[24] Bierbaum G, Götz F, Peschel A, Kupke T, van de Kamp M, Sahl HG. The biosynthesis of the lantibiotics epidermin, gallidermin, Pep5 and epilancin K7. Antonie Van Leeuwenhoek 1996;69(2):119—27.

[25] Kim PI, Sohng JK, Sung C, Joo HS, Kim EM, Yamaguchi T, et al. Characterization and structure identification of an antimicrobial peptide, hominicin, produced by *Staphylococcus hominis* MBBL 2-9. Biochem Biophys Res Commun 2010;399(2):133—8.

[26] da Costa TM, Morgado PG, Cavalcante FS, Damasco AP, Nouér SA, Dos Santos KR. Clinical and microbiological characteristics of hetero-resistant and vancomycin-intermediate *Staphylococcus aureus* from bloodstream infections in a Brazilian teaching hospital. PLoS One 2016;11(8):e0160506.

[27] Sashihara T, Kimura H, Higuchi T, Adachi A, Matsusaki H, Sonomoto K, et al. A novel lantibiotic, nukacin ISK-1, of *Staphylococcus warneri* ISK-1: cloning of the structural gene and identification of the structure. Biosci Biotechnol Biochem 2000;64(11):2420—8.

[28] Héchard Y, Ferraz S, Bruneteau E, Steinert M, Berjeaud JM. Isolation and characterization of a *Staphylococcus warneri* strain producing an anti-Legionella peptide. FEMS Microbiol Lett 2005;252(1):19—23.

[29] Verdon J, Berjeaud JM, Lacombe C, Héchard Y. Characterization of anti-*Legionella* activity of warnericin RK and delta-lysin I from *Staphylococcus warneri*. Peptides 2008;29(6):978—84.

[30] Cogen AL, Yamasaki K, Muto J, Sanchez KM, Crotty Alexander L, Tanios J, et al. Staphylococcus epidermidis antimicrobial delta-toxin (phenol-soluble modulin-gamma) cooperates with host antimicrobial peptides to kill group A *Streptococcus*. PLoS One 2010;5(1):e8557.

[31] Cogen AL, Yamasaki K, Sanchez KM, Dorschner RA, Lai Y, MacLeod DT, et al. Selective antimicrobial action is provided by phenol-soluble modulins derived from *Staphylococcus epidermidis*, a normal resident of the skin. J Investig Dermatol 2010;130(1):192—200.

[32] Kong HH, Oh J, Deming C, Conlan S, Grice EA, Beatson MA, et al. Temporal shifts in the skin microbiome associated with disease flares and treatment in children with atopic dermatitis. Genome Res 2012;22(5):850—9.

[33] Nakatsuji T, Chen TH, Narala S, Chun KA, Two AM, Yun T, et al. Antimicrobials from human skin commensal bacteria protect against *Staphylococcus aureus* and are deficient in atopic dermatitis. Sci Transl Med 2017;9(378).

[34] Otto M, Echner H, Voelter W, Götz F. Pheromone cross-inhibition between *Staphylococcus aureus* and *Staphylococcus epidermidis*. Infect Immun 2001;69(3):1957—60.

[35] Cogen AL, Nizet V, Gallo RL. Skin microbiota: a source of disease or defence? Br J Dermatol 2008;158(3):442—55.

[36] Paul GE, Booth SJ. Properties and characteristics of a bacteriocin-like substance produced by *Propionibacterium acnes* isolated from dental plaque. Can J Microbiol 1988;34(12):1344—7.

[37] Shu M, Wang Y, Yu J, Kuo S, Coda A, Jiang Y, et al. Fermentation of *Propionibacterium acnes*, a commensal bacterium in the human skin microbiome, as skin probiotics against methicillin-resistant *Staphylococcus aureus*. PLoS One 2013;8(2):e55380.

[38] Wieland Brown LC, Acker MG, Clardy J, Walsh CT, Fischbach MA. Thirteen posttranslational modifications convert a 14-residue peptide into the antibiotic thiocillin. Proc Natl Acad Sci USA 2009;106(8):2549—53.

[39] Christensen GJ, Scholz CF, Enghild J, Rohde H, Kilian M, Thürmer A, et al. Antagonism between *Staphylococcus epidermidis* and *Propionibacterium acnes* and its genomic basis. BMC Genom 2016;17:152.

[40] Allhorn M, Arve S, Brüggemann H, Lood R. A novel enzyme with antioxidant capacity produced by the ubiquitous skin colonizer *Propionibacterium acnes*. Sci Rep 2016;6:36412.

[41] Okayama Y. Oxidative stress in allergic and inflammatory skin diseases. Curr Drug Targets - Inflamm Allergy 2005;4(4):517—9.

[42] Gao Z, Tseng CH, Strober BE, Pei Z, Blaser MJ. Substantial alterations of the cutaneous bacterial biota in psoriatic lesions. PLoS One 2008;3(7):e2719.

[43] Liavonchanka A, Hornung E, Feussner I, Rudolph MG. Structure and mechanism of the *Propionibacterium acnes* polyunsaturated fatty acid isomerase. Proc Natl Acad Sci USA 2006;103(8):2576—81.

[44] Beppu F, Hosokawa M, Tanaka L, Kohno H, Tanaka T, Miyashita K. Potent inhibitory effect of trans9, trans11 isomer of conjugated linoleic acid on the growth of human colon cancer cells. J Nutr Biochem 2006;17(12):830—6.

[45] O'Shea M, Bassaganya-Riera J, Mohede IC. Immunomodulatory properties of conjugated linoleic acid. Am J Clin Nutr 2004;79(6 Suppl.):1199S—206S.

[46] Yamasaki M, Yanagita T. Adipocyte response to conjugated linoleic acid. Obes Res Clin Pract 2013;7(4):e235—42.

[47] Fujita K, Ichimasa S, Zendo T, Koga S, Yoneyama F, Nakayama J, et al. Structural analysis and characterization of lacticin Q, a novel bacteriocin belonging to a new family of unmodified bacteriocins of gram-positive bacteria. Appl Environ Microbiol 2007;73(9):2871—7.

[48] Ramsey MM, Freire MO, Gabrilska RA, Rumbaugh KP, Lemon KP. *Staphylococcus aureus* shifts toward commensalism in response to *Corynebacterium* species. Front Microbiol 2016;7:1230.

[49] Dominguez-Bello MG, Costello EK, Contreras M, Magris M, Hidalgo G, Fierer N, et al. Delivery mode shapes the acquisition and structure of the initial microbiota across multiple body habitats in newborns. Proc Natl Acad Sci USA 2010;107(26):11971—5.

[50] Bosch AATM, de Steenhuijsen Piters WAA, van Houten MA, Chu MLJN, Biesbroek G, Kool J, et al. Maturation of the infant respiratory microbiota, environmental drivers, and health consequences. A prospective cohort study. Am J Respir Crit Care Med 2017;196(12):1582—90.

[51] de Steenhuijsen Piters WA, Sanders EA, Bogaert D. The role of the local microbial ecosystem in respiratory health and disease. Philos Trans R Soc Lond B Biol Sci 2015;370(1675):20140294.

[52] van den Bergh MR, Biesbroek G, Rossen JW, de Steenhuijsen Piters WA, Bosch AA, van Gils EJ, et al. Associations between pathogens in the upper respiratory tract of young children: interplay between viruses and bacteria. PLoS One 2012;7(10):e47711.

[53] McCullers JA. Insights into the interaction between influenza virus and pneumococcus. Clin Microbiol Rev 2006;19(3):571—82.

[54] Bosch AA, Biesbroek G, Trzcinski K, Sanders EA, Bogaert D. Viral and bacterial interactions in the upper respiratory tract. PLoS Pathog 2013;9(1):e1003057.

[55] Mancini DA, Mendonça RM, Dias AL, Mendonça RZ, Pinto JR. Co-infection between influenza virus and flagellated bacteria. Rev Inst Med Trop Sao Paulo 2005;47(5):275—80.

[56] Tashiro M, Ciborowski P, Klenk HD, Pulverer G, Rott R. Role of *Staphylococcus* protease in the development of influenza pneumonia. Nature 1987;325(6104):536—7.

[57] Tashiro M, Ciborowski P, Reinacher M, Pulverer G, Klenk HD, Rott R. Synergistic role of staphylococcal proteases in the induction of influenza virus pathogenicity. Virology 1987;157(2):421—30.

[58] Regev-Yochay G, Trzcinski K, Thompson CM, Malley R, Lipsitch M. Interference between *Streptococcus pneumoniae* and *Staphylococcus aureus*: *In vitro* hydrogen peroxide-mediated killing by *Streptococcus pneumoniae*. J Bacteriol 2006;188(13):4996—5001.

[59] Veenhoven R, Bogaert D, Uiterwaal C, Brouwer C, Kiezebrink H, Bruin J, et al. Effect of conjugate pneumococcal vaccine followed by polysaccharide pneumococcal vaccine on recurrent acute otitis media: a randomised study. Lancet 2003;361(9376):2189—95.

[60] Kluytmans J, van Belkum A, Verbrugh H. Nasal carriage of *Staphylococcus aureus*: epidemiology, underlying mechanisms, and associated risks. Clin Microbiol Rev 1997;10(3):505—20.

[61] Kuehnert MJ, Kruszon-Moran D, Hill HA, McQuillan G, McAllister SK, Fosheim G, et al. Prevalence of *Staphylococcus aureus* nasal colonization in the United States, 2001-2002. J Infect Dis 2006;193(2):172—9.

[62] von Eiff C, Becker K, Machka K, Stammer H, Peters G. Nasal carriage as a source of *Staphylococcus aureus* bacteremia. Study Group. N Engl J Med 2001;344(1):11—6.

[63] Iwase T, Uehara Y, Shinji H, Tajima A, Seo H, Takada K, et al. Staphylococcus epidermidis Esp inhibits *Staphylococcus aureus* biofilm formation and nasal colonization. Nature 2010;465(7296):346—9.

[64] Zipperer A, Konnerth MC, Laux C, Berscheid A, Janek D, Weidenmaier C, et al. Human commensals producing a novel antibiotic impair pathogen colonization. Nature 2016;535(7613):511—6.

[65] Wilson MR, Zha L, Balskus EP. Natural product discovery from the human microbiome. J Biol Chem 2017;292(21):8546—52.

[66] Yan M, Pamp SJ, Fukuyama J, Hwang PH, Cho DY, Holmes S, et al. Nasal microenvironments and interspecific interactions influence nasal microbiota complexity and *S. aureus* carriage. Cell Host Microbe 2013;14(6):631—40.

[67] Heiniger N, Spaniol V, Troller R, Vischer M, Aebi C. A reservoir of *Moraxella catarrhalis* in human pharyngeal lymphoid tissue. J Infect Dis 2007;196(7):1080–7.

[68] Perez Vidakovics ML, Riesbeck K. Virulence mechanisms of *Moraxella* in the pathogenesis of infection. Curr Opin Infect Dis 2009;22(3):279–85.

[69] Khan MA, Northwood JB, Levy F, Verhaegh SJ, Farrell DJ, Van Belkum A, et al. Bro beta-lactamase and antibiotic resistances in a global cross-sectional study of *Moraxella catarrhalis* from children and adults. J Antimicrob Chemother 2010;65(1):91–7.

[70] Levy F, Walker ES. BRO beta-lactamase alleles, antibiotic resistance and a test of the BRO-1 selective replacement hypothesis in *Moraxella catarrhalis*. J Antimicrob Chemother 2004;53(2):371–4.

[71] Schaar V, de Vries SP, Perez Vidakovics ML, Bootsma HJ, Larsson L, Hermans PW, et al. Multicomponent *Moraxella catarrhalis* outer membrane vesicles induce an inflammatory response and are internalized by human epithelial cells. Cell Microbiol 2011;13(3):432–49.

[72] Florez C, Raab JE, Cooke AC, Schertzer JW. Membrane distribution of the *Pseudomonas* Quinolone Signal modulates outer membrane vesicle production in *Pseudomonas aeruginosa*. mBio 2017;8(4).

[73] Schaar V, Nordström T, Mörgelin M, Riesbeck K. Moraxella catarrhalis outer membrane vesicles carry β-lactamase and promote survival of *Streptococcus pneumoniae* and *Haemophilus influenzae* by inactivating amoxicillin. Antimicrob Agents Chemother 2011;55(8):3845–53.

[74] Krishnamurthy A, McGrath J, Cripps AW, Kyd JM. The incidence of *Streptococcus pneumoniae* otitis media is affected by the polymicrobial environment particularly *Moraxella catarrhalis* in a mouse nasal colonisation model. Microbes Infect 2009;11(5):545–53.

[75] Hol C, Van Dijke EE, Verduin CM, Verhoef J, van Dijk H. Experimental evidence for *Moraxella*-induced penicillin neutralization in pneumo-coccal pneumonia. J Infect Dis 1994;170(6):1613–6.

[76] Joice R, Yasuda K, Shafquat A, Morgan XC, Huttenhower C. Determining microbial products and identifying molecular targets in the human microbiome. Cell Metab 2014;20(5):731–41.

[77] Antunes LC, Finlay BB. A comparative analysis of the effect of antibiotic treatment and enteric infection on intestinal homeostasis. Gut Microb 2011;2(2):105–8.

[78] Antunes LC, Davies JE, Finlay BB. Chemical signaling in the gastrointestinal tract. F1000 Biol Rep 2011;3:4.

[79] Rooks MG, Garrett WS. Gut microbiota, metabolites and host immunity. Nat Rev Immunol 2016;16(6):341–52.

[80] Vogt SL, Peña-Díaz J, Finlay BB. Chemical communication in the gut: effects of microbiota-generated metabolites on gastrointestinal bacterial pathogens. Anaerobe 2015;34:106–15.

[81] Vogt SL, Finlay BB. Gut microbiota-mediated protection against diarrheal infections. J Travel Med 2017;24(Suppl._1):S39–43.

[82] Ferreira RB, Gill N, Willing BP, Antunes LC, Russell SL, Croxen MA, et al. The intestinal microbiota plays a role in *Salmonella*-induced colitis independent of pathogen colonization. PLoS One 2011;6(5):e20338.

[83] Antunes LC, Buckner MM, Auweter SD, Ferreira RB, Lolić P, Finlay BB. Inhibition of *Salmonella* host cell invasion by dimethyl sulfide. Appl Environ Microbiol 2010;76(15):5300–4.

[84] Macfarlane GT, Gibson GR. Carbohydrate fermentation, energy transduction and gas metabolism in the human large intestine. In: Mackie R, White B, editors. Gastrointestinal microbiology. Springer; 1997. p. 269–318.

[85] Sun Y, O'Riordan MX. Regulation of bacterial pathogenesis by intestinal short-chain Fatty acids. Adv Appl Microbiol 2013;85:93–118.

[86] Fukuda S, Toh H, Hase K, Oshima K, Nakanishi Y, Yoshimura K, et al. Bifidobacteria can protect from enteropathogenic infection through production of acetate. Nature 2011;469(7331):543–7.

[87] Van Immerseel F, De Buck J, Pasmans F, Velge P, Bottreau E, Fievez V, et al. Invasion of *Salmonella enteritidis* in avian intestinal epithelial cells in vitro is influenced by short-chain fatty acids. Int J Food Microbiol 2003;85(3):237–48.

[88] Zhang LS, Davies SS. Microbial metabolism of dietary components to bioactive metabolites: opportunities for new therapeutic interventions. Genome Med 2016;8(1):46.

[89] Schroeder BO, Bäckhed F. Signals from the gut microbiota to distant organs in physiology and disease. Nat Med 2016;22(10):1079–89.

[90] Derrien M, Veiga P. Rethinking diet to aid human-microbe symbiosis. Trends Microbiol 2017;25(2):100–12.

[91] Hsiao A, Ahmed AM, Subramanian S, Griffin NW, Drewry LL, Petri WA, et al. Members of the human gut microbiota involved in recovery from *Vibrio cholerae* infection. Nature 2014;515(7527):423–6.

[92] Donia MS, Fischbach MA. Human Microbiota. Small molecules from the human microbiota. Science 2015;349(6246):1254766.

[93] Theriot CM, Young VB. Microbial and metabolic interactions between the gastrointestinal tract and *Clostridium difficile* infection. Gut Microb 2014;5(1):86–95.

[94] Garrett WS. Cancer and the microbiota. Science 2015;348(6230):80–6.

[95] Donia MS, Cimermancic P, Schulze CJ, Wieland Brown LC, Martin J, Mitreva M, et al. A systematic analysis of biosynthetic gene clusters in the human microbiome reveals a common family of antibiotics. Cell 2014;158(6):1402–14.

[96] Coburn PS, Gilmore MS. The *Enterococcus faecalis* cytolysin: a novel toxin active against eukaryotic and prokaryotic cells. Cell Microbiol 2003;5(10):661–9.

[97] McWilliams V, Whiteside G, McKeage K. Linaclotide: first global approval. Drugs 2012;72(16):2167–75.

[98] Duquesne S, Destoumieux-Garzón D, Peduzzi J, Rebuffat S. Microcins, gene-encoded antibacterial peptides from enterobacteria. Nat Prod Rep 2007;24(4):708–34.

[99] Gonzalez DJ, Lee SW, Hensler ME, Markley AL, Dahesh S, Mitchell DA, et al. Clostridiolysin S, a post-translationally modified biotoxin from *Clostridium botulinum*. J Biol Chem 2010;285(36):28220–8.

[100] Nougayrède JP, Homburg S, Taieb F, Boury M, Brzuszkiewicz E, Gottschalk G, et al. *Escherichia coli* induces DNA double-strand breaks in eukaryotic cells. Science 2006;313(5788):848–51.

[101] Balskus EP. Colibactin: understanding an elusive gut bacterial genotoxin. Nat Prod Rep 2015;32(11):1534–40.

Chapter 13

Role of the Microbiome in Intestinal Barrier Function and Immune Defense

Aline Ignacio[1], Fernanda Fernandes Terra[1], Ingrid Kazue Mizuno Watanabe[2], Paulo José Basso[1] and Niels Olsen Saraiva Câmara[1]

[1]*Department of Immunology, Institute of Biomedical Sciences, University of São Paulo, São Paulo, Brazil;* [2]*Department of Medicine, Federal University of São Paulo, São Paulo, Brazil*

THE HUMAN MICROBIOTA

The human body, especially the mucosae, harbors complex communities of commensal microbes, including bacteria, fungi, viruses, protozoa, and archaea. This great number of microorganisms is collectively known as indigenous microbiota, and the genes codified by them are collectively called microbiome. Recent studies indicating the presence of bacteria in the placenta and meconium suggest that our colonization begins still in the feto-placental unit [1,2].

The disruption of indigenous microbiota, also named dysbiosis, may be characterized by the reduction of its number and/or diversity. Recently, dysbiosis has been associated to a myriad of intestinal and systemic diseases, such as inflammatory bowel disease (IBD), asthma, arthritis, cardiovascular diseases, and obesity. Such diseases can be triggered by both pathogenic and symbiotic microbes that can be harmful in certain circumstances.

Intestinal Microbiota: A Functional Organ

Along the gut, the bacterial density increases from the stomach toward the colon, while its diversity decreases from the lumen to the intestinal epithelia [3].

Looking for the mutualistic pattern maintenance, the intestine provides an auspicious environment for the microbes to live, with free access to nutrients, and controlled temperature. Meanwhile, the gut microbiota protects the host from potential pathogens and regulates the metabolism and nutrition, besides regulating the morphogenesis and physiology of the immune system. Many other nonconventional functions, such as adiposity control, bone mass formation and behavior are also performed by the microbiota.

Key Commensals and Pathogens

Most of the resident gut microbes live as commensals or in mutualism; however, a small fraction is potentially pathogenic, often called pathobionts, and can trigger some diseases when homeostasis is disrupted. *Clostridium difficile*, for example, is a spore-forming bacterium which composes the indigenous microbiota; however, once commensal diversity is reduced, its germination occurs and the release of toxins in the gut can lead to severe intestinal inflammation [4]. Therefore, besides protecting the host from exogenous pathogens, the microbiota must also control pathobiont growth [5].

Colonization Resistance and Pathogen Inhibition

Colonization resistance is the term used to describe the microbiota's capacity to limit the introduction of exogenous microorganisms and pathobiont expansion. Reduction in diversity and quantity of microbes in germ-free (GF) animals or by treating mice with antibiotics reveals increased susceptibility to a variety of infections, such as *C. difficile* [6], *Shigella flexneri* [7] and *Salmonella enterica* serovar Typhimurium [8]. In parallel, many studies have shown that microbiota transplantation is capable of rendering resistant animals into susceptible to pathogens, or even mimicking some disease phenotype [9,10].

Commensals can resist colonization by competing for nutritional requirements. Several microbes sharing the same niche present similar nutritional needs; therefore, the consumption of organic acids, amino acids, sugars, and other nutrients by indigenous microbiota can restrain competing pathogens. *Bacteroides thetaiotaomicron*, for example, limits *Citrobacter rodentium* by carbohydrate competition [11]. Microbiota-derived bioproducts can also directly control pathogen dissemination. Some short-chain fatty acids (SCFAs) produced during carbohydrate fermentation in the gut, such as butyrate and acetate, can control *Salmonella enterica* serovar Enteritidis and Typhimurium [12]. The indigenous microbiota secretes antimicrobial peptides, termed bacteriocins, which limit microorganisms of the same species or related ones. Some strains of commensal *Escherichia coli* are able to produce bacteriocins, which directly inhibit pathogenic enterohemorrhagic *Escherichia coli* (EHEC) [13] (Fig 13.1).

Reduction in oxygen tension [14] and pH alteration [15], are other known mechanisms developed by commensal bacteria to turn the environment more hostile to certain pathogens. Besides, the metabolism of bile acids by microbes plays an important role in colonization resistance to *C. difficile* [16].

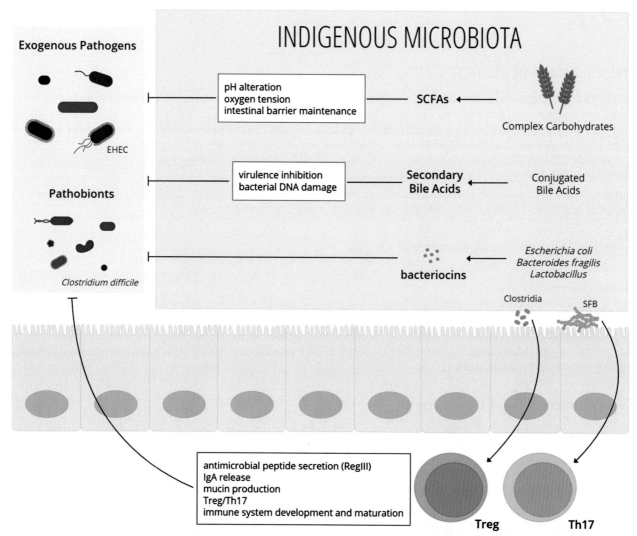

FIGURE 13.1 Microbiota—host homeostasis. Products of bacterial metabolism, such as short-chain fatty acids (SCFAs) and secondary bile acids act synergistically to avoid gut colonization through pH alteration, reduction in oxygen tension, virulence inhibition, and direct killing of some pathogens. The secretion of inhibitory substances such as bacteriocins by some commensals may target and inhibit pathobionts and pathogens. Indigenous microbiota facilitates host barrier maintenance and plays a central role in immune system maturation. Commensals, such as segmented filamentous bacteria (SFB) and various clostridia, can induce accumulation of Th17 and Treg, respectively. Secretion of antimicrobial peptides, such as RegIII, upregulation of mucus, and IgA release can also contribute to promote gut colonization resistance.

Despite all the strategies mentioned above, some pathogens developed evasion mechanisms that allow their maintenance in the host gut. *C. difficile* and *S. typhimurium*, for instance, exploit the metabolization of certain sugars produced by indigenous microbes, such as *B. thetaiotaomicron*, as energy source for their survival [17]. Other bacteria, such as EHEC, can sense the environment and change their metabolism to avoid commensal nutrient competition [18,19]. A study evaluating the colonization resistance to EHEC demonstrated that animals previously colonized with three strains of commensal *E. coli*, with different nutritional requirements are resistant to EHEC colonization, but when only one of the strains is inoculated the animals become susceptible [20].

METABOLITES AND PATHWAYS

Microbiota can produce a large amount of small bioactive molecules during the metabolism of diet, which are joined by endogenous metabolites, mucosal-derived macromolecules, and xenobiotics. Once in the bloodstream, these molecules link the microbiota to the rest of the human body. The metabolic axis host-microbe varies along the intestine, and the synthesized metabolites play a role in both host physiology and microbiota homeostasis, such as SCFAs, aromatic amino acids derivatives, choline, and bile acids.

Short-Chain Fatty Acids

Complex carbohydrates and, less extensively, amino acid fermentation by colonic microbes may yield SCFAs. Acetate, butyrate, and propionate are the most abundant SCFAs in the gut, and are sensed by G protein-coupled receptors, such as GPR41 and GPR43 [21]. While GPR41 is expressed by many different cells, GPR43 is observed in immune cells. Apart from being used as energy source to colonocytes, SCFAs regulate host energy utilization, control gut pH, and play an important role in maintaining intestinal barrier integrity [22,23].

Primary and Secondary Bile Acids

Bile acids, in turn, are mainly metabolized by *Lactobacillus* spp., *Bifidobacteria* spp., *Enterobacter* spp., *Bacteroides* spp., and *Clostridium* spp. [24]. Bile acids are synthesized in the liver and before being released into intestine, they are conjugated to enhance their surfactant activity. Most of these conjugated bile acids are reabsorbed in the ileum and go back to the liver; however, a small amount remains in the intestine, and is metabolized into secondary bile acids by gut microbes [25].

Deoxycholate and lithocholate are the main secondary bile acids, and they activate the nuclear farnesoid X receptor (FXR) that is expressed in intestinal and hepatic cells [26]. While modulating host lipid absorption and their own metabolism [27], bile acids can disrupt bacterial membrane [28], induce DNA damage [29], alter protein conformation [30], chelate iron [31] and regulate the expression of bacterial virulence factors [32] (Fig 13.1).

Amino Acid and Choline Metabolites

The catabolism of aromatic amino acids such as tryptophan by microorganisms of the intestinal microbiota generates mainly tryptamine and indole-3-acetate [33]. In intestinal epithelial cells (IECs), these molecules are capable of regulating the production of cytokines, antimicrobial peptides, and the development of regulatory T (Treg) cells, by signaling through the aryl hydrocarbon receptor(AhR) [34].

The metabolization of choline by different microbes generates trimethylamine, which is an important modulator of lipid metabolism and glucose homeostasis [35]. Changes in the composition of the microbiota, which in turn lead to the increase in choline metabolism, have been associated with cardiovascular and liver diseases.

COMMENSALS AND MUCOSAL IMMUNITY: THE CROSSTALK AGAINST PATHOGENS

The commensal microbiota is an important source of nutrients and vitamins, enables the digestion of complex polysaccharides, competes with pathogens for the same niche, and shapes the host immune system. The mechanisms by which the microbiota contributes to the development of the immune system include: (1) direct contact with host cells and (2) generation of molecules either through de novo synthesis, or through modification of food components or their metabolites [36].

Mucosal Anatomy

Among the four layers that compose the intestine (the mucosa, the submucosa, the muscularis, and the serosa), the mucosa is the nearest stratum to the lumen, consisting of a single layer of epithelial cells underlying a loose connective tissue denominated Lamina Propria (LP), which contains the majority of immune cells. Immune cells are located in organized lymphoid structures (gut-associated lymphoid tissues—GALT), such as Peyer patches (PPs), cryptopatches (CPs), and isolated lymphoid follicles (ILFs), or diffusely distributed. In mice, the development of CPs and ILFs occurs after birth, in a process influenced by dietary factors. The absence of AhR signaling pathway reduces the ILFs [37].

AhR ligands can be provided by the diet or through tryptophan metabolism by the microbiota [38], demonstrating that the microbiota can impact immune system development. In addition, the recognition of peptidoglycan from Gram-negative bacteria is also required to induce the establishing of ILFs, through binding to NOD1 (nucleotide-binding oligomerization domain containing 1) in IECs. In the small intestine (SI), ILFs generate commensal-reactive Ig-A-producing B cells essential for determining the microbiota diversity, and to prevent bacterial invasion and systemic spread. The maturation of ILFs into B cell clusters demands bacteria detection by toll-like receptors (TLRs) [39]. That way, absence of gut microbiota impacts mucosal immune system, which in turn can reciprocate on microbiota composition.

The Gut as an Immune Organ

IECs form a single cell layer that constitutes a physical barrier between the luminal content and the underlying immune cells. The expression of pattern recognition receptors (PRRs), including TLRs, nucleotide-binding domain leucine-rich repeat containing receptors (NLRs), RIG-I like receptors (RLRs), C-type lectin family, and AIM2-like receptors (ALRs) on IECs, allow the proper communication between the environment and the immune compartment. The expression and localization of TLRs and the expression of antimicrobial proteins by IECs are also influenced by bacterial colonization.

It has been shown that GF mice are deficient in defensins, while Gram-negative commensal microorganism *B. thetaiotaomicron* induces the secretion of the antimicrobial TLR-related protein RegIIIγ [40,41]. Collectively, those observations regarding tissues and cellular defects on GF mice support the idea that "normal" immune function demands the presence of the gut microbiota.

GPR43, also known as free fatty acid receptor 2 (FFAR2), is a SCFA's receptor expressed by IECs, monocytes, neutrophils, eosinophils, and lymphocytes, and it has been shown that its stimulation is necessary for the resolution of inflammatory responses. GPR43-deficient mice ($Gpr43^{-/-}$) showed exacerbation and unresolved inflammation in DSS-induced colitis model due to higher production of inflammatory molecules and increased immune cell infiltration in the gut. GF mice present none or low SCFAs levels, along with increased inflammation and lower resolution of inflammation, compared to colonized mice [42].

Flagellin and Pathogenic Bacteria

Pro-IL1β is the immature precursor of the proinflammatory cytokine IL-1β, produced by intestinal mononuclear phagocytes (iMPs) (macrophages and dendritic cells—DCs). The conversion of pro-IL1β into your active form requires the activation of caspase-1 via NLCR4 inflammasome, which is induced by the bacteria protein flagellin. The enteric pathogens *Pseudomonas aeruginosa* and *S. Typhimurium* express type III secretion system, and then introduce flagellin into the host cytosol, inducing the production of IL-1β. Commensal bacteria do not express type III secretion system, and in this way, do not induce the production of IL-1β. However, it has been shown that in steady state iMPs express pro-IL1β protein [43], indicating a mechanism by which pathogenic bacteria can be discriminated from commensals.

T-Cell Subsets

The microbiota also stimulates adaptive immunity, by inducing the differentiation of CD4+ T cells. The gut CD4+ T cells compartment includes four main T-cell subsets: Foxp3+ regulatory T cells (Treg), T-helper 1 (Th1), Th2, and Th17 cells, and it is believed that the distribution of the different subsets is a dynamic process, guided by microbiota composition [44].

Th17 cells play an important role in maintaining the epithelial barrier, and in host immune defense against extracellular pathogens of the mucosa surface. GF mice lack Th17 cells in the SI and colon LP but can acquire them after colonization with conventional microbiota. Although not all members of the microbiota can influence the differentiation of Th17 cells, the presence of segmented filamentous bacteria (SFB), typical of rodents and important for the immune system, increases the frequency of this subset. Moreover, the colonization with SFB along with Th17 differentiation are thought to be of homeostatic nature, and correlates with increased expression of genes associated with inflammation and antimicrobial defenses, resulting in enhanced resistance to the intestinal pathogen *C. rodentium* [45].

T Regulatory Cells (Treg)

Higher proportions of Foxp3+ Treg cells are found in the small and large intestinal LP. This subset can arise from the thymus, can be natural (nTreg), or can be peripherally induced (iTreg), being further subdivided based on the expression of RORγt transcription factor [46]. Several studies have demonstrated induction of iTreg following microbial colonization. The colonization of GF mice with a consortium of eight murine bacteria species (altered Schaedler flora—ASF), was able to promote the differentiation of iTreg in the colon, but not in the SI [47]. *Clostridia* members, notably, clusters IV and XIVa, have been described as efficient commensals for the induction of iTregs [48]. However, the precise role and mechanisms by which microbial antigens and metabolites regulate the differentiation of Foxp3+ Tregs subsets need to be better elucidated.

Bacteriocins

Lactic acid bacteria can produce antimicrobial peptides known as bacteriocins, with the ability to antagonize the growth of harmful microorganisms, and protect the host from infectious diseases (Fig 13.1) [49].

GUT MICROBIOTA MODULATION: A FOCUS ON THERAPIES

Some therapeutic benefits after use of probiotics, prebiotics, and synbiotics are described in Table 13.1. More recently, a new approach has also been highlighted in clinical trials, the Fecal Microbiota Transplantation (FMT), questing to transfer the whole microbial and substrate content from healthy intestine to sick patients [50].

Antibiotics

Antibiotics are dysbiosis inducers, commonly used to eliminate or decrease pathogenic strains in bacterial-mediated diseases. Although bactericidal and bacteriostatic antibiotics are heavily used in clinical practice, the commercially available drugs persist with low target specificity, and several beneficial strains of bacteria also decrease after their use. In addition, the overuse and misuse of antibiotics accelerate the process of resistance and treatment failure in primary care [51].

Although bacteria have resilience properties, after long periods of antibiotic treatment the intestinal microbiota becomes slightly different from that found before therapy [52]. Moreover, the use of antibiotics changes the gene expression and metabolism of the remaining bacteria, and consequently alters several biological processes of the host (e.g., immune responses, susceptibility to other infections, and gut permeability) [53]. For instance, microbiota depletion impairs the course of IBD by decreasing the availability of bacterial-derived TLR ligands in the gut, which in turn, reduce the immunological surveillance and intestinal repair [54]. Although antibiotics are highly recommended in many cases, alternative therapies should be investigated to improve the outcome of the diseases without impairing the maintenance of host homeostasis.

Probiotics

Most microbial species used in probiotic formulations are *Lactobacillus* (e.g., *reuteri*, *rhamnosus*, *casei*, *acidophilus*, *gasseri*) and *Bifidobacterium* (e.g., *bifidum*, *animalis*, *breve*, *lactis*), but also others such as *Streptococcus* spp., *Lactococcus* spp., *Escherichia coli* (strain Nissle), and *Clostridium* ssp. The yeast *Saccharomyces boulardii* is also commonly used as probiotic [55,56].

Probiotic strains control pathogens through production of antimicrobial substances (e.g., SCFAs, hydroperoxides, bacteriocins, and bile acids), competition for adhesion to the gut epithelium and for nutrients, and inhibition of bacterial toxin synthesis. The soluble mediators produced by probiotic strains and the release of cellular components of pathogenic bacteria (cellular wall components, DNA) increase the immune responses by enhancing immunoglobulin production, macrophage and lymphocyte activity, and production of proinflammatory cytokines [55].

For instance, *Lactobacillus* strains were described as early inducers of TNF-α (tumor necrosis factor α), IL-1β, IL-6, IL-8, and IL-12p70, and they also increase the bactericidal and phagocytosis capacity of human macrophages [57]. The use of probiotics by mothers during or after pregnancy also changes the microbiota composition of fetus and newborns, respectively [58,59], expanding the potential use of probiotics to modulate the dysbiotic microbiota before birth.

TABLE 13.1 Microbiota Modulation: Prebiotics, Probiotics, and Synbiotics

Classification/Ref	Composition/Dosage	Duration	Status/Disease	Outcome
Probiotic[1]	L. casei (10^8 CFU/mL)	12 months	Allergic asthma and/or rhinitis	Decreased number of rhinitis episodes
Probiotic[2]	B. longum 536 ($1–3 \times 10^{11}$ CFU)	8 weeks	Ulcerative colitis, proctitis	Clinical remission
Probiotic[3]	L. delbrueckiisubsp. Bulgaricus (NI), S. thermophilus (NI), B. animalis subsp. Lactis ($1,11 \times 10^7$), L. acidophilus ($1,11 \times 10^7$)	8 weeks	Type 2 diabetes	Decreased HbA1c and TNF-α levels
Probiotic[4]	L. rhamnosus GR-1 (10^9) and L. reuteri RC-14 (10^9)	4 weeks	Vulvovaginal candidiasis	Decreased number of yeast and symptoms
Probiotic[5]	Escherichia coli Nissle 1917 ($2.5–25 \times 10^9$ CFU)	12 weeks	Irritable bowel syndrome	Improved clinical signs
Prebiotic[6]	FOS (15 g)	3 weeks	Crohn disease	Increased Bifidobacteria ssp., IL-10+TLR2+TLR4+ DCs, decreased Crohn's disease index score
Prebiotic[7]	GOS (3.5–7 g)	12 weeks	Irritable bowel syndrome	Increased Bifidobacteria and decreased symptoms
Prebiotic[8]	Inulin (20 g)	3 weeks	Hypercholesterolemia	Decreased triglycerides and cholesterol levels
Prebiotic[9]	GOS/FOS (9:1 mix ratio; 0.4 g/100 mL)	12 months	Intestinal and respiratory infection	Decreased number of infectious cases
Synbiotic[10]	L. salivarius (2.10^9 CFU), FOS (10 g)	6 weeks	Healthy young individuals	Decreased E. coli, lipids, cytokines (TNF-α, IL-1β, IL-6), increased lactobacilli
Synbiotic[11]	L. acidophilus (2×10^9), L. casei (7×10^9), L. rhamnosus (1.5×10^9); L. bulgaricus (2×10^8), B. breve (2×10^{10}), B. longum (7×10^9), S. thermophilus (1.5×10^9), FOS (100 mg)	8 weeks	Type 2 diabetes	Decreased fasting plasma glucose levels
Synbiotic[12]	B. longum (2×10^{11}), FOS/Inulin(6 g, 1:1)	4 weeks	Ulcerative colitis	Reduction of inflammation
Synbiotic[13]	L. acidophilus (7×10^8), B. longum (4×10^8), B. infantis (3×10^8), L. rhamnosus (4×10^8), L. plantaris (3×10^8), L. casei (3×10^8), L. bulgaricus (3×10^8), B. breve (3×10^8), FOS (100 mg)	1 week	Necrotizing enterocolitis, sepsis, mortality	Reduced necrotizing enterocolitis severity
Synbiotic[14]	L. paracasei ssp. paracasei (10^8/100 mL), GOS (540 mg), FOS (61 mg)	6 months	Respiratory tract infections	Decreased number of infections

B., Bifidobacterium; DCs, dendritic cells; E., Escherichia; FOS, fructooligosaccharides; GOS, galactooligosaccharides; HbA1c, glycated hemoglobin A1c; L., Lactobacillus; NI, not informed; S., Streptococcus; TLR, Toll-like receptors.

[1]Giovannini M, et al. A randomized prospective double blind controlled trial on effects of long-term consumption of fermented milk containing Lactobacillus casei in pre-school children with allergic asthma and/or rhinitis. Pediatric Res 2007;62:215.

[2]Tamaki H, et al. Efficacy of probiotic treatment with Bifidobacterium longum 536 for induction of remission in active ulcerative colitis: a randomized, double-blinded, placebo-controlled multicenter trial. Dig Endosc 2016;28(1):67–74.

[3]Mazloom Z, Yousefinejad A, Dabbaghmanesh MH. Effect of probiotics on lipid profile, glycemic control, insulin action, oxidative stress, and inflammatory markers in patients with type 2 diabetes: a clinical trial. Iran J Med Sci 2013;38(1):38–43.

[4]Martinez RC, et al. Improved treatment of vulvovaginal candidiasis with fluconazole plus probiotic Lactobacillus rhamnosus GR-1 and Lactobacillus reuteri RC-14. Lett Appl Microbiol 2009;48(3):269–74.

[5]Kruis W, et al. A double-blind placebo-controlled trial to study therapeutic effects of probiotic Escherichia coli Nissle 1917 in subgroups of patients with irritable bowel syndrome. Int J Colorectal Dis 2012;27(4):467–74.

[6]Lindsay JO, et al. Clinical, microbiological, and immunological effects of fructooligosaccharide in patients with Crohn disease. Gut 2006;55(3):348–55.

[7]Silk DB, et al. Clinical trial: the effects of a trans-galactooligosaccharide prebiotic on fecal microbiota and symptoms in irritable bowel syndrome. Aliment Pharmacol Ther 2009;29(5):508–18.

[8]Causey JL, et al. Effects of dietary inulin on serum lipids, blood glucose, and the gastrointestinal environment in hypercholesterolemic men. Nutr Res 2000;20(2):191–201.

[9]Bruzzese E, et al. A formula containing galacto- and fructooligosaccharides prevents intestina and extra-intestinal infections: an observational study. Clin Nutr 2009;28(2):156–61.

[10]Rajkumar H, et al. Effect of Probiotic Lactobacillus salivarius UBL S22 and prebiotic fructo-oligosaccharide on serum lipids, inflammatory markers, insulin sensitivity, and gut bacteria in healthy young volunteers: a randomized controlled single-blind pilot study. J Cardiovasc Pharmacol Ther 2015;20(3):289–98.

[11]Asemi Z, et al. Effect of multispecies probiotic supplements on metabolic profiles, hs-CRP, and oxidative stress in patients with type 2 diabetes. Ann Nutr Metab 2013;63(1–2):1–9.

[12]Furrie E, et al. Synbiotic therapy (Bifidobacterium longum Synergy 1) initiates resolution of inflammation in patients with active ulcerative colitis: a randomized controlled pilot trial. Gut 2005;54(2):242–9.

[13]Nandhini LP, et al. Synbiotics for decreasing incidence of necrotizing enterocolitis among preterm neonates—a randomized controlled trial. J Matern Fetal Neonatal Med 2016;29(5):821–5.

[14]Szajewska H, et al. Effects of infant formula supplemented with prebiotics compared with synbiotics on growth up to the age of 12 mo: a randomized controlled trial. Pediatr Res 2017;81(5):752–8.

Prebiotics

Examples of prebiotics include the fructans (fructooligosaccharides— FOS— and inulin) and galactans (galactooligosaccharides—GOS). However, the carbohydrates are potential sources to be used as prebiotics. These bioactive compounds mainly act on *Lactobacillus* spp. and/or *Bifidobacterium* spp., but also on *Roseburia* spp., *Eubacterium* spp., and *Faecalibacterium* spp.

Inulin increases the proportion of the genus *Bifidobacteria*, and has been shown to improve both obesity- and diabetes-related complications [60]. On the other hand, both FOS and GOS support the growth of species of *Bacteroides*, *Lactobacilli*, *Enterobacteria*, *Streptococci*, and *Bifidobacteria*, while decreasing some strains of *Bacteroides* and *Candida*, and they have showed good potential to treat both *Listeria monocytogenes* and *S. typhimurium* infections, with some actions on colorectal cancer [61,62].

In addition to changing microbiota composition, prebiotics enhance digestion processing, synthesis of beneficial bioactive and antibiotic compounds, increase mineral bioavailability, and enhance the viability of epithelial cells in the colon by increasing SCFAs production [63,64]. GOS- and FOS-induced SCFAs have been shown to increase mucin production by intestinal epithelial cells, number of leukocytes in GALT and blood, IgA secretion and macrophage phagocytic activity [55,65]. Moreover, the use of FOS alone was sufficient to decrease Crohn Disease Activity Index in sick patients [66].

Synbiotics

Synbiotics increase good bacterial strains (e.g., *Bifidobacterium* spp. and *Lactobacillus* spp.) and production of bioactive compounds (e.g., SCFAs) [67]. They were created to increase probiotic survival and function, since the gastrointestinal tract can be a quite hostile environment. The use of synbiotics has been promising by decreasing the number of infections in postoperative patients, improving the outcome of IBD, and presenting anticancer effects [68]. The prophylactic use of a pharmaceutical formulation containing oligofructose-enriched inulin, *Lactobacillus rhamnosus,* and *Bifidobacterium lactis* Bb12, decreased cancer pathogenicity in colon cancer and polypectomized patients [69].

OTHER REPERCUSSIONS OF MUCOSAL IMMUNITY

About 10% of the host transcriptome is microbially regulated, especially genes involved in immunity, cell proliferation, and metabolism [70]. The colonization of GF mice with a single commensal bacterium resulted in increased transcription of genes involved in several intestinal functions, including nutrient absorption, mucosal barrier functionality, xenobiotic metabolism, angiogenesis, and postnatal intestinal maturation [71].

Gut microbes and their products can be recognized by a variety of receptors, including TLRs and NLRs. The pattern recognition of microbial molecules by the gut mucosal immunity regulates intestinal homeostasis and influences systemic immunity and the host metabolism. PRRs in the intestinal mucosa are not only involved in the elimination of pathogens but are also involved in host metabolic homeostasis. A variety of studies using global knockout models of TLRs, including TLR2 and TLR5, observed an increase in body weight or fat mass [72,73]. Surprisingly, the phenotypes in cell-specific knockout models are rather diverse. Loss of TLR5 from intestinal epithelial cells leads to altered gut microbiota, low-grade inflammation, and metabolic syndrome features [74]. The lack of TLR5 would allow intestinal epithelium breaching by bacteria, compromising the efficient recruitment of immune cells. On the other hand, the deletion of the signaling adaptor molecule of TLRs in IECs, Myd88 (Myeloid differentiation primary response 88), reduces fat mass accumulation and body weight and improves glucose metabolism [75].

The gut microbiota plays a critical role in arthritis. An elegant study revealed that intestinal leukocytes migrate to and from the intestine in steady state [76]. In addition, the fraction of gut-derived Th17 cells present in the spleen correlated with serum levels of autoantibodies, in a model of rheumatoid arthritis. Other studies found high levels of Th17 cells, and decreased levels of Tregs in blood of patients with active rheumatoid arthritis [77,78]. Since Th17 differentiation in intestinal mucosa is dependent on specific intestinal microbiota, exposure to disturbed gut microbiota may be critical in rheumatoid arthritis.

Increasing data also reveal that microbiota—host interactions drive brain development, function, and behavior. GF mice exhibit profound behavioral alterations and neuropathologies including psychiatric, neurodevelopmental, and neurodegenerative disorders [79]. Maturation of nervous system appears to be highly dependent on microbial colonization, during a specific time frame of development. GF mice show an exaggerated hypothalamic pituitary adrenal (HPA) stress response, which is normalized following monocolonization at 6 weeks, but not at 14 weeks [80].

Besides bacterial products and release of gut hormones, microbiota has access to the brain via an interconnection among primary afferent neurons, immune cells, and enteroendocrine cells. Subsets of vagal afferent nerve terminals, closely located to mucosal immune cells, are responsive to signaling molecules from these cells such as mast cells products [81]. Immune cell products can also modulate enteroendocrine cell functions, including cholecystokinin secretion [82].

BACTERIAL TRANSLOCATION: JUST A THEORETICAL POSSIBILITY OR A REAL CAUSE OF SYSTEMIC INFECTION

The intestinal barrier is a complex and dynamic system composed by several lines of defense against pathogens, including physical, chemical, and functional immunological barriers. In order to have access to the systemic circulation, microbes would have to pass through the luminal mucus, IgA, and antimicrobial defensins. Secondly, the epithelial barrier represents a major obstacle against invaders. Thirdly, the GIT is a major reservoir of macrophages, highly specialized in recognition and phagocytosis [83,84].

Other structures, such as blood vessels, smooth muscle layers, and the enteric nervous system, participate in the regulation of intestinal barrier. An additional clearance can be performed by the liver, where sinusoidal endothelial cells and Kupffer cells are responsive to direct stimulation by bacterial antigens [85]. Finally, in healthy conditions, circulating factors, peripheral blood monocytes, and tissue macrophages can neutralize microbial products.

The intestinal physical barrier is essentially composed by the mucus layer and the epithelial monolayer, which confers a selective permeability to the intestinal mucosa. A series of intercellular junctions seals the paracellular space, regulating the transmucosal flux [86]. Enteric pathogens can disrupt tight junctions by either altering cellular cytoskeleton, or by affecting tight junction proteins, via bacterial-derived proteases or biochemical alterations [87]. In addition, proinflammatory cytokines can also increase intestinal permeability.

Preclinical studies revealed that TNF-α signaling directly regulates tight junction functions. TNF-α increases intestinal permeability by both stimulating myosin light chain phosphorylation [88] and causing occludin endocytosis [89]. In case of tight junction disruption or direct epithelial cell damage, luminal content can cross the intestinal barrier in an unrestricted pathway.

Bacterial translocation (BT) is defined as translocation of bacteria and/or bacterial products (lipopolysaccharides [LPS], peptidoglycans, muramyl-dipeptides, and bacterial DNA) through the gut mucosa to normally sterile tissues such as the mesenteric lymph nodes and the internal organs [90]. BT may be a physiological phenomenon that occurs in healthy individuals without deleterious consequences [91]. It has been suggested that indigenous bacteria are continuously translocating in low number from the GIT, and they would be eliminated by the host reticuloendothelial system [92].

In nonhomeostatic conditions such as immunosuppression, bacterial overgrowth, and intestinal hypomotility, intestinal permeability may be compromised, resulting in a continuous process of BT, ultimately triggering long-term inflammatory process [93]. Dietary factors may also impair intestinal permeability, by altering tight junction proteins. High-fat diet reduced expression of claudin-1, claudin-3, occludin, and junctional adhesion molecule-1 in the intestinal mucosa [94].

Individuals with IBD have increased systemic levels of proinflammatory cytokines, which has been suggested to be due to elevated levels of LPS [95], bacterial DNA [96], endotoxin core antibodies [97] and LPS-binding protein [98]. A variety of intestinal and extraintestinal disorders have been associated with altered intestinal barrier and increased intestinal permeability, including IBS, celiac disease, allergies, arthritis, and metabolic diseases [99].

The majority of these associations were merely correlative, and there is little evidence relating barrier impairment to disease pathogenesis. Most in vivo studies involve assessment of the paracellular pathway, and when increased permeability is discovered, extrapolations to bacterial translocation are made [100]. Furthermore, animal studies have revealed that physiologic breaks in the barrier must occur, for homeostatic regulatory T-cell responses, necessary for mucosa protection from inflammation [101].

LIST OF ACRONYMS AND ABBREVIATIONS

AhR Aryl hydrocarbon receptor
ALR AIM2-like receptor
ASF Altered Schaedler flora
BT Bacterial translocation
CFU Colony-forming unit
CP Cryptopatch
DC Dendritic cell

DNA Deoxyribonucleic acid
DSS Dextran sulfate sodium
EHEC Enterohemorrhagic *Escherichia coli*
FFAR2 Free fatty acid receptor 2
FOS Fructooligosaccharide
FXR Farnesoid X receptor
GALT Gut-associated lymphoid tissue
GF Germ free
GIT Gastrointestinal tract
GOS galactooligosaccharide
GPR41 G-protein-coupled receptor 41
GPR43 G-protein-coupled receptor 43
HPA Hypothalamic pituitary adrenal
IBD Inflammatory Bowel Disease
IBS Inflammatory intestinal syndrome
IEC Intestinal epithelial cell
Ig Immunoglobulin
ILF Isolated lymphoid follicle
iMP Intestinal mononuclear phagocyte
iTreg Induced Treg
LP Lamina propria
LPS lipopolysaccharides
MyD88 Myeloid differentiation primary response 88
NLR Nucleotide-binding domain leucine-rich repeat containing receptor
NLRC4 NLR family CARD domain-containing protein 4
NOD1 Nucleotide-binding oligomerization domain containing 1
nTreg Natural Treg
PP Peyer patch
PRR Pattern recognition receptor
RLR RIG-I like receptor
RORγt RAR-related orphan receptor gamma t
SCFA Short-chain fatty acid
SFB Segmented filamentous bacterium
SI Small intestine
TGF-α Transforming growth factor-beta
Th T-helper cell
TLR Toll-like receptor
TNF-α Tumor necrosis factor α
Treg Regulatory T cell
WHO World Health Organization

REFERENCES

[1] Aagaard K, Ma J, Antony KM, Ganu R, Petrosino J, Versalovic J. The placenta harbors a unique microbiome. Sci Transl Med 2014;6(237):237ra65.
[2] Jiménez E, Marín ML, Martín R, et al. Is meconium from healthy newborns actually sterile? Res Microbiol 2008;159(3):187–93.
[3] Sekirov I, Russell SL, Antunes LC, Finlay BB. Gut microbiota in health and disease. Physiol Rev 2010;90(3):859–904.
[4] Leffler DA, Lamont JT. *Clostridium difficile* infection. N Engl J Med 2015;372(16):1539–48.
[5] Chow J, Tang H, Mazmanian SK. Pathobionts of the gastrointestinal microbiota and inflammatory disease. Curr Opin Immunol 2011;23(4):473–80.
[6] Theriot CM, Young VB. Interactions between the gastrointestinal microbiome and *Clostridium difficile*. Annu Rev Microbiol 2015;69:445–61.
[7] Osawa N, Mitsuhashi S. Infection of germeree mice with shigella flexneri 3A. Jpn J Exp Med 1964;34:77–80.
[8] Ferreira RB, Gill N, Willing BP, et al. The intestinal microbiota plays a role in *Salmonella*-induced colitis independent of pathogen colonization. PLoS One 2011;6(5):e20338.
[9] Willing BP, Vacharaksa A, Croxen M, Thanachayanont T, Finlay BB. Altering host resistance to infections through microbial transplantation. PLoS One 2011;6(10):e26988.
[10] Garrett WS, Punit S, Gallini CA, et al. Colitis-associated colorectal cancer driven by T-bet deficiency in dendritic cells. Cancer Cell 2009;16(3):208–19.

[11] Kamada N, Kim YG, Sham HP, et al. Regulated virulence controls the ability of a pathogen to compete with the gut microbiota. Science 2012;336(6086):1325—9.

[12] Gantois I, Ducatelle R, Pasmans F, et al. Butyrate specifically down-regulates salmonella pathogenicity island 1 gene expression. Appl Environ Microbiol 2006;72(1):946—9.

[13] Schamberger GP, Diez-Gonzalez F. Selection of recently isolated colicinogenic *Escherichia coli* strains inhibitory to *Escherichia coli* O157:H7. J Food Prot 2002;65(9):1381—7.

[14] Marteyn B, West NP, Browning DF, et al. Modulation of Shigella virulence in response to available oxygen in vivo. Nature 2010;465(7296):355—8.

[15] Cherrington CA, Hinton M, Pearson GR, Chopra I. Short-chain organic acids at ph 5.0 kill *Escherichia coli* and *Salmonella* spp. without causing membrane perturbation. J Appl Bacteriol 1991;70(2):161—5.

[16] Buffie CG, Bucci V, Stein RR, et al. Precision microbiome reconstitution restores bile acid mediated resistance to *Clostridium difficile*. Nature 2015;517(7533):205—8.

[17] Ng KM, Ferreyra JA, Higginbottom SK, et al. Microbiota-liberated host sugars facilitate post-antibiotic expansion of enteric pathogens. Nature 2013;502(7469):96—9.

[18] Fabich AJ, Jones SA, Chowdhury FZ, et al. Comparison of carbon nutrition for pathogenic and commensal *Escherichia coli* strains in the mouse intestine. Infect Immun 2008;76(3):1143—52.

[19] Pacheco AR, Curtis MM, Ritchie JM, et al. Fucose sensing regulates bacterial intestinal colonization. Nature 2012;492(7427):113—7.

[20] Maltby R, Leatham-Jensen MP, Gibson T, Cohen PS, Conway T. Nutritional basis for colonization resistance by human commensal *Escherichia coli* strains HS and Nissle 1917 against *E. coli* O157:H7 in the mouse intestine. PLoS One 2013;8(1):e53957.

[21] Kim MH, Kang SG, Park JH, Yanagisawa M, Kim CH. Short-chain fatty acids activate GPR41 and GPR43 on intestinal epithelial cells to promote inflammatory responses in mice. Gastroenterology 2013;145(2):396—406.e1-10.

[22] Samuel BS, Shaito A, Motoike T, et al. Effects of the gut microbiota on host adiposity are modulated by the short-chain fatty-acid binding G protein-coupled receptor, Gpr41. Proc Natl Acad Sci USA 2008;105(43):16767—72.

[23] Macfarlane GT, Macfarlane S. Fermentation in the human large intestine: its physiologic consequences and the potential contribution of prebiotics. J Clin Gastroenterol 2011;45(Suppl.):S120—7.

[24] Ridlon JM, Kang DJ, Hylemon PB. Bile salt biotransformations by human intestinal bacteria. J Lipid Res 2006;47(2):241—59.

[25] Dawson PA, Lan T, Rao A. Bile acid transporters. J Lipid Res 2009;50(12):2340—57.

[26] Swann JR, Want EJ, Geier FM, et al. Systemic gut microbial modulation of bile acid metabolism in host tissue compartments. Proc Natl Acad Sci USA 2011;108(Suppl. 1):4523—30.

[27] Lefebvre P, Cariou B, Lien F, Kuipers F, Staels B. Role of bile acids and bile acid receptors in metabolic regulation. Physiol Rev 2009;89(1):147—91.

[28] Fujisawa T, Mori M. Influence of various bile salts on beta-glucuronidase activity of intestinal bacteria. Lett Appl Microbiol 1997;25(2):95—7.

[29] Kandell RL, Bernstein C. Bile salt/acid induction of DNA damage in bacterial and mammalian cells: implications for colon cancer. Nutr Cancer 1991;16(3—4):227—38.

[30] Flahaut S, Frere J, Boutibonnes P, Auffray Y. Comparison of the bile salts and sodium dodecyl sulfate stress responses in *Enterococcus faecalis*. Appl Environ Microbiol 1996;62(7):2416—20.

[31] Sanyal AJ, Shiffmann ML, Hirsch JI, Moore EW. Premicellar taurocholate enhances ferrous iron uptake from all regions of rat small intestine. Gastroenterology 1991;101(2):382—9.

[32] Chatterjee A, Dutta PK, Chowdhury R. Effect of fatty acids and cholesterol present in bile on expression of virulence factors and motility of *Vibrio cholerae*. Infect Immun 2007;75(4):1946—53.

[33] Zelante T, Iannitti RG, Cunha C, et al. Tryptophan catabolites from microbiota engage aryl hydrocarbon receptor and balance mucosal reactivity via interleukin-22. Immunity 2013;39(2):372—85.

[34] Nguyen NT, Kimura A, Nakahama T, et al. Aryl hydrocarbon receptor negatively regulates dendritic cell immunogenicity via a kynurenine-dependent mechanism. Proc Natl Acad Sci USA 2010;107(46):19961—6.

[35] Martínez-del Campo A, Bodea S, Hamer HA, et al. Characterization and detection of a widely distributed gene cluster that predicts anaerobic choline utilization by human gut bacteria. mBio 2015;6(2).

[36] van de Pavert SA, Olivier BJ, Goverse G, et al. Chemokine CXCL13 is essential for lymph node initiation and is induced by retinoic acid and neuronal stimulation. Nat Immunol 2009;10(11):1193—9.

[37] Kiss EA, Vonarbourg C, Kopfmann S, et al. Natural aryl hydrocarbon receptor ligands control organogenesis of intestinal lymphoid follicles. Science 2011;334(6062):1561—5.

[38] Cella M, Colonna M. Aryl hydrocarbon receptor: linking environment to immunity. Semin Immunol 2015;27(5):310—4.

[39] Bouskra D, Brezillon C, Berard M, et al. Lymphoid tissue genesis induced by commensals through NOD1 regulates intestinal homeostasis. Nature 2008;456(7221):507—10.

[40] Round JL, Mazmanian SK. The gut microbiota shapes intestinal immune responses during health and disease. Nat Rev Immunol 2009;9(5):313—23.

[41] Cash HL, Whitham CV, Behrendt CL, Hooper LV. Symbiotic bacteria direct expression of an intestinal bactericidal lectin. Science 2006;313(5790):1126—30.

[42] Maslowski KM, Vieira AT, Ng A, et al. Regulation of inflammatory responses by gut microbiota and chemoattractant receptor GPR43. Nature 2009;461(7268):1282−6.

[43] Franchi L, Kamada N, Nakamura Y, et al. NLRC4-driven production of IL-1beta discriminates between pathogenic and commensal bacteria and promotes host intestinal defense. Nat Immunol 2012;13(5):449−56.

[44] Lu JT, Xu AT, Shen J, Ran ZH. Crosstalk between intestinal epithelial cell and adaptive immune cell in intestinal mucosal immunity. J Gastroenterol Hepatol 2017;32(5):975−80.

[45] Ivanov II, Atarashi K, Manel N, et al. Induction of intestinal Th17 cells by segmented filamentous bacteria. Cell 2009;139(3):485−98.

[46] Yadav M, Louvet C, Davini D, et al. Neuropilin-1 distinguishes natural and inducible regulatory T cells among regulatory T cell subsets in vivo. J Exp Med 2012;209(10):1713−22. s1−19.

[47] Geuking MB, Cahenzli J, Lawson MA, et al. Intestinal bacterial colonization induces mutualistic regulatory T cell responses. Immunity 2011;34(5):794−806.

[48] Atarashi K, Tanoue T, Shima T, et al. Induction of colonic regulatory T cells by indigenous Clostridium species. Science 2011;331(6015):337−41.

[49] Chikindas ML, Weeks R, Drider D, Chistyakov VA, Dicks LM. Functions and emerging applications of bacteriocins. Curr Opin Biotechnol 2018;49:23−8.

[50] Rossen NG, MacDonald JK, de Vries EM, et al. Fecal microbiota transplantation as novel therapy in gastroenterology: a systematic review. World Journal of Gastroenterology 2015;21(17):5359−71.

[51] Currie CJ, Berni E, Jenkins-Jones S, et al. Antibiotic treatment failure in four common infections in UK primary care 1991-2012: longitudinal analysis. BMJ 2014;349:g5493.

[52] Dethlefsen L, Relman DA. Incomplete recovery and individualized responses of the human distal gut microbiota to repeated antibiotic perturbation. Proc Natl Acad Sci USA 2011;108(Suppl. 1):4554−61.

[53] Tulstrup MV, Christensen EG, Carvalho V, et al. Antibiotic treatment affects intestinal permeability and gut microbial composition in wistar rats dependent on antibiotic class. PLoS One 2015;10(12):e0144854.

[54] Rakoff-Nahoum S, Paglino J, Eslami-Varzaneh F, Edberg S, Medzhitov R. Recognition of commensal microflora by toll-like receptors is required for intestinal homeostasis. Cell 2004;118(2):229−41.

[55] Markowiak P, Slizewska K. Effects of probiotics, prebiotics, and synbiotics on human health. Nutrients 2017;9(9).

[56] Patel R, DuPont HL. New approaches for bacteriotherapy: prebiotics, new-generation probiotics, and synbiotics. Clin Infect Dis 2015;60(Suppl. 2):S108−21.

[57] Rocha-Ramirez LM, Perez-Solano RA, Castanon-Alonso SL, et al. Probiotic lactobacillus strains stimulate the inflammatory response and activate human macrophages. J Immunol Res 2017;2017:4607491.

[58] Gueimonde M, Sakata S, Kalliomaki M, Isolauri E, Benno Y, Salminen S. Effect of maternal consumption of lactobacillus GG on transfer and establishment of fecal bifidobacterial microbiota in neonates. J Pediatr Gastroenterol Nutr 2006;42(2):166−70.

[59] Schultz M, Gottl C, Young RJ, Iwen P, Vanderhoof JA. Administration of oral probiotic bacteria to pregnant women causes temporary infantile colonization. J Pediatr Gastroenterol Nutr 2004;38(3):293−7.

[60] Cani PD, Possemiers S, Van de Wiele T, et al. Changes in gut microbiota control inflammation in obese mice through a mechanism involving GLP-2-driven improvement of gut permeability. Gut 2009;58(8):1091−103.

[61] Buddington KK, Donahoo JB, Buddington RK. Dietary oligofructose and inulin protect mice from enteric and systemic pathogens and tumor inducers. J Nutr 2002;132(3):472−7.

[62] Roberfroid M. Prebiotics: the concept revisited. J Nutr 2007;137(3 Suppl. 2):830S−7S.

[63] Tan J, McKenzie C, Potamitis M, Thorburn AN, Mackay CR, Macia L. The role of short-chain fatty acids in health and disease. Adv Immunol 2014;121:91−119.

[64] Singh SP, Jadaun JS, Narnoliya LK, Pandey A. Prebiotic oligosaccharides: special focus on fructooligosaccharides, its biosynthesis and bioactivity. Appl Biochem Biotechnol 2017;183(2):613−35.

[65] Scholtens PA, Alliet P, Raes M, et al. Fecal secretory immunoglobulin A is increased in healthy infants who receive a formula with short-chain galacto-oligosaccharides and long-chain fructo-oligosaccharides. J Nutr 2008;138(6):1141−7.

[66] Lindsay JO, Whelan K, Stagg AJ, et al. Clinical, microbiological, and immunological effects of fructo-oligosaccharide in patients with Crohn's disease. Gut 2006;55(3):348−55.

[67] Bull MJ, Plummer NT. Part 2: treatments for chronic gastrointestinal disease and gut dysbiosis. Integr Med 2015;14(1):25−33.

[68] Pandey KR, Naik SR, Vakil BV. Probiotics, prebiotics and synbiotics - a review. J Food Sci Technol 2015;52(12):7577−87.

[69] Rafter J, Bennett M, Caderni G, et al. Dietary synbiotics reduce cancer risk factors in polypectomized and colon cancer patients. Am J Clin Nutr 2007;85(2):488−96.

[70] Sommer F, Nookaew I, Sommer N, Fogelstrand P, Backhed F. Site-specific programming of the host epithelial transcriptome by the gut microbiota. Genom Biol 2015;16:62.

[71] Hooper LV, Wong MH, Thelin A, Hansson L, Falk PG, Gordon JI. Molecular analysis of commensal host-microbial relationships in the intestine. Science 2001;291(5505):881−4.

[72] Caricilli AM, Picardi PK, de Abreu LL, et al. Gut microbiota is a key modulator of insulin resistance in TLR 2 knockout mice. PLoS Biol 2011;9(12):e1001212.

[73] Vijay-Kumar M, Aitken JD, Carvalho FA, et al. Metabolic syndrome and altered gut microbiota in mice lacking Toll-like receptor 5. Science 2010;328(5975):228−31.

[74] Chassaing B, Ley RE, Gewirtz AT. Intestinal epithelial cell toll-like receptor 5 regulates the intestinal microbiota to prevent low-grade inflammation and metabolic syndrome in mice. Gastroenterology 2014;147(6):1363−1377 e17.

[75] Everard A, Geurts L, Caesar R, et al. Intestinal epithelial MyD88 is a sensor switching host metabolism towards obesity according to nutritional status. Nat Commun 2014;5:5648.

[76] Morton AM, Sefik E, Upadhyay R, Weissleder R, Benoist C, Mathis D. Endoscopic photoconversion reveals unexpectedly broad leukocyte trafficking to and from the gut. Proc Natl Acad Sci USA 2014;111(18):6696−701.

[77] Samson M, Audia S, Janikashvili N, et al. Brief report: inhibition of interleukin-6 function corrects Th17/Treg cell imbalance in patients with rheumatoid arthritis. Arthritis Rheum 2012;64(8):2499−503.

[78] Wang W, Shao S, Jiao Z, Guo M, Xu H, Wang S. The Th17/Treg imbalance and cytokine environment in peripheral blood of patients with rheumatoid arthritis. Rheumatol Int 2012;32(4):887−93.

[79] Sampson TR, Mazmanian SK. Control of brain development, function, and behavior by the microbiome. Cell Host Microbe 2015;17(5):565−76.

[80] Sudo N, Chida Y, Aiba Y, et al. Postnatal microbial colonization programs the hypothalamic-pituitary-adrenal system for stress response in mice. J Physiol 2004;558(Pt 1):263−75.

[81] Barbara G, Wang B, Stanghellini V, et al. Mast cell-dependent excitation of visceral-nociceptive sensory neurons in irritable bowel syndrome. Gastroenterology 2007;132(1):26−37.

[82] McDermott JR, Leslie FC, D'Amato M, Thompson DG, Grencis RK, McLaughlin JT. Immune control of food intake: enteroendocrine cells are regulated by CD4+ T lymphocytes during small intestinal inflammation. Gut 2006;55(4):492−7.

[83] Smith PD, Ochsenbauer-Jambor C, Smythies LE. Intestinal macrophages: unique effector cells of the innate immune system. Immunol Rev 2005;206:149−59.

[84] Smythies LE, Sellers M, Clements RH, et al. Human intestinal macrophages display profound inflammatory anergy despite avid phagocytic and bacteriocidal activity. J Clin Investig 2005;115(1):66−75.

[85] Shnyra A, Lindberg AA. Scavenger receptor pathway for lipopolysaccharide binding to Kupffer and endothelial liver cells in vitro. Infect Immun 1995;63(3):865−73.

[86] Shen L, Weber CR, Raleigh DR, Yu D, Turner JR. Tight junction pore and leak pathways: a dynamic duo. Annu Rev Physiol 2011;73:283−309.

[87] Berkes J, Viswanathan VK, Savkovic SD, Hecht G. Intestinal epithelial responses to enteric pathogens: effects on the tight junction barrier, ion transport, and inflammation. Gut 2003;52(3):439−51.

[88] Zolotarevsky Y, Hecht G, Koutsouris A, et al. A membrane-permeant peptide that inhibits MLC kinase restores barrier function in in vitro models of intestinal disease. Gastroenterology 2002;123(1):163−72.

[89] Marchiando AM, Shen L, Graham WV, et al. Caveolin-1-dependent occludin endocytosis is required for TNF-induced tight junction regulation in vivo. J Cell Biol 2010;189(1):111−26.

[90] Berg RD, Garlington AW. Translocation of certain indigenous bacteria from the gastrointestinal tract to the mesenteric lymph nodes and other organs in a gnotobiotic mouse model. Infect Immun 1979;23(2):403−11.

[91] Sedman PC, Macfie J, Sagar P, et al. The prevalence of gut translocation in humans. Gastroenterology 1994;107(3):643−9.

[92] Berg RD. Bacterial translocation from the gastrointestinal tract. Trends Microbiol 1995;3(4):149−54.

[93] Koh IH, Liberatore AM, Menchaca-Diaz JL, et al. Bacterial translocation, microcirculation injury and sepsis. Endocr Metab Immune Disord - Drug Targets 2006;6(2):143−50.

[94] Suzuki T, Hara H. Dietary fat and bile juice, but not obesity, are responsible for the increase in small intestinal permeability induced through the suppression of tight junction protein expression in LETO and OLETF rats. Nutr Metab 2010;7:19.

[95] Caradonna L, Amati L, Magrone T, Pellegrino NM, Jirillo E, Caccavo D. Enteric bacteria, lipopolysaccharides and related cytokines in inflammatory bowel disease: biological and clinical significance. J Endotoxin Res 2000;6(3):205−14.

[96] Gutierrez A, Frances R, Amoros A, et al. Cytokine association with bacterial DNA in serum of patients with inflammatory bowel disease. Inflamm Bowel Dis 2009;15(4):508−14.

[97] Pasternak BA, D'Mello S, Jurickova II, et al. Lipopolysaccharide exposure is linked to activation of the acute phase response and growth failure in pediatric Crohn's disease and murine colitis. Inflamm Bowel Dis 2010;16(5):856−69.

[98] Pastor Rojo O, Lopez San Roman A, Albeniz Arbizu E, de la Hera Martinez A, Ripoll Sevillano E, Albillos Martinez A. Serum lipopolysaccharide-binding protein in endotoxemic patients with inflammatory bowel disease. Inflamm Bowel Dis 2007;13(3):269−77.

[99] Konig J, Wells J, Cani PD, et al. Human intestinal barrier function in health and disease. Clin Transl Gastroenterol 2016;7(10):e196.

[100] Quigley EM. Leaky gut - concept or clinical entity? Curr Opin Gastroenterol 2016;32(2):74−9.

[101] Boirivant M, Amendola A, Butera A, et al. A transient breach in the epithelial barrier leads to regulatory T-cell generation and resistance to experimental colitis. Gastroenterology 2008;135(5):1612−1623 e5.

Chapter 14

The Cross Talk Between Bile Acids and Intestinal Microbiota: Focus on Metabolic Diseases and Bariatric Surgery

Jarlei Fiamoncini

Department of Food and Experimental Nutrition, School of Pharmaceutical Sciences, University of São Paulo, São Paulo, Brazil

INTRODUCTION

For a long time, bile acids (BA) have been known for their role in the digestion of dietary lipids. In the last 20 years, new functions have been attributed to them, putting these metabolites in the spotlight of research in different areas of physiology [1–3]. The popularization of metabolomics led to the development of high-throughput analytical methods, providing tools for the investigation of their metabolism. Now it is known that BA can directly modulate energy metabolism and immune function, by acting on specific signaling pathways, or indirectly by changing the composition of the intestinal microbiota [4,5].

Similarly, the intestinal microbiome has been long acknowledged for its functions, as part of the innate immune system, and also synthesizing vitamins and other bioactive compounds. We aim to provide in the next pages the basis to understand the complex relationship between BA and the intestinal microbiota and its impact on health.

BILE ACID PHYSIOLOGY AND METABOLISM

Bile acids are synthetized in the liver from cholesterol in a process that constitutes the major pathway for cholesterol degradation in humans. The first step for BA synthesis via the classical pathway is the hydroxylation at the C7 position of the steroid ring of cholesterol, catalyzed by the enzyme cholesterol 7α-hydroxylase (Cyp7a1), followed by modification of the steroid ring and side-chain shortening, and culminating in the synthesis of cholic acid. Alternatively, cholesterol can be hydroxylated at the C27 position of the side chain, by the enzyme mitochondrial sterol 27-hydroxylase (Cyp27a1), eventually leading to the synthesis of chenodeoxycholic acid by the acidic pathway [6,7].

Cholic acid and chenodeoxycholic acid are named primary BA. These BA are conjugated to taurine or glycine, by the enzyme bile acid CoA: amino acid *N*-acyltransferase (BAAT) [1]. While glycine is preferably used in humans, taurine is the most common conjugate of BA in rodents. After the synthesis, BA are stored in the gallbladder, making up the bile together with cholesterol, phospholipids, bilirubin, and water [2,3,8]. When nutrients reach the small intestine and are sensed by enteroendocrine cells, cholecystokinin is released and induces gallbladder contraction and secretion of the bile into the duodenum.

Bile promotes emulsification of dietary lipids, for digestion and absorption of fatty acids, cholesterol, fat-soluble vitamins, and other hydrophobic components of the diet [4,5,9]. Up to 95% of secreted BA are reabsorbed in the small intestine either via passive diffusion, uniporters or Na^+-coupled symporters (Asbt), and returned via the portal vein to the liver. This process is called enterohepatic circulation of BA. Despite efficient hepatic extraction of BA, a portion of it is not taken up in the first passage through the liver and thus reaches systemic circulation.

This spillover of BA into systemic circulation may be of crucial importance for the systemic effects of BA, as it causes in average a sixfold increase in the circulatory BA concentration, which can last for several hours after the meal. The magnitude of this increase as well as its time course are highly variable among different individuals, and it is related to gender, diet composition, and genetic variants [10]. After their uptake into hepatocytes, BA are most likely conjugated to taurine and glycine and stored in the gallbladder, closing a cycle that can be repeated several times a day.

Secondary Bile Acids

Bile acids that are not absorbed in the small intestine reach the large intestine and are metabolized by the microbiota, before being almost totally reabsorbed from the colon [2,3,11]. Deoxycholic, lithocholic, and ursodeoxycholic acid are the most common products of bacterial metabolism of primary BA in the human intestine and are thus named secondary BA. The 15 most common BA in humans are presented in Fig. 14.1, displaying the main structural differences between these BA, which are mostly related to the number, position, and orientation of the hydroxyl groups, attached to the steroid nucleus, as well as conjugation to taurine and glycine.

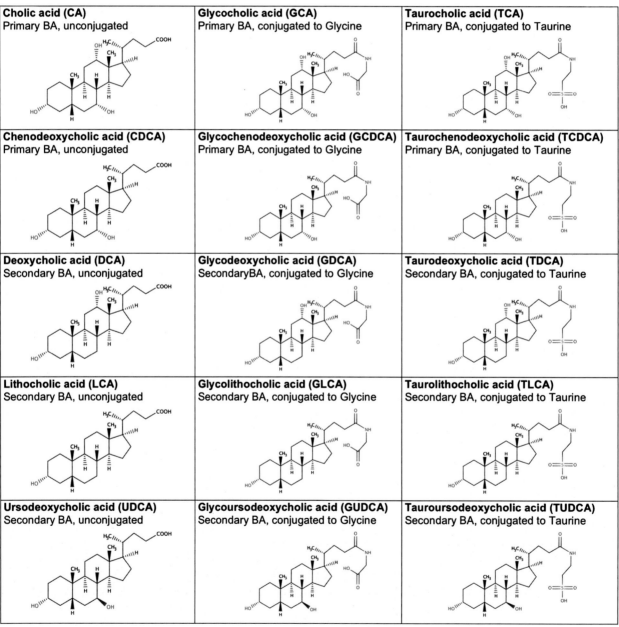

FIGURE 14.1   The most common BA found in humans.

MICROBIAL BIOTRANSFORMATION OF BILE ACIDS

The metabolization of BA and their conjugates by intestinal microorganisms involves their deconjugation from glycine and taurine, oxidation/epimerization, dehydroxylation, and esterification. These reactions can dramatically change physicochemical properties of the BA, affecting their absorption and microbial toxicity. The possible microbial biotransformations that cholic acid can undergo in the intestine, being converted into secondary BA, are depicted in Fig. 14.2.

Deconjugation

Bile salt hydrolases catalyze the deconjugation of BA from glycine and taurine and have been found in main bacterial genera of the intestinal microbiota [12]. Deconjugation of BA assists the microbial colonization of the gut and can provide sources of carbon, nitrogen, and sulfur to the bacteria [13–15]. The unconjugated BA have higher logP values, certainly affecting their reabsorption that no longer relies on transporters, as these molecules can flip-flop through the lipid bilayer.

Oxidation and Epimerization

The oxidation and epimerization of the hydroxyl groups of BA are catalyzed by hydroxysteroid dehydrogenases (HSDH) that promote the reversible oxidation of hydroxyl to oxo groups. Stereospecific oxidation followed by stereospecific reduction results in epimerization of the hydroxyl groups at C3, C7, and C12 and the generation of oxo-BA. This process involves two distinct HSDH that can be contained in the same bacteria or be catalyzed by distinct microorganisms [16].

7-α-Dehydroxylation

The 7-α-dehydroxylation is considered the most important reaction, in intestinal microbial metabolization of BA, being responsible for the conversion of cholic and chenodeoxycholic acids (primary BA), into deoxycholic and lithocholic acids (secondary BA), respectively [17]. This reaction can only happen after deconjugation of the substrates.

Esterification

A considerable amount of fecal BA is esterified and ethyl esters, as well as long-chain fatty acid esters and polyesters of BA, have been identified. The role of gut microbiota is not yet clear in the formation of these esters [18].

The importance of the intestinal bacteria in the conversion of primary into secondary BA has long been reported [11]. Effects of intestinal microbiota go further than changes in the composition, by the biotransformations described above. In fact, germ-free and antibiotic-treated mice have reduced fecal excretion of BA but increased BA pool as well as BA secretion and reabsorption from the intestine, due to modulation of the expression of the sodium-coupled intestinal transporter Asbt [19,20].

BILE ACID EFFECTS ON INTESTINAL MICROBIOTA

The amphipathic character that allows BA to solubilize dietary fats also grants these molecules antimicrobial properties. High concentrations of BA and their salts can dissolve membrane lipids and cause leakage of cell content, but also low concentration of BA may affect membrane permeability and fluidity, altering the function of membrane-bound proteins [21]. On top of their effects on membranes, BA can also induce the formation of RNA secondary structures and DNA damage [22]. Bile acids can also affect the folding of proteins and cause oxidative stress [21].

In parallel to their direct antimicrobial effects, BA control intestinal microbial communities via indirect effects that involve immune responses. The activation of FXR and GPBAR1 modulate the integrity of the intestinal barrier, the production of inflammatory mediators, and antimicrobial agents such as IL-18 and iNOS in the intestine [23–25].

The administration of BA to mice modulates the composition of gut microbiota in a similar fashion as a high-fat diet, leading to increase of Firmicutes at the expense of Bacteroidetes [26]. Lower concentrations of BA in the intestinal lumen favor gram-negative bacteria, including potential pathogens. On the other hand, increased BA levels in the intestine favor gram-positive members of the Firmicutes, including bacteria that cause the 7α-dehydroxylation of primary BA, generating toxic secondary BA [26,27].

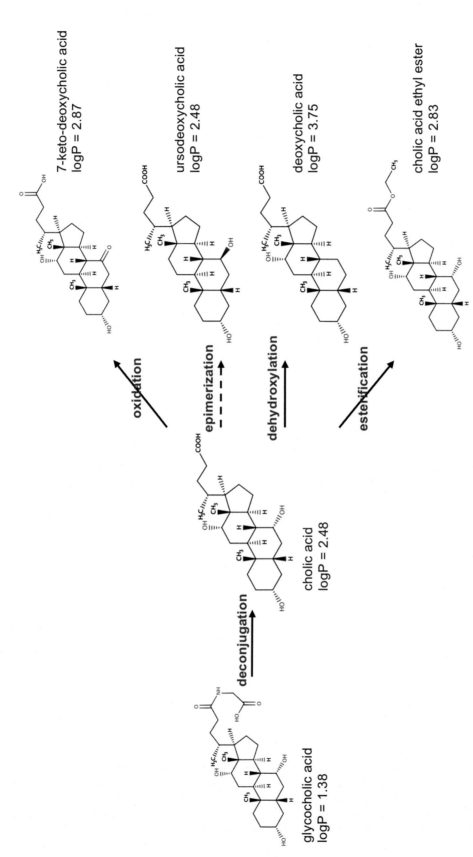

FIGURE 14.2 Microbial biotransformations that can affect glycocholic acid. The given example shows how intestinal microbiota can metabolize and convert a conjugated primary BA into secondary ones and how this affects logP. Structures were drawn and logP values calculated, using Marvin 17.29, 2017 from ChemAxon (http://www.chemaxon.com).

BILE ACIDS AS SIGNALING MOLECULES

Bile acids can activate at least five different nuclear receptors: farnesoid X receptor (FXR) [28], liver X receptor α (LXRα) [29], the constitutive androstane receptor (CAR, NR1H3) [30], pregnane X receptor (PXR, NR1H2) [31], and vitamin D receptor (VDR, NR1H1) [32]. They can also activate G-protein-coupled receptors (GPCR), including GPBAR1 (also known as M-BAR or TGR5) [33,34]. Although each receptor interacts with more than one BA, chenodeoxycholic acid is the strongest agonist of FXR, while secondary BA (deoxycholic and lithocholic acids) are more specific for GPBAR1 [36].

Since 1999 it is known that BA are ligands of FXR, which is highly expressed in the liver and intestine, modulating their own synthesis and transport [28,36,37]. FXR activation results in decreased CYP7A1 expression, and inhibition of BA synthesis through the neutral pathway [38,39]. In addition, FXR reduces the uptake and stimulates the excretion of BA by hepatocytes by controlling the expression of BA transporters in these cells [35]. In the intestine, FXR represses the expression of the apical sodium bile acid transporter Asbt and induces the expression of basolateral organic solute transporters (OSTα and OSTb).

FARNESOID X RECEPTOR, OBESITY, AND GLUCOSE METABOLISM

Another important intestinal action of FXR, confirmed in rodents, is the increased expression and release of fibroblast growth factor (FGF)-15 (FGF-19 in humans), which represses CYP7A1 expression in hepatocytes [40]. The activation of FXR recruits the short heterodimer partner (SHP) [1−3,36,37,41], whose actions toward HNF-4 or Foxo1, lead to decreased expression of gluconeogenic enzymes impacting glucose homeostasis [4,5,8,42,43]. FXR activation depends on the interaction between intestinal microbiota and BA, and both factors are subjected to dietary modulation, increasing the complexity of its regulation. In mice, it seems that FXR has beneficial effects when the animals feed on chow diet [44], whereas on high-fat diet this nuclear receptor promotes metabolic disease [45,46].

GPBAR1 was described as a membrane receptor for BA, thus leading to faster responses than those mediated through FXR [9,33,47−50]. In the small intestine L-cells, GPBAR1 modulates the release of glucagon-like peptide-1 (GLP1), which stimulates insulin secretion, suppresses appetite, and slows down gastrointestinal transit [34]. In mice, the administration of cholic acid increased the expression of enzymes involved in fatty acid oxidation, preventing obesity and insulin resistance. Bile acids elicited similar effects in human skeletal myocytes, via GPBAR1-mediated activation of deiodinase 2, that converts T_4 into T_3, the active thyroid hormone [4,5,51,52]. These findings sparkled the interest of using BA to target obesity and associated diseases [45,46].

BILE ACIDS HEALTH EFFECTS, BARIATRIC SURGERY, AND MICROBIOTA

The potential of BA as modulators of glucose homeostasis was first reported by Garg and Grundy who described glucose-lowering effects, induced by cholestyramine therapy [2,3,53]. After that, several studies with BA sequestrants reached the same conclusions, and differences in circulating BA levels were found between obese and diabetic patients, in comparison to lean and healthy individuals [4,5,54]. FXR activation can directly inhibit the expression of different gluconeogenic enzymes and increase insulin synthesis and secretion, while protecting pancreatic islets from lipotoxicity [1,7,8,55,56].

BARIATRIC INTERVENTIONS

Bariatric surgery is the most effective procedure for obesity treatment and induces long-term effects on gut microbiota, prioritizing Proteobacteria in detriment of Firmicutes, which parallels increases in BA pool [57]. Microbiota changes that follow Roux-en-Y gastric bypass surgery (RYGB) seem to be important for weight loss, as microbiota transplantation from RYGB surgery patients into mice regulates fat mass [58].

Bariatric surgery increases the levels of the incretin GLP-1, thus potentiating the actions of insulin in these patients after surgery, even before weight loss [59,60]. The levels of FGF-19 that are produced in the ileum, under control of FXR, are also increased after RYGB surgery, with reported positive health effects [58,61]. The changes in intestinal microbiota that lead to enhanced production of secondary BA might explain the effect of bariatric surgery on incretins. The activation of GPBAR1, as well as FXR, is at least part of the mechanism, underpinning the effects of the bariatric surgery on metabolic health.

CONCLUSIONS AND PERSPECTIVES

Reciprocal effects of BA and intestinal microbiota shape their composition and play a central role in health maintenance. The recent discovery of BA signaling, particularly via GPBAR1 and FXR, with related effects in numerous physiological processes, came as a promising target to treat metabolic diseases and deserves further studies.

REFERENCES

[1] Solaas K, Ulvestad A, Söreide O, Kase BF. Subcellular organization of bile acid amidation in human liver: a key issue in regulating the biosynthesis of bile salts. J Lipid Res 2000;41:1154–62.

[2] Zhou H, Hylemon PB. Bile acids are nutrient signaling hormones. Steroids 2014;86:62–8.

[3] de Aguiar Vallim TQ, Tarling EJ, Edwards PA. Pleiotropic roles of bile acids in metabolism. Cell Metab 2013;17:657–69.

[4] Müller VM, et al. Gut barrier impairment by high-fat diet in mice depends on housing conditions. Mol Nutr Food Res 2016;60:897–908.

[5] Zhang Y, Limaye PB, Renaud HJ, Klaassen CD. Effect of various antibiotics on modulation of intestinal microbiota and bile acid profile in mice. Toxicol Appl Pharmacol 2014;277:138–45.

[6] Marin JJG, Macias RIR, Briz O, Banales JM, Monte MJ. Bile acids in physiology, pathology and pharmacology. Curr Drug Metab 2015;17:4–29.

[7] Lefebvre P, Cariou B, Lien F, Kuipers F, Staels B. Role of bile acids and bile acid receptors in metabolic regulation. Physiol Rev 2009;89:147–91.

[8] Hofmann AF. Bile acids: the good, the bad, and the ugly. News Physiol Sci 1999;14:24–9.

[9] Liddle RA, Goldfine ID, Rosen MS, Taplitz RA, Williams JA. Cholecystokinin bioactivity in human plasma. Molecular forms, responses to feeding, and relationship to gallbladder contraction. J Clin Investig 1985;75:1144–52.

[10] Fiamoncini J, et al. Determinants of postprandial plasma bile acid kinetics in human volunteers. Aust J Pharm Gastrointest Liver Physiol 2017;313:G300–12.

[11] Ridlon JM, Kang D-J, Hylemon PB. Bile salt biotransformations by human intestinal bacteria. J Lipid Res 2006;47:241–59.

[12] Jones BV, Begley M, Hill C, Gahan CGM, Marchesi JR. Functional and comparative metagenomic analysis of bile salt hydrolase activity in the human gut microbiome. Proc Natl Acad Sci USA 2008;105:13580–5.

[13] Grill JP, Perrin S, Schneider F. Bile salt toxicity to some bifidobacteria strains: role of conjugated bile salt hydrolase and pH. Can J Microbiol 2000;46:878–84.

[14] Dussurget O, et al. Listeria monocytogenes bile salt hydrolase is a PrfA-regulated virulence factor involved in the intestinal and hepatic phases of listeriosis. Mol Microbiol 2002;45:1095–106.

[15] Carbonero F, Benefiel AC, Alizadeh-Ghamsari AH, Gaskins HR. Microbial pathways in colonic sulfur metabolism and links with health and disease. Front Physiol 2012;3:448.

[16] Gérard P. Metabolism of cholesterol and bile acids by the gut microbiota. Pathogens 2013;3:14–24.

[17] Hamilton JP, et al. Human cecal bile acids: concentration and spectrum. Am J Physiol Gastrointest Liver Physiol 2007;293:G256–63.

[18] Ridlon JM, Harris SC, Bhowmik S, Kang D-J, Hylemon PB. Consequences of bile salt biotransformations by intestinal bacteria. Gut Microb 2016;7:22–39.

[19] Out C, et al. Accepted manuscript. J Hepatol 2015:1–32. https://doi.org/10.1016/j.jhep.2015.04.030.

[20] Sayin SI, et al. Gut microbiota regulates bile acid metabolism by reducing the levels of tauro-beta-muricholic acid, a naturally occurring FXR antagonist. Cell Metab 2013;17:225–35.

[21] Begley M, Gahan CGM, Hill C. The interaction between bacteria and bile. FEMS Microbiol Rev 2005;29:625–51.

[22] Bernstein C, Bernstein H, Payne CM, Beard SE, Schneider J. Bile salt activation of stress response promoters in *Escherichia coli*. Curr Microbiol 1999;39:68–72.

[23] Inagaki T, et al. Regulation of antibacterial defense in the small intestine by the nuclear bile acid receptor. Proc Natl Acad Sci USA 2006;103:3920–5.

[24] Vavassori P, Mencarelli A, Renga B, Distrutti E, Fiorucci S. The bile acid receptor FXR is a modulator of intestinal innate immunity. J Immunol 2009;183:6251–61.

[25] Cipriani S, et al. The bile acid receptor GPBAR-1 (TGR5) modulates integrity of intestinal barrier and immune response to experimental colitis. PLoS One 2011;6:e25637.

[26] Islam KBMS, et al. Bile acid is a host factor that regulates the composition of the cecal microbiota in rats. Gastroenterology 2011;141:1773–81.

[27] Kakiyama G, et al. Modulation of the fecal bile acid profile by gut microbiota in cirrhosis. J Hepatol 2013;58:949–55.

[28] Makishima M, et al. Identification of a nuclear receptor for bile acids. Science 1999;284:1362–5.

[29] Song C, Hiipakka RA, Liao S. Selective activation of liver X receptor alpha by 6alpha-hydroxy bile acids and analogs. Steroids 2000;65:423–7.

[30] Saini SPS, et al. A novel constitutive androstane receptor-mediated and CYP3A-independent pathway of bile acid detoxification. Mol Pharmacol 2004;65:292–300.

[31] Staudinger JL, et al. The nuclear receptor PXR is a lithocholic acid sensor that protects against liver toxicity. Proc Natl Acad Sci USA 2001;98:3369–74.

[32] Makishima M, et al. Vitamin D receptor as an intestinal bile acid sensor. Science 2002;296:1313–6.

[33] Kawamata Y, et al. A G protein-coupled receptor responsive to bile acids. J Biol Chem 2003;278:9435–40.

[34] Katsuma S, Hirasawa A, Tsujimoto G. Bile acids promote glucagon-like peptide-1 secretion through TGR5 in a murine enteroendocrine cell line STC-1. Biochem Biophys Res Commun 2005;329:386–90.

[35] Fiorucci S, Distrutti E. Bile acid-activated receptors, intestinal microbiota, and the treatment of metabolic disorders. Trends Mol Med 2015;21:702–14.

[36] Parks DJ, et al. Bile acids: natural ligands for an orphan nuclear receptor. Science 1999;284:1365–8.

[37] Wang H, Chen J, Hollister K, Sowers LC, Forman BM. Endogenous bile acids are ligands for the nuclear receptor FXR/BAR. Mol Cell 1999;3:543–53.

[38] Goodwin B, et al. A regulatory cascade of the nuclear receptors FXR, SHP-1, and LRH-1 represses bile acid biosynthesis. Mol Cell 2000;6:517–26.

[39] Wang L, et al. Redundant pathways for negative feedback regulation of bile acid production. Dev Cell 2002;2:721−31.

[40] Inagaki T, et al. Fibroblast growth factor 15 functions as an enterohepatic signal to regulate bile acid homeostasis. Cell Metab 2005;2:217−25.

[41] Houten SM, Watanabe M, Auwerx J. Endocrine functions of bile acids. EMBO J 2006;25:1419−25.

[42] Yamagata K, et al. Bile acids regulate gluconeogenic gene expression via small heterodimer partner-mediated repression of hepatocyte nuclear factor 4 and Foxo1. J Biol Chem 2004;279:23158−65.

[43] Ma K, Saha PK, Chan L, Moore DD. Farnesoid X receptor is essential for normal glucose homeostasis. J Clin Investig 2006;116:1102−9.

[44] Lambert G, et al. The farnesoid X-receptor is an essential regulator of cholesterol homeostasis. J Biol Chem 2003;278:2563−70.

[45] Parséus A, et al. Microbiota-induced obesity requires farnesoid X receptor. Gut 2017;66:429−37.

[46] Wahlström A, Sayin SI, Marschall H-U, Bäckhed F. Intestinal crosstalk between bile acids and microbiota and its impact on host metabolism. Cell Metab 2016;24:41−50.

[47] Tagliacozzi D, et al. Quantitative analysis of bile acids in human plasma by liquid chromatography-electrospray tandem mass spectrometry: a simple and rapid one-step method. Clin Chem Lab Med 2003;41:1633−41.

[48] Griffiths WJ, Sjövall J. Bile acids: analysis in biological fluids and tissues. J Lipid Res 2010;51:23−41.

[49] Maruyama T, et al. Identification of membrane-type receptor for bile acids (M-BAR). Biochem Biophys Res Commun 2002;298:714−9.

[50] Alnouti Y, Csanaky IL, Klaassen CD. Quantitative-profiling of bile acids and their conjugates in mouse liver, bile, plasma, and urine using LC-MS/MS. J Chromatogr B Analyt Technol Biomed Life Sci 2008;873:209−17.

[51] Watanabe M, et al. Bile acids induce energy expenditure by promoting intracellular thyroid hormone activation. Nature 2006;439:484−9.

[52] Zietak M, Kozak LP. Bile acids induce uncoupling protein 1-dependent thermogenesis and stimulate energy expenditure at thermoneutrality in mice. Am J Physiol Endocrinol Metab 2015. https://doi.org/10.1152/ajpendo.00485.2015.

[53] Garg A, Grundy SM. Cholestyramine therapy for dyslipidemia in non-insulin-dependent diabetes mellitus. A short-term, double-blind, crossover trial. Ann Intern Med 1994;121:416−22.

[54] Staels B, Fonseca VA. Bile acids and metabolic regulation: mechanisms and clinical responses to bile acid sequestration. Diabetes Care 2009;32(Suppl. 2):S237−45.

[55] Popescu IR, et al. The nuclear receptor FXR is expressed in pancreatic beta-cells and protects human islets from lipotoxicity. FEBS Lett 2010;584:2845−51.

[56] Renga B, Mencarelli A, Vavassori P, Brancaleone V, Fiorucci S. The bile acid sensor FXR regulates insulin transcription and secretion. Biochim Biophys Acta 2010;1802:363−72.

[57] Graessler J, et al. Metagenomic sequencing of the human gut microbiome before and after bariatric surgery in obese patients with type 2 diabetes: correlation with inflammatory and metabolic parameters. Pharmacogenomics J 2013;13:514−22.

[58] Tremaroli V, et al. Roux-en-Y gastric bypass and vertical banded gastroplasty induce long-term changes on the human gut microbiome contributing to fat mass regulation. Cell Metab 2015;22:228−38.

[59] Laferrère B, et al. Incretin levels and effect are markedly enhanced 1 month after Roux-en-Y gastric bypass surgery in obese patients with type 2 diabetes. Diabetes Care 2007;30:1709−16.

[60] Salehi M, Prigeon RL, D'Alessio DA. Gastric bypass surgery enhances glucagon-like peptide 1-stimulated postprandial insulin secretion in humans. Diabetes 2011;60:2308−14.

[61] Sachdev S, et al. FGF 19 and bile acids increase following Roux-en-Y gastric bypass but not after medical management in patients with type 2 diabetes. Obes Surg 2016;26:957−65.

Block III

Established and Experimental Interventions

Chapter 15

Use of Probiotics in Inflammatory Bowel Disease

J. Plaza-Diaz[1,2], F.J. Ruiz-Ojeda[1,2], M.J. Arias-Tellez[3], P. Solis-Urra[4,5], M.J. Saez-Lara[2,6] and A. Gil[1,2]

[1]Department of Biochemistry and Molecular Biology II, School of Pharmacy, University of Granada, Granada, Spain; [2]Institute of Nutrition & Food Technology "Jose Mataix", Biomedical Research Center, University of Granada, Armilla, Spain; [3]Department of Nutrition, University of Chile, Santiago, Chile; [4]PROFITH "PROmoting FITness and Health through physical activity" research group, Department of Physical Education and Sport, Faculty of Sport Sciences, University of Granada, Granada, Spain; [5]IRyS Group, School of Physical Education, Physical Activity School, Pontificia Universidad Católica de Valparaíso, Valparaiso, Chile; [6]Department of Biochemistry and Molecular Biology I, School of Sciences, University of Granada, Granada, Spain

INFLAMMATORY BOWEL DISEASES

Inflammatory bowel disease (IBD) can be understood as a condition of disrupted physiology, microbiology, immunology, and genetics. IBD mainly includes Crohn disease (CD) and ulcerative colitis (UC), which are characterized by chronic inflammation of the gastrointestinal (GI) tract. Breast milk and other foods such as traditional fermented products are the primary sources of lactic acid bacteria (LAB) and *Bifidobacteria*, which subsequently colonize the GI tract and the feces of human subjects [1−3]. LAB and *Bifidobacteria* produce lactic acid as a major metabolic end-product of carbohydrate fermentation, and exhibit an increased tolerance to acidity. In addition to having important applications in the food industry, LAB and *Bifidobacteria* can have beneficial health effects, as adjuvants to decrease the intestinal microbiota imbalance, induced by the use of antibiotics or by pathological conditions, particularly IBD [1,4].

RELEVANT MICROBIAL POPULATIONS

Some studies have demonstrated that intestinal microbiome is an essential factor in the progression and development of IBD. In fact, antibiotic short-term treatment ameliorates intestinal inflammation. In IBD, the microbiome has a reduction in biodiversity, an important measure of the total number of the community species, including decreased Firmicutes and concomitant increase of Gammaproteobacteria, in both CD and UC [5].

The colon has two distinct mucous layers, a firmly attached inner mucous layer and an outer one of variable thickness. This contrasts with the small intestine, endowed with a single layer only. In a normal host, mucins, trefoil peptides, and immunoglobulin A make up the mucous layer, which acts to provide a barrier between the gut microbiota and intra-epithelial cells. Patients suffering from IBD are thought to have a compromised mucous layer, thus allowing luminal microflora to penetrate intraepithelial cells and drive inflammatory and proliferative processes. Mice deficient in mucins essential in maintaining intestinal mucosal integrity develop spontaneous colitis resembling human UC. Recent studies have demonstrated a higher abundance of *Fusobacterium varium* in patients with UC, which may in part be responsible for colonic mucosal erosion, as seen in mice models due to compromised mucosal integrity.

In biopsies obtained during colonoscopy of the rectum and terminal ileum, there was a positive correlation between a diagnosis of CD, and the presence of Pasteurellaceae (*Haemophilus* sp.), *Veillonella parvula*, Neisseriaceae (*Eikenella corrodens*), and *Fusobacteriaceae (Fusobacterium nucleatum)*. Additionally, the prevalence of Pasteurellaceae, *V. parvula*, and *Rothia mucilaginosa* correlated with deep ulcerations (ileal or colonic) seen during diagnostic colonoscopy. Dysbiosis within mucosa-associated bacteria may be responsible for new-onset CD [2−4].

Antibiotics remain the mainstay therapy in the treatment of septic complications of IBD, such as abscesses and wound infections. Yet wounds in the gut are associated with microenvironmental changes that allow the growth of microbial species that promote healing. Identification of wound-associated bacterial species showed that the abundance of anaerobic bacteria, in particular *Akkermansia* spp., increased substantially in early regenerative mucosa (between 2 and 4 days post injury) [6,7]. These findings suggest less enthusiasm with preoperative colon preparing, and more rigorous selection of postoperative antibiotic agents, in order not to harm beneficial species.

Key Molecules and Signaling Pathways in IBD

Impaired recognition and killing of commensal bacteria contribute to IBD development, and many of the identified IBD-susceptibility genes regulate host−microbial interactions. NOD2, which is an intracellular sensor of bacterial peptidoglycan, was identified as a susceptibility gene for CD, and CD-associated NOD2 mutations are associated with a loss of function of the protein. Mutations in regulatory genes such as autophagy-related gene 16-like 1 (ATG16L1) and immunity-related GTPase family M (IRGM 1) are also linked to CD, and autophagy dysfunction is associated with defects in bacterial killing. In addition, NOD2 and autophagy proteins regulate the function of Paneth cells, which release granules containing antimicrobial peptides in response to bacteria.

NOD2 is highly expressed in these cells, and some studies report diminished expression of α-defensins in individuals with CD-associated NOD2 mutations. Nod2$^{-/-}$ mice have impaired antimicrobial functions in Paneth cells, resulting in the accumulation of ileal bacteria, which might contribute to IBD pathogenesis. ATG16L1-mutant mice show abnormal granule formation in Paneth cells, which is also observed in patients with CD who have homozygous mutations in ATG16L1 [8].

In Vitro and in Vivo Models

In vitro studies principally involve intestinal porcine epithelial cells, CaCo-2 cells, human dendritic cells obtained from peripheral blood and umbilical cord blood, monocyte-derived dendritic cells, peripheral blood mononuclear cells, and intestinal T cells. In vivo studies involve animal models (mice, rats, and dogs) with chemical inflammation induction.

METABOLOME, TRANSCRIPTOME, AND PROTEOME

Metabolome

NMR spectroscopy uncovered a differential metabolite profile of lipoproteins and choline between CD and UC. At the same time, a general metabotype for IBD is postulated involving elevated biogenic amines, amino acids, and lipids [8,9]. Utilization of both of these techniques has implicated tryptophan and its metabolites in IBD severity. Identification of the relevant metabolic profiles may not only provide insight to the underlying pathophysiology but also lead us to important diagnostic and prognostic clinical assays.

Proteome

Utilizing antibody-marking, fluorescence-based flow cytometry, and the more recent time-of-flight mass cytometry (CyTOF), we are able to study single cell expression patterns of multiple proteins. By compiling and then comparing the levels of expression of many proteins across multiple cell types in disease and health, we can refine the molecular networks relevant to pathogenesis. CyTOF helped clarify the role of a novel gene in CD, through the differentiation of expression patterns in multiple immune cell types [9].

Interfaces With Antibiotics and Other Drugs

Treatment of suppurative complications of CD such as abscesses and fistulas includes drainage and antibiotic therapy, most often ciprofloxacin, metronidazole, or a combination of both. Data in UC mostly consists of trials evaluating ciprofloxacin, metronidazole, and rifaximin. Most trials did not show a benefit for the treatment of active UC with antibiotics, though two metaanalyses concluded that antibiotic therapy is associated with a modest improvement in clinical symptoms [10].

THERAPEUTIC PROTOCOLS

Current treatments of IBD first involve the induction of remission, which is followed by maintaining remission. Patients with an active disease are treated with topical or systemic 5-aminosalicylic acids, corticosteroids, or immunomodulators, such as azathioprine and 6-mercaptopurine, in addition to anti-TNF monoclonal antibodies. The efficacy of some probiotics in improving IBD patients' quality of life has been recently reported. There is evidence that commensal enteric bacteria and their products create a local environment that affects the course of IBD. These high bacterial concentrations in IBD patients are characterized by decreased numbers of LAB and *Bifidobacteria*, and increased numbers of *E. coli*, coliforms, and Bacteroides in the colon. In this sense, probiotics might increase intestinal biodiversity and improve the symptoms of IBD patients. Probiotics suppressing inflammation and/or activating innate immunity could be used within therapeutic strategies to restore the host gut microbiota [1,4].

Use of Probiotics in Ulcerative Colitis

Modulation of the intestinal microbiota can be performed either by antibiotics or by probiotics, but the former are not good candidates for chronic disease because of antibiotic resistance, potential side effects, and ecological concerns [1,11,12]. *Bifidobacteria*-fermented milk (BFM) supplementation may reduce exacerbations of UC through normalization of the intestinal microbiota and decreased luminal butyrate concentration, a key molecule in the remission of colitis. This reduction reflects the increased uptake or oxidation of short-chain fatty acids/SCFAs, by the improved colorectal mucosa.

Tumor necrosis factor alpha (TNF-α) exerts a pivotal role in the pathogenesis of active UC. Coculturing colonic biopsies from active UC with *Bifidobacterium longum* reduced the release of TNF-α and interleukin (IL)-8. It is well-known that the activation of nuclear factor kappa B (NF-κB) can regulate inflammatory cytokines such as TNF-α, IL-8, and IL-6. Immunohistochemistry staining of NF-κB p65 in colonic biopsies from active UC showed fewer NF-κB-positive cells when cocultured with either *B. longum* or dexamethasone, which indicates that *B. longum* can inhibit NF-κB activation in lamina propria cells.

The most studied probiotic in clinical trials is *L. rhamnosus* GG. After 6 to 12 months of treatment, the percentage of patients maintaining clinical remission was 91% and 85%, respectively, for the *Lactobacillus rhamnosus* GG group, 87% and 80% for the mesalazine group, and 94% and 84% for the combined treatment [13].

In children with distal active UC, rectal administration of *Lactobacillus reuteri* ATCC 55730 for 8 weeks in addition to standard oral mesalazine resulted in a significant decrease in the Mayo Disease Activity Index. All of the children on *L. reuteri* had a clinical response, whereas only 53% of the children on the placebo responded. Clinical remission was achieved in 31% of the *L. reuteri* group and in no children of the placebo group [14].

Administration of *B. longum* plus a prebiotic for a period of 1 month to patients with active UC improved the full clinical appearance of chronic inflammation. The proinflammatory cytokines, TNF-α and IL-1α, were significantly reduced, and *Bifidobacteria* determined by quantitative PCR increased 42-fold in the synbiotic group, but only 4.6-fold in the placebo group [15].

VSL#3 contains four strains of *Lactobacillus* (*Lactobacillus casei*, *Lactobacillus plantarum*, *Lactobacillus acidophilus*, and *Lactobacillus delbrueckii* subsp. *bulgaricus*), three of *Bifidobacterium* (*Bifidobacterium longum*, *Bifidobacterium breve*, and *Bifidobacterium infantis*), one of *Streptococcus salivarius* subsp. *thermophilus*, and cornstarch. VSL#3 is capable of colonizing the gut, and significantly decreases fecal pH in UC patients who are intolerant or allergic to 5-ASA. The combination is safe in adults with UC, and in the maintenance of remission in children. A metaanalysis totaling 1763 patients evaluated the effects of probiotics on remission and maintenance therapy in UC. The use of VSL#3 significantly increased remission rates in patients with active UC [13−15].

Use of Probiotics and Synbiotics in Crohn Disease

L. rhamnosus GG revealed no differences in endoscopic and clinical remission in CD. However, prebiotics and SCFA may contribute to the maintenance of colonic homeostasis [1,4]. Following synbiotic therapy, patients had improved clinical symptoms and clinical scores were significantly reduced. Also *B. longum* plus prebiotic were tested in active CD. The investigation evidenced that synbiotics can be developed into acceptable therapies for acute and active CD [16−19].

Fecal Microbiota Transplantation/FMT

Two systematic reviews of this topic, in 2012 and 2013, were predominantly comprised of case reports. These studies included IBD patients, both with and without comorbid *C. difficile* infections, and were limited in quantitative analysis. Short-term use of FMT looks beneficial for improvement in clinical symptoms, and endoscopic healing of patients with UC. Despite this benefit, there are still unanswered questions that require further research [20].

DIET, NUTRITION, AND EXERCISE AS OTHER LIFESTYLE REPERCUSSIONS
Diet and Nutrition

The consumption of milk, cheese, eggs, and red or processed meats has been associated with an increased risk of IBD. One of the explanations is high levels of cysteine, which can be used for the generation of hydrogen sulfide (H_2S), by sulfate-reducing bacteria of the intestine. H_2S has been shown to have inflammatory effects in the colon, including increased DNA damage [21−23]. The conventional regimen for IBD is a low-residue diet, but there is no evidence that such a diet is ideal for IBD. Other options are listed below.

- Semivegetarian diet. Reduced consumption of animal meats and derivatives, or more specifically a high-carbohydrate diet. It could have antiinflammatory properties to improve or maintain a clinical response in patients with CD.
- Diet low in FODMAPs. The FODMAPs (fermentable oligosaccharides, disaccharides, monosaccharides, and polyols) are a mix of short-chain carbohydrates, undigested in the small intestine, and fermented by colonic bacteria. They can cause an osmotic effect in patients with IBD, generating distention, bloating, pain, and diarrhea. The reduction of consumption of FODMAPs might help to reduce functional gastrointestinal symptoms in IBD.
- Gluten-free diet. It is used in the treatment for celiac disease. However, many symptoms related to gluten exposure in the general population are also common in IBD patients, diminishing their quality of life. There are few large studies analyzing this option.

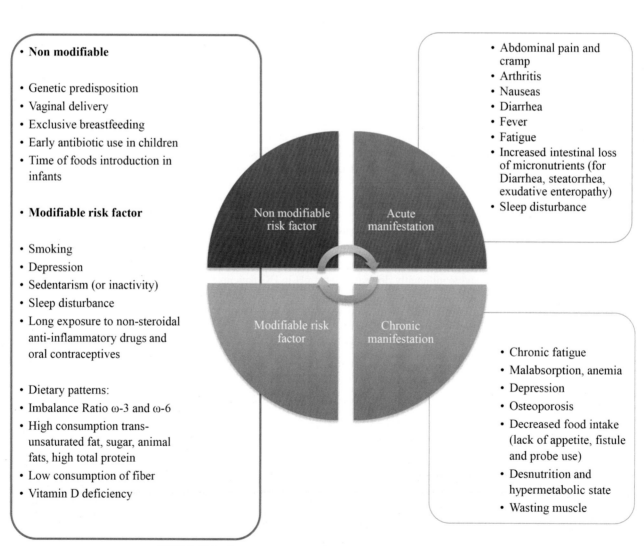

FIGURE 15.1 Consequences and risk factors in inflammatory bowel disease.

- Paleolithic diet. Includes a wide diversity of fruits, nuts, and vegetables; including wild-plant foods that contain high amounts of calcium and other minerals. Other features are lean meat, with low dairy, grain, sugar, and salt content. Again, clinical experience is rather limited [23−26].

Exercise, Physical Activity, and Lifestyle Repercussions

Sedentary lifestyle and inactivity are capable of provoking inflammation, and specifically IBD. Some mechanisms are postulated.

(1) The antiinflammatory effect of IL-6 released from the muscle in response to exercise, (2) The peroxisome proliferator−activated receptor gamma coactivator 1-alpha/PGC1α, likely reduces the activity of proinflammatory gene expression, and (3) Probably, the increase in gut microbial diversity.

Some evidence suggests that moderate aerobic exercise is more adequate than vigorous exercise as intensities over 70% VO$_2$max have shown negative effects due to the potential tendency to inflammation. Yoga, cycling, and walking are the most recommended. Exercise at 60% VO$_2$max, and less than 1 h duration, three times a week, enhances the quality of life and general health status, even counteracting some IBD complications. Benefits include nutritional, muscle and bone health, body composition, and even mood state [27−30]. In addition, patients with IBD exhibit higher incidence of colorectal cancer. Exercise and physical activity could contribute to the prevention of this risk [27−30] (Fig. 15.1).

REFERENCES

[1] Saez-Lara MJ, Gomez-Llorente C, Plaza-Diaz J, Gil A. The role of probiotic lactic acid bacteria and bifidobacteria in the prevention and treatment of inflammatory bowel disease and other related diseases: a systematic review of randomized human clinical trials. BioMed Res Int 2015;2015:505878.

[2] Gevers D, Kugathasan S, Denson LA, Vazquez-Baeza Y, Van Treuren W, Ren B, Schwager E, Knights D, Song SJ, Yassour M, Morgan XC, Kostic AD, Luo C, González A, McDonald D, Haberman Y, Walters T, Baker S, Rosh J, Stephens M, Heyman M, Markowitz J, Baldassano R, Griffiths A, Sylvester F, Mack D, Kim S, Crandall W, Hyams J, Huttenhower C, Knight R, Xavier R. The treatment-naive microbiome in new-onset Crohn's disease. Cell Host Microbe 2014;15:382−92.

[3] Plaza-Díaz J, Ruiz-Ojeda FJ, Vilchez-Padial LM, Gil A. Evidence of the anti-inflammatory effects of probiotics and synbiotics in intestinal chronic diseases. Nutrients 2017;9:E555.

[4] Ghouri YA, Richards DM, Rahimi EF, Krill JT, Jelinek KA, DuPont AW. Systematic review of randomized controlled trials of probiotics, prebiotics, and synbiotics in inflammatory bowel disease. Clin Exp Gastroenterol 2014;7:473−87.

[5] Naidoo K, Gordon M, Fagbemi AO, Thomas AG, Akobeng AK. Probiotics for maintenance of remission in ulcerative colitis. Cochrane Database Syst Rev 2011;12:CD007443.

[6] Derikx LA, Dieleman LA, Hoentjen F. Probiotics and prebiotics in ulcerative colitis. Best Pract Res Clin Gastroenterol 2016;30:55−71.

[7] Zeuthen LH, Fink LN, Frøkiaer H. Toll-like receptor 2 and nucleotide-binding oligomerization domain-2 play divergent roles in the recognition of gut-derived lactobacilli and bifidobacteria in dendritic cells. Immunology 2008;124:489−502.

[8] Santoru ML, Piras C, Murgia A, Palmas V, Camboni T, Liggi S, Ibba I, Lai MA, Orrù S, Loizedda AL, Griffin JL, Usai P, Caboni P, Atzori L, Manzin A. Cross sectional evaluation of the gut-microbiome metabolome axis in an Italian cohort of IBD patients. Sci Rep 2017;7:9523.

[9] Picoraro JA, LeLeiko NS. The omes of inflammatory bowel disease-a primer for clinicians. J Pediatr Gastroenterol Nutr 2017. https://doi.org/10.1097/MPG.0000000000001844.

[10] Nitzan O, Elias M, Peretz A, Saliba W. Role of antibiotics for treatment of inflammatory bowel disease. World J Gastroenterol 2016;22:1078−87.

[11] Ishikawa H, Matsumoto S, Ohashi Y, Imaoka A, Setoyama H, Umesaki Y, Tanaka R, Otani T. Beneficial effects of probiotic Bifidobacterium and galacto-oligosaccharide in patients with ulcerative colitis: a randomized controlled study. Digestion 2011;84:128−33.

[12] Groeger D, O'Mahony L, Murphy EF, Bourke JF, Dinan TG, Kiely B, Shanahan F, Quigley EM. Bifidobacterium infantis 35624 modulates host inflammatory processes beyond the gut. Gut Microb 2013;4:325−9.

[13] Zocco MA, dal Verme LZ, Cremonini F, Piscaglia AC, Nista EC, Candelli M, Novi M, Rigante D, Cazzato IA, Ojetti V, Armuzzi A, Gasbarrini G, Gasbarrini A. Efficacy of Lactobacillus GG in maintaining remission of ulcerative colitis. Aliment Pharmacol Ther 2006;23:1567−74.

[14] D'Incà R, Barollo M, Scarpa M, Grillo AR, Brun P, Vettorato MG, Castagliuolo I, Sturniolo GC. Rectal administration of Lactobacillus casei DG modifies flora composition and toll-like receptor expression in colonic mucosa of patients with mild ulcerative colitis. Dig Dis Sci 2011;56:1178−87.

[15] Fujimori S, Gudis K, Mitsui K, Seo T, Yonezawa M, Tanaka S, Tatsuguchi A, Sakamoto C. A randomized controlled trial on the efficacy of synbiotic versus probiotic or prebiotic treatment to improve the quality of life in patients with ulcerative colitis. Nutrition 2009;25:520−5.

[16] Schultz M, Timmer A, Herfarth HH, Sartor RB, Vanderhoof JA, Rath HC. Lactobacillus GG in inducing and maintaining remission of Crohn's disease. BMC Gastroenterol 2004;4:5.

[17] Chermesh I, Tamir A, Reshef R, Chowers Y, Suissa A, Katz D, Gelber M, Halpern Z, Bengmark S, Eliakim R. Failure of synbiotic 2000 to prevent postoperative recurrence of Crohn's disease. Dig Dis Sci 2007;52:385−9.

[18] Steed H, Macfarlane GT, Blackett KL, Bahrami B, Reynolds N, Walsh SV, Cummings JH, Macfarlane S. Clinical trial: the microbiological and immunological effects of synbiotic consumption — a randomized double-blind placebo-controlled study in active Crohn's disease. Aliment Pharmacol Ther 2010;32:872—83.

[19] Rolfe VE, Fortun PJ, Hawkey CJ, Bath-Hextall F. Probiotics for maintenance of remission in Crohn's disease. Cochrane Database Syst Rev 2006;4:CD004826.

[20] Colman RJ, Rubin DT. Fecal microbiota transplantation as therapy for inflammatory bowel disease: a systematic review and meta-analysis. J Crohns Colitis 2014;8:1569—81.

[21] Chiba M, Abe T, Tsuda H, Sugawara T, Tsuda S, Tozawa H, Fujiwara K, Imai H. Lifestyle-related disease in Crohn's disease: relapse prevention by a semi-vegetarian diet. World Journal of Gastroenterology 2010;16:2484—95.

[22] Eaton SB, Konner M. Paleolithic nutrition. A consideration of its nature and current implications. N Engl J Med 1985;312:283—9.

[23] Herfarth HH, Martin CF, Sandler RS, Kappelman MD, Long MD. Prevalence of a gluten-free diet and improvement of clinical symptoms in patients with inflammatory bowel diseases. Inflamm Bowel Dis 2014;20:1194—7.

[24] Lewis JD, Abreu MT. Diet as a trigger or therapy for inflammatory bowel diseases. Gastroenterology 2017;152:398—414.

[25] Prince AC, Myers CE, Joyce T, Irving P, Lomer M, Whelan K. Fermentable carbohydrate restriction (low fodmap diet) in clinical practice improves functional gastrointestinal symptoms in patients with inflammatory bowel disease. Inflamm Bowel Dis 2016;22:1129—36.

[26] Racine A, Carbonnel F, Chan SS, Hart AR, Bueno-de-Mesquita HB, Oldenburg B, van Schaik FD, Tjønneland A, Olsen A, Dahm CC, Key T, Luben R, Khaw KT, Riboli E, Grip O, Lindgren S, Hallmans G, Karling P, Clavel-Chapelon F, Bergman MM, Boeing H, Kaaks R, Katzke VA, Palli D, Masala G, Jantchou P, Boutron-Ruault MC. Dietary patterns and risk of inflammatory bowel disease in europe: results from the EPIC Study. Inflamm Bowel Dis 2016;22:345—54.

[27] Costa RJS, Snipe RMJ, Kitic CM, Gibson PR. Systematic review: exercise-induced gastrointestinal syndrome-implications for health and intestinal disease. Alimentary pharmacology & therapeutics 2017;46:246—65.

[28] Shephard RJ. The case for increased physical activity in chronic inflammatory bowel disease: a brief review. Int J Sports Med 2016;37:505—15.

[29] Bilski J, Mazur-Bialy A, Brzozowski B, Magierowski M, Zahradnik-Bilska J, Wojcik D, Magierowska K, Kwiecien S, Mach T, Brzozowski T. Can exercise affect the course of inflammatory bowel disease? Experimental and clinical evidence. Pharmacol Rep 2016;68:827—36.

[30] Cheifetz AS, Gianotti R, Luber R, Gibson PR. Complementary and alternative medicines used by patients with inflammatory bowel diseases. Gastroenterology 2017;152:415—29.

Chapter 16

Current Status of Fecal Microbiota Transplantation

J. Reygner[1] and N. Kapel[1,2]

[1]Faculté de pharmacie, Université Paris Descartes, Sorbonne Paris Cité, Paris, France; [2]Laboratoire de Coprologie Fonctionnelle, APHP, Hôpitaux Universitaires Pitié Salpêtrière-Charles Foix, Paris, France

INTRODUCTION

The intestinal microbiota is currently respected as a well-orchestrated organism, with a major role in the development of immunity and maintenance of health, and various pathologies are associated with dysbiosis. Fecal microbiota transplantation (FMT) aims to treat diseases associated with alteration of gut microbiota by introduction of a fecal suspension from a healthy donor into the gut of a receiver [1]. The goal of FMT is to restore the phylogenetic diversity, and homeostasis of the gut microbiota. The oldest account of FMT dates back to the 4th century when a Chinese physician named Ge Hong advised to consume fresh stool from a healthy neighbor when suffering from severe diarrhea. The first report in modern medical literature appeared in 1958, when Eiseman reported the successful use of FMT via fecal enema in four patients to treat severe pseudomembranous colitis, long before *Clostridium difficile* was identified as the causative agent of this condition.

This procedure has garnered significant attention in the last decade, in the face of the increasing prevalence, severity, and mortality related to *C. difficile* infection (CDI), and the growing interest in new therapeutics based on manipulation of the microbiota. Today, most cases of CDI are treated with metronidazole, vancomycin, or fidaxomicin, but recurrence ranges from 20% after the initial episode to 60% after multiple treatments. Following a first randomized clinical trial (RCT) [2], numerous studies have confirmed the efficacy of FMT, compared to conventional antibiotic treatment, for recurrent forms of *C. difficile* infection (rCDI). This procedure is thus recommended for this indication in the North American and European guidelines [3,4].

Since dysbiosis may be found in many other diseases, the use of FMT has been considered as a therapeutic option in non-CDI indications such as chronic inflammatory bowel diseases (IBD), irritable bowel syndrome (IBS), metabolic disorders, neuropsychiatric conditions including autism, or in antibiotic-resistant bacterial infections. Preliminary studies show encouraging results, but none of them allows recommendation to use FMT, except in a context of clinical research.

FECAL MICROBIOTA TRANSPLANTATION: REGULATORY ASPECTS

The regulatory status of the fecal microbiota as a medical treatment remains to be clarified, as FMT policy and legislation differ from country to country [5]. In 2012, the FDA classified human feces as a drug. However, FMT has not undergone traditional regulatory approval process of pharmaceutical products, with sequential testing leading to large phase III trials, assessing efficacy and safety prior to clinical use. The European Medicine Agency has not defined its position, leaving each country free to assign a qualification. In France, FMT is considered a drug by the National Agency for Medicine and Health Product Safety, placing the fecal material under the jurisdiction of the hospital pharmacy [5]. Thus production should be carried out under responsibility of pharmacists and storage within the pharmacy.

Other European countries such as Austria or Finland see FMT as a therapeutic intervention that should not be considered a pharmaceutical drug, and did not consider relevant to establish specific regulations. In those countries, doctors may perform FMT based on their own judgment without any approval from the drug authority [6,7].

FECAL MICROBIOTA TRANSPLANTATION PROCEDURE: PREPARATION, ADMINISTRATION

After a first RCT by van Nood [2], which has shown that FMT is highly effective for the treatment of rCDI, this procedure has gained popularity as clinical trials have consistently confirmed its high efficacy in resolving rCDI. No serious adverse events were reported, whereas incidence, severity, and mortality due to *C. difficile* increased. However, fecal microbiota transplantation is a complex intervention that involves multiple components, ranging from donor selection to methods of transplantation, and the clinical use of FMT remains difficult, because of uncertainty surrounding optimal methods of fecal bacterial community processing and administration.

FMT requires healthy stool donors, infrastructure for the processing and storage of donated stool, and a mechanism for introducing stool from the healthy donor into the infected patient's small intestine or colon. Up to now, there is no clear consensus on the optimal protocol for feces preparation before FMT, and practices vary among groups. However, some standards have emerged, allowing to propose a set of steps [8].

Donor Selection

Donor selection is a crucial step, which aims to reduce and prevent any adverse effects associated with fecal administration. Firstly potential donors are asked to provide detailed information about their medical history and lifestyle to identify any risk factor for transmission of infectious agents, that is, known HIV or viral hepatitis, illicit drug use, unprotected sex, and sex with high risk of HIV or hepatitis C virus acquisition. Gastrointestinal (GI) diseases such as inflammatory bowel diseases (IBD), irritable bowel syndrome (IBS), GI malignancy, or any other medical conditions or treatments that might impact the gut microbial community such as a recent international travel to areas at high risk for enteric infections or multiple drug resistant bacteria, recent antibiotic use, obesity, metabolic syndrome are equally questioned.

Secondly, they are submitted to a framework of screening including serologic and stool tests, with large variation according to guidelines published by different groups (Table 16.1) [2,9—11]. Donors may or may not be related to the recipient. Some authors favor abbreviated screening, if the medical history is reassuring, and if the donor is a family member or intimate to the recipient, while others favor extensive screening for all donors. No difference in effectiveness of FMT, in the recovery of patients suffering from rCDI, has been reported according this parameter, but there are still no RCT comparing the two categories of donor.

PREPARATION OF FMT

The donor should be notified to avoid any food to which a recipient may be allergic for 5 days prior to the procedure. On the day of donation, a second interview must be completed by the donor to ensure that no adverse event has occurred since screening [10—12].

The first RCTs of FMT used freshly donated stools [2,13], but results from subsequent studies have supported the equivalent efficacy of frozen stool preparation [14—18]. Considering potential logistic advantages of frozen fecal material, this option is now widely used, as it allows the development of fecal banks, and a faster management of the patient's treatment, without the delay related to the donor recruitment phase [19]. However, such banks are not suitable when FMT is considered as a drug which has to be prepared and stored within the hospital pharmacy [12].

Processing donated stool typically involves two steps: (1) homogenization of stools with diluent and (2) filtration to remove large particles [8]. A third step, with concentration of bacteria via centrifugation may be performed for the use of frozen capsules [20]. All procedures are performed at room temperature. It is highly recommended that fecal suspension should be prepared under a hood, because stool is considered as a level 2 biohazard. As a complete screening is performed prior to selection, this will mainly avoid the risk of crosscontamination and protect lab staff [21].

Whether feces should be prepared in an anaerobic environment to preserve anaerobic bacteria is still discussed. Most laboratories lack anaerobic chambers, and most successes have been obtained with aerobic preparation, so the use of a chamber under N_2 gas seems not to be absolutely required up to now [2,20]. Stools have to be diluted in an adapted volume of sterile diluent, according to the consistency of the fecal sample, that is, sterile water, 4% milk, preservative-free normal saline, or phosphate buffered saline for fresh FMT [22]. For frozen storage without risk of bacterial cell lysis, glycerol 10% —80% is advised [14—16].

After homogenization by hand or a dedicated blender, filtration of fecal solution is performed twice, using sterile gauze or stainless steel laboratory sieves, to remove debris and thus avoid clogging the administration equipment. Although stool

TABLE 16.1 Recommended Screening Tests for Stool Donors

	Bakken et al. [9]	van Nood et al. [2]	Sokol et al. [10] and French ANSM [12]	Cammarota et al. [11]
Blood (serologies)	HIV types 1 and 2 Ab HAV, HBV, HCV *Treponema pallidum*	HIV types 1 and 2 HTLV I and II HAV, HBV, HCV EBV *T. pallidum* *Entamoeba histolytica* *Strongyloides stercoralis*	HIV types 1 and 2 HTLV I and II HAV, HBV, HCV, HEV CMV[a] *T. pallidum* *S. stercoralis* *Trichinella species* *E. histolytica* EBV[b] *Toxoplasma gondii*[b]	HIV-1 and HVI-2 HTLV I and II HAV, HBV, HCV, HEV CMV EBV *T. pallidum* *E. histolytica* *S. stercoralis*
Stools	*Clostridium difficile* toxin B by PCR; Routine bacterial culture for enteric pathogens Fecal *Giardia* antigen Fecal *Cryptosporidium* antigen Acid-fast stain for *Cyclospora* and *Isospora* Ova and parasites *H. pylori* fecal antigen (for upper GI routes of FMT)	*C. difficile* (toxin ELISA and culture or PCR) Routine bacterial culture for enteric pathogens Parasitological evaluation according to local standards (triple feces test or PCR)	Adenovirus, Astrovirus[b] Calcivirus (Norovirus, Sapovirus)[b] Picornavirus (Enterovirus, Aichi virus)[b] Bacterial culture for: *C. difficile* (toxin ELISA and culture or PCR) *Listeria monocytogenes* *Vibrio cholerae/V. parahemolyticus*,[b] *Salmonella, Shigella, Yersinia,* *Campylobacter* sp., Multiresistant-drug bacteria, *S. stercoralis,* *Cryptosporidium* species, *Cyclospora* species, *E. histolytica,* *Giardia intestinalis,* *Isospora* species Microsporidia *Blastocystis hominis*[b] *Dientamoeba fragilis*[b]	*Helicobacter pylori* fecal antigen Rotavirus *C. difficile* (toxin ELISA and culture or PCR) Bacterial culture for: *Salmonella,* *Shigella,* *Campylobacter,* *Escherichia coli* O157 H7, *Yersinia,* *V. cholera,* *L. monocytogenes,* Multiresistant-drug bacteria, Norovirus, Protozoa (including *Blastocystis hominis, Giardia lamblia,* and *Cryptosporidium parvum*), and helminths. Fecal occult blood testing *Isospora* Microsporidia

[a] to exclude serodiscordance with the recipient.
[b] these recommendations have been proposed in the case of clinical trials by the French National Agency for Medicines and Health Products Safety 2016.

weight is not a highly representative measure of the amount of microbiota, due to the large interindividual variability of microorganism load and water content in feces, most studies use ~50 g of stool for FMT. However, a wide range of masses has been reported, and recent studies pointed out satisfactory results in the treatment of rCDI with as little as 30 g of fecal material [11,22].

When FMT is performed with fresh stools, administration should be performed within 6 h after bowel movement, to preserve sample from any alteration [10,11]. Administration of fresh stools is being gradually abandoned in favor of frozen stools. Frozen samples must be properly labeled, to be traceable as for blood. Stool preparations should be stored as aliquots for single use in order to avoid repeated freezing/thawing of the fecal suspension. The aliquots must be stored at $-80°C$, and storage at $-20°C$ should be avoided as some enzymes may still be active and degrade bacteria [11]. There is currently no codified protocol for thawing (gradual thawing or use of a 37°C water bath); however, consensually the fecal suspension should be administered within 6 h after thawing.

Delivery of FMT

The treatment sequence for rCDI usually involves three steps: (1) oral antibiotic therapy with vancomycin, (2) bowel preparation, and (3) delivery of the fecal suspension [10]. However, there is no general agreement on how best to prepare the recipient prior to FMT. Bowel lavage is known to significantly reduce the luminal bacterial load [23], but whether it should be performed prior to FMT is uncertain. It is typical to stop anti-CDI antibiotics 24—48 h prior to the FMT procedure; however, the impact of this timing on FMT success has not been rigorously evaluated nor is there robust evidence to support which anti-CDI antibiotic should be used prior to FMT or at which dose.

There is also no general agreement on the best approach for delivering fecal microbiota or for optimal volume. The FMT dose can be administered via the upper or lower GI tract. Upper GI delivery methods include nasogastric tubes, nasoduodenal tubes, and oral capsules, while lower GI delivery methods include mainly colonoscopy and enema. A nasogastric tube has been used in the first RCTs [2,16] and in some further studies, namely in elderly and immuno-compromised patients [24,25]. It requires less patient preparation, time, and inconvenience, and is technically easier to perform than colonoscopy [16,24,25]. However, this method may carry the risk of vomiting and aspirating fecal material; so a gradual administration at a flow rate of 50 mL/2—3 min should be performed, in order to avoid the risk of reflux [8].

Moreover, the engraftment of the microbiota at the colonic level remains questionable. Colonoscopy, which allows delivery of a large volume of fecal suspension (100—1000 mL) throughout the entire colon has been widely used [16,18,26—29]. Interestingly, it allows simultaneous inspection of the colon mucosa and determination of preferential sites for infusing sufficient amounts of donor feces. However, manipulation with colonoscopy through an inflamed colon can be difficult and dangerous.

Studies directly comparing delivery methods are scarce. A recent metaanalysis reported a significant difference between lower GI and upper GI delivery of FMT with levels of clinical resolution of 95% (95%CI 92%—97%) versus 88% (95%CI 82%—94%), respectively ($P = .02$) [30]. Gundacker et al. confirmed that nasogastric delivery of FMT was less effective than lower endoscopic delivery. However, when patients were stratified by illness severity, nasogastric delivery achieved similar cure rates in healthier individuals, whereas lower endoscopic delivery was preferred for relatively ill individuals [31]. Although this study was monocentric and the post-FMT follow-up was short (15 days), it has the advantage of including patients with different degrees of severity of the disease. It is worth noting that a time-varying analysis has suggested that patients treated by the upper route were two to three times more likely to face clinical failure than those treated by the lower route after 30 and 90 days post-FMT follow-up, respectively [32].

FMT by retention enema and rectal tube instillation continues to be popular due to its simplicity, low risk, and low cost [15,33,34]. Retention enema is safe, simple, and inexpensive. However, a retention time of at least 2 h is required for better efficacy. Moreover, the theory that rectal repopulation of the microbiota will lead to proximal colonization has yet to be validated and diminished rectal sphincter tone in the elderly may compromise retention of the FMT infusate and decrease success. The main drawback of this technique is that it usually requires more than one administration. In the study of Lee et al. [15], the cure rate was 50.5% after the first procedure, and 85.6% after several administrations.

Recently protocols using capsules of frozen (2 × 15 capsules, 1 day apart) or lyophilized material have reported cure rates similar to the ones achieved with other FMT procedures, that is, ~90% [20,35,36,54]. These results have been confirmed in a recent RCT, which reported that FMT via oral capsules was not inferior to delivery by colonoscopy in patients with rCDI, for preventing recurrent infection over 12 weeks of follow-up [26].

CURRENT INDICATION: TREATMENT OF RECURRENT FORMS OF *CLOSTRIDIUM DIFFICILE* INFECTIONS

According to recommendations, the first-line treatment for CDI remains anti-*C difficile* antibiotics (metronidazole, vancomycin, or fidaxomicin). This treatment is typically successful in relieving CDI symptoms, but as many as 20% of CDI patients face recurrence, and the risk of rCDI increases with each subsequent CDI episode. Molecular analyses of intestinal microbiota in patients with rCDI demonstrate lower fecal microbiota diversity compared to healthy subjects, which supports the role of FMT in the treatment of rCDI.

In 2013, a Dutch team reported the first RCT of duodenal infusion of donor fresh feces, which definitively established the remarkable efficacy of FMT [2]. In fact, the trial was stopped early after an interim analysis which demonstrated superiority of FMT compared to vancomycin alone or associated with bowel lavage. Since that first RCT, enthusiasm for FMT has increased, and cumulative experience from RCT and prospective case series showed that FMT is effective (80% −90%) when used to treat rCDI, after one or sequential infusions, without severe adverse effects.

An important barrier to FMT includes the limited time window to recruit and screen a suitable donor and prepare the material, and this leads to using frozen stool material. The elderly are especially vulnerable to CDI and rCDI. Recent studies have shown safety and equivalent cure rate of FMT in this population, as well as in immunocompromised patients, with or without significant comorbidities, allowing the use of FMT as an emerging treatment option. Major results about the use of FMT in the treatment of rCDI (RCT and prospective case series) are reported in Table 16.2.

However there are still many open questions to be addressed, mainly related to the mechanism of action of FMT: Is the therapeutic action related to the whole microbiota, or only to a few bacteria having special properties such as those with 7α-dehydroxylase activity, which can metabolize primary to secondary bile acids [37]? Is it necessary to deliver living bacteria? One study reported a successful treatment of 5 patients with FMT, using sterile fecal filtrates free of bacteria, with a 6 months of follow-up [38]. Are the bacteria the definitive therapeutic agents, or could other components such as bacteriophages be active against rCDI [39]?

FMT BEYOND *C. DIFFICILE*: PERSPECTIVES

FMT and Inflammatory Bowel Disease

Dysbiosis, with developing evidence of fungi and viruses, contributes to development of Crohn disease (CD), ulcerative colitis (UC), and pouchitis. Characteristic compositional changes observed in patients with IBD include decrease in bacterial diversity, with expansion of putative aggressive groups (such as Proteobacteria, *Fusobacterium* species, and *Ruminococcus gnavus*) combined with decreases in protective groups (such as *Bifidobacterium* species, *Roseburia*, and *Faecalibacterium prausnitzii*), causing alteration of the metagenome and production of metabolites, namely short-chain fatty acids [40].

The question of FMT in these indications was therefore quickly raised. Although trials were reported as early as the end of the 1980s, studies remain limited, including small cohorts with limited follow-up. Variability is wide in FMT routine (type of donors, antibiotic pretreatment, bowel lavage, mode of administration, number and frequency of infusions) as well as corresponding results. To date, only three RCTs have been reported in UC patients, showing conflicting results, that is, two positives and one negative [41−43].

The evidence for FMT in CD is even sparser, and additional well-designed controlled studies are needed. However, encouraging results are available, with clinical response rates of 50%−60%. Of note, in the only CD cohort study to report endoscopic outcomes, no patient experienced endoscopic remission [44]. However, a recent metaanalysis suggested a significant efficacy in attaining clinical remission in the most common phenotypes of IBD, with 201/555 patients (36%) with UC, 42/83 (50.5%) with CD, and 5/23 (21.5%) with pouchitis achieving clinical remission [45].

These results are promising, but the success rate does not reach that observed for rCDI. Moreover, long-term durability and safety remain unclear, so there is no evidenced-based recommendation for the use of FMT in UC. For both CD and UC, all studies point out the importance of donor selection, which is today only based on microbiological examinations, to ensure the absence of risk of transmission of pathogens.

FMT and Irritable Bowel Syndrome

IBS is a common disorder characterized by chronic abdominal pain and discomfort, associated with alterations of bowel habits, in the absence of a demonstrable pathology. Despite considerable research efforts, its treatment remains a significant challenge. During the past years, intestinal dysbiosis has been closely linked to the pathophysiology of IBS [46].

TABLE 16.2 Major RCTs and Prospective Studies on FMT for rCDI Treatment

Study	Delivery Mode	Special Feature	Number of Patients	Main Results	References
Randomized clinical trial	Upper route	Vancomycin Vancomycin + bowel lavage FMT with fresh stools	13 13 16	End point: no diarrhea and no relapse after 10 weeks of follow-up Vancomycin group: cure rate 31% Vancomycin + bowel lavage group: cure rate 23% FMT group: cure rate 81% after one FMT and 94% after a second infusion	[2]
	Colonoscopy	Vancomycin FMT with fresh stools	19 20	Cure rate 26% Cure rate 90%	[13]
	Upper route Colonoscopy	Frozen stool (storage 29–156 days at −80°C)	10 10	End point: no diarrhea and no relapse 8 weeks after FMT Upper route: cure rate 60% Colonoscopy: cure rate 80%	[16]
	Colonoscopy	Fresh stool Frozen stool Lyophilized stool	25 24 23	Overall cure rate: 87% after 8 weeks of follow-up fresh stool: 100%; frozen stool 83.3%; lyophilized stool 78%	[18]
	Colonoscopy	Donor stools (= heterologous FMT) Patient's own stool (= autologous FMT)	22 24	Cure rate = 90.9% Cure rate = 62.5%	[27]
	Colonoscopy Oral capsules		59 57	End point: no rCDI 12 weeks after FMT Cure rate: 96.2% in both groups	[26]
	Enema	Refractory and recurrent CDI Fresh stool Frozen stool (storage < 30 days at −20°C)	111 108	End point: no diarrhea and no relapse at week 13 Cure rate 85.6% (fresh stool) vs. 90.7% (frozen stool)	[15]

Number of FMT	Fresh stools	Frozen stools
1	56	57
2	22	75
3 à 5	12	13
>5	5	4
Total	95	98

	Route	Description	Number	Outcome	Reference
Prospective studies	Upper route	Elderly patients (>65 years) Jejunal and colonic administration	29	Cure rate 93.1%	[24]
	Upper route Colonoscopy	Hematopoietic stem cell transplantation recipients with CDI	7	Cure rate 85.7% and 100% after the first FMT and second FMT, respectively	[25]
	Upper route Colonoscopy	Enteroscopy and Colonoscopy in a single session	27	Cure rate 100% after 12 weeks of follow-up	[29]
	Colonoscopy	Elderly patients (mean age: 83.7 years)	4	Cure rate 100% for up to 3 months without any adverse events	[28]
	Colonoscopy	First use of frozen material	43	Cure rate 86% after 2 months of follow-up Similar efficacy in patients that had underlying IBD ($\sim$30%)	[14]
	Oral delivery	Capsules containing frozen stool	180	Cure rate 82% after the first dose (2×15 capsules at 1 day apart), 91% after the second dose and 93% after the third dose	[35]
	Oral delivery	Freeze-dried encapsulated preparation of standardized fecal microbiota	49	Cure rate 88% A single administration of the lowest dose ($2.1–2.5 \times 10^{11}$ bacteria) in a 2 or 3 capsules was highly successful	[36]
	Enema	Recurrent, severe, or complicated CDI in elderly patients	146	Cure rate 82.9% and 95.9% after the first FMT and second FMT, respectively	[33]

It has been demonstrated that diversity of the microbial population is reduced, the proportion of specific bacterial groups is altered, and the degree of variability in the microbiota composition is different in IBS patients when compared with healthy subjects. Furthermore, a higher degree of temporal instability of the microbiota among IBS patients has been detected [47]. FMT has thus been suggested as a potential treatment; however, results are far from conclusive yet. A first case series has been published in 1989, demonstrating approximately 50% relief in symptoms. However, the study also included IBD and CDI patients, without any distinction between these different diseases [48]. The same team has then reported the successful treatment of more than 300 IBS patients, with diarrheic phenotype [49].

Another single-center study has reported promising results with refractory disease: 69% of the patients experienced a resolution or improvement in their IBS symptoms. However, patient-specific long-term treatment goals were only achieved in 46% of the subjects [50].

FMT and Decolonization of Antibiotic-Resistant Bacteria in the Gastrointestinal Tract

Selective pressures created by the clinical and agricultural use and misuse of antibiotics have resulted in the development of multidrug resistant bacteria. Several case studies have suggested the potential efficacy of FMT for the decolonization of the digestive tract of patients with multiresistant bacteria, regardless of the immune status. Recently, Bilinski et al. reported the successful decolonization of 60% (15/25) of the patients at 1 month after FMT, and more frequently in those that had no preprocedural use of antibiotics (79% vs. 36%, $P < .05$) [51]. In a recent French study with digestive tract colonization by carbapenem-resistant *Enterobacteriaceae* (CRE, n = 6) or vancomycin-resistant *Enterococci* (VRE, n = 2), FMT administered nasoduodenally allowed the eradication of carriage in 2/6 patients with CRE and 1/2 patients with VRE at 3 month follow-up, with no noticeable side effects [52].

It should be underscored that the delivery of FMT via enema has not demonstrated the ability to decolonize VRE from the gut, suggesting that the route of administration represents an important factor in efficacy [53].

FMT and Other Indications

The development of high-throughput microbial sequencing has enabled the identification of a plethora of diseases (liver disease, metabolic syndrome, autism, rheumatism, and others), associated with underlying gastrointestinal microbiota dysbiosis. Interest has thus grown in the potential of microbial manipulation through the use of FMT. However, these studies remain experimental.

CONCLUSION

Since 2013 and the study by van Nood [2], FMT has revolutionized the treatment of rCDI and it is now recognized and recommended as a highly efficacious and safe treatment method for rCDI. Moreover, the capacity of FMT to restore a dysbiotic microbiota has prompted various researches into its use as a therapeutic agent for other diseases associated with gastrointestinal microbial dysbiosis. However, a lot of work remains to be done to develop international recommendations and to standardize key methodological aspect of the FMT intervention such as donor selection, FMT preparation, receiver preparation, and follow-up to facilitate reporting and implementation in clinical practice in various clinical situations.

LIST OF ACRONYMS AND ABBREVIATIONS

CD Crohn disease
CDI *Clostridium difficile* infection
CRE Carbapenem-resistant Enterobacteriaceae
FMT Fecal microbiota transplantation
GI Gastrointestinal
IBD Inflammatory bowel disease
IBS Irritable bowel syndrome
rCDI Recurrent forms of *Clostridium difficile* infection
RCT Randomized controlled trial
UC Ulcerative colitis
VRE Vancomycin-resistant *Enterococci*

REFERENCES

[1] Smits LP, Bouter KE, de Vos WM, Borody TJ, Nieuwdorp M. Therapeutic potential of fecal microbiota transplantation. Gastroenterology 2013;145:946−53.

[2] van Nood E, Vrieze A, Nieuwdorp M, Fuentes S, Zoetendal EG, de Vos WM, Visser CE, Kuijper EJ, Bartelsman JF, Tijssen JG, Speelman P, Dijkgraaf MG, Keller JJ. Duodenal infusion of donor feces for recurrent *Clostridium difficile*. N Engl J Med 2013;368:407−15.

[3] Debast SB, Bauer MP, Kuijper EJ. European Society of clinical Microbiology and infectious diseases: update of the treatment guidance document for *Clostridium difficile* infection. Clin Microbiol Infect 2014;20Suppl(2):1−26.

[4] Surawicz CM, Brandt LJ, Binion DG, Ananthakrishnan AN, Curry SR, Gilligan PH, McFarland LV, Mellow M, Zuckerbraun BS. Guidelines for diagnosis, treatment, and prevention of *Clostridium difficile* infections. Am J Gastroenterol 2013;108:478−98.

[5] Vyas D, Aekka A, Vyas A. Fecal transplant policy and legislation. World J Gastroenterol 2015;21:6−11.

[6] Kump PK, Krause R, Allerberger F, Högenauer C. Faecal microbiota transplantation-the Austrian approach. Clin Microbiol Infect 2014;20:1106−11.

[7] Lahtinen P, Mattila E, Anttila VJ, Tillonen J, Teittinen M, Nevalainen P, Salminen S, Satokari R, Arkkila P. Faecal microbiota transplantation in patients with *Clostridium difficile* and significant comorbidities as well as in patients with new indications: a case series. World Journal Gastroenterol 2017;23:7174−84.

[8] Kapel N, Thomas M, Corcos O, Mayeur C, Barbot-Trystram L, Bouhnik Y, Joly F. Practical implementation of faecal transplantation. Clin Microbiol Infect 2014;20:1098−105.

[9] Bakken JS, Borody T, Brandt LJ, Brill JV, Demarco DC, Franzos MA, Kelly C, Khoruts A, Louie T, Martinelli LP, Moore TA, Russell G, Surawicz C. Fecal Microbiota Transplantation Workgroup. Treating *Clostridium difficile* infection with fecal microbiota transplantation. Clin Gastroenterol Hepatol 2011;9:1044−9.

[10] Sokol H, Galperine T, Kapel N, Bourlioux P, Seksik P, Barbut F, Scanzi J, Chast F, Batista R, Joly F, Joly AC, Collignon A, Guery B, Beaugerie L. French group of faecal microbiota transplantation (FGFT). Faecal microbiota transplantation in recurrent *Clostridium difficile* infection: recommendations from the French group of faecal microbiota transplantation. Dig Liver Dis 2016;48:242−7.

[11] Cammarota G, Ianiro G, Tilg H, Rajilić-Stojanović M, Kump P, Satokari R, Sokol H, Arkkila P, Pintus C, Hart A, Segal J, Aloi M, Masucci L, Molinaro A, Scaldaferri F, Gasbarrini G, Lopez-Sanroman A, Link A, de Groot P, de Vos WM. Högenauer C,Malfertheiner P, Mattila E, Milosavljević T, Nieuwdorp M, Sanguinetti M, Simren M, Gasbarrini a; European FMT working group. European consensus conference on faecal microbiota transplantation in clinical practice. Gut 2017;66:569−80.

[12] ANSM. La transplantation de microbiote fécal et son encadrement dans les essais cliniques. Novembre 2016.

[13] Cammarota G, Masucci L, Ianiro G, Bibbò S, Dinoi G, Costamagna G, Sanguinetti M, Gasbarrini A. Randomised clinical trial: faecal microbiota transplantation by colonoscopy vs. vancomycin for the treatment of recurrent *Clostridium difficile* infection. Aliment Pharmacol Ther 2015;41:835−43.

[14] Hamilton MJ, Weingarden AR, Sadowsky MJ, Khoruts A. Standardized frozen preparation for transplantation of fecal microbiota for recurrent *Clostridium difficile* infection. Am J Gastroenterol 2012;107:761−7.

[15] Lee CH, Steiner T, Petrof EO, Smieja M, Roscoe D, Nematallah A, Weese JS, Collins S, Moayyedi P, Crowther M, Ropeleski MJ, Jayaratne P, Higgins D, Li Y, Rau NV, Kim PT. Frozen vs fresh fecal microbiota transplantation and clinical resolution of diarrhea in patients with recurrent *Clostridium difficile* infection: a randomized clinical trial. J Am Med Assoc 2016;315:142−9.

[16] Youngster I, Sauk J, Pindar C, Wilson RG, Kaplan JL, Smith MB, Alm EJ, Gevers D, Russell GH, Hohmann EL. Fecal microbiota transplant for relapsing *Clostridium difficile* infection using a frozen inoculum from unrelated donors: a randomized, open-label, controlled pilot study. Clin Infect Dis 2014;58:1515−22.

[17] Tang G, Yin W, Liu W. Is frozen fecal microbiota transplantation as effective as fresh fecal microbiota transplantation in patients with recurrent or refractory *Clostridium difficile* infection: a meta-analysis? Diagn Microbiol Infect Dis 2017;88:322−9.

[18] Jiang ZD, Ajami NJ, Petrosino JF, Jun G, Hanis CL, Shah M, Hochman L, Ankoma-Sey V, DuPont AW, Wong MC, Alexander A, Ke S, DuPont HL. Randomised clinical trial: faecal microbiota transplantation for recurrent *Clostridium difficile* infection - fresh, or frozen, or lyophilised microbiota from a small pool of healthy donors delivered by colonoscopy. Aliment Pharmacol Ther 2017;45:899−908.

[19] Costello SP, Tucker EC, La Brooy J, Schoeman MN, Andrews JM. Establishing a fecal microbiota transplant service for the treatment of *Clostridium difficile* infection. Clin Infect Dis 2016;62:908−14.

[20] Youngster I, Russell GH, Pindar C, Ziv-Baran T, Sauk J, Hohmann EL. Oral, capsulized, frozen fecal microbiota transplantation for relapsing Clostridium difficile infection. J Am Med Assoc 2014;312:1772−8.

[21] Brandt LJ, Aroniadis OC. An overview of fecal microbiota transplantation: techniques, indications, and outcomes. Gastrointest Endosc 2013;78:240−9.

[22] Kassam Z, Lee CH, Yuan Y, Hunt RH. Fecal microbiota transplantation for *Clostridium difficile* infection: systematic review and meta-analysis. Am J Gastroenterol 2013;108:500−8.

[23] Jalanka J, Salonen A, Salojärvi J, Ritari J, Immonen O, Marciani L, Gowland P, Hoad C, Garsed K, Lam C, Palva A, Spiller RC, de Vos WM. Effects of bowel cleansing on the intestinal microbiota. Gut 2015;64:1562−8.

[24] Girotra M, Garg S, Anand R, Song Y, Dutta SK. Fecal microbiota transplantation for recurrent *Clostridium difficile* infection in the elderly: long-term outcomes and microbiota changes. Dig Dis Sci 2016;61:3007−15.

[25] Webb BJ, Brunner A, Ford CD, Gazdik MA, Petersen FB, Hoda D. Fecal microbiota transplantation for recurrent *Clostridium difficile* infection in hematopoietic stem cell transplant recipients. Transpl Infect Dis 2016;18:628−33.

[26] Kao D, Roach B, Silva M, Beck P, Rioux K, Kaplan GG, Chang HJ, Coward S, Goodman KJ, Xu H, Madsen K, Mason A, Wong GK, Jovel J, Patterson J, Louie T. Effect of oral capsule- vs colonoscopy-delivered fecal microbiota transplantation on recurrent *Clostridium difficile* infection: a randomized clinical trial. J Am Med Assoc 2017;318:1985–93.

[27] Kelly CR, Khoruts A, Staley C, Sadowsky MJ, Abd M, Alani M, Bakow B, Curran P, McKenney J, Tisch A, Reinert SE, Machan JT, Brandt LJ. Effect of fecal microbiota transplantation on recurrence in multiply recurrent Clostridium difficile infection: a randomized trial. Ann Intern Med 2016;165:609–16.

[28] Bamba S, Nishida A, Imaeda H, Inatomi O, Sasaki M, Sugimoto M, Andoh A. Successful treatment by fecal microbiota transplantation for Japanese patients with refractory Clostridium difficile infection: a prospective case series. J Microbiol Immunol Infect November 5, 2017;1182(17):30235–9. https://doi.org/10.1016/j.jmii.2017.08.027. S1684.

[29] Dutta SK, Girotra M, Garg S, Dutta A, von Rosenvinge EC, Maddox C, Song Y, Bartlett JG, Vinayek R, Fricke WF. Efficacy of combined jejunal and colonic fecal microbiota transplantation for recurrent *Clostridium difficile* Infection. Clin Gastroenterol Hepatol 2014;12:1572–6.

[30] Quraishi MN, Widlak M, Bhala N, Moore D, Price M, Sharma N, Iqbal TH. Systematic review with meta-analysis: the efficacy of faecal microbiota transplantation for the treatment of recurrent and refractory *Clostridium difficile* infection. Aliment Pharmacol Ther 2017;46:479–93.

[31] Gundacker ND, Tamhane A, Walker JB, Morrow CD, Rodriguez JM. Comparative effectiveness of faecal microbiota transplant by route of administration. J Hosp Infect 2017;96:349–52.

[32] Furuya-Kanamori L, Doi SA, Paterson DL, Helms SK, Yakob L, McKenzie SJ, Garborg K, Emanuelsson F, Stollman N, Kronman MP, Clark J, Huber CA, Riley TV, Clements AC. Upper versus lower gastrointestinal delivery for transplantation of fecal microbiota in recurrent or refractory Clostridium difficile infection: a collaborative analysis of individual patient data from 14 studies. J Clin Gastroenterol 2017;51:145–50.

[33] Agrawal M, Aroniadis OC, Brandt LJ, Kelly C, Freeman S, Surawicz C, Broussard E, Stollman N, Giovanelli A, Smith B, Yen E, Trivedi A, Hubble L, Kao D, Borody T, Finlayson S, Ray A, Smith R. The long-term efficacy and safety of fecal microbiota transplant for recurrent, severe, and complicated *Clostridium difficile infection* in 146 elderly Individuals. J Clin Gastroenterol 2016;50:403–7.

[34] Baro E, Galperine T, Denies F, Lannoy D, Lenne X, Odou P, Guery B, Dervaux B. Cost-effectiveness analysis of five competing strategies for the management of multiple recurrent community-onset *Clostridium difficile* infection in France. PLoS One 2017;12(1):e0170258.

[35] Youngster I, Mahabamunuge J, Systrom HK, Sauk J, Khalili H, Levin J, Kaplan JL, Hohmann EL. Oral, frozen fecal microbiota transplant (FMT) capsules for recurrent *Clostridium difficile* infection. BMC Med 2016;14(1):134.

[36] Staley C, Hamilton MJ, Vaughn BP, Graiziger CT, Newman KM, Kabage AJ, Sadowsky MJ, Khoruts A. Successful resolution of recurrent Clostridium difficile infection using freeze-dried, encapsulated fecal microbiota; pragmatic cohort study. Am J Gastroenterol 2017;112:940–7.

[37] Thanissery R, Winston JA, Theriot CM. Inhibition of spore germination, growth, and toxin activity of clinically relevant *C. difficile* strains by gut microbiota derived secondary bile acids. Anaerobe 2017;45:86–100.

[38] Ott SJ, Waetzig GH, Rehman A, Moltzau-Anderson J, Bharti R, Grasis JA, Cassidy L, Tholey A, Fickenscher H, Seegert D, Rosenstiel P, Schreiber S. Efficacy of sterile fecal filtrate transfer for treating patients with *Clostridium difficile* infection. Gastroenterology 2017;152:799–811.

[39] Nale JY, Spencer J, Hargreaves KR, Buckley AM, Trzepiński P, Douce GR, Clokie MR. Bacteriophage combinations significantly reduce *Clostridium difficile* growth in vitro and proliferation in vivo. Antimicrob Agents Chemother 2016;60:968–81.

[40] Sartor RB, Wu GD. Roles for intestinal bacteria, viruses, and fungi in pathogenesis of inflammatory bowel diseases and therapeutic approaches. Gastroenterology 2017;152:327–39.

[41] Moayyedi P, Surette MG, Kim PT, Libertucci J, Wolfe M, Onischi C, Armstrong D, Marshall JK, Kassam Z, Reinisch W, Lee CH. Fecal microbiota transplantation induces remission in patients with active ulcerative colitis in a randomized controlled trial. Gastroenterology 2015;149:102–9.

[42] Rossen NG, Fuentes S, van der Spek MJ, Tijssen JG, Hartman JH, Duflou A, Löwenberg M, van den Brink GR, Mathus-Vliegen EM, de Vos WM, Zoetendal EG, D'Haens GR, Ponsioen CY. Findings from a randomized controlled trial of fecal transplantation for patients with ulcerative colitis. Gastroenterology 2015;149:110–8.

[43] Paramsothy S, Kamm MA, Kaakoush NO, Walsh AJ, van den Bogaerde J, Samuel D, Leong RWL, Connor S, Ng W, Paramsothy R, Xuan W, Lin E, Mitchell HM, Borody TJ. Multidonor intensive faecal microbiota transplantation for active ulcerative colitis: a randomised placebo-controlled trial. Lancet 2017;389:1218–28.

[44] Vermeire S, Joossens M, Verbeke K, Wang J, Machiels K, Sabino J, Ferrante M, Van Assche G, Rutgeerts P, Raes J. Donor species richness determines faecal microbiota transplantation success in inflammatory bowel disease. J Crohns Colitis 2016;10:387–94.

[45] Paramsothy S, Paramsothy R, Rubin DT, Kamm MA, Kaakoush NO, Mitchell HM, Castaño-Rodríguez N. Faecal microbiota transplantation for inflammatory bowel disease: a systematic review and meta-analysis. J Crohns Colitis 2017;11:1180–99.

[46] Dlugosz A, Winckler B, Lundin E, Zakikhany K, Sandström G, Ye W, Engstrand L, Lindberg G. No difference in small bowel microbiota between patients with irritable bowel syndrome and healthy controls. Sci Rep 2015;5:8508.

[47] Distrutti E, Monaldi L, Ricci P, Fiorucci S. Gut microbiota role in irritable bowel syndrome: new therapeutic strategies. World J Gastroenterol 2016;22:2219–41.

[48] Borody TJ, George L, Andrews P, Brandl S, Noonan S, Cole P, Hyland L, Morgan A, Maysey J, Moore-Jones D. Bowel-flora alteration: a potential cure for inflammatory bowel disease and irritable bowel syndrome? Med J Aust 1989;150:604.

[49] Borody TJ, Paramsothy S, Agrawal G. Fecal microbiota transplantation: indications, methods, evidence, and future directions. Curr Gastroenterol Rep 2013;15:337.

[50] Pinn DM, Aroniadis OC, Brandt LJ. Is fecal microbiota transplantation the answer for irritable bowel syndrome? A single-center experience. Am J Gastroenterol 2014;109:1831–2.

[51] Bilinski J, Grzesiowski P, Sorensen N, Madry K, Muszynski J, Robak K, Wroblewska M, Dzieciatkowski T, Dulny G, Dwilewicz-Trojaczek J, Wiktor-Jedrzejczak W, Basak GW. Fecal microbiota transplantation in patients with blood disorders inhibits gut colonization with antibiotic-resistant bacteria: results of a prospective, single-center study. Clin Infect Dis 2017;65:364–70.

[52] Davido B, Batista R, Michelon H, Lepainteur M, Bouchand F, Lepeule R, Salomon J, Vittecoq D, Duran C, Escaut L, Sobhani I, Paul M, Lawrence C, Perronne C, Chast F, Dinh A. Is faecal microbiota transplantation an option to eradicate highly drug-resistant enteric bacteria carriage? J Hosp Infect 2017;95:433–7.

[53] Sohn KM, Cheon S, Kim YS. Can fecal microbiota transplantation (FMT) eradicate fecal colonization with vancomycin-resistant enterococci (VRE)? Infect Control Hosp Epidemiol 2016;37:1519–21.

[54] Cheminet G, Kapel N, Bleibtreu A, Sadou-Yaye H, Bellanger A, Duval X, Joly F, Fantin B, de Lastours V and the French Group for Fecal Microbiota Transplantation (GFTF). Fecal microbiota transplantation with frozen capsules for relapsing Clostridium difficile infections: the first experience from 15 consecutive patients in France. J Hosp Infect. 2018 Jul 12. pii: S0195-6701(18)30369-4. https://doi.org/10.1016/j.jhin.2018.07.00

Chapter 17

Effects of Probiotics on Improvement of Metabolic Factors in Pregnant Women: A Metaanalysis of Randomized Placebo-Controlled Trials

Jing Sun[1], Damodar Gajurel[3], Nicholas Buys[3] and Chenghong Yin[2]

[1]*School of Medicine, Griffith University, Gold Coast, QLD, Australia;* [2]*Beijing Obstetrics and Gynecology Hospital, Capital Medical University, Beijing, China;* [3]*Menzies Health Institute Queensland, Griffith University, Gold Coast, QLD, Australia*

INTRODUCTION

Pregnancy and childbirth are momentous events in the lives of women. Preventing or minimizing pregnancy-related complications is therefore a major priority for the healthcare system. Common and far-reaching complications of pregnancy are related to metabolic abnormalities, the prevalence of which has significantly increased in recent years [1]. Lifestyle changes, in particular, have resulted in an increase in body weight and decrease in physical activities, which has had a detrimental impact on pregnancy, due to the ensuing disturbed metabolic homeostasis.

Obesity is a significant global health problem, with more women entering into pregnancy in this state [2]. In 2014, 40% of women aged 18 years and over were overweight (BMI $\geq$ 25 kg/m^2) and 15% were obese (BMI $\geq$ 30 kg/m^2). More than half of all pregnant women in the United States are considered obese, with 8% considered extremely obese [3]. Both overweight and obesity have shown a marked increase over the past four decades [4]. Obesity and diabetes mellitus, which often coexist, are common presentations during pregnancies, with an estimated one-third of all pregnant women being affected [5].

Obesity and Diabetes in Pregnancy

Obese women are at an increased risk for maternal complications including gestational diabetes mellitus (GDM), pre-eclampsia, and thromboembolic events. In addition, they may experience obstetric complications such as fetal macrosomia, stillbirth, shoulder dystocia, caesarean delivery, and preterm delivery [1]. Pregnancy-induced production of cytokines, due to the excess body weight, including leptin, resistin, interleukin-6(IL-6), together with low physical activity, is associated with abnormal glucose homeostasis, insulin resistance, and increased systemic inflammation. Maternal hyperinsulinism and glycemic status typically result in preterm delivery, higher rate of caesarean section, the development of pregnancy-induced hypertension (PIH), and GDM [6].

Gestational Diabetes Mellitus

GDM has now reached epidemic and pandemic proportions in Western society, affecting between 1% and 14% of pregnancies, depending on the screening and diagnostic methods employed [2,7]. GDM is defined as the development of glucose intolerance, that for the first time occurs during pregnancy [8] and is associated with adverse maternal and infant health outcomes during gestation, childbirth, and postpartum, as well as preeclampsia and increased risk of infection throughout pregnancy. It is related to several adverse pregnancy outcomes, including shoulder dystocia, neonatal hypoglycemia, respiratory distress, hypocalcemia, and macrosomia.

A several-fold increased risk of the mother developing type 2 diabetes (T2DM) postpartum has also been reported. Additionally, infant morbidity includes the risk of fetal malformations and diabetic fetopathy, which may cause macrosomia and subsequent mechanical complications during labor. Studies also suggest that children born to mothers with GDM have an increased risk of diabetes mellitus and metabolic dysfunction later in life [9].

Oxidative Stress

Due to enhanced weight during the mid-pregnancy period and physiological alterations in lipoprotein levels resulting from hormonal changes, pregnancy is associated with elevated levels in lipid profiles [10,11]. In addition, the increased oxygen requirement, systemic release of placental factors, and reduced scavenging power of antioxidants during pregnancy may result in increased susceptibility to oxidative stress. Maternal dyslipidemia and oxidative stress are associated with several complications including GDM, preeclampsia, and intrauterine growth restriction (IUGR) [12].

Therapeutic Options in Gestational Diabetes Mellitus

GDM management is currently based on the modification of maternal diet, and possibly insulin therapy. Pharmacotherapy may result in significant side effects including abdominal discomfort, dizziness, diarrhea, and hypoglycemia. It has recently been suggested that probiotics may ameliorate some of the adverse metabolic effects of diabetes, and provide an alternative option for women with GDM [9]. The regular intake of probiotics can beneficially impact on the composition of the gut microbiota [13,14], resulting in improved glucose homeostasis, attenuation of inflammation, regulation of insulin production, maintenance of the integrity of the gastrointestinal lining, and the harvesting of nutrients from the host diet [15,16].

Literature Search

This review protocol was registered at Prospero International Prospective Register of Systematic Reviews (CRD42017068789). Relevant studies published in English from January 1, 1987, to June 30, 2017, were accessed by the researchers from PubMed, Scopus, CINHAL, Web of Science and Cochrane Library. Articles were also identified through reference lists. In articles where the data was not full, reported contact was made with the authors to ascertain if the data was available. Key words used in the search process included: "Lactobacillus and pregnancy," "Bifidobacterium and pregnancy," "Saccharomyces and pregnancy," "yogurt and pregnancy," "probiotics and pregnancy," "probiotics and gestational diabetes," "probiotics and biomedical factors in pregnancy," "probiotics and glucose in pregnancy," "probiotics and insulin in pregnancy," "probiotics and HOMA in pregnancy," "probiotics and HbA1c in pregnancy," "probiotics and lipids in pregnancy," "probiotics and cholesterol in pregnancy," "probiotics and LDL in pregnancy," "probiotics and BMI in pregnancy," "probiotics and body weight in pregnancy," "probiotics and HDL in pregnancy," "probiotics and triglyceride in pregnancy," "probiotics and sugar in pregnancy," and "probiotics and blood sugar in pregnancy."

All retrieved articles were downloaded into EndNote and duplicates removed. Articles were then screened based on title and abstract, and the full texts of the remaining articles obtained. Two researchers independently read, evaluated, and extracted data, and resolved any differences through dialogue. Preferred Reporting Items for Systematic Reviews and Meta-Analysis (PRISMA) guidelines were utilized [17]. The final metaanalysis thereby consisted of 16 articles, 12 of which were assessed as high quality and 4 as moderate quality. Included studies are summarized in Fig. 17.1, Tables 17.1 and 17.2.

Study Descriptions

The final sample drawn from the 16 randomized controlled trials (RCTs) was 1339 subjects, with 669 in the probiotic group and 670 in the control group (see Table 17.2). Among these studies, four were single-blinded [18−21] and the 12 were double-blinded [6,12,22−31]. Studies were conducted in three countries: Finland, Iran, and Ireland. Participant characteristics included six studies with GDM [24,25,27−30], and two studies with obesity [22,23]. Most of the studies measured glucose, HOMA-IR, insulin, weight/body mass index (BMI) and lipid profile (total cholesterol, LDL, HDL, and Triglycerides). Some measured very low-density lipoprotein (VLDL), markers of oxidative stress (total antioxidant capacity (TAC), glutathione (GSH)), and HbA1c. Table 17.2 summarizes the characteristics for each study.

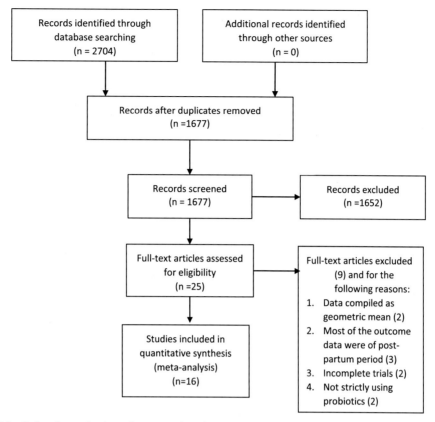

FIGURE 17.1 Preferred reporting items for systematic reviews and metaanalysis (PRISMA) flow chart of the included studies.

Effects on fasting Glucose

There were 12 studies with 994 participants reporting the effects of probiotics on fasting glucose (see Fig 17.2A). The average reduction of glucose was -0.32 mmol/L (range from -0.07 to -1.18) in the probiotics group compared with -0.18 mmol/L (range from 0.07 to -0.91) in the control group. The pooled effect was -0.15 mmol/L (95% CI, -0.27, -0.02; $P < .05$; $I^2 = 90\%$, $P < .01$) (Table 17.3).

Probiotics resulted in decreased glucose only in trials using multiple types of bacteria (-0.21 mmol/L (95% CI -0.37, -0.05; $P < .01$; $I^2 = 93\%$, $P < .01$) located outside Europe (-0.21 mmol/L (95% CI, -0.39, -0.04; $P < .01$; $I^2 = 92\%$, $P < .01$), as well as in those using probiotics without synbiotics, with a pooled mean difference of -0.15 mmol/L (95% CI -0.29, -0.01; $P < .01$; $I^2 = 92\%$, $P < .01$) compared with those combining synbiotics with probiotics. No significant effect was seen in impaired glucose tolerance.

TABLE 17.1 Inclusion criteria of included studies

PICO Step	Criteria
Population	1. Trial participants must be aged 18 or older; 2. Trial participants must be pregnant women.
Intervention	1. Trials must be randomized placebo-controlled studies; 2. All intervention groups used probiotics as the intervention; 3. Where there were multiple groups in a study, only the probiotics and placebo groups were included.
Comparison	1. All trials used a control group (placebo).
Outcome	1. Outcome variables include glucose, insulin, C-peptide, HOMA-IR, HOMA-B, HbA1c, QUICKI, total cholesterol, LDL, HDL, triglyceride, VLDL, weight, BMI, TAC, GSH.

TABLE 17.2 Characteristics of Included Studies

Studies	Participants (Number of P/C)	Design & Location	Age (P/C) (Mean [SD]) (years)	Probiotic (week)	Probiotic Strains	Dose (CFU/ G And/or Quantity/day)	Measured Outcome	Quality of Studies Assessed by PEDro Tool
[22]	Obese: BMI >30 (63/75)	DB,PC,P Ireland	31.4 (5.0)/31 (5.2)	Capsule (4)	Lactobacillus salivarius	1×10^9	Glucose, insulin, HOMA-IR, C-peptide, cholesterol, HDL, LDL, TG, and CRP	10, high
[24]	GDM/IGT (74/75)	DB,PC,P Ireland	33.5 (5.0)/32.6 (4.5)	Capsule (diagnosis to delivery)	L. salivarius	1×10^9	Glucose, Insulin, HOMA-IR, C-peptide, cholesterol, HDL, LDL, TG, cholesterol: HDL ratio, and LDL-HDL ratio	10, high
[23]	Obese: BMI 30–39.9 (52/58)	DB,PC,P Ireland	N/A	Capsule (4)	L. salivarius	1×10^9	Glucose, insulin, HOMA-IR, C-peptide, cholesterol, HDL, LDL, TG, and CRP	10, high
[25]	GDM (30/30)	DB,PC,P Iran	31.8 (6.0)/29.7 (4.0)	Capsule (6)	L. acidophilus, Lactobacillus casei, Bifidobacterium bifidum	2×10^9 each	Glucose, insulin, HOMA-IR, HOMA-B, QUICKI, cholesterol, HDL, LDL, TG, and VLDL	10, high
[26]	Normal pregnancy (30/30)	DB,PC,P Iran	27.1 (5.1)/28.4 (5.3)	Capsule (12)	Lactobacillus acidophilus, L. casei, B. bifidum	2×10^9 each	Glucose, Insulin, HOMA-IR, HOMA-B, QUICKI, cholesterol, HDL, LDL, TG, VLDL, hs-CRP, NO, TAC, GSH, and MDA	10, high
[27]	GDM (41/41)	DB,PC,P Iran	32.4 (3.1)/31.9 (4.0)	Capsule (8)	Streptococcus thermophilus, Bifidobacterium breve, Bifidobacterium longum, Bifidobacterium infantis, L. acidophilus, Lactobacillus plantarum, Lactobacillus paracasei, and Lactobacillus delbrueckii	112.5×10^9 of 8 strains in one capsule	Glucose, insulin, HOMA-IR, HbA1c, hs-CRP, IL-10, IFN-c, IL-6, and TNF-α	10, high
[28]	GDM (29/27)	DB, PC, P Iran	28.14(6.2)/26.48 (5.2)	Capsule (8)	L. acidophilus, Bifidobacterium, S. thermophilus, Lactobacillus delbrueckii bulgaricus	10^9 of each	Glucose, insulin, HOMA-IR, and QUICKI	10, high
[29]	GDM (30/30)	DB, PC, P Iran	28.8 (5.4)/27.8 (3.7)	Capsule (6)	L. acidophilus, L. casei, B. bifidum	2×10^9 each	Glucose, hs-CRP, NO, TAC, GSH, MAD, MAD: TAC ratio	10, high
[21]	Normal pregnancy (37/33)	SB, C, P Iran	24.2 (3.3)/25.7 (3.1)	Yoghurt (9)	S. thermophilus, L. bulgaricus, L. acidophilus LA5, and Bifidobacterium lactis B B12	1×10^7 in 200 g yoghurt	Glucose, insulin, HOMA-IR, SBP, and DBP	6, medium

Ref	Population (n)	Design, Country	BMI	Form (weeks)	Intervention	Dose	Outcomes	Score, quality
[20]	Normal pregnancy (37/33)	SB, C, P Iran	24.2 (3.3)/ 25.7 (3.1)	Yoghurt (9)	S. thermophilus, L. bulgaricus, and B. lactis B B12	1×10^7 each in 200 g yoghurt	Cholesterol, HDL, LDL, TG, and cholesterol: HDL ratio	6, medium
[18]	Normal pregnancy (37/33)	SB, C, P Iran	24.2(3.3)/ 25.72 (3.1)	Yoghurt (9)	S. thermophilus, L. bulgaricus, L. acidophilus LA5, and Bifidobacterium lactis B B12	1×10^7 each in 200 g yoghurt	h-CRP TNF-alpha	6, medium
[30]	GDM (35/35)	DB, PC, P Iran	28.5 (5.8)/ 28.7 (3.4)	Capsule (6)	Bifidobacterium animalis, L. acidophilus, L. casei, B. bifidum and Inulin	2×10^9 each 800 mg	Glucose, insulin, HOMA-IR, HOMA-B, QUICKI, cholesterol, HDL, LDL, VLDL, and triglycerides	10, high
[31]	Normal pregnancy (85/85)	DB, PC, P Finland	29.7 (4.1)/ 30.2 (5.0)	Capsule <17 –52 weeks	Lactobacillus rhamnosus GG B. lactis Bb12; (diet/probiotics)	1×10^{10} each	Glucose, HbA1C, insulin, HOMA and QUICKI	10, high
[12]	Normal pregnancy (26/26)	DB, PC, P Iran	26.9 (6.3)/29 (4.6)	Capsule (9 weeks)	Lactobacillus sporogenes (1×10^7 CFU) and 0.04 g inulin as the prebiotic per g per day	$18 \times 10^7/18$ g package	TG, TC, HDL, VLDL, LDL, TAC, GSH	9, high
[12]	Normal pregnancy (26/26)	DB, PC, P Iran	26.4 (6.3)/29 (4.6)	Capsule (9 weeks)	L. sporogenes and 0.04 g inulin as prebiotic with 0.38 g isomalt, 0.36 g sorbitol and 0.05 g stevia as sweetener per 1 g per day	$18 \times 10^7/18$ g package	FPG, HOMA-IR, HOMA-B, QUICKI, and hs-CRP	9, high
[19]	Normal pregnancy (37/33)	SB, C, P Iran	24.2 (3.3)/ 25.7 (3.1)	Yoghurt (9 weeks)	Streptococcus thermophilus, Lactobacillus bulgaricus, Lactobacillus acidophilus LA5, and Bifidobacterium lactis B B12	10^7 each in 200 g yoghurt	TAC, GSH, GPx, GR, 8-oxo-G	6, medium

8-oxo-G, serum 8-oxo-7,8-dihydroguanine; BMI, body-mass index; C, controlled; CFU, colony-forming units; DB, double blinded; DBP, diastolic blood pressure; GDM, gestational diabetes mellitus patients; GPx, erythrocyte glutathione peroxidase; GR, glutathione reductase; GSH, plasma glutathione; HbA1c, glycosylated haemoglobin; HDL, high density lipoprotein; HOMA-B, homeostatic model assessment-beta cell function; HOMA-IR, homeostasis model of assessment-insulin resistance; hs-CRP, high sensitivity C-reactive protein; IGT, impaired glucose tolerance; LDL, low density lipoprotein; MDA, malondialdehyde; NO, nitric oxide; P, parallel; P/C, probiotic group/control group; PC, placebo controlled; PEDro, The Physiotherapy Evidence Database tool; QUICKI, quantitative insulin sensitivity; SBP, systolic blood pressure; SD, standard deviation; TAC, total anti-oxidant capacity; TC, total cholesterol; TG, triglyceride; VLDL, very low density lipoprotein.

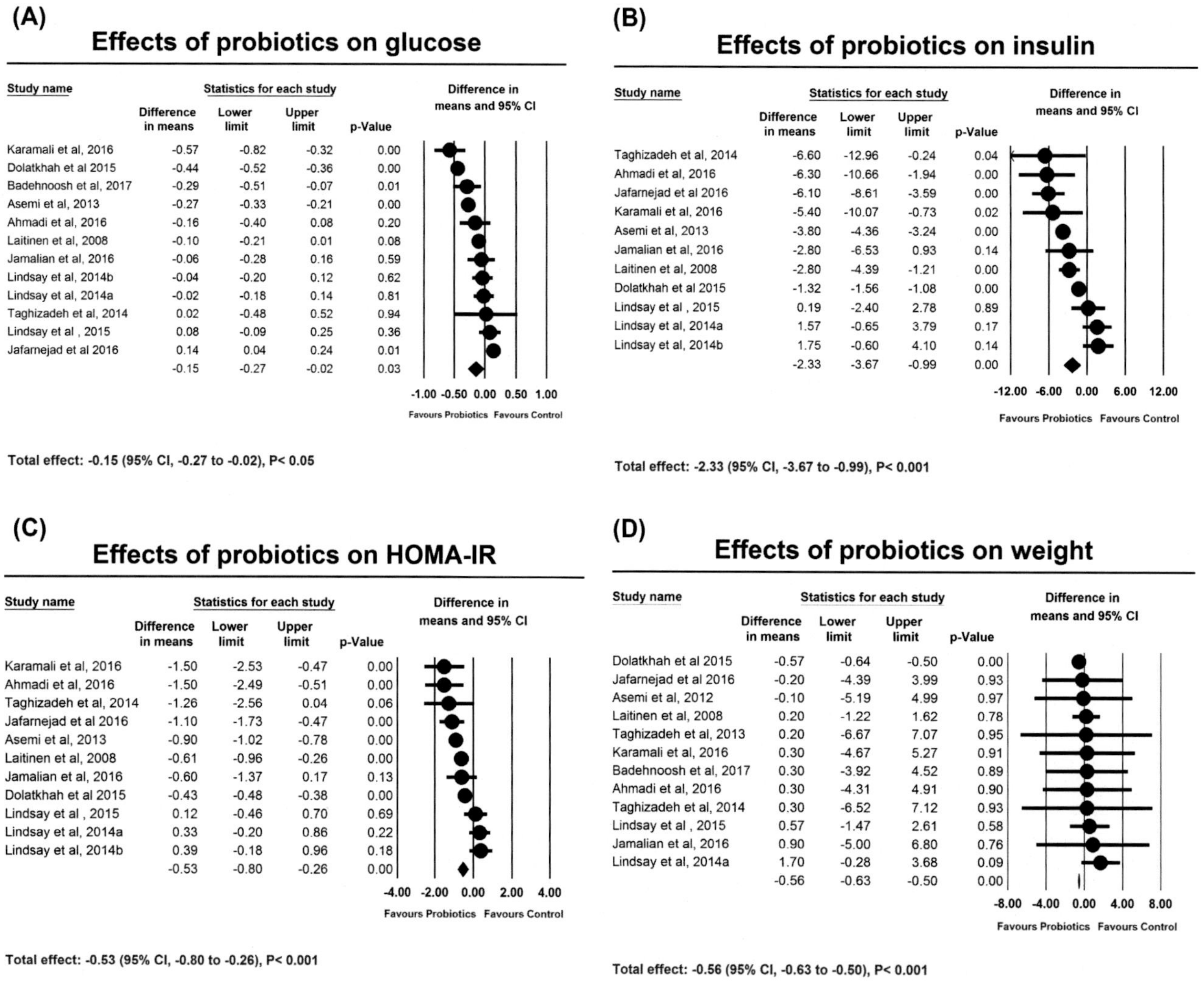

(A) Effects of probiotics on glucose

Study name	Difference in means	Lower limit	Upper limit	p-Value
Karamali et al, 2016	-0.57	-0.82	-0.32	0.00
Dolatkhah et al 2015	-0.44	-0.52	-0.36	0.00
Badehnoosh et al, 2017	-0.29	-0.51	-0.07	0.01
Asemi et al, 2013	-0.27	-0.33	-0.21	0.00
Ahmadi et al, 2016	-0.16	-0.40	0.08	0.20
Laitinen et al, 2008	-0.10	-0.21	0.01	0.08
Jamalian et al, 2016	-0.06	-0.28	0.16	0.59
Lindsay et al, 2014b	-0.04	-0.20	0.12	0.62
Lindsay et al, 2014a	-0.02	-0.18	0.14	0.81
Taghizadeh et al, 2014	0.02	-0.48	0.52	0.94
Lindsay et al , 2015	0.08	-0.09	0.25	0.36
Jafarnejad et al 2016	0.14	0.04	0.24	0.01
	-0.15	-0.27	-0.02	0.03

Difference in means and 95% CI: -1.00 -0.50 0.00 0.50 1.00 — Favours Probiotics / Favours Control

Total effect: -0.15 (95% CI, -0.27 to -0.02), P< 0.05

(B) Effects of probiotics on insulin

Study name	Difference in means	Lower limit	Upper limit	p-Value
Taghizadeh et al, 2014	-6.60	-12.96	-0.24	0.04
Ahmadi et al, 2016	-6.30	-10.66	-1.94	0.00
Jafarnejad et al 2016	-6.10	-8.61	-3.59	0.00
Karamali et al, 2016	-5.40	-10.07	-0.73	0.02
Asemi et al, 2013	-3.80	-4.36	-3.24	0.00
Jamalian et al, 2016	-2.80	-6.53	0.93	0.14
Laitinen et al, 2008	-2.80	-4.39	-1.21	0.00
Dolatkhah et al 2015	-1.32	-1.56	-1.08	0.00
Lindsay et al , 2015	0.19	-2.40	2.78	0.89
Lindsay et al, 2014a	1.57	-0.65	3.79	0.17
Lindsay et al, 2014b	1.75	-0.60	4.10	0.14
	-2.33	-3.67	-0.99	0.00

Difference in means and 95% CI: -12.00 -6.00 0.00 6.00 12.00 — Favours Probiotics / Favours Control

Total effect: -2.33 (95% CI, -3.67 to -0.99), P< 0.001

(C) Effects of probiotics on HOMA-IR

Study name	Difference in means	Lower limit	Upper limit	p-Value
Karamali et al, 2016	-1.50	-2.53	-0.47	0.00
Ahmadi et al, 2016	-1.50	-2.49	-0.51	0.00
Taghizadeh et al, 2014	-1.26	-2.56	0.04	0.06
Jafarnejad et al 2016	-1.10	-1.73	-0.47	0.00
Asemi et al, 2013	-0.90	-1.02	-0.78	0.00
Laitinen et al, 2008	-0.61	-0.96	-0.26	0.00
Jamalian et al, 2016	-0.60	-1.37	0.17	0.13
Dolatkhah et al 2015	-0.43	-0.48	-0.38	0.00
Lindsay et al , 2015	0.12	-0.46	0.70	0.69
Lindsay et al, 2014a	0.33	-0.20	0.86	0.22
Lindsay et al, 2014b	0.39	-0.18	0.96	0.18
	-0.53	-0.80	-0.26	0.00

Difference in means and 95% CI: -4.00 -2.00 0.00 2.00 4.00 — Favours Probiotics / Favours Control

Total effect: -0.53 (95% CI, -0.80 to -0.26), P< 0.001

(D) Effects of probiotics on weight

Study name	Difference in means	Lower limit	Upper limit	p-Value
Dolatkhah et al 2015	-0.57	-0.64	-0.50	0.00
Jafarnejad et al 2016	-0.20	-4.39	3.99	0.93
Asemi et al, 2012	-0.10	-5.19	4.99	0.97
Laitinen et al, 2008	0.20	-1.22	1.62	0.78
Taghizadeh et al, 2013	0.20	-6.67	7.07	0.95
Karamali et al, 2016	0.30	-4.67	5.27	0.91
Badehnoosh et al, 2017	0.30	-3.92	4.52	0.89
Ahmadi et al, 2016	0.30	-4.31	4.91	0.90
Taghizadeh et al, 2014	0.30	-6.52	7.12	0.93
Lindsay et al , 2015	0.57	-1.47	2.61	0.58
Jamalian et al, 2016	0.90	-5.00	6.80	0.76
Lindsay et al, 2014a	1.70	-0.28	3.68	0.09
	-0.56	-0.63	-0.50	0.00

Difference in means and 95% CI: -8.00 -4.00 0.00 4.00 8.00 — Favours Probiotics / Favours Control

Total effect: -0.56 (95% CI, -0.63 to -0.50), P< 0.001

FIGURE 17.2   Forest plots: Effects of probiotics on: (A) Fasting glucose, (B) insulin, (C) HOMA, (D) body weight.

TABLE 17.3 Overall Mean Difference for Probiotic Intervention on Metabolic Parameters Outcomes

Variable	Mean Difference (95% CI)	P-value	I-square (I^2)
Glucose	−0.15 (−0.27, −0.02)	<0.05	90.94%
Insulin	−2.33 (−3.67, −0.99)	<0.001	90.45%
HOMA-IR	−0.53 (−0.80, −0.26)	<0.001	88.45%
Weight	−0.56 (−0.63, −0.50)	<0.001	0%
Triglyceride	−0.23 (−0.42, −0.04)	0.02	74.37%
VLDL	−0.17 (−0.26, −0.07)	<0.001	0%
TAC	116.50 (3.57, 229.43)	0.04	80.29%
GSH	52.59 (18.72, 86.47)	<0.001	0%

CI, confidence interval; *GSH*, glutathione; *HOMA-IR*, Homeostatic model assessment-insulin resistance; *TAC*, total anti-oxidation capacity; *VLDL*, very low density lipo-protein.

Effects on Insulin

There were 11 studies with 915 participants reporting the effects of probiotics on fasting insulin (see Fig 17.2B). The average reduction of insulin was -0.20 mU/L (range from -2.50 to 1.78) in the probiotic group and 2.67 mU/L (range from -1.03 to 6.34) in the control group. The pooled mean difference was -2.33 (95% CI -3.67, -0.99; $P < .001$; $I^2 = 90\%$, $P < .01$). Probiotics resulted in decreased insulin in participants using multiple bacterial species ($-0.3.64$ mU/L (95% CI -5.23, -2.05; $P < .01$; $I^2 = 93\%$ $P < .01$)), with impaired glucose tolerance (-2.83 (95% CI -5.49, -0.16; $P < .05$; $I^2 = 83\%$, $P < .01$)), and located outside Europe-4.02 (95% CI -5.79, -2.24; $P < .01$; $I^2 = 93\%$, $P < .01$). Both probiotic alone (-1.92 (95% CI -3.31, -0.53; $P < .01$; $I^2 = 92\%$, $P < .01$), and with synbiotic (-6.40 (95% CI -9.99, -2.80; $P < .01$; $I^2 = 0\%$, $P = .94$)) had significant decreasing effect (Fig. 17.2).

Effects on Homoeostasis Model Assessment–Estimated Insulin Resistance (HOMA–IR)

There were 11 studies with 915 participants reporting the effects of probiotics on HOMA–IR (see Fig 17.2C). The average reduction of HOMA–IR was -0.17 (range from -0.6 to 0.32) in the probiotic group and 0.47 (range from -0.42 to 1.13) in the control group. The pooled effect was -0.53 (-0.80, -0.26; $P < .001$; $I^2 = 88\%$, $P < .01$). There was significant decreasing effect on HOMA–IR in participants using multiple bacterial species(-0.81 (95% CI -1.11, -0.50; $P < .01$; $I^2 = 90\%$ $P < .01$), with impaired glucose tolerance (-0.61 (95% CI -1.12, -0.10; $P < .01$; $I^2 = 75\%$, $P < .01$)), and located outside Europe (-0.88 (95% CI -1.23, -0.54; $P < .01$; $I^2 = 9\%$, $P < .01$)). Both probiotic alone (-0.45 (95% CI -0.73, -0.16; $P < .01$; $I^2 = 90\%$, $P < .01$) or with synbiotic (-1.41 (95% CI -2.20, -0.63; $P < .01$; $I^2 = 0\%$, $P = .77$)) had significant decreasing effect.

Effects on Body Weight

There were 12 studies with 980 participants reporting the effects of probiotics on body weight (see Fig 17.2D). The average increase in weight in the probiotic group was 5.37 kg (range from 1.65 to 15) and 5.04 kg (range from 2.22 to 14.8) in the control group. The pooled effect was -0.56 kg (95% CI -0.63, -0.50; $P < .001$; $I^2 = 0\%$, $P = .70$). There was significant decreasing effect on body weight with the use of multiple bacterial species (-0.57 kg (95% CI -0.64, -0.54; $P < .01$; $I = 0\%$, $P > .05$)), participants with impaired glucose tolerance (-0.57 (95% CI -0.64, -0.54; $P < .01$; $I = 0\%$, $P > .05$)), located in Europe (-0.57 (95% CI -0.64, -0.50; $P < .01$; $I = 0\%$, $P > .05$)), and those using probiotic alone (-0.56 (95% CI -0.63, -0.50; $P < .01$; $I = 0\%$, $P > .05$)), without synbiotic.

Effects on Triglycerides

There were eight studies with 636 participants reporting the effects of probiotics on triglycerides (see Fig 17.3A). The pooled effect was -0.23 (95% CI -0.42, -0.04; $P = .02$; $I^2 = 74\%$, $P < .01$). There was significant decreasing effect of probiotics on triglycerides with multiple bacterial species (0.37 mmol/L (95% CI -0.42, 0.31; $P < .01$; $I^2 = 0\%$, $P > .05$), and in those located outside Europe (0.37 (95% CI -0.42, 0.31; $P < .01$; $I^2 = 0\%$, $P > .05$)) (Fig. 17.3).

Effects on VLDL, TAC, and GSH

There were four studies with a total of 242 participants reporting the effects of probiotics on VLDL (see Fig 17.3B). There was statistically significant pooled effect with -0.17 mmol/L (95% CI -0.26, -0.07; $P < .001$; $I^2 = 0\%$, $P > .05$). Similarly, three studies with 190 participants looked at the effect of probiotics on TAC (see Fig 17.3C). There was statistically significant pooled effect with 116.50 mmol/L (95% CI 3.57, 229.43; $P < .05$); ($I^2 = 80\%$, $P < .01$). Likewise, three studies with total 190 participants reported the effect of probiotics on GSH (see Fig 17.3D). There was a statistically significant pooled effect of 52.59 μmol/L (95% CI 18.72, 86.47; $P < .001$; $I^2 = 0\%$, $P > .05$).

COMMENTS

Significant reductions in insulin and HOMA–IR but not glucose were found among participants with gestational diabetes, indicating that consumption of probiotics has more significant effects on reducing glucose levels during impaired glucose tolerance (IGT). Animal studies have also shown similar findings [32]. Possible mechanisms include the effect of

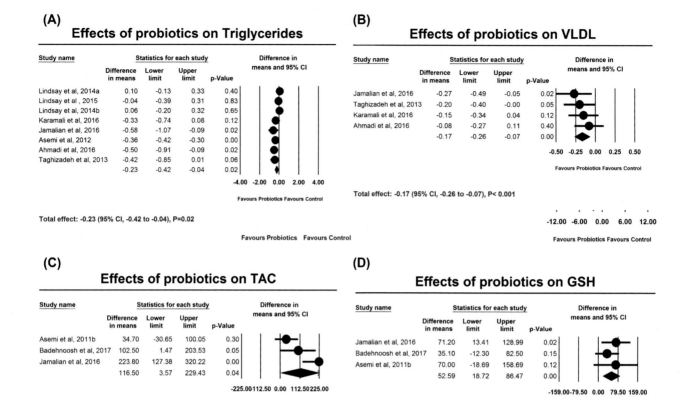

FIGURE 17.3 Effects of probiotics on: (A) Triglycerides, (B) VLDL, (C) TAC, (D) GSH.

probiotics on intestinal microbiota, and production of short-chain fatty acids (SCFA), which have a beneficial effect on glucose metabolism [33].

Probiotics containing *Lactobacillus acidophilus* and *Lactobacillus casei* are effective in ameliorating streptozotocin-induced diabetes through inhibiting its progression by suppressing glucose intolerance and blood glucose, and sustaining insulin levels. Probiotic use may therefore have an antidiabetic impact, by protecting pancreatic β-cells from damage [34]. Our analysis demonstrated that TAC and GSH, both antioxidants, were significantly increased, which in turn may indirectly impact insulin levels and glucose homoeostasis [35]. This effect with GDM participants, as stated earlier, may relate to improved composition of intestinal microflora, thereby delaying or preventing elevations in insulin level. It may also act by assisting GDM participants with their glucose homoeostasis.

Our findings indicate that a number of factors may account for the effects of probiotics on GDM participants [36]. Consumption of probiotics containing multiple strains resulted in a statistically significant reduction in glucose, insulin, and HOMA−IR levels. In contrast, trials with single strains did not have a significant effect. A possible explanation may be that multiple strains regulate multiple systems, through intestinal SCFA production and other mechanisms. Further research is required in this area due to the small number of studies with single bacterial strains.

Implications for Treatment and Future Research

The findings of this metaanalysis suggest that probiotics may play a key role in the prevention and treatment of gestational diabetes in pregnancy. However, further research relating to the optimization of the type, dose, and number of microbial organisms needs to be carried out. Furthermore, RCTs involving both short-term (as trials in this review) and long-term interventions with single versus multiple strains of probiotic regimens are required.

It is possible that probiotics can be used as an alternative method for glycemic control in mild and moderate cases of gestational diabetes, without harming the baby. The finding of statistically significant effects of probiotics on triglyceride, VLDL, and oxidative stress relieving factors has to be researched further with higher number of participants.

REFERENCES

[1] Ryckman KK, Borowski KS, Parikh NI, Saftlas AF. Pregnancy complications and the risk of metabolic syndrome for the offspring. Curr Cardiovasc Risk Rep 2013;7:217—23.

[2] Correa PJ, Vargas JF, Sen S, Illanes SE. Prediction of gestational diabetes early in pregnancy: targeting the long-term complications. Gynecol Obstet Investig 2014;77:145—9.

[3] Flegal KM, Carroll MD, Kit BK, Ogden CL. Prevalence of obesity and trends in the distribution of body mass index among US adults, 1999-2010. J Am Med Assoc 2012;307:491—7. Epub 2012/01/19.

[4] WHO. Global Health Observatory (GHO) data on overweight and obesity. 2017.

[5] Gunatilake RP, Perlow JH. Obesity and pregnancy: clinical management of the obese gravida. Am J Obstet Gynecol 2011;204:106—19.

[6] Taghizadeh M, Asemi Z. Effects of synbiotic food consumption on glycemic status and serum hs-CRP in pregnant women: a randomized controlled clinical trial. Hormones 2014;13:398—406.

[7] Karcaaltincaba D, Kandemir O, Yalvac S, Güvendag-Guven S, Haberal A. Prevalence of gestational diabetes mellitus and gestational impaired glucose tolerance in pregnant women evaluated by National Diabetes Data Group and Carpenter and Coustan criteria. Int J Gynecol Obstet 2009;106:246—9.

[8] Hadar E, Hod M. Gestational diabetes and pregnancy outcome: do we need an update on diagnostic criteria? Nutr Metabol Cardiovasc Dis 2009;19:75—6. Epub 2009/02/03.

[9] Taylor BL, Woodfall GE, Sheedy KE, O'Riley ML, Rainbow KA, Bramwell EL, Kellow NJ. Effect of probiotics on metabolic outcomes in pregnant women with gestational diabetes: a systematic review and meta-analysis of randomized controlled trials. Nutrients 2017;9:461.

[10] Rössner S, Öhlin A. Pregnancy as a risk factor for obesity: lessons from the Stockholm pregnancy and weight development study. Obesity 1995;3.

[11] Basaran A. Pregnancy-induced hyperlipoproteinemia: review of the literature. Reprod Sci 2009;16:431—7.

[12] Taghizadeh M, Hashemi T, Shakeri H, Abedi F, Sabihi SS, Alizadeh SA, Asemi Z. Synbiotic food consumption reduces levels of triacylglycerols and VLDL, but not cholesterol, LDL, or HDL in plasma from pregnant women. Lipids 2014;49:155—61. Epub 2013/11/26.

[13] Sanders ME. Probiotics: definition, sources, selection, and uses. Clin Infect Dis 2008;46:S58—61.

[14] Lye H-S, Kuan C-Y, Ewe J-A, Fung W-Y, Liong M-T. The improvement of hypertension by probiotics: effects on cholesterol, diabetes, renin, and phytoestrogens. Int J Mol Sci 2009;10:3755—75.

[15] Koren O, Goodrich JK, Cullender TC, Spor A, Laitinen K, Bäckhed HK, Gonzalez A, Werner JJ, Angenent LT, Knight R. Host remodeling of the gut microbiome and metabolic changes during pregnancy. Cell 2012;150:470—80.

[16] Zhang H, DiBaise JK, Zuccolo A, Kudrna D, Braidotti M, Yu Y, Parameswaran P, Crowell MD, Wing R, Rittmann BE. Human gut microbiota in obesity and after gastric bypass. Proc Natl Acad Sci USA 2009;106:2365—70.

[17] Moher D, Liberati A, Tetzlaff J, Altman DG. Preferred reporting items for systematic reviews and meta-analyses: the PRISMA statement. J Clin Epidemiol 2009;62:1006—12. Epub 2009/07/28.

[18] Asemi Z, Jazayeri S, Najafi M, Samimi M, Mofid V, Shidfar F, Foroushani AR, Shahaboddin ME. Effects of daily consumption of probiotic yoghurt on inflammatory factors in pregnant women: a randomized controlled trial. Pakistan J Biol Sci 2011;14:476—82. Epub 2011/09/23.

[19] Asemi Z, Jazayeri S, Najafi M, Samimi M, Mofid V, Shidfar F, Shakeri H, Esmaillzadeh A. Effect of daily consumption of probiotic yogurt on oxidative stress in pregnant women: a randomized controlled clinical trial. Ann Nutr Metabol 2012;60:62—8.

[20] Asemi Z, Samimi M, Tabasi Z, Talebian P, Azarbad Z, Hydarzadeh Z, Esmaillzadeh A. Effect of daily consumption of probiotic yoghurt on lipid profiles in pregnant women: a randomized controlled clinical trial. J Matern Fetal Neonatal Med 2012;25:1552—6.

[21] Asemi Z, Samimi M, Tabassi Z, Naghibi Rad M, Rahimi Foroushani A, Khorammian H, Esmaillzadeh A. Effect of daily consumption of probiotic yoghurt on insulin resistance in pregnant women: a randomized controlled trial. Eur J Clin Nutr 2013;67:71—4.

[22] Lindsay K, Maguire O, Smith T, Brennan L, McAuliffe F. A randomized controlled trial of probiotics to reduce maternal glycaemia in obese pregnancy. Am J Obstet Gynecol 2014;210:S342.

[23] Lindsay KL, Kennelly M, Culliton M, Smith T, Maguire OC, Shanahan F, Brennan L, McAuliffe FM. Probiotics in obese pregnancy do not reduce maternal fasting glucose: a double-blind, placebo-controlled, randomized trial (Probiotics in Pregnancy Study). Am J Clin Nutr 2014;99:1432—9.

[24] Lindsay KL, Brennan L, Kennelly MA, Maguire OC, Smith T, Curran S, Coffey M, Foley ME, Hatunic M, Shanahan F, McAuliffe FM. Impact of probiotics in women with gestational diabetes mellitus on metabolic health: a randomized controlled trial. Am J Obstet Gynecol 2015;212.

[25] Karamali M, Dadkhah F, Sadrkhanlou M, Jamilian M, Ahmadi S, Tajabadi-Ebrahimi M, Jafari P, Asemi Z. Effects of probiotic supplementation on glycaemic control and lipid profiles in gestational diabetes: a randomized, double-blind, placebo-controlled trial. Diabetes Metabol 2016;42:234—41.

[26] Jamilian M, Bahmani F, Vahedpoor Z, Salmani A, Tajabadi-Ebrahimi M, Jafari P, Hashemi Dizaji S, Asemi Z. Effects of probiotic supplementation on metabolic status in pregnant women: a randomized, double-blind, placebo-controlled trial. Arch Iran Med 2016;19:687—92.

[27] Jafarnejad S, Saremi S, Jafarnejad F, Arab A. Effects of a multispecies probiotic mixture on glycemic control and inflammatory status in women with gestational diabetes: a randomized controlled clinical trial. Journal Nutr Metab 2016;2016:5190846. Epub 2016/07/19.

[28] Dolatkhah N, Hajifaraji M, Abbasalizadeh F, Aghamohammadzadeh N, Mehrabi Y, Abbasi MM, Mesgari Abbasi M. Is there a value for probiotic supplements in gestational diabetes mellitus? A randomized clinical trial. J Health Popul Nutr 2015;33:1—8.

[29] Badehnoosh B, Karamali M, Zarrati M, Jamilian M, Bahmani F, Tajabadi-Ebrahimi M, Jafari P, Rahmani E, Asemi Z. The effects of probiotic supplementation on biomarkers of inflammation, oxidative stress and pregnancy outcomes in gestational diabetes. J Matern Fetal Neonatal Med 2017:1—9. Epub 2017/03/23.

[30] Ahmadi S, Jamilian M, Tajabadi-Ebrahimi M, Jafari P, Asemi Z. The effects of synbiotic supplementation on markers of insulin metabolism and lipid profiles in gestational diabetes: a randomised, double-blind, placebo-controlled trial. Br J Nutr 2016;116:1394—401.

[31] Laitinen K, Poussa T, Isolauri E. Probiotics and dietary counselling contribute to glucose regulation during and after pregnancy: a randomised controlled trial. Br J Nutr 2009;101:1679−87.

[32] Al-Salami H, Butt G, Fawcett JP, Tucker IG, Golocorbin-Kon S, Mikov M. Probiotic treatment reduces blood glucose levels and increases systemic absorption of gliclazide in diabetic rats. Eur J Drug Metab Pharmacokinet 2008;33:101−6.

[33] Sadrzadeh-Yeganeh H, Elmadfa I, Djazayery A, Jalali M, Heshmat R, Chamary M. The effects of probiotic and conventional yoghurt on lipid profile in women. Br J Nutr 2010;103:1778−83.

[34] Yadav H, Jain S, Sinha PR. Oral administration of dahi containing probiotic Lactobacillus acidophilus and Lactobacillus casei delayed the progression of streptozotocin-induced diabetes in rats. J Dairy Res 2008;75:189−95.

[35] Rajput IR, Li YL, Xu X, Huang Y, Zhi WC, Yu DY, Li W. Supplementary effects of Saccharomyces boulardii and Bacillus subtilis B10 on digestive enzyme activities, antioxidation capacity and blood homeostasis in broiler. Int J Agric Biol 2013;15:231−7.

[36] Gomes AC, Bueno AA, de Souza RGM, Mota JF. Gut microbiota, probiotics and diabetes. Nutr J 2014;13:60.

FURTHER READING

[1] Cheung NW. The management of gestational diabetes. Vasc Health Risk Manag 2009;5:153−64. Epub 2009/05/14.

[2] Seshiah V, Balaji V, Balaji MS, Paneerselvam A, Arthi T, Thamizharasi M, Datta M. Gestational diabetes mellitus manifests in all trimesters of pregnancy. Diabetes Res Clin Pract 2007;77:482−4.

[3] Katon J, Maynard C, Reiber G. Attempts at weight loss in US women with and without a history of gestational diabetes mellitus. Womens Health Issues 2012;22:e447−53.

[4] Dugoua JJ, Machado M, Zhu X, Chen X, Koren G, Einarson TR. Probiotic safety in pregnancy: a systematic review and meta-analysis of randomized controlled trials of Lactobacillus, Bifidobacterium, and Saccharomyces spp. J Obstet Gynaecol Can 2009;31:542−52.

[5] Verhagen AP, de Vet HC, de Bie RA, Kessels AG, Boers M, Bouter LM, Knipschild PG. The Delphi list: a criteria list for quality assessment of randomized clinical trials for conducting systematic reviews developed by Delphi consensus. J Clin Epidemiol 1998;51:1235−41.

[6] Meta-analysis BC. Comprehensive meta-analysis. Englewood, NJ: Biostat. Available from: https://www.meta-analysis.com/.

[7] Egger M, Smith GD, Schneider M, Minder C. Bias in meta-analysis detected by a simple, graphical test. BMJ 1997;315:629−34.

[8] Sun J, Buys NJ. Glucose-and glycaemic factor-lowering effects of probiotics on diabetes: a meta-analysis of randomised placebo-controlled trials. Br J Nutr 2016;115:1167−77.

Chapter 18

Current Options for Fecal Transplantation in *Clostridium difficile* Infection

Nathaniel Aviv Cohen[1,2] and Nitsan Maharshak[1,2]

[1]IBD Center and Bacteriotherapy Clinic, Department of Gastroenterology and Liver Diseases, Tel Aviv Medical Center, Tel Aviv, Israel;
[2]Sackler Faculty of Medicine, Tel Aviv University, Tel Aviv, Israel

INTRODUCTION

Clostridium difficile, a gram-positive spore forming anaerobe and a member of the Clostridiaceae family of the Firmicutes phylum, is the most common cause of hospital associated diarrhea. The incidence of *Clostridium difficile* infection (CDI) has been increasing over the past 2 decades, with resultant increases in mortality, hospital stays, and medical costs [1]. Particular populations at higher risk include those older than 65 years, patients with comorbidities such as malignancy, inflammatory bowel disease and diabetes mellitus, recent hospitalizations, residency in long-term care facilities, and medication exposure such as antibiotics, proton pump inhibitors, and chemotherapy [2]. During the last decade, it has become clear that a healthy gut microbiome has a crucial role in protection against CDI infection.

The first-line therapeutic intervention against CD includes antibiotics (metronidazole or vancomycin), which are effective in up to 75% of the patients. However, recurrent CDI episodes, not responsive to antibiotics, occur often, and mortality rates reach up to 15%. In these patients, fecal microbial transplantation (FMT) is extremely effective and fast acting, for the treatment and prevention of recurrent CDI (rCDI), by restoration of the gut microbiome. This therapy is now approved for patients with rCDI, underscoring the importance of a normal intestinal microbiome for protection against CDI.

RELEVANT MICROBIAL POPULATIONS

The diversity of the healthy gut microbiota is important for the prevention of colonization and proliferation of pathogenic microbes. Patients who suffer from CDI are characterized by reduced enteric microbial diversity, which is more prominent in patients with recurrent and severe CDI, and also by the loss of multiple bacterial groups and their functions. At the phyla level, Firmicutes/Bacteroidetes ratio is decreased, leading to decreased proportion of multiple bacterial groups that belong to the Firmicutes phylum, such as *Ruminococceae*, *Lachnospiraceae*, and *Clostridium* clusters IV and XIVa [3], and additional bacterial groups from other phyla such as *Alistipes*, *Bacteroides*, and *Barnesiella* genera.

Other than *C. difficile*, there is an increase in bacterial groups that are associated with inflammation and belong to the Proteobacteria phylum, such as *Klebsiella* and *Escherichia/Shigella* genera [4]. Patients with recurrent CDI show a progressive decrease in diversity with each recurrence and also significantly fewer members of the *Clostridiales* family and *Collinsella* genus [5].

In addition, it appears that the microbiome plays a role in determining the virulence of *C. difficile*. Patients who are asymptomatic carriers show a lower proportion of *Proteobacteria* and higher proportion of *Bifidobacteria*, compared to patients with clinical infection, who show increased Enterobacteriaceae, with a decrease in Enterococcaceae [6].

Moreover, patients infected with the hyper-virulent ribotype 027, which is associated with severe infection and increased mortality, show elevations in Peptostreptococcaceae, *Ruminococcus bromii*, and *Enterococcus* sp. 1, with

Microbiome and Metabolome in Diagnosis, Therapy, and other Strategic Applications. https://doi.org/10.1016/B978-0-12-815249-2.00018-X

decreases in some *Bacteroides* species and Enterobacteriaceae 1, 2 and 3 and a significant reduction in *Bifidobacterium longum* genus [7]. Indeed, *Bifidobacterium longum* has been shown to be associated with *C. difficile* negativity.

CRUCIAL MOLECULES AND PATHWAYS

It is thought that the success of FMT in CDI patients is multidimensional and results from the restoration of a healthy microbial composition, and the metabolic ramifications of those microbial changes, on the various stages of the *C. difficile* life cycle.

Clostridium difficile Life Cycle

The exposed individual ingests *C. difficile* in both vegetative and spore forms. The acidic environment of the stomach generally clears vegetative cells, while the spores survive and can germinate in the duodenum. However, local bile salts, the host immune response, and the normal gut microbiome, in part through the production of short-chain fatty acids (SCFAs) and nutrient limitation, limit germination and colonization. In patients with CDI these primary defenses are breached, allowing germination from spores to vegetative cells and host colonization in the large intestine, where *C. difficile* produces its toxins that result in clinical infection. Restoring the normal microbiome through FMT prevents this process via a number of mechanisms.

Prevention of Colonization

The Role of Secondary Bile Salts

In order for *C. difficile* spores to germinate, the appropriate host environment and favorable nutrient conditions are required. The germinant receptor, CspC, has been identified in *C. difficile* spores, and this receptor is stimulated by cholic acid class bile acids [8]. Primary bile acids e.g., taurocholic acid, promote the germination of *C. difficile* and the secondary bile acids, such as deoxycholic acid, inhibit *C. difficile* growth.

This is supported by the observation that in patients with rCDI, there are high levels of primary bile acids, while post-FMT there are mainly secondary bile acids, similar to in healthy controls [9]. Indeed, bile acids in concentrations found in *C. difficile* patients, prior to FMT, induce *C. difficile* isolate growth, whereas bile acids in concentrations following FMT inhibit isolate growth [10]. Further, in the gut, bile salt hydrolases (BSH) are partially responsible for converting primary to secondary bile acids.

In patients with CDI who undergo FMT, BSH-producing organisms are enriched and are negatively correlated to taurocholic acid levels and positively correlated to deoxycholic levels [11]. Additionally, specific bacteria such as *Clostridium scindens*, which is a bile acid 7 alpha-dehydroxylating gut bacteria, when administered to *C. difficile* patients, promotes resistance to infection [12].

Colonization Resistance and Nutrient Limitation

Resident gut microbiota compete with *C. difficile* for nutrients and space and interfere with its ability to proliferate and colonize the host's gastrointestinal tract. *C. difficile* is able to ferment many amino acids and carbohydrates, and in vitro these nutrients are used to stimulate *C. difficile* growth. Micronutrients such as monomeric glucose and *N*-acetylglucosamine, released by intestinal microorganisms, are more accessible to *C. difficile* when the host microbial diversity is reduced. In addition to stimulating *C. difficile* growth, these nutrients also play a role in the virulence of *C. difficile*, by stimulating the production of toxins, the primary virulence factors of *C. difficile*.

The synthesis of toxins is dependent on the activity of TcdR, an initiation factor of RNA polymerase. CodY, a global regulatory protein, monitors nutrient availability and represses TcdR activity [13]. Additionally, other regulatory proteins such as CcpA, a carbon catabolite control protein, respond to available nutrients and regulate toxin gene expression [14]. This suggests a link between metabolism of various nutrients and *C. difficile* virulence.

A successful FMT in patients with rCDI leads to engraftment of the donor's microbiota, with similar diversity and compositional profile as that of the healthy donor [15–17]. Post-FMT, colonization resistance is achieved, by the competition and synergistic cross talk of the healthy gut bacteria with each other, in the provision of space and nutrients within the gut environment. Competition for the micronutrients with other microbes may limit the growth of *C. difficile*, dampen the inflammatory response to *C. difficile* toxins, and eradicate it.

The Antiinflammatory Effect of the Normal Microbiome

The toxins released by *C. difficile* namely, TcdA and TcdB result in an inflammatory response. These toxins disrupt essential cellular signaling pathways, resulting in the weakening of epithelial cell tight junctions and apoptotic death of colonic epithelial cells. The initiation of this inflammatory process results in the release of multiple cytokines and the local and systemic inflammatory response noted in patients with CDI. Indeed, the intensity of this cytokine release, in particular high levels of IL-8 and CXCL5, correlates with worse outcomes [18].

The innate immune system uses toll-like receptors (TLRs), and adapter modules such as MyD88, to continually sample luminal contents and thus detect bacterial derived molecules, such as lipopolysaccharides and flagellin [19]. This continuous detection of normal microbial molecules allows for homeostasis, by accumulation of microbe-induced regulatory T cells (iTregs), and suppression of proinflammatory T helper cells (Th17) in the lamina propria. The induction of iTregs results in the release of antiinflammatory cytokines, such as TGF-β and IL-10, which further reduce inflammation [20].

In patients with severe systemic inflammatory response syndrome (SIRS), FMT rapidly stabilized their hemodynamic parameters [21]. This indicates that possibly certain elements of the healthy donor microbiome, such as butyrate, induce rapid dampening of the host immune response, and allow for patient stabilization and time for gut repair. The normal microbiota can induce secretion of IgA that protects from invading pathogens [22]. It is proposed that FMT, by restoring the host microbiome, balances the immune response to *C. difficile*, and both reduces inflammation and clears the infection.

Normal Microbiota Produces Bactericidal Metabolites

The normal gut microbiota produces multiple peptides, with both bactericidal and bacteriostatic antimicrobial effects. An example of a narrow-spectrum bactericidal peptide is Thuricin CD, which is produced by *Bacillus thuringiensis* DPC 6431. Thuricin CD consists of 2 distinct peptides, which act synergistically to kill multiple CD isolates including ribotype 027, which is responsible for recent hospital outbreaks [23]. An additional group of bacteriocins are the lantibiotics, lacticin 3147 and nisin, produced by *Lactococcus lactis*. These are active against CD, but have a broader spectrum of action, particularly towards gram-positive bacteria [24]. There are numerous other bacteriocins, e.g., LFF571 and Actagardine A, produced by bacteria in the normal gut microbiome. The combination of these peptides, which play a role in resisting infection with *C. difficile*, and post-FMT, most probably play a role in clearing the pathogen.

Short-Chain Fatty Acid Metabolism and Gut Epithelial Integrity

The multiple metabolic and trophic functions of the healthy gut microbiome include the metabolization of carbohydrates into SCFAs such as butyrate, propionate, and acetate. SCFAs decrease gut permeability and increase the production of antiinflammatory molecules, particularly butyrate, which is also the preferred source of energy of enterocytes [25]. Indeed, butyrate producing bacteria, such as members of the *Clostridium* clusters IV and XIVa, are absent or significantly depleted in patients with CDI. This has lead to the hypothesis that SCFAs may play a role in resisting CDI, although currently, the results of in vivo studies do not support this assumption [26].

Transfer of Bacteriophages

The fecal matter transplanted during FMT includes bacteriophages, which are viruses that infect and replicate within bacteria. A successful FMT is associated with higher richness of *Caudovirales* taxa of the bacteriophages. Post-FMT enriched virome and bacterial community were associated with recurrence free cure of CDI, whereas patients with only enriched bacterial community had recurrence of CDI [27]. Interestingly, infusion of bacteria-free fecal filtrate, containing bacteriophages, from a healthy donor, cured CDI in a recent publication [28]. These studies show that the viral component of the FMT may have an important part in success, by shaping the microbial community.

INTERFACES WITH ANTIBIOTICS AND OTHER DRUGS

Besides exposure to antibiotics, which directly impact the microbiome and its function, additional drugs may interfere with the gut microbiome and serve as a risk factor for CDI. Proton pump inhibitors (PPIs), which are widely used and reduce acid secretion, increase the risk of CDI. These drugs enable the survival of ingested *C. difficile* vegetative forms. In addition, PPIs are associated with reduction of bacterial diversity and overrepresentation of multiple oral bacteria in the fecal microbiome, such as *Rothia*, *Enterococcus*, *Streptococcus*, *Staphylococcus*, and *Escherichia coli* [29]. These changes are in keeping with known changes that predispose to CDI. Antidepressants, statins, chemotherapy agents and corticosteroids, also influence the gut microbiota [30–32]. However, only corticosteroids have been associated with increased risk of CDI [31] probably through their immunosuppressive effects or their association with significant comorbidities.

DIAGNOSTIC TESTS

Traditionally, the diagnosis of CDI is based on clinical features and laboratory tests. The presence of moderate to severe diarrhea or ileus is required, with the addition of an enzyme-linked immunoassay (ELISA) positive for *C. difficile* toxins, and antigen glutamate dehydrogenase (GDH). Discrepant ELISA tests require the confirmation of toxigenic *C. difficile,* by polymerase chain reaction (PCR). Endoscopy may prove useful in cases where an alternative diagnosis is suspected, if symptoms persist despite adequate treatment, or if there is ileus or fulminant colitis without diarrhea, and pseudomembranes are sought to confirm the diagnosis. The introduction of an immunochromatographic test, which simultaneously checks for toxins and GDH antigen, has allowed for more rapid and accurate diagnosis of CDI [33].

THERAPEUTIC PROTOCOLS

Treatment of *Clostridium difficile* Infection

Metronidazole and vancomycin are the mainstay of management strategies, and are effective in treating up to 75% of the patients at their initial infection; however, following a recurrence, these medications have limited efficacy, and are associated with high recurrence rates, of up to 65% [34]. Historically, alternative treatment methods and strategies have been sought (Table 18.1). However, most of these interventions were only mildly successful. Fidaxomicin has been recently approved for CDI eradication, and compared to vancomycin is associated with a similar efficacy, but with lower recurrence rates [35].

Fecal Microbial Transplantation

Fecal microbial transplantation (FMT) has been found to be extremely effective and safe, for the treatment of CDI, and has been approved for the treatment of recurrent (more than three episodes) or resistant CDI (rCDI) in many countries. A randomized, controlled trial comparing duodenally delivered FMT to vancomycin, has been stopped due to a significant response rate of 81% in the FMT group, compared to 31% in the vancomycin group [16]. In addition, a large systematic review including 536 patients with rCDI, treated with FMT, showed a cure rate of 87% [36].

TABLE 18.1 Alternative and Experimental Treatment Strategies for *Clostridium difficile* Infection

Treatment	Detail	Evidence
Vancomycin tapering and pulse therapy	Patients treated for prolonged period with gradually reducing dose. Thought to suppress formation of *C. difficile* spores, restore microbiome	Cures 60%–80% of patients. Associated with less recurrence than fixed doses
Rifaximin	Semi-synthetic, non-absorbable antibiotic	Prevents recurrence, immediately after vancomycin
Ursodeoxycholic acid	Inhibitory to the germination and vegetative growth of *C. difficile*	Minimal experience, long term treatment may cure CDI
Intravenous immune globulin (IVIG)	Administered in addition to antibiotics	Mixed results, more investigation needed
Monoclonal antibodies targeting *C. difficile* toxins	Administered in combination with standard antibiotics. Targets *C. difficile* toxin A and B	Reduces the rate of recurrence of CDI
Vaccination against Toxin A + B	Chimeric toxin vaccine	Protection against numerous strains, prevented relapse
Sanofi Pasteur anti-toxin vaccine	Bivalent formalin-inactivated vaccine against toxins A + B	Seroconversion in 75% of patients

The long-term durability of FMT was shown in a large multicenter trial, which revealed that over 90% of patients remain recurrence free, post treatment [37]. Indeed, 4 year recurrence rates of CDI post FMT, as low as 2.4% have been reported [38].

Fecal Microbial Transplantation Protocol

Selecting a Donor

An important aspect of FMT is donor selection. Fecal material is medically classified as human tissue, and as such screening of potential donors with, or at risk of contracting medical conditions, such as human immunodeficiency virus (HIV), hepatitis B or C and other sexually transmitted infections is required. Potential donors need to complete comprehensive questionnaires detailing their health history. Table 18.2 lists the various screening questions and tests donors have to undergo.

The benefits of using a young, rigorously screened and unrelated donor, may include reducing the risk of transplanting a similar susceptible microbiome from a related donor, and reducing the burden of finding a donor from the family. Donors with a proven history can thus be used.

The Personalized Fecal Microbial Transplantation Procedure

Saline solution is added to the fresh stool, and blended to create a homogenous liquid, which is then filtered to remove any particulate matter. FMT can be performed using either fresh, or frozen fecal fluid to which glycerol is added as a bacterial cryoprotector. Frozen fecal fluid can be stored at $-80°C$ for at least 2 years, and then thawed and used. The fecal solution can be administered via a number of different routes i.e., nasoduodenal, transgastroscopic, transcolonoscopic, enema based or by capsules. All the routes are effective for the treatment of rCDI. Additional benefits of administration via colonoscopy and enemas are the ability to assess disease severity at the time of the procedure, and superior cost effectiveness, respectively.

FMT FOR MAINSTREAM USE, AND THE FUTURE OF FECAL-BASED THERAPIES

Currently the most common method of performing FMT is infusion by endoscopy via the upper or lower GI tract. These methods limit feasibility of treating all patients with CDI, due to contraindications to the procedures or patient preferences. These limitations have in part been addressed by the introduction of frozen fecal capsules (single donor), or serially manufactured nonfrozen capsules (pool of donors), which are equally effective as regular FMT. There is still a need to adapt this treatment, to make it even more accessible, standardized, and tolerable.

Stool-Substitute Therapies

These therapies use defined, standardized preparations of stool-derived products, retaining the compositional, metabolic and transcriptomic properties of fecal communities. In 1989 10 CDI patients received a consortium of 10 bacterial strains alone and showed a 60% cure rate [39]. More recently, a stool substitute preparation of 33 purified bacterial strains, derived from a single healthy donor, was used to cure 2 patients with rCDI [40]. These stool-substitute therapies are highly standardized, safe and tolerable, and can be commercially distributed, allowing a greater number of patients to benefit.

Use of Sterile Filtrates From Donor Stool

Sterile fecal filtrate transfer has been effective in CDI patients. This sterile fecal filtrate contains the metabolic components of the donor sample including the virome, but without any of the bacteria. In patients who responded to this treatment, the analysis of virus-like particles postinfusion, showed a complex signature of bacteriophages, indicating that these could have a role in treating CDI [28]. Sterile fecal filtrates have a potentially better safety profile, particularly among severely immunocompromised patients.

Bacteriophage-Based Therapies

Bacteriophages target specific bacteria and can be both bactericidal and bacteriostatic, by hampering bacterial metabolic function. They are strain specific, without deleterious effects on the gut microbiota. In mouse models, specific strains of

TABLE 18.2 Typical Exclusion Criteria and Screening Tests for Stool Donors

Exclusion Criteria	Screening Tests
Personal factors Blood-borne disease risk, Recent travel to high risk/diarrhea endemic areas, Recent hospitalization, Smoking, Illicit drug use, Sexual behavior, Incarceration, Tattoos/piercings, Exposure to HIV/viral hepatitis, Needle-stick injury, Receipt of blood products, Recent vaccinations	**Stool tests** *Clostridium difficile*, Parasites/ova, *Giardia lamblia*, *Cryptosporidium parvum*, Vancomycin resistant *Enterococci*, Methicillin resistant *Staphylococcus aureus*, Enterohemorrhagic *E coli*, *Yersinia enterocolitica*, *Campylobacter jejuni*, *Salmonella typhi*, *Shigella dysenteriae*, Extended spectrum β lactamase producing organisms, Fecal calprotectin
Medications Recent antibiotic exposure, Probiotics, Laxatives, Antineoplastic drugs, Immunosuppressive drugs (calcineurin inhibitors, exogenous glucocorticoids etc.), Chronic ongoing medication therapy, Over the counter aids for digestion, Drug abuse/illicit drugs	**Blood tests** Human immunodeficiency virus, Hepatitis B and C sorology, *Treponema pallidum*, Human T lymphotrophic virus, *Entamoeba histolytica*, *Strongyloides stercoralis*, Adenovirus, norovirus and rotavirus, Liver enzymes and liver function tests, Full blood count, General chemistry and renal function tests, Cholesterol and triglycerides, Thyroid function tests, Autoimmune serology (ANA, ANCA, celiac screen, ASCA)

Past medical history Communicable diseases, GI conditions (IBS, IBD, chronic diarrhea/constipation, GI surgery etc.), Systemic autoimmunity (multiple sclerosis, connective tissue disease etc.), Allergy/atopy (asthma, eczema etc.), Chronic pain syndromes (fibromyalgia, chronic fatigue syndrome), Ingestion of recipient's allergens, Metabolic syndrome (obesity, poor OGTT etc.), Diabetes Mellitus, Malignancy, Polyposis, Variant Creutzfeld-Jakob disease risk, Psychiatric/neurological disease

bacteriophages used in combination, resulted in reduced *C. difficile* colonization. In a challenge model, they prophylactically delayed the onset of symptoms, compared to untreated mice [41]. This has also been previously shown in in vitro models, in which a bacteriophage prophylaxis regime was effective and disrupted the production of bacterial toxins [42].

In a human colon model, the use of a specific bacteriophage significantly reduced the burden of *C. difficile* bacteria and toxins, with no detrimental effect on commensal bacterial populations [43]. Consequently they may have a prophylactic role in preventing CDI in susceptible patients. There are, however, some concerns regarding the transfer of bacterial genes, encoding virulence factors or antibiotic resistance.

Bacterial Spore—Based Therapies

Many members of the beneficial bacteria, thought to be important for the protection against CDI, belong to the Clostridiaceae family. Some of these bacteria are spore forming, and can become activated in the gastrointestinal tract. Exposure of the fecal sample to harsh conditions, such as ethanol, will kill most bacteria, but bacterial spores will survive. Such therapy will be more specific and safe, compared to regular FMT. Currently, a phase III trial investigating the efficacy of SER-109, a combination of 50 bacterial spores prepared from healthy donors is underway, after this treatment was shown to prevent CDI recurrence, and restore a healthy gut microbiome [44].

Probiotics for the Treatment of CDI

The probiotic yeast *Saccharomyces boulardii* has shown immune modulating effects, with the inhibition of CD toxin-A receptor binding, thus preventing downstream activation of MAP kinases, IL-8 production, fluid secretion and intestinal inflammation [45]. Recent studies have shown that timely probiotic use may prevent CDI in patients receiving antibiotics [46]. Further studies are required to determine the correct dosage, timing, and formulation.

CONCLUSION

The use of FMT in CDI has a sound biological basis; however, mechanisms of actions are yet to be fully understood. Further investigation of these mechanisms will lead to safer, more standardized, and more specific stool component based management, with a role in both prophylaxis and treatment of this significant medical condition.

REFERENCES

[1] Khanna S, Pardi DS. The growing incidence and severity of *Clostridium difficile* infection in inpatient and outpatient settings. Expet Rev Gastroenterol Hepatol 2010;4:409–16.

[2] McDonald LC, Owings M, Jernigan DB. *Clostridium difficile* infection in patients discharged from US short-stay hospitals, 1996-2003. Emerg Infect Dis 2006;12:409–15.

[3] Antharam VC, Li EC, Ishmael A, et al. Intestinal dysbiosis and depletion of butyrogenic bacteria in *Clostridium difficile* infection and nosocomial diarrhea. J Clin Microbiol 2013;51:2884–92.

[4] Milani C, Ticinesi A, Gerritsen J, et al. Gut microbiota composition and *Clostridium difficile* infection in hospitalized elderly individuals: a metagenomic study. Sci Rep 2016;6:25945.

[5] Allegretti JR, Kearney S, Li N, et al. Recurrent *Clostridium difficile* infection associates with distinct bile acid and microbiome profiles. Aliment Pharmacol Ther 2016;43:1142–53.

[6] Zhang L, Dong D, Jiang C, Li Z, Wang X, Peng Y. Insight into alteration of gut microbiota in *Clostridium difficile* infection and asymptomatic *C. difficile* colonization. Anaerobe 2015;34:1–7.

[7] Skraban J, Dzeroski S, Zenko B, Mongus D, Gangl S, Rupnik M. Gut microbiota patterns associated with colonization of different *Clostridium difficile* ribotypes. PLoS One 2013;8:e58005.

[8] Francis MB, Allen CA, Shrestha R, Sorg JA. Bile acid recognition by the *Clostridium difficile* germinant receptor, CspC, is important for establishing infection. PLoS Pathogens 2013;9:e1003356.

[9] Weingarden AR, Chen C, Bobr A, et al. Microbiota transplantation restores normal fecal bile acid composition in recurrent *Clostridium difficile* infection. Am J Physiol Gastrointest Liver Physiol 2014;306:G310–9.

[10] Weingarden AR, Dosa PI, DeWinter E, et al. Changes in colonic bile acid composition following fecal microbiota transplantation are sufficient to control *Clostridium difficile* germination and growth. PLoS One 2016;11:e0147210.

[11] Mullish B, McDonald J, Kao D, et al. OC-063 Gut microbiota-host bile acid metabolism interactions in *Clostridium difficile* infection: the explanation for the efficacy of faecal microbiota transplantation? Gut 2017;66:A33–4.

[12] Buffie CG, Bucci V, Stein RR, et al. Precision microbiome reconstitution restores bile acid mediated resistance to *Clostridium difficile*. Nature 2015;517:205–8.

[13] Dineen SS, Villapakkam AC, Nordman JT, Sonenshein AL. Repression of *Clostridium difficile* toxin gene expression by CodY. Mol Microbiol 2007;66:206–19.

[14] Antunes A, Martin-Verstraete I, Dupuy B. CcpA-mediated repression of *Clostridium difficile* toxin gene expression. Mol Microbiol 2011;79:882–99.

[15] Seekatz AM, Aas J, Gessert CE, et al. Recovery of the gut microbiome following fecal microbiota transplantation. mBio 2014;5:e00893–14.

[16] van Nood E, Vrieze A, Nieuwdorp M, et al. Duodenal infusion of donor feces for recurrent *Clostridium difficile*. N Engl J Med 2013;368:407–15.

[17] Khanna S, Jones C, Jones L, Bushman F, Bailey A. Increased microbial diversity found in successful versus unsuccessful recipients of a next-generation FMT for recurrent *Clostridium difficile* infection. Open Forum Infect Dis 2015;2:759.

[18] El Feghaly RE, Bangar H, Haslam DB. The molecular basis of *Clostridium difficile* disease and host response. Curr Opin Gastroenterol 2015;31:24–9.

[19] Chieppa M, Rescigno M, Huang AY, Germain RN. Dynamic imaging of dendritic cell extension into the small bowel lumen in response to epithelial cell TLR engagement. J Exp Med 2006;203:2841–52.

[20] Curotto de Lafaille MA, Lafaille JJ. Natural and adaptive foxp3+ regulatory T cells: more of the same or a division of labor? Immunity 2009;30:626–35.

[21] Weingarden AR, Hamilton MJ, Sadowsky MJ, Khoruts A. Resolution of severe *Clostridium difficile* infection following sequential fecal microbiota transplantation. J Clin Gastroenterol 2013;47:735–7.

[22] Macpherson AJ, Uhr T. Induction of protective IgA by intestinal dendritic cells carrying commensal bacteria. Science 2004;303:1662–5.

[23] Rea MC, Sit CS, Clayton E, et al. Thuricin CD, a posttranslationally modified bacteriocin with a narrow spectrum of activity against *Clostridium difficile*. Proc Natl Acad Sci USA 2010;107:9352–7.

[24] Rea MC, Alemayehu D, Ross RP, Hill C. Gut solutions to a gut problem: bacteriocins, probiotics and bacteriophage for control of *Clostridium difficile* infection. J Med Microbiol 2013;62:1369–78.

[25] Wong JM, de Souza R, Kendall CW, Emam A, Jenkins DJ. Colonic health: fermentation and short chain fatty acids. J Clin Gastroenterol 2006;40:235–43.

[26] Rolfe RD. Role of volatile fatty acids in colonization resistance to *Clostridium difficile*. Infect Immun 1984;45:185–91.

[27] Zuo T, Wong SH, Lam K, et al. Bacteriophage transfer during faecal microbiota transplantation in *Clostridium difficile* infection is associated with treatment outcome. Gut 2018;67(4):634–43.

[28] Ott SJ, Waetzig GH, Rehman A, et al. Efficacy of sterile fecal filtrate transfer for treating patients with *Clostridium difficile* infection. Gastroenterology 2017;152:799–811, e7.

[29] Imhann F, Bonder MJ, Vich Vila A, et al. Proton pump inhibitors affect the gut microbiome. Gut 2016;65:740–8.

[30] Fuereder T, Koni D, Gleiss A, et al. Risk factors for *Clostridium difficile* infection in hemato-oncological patients: a case control study in 144 patients. Sci Rep 2016;6:31498.

[31] Haran JP, Bradley E, Howe E, Wu X, Tjia J. Medication exposure and risk of recurrent *Clostridium difficile* infection in community-dwelling older people and nursing home residents. J Am Geriatr Soc 2017;66(2):333—8.

[32] Imhann F, Vich Vila A, Bonder MJ, et al. The influence of proton pump inhibitors and other commonly used medication on the gut microbiota. Gut Microbes 2017;8:351—8.

[33] Samra Z, Madar-Shapiro L, Aziz M, Bishara J. Evaluation of a new immunochromatography test for rapid and simultaneous detection of *Clostridium difficile* antigen and toxins. Isr Med Assoc J 2013;15:373—6.

[34] McFarland LV, Surawicz CM, Rubin M, Fekety R, Elmer GW, Greenberg RN. Recurrent *Clostridium difficile* disease: epidemiology and clinical characteristics. Infect Control Hosp Epidemiol 1999;20:43—50.

[35] Louie TJ, Miller MA, Mullane KM, et al. Fidaxomicin versus vancomycin for *Clostridium difficile* infection. N Engl J Med 2011;364:422—31.

[36] Cammarota G, Ianiro G, Gasbarrini A. Fecal microbiota transplantation for the treatment of *Clostridium difficile* infection: a systematic review. J Clin Gastroenterol 2014;48:693—702.

[37] Brandt LJ, Aroniadis OC, Mellow M, et al. Long-term follow-up of colonoscopic fecal microbiota transplant for recurrent *Clostridium difficile* infection. Am J Gastroenterol 2012;107:1079—87.

[38] NICE NIfHaCE. Fecal microbiota transplantation for recurrent clostridium difficile infection. 2014.

[39] Tvede M, Rask-Madsen J. Bacteriotherapy for chronic relapsing *Clostridium difficile* diarrhoea in six patients. Lancet 1989;1:1156—60.

[40] Petrof EO, Gloor GB, Vanner SJ, et al. Stool substitute transplant therapy for the eradication of *Clostridium difficile* infection: 'RePOOPulating' the gut. Microbiome 2013;1:3.

[41] Nale JY, Spencer J, Hargreaves KR, et al. Bacteriophage combinations significantly reduce *Clostridium difficile* growth in vitro and proliferation in vivo. Antimicrob Agents Chemother 2016;60:968—81.

[42] Meader E, Mayer MJ, Gasson MJ, Steverding D, Carding SR, Narbad A. Bacteriophage treatment significantly reduces viable *Clostridium difficile* and prevents toxin production in an in vitro model system. Anaerobe 2010;16:549—54.

[43] Meader E, Mayer MJ, Steverding D, Carding SR, Narbad A. Evaluation of bacteriophage therapy to control *Clostridium difficile* and toxin production in an in vitro human colon model system. Anaerobe 2013;22:25—30.

[44] Khanna S, Pardi DS, Kelly CR, et al. A novel microbiome therapeutic increases gut microbial diversity and prevents recurrent *Clostridium difficile* infection. J Infect Dis 2016;214:173—81.

[45] Chen X, Fruehauf J, Goldsmith JD, et al. *Saccharomyces boulardii* inhibits EGF receptor signaling and intestinal tumor growth in Apc(min) mice. Gastroenterology 2009;137:914—23.

[46] Shen N, Maw A, Tmanova L, et al. Timely use of probiotics in hospitalized adults prevents *Clostridium difficile* infection: a systematic review with meta-regression analysis. Gastroenterology 2017;152:1889—900.

Chapter 19

Targeted Delivery of Bacteriophages to the Gastrointestinal Tract and Their Controlled Release: Unleashing the Therapeutic Potential of Phage Therapy

Danish J. Malik

Chemical Engineering Department, Loughborough University, Loughborough LE11 3TU, United Kingdom

INTRODUCTION

Bacteriophages are the smallest entity in the human microbial ecosystem and colonize all niches of the human body, including those previously thought to be sterile [1]. Phages are viruses that infect bacteria, including antibiotic-resistant ones. They exert significant selective pressure on their bacterial hosts and undoubtedly impact on human health through their interactions with the host microbiome [1]. Bacteriophages are inherent constituents of the human microbiome and are ubiquitous in the gut (estimated at 10^9 phage per gram of faeces), with the large intestine harboring the most densely populated microbial ecosystem (estimated at 10^{13}–10^{14} microbial cells per gram of faecal matter) [2]. Like the bacterial microbiome, the human intestinal virome is highly personalized, dynamic, and may be perturbed and modulated, e.g., through changes in diet or during disease [3–6].

Phages in the human gut appear to predominantly favor a temperate life cycle [1,3,7]. Gut phages encode a rich diversity of functional genes, through which lysogeny may confer beneficial traits to the bacterial host. There is considerable interpersonal gut virome diversity, suggesting that it is highly personalized, along with evidence of considerable intrapersonal virome stability within the human gut [3,5]. Similar to the concept of a "healthy gut microbiome," where a similar set of bacteria present across healthy individuals confers functions that promote human health, a "healthy gut phageome" may possess a core common set of phages shared among healthy individuals [1]. Individuals suffering from gut-related diseases display discordant gut viromes, including those suffering from obesity, inflammatory bowel disease, ulcerative colitis, Crohn's disease, etc [8]. It is unclear whether this is reflective of the different bacterial microbiomes or of other contexts.

Mucus-Associated Phages

The mucosa covering the gastrointestinal epithelial cells is a highly microbe-rich environment, with associated diversity in the bacteriophage population, due to weak binding interactions between phage capsid proteins and mucin glycoproteins, which trap phages and bacteria in structured mucin networks [9]. Mucus adherent phage may provide a nonhost-derived immunity [9]. The spatial distribution of bacterial hosts and their physiological status within the mucus are important in terms of phage access and replication therein. Low concentration of bacteria and high bacterial diversity are problematic with regard to phage propagation, given their narrow host range, and concentration-dependent replication kinetics following the mass action law [10].

Safe Alternatives for Current Antibiotic Options

The beneficial effects arising from the diversity of the gut microbiome is brought into sharp focus when broad-spectrum antibiotics are used, resulting in gut microbiome dysbiosis, e.g., *Clostridium difficile* infections. There is a pressing need to develop species specific, narrow-spectrum antibiotics, which can be employed to manipulate the microbiome for beneficial health outcomes. Examples of such approaches include programmable sequence-specific antimicrobials, using RNA-guided nuclease Cas9 delivered using phagemids [11].

Bioengineered Avirulent Bacteria

Cas9 reprogrammed to target virulence genes, using bacteriophages as the delivery system, was recently shown to kill virulent, but not avirulent, *Staphylococcus aureus*. Such CRISPR-Cas9 antimicrobials could be deployed to decolonize patients harboring antibiotic-resistant gastrointestinal pathogenic bacteria, e.g., *Enterococci*, other *Enterobacteriaceae*, and toxigenic *Clostridium*. Nontargeted bacteria are free to occupy niches, without undue disturbance of the gut microbiota. As the phagemid does not replicate in the bacteria, targeted delivery of high concentrations of phagemid is needed at the site of infection. The technology could easily be adapted to remove expression of antibiotic resistance or virulence genes, without causing death of the bacterium, using dCas9 the nuclease-defective version of Cas9.

Available Phage Families

The focus of recent phage therapy approaches is on the use of lytic tailed phages, all of which belong to the Caudovirales and include the *Myoviridae*, *Siphoviridae*, and *Podoviridae* families [12]. Members of the Caudovirales have an icosahedral capsid head that contains double-stranded DNA (15−500 Kbp) and a tail with surface receptor proteins, which interact with surface features on the host bacterium [13]. Successful phage therapy requires interaction between the phage and the bacterium, resulting in adsorption of the phage to the host bacterium, followed by injection of the phage DNA.

The pharmacodynamics of this process has been modeled using mathematical models based on colloidal interactions [14−16]. On infection, the phage replication cycle ensues, culminating, after a short latency period, in cell lysis and the liberation of multiple phage virions (with a burst size that is typically between 10 and 100 [17]). Gill and Hyman [18] and Weber-Dabrowska et al. [19] provide an overview of key considerations related to phage choice, isolation, and purification for phage therapy. Phage therapy practice relies on the isolation of naturally occurring phages abundantly found in the environment [20].

Phage Laboratory Processing

Typically, phages are isolated from the environment and screened against commonly occurring pathogenic bacterial strains (to identify host ranges) and then evaluated using in vitro and in vivo animal models. The limitation of host ranges is overcome through the use of phage cocktails to ensure sufficient coverage of commonly occurring strains, and the use of phage mixtures targeting different receptors reduces the probability of encountering phage-resistant bacterial mutants. Phages are generally manufactured using standard fermentation process technology. In brief, host bacteria are grown in liquid culture in a bioreactor. During the log growth phase, phages are added to the bioreactor to infect the bacteria.

Incubation of phages with bacteria results in phage adsorption to the bacteria, infection, and following a short lag period, release of bacteriophage virions. The resulting lysate contains the product (the amplified phage) along with bacterial debris and residual fermentation media. Removal of cellular debris is typically done using centrifugation and/or filtration. Ion exchange, gel filtration, etc., can be used to further purify the bacteriophages, e.g., for the removal of endotoxin from Gram negative bacteria. Phages may then be resuspended in simple saline or buffer and stored under refrigerated conditions or processed further, e.g., spray-dried to improve storage shelf life or encapsulated in micro- or nanoparticle formulations.

Mechanism of Action

Phages are unique antimicrobials in that in the presence of host bacteria, they are able to increase their numbers by infecting the bacteria and producing virion progeny, while minimally affecting the overall microbiota and body tissues. Phages carrying polysaccharide depolymerases in their structure may be able to disrupt bacterial defensive biofilms [21−23]. For example, *Enterococcus* and *Staphylococcus* phages capable of destroying biofilms have been reported [24,25].

In addition to their potential as human biotherapeutics, phages are being developed for agricultural use to rid the environment and domestic animals of pathogens, which could contaminate the food supply chain [26], in aquaculture for the treatment of fish pathogens [27], and for the control of infections in intensively farmed poultry [28,29]. Recent advances in molecular biology and sequencing technology have improved our basic understanding of how bacteriophages interact with bacteria and have opened new possibilities for utilizing phages, including genetically engineered phages, for potential therapeutic and diagnostic applications [13,30].

Gastrointestinal Bacterial Infections

The use of phage therapy is a particularly promising alternative to broad-spectrum antibiotic treatment for acute enteric infections because typically intestinal concentration of infecting bacteria is high, the causative agent and strain may be suitably diagnosed using rapid diagnostic tools, and application of phage therapy with a sufficiently high initial phage dose would promote rapid in situ phage multiplication and decrease in the host bacterial population [14,31–33]. In conditions where lytic dynamics tend to dominate, phage therapy may be successful to treat acute infections in the gut [34].

Enteric infections worldwide are typically caused by pathogens such as *C. difficile*, *Escherichia coli*, *Salmonella* spp., *Vibrio cholerae* [35]. Any such pathogen is potentially a target for phage therapy; however, there are major barriers to be overcome in terms of understanding the dynamics of the interaction between phage and bacteria in the gut environment and in terms of the logistics of delivering a stable defined product to the infection site [34].

ENCAPSULATION OF BACTERIOPHAGES

Developing formulations that incorporate bacteriophages for therapeutic applications requires an appreciation of the chemical and physical stresses, where phages may encounter both during processing and during storage once formulated. Phage inactivation and long-term reduction in phage titer on storage are highly undesirable. Accurate loading of phages per particle requires particle monodispersity, which is rarely achieved in practice; however, some techniques are considerably better than others. The particle morphology should be without deformities, and the particles should not aggregate or uncontrollably fuse together with material in the surrounding environment.

A polymer or lipid may be used to coat an existing structure containing the phage. For example, Murthy and Engelhardt [36] sprayed phage on dried skimmed milk and then encapsulated them in a lipid coating. Alternatively, phages may be incorporated in the formulation within the droplet, which on drying or cross-linking results in phage entrapment in the particle core. Depending on the technique used, polymer gelling may be part of the technique or it may be a process occurring downstream. There are many techniques and processes that may be used for stabilizing, immobilizing, and encapsulating phages. The most common methods are spray drying, spray freeze drying, freeze drying, extrusion dripping methods, emulsion, and polymerization techniques.

Phage In Vitro Stability

Phages are protein structures, and they are therefore susceptible to factors known to denature proteins. These include exposure to organic solvents [37,38], high temperatures [39], pH [39,40], ionic strength [40], and interfacial effects. Additionally, mechanical stresses during formulation or encapsulation, including shear stresses during mixing and agitation, atomization during spraying [41], and desiccation stresses during drying [42], need careful consideration. Drying and lyophilization of bacteriophages, with suitable carriers and bulking agents to improve storage life, have been the focus of numerous studies.

A number of encapsulation techniques have been used, followed by drying [43] or incorporating solvent evaporation, e.g., during electrospinning [44], to improve phage storage. One of the most common and successful modes of long-term preservation of bacteriophages is storage at 4°C in trypticase soy agar and brain heart infusion broth, whereas storage at −80°C requires 50% glycerol as a cryoprotectant [45] or freeze drying with excipients (e.g., sucrose or trehalose) as lyophilization and cryoprotectants [46,47]. Ackermann [47] observed that phage titers for one of the largest collections of tailed phages typically tended to drop by 1 log over the course of a year, but then remained fairly stable, although there were many individual variations.

FREEZE DRYING OF BACTERIOPHAGES

Freeze drying (lyophilization) is routinely used in the pharmaceutical industry [48] to dry proteins, vaccines, peptides, or colloidal carriers such as liposomes, nanoparticles, and nanoemulsions [49]. Freeze-dried material typically results in a cake that needs further processing, e.g., milling to achieve fine particles suitable for loading in dry powder inhalers (DPI) [50].

Articles on phage and virus freeze drying have mainly focused on evaluation of formulations to stabilize phage for storage [42,47,51−53]. A number of publications have shown that addition of amino acids, e.g., sodium glutamate [51], peptides, peptone [46], and proteins such as casein and lactoferrin [54], improves phage viability on freeze drying and following rehydration. Literature suggests that disaccharides, e.g., lactose [54], sucrose [42,52,53,55], and trehalose [41,42,52,56,57], improve phage survival during freezing and subsequent lyophilization. Rapid rates of freezing at relatively low temperatures ($<-20°C$) have been found to result in better phage survival compared with slow freezing. This has been attributed to having less time for osmotic damage to occur [46].

SPRAY DRYING OF BACTERIOPHAGES

Spray drying processes atomize a liquid containing dissolved solids, converting it into a fine mist, which is contacted with a hot dry gas, typically hot dry air, inside a drying chamber. Due to the high surface area to volume ratio of the very small atomized droplets, the solvent rapidly evaporates with each droplet, forming a particle comprising the nonvolatile components (bacteriophage and excipients) along with a small amount of residual moisture. The particles are separated from the airstream exiting the dryer, and cyclones or bag filters are typically used for this. Spray drying is a scalable industrial process technology and is widely used to produce fine powders for pharmaceutical applications, including pulmonary delivery via DPI [58]. Dry powders are favored for respiratory drug delivery because they show relatively long storage stability without requiring refrigeration [59,60], and DPI are simple to use and do not require regular cleaning and disinfection [61]. This is a particularly interesting alternative for phages targeting respiratory infections.

Other Technical Considerations

Phages have been shown not to be able to withstand high shear stress for too long. The nebulization process has been shown to result in loss of phage titer [41,62]. Phages are also sensitive to thermal stresses and partially lose activity at temperatures higher than 60°C [46,63]. Low spray drying temperatures ($\sim$40°C outlet air temperature) result in higher phage titer [64], with high loss at higher temperatures [41,62,65,66]. Literature on spray drying of phage suspensions typically includes excipients in the formulation to protect the phage from desiccation and thermal stress.

Trehalose is the most frequently reported excipient used in the spray drying of phages, and it has been shown to result in spray-dried powders with high phage titer and good storage stability [41,64,66,67]. Trehalose has low toxicity and has been shown to protect biological material, including proteins, probiotics, and vaccines, against desiccation and thermal stress. Crowe et al. [68] suggested that the efficacy of trehalose as an excipient in drying is partly not only because of its high glass transition temperature at all water contents but also because it binds residual water left over from the drying process to form a dihydrate, which might otherwise participate in lowering the glass transition temperature to below ambient [67].

Different phage strains, formulated and spray-dried under identical conditions, showed significant differences in the resulting phage titer, suggesting the need for individually formulating each phage to be used in a phage cocktail [62,67]. A number of studies have reported spray-dried phage powders to have a suitable mass median aerosolized diameter for pulmonary delivery of phage to treat respiratory infections [41,66−70].

METHODS USED FOR PHAGE ENCAPSULATION IN MICRO- AND NANOPARTICLE SOLID DOSAGE FORMS

Bacteriophages may be encapsulated in protective micro- and nanoparticles to overcome adverse storage and physiological conditions *en route* to delivering the phage load to modulate the gut microbiome or to treat gut infections. Controlled release and sustained release strategies may be achieved using a diverse array of strategies. These include systems based on diffusion-controlled release (e.g., solvent diffusion−based osmotic pumps), matrix dissolution and erosion-controlled systems, and ion-exchange swelling-based systems. Phage compatible formulation and encapsulation processes need to be carefully designed to prevent damage to viral capsid and DNA/RNA components, and stabilization of the structure of viral capsid and tail proteins to prevent loss of phage viability during manufacturing operations. It is important that the carrier encapsulating the phage is able to withstand adverse environmental conditions, for the duration of exposure, and is capable of delivering the bacteriophage at a suitable high dose to the relevant site where host bacteria are present [71−82].

Phage Survival in the Upper Gastrointestinal Tract

Polymer phage encapsulation literature has focused largely on gastrointestinal infections. Oral administration of bacteriophage for human or animal use requires careful consideration of a number of factors, including the acidic pH of the stomach, digestive enzymes (pepsin, proteases, lipases, amylase, and trypsinogen), bile salts, pancreatic juices, residence time in different intestinal compartments (duodenum, jejunum, ileum), and phage permeability into the mucosal lining where the infection may reside.

Acid pH

Overcoming phage susceptibility to gastric acid is a major concern, as with other live agents (probiotics, synbiotics). It has been addressed in some studies through administration of antacids, e.g., sodium bicarbonate, prior to oral phage treatment or co-encapsulation of calcium carbonate [83]. Previous efforts targeting *Salmonella* in the food chain have attributed to poor phage stability in the gastrointestinal tract and the insufficient in vivo efficacy [84]. *Salmonella* Felix O1 phage (belonging to the *Myoviridae* family) was shown here to be highly sensitive to acidic pH. Similar results have been reported previously for phage Felix O1, other *Salmonella* phages, as well as *S. aureus* bacteriophage K (also a myovirus), and for *E. coli*–specific phages [74,77,85–87]. In contrast, Felix O1 microencapsulated in alginate beads has previously been shown to survive and amplify in the gastrointestinal tract of pigs [84].

Gastrointestinal Motility, Osmotic Gradient

Carriers may be designed to respond to specific pH, which differs from the stomach (pH 1–3) to the small intestine (pH 5.5–6.5) and the colon (pH 6.5–7.2) [43,72]. Gastric emptying rates are another important factor. In humans gastric emptying of small microcapsules (less than 2 mm) is rapid (~ 0.5–1.5 h) and is not greatly affected by the digestive state of the individual [88,89]. The gastrointestinal tract may be modeled as a distinct set of compartments (duodenum, jejunum, ileum, etc.), with each compartment having different pH and transit times [90–93]. Moreover, considering, for instance, the treatment of infectious diarrhea, the osmotic gradient between the epithelium and the colon lumen decreases, resulting in fluid leakage and shorter transit times [94]. Changes in pH have been observed as well, which can directly affect the microbial population and transit times [95,96].

Phage Encapsulation Using Membrane Microarrays

Membrane emulsification (ME) using micropore arrays is a process that allows formation of uniform drops by injecting a dispersed phase liquid through a microporous membrane into the continuous phase. Alternatively, a preemulsified mixture of the dispersed and continuous phase may be repeatedly injected through the membrane (Fig. 19.1). To encapsulate bacteriophages, the formed drops, typically composed of a mixture of the wall-forming materials, solvent(s), and phages, could be solidified under mild agitation using various solidification reactions or processes, such as free-radical

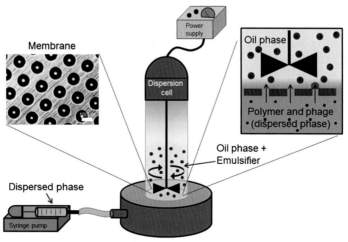

FIGURE 19.1 Schematic representation of the membrane emulsification system used for microencapsulation of bacteriophages.

polymerization, polycondensation, ionotropic/thermal gelation, cooling crystallization, and molecular or particle self-assembly, triggered by solvent evaporation [97].

The advantages of ME over standard emulsification procedures, using high-pressure valve homogenizers or rotor—stator devices, are in higher drop size uniformity and lower energy inputs and applied shear, which can be useful to preserve phage integrity. It is well known that bacteriophages are sensitive to mechanical shearing [41,62]. In direct ME, the shear rate on the membrane surface is $\sim 10^3 - 10^4 \, s^{-1}$; however, uniform drops may be obtained without any shear by spontaneous droplet formation driven by Laplace pressure gradients [98]. The shear rate in high-shear mixers and colloid mills is typically $\sim 10^5 \, s^{-1}$ and can exceed $10^7 \, s^{-1}$ in microfluidizers.

The most common membranes used in ME are Shirasu porous glass and microsieve metallic membranes [99]. Shear on the membrane surface, required for drop detachment, can be generated in various ways including (1) using a paddle stirrer placed above the membrane surface [100]; (2) rotating membrane [101]; and (3) oscillating membrane [102]. In the oscillating ME system, the tubular membrane can oscillate tangentially, clockwise or counterclockwise [103], or radially upward and downward [102], with frequencies ranging from 10 to 90 Hz.

Microfluidic Microencapsulation

Fabrication of bacteriophage-encapsulated microparticles, using microfluidic technology, has recently been shown to permit precise control over the particle size, phage dose per particle, and tailoring of the release profile for targeted delivery, and controlled release of phages [104]. The low controlled shear in the microfluidic droplet generation unit potentially allows 100% phage encapsulation and production of highly uniform droplets. The significant potential of microfluidic microencapsulation for phages is relatively unexplored, with only two previous published papers to date, including one from our group [104,105]. Little consideration has previously been given to the control of the particle size distribution, the phage loading per particle, and the resulting heterogeneity of the release kinetics from each particle [74,75,77,85,106]. A key advantage of the microfluidic microcapsule fabrication process is the precision with which uniform small microparticles (mean size 10—100 μm) containing encapsulated phage may be prepared under low shear conditions.

Droplet generation in a microfluidic device results in each droplet being produced at the same position. Disruptive forces of the same magnitude act on each individual droplet, and these forces become negligible after pinch-off (Fig. 19.2). In conventional bulk emulsification systems, the droplets are continuously being subjected to shear, which typically varies widely across the emulsification or homogenization unit [107—109]. Continuous excessive exposure to shear results in the droplets breaking further into smaller droplets, resulting in emulsions with a high degree of polydispersity [110]. The microfluidic fabrication process provides a highly controlled uniform shear environment, resulting in microparticles with low polydispersity [104].

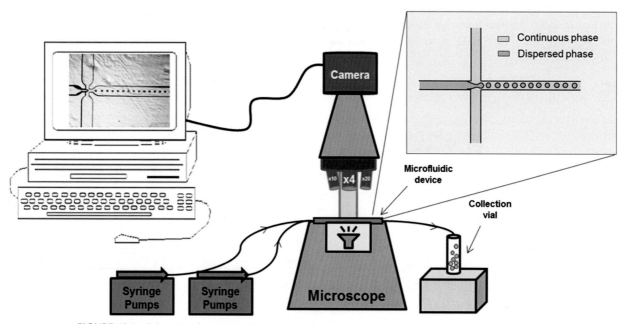

FIGURE 19.2 Schematic of experimental setup to produce water-in-oil emulsion for phage microencapsulation.

CONCLUSIONS

Exploiting the potential of bacteriophages to selectively manipulate the microbiome is an exciting future prospect. There is, however, a pressing need to develop solid dosage forms to precisely deliver high concentrations of phages in different regions of the gut. Repeated frequent dosing of phage may be needed but would be impractical. Phage susceptibility to environmental stresses in the gut is an important limitation. Encapsulation processes may be low cost and simple, such as spray drying, or highly advanced such as those based on microfluidic fabrication of highly uniform microcapsules. Freeze drying of microencapsulated phages may allow improvement in product storage shelf life. Encapsulation technologies have the potential to enable precise control over phage loading and production of highly uniform microparticles. Using different polymer formulations would enable high level of innovation, e.g., by employing a multitude of different triggers for phage release at the target site, including light, temperature, pH, enzymes, etc. Furthermore, the fabrication of structured multiple emulsions, e.g., composed of concentric onion-like shells around the core drop, may allow production of complex microcapsules for co-encapsulation and simultaneous or sequential release of several different phages encapsulated in individually optimized formulations, which would allow evaluation of novel strategies for gut microbiome modulation.

REFERENCES

[1] Barr JJ. A bacteriophages journey through the human body. Immunol Rev 2017;279:106−22.
[2] Marchesi J, Shanahan F. The normal intestinal microbiota. Curr Opin Infect Dis 2007;20:508−13.
[3] Minot S, Bryson A, Chehoud C, Wu GD, Bushman FD, Lewis JD. Rapid evolution of the human gut virome. Proc Natl Acad Sci U S A 2013;110(30):12450−5.
[4] Hooper LV, Littman DR, Macpherson AJ. Interactions between the microbiota and the immune system. Gut Microbiota June 2012;336:1268−73.
[5] Reyes A, Haynes M, Hanson N, Angly FE, Heath AC, Rohwer F, et al. Viruses in the faecal microbiota of monozygotic twins and their mothers. Nature 2010;466(7304):334−8.
[6] Minot S, Sinha R, Chen J, Li H, Keilbaugh SA, Wu GD, et al. The human gut virome: inter-individual variation and dynamic response to diet. Genome Res 2011;21(10):1616−25.
[7] Knowles B, Silveira CB, Bailey BA, Barott K, Cantu VA, Cobian-Guëmes AG, et al. Lytic to temperate switching of viral communities. Nature 2016;531(7595):466−70.
[8] Marques TM, Cryan JF, Shanahan F, Fitzgerald GF, Ross RP, Dinan TG, et al. Gut microbiota modulation and implications for host health: dietary strategies to influence the gut-brain axis. Innov Food Sci Emerg Technol 2014;22:239−47.
[9] Barr JJ, Auro R, Sam-Soon N, Kassegne S, Peters G, Bonilla N, et al. Subdiffusive motion of bacteriophage in mucosal surfaces increases the frequency of bacterial encounters. Proc Natl Acad Sci 2015;112(44):13675−80.
[10] Malik DJ, Sokolov IJ, Vinner GK, Mancuso F, Cinquerrui S, Vladisavljevic GT, et al. Formulation, stabilisation and encapsulation of bacteriophage for phage therapy. Adv Colloid Interface Sci May 2017;249:100−33.
[11] Bikard D, Euler C, Jiang W, Nussenzweig PM, Gregory W, Duportet X, et al. Development of sequence-specific antimicrobials based on programmable CRISPR-Cas nucleases. Nat Biotechnol 2015;32(11):1146−50.
[12] Wittebole X, De Roock S, Opal SM. A historical overview of bacteriophage therapy as an alternative to antibiotics for the treatment of bacterial pathogens. Virulence 2014;5(1):226−35.
[13] Young R, Gill JJ. Phage therapy redux — what is to be done? Science 2015;350(6265):1163−4.
[14] Levin BR, Bull JJ. Population and evolutionary dynamics of phage therapy. Nat Rev Microbiol 2004;2(2):166−73.
[15] Cairns BJ, Timms AR, Jansen VAA, Connerton IF, Payne RJH. Quantitative models of in vitro bacteriophage-host dynamics and their application to phage therapy. PLoS Pathog 2009;5(1):1−10.
[16] Bull JJ, Gill JJ. The habits of highly effective phages: population dynamics as a framework for identifying therapeutic phages. Front Microbiol November 2014;5:1−12.
[17] Hadas H, Einav M, Fishov I, Zaritsky A. Baleriophage T4 development depends on the physiology of its host Escherichia coli. Microbiology 1997;143:179−85.
[18] Gill JJ, Hyman P. Phage choice, isolation and preparation for phage therapy. Curr Pharm Biotechnol 2010;11:2014.
[19] Weber-Dabrowska B, Jonczyk-Matysiak E, Zaczek M, Lobocka M, Lusiak-Szelachowska M, Gorski A. Bacteriophage procurement for therapeutic purposes. Front Microbiol August 2016;7:1−14.
[20] Mattila S, Ruotsalainen P, Jalasvuori M. On-demand isolation of bacteriophages against drug-resistant bacteria for personalized phage therapy. Front Microbiol November 2015;6:1−7.
[21] Sutherland IW, Hughes KA, Skillman LC, Tait K. The interaction of phage and biofilms. FEMS Microbiol Lett 2004;232(1):1−6.
[22] Cornelissen A, Ceyssens PJ, T'Syen J, van Praet H, Noben JP, Shaburova OV, et al. The t7-related pseudomonas putida phage φ15 displays virion-associated biofilm degradation properties. PLoS One 2011;6(4).
[23] Pires DP, Oliveira H, Melo LDR, Sillankorva S, Azeredo J. Bacteriophage-encoded depolymerases: their diversity and biotechnological applications. Appl Microbiol Biotechnol 2016;100(5):2141−51.
[24] Gutiérrez D, Vandenheuvel D, Martínez B, Rodríguez A, Lavigne R, García P. Two phages, phiIPLA-RODI and phiIPLA-C1C, lyse mono- and dual-species Staphylococcal biofilms. Appl Environ Microbiol 2015;81(10):3336−48.

[25] Khalifa L, Brosh Y, Gelman D, Coppenhagen-Glazer S, Beyth S, Poradosu-Cohen R, et al. Targeting *Enterococcus faecalis* biofilms with phage therapy. Appl Environ Microbiol 2015;81(8):2696−705.

[26] Smith HW, Huggins MB, Shaw KM. The control of experimental *Escherichia coli* diarrhoea in calves by means of bacteriophages. J Gen Microbiol 1987;133(5):1111−26.

[27] Nakai T, Sugimoto R, Park K-H, Matsuoka S, Mori K, Nishioka T, et al. Protective effects of bacteriophage on experimental Lactococcus garvieae infection in yellowtail. Dis Aquat Organ 1999;37:33−41.

[28] Smith HW, Huggins MB. Effectiveness of phages in treating experimental *Escherichia coli* diarrhoea in calves, piglets and lambs. J Gen Microbiol 1983;129(8):2659−75.

[29] Atterbury RJ, Van Bergen MAP, Ortiz F, Lovell MA, Harris JA, De Boer A, et al. Bacteriophage therapy to reduce Salmonella colonization of broiler chickens. Appl Environ Microbiol 2007;73(14):4543−9.

[30] Peltomaa R, López-Perolio I, Benito-Peña E, Barderas R, Moreno-Bondi MC. Application of bacteriophages in sensor development. Anal Bioanal Chem 2016;408(7):1805−28.

[31] Barrow PA, Soothill JS. Bacteriophage therapy and prophylaxis: rediscovery and renewed assessment of potential. Trends Microbiol 1997;5(7):268−71.

[32] Galtier M, De Sordi L, Maura D, Arachchi H, Volant S, Dillies MA, et al. Bacteriophages to reduce gut carriage of antibiotic resistant uropathogens with low impact on microbiota composition. Environ Microbiol 2016;18(7):2237−45.

[33] Merril CR, Scholl D, Adhya SL. The prospect for phage therapy in western medicine. Nat Rev Drug Discov 2003;2:489−97.

[34] Sarker SA, Sultana S, Reuteler G, Moine D, Descombes P, Charton F, et al. Oral phage therapy of acute bacterial diarrhea with two coliphage preparations: a randomized trial in children from Bangladesh. EBioMedicine 2016;4:124−37.

[35] Viswanathan VK, Hodges K, Hecht G. Enteric infection meets intestinal function: how bacterial pathogens cause diarrhoea. Nat Rev Microbiol 2009;7(2):110−9.

[36] Murthy K, Engelhardt R. US 2009/0130196 A1: bacteriophage composition. 2009. p. 1−12.

[37] Puapermpoonsiri U, Spencer J, Van Der Walle CF. A freeze-dried formulation of bacteriophage encapsulated in biodegradable microspheres. Eur J Pharm Biopharm 2009;72(1):26−33.

[38] Lee SW, Belcher AM. Virus-based fabrication of micro- and nanofibers using electrospinning. Nano Lett 2004;4(3):387−90.

[39] Briers Y, Miroshnikov K, Chertkov O, Nekrasov A, Mesyanzhinov V, Volckaert G, et al. The structural peptidoglycan hydrolase gp181 of bacteriophage phiKZ. Biochem Biophys Res Commun 2008;374(4):747−51.

[40] Knezevic P, Obreht D, Curcin S, Petrusic M, Aleksic V, Kostanjsek R, et al. Phages of *Pseudomonas aeruginosa*: response to environmental factors and in vitro ability to inhibit bacterial growth and biofilm formation. J Appl Microbiol 2011;111(1):245−54.

[41] Leung SSY, Parumasivam T, Gao FG, Carrigy NB, Vehring R, Finlay WH, et al. Production of inhalation phage powders using spray freeze drying and spray drying techniques for treatment of respiratory infections. Pharm Res (N Y) 2016;33(6):1486−96.

[42] Dini C, de Urraza PJ. Effect of buffer systems and disaccharides concentration on *Podoviridae* coliphage stability during freeze drying and storage. Cryobiology 2013;66(3):339−42.

[43] Stanford K, Niu YD, Johnson R. Oral delivery systems for encapsulated bacteriophages targeted at *Escherichia coli* O157: H7 in feedlot cattle. J Food Prot 2010;73(7):1304−12.

[44] Korehei R, Kadla J. Incorporation of T4 bacteriophage in electrospun fibres. J Appl Microbiol 2013;114(5):1425−34.

[45] Clark WA. Comparison of several methods for preserving bacteriophages. Appl Environ Microbiol 1962;10:466−71.

[46] Davies JD, Kelly MJ. The preservation of bacteriophage H1 of Corynebacterium ulcerans U103 by freeze-drying. J Hyg 1969;67(4):573−83.

[47] Ackermann H-W, Tremblay D, Moineau S. Long-term bacteriophage preservation. WFCC Newsl 2004;38:35−40.

[48] Wang W. Lyophilization and development of solid protein pharmaceuticals. Int J Pharm 2000;203:1−60.

[49] Abdelwahed W, Degobert G, Stainmesse S, Fessi H. Freeze-drying of nanoparticles: formulation, process and storage considerations. Adv Drug Deliv Rev 2006;58(15):1688−713.

[50] Telko MJ, Hickey AJ. Dry powder inhaler formulation. Respir Care 2005;50(9):1209−27.

[51] Engel HWB, Smith D, Berwald LG. The preservation of mycobacteriophages by means of freeze drying. Am Rev Respir Dis 1974;109:561−6.

[52] Merabishvili M, Vervaet C, Pirnay JP, de Vos D, Verbeken G, Mast J, et al. Stability of *Staphylococcus aureus* phage ISP after freeze-drying (lyophilization). PLoS One 2013;8(7):1−7.

[53] Malenovská H. The influence of stabilizers and rates of freezing on preserving of structurally different animal viruses during lyophilization and subsequent storage. J Appl Microbiol 2014;117(6):1810−9.

[54] Golshahi L, Lynch KH, Dennis JJ, Finlay WH. In vitro lung delivery of bacteriophages KS4-M and φKZ using dry powder inhalers for treatment of Burkholderia cepacia complex and *Pseudomonas aeruginosa* infections in cystic fibrosis. J Appl Microbiol 2010;110(1):106−17.

[55] Puapermpoonsiri U, Ford SJ, van der Walle CF. Stabilization of bacteriophage during freeze drying. Int J Pharm 2010;389(1−2):168−75.

[56] Dai M, Senecal A, Nugen SR. Electrospun water-soluble polymer nanofibers for the dehydration and storage of sensitive reagents. Nanotechnology 2014;25(22):225101.

[57] Colom J, Cano-Sarabia M, Otero J, Cortés P, Maspoch D, Llagostera M. Liposome-encapsulated bacteriophages for enhanced oral phage therapy against *Salmonella* spp. Appl Environ Microbiol 2015;81(14):4841−9.

[58] Hoe S, Ivey JW, Boraey MA, Shamsaddini-Shahrbabak A, Javaheri E, Matinkhoo S, et al. Use of a fundamental approach to spray-drying formulation design to facilitate the development of multi-component dry powder aerosols for respiratory drug delivery. Pharm Res (N Y) 2014;31(2):449−65.

[59] Chew NYK, Chan HK. The role of particle properties in pharmaceutical powder inhalation formulations. J Aerosol Med 2002;15:325−30.

[60] Klingler C, Müller BW, Steckel H. Insulin-micro- and nanoparticles for pulmonary delivery. Int J Pharm 2009;377(1−2):173−9.

[61] Weers J. Inhaled antimicrobial therapy - barriers to effective treatment. Adv Drug Deliv Rev 2015;85:24−43.

[62] Vandenheuvel D, Singh A, Vandersteegen K, Klumpp J, Lavigne R, Van Den Mooter G. Feasibility of spray drying bacteriophages into respirable powders to combat pulmonary bacterial infections. Eur J Pharm Biopharm 2013;84(3):578−82.

[63] Jończyk E, Kłak M, Miedzybrodzki R, Górski A. The influence of external factors on bacteriophages−review. Folia Microbiol 2011;56(3):191−200.

[64] Leung SSY, Parumasivam T, Gao FG, Carrigy NB, Vehring R, Finlay WH, et al. Effects of storage conditions on the stability of spray dried, inhalable bacteriophage powders. Int J Pharm 2017.

[65] Walbeck AK. Methods for drying bacteriophage and bacteriophage-containing compositions, the resulting dry compositions, and methods of use. US20138501453B2. 2013. p. 1.

[66] Matinkhoo S, Lynch KH, Dennis JJ, Finlay WH, Vehring R. Spray-dried respirable powders containing bacteriophages for the treatment of pulmonary infections. J Pharm Sci 2011;100(12):5197−205.

[67] Vandenheuvel D, Meeus J, Lavigne R, Van Den Mooter G. Instability of bacteriophages in spray-dried trehalose powders is caused by crystallization of the matrix. Int J Pharm 2014;472(1−2):202−5.

[68] Crowe LM, Reid DS, Crowe JH. Is trehalose special for preserving dry biomaterials? Biophys J 1996;71(4):2087−93.

[69] Steckel H, Bolzen N. Alternative sugars as potential carriers for dry powder inhalations. Int J Pharm 2004;270(1−2):297−306.

[70] Vandenheuvel D, Singh A, Vandersteegen K, Klumpp J, Lavigne R, Van den, Mooter G. Feasibility of spray drying bacteriophages into respirable powders to combat pulmonary bacterial infections. Eur J Pharm Biopharm 2013;84(3):578−82.

[71] Ibekwe VC, Fadda HM, McConnell EL, Khela MK, Evans DF, Basit AW. Interplay between intestinal pH, transit time and feed status on the in vivo performance of pH responsive ileo-colonic release systems. Pharm Res (N Y) 2008;25(8):1828−35.

[72] McConnell EL, Fadda HM, Basit AW. Gut instincts: explorations in intestinal physiology and drug delivery. Int J Pharm December 2008;364(2):213−26.

[73] Esteban PP, Alves DR, Enright MC, Bean JE, Gaudion A, Jenkins ATA, et al. Enhancement of the antimicrobial properties of bacteriophage-K via stabilization using oil-in-water nano-emulsions. Biotechnol Prog 2014;30(4):932−44.

[74] Ma Y, Pacan JC, Wang Q, Xu Y, Huang X, Korenevsky A, et al. Microencapsulation of bacteriophage Felix O1 into chitosan-alginate microspheres for oral delivery. Appl Environ Microbiol 2008;74(15):4799−805.

[75] Tang Z, Huang X, Sabour PM, Chambers JR, Wang Q. Preparation and characterization of dry powder bacteriophage K for intestinal delivery through oral administration. LWT Food Sci Technol (Lebensmittel-Wissenschaft-Technol) 2015;60(1):263−70.

[76] Colom J, Cano-Sarabia M, Otero J, Aríñez-Soriano J, Cortés P, Maspoch D, et al. Microencapsulation with alginate/CaCO$_3$: a strategy for improved phage therapy. Sci Rep January 2017;7:41441.

[77] Ma Y, Pacan JC, Wang Q, Sabour PM, Huang X, Xu Y. Enhanced alginate microspheres as means of oral delivery of bacteriophage for reducing *Staphylococcus aureus* intestinal carriage. Food Hydrocoll March 2012;26(2):434−40.

[78] Lotfipour F, Mirzaeei S, Maghsoodi M. Preparation and characterization of alginate and psyllium beads containing *Lactobacillus acidophilus*. ScientificWorldJournal 2012;2012:680108.

[79] Dini C, Islan GA, de Urraza PJ, Castro GR. Novel biopolymer matrices for microencapsulation of phages : enhanced protection against acidity and protease activity. Macromol Biosci 2012;12:1200−8.

[80] Samtlebe M, Ergin F, Wagner N, Neve H, Küçükçetin A, Franz CMAP, et al. Carrier systems for bacteriophages to supplement food systems : encapsulation and controlled release to modulate the human gut microbiota. LWT Food Sci Technol (Lebensmittel-Wissenschaft-Technol) 2016;68:334−40.

[81] Hathaway H, Alves DR, Bean J, Esteban PP, Ouadi K, Sutton JM, et al. Poly (N-isopropylacrylamide- co -allylamine) (PNIPAM- co -ALA) nanospheres for the thermally triggered release of bacteriophage K. Eur J Pharm Biopharm 2015;96:437−41.

[82] Bean JE, Alves DR, Laabei M, Esteban PP, Thet NT, Enright MC, et al. Triggered release of bacteriophage K from agarose/hyaluronan hydrogel matrixes by *Staphylococcus aureus* virulence factors. Chem Mater 2014;26:7201−8.

[83] Jamalludeen N, Johnson RP, Shewen PE, Gyles CL. Evaluation of bacteriophages for prevention and treatment of diarrhea due to experimental enterotoxigenic *Escherichia coli* O149 infection of pigs. Vet Microbiol 2009;136:135−41.

[84] Wall SK, Zhang J, Rostagno MH, Ebner PD. Phage therapy to reduce preprocessing *Salmonella* infections in market-weight swine. Appl Environ Microbiol 2010;76(1):48−53.

[85] Kim S, Jo A, Ahn J. Application of chitosan-alginate microspheres for the sustained release of bacteriophage in simulated gastrointestinal conditions. Int J Food Sci Technol 2015;50(4):913−8.

[86] Albino LAA, Rostagno MH, Hungaro HM, Mendonca RCS. Isolation, characterization and application of bacteriophages for *Salmonella* spp. biocontrol in pigs. Foodborne Pathog Dis 2014;11(8):602−9.

[87] Saez AC, Zhang J, Rostagno MH, Ebner PD. Direct feeding of microencapsulated bacteriophages to reduce salmonella colonization in pigs. Foodborne Pathog Dis 2011;8(12):1269−74.

[88] Davis SS, Hardy JG, Fara JW. Transit of pharmaceutical dosage forms through the small intestine. Gut 1986;27(8):886−92.

[89] Davis SS, Illum L, Hinchcliffe M. Gastrointestinal transit of dosage forms in the pig. J Pharm Pharmacol 2001;53(1):33−9.

[90] Uriot O, Galia W, Awussi AA, Perrin C, Denis S, Chalancon S, et al. Use of the dynamic gastro-intestinal model TIM to explore the survival of the yogurt bacterium *Streptococcus thermophilus* and the metabolic activities induced in the simulated human gut. Food Microbiol 2016;53:18−29.

[91] Philip AK, Philip B. Colon targeted drug delivery systems: a review on primary and novel approaches. Oman Med J 2010;25(2):79−87.

[92] Choonara BF, Choonara YE, Kumar P, Bijukumar D, du Toit LC, Pillay V. A review of advanced oral drug delivery technologies facilitating the protection and absorption of protein and peptide molecules. Biotechnol Adv 2014;32(7):1269−82.

[93] Amidon S, Brown JE, Dave VS. Colon-targeted oral drug delivery systems: design trends and approaches. AAPS PharmSciTech 2015;16(4):731−41.

[94] Van Citters GW, Lin HC. Ileal brake: neuropeptidergic control of intestinal transit. Curr Gastroenterol Rep 2006;8(5):367−73.

[95] Hua S, Marks E, Schneider JJ, Keely S. Advances in oral nano-delivery systems for colon targeted drug delivery in inflammatory bowel disease: selective targeting to diseased versus healthy tissue. Nanomedicine 2015;11(5):1117−32.

[96] Malik DJ, Sokolov IJ, Vinner GK, Mancuso F, Cinquerrui S, Vladisavljevic GT, et al. Formulation, stabilisation and encapsulation of bacteriophage for phage therapy. Adv Colloid Interface Sci May 2017:1−34.

[97] Vladisavljevic GT, Williams RA. Recent developments in manufacturing emulsions and particulate products using membranes. Adv Colloid Interface Sci 2005;113(1):1−20.

[98] Kukizaki M. Shirasu porous glass (SPG) membrane emulsification in the absence of shear flow at the membrane surface: influence of surfactant type and concentration, viscosities of dispersed and continuous phases, and transmembrane pressure. J Memb Sci 2009;327(1−2):234−43.

[99] Vladisavljević GT, Kobayashi I, Nakajima M. Production of uniform droplets using membrane, microchannel and microfluidic emulsification devices. Microfluid Nanofluidics February 18, 2012;13(1):151−78.

[100] Kosvintsev SR, Gasparini G, Holdich RG, Cumming IW, Stillwell MT. Liquid-liquid membrane dispersion in a stirred cell with and without controlled shear. Ind Eng Chem Res 2005;44:9323−30.

[101] Vladisavljević GT, Williams RA. Manufacture of large uniform droplets using rotating membrane emulsification. J Colloid Interface Sci 2006;299(1):396−402.

[102] Holdich RG, Dragosavac MM, Vladisavljević GT, Kosvintsev SR. Membrane emulsification with oscillating and stationary membranes. Ind Eng Chem Res 2010;49(8):3810−7.

[103] Silva PS, Dragosavac MM, Vladisavljevic GT, Bandulasena HCH, Holdich RG, Stillwell M, et al. Azimuthally oscillating membrane emulsification for controlled droplet production. AIChE J 2015;61(11):3607−15.

[104] Vinner GK, Vladisavljevi GT, Clokie MRJ, Malik DJ. Microencapsulation of *Clostridium difficile* specific bacteriophages using microfluidic glass capillary devices for colon delivery using pH triggered release. PLoS One 2017;12(10):1−27.

[105] Boggione DMG, Batalha LS, Gontijo MTP, Lopez MES, Teixeira AVNC, Santos IJB, et al. Evaluation of microencapsulation of the UFV-AREG1 bacteriophage in alginate-Ca microcapsules using microfluidic devices. Colloids Surf B Biointerfaces 2017;158:182−9.

[106] Tang Z, Huang X, Baxi S, Chambers JR, Sabour PM, Wang Q. Whey protein improves survival and release characteristics of bacteriophage Felix O1 encapsulated in alginate microspheres. Food Res Int July 2013;52(2):460−6.

[107] Jain D, Panda AK, Majumdar DK. Eudragit S100 entrapped insulin microspheres for oral delivery. AAPS PharmSciTech January 2005;6(1):E100−7.

[108] Surh J, Mun S, McClements DJ. Preparation and characterization of water/oil and water/oil/water emulsions containing biopolymer-gelled water droplets. J Agric Food Chem 2007;55(1):175−84.

[109] Sağlam D, Venema P, de Vries R, Sagis LMC, van der Linden E. Preparation of high protein micro-particles using two-step emulsification. Food Hydrocoll July 2011;25(5):1139−48.

[110] Kong T, Wu J, To M, Wai Kwok YK, Cheung SH, Wang L. Droplet based microfluidic fabrication of designer microparticles for encapsulation applications. Biomicrofluidics 2012;6(3). https://doi.org/10.1063/1.4738586.

FURTHER READING

[1] Cox CS, Harris WJ, Lee J. Viability and electron microscope studies of phages T3 and T7 subjected to freeze-drying, freeze-thawing and aerosolization. J Gen Microbiol 1974;81(1):207−15.

[2] Shi LE, Zheng W, Zhang Y, Tang ZX. Milk-alginate microspheres: protection and delivery of *Enterococcus faecalis* HZNU P2. LWT Food Sci Technol (Lebensmittel-Wissenschaft-Technol) 2016;65:840−4.

[3] Watanabe R, Matsumoto T, Sano G, Ishii Y, Tateda K, Sumiyama Y, et al. Efficacy of bacteriophage therapy against gut-derived sepsis caused by *Pseudomonas aeruginosa* in mice. Antimicrob Agents Chemother 2007;51(2):446−52.

Chapter 20

The Unknown Effect of Antibiotic-Induced Dysbiosis on the Gut Microbiota

Aleksandr Birg, Nathaniel L. Ritz and Henry C. Lin[a]
Medicine Service, New Mexico VA Health Care System and the Division of Gastroenterology and Hepatology, University of New Mexico,
Albuquerque, NM, United States

INTRODUCTION

In recent years, evaluation of changes in the gut microbial community after antibiotic exposure has gained traction as an important health topic. While antibiotic use has proven to be extremely beneficial in combatting infectious disease, the classic vignette has been that microbes are costly and deleterious to life and should be eliminated at every opportunity. This stance has been reversed in recent years, with our growing knowledge of microbial communities and their role in promoting host health and homeostasis. Therefore, the paradoxical widespread and frequent overuse of antibiotics, in hospitals, agriculture, and the food industry, can be counterproductive to the health of the microbiota and subsequently to the health of the host. Increasing focus has been given, to evaluate how the intestinal microbiome is impacted by antibiotic use, both acutely and long term. Following the use of antibiotics, the beneficial properties of the gut microbiota can be lost and can promote the evolution of antibiotic-resistant pathogens.

The intestinal microbiota is a densely populated microbial ecosystem, comprised of trillions of microbes belonging to hundreds of species [1]. The composition of the gut microbiota is altered by numerous environmental factors, including the general health of the host, diet, lifestyle, and antibiotic use. Gastrointestinal microbial colonization increases both from proximal to distal gut (stomach 10^1 microbial cells/g contents, small intestine 10^{3-7} cells/g, and colon 10^{12} cells/g), and within the intestinal lumen, from lateral (regions adherent and adjacent to host tissues) to medial [2].

In healthy individuals, there is great microbial richness and diversity, both within individuals (alpha diversity) and between individuals (beta diversity). Despite this high diversity at the species level, the microbiota consists of only a few bacterial phyla, with greater than 90% belonging to Bacteroidetes, Firmicutes, Actinobacteria, and Proteobacteria [3,4]. Verrucomicrobia, Fusobacteria, and Cyanobacteria combine for most of the remaining bacteria, in low numbers.

Bacteroidetes are gram-negative bacteria that ferment polysaccharides and otherwise indigestible carbohydrates and produce short-chain fatty acids (SCFAs) that have many beneficial effects in the gut. Firmicutes are primarily composed of gram-positive bacteria that range from mutualistic symbionts (e.g., *Lactobacillus* spp., *Clostridium* clusters IV-XIVa) to pathobionts (e.g., *Clostridium difficile*, *Clostridium perfringens*). Actinobacteria are gram-positive bacteria that are usually beneficial (e.g., *Bifidobacterium* spp.) and commonly used as probiotics. Proteobacteria are a diverse group of gram-negative bacteria that tend to increase in number during dysbioses (e.g., Enterobacteriaceae family) [5].

While bacteria make up the vast majority of microbial biomass in the gut and are the most commonly studied, it is important to note that there are also eukaryotes (mostly yeasts), viruses (mostly bacteriophage), and archaea (mostly methanogens) present with distinct roles within the microbial community. Following perturbation, the proportion and representation of these groups can be thrown into a state of imbalance, or dysbiosis, that can cause a loss of functionality and lead to disease.

---

a. Senior author.

Microbiome and Metabolome in Diagnosis, Therapy, and other Strategic Applications. https://doi.org/10.1016/B978-0-12-815249-2.00020-8

BACTERIAL BIOFILMS

There has been a growing appreciation for the complexity of the interrelationships, among the large consortia of microbes of the gut. A prominent example of this is seen when microbes form biofilms by adhering to surfaces, aggregate, and become embedded in extracellular matrices. These systems provide defense against antimicrobials and allow direct access for microbial nutrient transfer. Biofilms have been shown to be beneficial for the host, by aiding the digestion of luminal contents and creating a barrier to prevent the colonization of pathogens [6]. In fact, colonization resistance is one of the most significant defensive mechanisms a healthy, intact microbiota provides. However, biofilms can also behave pathologically, with estimates that 65% of infections were associated with biofilm production [7].

ANTIBIOTIC ADMINISTRATION

Antibiotics can alter the microbiota, by selecting for antibiotic-resistant species. Resistance genes spread through vertical (i.e., resistant mutation passed down from parent to offspring) or horizontal (i.e., transformation or conjugation) transmission. Following antibiotic use, resistant organisms and mobile elements can remain within the population, for an extended period of time. Studies have reported that resistant species can persist 3−4 years, after patients were originally treated with antibiotics [4,5].

The antibiotic spectrum, dose and duration, pharmacokinetic and pharmacodynamic properties, and route of administration, all contribute to how the microbiota is affected [8]. Different classes of antibiotics will not only affect different microbes but also have different host absorption and clearance properties, which impact the distribution and excretion of the compounds in relation to the microbiota, where the site of greatest impact is within the distal gut.

ANTIBIOTIC SPECTRUM

Oral administration of broad-spectrum and poor absorption antibiotics have a greater impact on the gut microbiota, compared to narrow-spectrum and high absorption antibiotics [9]. The spectrum of activity for an antibiotic can play an important role in pathobiont (any organism with the potential to become pathological) overgrowth and/or acute pathogen infection. This scenario is exemplified following a narrow-spectrum antibiotic treatment, which primarily targets anaerobes, resulting in a vastly increased risk of overgrowth by facultative anaerobes *Enterococcus* and *Streptococcus*; this event is not seen following the use of broad-spectrum antibiotics [10].

BACTERIAL DIVERSITY

Multiple studies testing the effects on antibiotics have shown that while total microbial numbers may return to normal after a 1−2 week period, the overall diversity significantly decreases, and certain bacterial taxa may be permanently lost. Additionally, colonization resistance to common nosocomial pathogens was lost, until microbiota diversity was restored [11−14]. The burgeoning replacement bacterial populations within the microbiota varied across studies and depended on the initial colonization of microbes still present after cessation of the antibiotic treatment.

PROTEOBACTERIA OVERGROWTH

Worrisomely, bacteria of the phylum Proteobacteria frequently expanded following antibiotic treatments. One explanation for the Proteobacteria bloom in dysbioses is that both intestinal inflammation and antibiotic treatment increase epithelial oxygenation in the colon, creating a more aerobic atmosphere, favoring the facultative over obligate anaerobes [15]. These include major pathogens, some of them multiresistant to antibiotics, such as *Salmonella*, *Escherichia coli*, *Yersinia*, *Pseudomonas*, *Klebsiella*, *Shigella*, *Proteus*, *Enterobacter*, and *Serratia*.

ANTIBIOTIC-INDEPENDENT LOSS OF DIVERSITY

The loss of microbiome diversity has been reported in a multitude of diseases, impacting the gastrointestinal system, and parallels changes seen with antibiotics. Inflammatory bowel disease has been studied extensively, showing a loss of diversity in both ulcerative colitis and Crohn's disease patients [16]. Variable results in regard to alpha diversity, in inflammatory bowel disease, have been reported in the literature, with some studies showing significant change in phyla, particularly Proteobacteria and Bacteroidetes, while other studies show no notable change [16].

Irritable bowel syndrome also shows a loss of diversity in the diarrhea subtype, via sequencing of the stool [17]. Reduced diversity of the microbiome was also noted in *Clostridium difficile*—infected patients, with symptoms of acute diarrhea [18]. Interestingly, this trend in loss of microbial diversity has been reported in diseases outside the gut lumen. Patients evaluated with chronic liver disease were also shown to have a loss of gut bacterial diversity [19].

BACTERIAL CROSS TALK

The changes seen in bacterial diversity during a disease process emphasize the communication and interplay between the host and gut microbiota. With similar bacterial community changes seen during antibiotic use, any administration of antibacterial therapy needs to be carefully evaluated, for potential impact on the host [20]. It is also important to emphasize that antibiotics have been used extensively to treat a multitude of gastrointestinal infections and diseases, through the years.

ANTIBIOTIC THERAPY OF INTESTINAL DYSBIOSIS

More recently, antibiotics have been applied to treat dysbiosis [21]. The rationale here is to treat overgrowth of bacteria in the microbiome, as one would treat overgrowth of bacteria in the setting of an infection. While there is some comparability between these two types of overgrowth, the desired outcomes are drastically different.

ANTIBIOTICS IN URINARY INFECTION

Urinary tract infections are a common disease process that is often treated with antibiotics for abnormal changes in urinary microbiome that can cause symptoms to the host [22]. With recent studies evaluating urine via genetic sequencing, it has been determined that the bladder lumen is associated with a specific microbiome (as opposed to being a sterile environment). Changes to that microbiome can lead to an overgrowth of a certain species that can cause symptoms of a urinary infection [22].

THE URINARY MICROBIOME

The bacterial density of a healthy urine microbiome is magnitudes lower than that of the gut, usually ranging from 10^2 to 10^5 colony-forming units per milliliter, compared to 10^{12} colony-forming units per milliliter in intestines [22]. Urine cultures from urinary tract infections usually grow only one species, most commonly *E. coli* that is responsible for symptoms. These can typically be treated with narrow-coverage antibiotics [22]. Symptom improvement is seen after treatment with antibiotics, which curb the growth of the uropathogen.

URINARY VERSUS GASTROINTESTINAL DYSBIOSIS

The gut microbiota is a more complex system, with a significantly higher number of species present, compared to the urinary tract system; therefore changes that occur will impact host physiology differently [22]. When dealing with the gastrointestinal system, symptomatic dysbiosis does not occur due to alterations in a single species or a specific collection of species [21]. Treatment with narrow-range antibiotics will not yield the same results of symptom resolution, in patients with intestinal dysbiosis, as it would in patients with urinary tract infections.

Broad-spectrum antibiotics have been shown to be successful, in diminishing gastrointestinal symptoms in the setting of dysbiosis [23]. Many recent efforts have gone into the reasoning, behind the beneficial effects of the broad-spectrum antimicrobials. Outside of the usual bactericidal and bacteriostatic activity of antibiotics, additional properties must be involved in the exact effects on the altered microbiome, in order to facilitate restoration of the normal function [21].

NON-ANTIBIOTIC MANAGEMENT OF DYSBIOSIS

There has been a revitalization of efforts, delving into alternative therapies to improve symptoms of dysbiosis. As antibiotics have been shown to cause loss of gut bacterial diversity that can lead to further deleterious changes, advantageous properties of probiotics and prebiotics have been acknowledged as useful therapies.

PROBIOTICS

Probiotics are live organisms that provide benefits to the host after consumption [24]. Probiotics have been used to treat liver injury and hepatic inflammation, in the event of intestinal bacterial overgrowth, by maintaining permeability of tight junctions and preventing pathogen translocation [25]. In addition, JAMA meta-analysis has shown that probiotics can reduce antibiotic-associated diarrhea by up to 40%, while a review by Cochrane has shown that probiotic use in *Clostridium difficile* colitis will reduce diarrhea by up to 64% [26]. Single-strain probiotics have been shown to increase bacterial gut diversity as well [19].

PREBIOTICS

Prebiotics are nondigestible food ingredients that propagate beneficial microbial activity of the microbiome [24]. Prebiotics are usually oligosaccharides that provide nutrients selectively to intestinal symbionts [27]. Prebiotics, in themselves, have to work through probiotic strains, to provide benefits to the host and alter probiotic bacteria to control pathogens [25]. However, it is important to note that as of this time, probiotics are not regularly recommended for symptom control, except for known cases of antibiotic-induced diarrhea [28].

FECAL TRANSPLANTS

Fecal microbiota transplants (FMT) are another method developed to help control altered microbial changes [29]. The exact mechanism as to how FMT works is not well understood and needs further research [30]; however, FMT have shown to be very effective, for treating altered microbiota in the setting of *C. difficile* colitis [29]. Despite probiotics, prebiotics, and FMT showing great potential benefit to treat antibiotic alteration in the microbiome, there are still significant unknowns involved, which do not translate to regular use in medicine [25]. It has been reported that single-strain probiotics have very limited success in treating intestinal dysbiosis [31]. The lack of knowledge into long-term effects keeps these alternative therapies from being considered viable treatment options, for severe conditions that occur with antibiotic use (i.e., *C. difficile* colitis) [32].

ECOSYSTEM EVALUATION

Future research into changes seen in the gut microbiota will focus less on individual species that can be observed through culture and will need to emphasize the role of larger community variations. Culturing the gut microbiota is no longer considered a viable option, to truly understand those changes, as only 80% are culturable, and many genera are too difficult to grow under standard conditions [33,34]. Important and useful concepts, such as alpha and beta diversity, cannot be evaluated with the use of culturing and lead to lack of complete understanding of the changes that take place [35].

MULTIOMICS INVESTIGATION

When considering the complex and evolving changes, occurring in an intestinal environment after antibiotic exposure, interactions between the host and microbiota require more involved strategies [25]. Several "omic" technologies have emerged to the forefront, to study how the microbial community and microbial metabolites are affected, including changes by antibiotic disruption [25].

DIAGNOSTIC OPTIONS

As more of these techniques are being incorporated into everyday practice, the cost becomes more favorable. This allows for easier access and future integration into personalized medicine [25]. The current "gold standard" for studying microbial communities is 16S rRNA next-generation sequencing (also known as high-throughput sequencing) that allows for genetic definition of pathogens, rather than relying on their morphology [34]. In addition to genetic sequencing to analyze gut microbiome, several other techniques have emerged to study functional changes such as metagenomics and metatranscriptomics [34]. Another mechanism to study disruption in microbiome is evaluating metabolites in a technique called metabolomics, which is used to detect microbial products in urine, blood, saliva, and feces [25]. As we learn more about metagenomics and metabolomics and understand the role of microbes in human health and disease, future therapies can be based around the microbiome as well [27].

ONGOING STUDIES

Evaluating human data regarding alterations of the microbiome by antibiotics is difficult, lacking in human trials [25]. The major question in the coming years will be to determine the exact changes in the gut microbiota that will cause a disruption in function, and what specific parameters will be responsible for diseases and symptoms.

ACKNOWLEDGMENTS

This study is supported by the Winkler Bacterial Overgrowth Research Fund.

DISCLAIMER

Dr. Lin has patent rights in related area.

REFERENCES

[1] Rajilić-Stojanović M, de Vos WM. The first 1000 cultured species of the human gastrointestinal microbiota. FEMS Microbiol Rev 2014;38:996−1047.

[2] Sekirov I, Russell SL, Antunes LCM, Finlay BB. Gut microbiota in health and disease. Physiol Rev 2010;90:859−904.

[3] Jandhyala SM, Talukdar R, Subramanyam C, Vuyyuru H, Sasikala M, Nageshwar Reddy D. Role of the normal gut microbiota. World J Gastroenterol 2015;21:8787−803.

[4] Kim S, Covington A, Pamer EG. The intestinal microbiota: antibiotics, colonization resistance, and enteric pathogens. Immunol Rev 2017;279:90−105.

[5] Looft T, Johnson TA, Allen HK, et al. In-feed antibiotic effects on the swine intestinal microbiome. Proc Natl Acad Sci U S A 2012;109:1691−6.

[6] Bollinger RR, Barbas AS, Bush EL, Lin SS, Parker W, Parker W. Biofilms in the normal human large bowel: fact rather than fiction. Gut 2007;56:1481−2.

[7] Potera C. Forging a link between biofilms and disease. Science 1999;283(1837):1839.

[8] Jernberg C, Lofmark S, Edlund C, Jansson JK. Long-term impacts of antibiotic exposure on the human intestinal microbiota. Microbiology 2010;156:3216−23.

[9] Levison ME, Levison JH. Pharmacokinetics and pharmacodynamics of antibacterial agents. Infect Dis Clin North Am 2009;23:791−815.

[10] Taur Y, Xavier JB, Lipuma L, et al. Intestinal domination and the risk of bacteremia in patients undergoing allogeneic hematopoietic stem cell transplantation. Clin Infect Dis 2012;55:905−14.

[11] Lewis BB, Buffie CG, Carter RA, et al. Loss of microbiota-mediated colonization resistance to *Clostridium difficile* infection with oral vancomycin compared with metronidazole. J Infect Dis 2015;212:1656−65.

[12] Isaac S, Scher JU, Djukovic A, et al. Short- and long-term effects of oral vancomycin on the human intestinal microbiota. J Antimicrob Chemother 2017;72:128−36.

[13] Buffie CG, Jarchum I, Equinda M, et al. Profound alterations of intestinal microbiota following a single dose of clindamycin results in sustained susceptibility to *Clostridium difficile*-Induced colitis. Infect Immun 2012;80:62−73.

[14] Ubeda C, Taur Y, Jenq RR, et al. Vancomycin-resistant Enterococcus domination of intestinal microbiota is enabled by antibiotic treatment in mice and precedes bloodstream invasion in humans. J Clin Invest 2010;120:4332−41.

[15] Litvak Y, Byndloss MX, Tsolis RM, Bäumler AJ. Dysbiotic Proteobacteria expansion: a microbial signature of epithelial dysfunction. Curr Opin Microbiol 2017;39:1−6.

[16] Hansen R, Russell RK, Reiff C, et al. Microbiota of de-novo pediatric IBD: increased *Faecalibacterium prausnitzii* and reduced bacterial diversity in Crohn's but not in ulcerative colitis. Am J Gastroenterol 2012;107:1913−22.

[17] Rajilić-Stojanović M, Jonkers DM, Salonen A, et al. Intestinal microbiota and diet in IBS: causes, consequences, or epiphenomena? Am J Gastroenterol 2015;110:278−87.

[18] Chang JY, Antonopoulos DA, Kalra A, et al. Decreased diversity of the fecal microbiome in recurrent *Clostridium difficile*−associated diarrhea. J Infect Dis 2008;197:435−8.

[19] Chen Y, Yang F, Lu H, et al. Characterization of fecal microbial communities in patients with liver cirrhosis. Hepatology 2011;54:562−72.

[20] Dethlefsen L, Huse S, Sogin ML, Relman DA. The pervasive effects of an antibiotic on the human gut microbiota, as revealed by deep 16S rRNA sequencing. PLoS Biol 2008;6:e280.

[21] Walker AW, Lawley TD. Therapeutic modulation of intestinal dysbiosis. Pharmacol Res 2013;69:75−86.

[22] Brubaker L, Wolfe AJ. The female urinary microbiota, urinary health and common urinary disorders. Ann Transl Med 2017;5. https://doi.org/10.21037/atm.2016.11.62.

[23] Vemuri RC, Gundamaraju R, Shinde T, Eri R. Therapeutic interventions for gut dysbiosis and related disorders in the elderly: antibiotics, probiotics or faecal microbiota transplantation? Benef Microbes 2017;8:179−92.

[24] Mayer EA, Savidge T, Shulman RJ. Brain−gut microbiome interactions and functional bowel disorders. Gastroenterology 2014;146:1500−12.

[25] Khalsa J, Duffy LC, Riscuta G, Starke-Reed P, Hubbard VS. Omics for understanding the gut-liver-microbiome axis and precision medicine. Clin Pharmacol Drug Dev 2017;6:176–85.

[26] Bo L, Li J, Tao T, et al. Probiotics for preventing ventilator-associated pneumonia. In: Deng X, editor. Cochrane database of systematic reviews. Chichester, UK: John Wiley & Sons, Ltd.; 2014. CD009066.

[27] Candela M, Maccaferri S, Turroni S, Carnevali P, Brigidi P. Functional intestinal microbiome, new frontiers in prebiotic design. Int J Food Microbiol 2010;140:93–101.

[28] Acree M, Davis AM. Acute diarrheal infections in adults. JAMA 2017;318. https://doi.org/10.1053/j.gastro.2016.12.039.

[29] Wischmeyer PE, Mcdonald D, Knight R. Role of the microbiome, probiotics, and 'dysbiosis therapy' in critical illness. Curr Opin Crit Care 2016;22(4):347–53. https://doi.org/10.1097/MCC.0000000000000321.

[30] Shetty SA, Hugenholtz F, Lahti L, Smidt H, De Vos WM. Intestinal microbiome landscaping: insight in community assemblage and implications for microbial modulation strategies. FEMS Microbiol Rev 2017;45:182–99.

[31] Difficile C. Role of the intestinal microbiota in resistance to colonization by *Clostridium difficile*. Gastroenterology 2014;146:1547–53.

[32] Cox LM, Blaser MJ. Antibiotics in early life and obesity. Nat Rev Endocrinol 2014;11:182–90.

[33] De La Cochetière MF, Durand T, Lepage P, Bourreille A, Galmiche JP, Doré J. Resilience of the dominant human fecal microbiota upon short-course antibiotic challenge. J Clin Microbiol 2005;43:5588–92.

[34] Keeney KM, Yurist-Doutsch S, Arrieta M-C, Finlay BB. Effects of antibiotics on human microbiota and subsequent disease. Annu Rev Microbiol 2014;68. https://doi.org/10.1146/annurev-micro-091313-103456.

[35] Bokulich NA, Chung J, Battaglia T, et al. Antibiotics, birth mode, and diet shape microbiome maturation during early life. Sci Transl Med 2016;8. http://stm.sciencemag.org.libproxy.unm.edu/content/scitransmed/8/343/343ra82.full.pdf.

Chapter 21

The Myth and Therapeutic Potentials of Postbiotics

Hooi Ling Foo[1,2], Teck Chwen Loh[3,4], Nur Elina Abdul Mutalib[1] and Raha Abdul Rahim[2,5]

[1]Department of Bioprocess Technology, Faculty of Biotechnology and Biomolecular Sciences, Universiti Putra Malaysia, 43400 UPM Serdang, Selangor Darul Ehsan, Malaysia; [2]Institute of Bioscience, Universiti Putra Malaysia, 43400 UPM Serdang, Selangor Darul Ehsan, Malaysia; [3]Department of Animal Science, Faculty of Agriculture, Universiti Putra Malaysia, 43400 UPM Serdang, Selangor Darul Ehsan, Malaysia; [4]Institute of Tropical Agriculture and Food Security, Universiti Putra Malaysia, 43400 UPM Serdang, Selangor Darul Ehsan, Malaysia; [5]Department of Cell and Molecular Biology, Faculty of Biotechnology and Biomolecular Sciences, Universiti Putra Malaysia, 43400 UPM Serdang, Selangor Darul Ehsan, Malaysia

INTRODUCTION

The gut microbiota prevents the conversion of procarcinogenic substances, induces the production of essential vitamins such as vitamin B and K2, and promotes degradation of bile acids and digestion of nutrients [1]. A healthy gut determines a functional and developed gut immune system, thus contributing to the overall health of the host. Commensal bacteria act synergistically with the intestinal mucosa to form a barrier against pathogens [2].

Probiotics in the gastrointestinal tract (GIT) perform competitive exclusion with pathogenic microorganisms for the site of adhesion at intestinal wall, thus promoting intestinal epithelial cells survival, barrier function, and protective immune responses. They also neutralize enterotoxins and lipopolysaccharides (LPS) produced by pathogens, and some possess bactericidal and bacteriostatic activity [3,4].

Since the ban of antibiotics as growth promoters in livestock in 2006, probiotics and other natural feed additives are of utmost interest to the industry to provide the same beneficial effect as antibiotics [5].

Although they come with many advantages, probiotics also confer antagonistic effects to the host. Some disseminate antibiotic resistance encoded by the genes of the genomes or plasmids, which is transferable between organisms [6]. Furthermore, probiotic activity is dependent on its metabolic activity [7]. Therefore, any probiotics fed or ingested must be viable in sufficient amount when they reach the small and large intestines [8].

In clinical applications of probiotics, it is very difficult to determine the bioavailability of bacteria. The biosafety concern is also prevalent, regardless of the nonpathogenic and safe status given to the probiotics [9]. In addition, the action of probiotics on intestinal mucosa can require extensive biopsy sampling from the patients, provided with given acceptability and consent during clinical trials [10]. Administration of live probiotics to patients with severe acute inflammation is discouraged because of the increased risks of mortality [8,11].

Alternatives Without Live Bacteria

Soluble factors, metabolites, bacteriocins, or bacteria-free solutions derived from probiotics have been reported to give similar beneficial effects to live cell counterparts, minus the hassle and biosafety concerns. Takano [12] named these metabolites as biogenics, food components derived from microbial activity, which provide health benefits, without involving intestinal microflora. According to Tsilingiri and Rescigno [7,13], postbiotics are any factors resulting from the metabolic activity of a probiotic, or any released molecules capable of conferring beneficial effects to the host, in a direct or indirect way.

Recently, proteobiotics are referred specifically to the bioactive molecules derived from probiotics, which downregulate genes involved in adhesion to and invasion of epithelial cells through interference with cell-to-cell communication [14].

Microbiome and Metabolome in Diagnosis, Therapy, and other Strategic Applications. https://doi.org/10.1016/B978-0-12-815249-2.00021-X

KEY POSTBIOTIC PRODUCER CELLS

Probiotics mostly belong to lactic acid bacteria (LAB), with few exceptions such as yeasts [15] or Gram-negative bacteria such as *Escherichia coli* Nissle 1917 [16]. LAB are bacteria with the ability to metabolize carbohydrate (lactose and other sugars) to lactic acid via lactic acid fermentation. They are mostly facultative or obligatory anaerobes, which thrive well in anaerobic conditions. Industrially important genera include *Streptococcus*, *Lactobacillus*, *Lactococcus*, *Leuconostoc*, *Bifidobacteria*, and *Pediococcus*.

Lactobacillus is one of the main genera of LAB and plays an important role in balancing the gut microflora of animals. These bacilli can be either homo- or heterofermentative bacteria and are widely utilized in dairy and vegetable fermentation. In the human gut microflora, lactobacilli constitute a small niche in the ecosystem. *Lactobacillus acidophilus*, *Lactobacillus casei*, and *Lactobacillus plantarum*, for example, are allochthonous and persist in GIT via continuous administration or consumption [17].

Lactobacilli are resistant to harsh actions of bile salts and gastric acids. As probiotics, when administered to the hosts, lactobacilli compete with pathogens for binding to receptors, nutrients, and colonization of GIT [18] and to promote intestinal epithelial cell growth, survival, and barrier function [19,20].

Moghadam et al. [21] reported that six strains of *L. plantarum* (UL4, TL1, RS5, RI11, RG14, and RG11), isolated from Malaysian foods [22,23], harbored two classes of bacteriocin structural genes, namely plantaricin W and plantaricin EF. This indicates that these lactobacilli strain isolates are novel bacteriocin producers, exhibiting broad inhibitory activity [24,25]. According to Tai et al. [26], the simultaneous presence of two classes of bacteriocin structural genes within a probiotic cell has not been reported elsewhere.

Despite these six lactobacilli strains being highly similar at genomic level, random amplified polymorphic DNA analysis and short intergenic spacer region at 16S–23S region analysis showed distinct strain discrimination.

plw and *plnEF* loci belong to two different classes of bacteriocin molecules. Plantaricin W and plantaricin EF belong to class I and class II bacteriocins, respectively. Interestingly, UL4-*plnEF* locus is a composite from lactobacilli strain JDM1-*plnEF* locus and J51-*plnEF* locus in terms of genetic composition and organization [26].

POSTBIOTIC PRODUCTION

In most studies, the production of postbiotics was performed by separating the extracellular bioactive metabolites from producer cells by centrifugation and ultrafiltration techniques [24,27,28]. However, the postbiotics harvested may differ in terms of the concentrations of proteins or peptides, organic acids, and bacteriocins, depending on the media composition [28–31].

In our study, the postbiotic production was performed according to Foo et al. [31]. The stock culture of the selected probiotic strain was revived twice in the Man Rogosa Sharpe (MRS) broth and incubated for 48 and 24 h for each reviving step as a standing culture at 30°C. All the probiotic strains were grown anaerobically at 30°C unless stated otherwise.

The overnight culture was then streaked onto MRS agar and incubated again until a single colony formed on the agar plate which took up to 48 h. A single colony was picked, inoculated in MRS broth, and incubated overnight. An inoculum size of 1%–2% (v/v) from the overnight culture was again transferred and incubated overnight.

Metabolites were subsequently collected by centrifugation at $12,000 \times g$ for 10 min at 4°C to separate the postbiotics from the producer cells. The postbiotics were then filtered [30] prior to storage at 4°C until further use. Depending on the subsequent application, postbiotics can also be prepared in dry form via spray drying procedure [25].

Postbiotics can also be produced in conditioned media [28], incubated in eukaryotic cell culture medium for few hours. Probiotics isolated from food-based products may have different nutrient requirements to produce postbiotics. Leroy and Devust reported that probiotic *Lactobacillus sakei* CTC 494, isolated from fermented sausages, thrive well in meat-based peptone; thus the strains are well-adjusted in meat-like environment [32].

Carbon and nitrogen sources of the media composition of probiotics cultivation exhibited profound effects on the bacteriocin inhibitory activity of the postbiotic metabolites produced by *L. plantarum* I-UL4 [29]. Maximum bacteriocin inhibitory activity was achieved, when glucose and yeast extract were added to the modified media as the sole carbon and nitrogen sources, respectively. The addition of meat extract and peptone decreases the bacteriocin inhibitory activity of the UL4 postbiotic metabolites.

The addition of polysorbate 80 (Tween 80) has also been reported to affect the growth and production of bacteriocin by LAB [33,34]. Tween 80 incorporates oleic acid to the cell membrane and hence increases its fluidity to protect the bacteria from harsh environment conditions [35], enhancing nutrient uptake [36]. Similar findings were reported by Tan et al. [30], whereby the addition of Tween 80 to the reconstituted MRS media would greatly affect the proteinaceous postbiotic metabolites (PPM) production by *L. plantarum* I-UL4 [30].

In cytotoxicity assay against breast cancer cell lines, MCF-7 cells treated with the PPM by *L. plantarum* I-UL4, cultivated in reconstituted MRS with Tween 80, exhibited higher percentage of apoptosis [30].

CHARACTERISTICS AND FUNCTIONALITY OF POSTBIOTICS

Bacteriocin is a ribosomally synthesized small, heat-stable, cationic, amphiphilic, antimicrobial peptide ($\sim$5–20 kDa), produced by a wide variety of bacterial species with cognate immune system for the host self-protection [37].

Most bacteriocins have narrow spectrum of activity, thereby only inhibit the growth of closely related species to the producer strain. However, there are reports of bacteriocins with broad inhibition spectrum activity against both Gram-positive and Gram-negative bacteria, such as nisin [38].

A single LAB strain or metabolite from one LAB isolate showed inhibitory effect against pathogenic microorganisms at a similar level among these *L. plantarum* isolates [23], indicating that these isolates are bacteriocinogenic LAB. Nevertheless, as a cocktail of postbiotics, the combinations of metabolites showed higher level of inhibitory action [39,40]. The combination of postbiotic metabolites possesses broad inhibitory spectrum against both Gram-positive and Gram-negative bacteria, such as *Pediococcus acidilactici*, *Escherichia coli*, *Listeria monocytogenes*, *Salmonella typhimurium*, and vancomycin-resistant *enterococci* [39,40].

The bacteriocins produced by the isolates belong to class I and class II plantaricins, a lantibiotic and nonlantibiotic bacteriocin, respectively [21]. Additionally, the bacteriocins produced from TL1, RS5, RG14, RG11, RI11, and I-UL4 strains also have broad inhibition spectrum activity against other microorganisms [39,40], with varying degree of inhibition. Hence, these strains are good candidates as antimicrobial agents against food-spoiling microorganisms in food preservation, in animal feeds to replace antibiotics, and other applications.

Postbiotics derived from lactobacilli have the flexibility to be utilized in both liquid and spray-dried powder forms, with comparable effectiveness (Fig. 21.1).

THERAPEUTIC POSSIBILITIES OF POSTBIOTICS

Malignant Diseases

Postbiotics from *L. casei* strains have been tested to prevent the development and the relapse of colorectal tumor and bladder cancer [41,42]. In addition, postbiotic metabolites produced by the six strains of *L. plantarum* have the capability to exert cytotoxic effects on an array of human cancer cell lines [43], namely breast (MCF-7), colorectal (HT 29), cervical (HeLa), liver (Hep G2), and leukemia HL-60 and K-562.

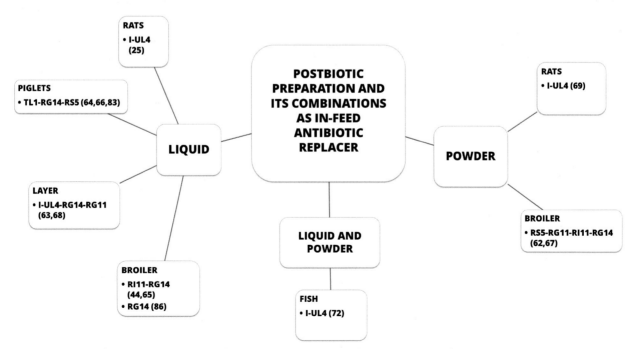

FIGURE 21.1 Applications of liquid and spray-dried postbiotic combinations as in-feed antibiotic replacer.

The postbiotic metabolites, containing bacteriocins, vitamin B, and organic acids, have the ability to substantially reduce cancer cell viability, inducing cell death via apoptotic pathway [44]. These metabolites do not exert the afore-mentioned effects on blood and normal human cell lines [30,44,45].

On MCF-7 cells, UL4 metabolites cause morphological changes such as chromatin condensation, cell shrinkage, membrane blebbing, and the formation of apoptotic bodies [43], the favorable (apoptotic) cell death mechanism that does not cause inflammatory reactions.

Anticancer properties harbored by microorganisms are one of the probiotic characteristics listed by FAO and WHO for the selection criteria of probiotic species [46]. Nisin [47] and its derivatives [48], as well as plantaricin [49—51], have been demonstrated to possess cytotoxic effect on human cancer cell lines. The cytotoxicity of bacteriocins toward cancer cells is due to cell surface variations present on the cancer cells.

Cancer cells are predominantly negatively charged because of the high level of anionic plasma membrane proteins and lipids on their membranes [52]. As a result, cationic bacteriocins or antimicrobial peptides preferentially bind to the negatively charged cells and readily permeabilize them [50] or induce apoptosis [30,49].

Additional Anticancer Molecules

The cytotoxic effect of postbiotic is not solely contributed by bacteriocins. In fact, some reports on bioactive metabolites with undetermined compositions produced by LAB have been demonstrated to possess anticancer properties against various human cancer cell lines [53,54], but yet proved to be harmless on primary human cell lines [44,55,56] (Table 21.1).

POTENTIAL OF POSTBIOTIC INTERFACES WITH ANTIBIOTICS AND OTHER DRUGS

Antibiotic Replacement

Postbiotics derived from *L. plantarum*, isolated from Malaysian fermented foods, have vast potential to fill in the gaps as one of the promising substitutes for antibiotic growth promoter in animals, as well as in general indications for antibiotics

TABLE 21.1 Cytotoxicity Studies of Postbiotics Against Human Cancer Cells
A) Cytotoxicity effects of bacteriocin

Bacteriocin	Producer (Ref)	Human Cancer Cell Lines
Nisin	*Lactococcus lactis* [47]	Head and neck squamous cell carcinoma (HNSCC)
Nisin ZP	*L. lactis* [48]	HNSCC
Plantaricin A	*Lactobacillus plantarum* C11 [49—51]	Jurkat (T cell leukemia), GH4 (pituitary cells)

B) Cytotoxicity effects of postbiotics with unspecified composition

Producer (Ref)	Cancer Cell Lines
Bifidobacterium breve [85]	LBR2 Burkitt's lymphoma
	MDA-MB-231 breast cancer
	PANC-1 pancreatic cancer
Bacillus *polyfermenticus* [28]	HT29 (colon), DLD-1 (colon) Caco-2 (colon)
Lactobacillus acidophilus LA102 *Lactobacillus casei* LC232 [55]	Caco-2, HRT-18 (rectum)
L. plantarum PCS20, PCS22, PCS25, and PCS26 [56]	Caco-2, melanoma cells
Lactobacillus helveticus R389 [53]	Murine breast cancer
Lactobacillus pentosus ITA23 *Lactobacillus acidipiscis* ITA44 [88]	MDA-MB-231 (breast) Hep G2 (liver)
Pediococcus pentosaceus 48.P3.3 *L. plantarum* 154.P7.2 *Weissella confusa* 163.P5.8 [54]	Caco-2
L. plantarum TL1, RS5, RG14, RG11, RI11, and I-UL4 [44]	MCF-7 (breast), HT 29, HeLa (cervical), Hep G2, HL-60 (leukemia), K-562 (leukemia)

(antibiotic growth promoter) [57−59]. Studies have been performed in broiler and layer chickens, pigs, and fish (Table 21.2). Postbiotics cocktails have been formulated to give maximal benefits.

Bacteriocins and organic acids inhibit the growth of *Enterobacteriaceae* and other pathogenic microorganisms, albeit in an extreme niche of GIT [37]. As a result, they synergistically modulate the composition of the gut microbiota in livestock [61−63,65,68] and influence the gut immune system [8,69].

Improvement of Gut Microenvironment

The organic acids act as acidifying agents, whereby commensal LAB thrives well in low pH environment, while the *Enterobacteriaceae* are susceptible in acidic environment. Bacteriocins too play a role in inhibiting the growth of pathogenic bacteria, thus contributing to the reduction of pathogenic bacteria populations and nullifying the effects of toxins produced by the pathogens [61]. Subsequently, the gut microflora is colonized with commensal LAB, improving the growth of the host animals.

The enhancement of volatile fatty acids (VFA) and short-chain fatty acids (SCFA) is also noted for microbial fermentation in colon, evidencing reduction of pathogens in GIT [68,70]. VFA have bacteriostatic properties, thus further decreasing the activity and number of pathogens [64].

Inhibition of Hypercholesterolemia

The increased commensal LAB population in the gut [31,60,64] inhibits the cholesterol enzyme synthesis in the host, assists in the elimination of cholesterol in faeces, and utilizes the circulating cholesterol for the synthesis of bacterial cell wall. It also enhances bile salt deconjugation and excretion, which in turn will stimulate the signal in the host to utilize cholesterol to synthesize more bile salts [71,72].

POTENTIAL APPLICATION OF POSTBIOTIC IN DISEASE PREVENTION

Intestinal Infections

Postbiotics containing bacteriocins have been reported to reduce the incidence of diarrhea in piglets, thus reducing the risk of mortality [58,65,70]. In aquaculture, disease outbreaks caused by pathogens such as *Aeromonas hydrophila* were reduced, when postbiotic UL4 was added to the fish feed [67,73−75] (Table 21.3).

TABLE 21.2 Postbiotic Metabolites by *Lactobacillus plantarum* Strains as Feed Supplements

Postbiotic (Ref)	Animal Model	Description
I-UL4 [24,25,31,60]	Rats	Reduce *Enterobacteriaceae* Reduce plasma cholesterol
Cocktails of RS5, RG14, and RG11 [61,62]	Broiler chickens	Improve overall growth performance Increase faecal LAB population Increase faecal volatile fatty acid (VFA) Increase intestinal villi height and crypt depth Reduce *Enterobacteriaceae* and pH of faeces Improve bile salt deconjugation
Cocktails of RG14, RI11, and RG11 [63,64]	Layer hens	Improve hens' day egg production Increase faecal LAB counts, VFA, and villi height Reduce *Enterobacteriaceae* and pH of the faeces Reduce cholesterol in plasma and yolk
Cocktails of TL1, RG14, and RS5 [65,66]	Piglets	Reduce diarrhea incidence Improve average daily gain and feed intake Increased in faecal LAB, SCFA, villi height, and crypt depth Reduced *Enterobacteriaceae* and pH of the faeces Improved protein digestibility
I-UL4 [67]	Fish	Increase overall growth and survival rate Reduced *Enterobacteriaceae* and *Aeromonas hydrophila* Higher LAB in faecal samples

LAB, lactic acid bacteria; *SCFA*, short-chain fatty acids.

TABLE 21.3 Postbiotics in Disease Prevention Reported in Human and Animal Models

Postbiotic (Ref)	Description
Lactocepin [76]	Serine protease from *Lactobacillus paracasei*, inactivates IP-10, lymphocyte recruiting chemokine produced by epithelial cells
p40 [19]	Produced by *Lactobacillus rhamnosus* GG (LGG), ameliorates cytokine-induced apoptosis in intestinal epithelial cells
Bacteriocin from *Lactobacillus plantarum* I-UL4 [67]	Prevents outbreaks by *Aeromonas hydrophila* in fishes, increased survival of *Tilapia*
Postbiotic combination from *L. plantarum* TL1, RG14, and RS5 [65,70]	Reduces mortality in piglets due to diarrhea and gut-associated diseases
Polyphosphates (polyP) [77]	Produced by *Lactobacillus brevis* SBC8803, enhance epithelial barrier and homeostasis through integrin–p38 MAPK pathway
S layer protein A [78]	Produced by *Lactobacillus acidophilus* NCFM, modulates immature dendritic and T cell functions
Polysaccharide A [79]	Produced by *Bacteroides fragilis*, protects animals from *Helicobacter hepaticus* colitis
Cocktail of supernatants of *Bifidobacterium breve* and *Streptococcus thermophilus* [80,81]	Antiinflammatory properties after intestinal transport enhance Th1 response and intestinal barrier function in mice
Lactobacillus casei DG soluble factors [11]	Reduce the inflammatory mucosal response in model of postinfectious irritable bowel syndrome
Secreted factors from LGG [82]	Protect colonic smooth muscle cells during LPS-induced inflammation
STp peptide from *L. plantarum* BMCM 12 [83]	Induces cytokines and skin-homing profile on stimulated T-cells (IBD), biomarker of gut homeostasis
Proteobiotics [14]	Molecules from *L. acidophilus* La-5 prevent enteric colibacillosis in pigs by enterotoxigenic *E. coli*

IBD, inflammatory bowel disease; *LPS*, lipopolysaccharides.

Inflammation and Immune Response

Postbiotic metabolites ranging from enzymes [76], peptides [83], proteins [19,78], polysaccharides [79], polyphosphates [77], bacteriocins [67], cocktails of postbiotics [65,66,80,81], and some with unspecified composition [11,14,82] have been reported to affect the modulation of gut inflammatory reactions and immune system, as summarized in Table 21.3.

THE WAY FORWARD OF POSTBIOTIC APPLICATIONS

There are other strategic and efficient ways to use the postbiotics to meet the demand and requirements in pharmaceutical, nutraceutical, and food industry, as summarized in Fig. 21.2.

Postbiotics as Alternative/Complementary/Integrative Treatment for Cancers

As reported by Ma et al. [28], the metabolites produced by the probiotic bacteria, *Bacillus polyfermenticus*, can stimulate the ErbB2 and ErbB3 protein inhibition. It has been reported that in anticancer therapy, targeting aberrant growth receptors, namely tyrosine kinase ErbB receptor family, is a relevant treatment strategy [84]. Overexpression of ErbB2 is predominantly observed in human lung, bladder, breast, and colon tumors, and ErbB3 expression is significantly correlated to the overexpression of ErbB2 in cancer tissues [84]. Therefore, postbiotics can be potentially employed to induce the killing of targeted cancer cells as an alternative to the available cytotoxic drugs, which are frequently accompanied by unfavorable side effects. It is a very feasible approach, especially for those in remission after chemotherapy [85].

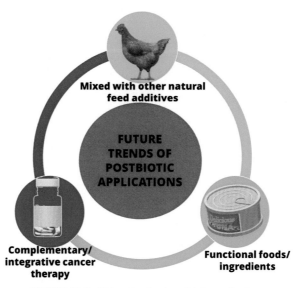

FIGURE 21.2 Future trends of postbiotic applications.

Postbiotics as Antioxidants

Some LAB with probiotic characteristics have been reported to possess antioxidant activity [86]. Reactive oxygen species (ROS) imbalance can lead to the development of cancer, cardiovascular disease, neurological disorders, and inflammation [87]. There are reports of metabolites of probiotics exerting antioxidant activities [85,88–92]. Zhang et al. [90] reported that the exopolysaccharide of *L. plantarum* C88 induces antioxidant effects such as scavenging of ROS, upregulation of enzymatic and nonenzymatic antioxidant activities, and reduction of lipid peroxidation.

LIST OF ACRONYMS AND ABBREVIATIONS

AGP Antibiotic growth promoter
BWG Body weight gain
FAO Food and Agriculture Organization
FCR Feed conversion ratio
GHR Growth hormone receptor
GIT Gastrointestinal tract
IBD Inflammatory bowel disease
IGF1 Insulin-like growth factor 1
IL Interleukin
ITS Intergenic spacer
LAB Lactic acid bacteria
LITAF LPS-induced tumor necrosis factor α factor
LPS Lipopolysaccharides
mRNA Messenger ribonucleic acid
MRS de Man Rogosa Sharpe
OAB Organic acid blend
PPM Proteinaceous postbiotic metabolites
RAPD Random amplified polymorphism DNA
ROS Reactive oxygen species
SCFA Short-chain fatty acids
VFA Volatile fatty acid
VRE Vancomycin-resistant Enterococci
WHO World Health Organization

REFERENCES

[1] Jonkers D, Stockbrügger R. Probiotics in gastrointestinal and liver diseases. Aliment Pharmacol Ther 2007;26(Suppl. 2):133—48.

[2] Collado MC, Grześkowiak SS. The role of microbiota and probiotics on the gastrointestinal health: prevention of pathogen infections. Bioactive Food as Dietary Interventions for Liver and Gastrointestinal Disease. 2013. p. 201—13.

[3] Guarner F, Schaafsma GJ. Probiotics. Int J Food Microbiol 1998;39(3):237—8.

[4] Morelli L, Capurso L. FAO/WHO guidelines on probiotics: 10 years later. J Clin Gastroenterol 2012;46:S1—2.

[5] European Commission. Ban on antibiotics as growth promoters in animal feed enters into effect. December 22, 2005. Available at: http://europa.eu/rapid/press_release_IP-05-1687_en.htm.

[6] Marteau P, Shanahan F. Basic aspects and pharmacology of probiotics: an overview of pharmacokinetics, mechanisms of action and side-effects. Bailliere's Best Pract Res Clin Gastroenterol 2003;17(5):725—40.

[7] Tsilingiri K, Rescigno M. Postbiotics: what else? Benef Microbes 2013;4(1):101—7.

[8] Tsilingiri K, Barbosa T, Penna G, Caprioli F, Sonzogni A, Viale G, Rescigno M. Probiotic and postbiotic activity in health and disease: comparison on a novel polarised *ex-vivo* organ culture model. Gut 2012;61(7):1007—15.

[9] Yan F, Polk DB. Characterization of a probiotic-derived soluble protein which reveals a mechanism of preventive and treatment effects of probiotics on intestinal inflammatory diseases Fang. Gut Microb 2012;3(1):25—8.

[10] Compare D, Rocco A, Coccoli P, Angrisani D, Sgamato C, Iovine B, Salvatore U, Nardone G. Lactobacillus casei DG and its postbiotic reduce the inflammatory mucosal response: an ex-vivo organ culture model of post-infectious irritable bowel syndrome. BMC Gastroenterol 2017;17(1):53.

[11] Besselink MG, van Santvoort HC, Buskens E, Boermeester MA, van Goor H, Timmerman HM, Nieuwenhuijs VB, Schaapherder FM, Dejong CHC, Wahab PJ, van Laarhoven CJHM, van der Harst E, van Eijck CHJ, Cuesta MA, Akkermans LMA, Gooszen HG. Probiotic prophylaxis in predicted severe acute pancreatitis: a randomised, double-blind, placebo-controlled trial. Lancet 2008;371(9613):651—9.

[12] Yamamoto N, Takano T. Antihypertensive peptides derived from milk proteins. Nahrung 1999;43(3):159—64.

[13] Gosálbez L, Ramón D. Probiotics in transition: novel strategies. Trends Biotechnol 2015;33(4):195—6.

[14] Nordeste R, Tessema A, Sharma S, Kovač Z, Wang C, Morales R, Griffiths MW. Molecules produced by probiotics prevent enteric colibacillosis in pigs. BMC Vet Res 2017;13:335.

[15] Palma ML, Zamith-Miranda D, Martins FS, Bozza FA, Nimrichter L, Montero-Lomeli M, Marques Jr ETA, Douradinha B. Probiotic *Saccharomyces cerevisiae* strains as biotherapeutic tools: is there room for improvement? Appl Biochem Biotechnol 2015;99:6563—70.

[16] Brader P, Stritzker J, Riedl CC, Zanzonico P, Cai S, Burnazi EM, Ghani ER, Hricak H, Szalay AA, Fong Y, Blasberg R. *Escherichia coli* Nissle 1917 facilitates tumor detection by positron emission tomography and optical imaging. Clin Cancer Res 2008;14(8):2295—303.

[17] Walter J. Ecological role of lactobacilli in the gastrointestinal tract: implications for fundamental and biomedical research. Appl Environ Microbiol 2008;74(16):4985—96.

[18] Candela M, Perna F, Carnevali P, Vitali B, Ciati R, Gionchetti P, Rizzello F, Campieri M, Brigidi P. Interaction of probiotic *Lactobacillus* and *Bifidobacterium* strains with human intestinal epithelial cells: adhesion properties, competition against enteropathogens and modulation of IL-8 production. Int J Food Microbiol 2008;125(3):286—92.

[19] Yan F, Cao H, Cover TL, Whitehead R, Washington MK, Polk DB. Soluble proteins produced by probiotic bacteria regulate intestinal epithelial cell survival and growth. Gasteoenterology 2007;132(2):562—75.

[20] Ewaschuk JB, Diaz H, Meddings L, Diederichs B, Dmytrash A, Backer J, Langen ML, Madsen KL. Secreted bioactive factors from Bifidobacterium infantis enhance epithelial cell barrier function. Am J Physiol Gastrointest Liver Physiol 2008;8:1025—34.

[21] Moghadam MS, Foo HL, Leow TC, Rahim RA, Loh TC. Novel bacteriocinogenic *Lactobacillus plantarum* strains and their differentiation by sequence analysis of 16S rDNA, 16S-23S and 23S-5S intergenic spacer regions and Randomly Amplified Polymorphic DNA analysis. Food Technol Biotechnol 2010;48(4):476—83.

[22] Lim YS, Foo HL, Raha AR, Loh TC, Bejo MH, Rusul GRA. The probiotic characteristics of *Lactobacillus plantarum* strains isolated from local foods. In: Proceeding of the 28th symposium of Malaysian society for microbiology, Melaka, Malaysia; 2006. p. 88—9.

[23] Foo HJ, Lim YZ, Rusul G. Isolation of bacteriocin producing lactic acid bacteria from Malaysian fermented food, Tapai. In: Proceedings of the 11th World Congress of Food Science and Technology, Seoul, Korea; 2001. p. 185. Available from: https://scholar.google.com/scholar?cluster=8454840520857765129.

[24] Foo HL, Loh TC, Lai PW, Lim YZ, Kufli CN, Rusul G. Effects of adding *Lactobacillus plantarum* I-UL4 metabolites in drinking water of rats. Pakistan J Nutr 2003;2(5):283—8.

[25] Loh TC, Lee TM, Foo HL, Law FL, Rajion MA. Growth performance and fecal microflora of rats offered metabolites from lactic acid bacteria. J Appl Anim Res 2008;34(1):61—4.

[26] Tai HF, Foo HL, Abdul Rahim R, Loh TC, Abdullah MP, Yoshinobu K. Molecular characterisation of new organisation of plnEF and plw loci of bacteriocin genes harbour concomitantly in *Lactobacillus plantarum* I-UL4. Microb Cell Fact 2015;14(1):89.

[27] Dinić M, Lukić J, Djokić J, Milenković M, Strahinić I, Golić N, Begović J. *Lactobacillus fermentum* postbiotic-induced autophagy as potential approach for treatment of acetaminophen hepatotoxicity. Front Microbiol April 2017;8:1—10.

[28] Ma EL, Choi YJ, Choi J, Pothoulakis C, Rhee SH, Im E. The anti-cancer effect of probiotic *Bacillus polyfermenticus* on human colon cancer cells is mediated through ErbB2 and ErbB3 inhibition. Int J Cancer 2015;347(6224):882—6.

[29] Ooi MF, Mazlan N, Foo HL, Loh TC, Mohamad R, Rahim RA, Ariff A. Effects of carbon and nitrogen sources on bacteriocin-inhibitory activity of postbiotic metabolites produced by *Lactobacillus plantarum* I-UL4. Malays J Microbiol 2015;11:176—84.

[30] Tan HK, Foo HL, Loh TC, Banu N, Alitheen M, Abdul Rahim R. Cytotoxic effect of proteinaceous postbiotic metabolites produced by *Lactobacillus plantarum* I-UL4 cultivated in different media composition on MCF-7 breast cancer cell. Malays J Microbiol 2015;11:207−14.

[31] Foo HL, Loh TC, Law FL, Lim YZ, Kufli CN, Rusul G. Effects of feeding *Lactobacillus plantarum* I-UL4 isolated from Malaysian tempeh on growth performance, faecal flora and lactic acid bacteria and plasma cholesterol concentrations in postweaning rats. Food Sci Biotechnol 2003;12(4).

[32] De Vuyst L, Leroy F. Bacteriocins from lactic acid bacteria: production, purification, and food applications. J Mol Microbiol Biotechnol 2007;13(4):194−9.

[33] Saraniya A, Jeevaratnam K. Optimization of nutritional and non-nutritional factors involved for production of antimicrobial compounds from *Lactobacillus pentosus* SJ65 using response surface methodology. Brazilian J Microbiol 2014;45(1):81−8.

[34] Malheiros PS, Sant'anna V, Todorov SD, Franco BDGM. Optimization of growth and bacteriocin production by *Lactobacillus sakei* subsp. *sakei*2a. Braz J Microbiol 2015;46(1):825−34.

[35] Li J, Zhang L, Du M, Han X, Yi H, Guo C. Effect of Tween series on growth and *cis*-9, *trans*-11 conjugated linoleic acid production of *Lactobacillus acidophilus* F0221 in the presence of bile salts. Int J Mol Sci 2011;80:9138−54.

[36] Hayek SA, Ibrahim SA. Current limitations and challenges with lactic acid Bacteria: a review. Food Nutr Sci 2013:73−87.

[37] Dobson A, Cotter PD, Paul Ross R, Hill C. Bacteriocin production: a probiotic trait? Appl Environ Microbiol 2012;78(1):1−6.

[38] Ruhr E, Sahl H-G. Mode of action of the peptide antibiotic nisin and influence on the membrane potential of whole cells and on cytoplasmic and artificial membrane vesicles. Antimicrob Agents Chemother 1985;27(5):841−5.

[39] Thanh NT, Chwen LT, Foo HL, Hair-Bejo M, Azhar K. Inhibitory activity of metabolites produced by strains of *Lactobacillus plantarum* isolated from Malaysian fermented food. Int J Probiotics Prebiotics 2010;5(1):37−43.

[40] Choe DW, Foo HL, Loh TC, Hair-Bejo M, Awis QS. Inhibitory property of metabolite combinations produced from *Lactobacillus plantarum* strains. Pertanika J Trop Agric Sci 2013;36(1):79−88.

[41] Ishikawa H, Akedo I, Otani T, Suzuki T, Nakamura T, Takeyama I, et al. Randomized trial of dietary fiber and *Lactobacillus casei* administration for prevention of colorectal tumors. Int J Cancer 2005;116(5):762−7.

[42] Nanno M, Kato I, Kobayashi T, Shida K. Biological effects of probiotics: what impact does *Lactobacillus casei shirota* have on us? Int J Immunopathol Pharmacol 2011;24(1):45S−50S.

[43] Kareem KY, Loh TC, Foo HL, Asmara SA, Akit H, Abdulla NR, Ooi MF. Carcass, meat and bone quality of broiler chickens fed with postbiotic and prebiotic combinations. Int J Probiotics Prebiotics 2015;10(1):23−30.

[44] Foo HL, Loh TC, Chuah LO, Noorjahan Banu Alitheen RAR. Tumour cytotoxic agent and methods thereof. 2016. US 2016/0030492.

[45] Nishio K, Ma J, Arora S, Kaur S, Kaur S. Bacteriocins as potential anticancer agents. Front Pharmacol 2015;6:272.

[46] FAO/WHO (Food, Agriculture Organization/World Health Organization). Report of a joint FAO/WHO expert consultation on evaluation of health and nutritional properties of probiotics in food including powder milk with live lactic acid bacteria. Cordoba, Agentina. October 2001. p. 1−4. Available from:, http://www.fao.org/3/a-a0512e.pdf.

[47] Joo NE, Ritchie K, Kamarajan P, Miao D, Kapila YL. Nisin, an apoptogenic bacteriocin and food preservative, attenuates HNSCC tumorigenesis via CHAC1. Cancer Med 2012;1(3):295−305.

[48] Kamarajan P, Hayami T, Matte B, Liu Y, Danciu T, Ramamoorthy A, Woorden F, Sunil K, Kapila Y, Nisin ZP. A bacteriocin and food preservative, inhibits head and neck cancer tumorigenesis and prolongs survival. PLoS One 2015;10(7):e0131008.

[49] Zhao H, Sood R, Jutila A, Bose S, Fimland G, Nissen-Meyer J, Kinnunen PVJ. Interaction of the antimicrobial peptide pheromone Plantaricin A with model membranes: Implications for a novel mechanism of action. Biochim Biophys Acta 2006;1758:1461−74.

[50] Sand SL, Haug TM, Nissen-Meyer J, Sand O. The bacterial peptide pheromone plantaricin A permeabilizes cancerous, but not normal, rat pituitary cells and differentiates between the outer and inner membrane leaflet. J Membr Biol 2007;216(2−3):61−71.

[51] Sand SL, Nissen-Meyer J, Sand O, Haug TM, Plantaricin A. A cationic peptide produced by *Lactobacillus plantarum*, permeabilizes eukaryotic cell membranes by a mechanism dependent on negative surface charge linked to glycosylated membrane proteins. Biochim Biophys Acta biomembr 2013;1828(2):249−59.

[52] Riedl S, Rinner B, Asslaber M, Schaider H, Walzer S, Novak A, Lohner K, Zweytick D. In search of a novel target- Phosphatidylserine exposed by non-apoptotic tumor cells and metastases of malignancies with poor treatment efficacy. Biochim Biophys Acta 2011;1808:2638−45.

[53] de LeBlanc A, de M, Matar C, Thériault C, Perdigón G. Effects of milk fermented by *Lactobacillus helveticus* R389 on immune cells associated to mammary glands in normal and a breast cancer model. Immunobiology 2005;210(5):349−58.

[54] Er S, Koparal AT, Kivanç M. Cytotoxic effects of various lactic acid bacteria on Caco-2 cells. Turkish J Biol 2015;39(1):23−30.

[55] Awaisheh SS, Obeidat MM, Al-Tamimi HJ, Assaf AM, EL-Qudah JM, Al-Khaza'leh JM, Rahahleh RJ. In vitro cytotoxic activity of probiotic bacterial cell extracts against Caco-2 and HRT-18 colorectal cancer cells. Milk Dairy Prod Hum Nutr 2016;69:33−7.

[56] Dimitrovski D, Cenci A, Winkelhausen E, Langerholc TZ. *Lactobacillus plantarum* extracellular metabolites: in vitro assessment of probiotic effects on normal and cancerogenic human cells. Int Dairy J 2014;39:293−300.

[57] Yirga H. The use of probiotics in animal nutrition. J Probiotics Heal 2015;3(2).

[58] Cho JH, Zhao PY, Kim IH. Probiotics as a dietary additive for pigs: a review. J Anim Vet Adv 2011;10:2127−34.

[59] S. 619 — 111th Congress: Preservation of Antibiotics for Medical Treatment Act of 2009. www.GovTrack.us. 2009. November 19, 2017 https://www.govtrack.us/congress/bills/111/s619; 2009 Available from: https://www.gpo.gov/fdsys/pkg/BILLS-111s619is/pdf/BILLS-111s619is.pdf.

[60] Loh TC, Chong SW, Foo HL, Law FL. Effects on growth performance, faecal microflora and plasma cholesterol after supplementation of spray-dried metabolite to postweaning rats. Czech J Anim Sci 2009;54(1):10−6.

[61] Thanh NT, Loh TC, Foo HL, Hair-Bejo M, Azhar K. Effects of feeding metabolite combinations produced by *Lactobacillus plantarum* on growth performance, faecal microbial population, small intestine villus height and faecal volatile fatty acids in broilers. Br Poult Sci 2009;50(3):298−306.

[62] Loh TC, Thanh NT, Foo HL, Hair-Bejo M. *Lactobacillus plantarum* on plasma and breast meat lipids in broiler chickens birds. Braz J Poult Sci 2013;15(4):307−16.

[63] Choe DW, Loh TC, Foo HL, Hair-Bejo M, Awis QS. Egg production, faecal pH and microbial population, small intestine morphology, and plasma and yolk cholesterol in laying hens given liquid metabolites produced by *Lactobacillus plantarum* strains. Br Poult Sci 2012;53(1):106−15.

[64] Loh T, Choe D, Foo H, Sazili A, Bejo M. Effects of feeding different postbiotic metabolite combinations produced by *Lactobacillus plantarum* strains on egg quality and production performance, faecal parameters and plasma cholesterol in laying hens. BMC Vet Res 2014;10(1):149.

[65] Loh TC, Van TT, Foo HL, Bejo MH. Effects of different levels of metabolite combination produced by *Lactobacillus plantarum* on growth performance, diarrhoea, gut environment and digestibility of postweaning piglets. J Appl Anim Res 2013;41(2):200−7.

[66] Thu TV, Loh TC, Foo HL, Yaakub H, Bejo MH. Effects of liquid metabolite combinations produced by *Lactobacillus plantarum* on growth performance, faeces characteristics, intestinal morphology and diarrhoea incidence in postweaning piglets. Trop Anim Health Prod 2011;43(1):69−75.

[67] Foo HL, Loh TC, Karunakaramoorthy A, Shamsudin MN, Rahim RA. Metabolites in animal feed. 2013. US 2013/0108600 A1.

[68] Loh TC, Thanh NT, Foo HL, Hair-Bejo M, Azhar BK. Feeding of different levels of metabolite combinations produced by *Lactobacillus plantarum* on growth performance, fecal microflora, volatile fatty acids and villi height in broilers. Anim Sci J 2010;81(2):205−14.

[69] Kareem KY, Loh TC, Foo HL, Akit H, Samsudin AA. Effects of dietary postbiotic and inulin on growth performance, IGF1 and GHR mRNA expression, faecal microbiota and volatile fatty acids in broilers. BMC Vet Res 2016;12(1):163.

[70] Thu TV, Chwen LT, Foo HL, Halimatun Y, Bejo MH. Effects of metabolite combinations produced by *Lactobacillus plantarum* on plasma cholesterol and fatty acids in piglets. Am J Anim Vet Sci 2010;5(4):233−6.

[71] Lee D, Jang S, Baek E, Kim M, Lee K, Shin H, et al. Lactic acid bacteria affect serum cholesterol levels, harmful fecal enzyme activity, and fecal water content. Lipids Health Dis 2009;8(1):21.

[72] Stojanovska L. Cholesterol reduction mechanisms and fatty acid composition of cellular membranes of probiotic Lactobacilli and Bifidobacteria. J Funct Foods 2014;9:295−305.

[73] West NP, Horn PL, Pyne DB, Gebski VJ, Lahtinen SJ, Fricker PA, et al. Probiotic supplementation for respiratory and gastrointestinal illness symptoms in healthy physically active individuals. Clin Nutr 2014;33:581−7.

[74] Balcázar JL, Decamp O, Vendrell D, De Blas I, Ruiz-Zarzuela I. Health and nutritional properties of probiotics in fish and shellfish. Microb Ecol Health Dis 2006;18(2):65−70.

[75] Ligaarden SC, Axelsson L, Naterstad K, Lydersen S, Farup G. A candidate probiotic with unfavourable effects in subjects with irritable bowel syndrome: a randomised controlled trial. BMC Gastroenterol 2010;10(16).

[76] von Schillde M-A, Hö Rmannsperger G, Weiher M, Alpert C-A, Hahne H, Bäuerl C, van H, Steidler L, Hrncir T, Pérez-Martínez G, Kuster B, Haller D, Shin HS, Chung MJ, Kim JE, Lee KO, Ha NJ. Lactocepin secreted by *Lactobacillus* exerts anti-inflammatory effects by selectively degrading proinflammatory chemokines. Cell Host Microbe 2012;11:387−96.

[77] Segawa S, Fujiya M, Konishi H, Ueno N, Kobayashi N, Shigyo T, Kohgo Y. Probiotic-derived polyphosphate enhances the epithelial barrier function and maintains intestinal homeostasis through integrin−p38 MAPK pathway. PLoS One 2011;6(8):e23278.

[78] Konstantinov SR, Smidt H, De Vos WM, Bruijns SCM, Singh SK, Valence F, Molle D, Lortal S, Altermann E, Klaenhammer TR, van Kooyk Y. S layer protein A of *Lactobacillus acidophilus* NCFM regulates immature dendritic cell and T cell functions. Proc Natl Acad Sci 2008;105(49):19474−9.

[79] Mazmanian SK, Round JL, Kasper DL. A microbial symbiosis factor prevents intestinal inflammatory disease. Nature 2008;453.

[80] Menard S. Lactic acid bacteria secrete metabolites retaining anti-inflammatory properties after intestinal transport. Gut 2004;53(6):821−8.

[81] Menard S, Laharie D, Asensio C, Vidal-Martinez T, Candalh C, Rullier A, Mégraud F, Matsiak-Budnik F, Heyman M. Bifidobacterium breve and *Streptococcus thermophilus* secretion products enhance T helper 1 immune response and intestinal barrier in mice. Exp Biol Med 2005;230:749−56.

[82] Cicenia A, Santangelo F, Gambardella L, Pallotta L, Iebba V, Scirocco A, Marignani M, Tellan G, Carabotti M, Corazziari ES, Schippa S, Severi S. Protective role of postbiotic mediators secreted by *Lactobacillus rhamnosus* GG versus lipopolysaccharide-induced damage in human colonic smooth muscle cells. J Clin Gastroenterol 2016;50:S140−4.

[83] Bernardo D, Sá Nchez B, Al-Hassi HO, Mann ER, Urdaci MC, Knight SC, Margolles A. Microbiota/host crosstalk biomarkers: regulatory response of human intestinal dendritic cells exposed to *Lactobacillus* extracellular encrypted peptide. PLoS One 2012;7(5).

[84] Kamath S, Buolamwini JK. Targeting EGFR and HER-2 receptor tyrosine kinases for cancer drug discovery and development. Med Res Rev 2006;26(5):569−94.

[85] Prosekov A, Dyshlyuk L, Milentyeva I, Sukhih S, Babich O, Ivanova S, Pavskyi V, Shishin M, Matskova L. Antioxidant, antimicrobial and antitumor activity of bacteria of the genus *Bifidobacterium*, selected from the gastrointestinal tract of human. Integr Mol Med 2015;2(5):295−303.

[86] Lin MY, Yen CL. Antioxidative ability of lactic acid bacteria. J Agric Food Chem 1999;47(4):1460−6.

[87] Sayre LM, Perry G, Smith MA. Oxidative stress and neurotoxicity. Chem Res Toxicol 2008;21(1):172−88.

[88] Shokryazdan P, Jahromi MF, Bashokouh F, Idrus Z, Liang JB. Antiproliferation effects and antioxidant activity of two new *Lactobacillus* strains. Braz J Food Technol 2017;21:e2016064.

[89] Xing J, Wang G, Zhang Q, Liu X, Gu Z, Zhang H, Chen YQ, Chen W. Determining antioxidant activities of lactobacilli cell-free supernatants by cellular antioxidant assay: a comparison with traditional methods. PLoS One 2015;10(3):e0119058.

[90] Zhang L, Liu C, Li D, Zhao Y, Zhang X, Zeng X, Yang Z, Li S. Antioxidant activity of an exopolysaccharide isolated from *Lactobacillus plantarum* C88. Int J Biol Macromol 2013;54(1):270−5.

[91] Rosyidah MR, Loh TC, Foo HL, Cheng XF, Bejo MH. Effect of feeding metabolites and acidifier on growth performance, faecal characteristics and microflora in broiler chickens. J Anim Vet Adv 2011;10:2758−64.

[92] Kareem KY, Loh TC, Foo HL, Samsudin AA, Akit H. Influence of postbiotic RG14 and inulin combination on cecal microbiota, organic acid concentration, and cytokine expression in broiler chickens. Poult Sci 2016;96(4):966−75.

Block IV

Diagnostic and Therapeutical Applications

Chapter 22

The Microbiome and Metabolome in Metabolic Syndrome

Rigoberto Pallares-Méndez[1] and Carla Fernández-Reynoso[2]

[1]Hospital Universitario "Dr. José Eleuterio Gonzalez" Universidad Autónoma de Nuevo León, Monterrey, México; [2]Centro de Estudios Universitarios Xochicalco, Tijuana, México

The gut microbiota is individually influenced by the environment of each portion of the gastrointestinal tract. The colonic microbiota is mainly regulated by the degradation of nondigestible carbohydrates. Microbiota functions include polysaccharide digestion, immune system enhancement, vitamin synthesis, and fat storage. The microbiome varies according to the individual dietary behavior, cultural habits, previous surgeries, antibiotic (AB) use, and genetic inheritance, giving the host a specific symbiotic functioning [1].

RELEVANT MICROBIAL POPULATIONS

Bacterial colonization starts after birth, reaching a more stable state at 3 years of age, when the microbiota is more similar to that from the adult host. Even so, a high intraindividual variability exists and a core gut microbiota has been identified. Around 90% of gut microbiota is constituted by Firmicutes (mostly Gram-positive), Bacteroidetes (mostly Gram-negative), Actinobacteria, and Proteobacteria [2]. The genes of the microbiome encode proteins for host survival that are not encoded by the human genome, the microbiome being considered as a "forgotten organ" [2,3]. Up to 1057 bacterial species have been recognized in the gut microbiota, which correspond to the Bacteria domain, the principal one, in addition to Archaea and Eukarya (or Eukaryota) [4,5].

Clostridium spp., *Streptococcus* spp., and coliforms are the dominant phylogenetic groups in the small intestine. The microbial composition of the small intestine is regulated by the uptake of simple carbohydrates (glucose, mannose, fructose) through the phosphotransferase system (PTS), carbohydrate fermentation, and amino acid metabolism. The distal gut microbiota is mainly regulated by the fermentation of nondigestible carbohydrates, mainly polysaccharides. This microbial composition regulates short-chain fatty acid (SCFA) absorption, liver processing, and adipocyte deposition, through an interaction between gut microbiota and host genes, supporting the evidence of the mechanisms for increased energy absorption and storage observed on obese hosts [1,6].

Firmicutes represent between 50% and 80% of the gut microbiota in healthy individuals, being the most diverse and abundant phylum. High bacterial diversity is associated with a healthy status, whereas low bacterial diversity is associated with obesity and inflammatory diseases. Obese phenotypes show a lower relative representation of Bacteroidetes and higher relative representation of Firmicutes than their lean counterparts. It has been observed in animal models that weight reduction after diet-induced obesity restores the relative representation of gut microbiota, increasing the relative Bacteroidetes abundance and decreasing that of the Firmicutes [7].

IN VIVO MODELS

Microbiota transfer from Western diet (high in saturated and unsaturated fat + sucrose) fed mice to germ-free (GF) mice promotes the obesity phenotype and increases weight and adipose tissue when compared with mice fed with a carbohydrate rich diet. When assessing if the changes on gut microbiota really had an effect on the host adiposity, initial carbohydrate-fed mice were given a Western diet and then changed to a modified carbohydrate-reduced or fat-reduced Western diet, showing a reversal of adipose tissue mass, a reduction of fat deposition, and an increase of the relative Bacteroidetes representation [7]. GF mice fed with high-fat diet (HFD) with less sucrose, a common food additive of Western diet, are partially protected from obesity.

During obesity the microbiota adapts to a high sucrose intake by enhancing the PTS to regulate simple carbohydrate uptake and metabolize them to SCFAs [5,8]. Western diet decreases microbiota diversity and increases "glycoside hydrolases". On the other hand, rich fiber diet increases microbiota diversity. Enriched metabolic pathways on mice fed with a Western Diet are PTS, fructose and mannose metabolism, glycolysis/gluconeogenesis, and glutamate metabolism. Depleted metabolic pathways are ATP-binding cassete transporters (ABC) transporters, bacterial chemotaxis, and bacterial motility proteins [7].

When mice were fed with lard versus fish oil—rich diets, as expected, lard-fed mice had higher fasting insulin, glucose levels, impaired insulin sensitivity, and gained more weight. Lard-fed mice have an increased potential to activate Toll-like receptor (TLR) 2 and TLR4, as HFDs increase intestinal absorption of bacterial products, such as lipopolysaccharide (LPS), which may be enough to promote inflammation.

When the TLR molecule MyD88 was knocked out, mice had protection against diet-induced obesity and showed decreased adipocyte size, lower fasting insulin levels, and better insulin sensitivity. Regarding adipose tissue inflammation, MyD88−/− mice showed less macrophage and leukocyte abundance in adipose tissue. On the other hand, expression of CCL2, a macrophage mediator, was increased in white adipose tissue of lard-fed conventional mice. Adipose tissue exposed to serum from lard-fed mice increases expression of TNF-α, supporting the fact that microbial factors promote an inflammatory state.

When AB-sterilized mice were transplanted with the gut microbiota of fish oil—fed or lard-fed mice and subsequently fed lard diet, those who were recipients of fish oil diet showed less adipose tissue inflammation and less weight gain [8]. Even with an uptake of 29% more calories, GF mice have 40% less fat in comparison with conventional mice and show resistance to obesity induced by a high-fat or high-sugar Western diet [6,7]. Furthermore, GF mice show protection to diet-induced glucose intolerance and insulin resistance [9]. Resistance to obesity is associated with a more rapid fatty acid metabolism, through a shift of skeletal muscle and liver activity, increased mitochondrial fatty acid oxidation [7], a trend toward lipid use as energy source, and a decrease of carbohydrate usage, supported by a lower respiratory quotient [10].

Fecal transplant from conventional mice to GF mice causes 57% corporal fat increase, insulin resistance, and higher hepatic triglyceride levels [11−13]. Microbiota of women, who had a previous Roux-en-Y gastric bypass (RYGB), vertical banding gastroplasty (VBG), or were obese, was transferred to GF mice. Mice recipient to RYGB or VBG stools showed 43% and 26% less body fat, respectively, in comparison with mice recipient to obese microbiota. Also, mice who received RYGB had an increased lean body mass [10].

It is concluded that gut microbiota regulates host weight, metabolism signaling, and energy harvesting. The microbiome after RYGB showed enhanced sugar metabolism and glycolysis, which point to metabolic pathways of phosphoglycerate, acetoacetate, SCFAs, and PTSs. On the other hand, after VBG, enhanced amino acid uptake and glyoxylate metabolism occurred, evidencing in both groups an increased energy production through amino acids and acetate. Metabolomic biomarkers for obesity such as SCFAs (acetate, propionate, butyrate) have been found decreased in RYGB and VBG, pointing toward decreased energy harvest [10].

Fecal transplant from obese hosts to GF mice is consistently associated with increased adiposity and metabolic disturbances. Saturated fats, processed sugars, and a low fiber intake change the bacterial representation within the gastrointestinal tract, which alter energy harvesting, SCFA usage, and oligonutrient fermentation, changing the interplay between the pathways of metabolic homeostasis. These findings support causality that a lack of fiber and a poor microbial diversity contribute to the obesity phenotype and metabolic alterations.

METABOLOME AND PROTEOME

Amino Acid Metabolism

Branched-chain amino acids (BCAAs) (isoleucine, leucine, and valine) and aromatic amino acids (tyrosine and phenyl-alanine) have a strong predictive value for insulin resistance and type 2 diabetes (T2D) [14,15]. Some amino acids can promote insulin resistance (IR) by disrupting insulin signaling in skeletal muscle. At the same time insulin resistance can perpetuate a catabolic state, impairing the uptake or release of BCAAs in skeletal muscle, working as a vicious cycle, as BCAAs are known to trigger β-cell insulin secretion and promote hyperinsulinemia and insulin depletion. However, this process seems not to be enough to prevent hyperglycemia [15]. Alpha-hydroxybutyrate (α-HB) is an important by-product of alpha-ketobutyrate (α-KB), which is required for cysteine biosynthesis and ultimately leads to glutathione synthesis [16].

Indeed α-HB accumulates when α-KB synthesis is greater than its catabolism and has been related to IR and oxidative stress. Cysteine assumes a major role when increased glutathione synthesis is required to counteract the damage done by reactive oxygen species (ROS), as it is known to be a glutathione precursor and has been found related to insulin resistance [16]. This metabolic network suggests that cysteine is an important meeting point between some ketoacids, such as α-HB and α-KB, and oxidative stress damage regulators, such as glutathione. In addition, 2-aminoadipic acid, a lysine intermediary, shows an increased risk (4.5-fold) of developing T2D, independently of fasting levels of glucose. When added as supplement in animal models, it has shown to decrease glucose levels, acting as a modulator of insulin secretion [15,17].

β-Hydroxypyruvate, a glycine intermediary, regulates myenteric neurons, reducing pancreatic islet insulin content on mice [18]. The Takayama Study has demonstrated that BCAA supplementation could be used as a therapeutic measure in subjects prone to insulin resistance and normal weight, as they have been associated with a decrease on diabetes risk and improvement of glucose control. Leucine was the most important BCAA in relationship with insulin release [19]. The amount of BCAAs in serum does not strictly correlate with its intake. In obese subjects, the increased BCAA concentration was explained by an impairment of its catabolism and clearance due to a decrease of the Branched chain ketoacid dehydrogenase (BCKD) complex function, the same mitochondrial enzyme involved in α-KB synthesis [19]. When insulin resistance and diabetes occur, mitochondrial function for BCAA catabolism gets impaired; the by-products of BCAAs thereby affect mitochondrial oxidation of glucose. In addition, the large neutral amino acid transporter 1 receptor gets impaired, and the entrance of BCAAs into the cell does not occur in a suitable fashion, contributing to the high serum concentrations of BCAAs [14].

Lipid Metabolism

Phospholipids, sphingomyelins, and triglycerides with low carbon number saturation and low double-bond content, such as palmitate and palmitoleate, have been associated with insulin resistance and T2D, whereas those with greater carbon number saturation and more double bonds have shown an inverse relationship. Lysophosphatidylcholines and lysophosphatidylethanolamines have been found to be increased in subjects with T2D [20]. Sphingomyelin synthesis reduction is associated with ROS and impairment of insulin secretion. Free fatty acids (FFAs) impair insulin action via oxidative stress, inflammation, mitochondrial dysfunction, and accumulation of diacylglycerol and ceramides.

Lipid signaling transduces glucose-mediated insulin release through G-coupled proteins on β-cell islets. Lineoleoylglycerophosphocholine (L-GPC) is inversely related with incident dysglycemia. Low L-GPC and high α-HB increase glucose-mediated insulin secretion and are strong predictors of glucose intolerance, with a similar power to the 2-h glucose challenge [21].

There is a close relationship between BCAAs and SCFAs. Both have been found to be increased in obesity. BCAAs stimulate SCFAs production, and SCFAs with straight chain affect the gut microbiota, which is responsible for production from carbohydrate fermentation and BCAAs. This suggests a feedback mechanism between gut microbiota and energetic homeostasis [22]. Amino acids also modulate the signaling pathways of gut microbiota, which contribute to epithelial cell integrity and immune system performance, by modulating epithelial growth [23].

The increase of FFAs and leptin stimulation of SCFAs through G-coupled proteins are all mechanisms that inhibit lipolysis. Hyperinsulinemia fails to regulate intracellular responses on adipocytes in the obese subjects, which are known to have a blunted response to the antilipolytic effect of insulin, shifting energy obtainment to FFAs and increasing hepatic lipolysis, as well as promoting a defective β-oxidation in liver and skeletal muscle, reflected by the increase of acylcarnitines on serum and urine [16,24]. The high amounts of FFAs that go through β-oxidation overwhelm the citric acid cycle and promote acetyl-CoA accumulation, leading to short-chain acylcarnitines synthesis, activating inflammatory pathways that cause insulin resistance and ROS [21,22,25].

Carbohydrate Metabolism

1,5-Anhydroglucitol is a novel metabolite that assesses short-term glucose control for 5−7 days and could be used as a diagnostic and follow-up tool. It has been found decreased in saliva of T2D patients and increased in urine [26]. It decreases during hyperglycemia and increases during normal glucose homeostasis. In urine it shows a different pattern, as it increases in dysglycemia, as its reabsorption in kidneys gets impaired, when the glucose threshold is exceeded [24] (Table 22.1).

TABLE 22.1 Important Metabolomic Cornerstones

Amino acid metabolism	• ↑BCAAs (iso, leu, and val) • ↑Aminoadipic acids (Thyr, Phe)	• Promote β-cell insulin secretion. • Promote glucose oxidation [19]. • Improved glucose control [19]. • Promote hyperinsulinemia [14,15].
	• ↑Alpha-hydroxybutyrate • β-Hydroxypyruvate	• Product of α-KB [16]. • Substrate for cysteine, which leads to glutathione synthesis [16]. • Related to insulin resistance [16]. • Oxidative stress regulators [16]. • Regulate myenteric neurons reducing pancreatic islet insulin [18].
Lipid metabolism	• ↑Lysophosphatidylcholines and lysophosphatidylethanolamines • ↑FFAs	• Increased in T2D [20]. • Impair insulin action [20].
Carbohydrate metabolism	• 1,5-Anhydroglucitol	• Assesses short-term glucose control for 5–7 days [26]. • Biomarker of glucose control [24]. • Diagnostic and follow-up tool [24].

FFAs, Free fatty acids.

CRUCIAL MOLECULES AND PATHWAYS

An inverse relationship has been observed in the microbiome load of obese subjects and insulin resistance, as lean subjects show a major DNA bacterial load and bacterial diversity [27]. Fecal transplant from lean, metabolically healthy subjects to those with insulin resistance increases insulin sensitivity and microbial diversity, evidencing that the gut microbiota plays an important role in metabolism and energetic homeostasis.

The LPS-CD14 Complex Induces NF-κB

Chylomicrons take and transport bacterial LPS toward target tissues and promote inflammation on the host, thereby insulin resistance. Lean individuals have low concentrations of LPS, whereas obese individuals show "metabolic endotoxemia" [28]. LPS crosses the gastrointestinal mucosa through tight junctions by chylomicron uptake or through a TLR4-dependent mechanism. After LPS is taken by chylomicrons, it reaches to the liver and adipose tissue through systemic circulation, promoting an innate immune response by activation of CD14 receptors. Activation of macrophages via LPS—LPS-binding protein (LPS-LBP) to surface CD14 receptors promotes expression of TNF-α, IL-1, and IL-6, which trigger signal transduction for the expression of genes that serve to encode inflammatory effects, such as NF-κB and activator protein-1. In time, more TNF-α and inflammation result [29].

When mice under continuous LPS infusion were fed a 72% HFD, endotoxemia increased by 2.7-fold when compared with normal fed mice. Endotoxemia also increased with 40% fat, showing a 1.4-fold increase [28], supporting the notion that an HFD promotes LPS transport [28,29]. AB-treated mice showed a blunted shift of IL-1 and TNF-α, lower LPS concentration, lower glucose levels, and lower adipocyte hypertrophy, suggesting better epithelial barrier function in the gut, as tight junction proteins such as ZO-1 and occludin were increased [30].

The role of CD14 was assessed with wild-type and CD14 knock-out mice under a continuous LPS infusion; IL-6, IL-1, plasminogen activator inhibitor-1, and TNF-α were all increased in wild-type mice, with increased weight, subcutaneous and visceral fat deposition, and dysglycemia. These effects support the direct effect on insulin resistance of the interaction between CD14 and LPS [28].

The Influence of CD4 T Cells Through RORγt Expression

Insulin resistance induced through an HFD is associated with reduced T-helper 17 (Th17) cells positive to IL-17, as well as expression of retinoic acid receptor-related orphan receptor gamma T (RORγt). Differentiation of Th17 cells is mediated by RORγt, which in time is induced by cytokines TGFβ-1, IL-6, IL-1β, IL-21, and IL-23 [31]. An HFD-induced dysbiotic microbiota downregulates Th17 production, decreasing IL-17 and Treg cells. These changes induce glucose intolerance and insulin resistance [32]. Th17 cells have been found decreased in ileum, whereas on mesenteric adipose tissue and liver they increased, consistent with a higher grade inflammation in liver and adipose tissue than small bowel.

The decrease of Th17 cells on HFD-fed mice follows the decrease of IL17 and number of RORγt-expressing CD4 T cells and has a direct relationship with T2D development. When evaluating RORγt−/− mice on normal diet, glucose intolerance and insulin resistance were increased, even when microbiota of RORγt−/− mice was transferred to Rag1−/− mice (no B or T cells). Furthermore, these mice gained more weight than wild-type mice did [31]. No insulin resistance or glucose intolerance was developed in mice with upregulated function of Th17 cells, even when fed HFD [31].

On transfer of Th17 cells to mice fed HFD, a reconstitution of the microbiota was observed, increasing Bacteroidetes and decreasing Firmicutes, ameliorating weight gain and metabolic disturbances [32]. Dysbiosis precedes Th17 dysfunction and has a direct relationship with metabolic disruptions. Moreover, IL-17 proves a key factor for intestinal immunity, intestinal bacterial profile, and epithelial barrier function in small bowel.

The Farnesoid X-Activated Receptor and the Bile Acid "Cross Talk"

Bile acids have important endocrine functions via G protein–coupled receptors (GPCRs). Primary biliary acids such as cholic acid and chenodeoxycholic acid (CDCA) account for the most abundant constituents of the biliary acid pool. The gut microbiota modifies the bile acid pool in composition and amount. Primary bile acids are transformed to secondary by the gut microbiota, such as deoxycholic acid and lithocholic acid. Bile acids regulate their own production through a complex negative feedback mechanism. Main signaling targets are the membrane bile acid receptor as a ligand for GPCRs and the nuclear bile acid receptor, known as the farnesoid X-activated receptor (FXR).

TGR-5, a GPCR involved in biliary acid signaling, has been related to immunomodulatory functions of bile acids. TGR-5 modulates metabolism through incretin signaling and mitochondrial homeostasis. FXR modulates the negative feedback mechanism that governs bile acid production from cholesterol. Bile acids activate FXR and induce small heterodimer partner (SHP) in the hepatocyte nucleus, increasing expression of FGF15/19 in the ileum. Both systems suppress CYP7A1 activation, the bile acid rate-limiting enzyme [33]. All bile acids are FXR agonists, except for murocholic acid (MCA) and ursodeoxycholic acid, which are potent antagonists. On the other hand, CDCA is the most potent agonist of FXR [34].

Intestinal microbiota may have an important role on regulation of FGF15/19 expression in ileum and CYP7A1 in liver through FXR-dependent mechanisms. Gut microbiota promotes deconjugation, dehydrogenation, and dehydroxylation of primary bile acids in ileum and colon, increasing biliary acid diversity through biosynthesis of secondary bile acids [35]. GF mice show a higher CYP7A1 activity and FXR antagonism through increased MCA. In CONV-R mice (recolonized animals), a reduction in MCA resulted in increased FXR activity, explaining the lower bile acid pool [35]. Microbiota of CONV-R mice influenced an increase on FXR action through target molecules that inhibit CYP7A1. Elevation of these molecules, such as SHP and FGF15/19, was observed in ileum.

FXF−/− mice showed a decrease of SHP activation and FGF15 signaling in ileum, proving that FXF is necessary for the expression of these signal molecules. When GF and CONV-R mice are treated with FXR agonists, FGF15 expression increases in ileum and suppresses hepatic CYP7A1. These associations denote that presence of gut microbiota regulates secondary bile acid metabolism and higher MCAs in GF mice, reducing the expression of FXR genes in ileum [35,36].

FXR takes part in the metabolic interplay. When bile acid sequestrants are given, bile acids exit the system through stool, thus cholesterol conversion to bile acids gets stimulated. This increases low-density lipoprotein expression in liver to promote an avid uptake of substrate. FXR activation enhances triglyceride clearance through higher LPL activity, which is in time favored by the expression of PPAR-α and ApoCII and the decrease of ApoCIII [37]. FXR seems to play a role in glucose homeostasis. FXR-deficient mice show impaired intestinal glucose absorption. TGR-5 has been proved to regulate Glucose-like peptide-1 (GLP-1) and thus takes part in glucose homeostasis. In CONV-R mice, a decrease of biliary acid pool fails to activate TGR-5, leading to decreased of GLP-1 activation through GPCRs, impacting glucose homeostasis [34,37].

FXR also modulates metabolic homeostasis through ceramide metabolism. As said before, high ceramide concentrations have been related with inflammation and metabolic disease. Observations show that heterozygous mice to dihydroceramide desaturase relate to lower ceramide concentrations and better insulin sensitivity. Lack of ceramide synthase shows less ceramide concentrations and is associated with HFD-induced obesity protection and better glucose tolerance. FXR could control the expression of genes for ceramide biosynthesis [38].

Mice under FXR antagonists (MCA) show decreased ceramide levels. On the contrary, FXR agonists induce ceramide metabolism and increased levels, thus greater inflammation. FXR activation then increases ceramide gene activation for sphingomyelin phosphodiesterase and serine palmitoyltransferase. High ceramide levels damage mitochondria, producing increased permeability and inducing apoptosis [20,38] (Fig. 22.1).

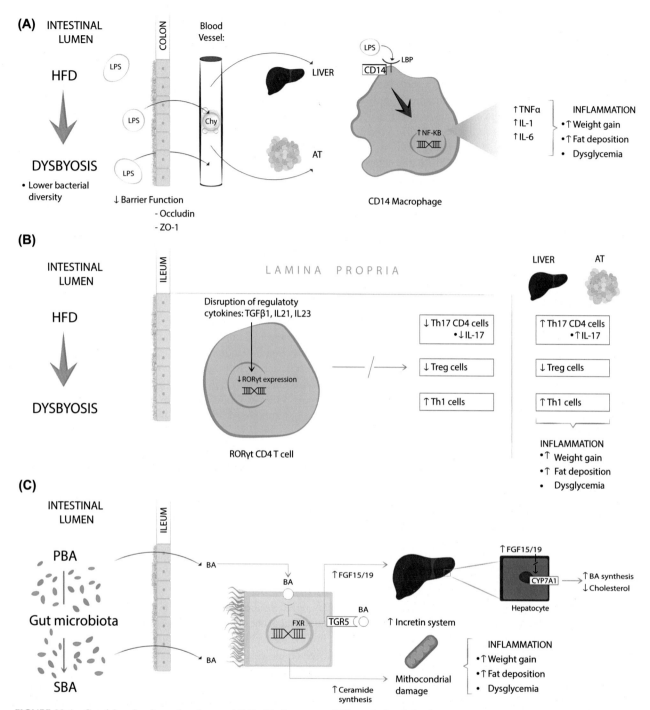

FIGURE 22.1 Crucial molecules and pathways. (A) Dysbiosis promotes LPS uptake by chylomicrons transported to liver and adipose tissue. This activates CD14 receptors on macrophages through LPB, which in time activates NF-κB, increasing inflammatory cytokines TNF-a, IL-1, IL6, and promoting inflammation, which triggers weight gain, fat deposition, and dysglycemia. (B) Dysbiosis affects immune cell differentiation. An altered microbiota is associated with disruption of TTGFβ-1, IL21, IL23 signaling, which causes a reduced expression of RORγt. This promotes less differentiation of TH17 CD4 cells and Treg cells in small intestine *lamina propria* and an increase of Th17 CD4 cells in adipose tissue and liver, resulting in a higher-grade inflammation in these target tissues, which in time promotes weight gain, fat deposition, and dysglycemia. (C) A diverse healthy microbiota promotes a larger and more diverse bile acid pool. Bile acids activate TGR-5 on the cell surface and FXR in the nucleus, which increase FGF15/19 expression and precipitate CYP7A1 inhibition, decreasing BA synthesis and cholesterol utilization. FXR activation promotes ceramide synthesis, which causes mitochondrial damage promoting inflammation, weight gain, fat deposition, and dysglycemia. Bile acids also increase the incretin system tone, promoting satiety and glucose tolerance. *AT*, adipose tissue; *BA*, bile acids; *Chy*, chylomicrons; *FXR*, farnesoid X-activated receptor; *HFD*, high-fat diet; *LBP*, lipopolysaccharide-binding protein; *LPS*, lipopolysaccharide; *PBA*, primary bile acids; *SBA*, secondary bile acids; *TLR 4*, Toll-like receptor 4.

INTERFACES WITH ANTIBIOTICS

AB administration promotes dysbiosis and affects up to 30% of the gut bacteria. Once ABs cease, these bacteria hardly return to the previous state. Dysbiosis impacts on gene expression, metabolism, and immunity [39]. A reduction of microbial-associated molecular patterns (MAMPS) has been related to AB use. When the MAMPS epithelial cell interaction is decreased, an altered NOD1 and TLR stimulation results, which affects lymphoid tissue development, T-cell differentiation, and cytokine release [39,40]. *Bacteroides fragilis* and *Clostridium* promote Treg cell differentiation in mice through TGF-β expression from intestinal epithelial cells. On the other hand, segmented filamentous bacteria promote proinflammatory TH17 cell regulation [40]. This was also observed when vancomycin administration, which kills Gram-positive bacteria, reduced Treg cells on *lamina propria*, supporting the role of Gram-positive bacteria in Treg differentiation [41].

Metronidazole causes thinning of the mucin layer on the gut, by decreasing Muc-2 protein expression, a major component of this layer. A thin mucin layer promotes more bacterial contact with the epithelium, which stimulates the innate immune system response and favors a proinflammatory state. Some components of the mucin layer decrease, such as antimicrobial peptides, i.e., C-type lectins and defensins, impairing the first-line response to pathogens on the gut epithelium [40].

The interaction between NOD1 receptors and bacterial peptidoglycan decreases under ABs. NOD1 receptors promote adaptive lymphoid tissue differentiation in response to bacterial cell wall recognition. This signal disruption impairs neutrophil ability to respond to pathogens [39]. Furthermore, TLR2/4 expression in macrophages decreases under AB treatment, inducing a poor response to LPS [40].

Broad-spectrum ABs reduce the Bacteroidetes population and increase Firmicutes. Although ABs target specific bacteria, other species could be affected, influencing metabolite interchange and waste products of the gut microbiota [40]. Perez-Cobas et al. [42] showed a shift from Firmicutes to Bacteroidetes abundance at day 14 of AB treatment. After 40 days of AB discontinuation, a tendency of recovery of the previous pattern was observed; however, the original pattern could not fully recover due to "bacterial resilience". Interestingly, at day 11 of AB treatment, the bacterial diversity was abruptly decreased with a subsequent Gram-positive regrowth. A reduction of enzymes necessary for glycolysis, Tricarboxylic acid (TCA) cycle, and glutamate metabolism was observed even after AB cessation.

This energy pathway disruption affects carbohydrate fermentation, which gets impaired and affects SCFA production, an important energy source for colonocytes. Butyrate, the most important, regulates epithelial functions such as tight junction assembly and mucin expression [40]. Butyrate protects the colon from inflammation. Mechanisms include inhibition of monocyte chemotactic protein expression and expression of PGE2, which inhibits TNF-α [40]. Disruption of these interactions promotes inflammation and could trigger a response through LPS and TNF-α.

THERAPEUTICAL PROTOCOLS AND DISEASE PREVENTION

Kobyliak et al. [43] thoroughly revised the effects of probiotics on host metabolism. Findings agree with decreased body weight gain and fat accumulation,as well as improved glucose metabolism and insulin sensitivity in HFD-fed models, both with *Lactobacillus* and *Bifidobacterium* strains, alone or combined. Administration of oligofructose with HFD increased *Akkermansia muciniphila*. Besides the beneficial effects previously mentioned, decreased metabolic endotoxemia and adipose tissue inflammation occurred through an enhanced mucus layer function on the gut. *A. muciniphila* has been found decreased in diabetic mice. Restoration promotes glucose-6-phosphatase expression, which correlates with reduced fasting plasma glucose. Furthermore, *A. muciniphila* enhances lipid oxidation, supported by increased mRNA expression of key enzymes, such as carnitine palmitoyltransferase-1 and PPAR-γ. These observations were only found with viable bacteria, not with heat-killed material [44].

It remains to be elucidated if these beneficial effects were due to administration of oligofructose, which has been found to increase goblet cell numbers and mucus content optimal to *A. muciniphila* growth, as *A. muciniphila* does not to grow in oligofructose-enriched media. Controversy still exists, as mucin-degrading bacteria have also been found deleterious for the mucus barrier function [45].

Mixed probiotics have shown to protect HFD-fed mice from obesity and hepatic steatosis. Glucose tolerance, decreased insulin levels, and improved Homeostatic model assessment - Insulin resistance (HOMA-IR) also resulted. With regard to Non-alcoholic fatty liver disease (NAFLD), mixed probiotics decreased triglycerides in hepatocytes and increased hepatic Natural killer T cells (NKT) cell count, contributing to diminished hepatic inflammation. Probiotic lipid load and antigens could take part in NKT stimulation and function, which improves NAFLD, although the interplay between NKT cells in NAFLD is still controversial [46].

Probiotic administration modulates the incretin system. HFD-fed mice show protection from obesity and glucose intolerance when given VSL#3 (mixed probiotics). They can also decrease triglyceride, FFAs, and hepatic steatosis. Mixed probiotic administration modifies weight gain, adiposity, and insulin sensitivity through appetite suppression, increasing GLP-1 expression from L cells. Butyrate levels in the gut are benefitted through higher abundance of butyrate-producing bacteria (*Bifidobacteria*), demonstrated by an increase of butyrate kinase gene expression (*buk*).

During L cell treatment with butyrate, interestingly, increased GLP-1 synthesis (*Gcg*, *Pcsk1*) and secretion (*Sclc5a1*) genes were observed [47]. VSL#3 mice showed an increase of the satiety gene proopiomelanocortin, in addition to GLP-1, which also regulates appetite. These interactions could regulate this microbiota neural cross talk, with reduced appetite leading to less weight gain, decreased adiposity, improved inflammation and insulin resistance, and hence a healthier microbiota.

Prebiotics can also influence the incretin system, as GLP-1 and GLP-2 are increased in oligofructose-treated models. Oligofructose enhances *Bifidobacteria* growth, which promotes the mechanisms described before [48]. Moreover, prebiotics have been associated with a decrease in the endocannabinoid tone, providing an enhanced gut barrier function [49]. The endocannabinoid system has been related to modulation of gut permeability, endotoxemia, and weight gain.

Prebiotic-treated (oligofructose) ob/ob mice show a decrease of endocannabinoid tone. N-arachidonoylethanolamine, an endogenous cannabinoid diminishes, and endocannabinoid-degrading enzymes, such as fatty acid amide hydrolase, show higher expression. The regulation of gut microbiota through prebiotics decreases CB1 receptor expression in colon. Furthermore, CB1r regulates endotoxemia, as treatment with CB1r antagonist (OB-SR) shows an increased expression of tight junction proteins ZO-1 and occludin and decreases LPS levels [50].

Fecal microbiota transplant (FMT) from lean donors to obese T2D subjects promotes bacterial diversity and butyrate synthesis by butyrate-producing bacteria (*Roseburia intestinalis* and *Eubacterium hallii*). An improvement of median rate of glucose disappearance in obese FMT receptors was registered [27]. These results support the effects of butyrate on gut barrier function, inflammation, and incretin signaling described before [40,47]. Further information is needed with regard to the long-lasting effects of FMT from healthy donors in obesity or T2D, as the available data concern mainly subjects with resistant *Clostridium difficile* infection.

DIET, NUTRITION, EXERCISE, AND OTHER LIFESTYLE REPERCUSSIONS

The increase of the *Bacteroides* genus has been associated with low plant-based diets. On the other hand, *Prevotella* and *Xylanibacter*, which are associated with rich plant-based diets, contain genes for the hydrolysis of xylan and cellulose, essential fibers for the metabolic activity of the gut microbiota. This diet behavior influences the microenvironment and the barrier function of the gut. Rich fiber diets enhance SCFA production, essential for energy harvesting and colonocyte metabolism, modulating the microbiome through pH modifications, which prevent pathogenic bacteria colonization and promote Firmicutes growth [51].

Greater intestinal microbial diversity has been reported in high-performance athletes, followed by regular adults with a body mass index (BMI) less than $25 \, kg/m^2$ and lower in patients with a BMI higher than $28 \, kg/m^2$, identifying an improvement related to protein intake, which was higher in high-performance athletes [52]. Resistance training combined with aerobic activity in T2D leads to a reduction of IL-1β, IL-6, TNF-α, and IFN-γ, as well as an increase in IL-4 and IL-10, which act as inflammation modulators. Expression of IL-10 in intestinal lymphocytes showed to be enhanced after exercise. This evidences that exercise can modify the gut microbiota, as well as diet-microbe-host and neuroendocrine and neuroimmune interactions. A combination of anaerobic and resistance exercise, decreasing IL-1β, IL-6, TNF-α, and IFN-γ, which translates to decrease of inflammation, correlates with a decrease of HOMA-IR and insulin levels [52,53].

Diet-restricted exercising mice were associated with higher Firmicutes and lower Bacteroidetes in fecal samples [54]. Diet and exercise can modify the gut microbiota; observations have shown a positive correlation of leptin with *Bifidobacterium* and *Lactobacillus*, whereas *Clostridium*, *Bacteroides*, and *Prevotella* show a negative correlation. Ghrelin has been negatively correlated with *Bifidobacterium* and *Lactobacillus*, whereas positively correlated with *Bacteroides* and *Prevotella* [54]. Higher levels of butyrate-producing bacteria, such as *Bifidobacteria, Clostridium*, and *Lactobacillus,* are also documented, either by fiber fermentation or cross-feeding interactions between different bacteria. Mucin maintenance is also important, as the balance of mucin-degrading bacteria such as *Prevotella* or *A. muciniphila* affects gut permeability and modifies the inflammatory and metabolic response of the host [54].

Subjects between 18 and 50 years were given a fiber preparation and then fresh stool samples were obtained. A higher fiber utilizing capacity of *Prevotella*, with enhanced SCFA production, was noticed compared with *Bacteroides*. Affinity to fiber fermentation according to the microbiota enterotype could be a future approach, promoting metabolic health through SCFA functions, such as gut barrier regulation, modulation of inflammatory stimulus, and appetite suppression [55].

The amount of fiber ingested in the diet affects the thickness of the mucus layer of the intestine. The greater the thickness, the more bacterial diversity and better barrier protection [45]. In gnotobiotic mice colonized by human synthetic microbiota, a fiber-rich diet induced colonization by bacteria that utilize fiber polysaccharides for energy, enhancing the gut barrier function. The fiber-free diet was associated with mucin-degrading bacteria, degrading mucus components as energy source and thinning the mucus barrier, hence increasing the gut epithelium susceptibility to inflammation and pathogens. In this study, prebiotics did not turn out to have protective capacity against bacterial erosions of the mucus layer, although they can help to promote growth of helpful bacteria. Pure polysaccharides could be employed to increase the mucus protective capacity and the gut barrier function [45].

FINAL CONSIDERATIONS

Some bacterial strains have been associated with appetite-regulating hormones. *Bifidobacterium* and *Lactobacillus* have shown a positive correlation with leptin and a negative correlation with ghrelin [54]. This observation may be associated to the effects of butyrate regarding appetite suppression, either by increased incretin tone or by a direct effect of butyrate on these appetite-regulating hormones [47,56].

Transfer of fecal microbiota from lean donors to T2D promotes insulin sensitivity and enhances glucose metabolism through butyrate-producing mechanisms, which protect the gut mucus barrier integrity, reduce inflammation, and enhance the incretin system [27]. A combination of aerobic and resistance-training exercise has shown to reduce inflammatory cytokines and increase inflammatory modulators, conducting to less inflammation that correlates with a decrease of HOMA-IR [52].

ACKNOWLEDGMENTS

Many thanks to Heidy Maccioni for the help with the diagram design.

REFERENCES

[1] Marchesi JR, Adams DH, Fava F, Hermes GDA, Hirschfield GM, Hold G, et al. The gut microbiota and host health: a new clinical frontier. Gut 2016;65(2):330−9. Available from: http://gut.bmj.com/lookup/doi/10.1136/gutjnl-2015-309990.

[2] D'Argenio V, Salvatore F. The role of the gut microbiome in the healthy adult status. Clin Chim Acta 2015;451:97−102. Available from: http://linkinghub.elsevier.com/retrieve/pii/S0009898115000170.

[3] O'Hara AM, Shanahan F. The gut flora as a forgotten organ. EMBO Rep 2006;7(7):688−93. Available from: http://embor.embopress.org/cgi/doi/10.1038/sj.embor.7400731.

[4] Rajilić-Stojanović M, de Vos WM. The first 1000 cultured species of the human gastrointestinal microbiota. FEMS Microbiol Rev 2014;38(5):996−1047.

[5] Fleissner CK, Huebel N, Abd El-Bary MM, Loh G, Klaus S, Blaut M. Absence of intestinal microbiota does not protect mice from diet-induced obesity. Br J Nutr 2010;104(6):919−29.

[6] Turnbaugh PJ, Ley RE, Mahowald M a, Magrini V, Mardis ER, Gordon JI. An obesity-associated gut microbiome with increased capacity for energy harvest. Nature 2006;444(7122):1027−31. Available from: http://www.ncbi.nlm.nih.gov/pubmed/17183312.

[7] Turnbaugh PJ, Bäckhed F, Fulton L, Gordon JI. Diet-induced obesity is linked to marked but reversible alterations in the mouse distal gut microbiome. Cell Host Microbe 2008;3(4):213−23.

[8] Caesar R, Tremaroli V, Kovatcheva-Datchary P, Cani PD, Bäckhed F. Crosstalk between gut microbiota and dietary lipids aggravates WAT inflammation through TLR signaling. Cell Metab 2015;22(4):658−68.

[9] Piya MK, McTernan PG, Kumar S. Adipokine inflammation and insulin resistance: the role of glucose, lipids and endotoxin. J Endocrinol 2013;216(1):T1−15. Available from: http://www.ncbi.nlm.nih.gov/pubmed/23160966.

[10] Tremaroli V, Karlsson F, Werling M, Ståhlman M, Kovatcheva-Datchary P, Olbers T, et al. Roux-en-Y gastric bypass and vertical banded gastroplasty induce long-term changes on the human gut microbiome contributing to fat mass regulation. Cell Metab 2015;22(2):228−38.

[11] Bäckhed F, Ding H, Wang T, Hooper LV, Koh GY, Nagy A, et al. The gut microbiota as an environmental factor that regulates fat storage. Proc Natl Acad Sci U S A 2004;101(44):15718−23.

[12] Bäckhed F, Manchester JK, Semenkovich CF, Gordon JI. Mechanisms underlying the resistance to diet-induced obesity in germ-free mice. Proc Natl Acad Sci U S A 2007;104(3):979−84.

[13] Ley RE, Bäckhed F, Turnbaugh P, Lozupone C a, Knight RD, Gordon JI. Obesity alters gut microbial ecology. Proc Natl Acad Sci U S A 2005;102(31):11070−5.

[14] Pallares-Méndez R, Aguilar-Salinas CA, Cruz-Bautista I, Del Bosque-Plata L. Metabolomics in diabetes. A Review Ann Med 2016;48(1−2).

[15] Guasch-Ferré M, Hruby A, Toledo E, Clish CB, Martínez-González MA, Salas-Salvadó J, et al. Metabolomics in prediabetes and diabetes: a systematic review and meta-analysis. Diabetes Care 2016;39(5):833−46.

[16] Gall WE, Beebe K, Lawton KA, Adam KP, Mitchell MW, Nakhle PJ, et al. A-hydroxybutyrate is an early biomarker of insulin resistance and glucose intolerance in a nondiabetic population. PLoS One 2010;5(5).

[17] Wang TJ, Ngo D, Psychogios N, Dejam A, Larson MG, Vasan RS, et al. 2-Aminoadipic acid is a biomarker for diabetes risk. J Clin Invest 2013;123(10):4309−17.

[18] Zhang S, Wang S, Puhl MD, Jiang X, Hyrc KL, Laciny E, et al. Global biochemical profiling identifies β-hydroxypyruvate as a potential mediator of type 2 diabetes in mice and humans. Diabetes 2015;64(4):1383−94.

[19] Nagata C, Nakamura K, Wada K, Tsuji M, Tamai Y, Kawachi T. Branched-chain amino acid intake and the risk of diabetes in a Japanese community: the Takayama study. Am J Epidemiol 2013;178(8):1226−32. Available from: http://www.ncbi.nlm.nih.gov/pubmed/24008908.

[20] Xu F, Tavintharan S, Sum CF, Woon K, Lim SC, Ong CN. Metabolic signature shift in type 2 diabetes mellitus revealed by mass spectrometry-based metabolomics. J Clin Endocrinol Metab 2013;98(6).

[21] Ferrannini E, Natali A, Camastra S, Nannipieri M, Mari A, Adam KP, et al. Early metabolic markers of the development of dysglycemia and type 2 diabetes and their physiological significance. Diabetes 2013;62(5):1730−7.

[22] Zheng X, Qiu Y, Zhong W, Baxter S, Su M, Li Q, et al. A targeted metabolomic protocol for short-chain fatty acids and branched-chain amino acids. Metabolomics 2013;9(4):818−27.

[23] De Vadder F, Kovatcheva-Datchary P, Goncalves D, Vinera J, Zitoun C, Duchampt A, et al. Microbiota-generated metabolites promote metabolic benefits via gut-brain neural circuits. Cell 2014;156(1−2):84−96.

[24] Suhre K, Meisinger C, Döring A, Altmaier E, Belcredi P, Gieger C, et al. Metabolic footprint of diabetes: a multiplatform metabolomics study in an epidemiological setting. PLoS One 2010;5(11):e13953. Available from: http://www.pubmedcentral.nih.gov/articlerender.fcgi?artid=2978704&tool=pmcentrez&rendertype=abstract.

[25] Batch BC, Shah SH, Newgard CB, Turer CB, Haynes C, Bain JR, et al. Branched chain amino acids are novel biomarkers for discrimination of metabolic wellness. Metabolism 2013;62(7):961−9.

[26] Mook-Kanamori DO, El-Din Selim MM, Takiddin AH, Al-Homsi H, Al-Mahmoud K a S, Al-Obaidli A, et al. 1,5-anhydroglucitol in saliva is a noninvasive marker of short-term glycemic control. J Clin Endocrinol Metab 2014;99(3):479−83.

[27] Vrieze A, Van Nood E, Holleman F, Salojärvi J, Kootte RS, Bartelsman JFWM, et al. Transfer of intestinal microbiota from lean donors increases insulin sensitivity in individuals with metabolic syndrome. 913.e7−916.e7. Gastroenterology 2012;143(4). https://doi.org/10.1053/j.gastro.2012.06.031. Erratum in Gastroenterology 2013:250

[28] Cani PD, Amar J, Iglesias MA, Poggi M, Knauf C, Bastelica D, et al. Metabolic endotoxemia initiates obesity and insulin resistance. Diabetes 2007;56(7):1761−72. Available from: http://diabetes.diabetesjournals.org/content/56/7/1761.abstract.

[29] Laugerette F, Vors C, Peretti N, Michalski MC. Complex links between dietary lipids, endogenous endotoxins and metabolic inflammation. Biochimie 2011;93(1):39−45.

[30] Cani PD, Bibiloni R, Knauf C, Waget A, Neyrinck AM, Delzenne NM, et al. Changes in gut microbiota control metabolic endotoxemia-induced inflammation in high-fat diet-induced obesity and diabetes in mice. Diabetes 2008;57(6):1470−81.

[31] Garidou L, Pomié C, Klopp P, Waget A, Charpentier J, Aloulou M, et al. The gut microbiota regulates intestinal CD4 T cells expressing RORgt and controls metabolic disease. Cell Metab 2015;22(1):100−12.

[32] Hong C-P, Park A, Yang B-G, Yun CH, Kwak M-J, Lee G-W, et al. Gut-specific delivery of T-helper 17 cells reduces obesity and insulin resistance in mice. Gastroenterology 2017;152(8):1998−2010. Available from: http://linkinghub.elsevier.com/retrieve/pii/S0016508517301804.

[33] Thomas C, Pellicciari R, Pruzanski M, Auwerx J, Schoonjans K. Targeting bile-acid signalling for metabolic diseases. Nat Rev Drug Discov 2008;7(8):678−93. Available from: http://www.nature.com/doifinder/10.1038/nrd2619.

[34] NIE Y, YAN X. Cross-talk between bile acids and intestinal microbiota in host metabolism and health. J Zhejiang Univ B 2015;16:436−46. Available from: http://link.springer.com/article/10.1631/jzus.B1400327.

[35] Sayin SI, Wahlström A, Felin J, Jäntti S, Marschall HU, Bamberg K, et al. Gut microbiota regulates bile acid metabolism by reducing the levels of tauro-beta-muricholic acid, a naturally occurring FXR antagonist. Cell Metab 2013;17(2):225−35.

[36] Prawitt J, Caron S, Staels B. Bile acid metabolism and the pathogenesis of type 2 diabetes. Curr Diabetes Rep 2011;11(3):160−6.

[37] Staels B, Fonseca VA. Bile acids and metabolic regulation: mechanisms and clinical responses to bile acid sequestration. Diabetes Care 2009;32.

[38] Gonzalez FJ, Jiang C, Patterson AD. An intestinal microbiota−farnesoid X receptor axis modulates metabolic disease. Gastroenterology 2016;151:845−59.

[39] Francino MP. Antibiotics and the human gut microbiome: dysbioses and accumulation of resistances. Front Microbiol 2016;12.

[40] Willing BP, Russell SL, Finlay BB. Shifting the balance: antibiotic effects on host-microbiota mutualism. Nat Rev Microbiol 2011;9:233−43.

[41] Atarashi K, Tanoue T, Shima T, Imaoka A, Kuwahara T, Momose Y, et al. Induction of colonic regulatory T cells by indigenous *Clostridium* species. Science 2011;331(6015):337−41. Available from: http://www.sciencemag.org/cgi/doi/10.1126/science.1198469.

[42] Perez-Cobas AE, Gosalbes MJ, Friedrichs A, Knecht H, Artacho A, Eismann K, et al. Gut microbiota disturbance during antibiotic therapy: a multi-omic approach. Gut 2012;62(11):1−11. Available from: http://www.ncbi.nlm.nih.gov/pubmed/23236009.

[43] Kobyliak N, Conte C, Cammarota G, Haley AP, Styriak I, Gaspar L, et al. Probiotics in prevention and treatment of obesity: a critical view. Nutr Metab (Lond) 2016;13(1):14. Available from: http://www.nutritionandmetabolism.com/content/13/1/14.

[44] Everard A, Belzer C, Geurts L, Ouwerkerk JP, Druart C, Bindels LB, et al. Cross-talk between *Akkermansia* muciniphila and intestinal epithelium controls diet-induced obesity. Proc Natl Acad Sci U S A 2013;110(22):9066−71. Available from: http://www.pnas.org/cgi/content/long/110/22/9066.

[45] Desai MS, Seekatz AM, Koropatkin NM, Kamada N, Hickey CA, Wolter M, et al. A dietary fiber-deprived gut microbiota degrades the colonic mucus barrier and enhances pathogen susceptibility. Cell 2016;167(5). 1339.e21−1353.e21.

[46] Liang S, Webb T, Li Z. Probiotic antigens stimulate hepatic natural killer T cells. Immunology 2014;141(2):203−10.

[47] Yadav H, Lee JH, Lloyd J, Walter P, Rane SG. Beneficial metabolic effects of a probiotic via butyrate-induced GLP-1 hormone secretion. J Biol Chem 2013;288(35):25088−97.

[48] Cani PD, Osto M, Geurts L, Everard A. Involvement of gut microbiota in the development of low-grade inflammation and type 2 diabetes associated with obesity. Gut Microbes 2012;3(4):279−88.

[49] Geurts L, Neyrinck AM, Delzenne NM, Knauf C, Cani PD. Gut microbiota controls adipose tissue expansion, gut barrier and glucose metabolism: novel insights into molecular targets and interventions using prebiotics. Benef Microbes 2014;5(1):3−17.

[50] Muccioli GG, Naslain D, B??ckhed F, Reigstad CS, Lambert DM, Delzenne NM, et al. The endocannabinoid system links gut microbiota to adipogenesis. Mol Syst Biol 2010;6.

[51] Jeffery IB, O'Toole PW. Diet-microbiota interactions and their implications for healthy living. Nutrients 2013;5(1):234−52.

[52] O'Sullivan O, Cronin O, Clarke SF, Murphy EF, Molloy MG, Shanahan F, et al. Exercise and the microbiota. Gut Microb 2015;6(2):131−6.

[53] Balducci S, Zanuso S, Nicolucci A, Fernando F, Cavallo S, Cardelli P, et al. Anti-inflammatory effect of exercise training in subjects with type 2 diabetes and the metabolic syndrome is dependent on exercise modalities and independent of weight loss. Nutr Metab Cardiovasc Dis 2010;20(8):608−17.

[54] Queipo-Ortuño MI, Seoane LM, Murri M, Pardo M, Gomez-Zumaquero JM, Cardona F, et al. Gut microbiota composition in male rat models under different nutritional status and physical activity and its association with serum leptin and ghrelin levels. PLoS One 2013;8(5).

[55] Chen T, Long W, Zhang C, Liu S, Zhao L, Hamaker BR. Fiber-utilizing capacity varies in *Prevotella*- versus *Bacteroides*-dominated gut microbiota. Sci Rep 2017;7(1):1−7. Springer US.

[56] Lin HV, Frassetto A, Kowalik EJ, Nawrocki AR, Lu MM, Kosinski JR, et al. Butyrate and propionate protect against diet-induced obesity and regulate gut hormones via free fatty acid receptor 3-independent mechanisms. PLoS One 2012;7(4):1−9.

Chapter 23

The Gut Microbiota as a Therapeutic Approach for Obesity

Trevor O. Kirby, Emily K. Hendrix and Javier Ochoa-Repáraz

Department of Biology, Eastern Washington University, Cheney, WA 99004, United States

INTRODUCTION: THE GUT MICROBIOME AND OBESITY

In 2016 the World Health Organization estimated 1.9 billion adults were overweight, among which 650 millions were considered obese [1]. A comprehensive global study conducted by the NCD Risk Factor Collaboration showed that the rate of obesity in men increased from 3.2% to 10.8% between 1975 and 2014 (a 3.4-fold increase). For women, it increased from 6.4% to 14.9% (a 2.3-fold increase) [2]. Incidence rates of obesity have nearly tripled since 1975 [1]. The obesity rate in children increased from 4.2% to 6.7% between 1990 and 2010 (a 1.6-fold increase) [3]. In 2016, over 340 million children and adolescents aged 5−19 were obese [1].

Defining obesity remains controversial, according to the American Obesity Treatment Association [4]. Predominantly a Body Mass Index (BMI) above 30 kg/m^2 is considered obese [5]. However, such a simplistic definition can be problematic. A BMI of 30 kg/m^2 brought on by large muscle mass with little fatty tissue is substantially different from a BMI >30 kg/m^2 with reduced muscle mass and more fatty tissue [4]. Obesity triggers are also relatively convoluted. It is now accepted that genetics, diet, environment, and physical activity play distinct roles in the development of obesity [5−7].

An emerging environmental factor associated with a number of human diseases of the metabolic, neural, and immune system is the microbiota, the sum of all microbial organisms including bacteria, archaea, viruses, as well as fungi and other eukaryotic microorganisms that live on and within us. The microbiome is considered the sum of the biotic, living forms and their products, and abiotic habitat within an environment or host. Generally, a healthy microbiome tends to be rich in diversity and functionally complex [8]. Recent evidences suggest, however, that the obese microbiome shows deviations from the normal characteristics seen in a "healthy" microbiome [9,10].

In this review chapter, these deviations and their potential causations are addressed, but it is important to note that simple signatures of obesity, other than at the phylum level, are difficult to detect in the gut microbiome [8,11]. Therefore, it is more important to identify potential factors that cause functional alterations in the obese gut microbiome.

THE MULTIFACTORIAL ASSOCIATIONS BETWEEN THE MICROBIOTA AND THE HOST

Can We Identify an Obese Microbiota?

The technological advances in molecular biology, genetics and bioinformatics allow us these days, to look at the intestinal microbiota as a complex ecosystem, where hundreds of different species that belong to the three domains (bacteria, archaea, and eukaryotes) interact between each other, the host and also environmental factors. Although we can describe the individual components of the microbiota, and even anticipate potential functionality, many questions arise from the analysis of the 16S ribosomal DNA (rDNA) sequences, or from whole-genomes.

An increase in the Firmicutes to Bacteroidetes (F/B) ratio was shown in mice fed with westernized diet [12]. Germ-free mice show increased resistance to glucose intolerance than conventionally housed mice [13]. Furthermore, the transfer of fecal content from obese mice results in the acquisition of an obese phenotype, which is not observed when mice receive fecal content from lean mice [14]. After feeding animals with high-fat diets (HFD), early studies showed that the alterations of the microbiota with antibiotics (ampicillin and neomycin) had an impact on type 2 diabetes and obesity [15]. Antibiotics were administered through the duration of the experiment (4 weeks).

When compared to mice on standard diet, HFD-fed mice treated with antibiotics showed a lower degree of intestinal barrier epithelium disruption, reduced endotoxemia, and proinflammatory markers and those indicative of oxidative stress. The treatment reduced glucose intolerance and weight gain. In contrast, a more recent study in HFD-fed mice showed that penicillin G from birth to the end of the experiment (32 weeks) was followed by exacerbated glucose intolerance and increased homeostasis model assessment (HOMA-IR) index and body weight [16].

The concentrations of IL-6 in serum, a proinflammatory cytokine, were increased in female mice, but not male. Leptin was increased in mice treated with penicillin G regardless of sex, whereas ghrelin levels were decreased in both sexes [17]. Despite the differences in the experimental methods and antibiotic use, the studies shown above indicate an association of microbiota perturbations with weight gain/loss dependent on diet. Combined with the germ-free studies, they suggest a functional link between the microbiota and obesity.

Human studies are showing a more complex scenario. The changes in F/B ratio predicted in the murine models are not necessarily observed in human studies [18]. In this regard, the structure of the microbiota alpha diversity was reduced in obese versus lean individuals [18], although other studies suggest no differences in diversity and evenness [11,19]. High interindividual variability affects interpretation of the results. Variability might be due to the sample sizes, but also to the difficulties of integrating all factors, including genetic and environmental that merge in the gut microbiota, and could impact the already complex interactions that define the composition and structure of the intestinal ecosystem [20].

A recent metaanalysis compared the composition and structure of obese and lean microbiotas [21]. A significant association between richness, evenness, diversity, and obesity emerged. However, the statistical approach failed to successfully separate between obese and lean individuals based only on the microbiota. Sample sizes were not sufficient to conclude differences on alpha diversity [21].

Initial studies suggested the possibility of defining clusters of populations, or enterotypes, based on the similarities in structure and function of the microbiota [22]; however, later studies indicated the potential limitations of such approach [23]. Significant modifications are rapidly observed in the microbiota as a consequence of changes in environmental factors, such as diet [24,25].

Genetics may also contribute to dissimilarities, as well as to microbiota similarities between individuals [18,26–28], that would add up to shared factors such as diet or housing. Studies in monozygotic and dizygotic twins suggest a partial genetic dependency of the interindividual microbiota diversity [29,30]. Goodrich et al. compared the microbiota of 171 monozygotic and 245 dizygotic twin pairs, for composition, richness, and BMI [30]. Investigators confirmed the partial effect of genetics on the composition of the microbiota and also linked the heritability to specific taxa, such as *Christensenellaceae*.

The presence of *Christensenellaceae* was associated with a lean phenotype in germ-free mice transplanted with human obese or lean fecal contents. *Christensenellaceae* are gram-negative, nonspore forming, strict anaerobes. Transplantation of stool samples from a donor with undetectable *Christensenella* combined with live *Christensenella minuta*, a member of the family with identified genome [31], changed the overall structure of the microbiota and remarkably also reduced weight gain and percentage of adiposity in recipient mice [30].

Proposed Cellular and Molecular Mechanisms for the Potential Link Between Microbiota and Obesity

Inflammation and inflammatory cytokines are observed in obese individuals. Features of the Western diet such as saturated fatty acids, red meats, sweetened drinks, and increased salt intake lead to proinflammatory responses. According to a study by Swank and Goodwin, HFD could contribute to increased production of storage lipids and cholesterol, causing a reduction in membrane fluidity. As a result, this could lead to potential capillary obstruction and bring about inflammation [32,33]. High sugar uptakes could increase insulin production which would then upregulate the biosynthetic pathways of arachidonic acid and its proinflammatory derivatives [32]. High salt intake is linked to potential induction of proinflammatory IL-17-producing T helper-17 (Th17) cells and related cytokines, which could have autoimmunological impacts [32,34,35].

Intestinal Dysbiosis

Inflammation and dysbiosis are closely linked with increase in gut endotoxin/lipopolysaccharide presence. Regulatory T (Treg) cell populations become defective and the aryl hydrocarbon receptors and proinflammatory Th17 cells are activated [15,32,36]. The imbalance in gut microbiota can lead to low-grade endotoxemia as well as chronic intestinal and systemic inflammation. Systemic inflammation increases the risk of chronic inflammation [32].

Anti-inflammatory Nutrients and Obesity

Polyphenols have been known to be anti-inflammatory, immunomodulatory, antiangiogenic as well as exhibit antiviral properties [37,38]. How they confer anti-inflammatory effects are based on their chemical structure [32,39]. Vitamin D_3 is hydroxylated by the liver to calcidiol by the P450 enzymes CYP27A1 or CYP2R1. Then calcidiol is activated in the kidney by CYP27B1 to calcitriol. Calcitriol is the active form of vitamin D, which can then be used in other cellular pathways [32,40]. Thiolic compounds, such as α-lipoic acid, stabilize the blood—brain barrier's integrity and stimulate cyclic adenosine monophosphate production, as well as the activity of protein kinase A [32,41].

Diets rich in complex carbohydrates increase the F/B ratio, due to the fact that Bacteroidetes groups tend to metabolize complex glycans. This would increase the production of butyrate from bacterial metabolism, which downregulates the activation of the NF-kB pathway, giving an overall anti-inflammatory response. The Western diet instead provides more substrates to Firmicutes. One such group of Firmicutes, Mollicutes, is more adept at energy extraction and harvesting and can, in turn, increase caloric intake [32]. The increase in Firmicutes has been associated with pro-inflammatory responses.

Studying the Gut Metabolome in Obesity

Metabolites are largely formed as products and by-products of genomic regulation, but they are also derived from diet and medicinal or toxic compounds [42]. The gut microbiome plays a critical role in nutrient digestion, including synthesis of micronutrients, catabolism of various toxicants, and production of short-chain fatty acids (SCFAs) [43,44].

Bariatric Surgery

High levels of SCFA are correlated with high diversity of microbial communities in the gut but are also found to be elevated in overweight and obese individuals. Through microbial fermentation, SCFAs create a caloric surplus [45]. However, propionate, also a SCFA, has been shown to stimulate gluconeogenesis in the gut following gastric bypass surgery. An increase in propionate reduces weight gain and is also correlated with an increased defense against obesity related to diet [46]. Patients who underwent gastric bypass surgery, rather than laparoscopic band surgery, have higher diversity in their gut microbiota that is associated with a healthy gut. Though diversity is largely increased versus patients who underwent laparoscopic band surgery, the gut microbiota of gastric bypass patients is distinct from normal and obese individuals, which may present future challenges [46].

Metabolic Markers Related to Carbohydrate and Amino Acid

Bacteroidetes digest carbohydrates via a series of metabolic pathways [44]. They also possess genes that encode for glycan-foraging enzymes to hydrolyze starch and degrade host-derived glycoconjugates and glycosaminoglycans that include hyaluronic acid, mucins, cellulose, and heparin. The fermentation of dietary carbohydrates will result in SCFAs [44,47].

According to the American Heart Association, metabolic syndrome affects about 23% of adults, presenting as abdominal obesity, triglyceride levels of >150 mg/dL and fasting glucose levels of $\geq$100 mg/dL in blood, and hypertension [48]. Shifts in microbial diversity promote dysbiosis of the gut and can trigger inflammation, lipid accumulation, and insulin insensitivity [49]. Schäfer et al. (2013) noted that in mice genetically predisposed to obesity, various derivatives of phosphatidylcholine were significantly lower than in lean mice, whereas serine, glycine, and arginine were higher. These metabolites were genetically linked to higher expression of adipocytes [50].

Particularly, trimethylamine-N-oxide was shown to modulate metabolic disease phenotypes and was also able to accurately predict obesity, impaired glucose tolerance, and behavior [51]. Methyl succinate, asparagine, urate, kynurenic acid, glycine, glutamic acid, and serine were associated with BMI, whereas other metabolites were successful in predicting weight gain after 5 years [52].

MICROBIOTA-BASED THERAPEUTIC APPROACHES

Antibiotics and Obesity

Antibiotics have been a part of human life and diet as far back as 350 CE. Human remains found in Sudan around this time exhibited traces of tetracycline. It was not until the 20th century that antibiotics have been used specifically as a chemotherapeutic agents to treat bacterial infections [53], reducing mortality up to 75%. However, over usage of antibiotics may have adverse effects [54], such as colitis by *Clostridium difficile* [55]. Broad-spectrum antibiotics cause structural changes in the gut microbiome that leave individuals more susceptible to *C. difficile* and other enteric pathogens [56].

Children exposed to antibiotics in the first 3 years of life showed lower gut microbe diversity [57]. Furthermore, the gut microbiome structure was less stable [57]. Increased usage of antibiotics has been associated with obesity and other metabolically associated diseases [58].

Phage Therapy

Bacteriophage (phage) therapy for bacterial infection is an option that bypasses overexposure to antibiotics. Phages are bacteria-specific viruses that target individual host species. Although initially introduced in 1919 [59], phage therapy is relatively new and therefore the safety of this therapeutic is still under evaluation. The oral delivery of phages, however, is considered to be a safe route of administration [60,61]. Interestingly, the usage of phages to bypass the intestinal epithelium, reaching the gut-associated lymphoid tissue as well as the blood stream directly, has also been contemplated [62].

Bacterial translocation is generally mediated by overgrowth of bacterial populations in the small intestine, increased intestinal permeability, and immune system deficiencies [63]. It is a normal process potentially crucial for immune tolerance concerning resident microbiota [64]; however, it can also have adverse effects associated with inflammation and endotoxemia [65]. Obesity has been proposed to have bacterial 16S rDNA sequences that are detectable in the blood [66]. The generation of phages specific for these bacterial translocations could be a novel therapy for obesity.

Of the intestinal virome, bacteriophages are the most predominant viral particle [67,68]. Their relative importance in disease etiology should be investigated further because they have been associated with several dysbioses [69] including inflammatory bowel disease [70].

Moreover, antibiotic resistance genes have been located in phage genomes [71]. Horizontal gene transfer from lysogenic phage to their bacterial hosts may also generate further complications, with pathogenic bacteria rendering them obsolete [72]. Conversely, lysogenic phage cocktails, such as one proposed by Regeimbal et al. for *Acinetobacter baumannii*, have been proposed as viable options that bypass the challenges of horizontal gene transfer [73].

Naturally Occurring Dietary Compounds, Obesity, and the Microbiome

Obesogenic microbiome patterns that differ from nonobese individuals have been outlined [45,74–76]. The F/B ratio can be diet-influenced independently from obesity [77]. Another aspect of the obese microbiome is reduction in bacterial diversity [77,78].

Western diets with high levels of carbohydrates and saturated fats can lead to chronic inflammation due to increased adiposity [79]. Chlorogenic acid is a phenolic acid that can be found in coffee and tea [80] with anti-inflammatory and antioxidant properties [81]. More importantly, chlorogenic acid has been associated with anti-obesity properties [82]. By preventing hepatic glucose-6-phosphatase activity, the release of glucose is inhibited [83]. Furthermore, chlorogenic acid can inhibit glucose-6-phosphate translocase, which would prevent glucose absorption in the small intestine [83]. Chlorogenic acid undergoes biotransformation in the gut microbiome and in turn induces selective changes, including bactericidal activity against species such as *Escherichia coli* [82,84].

Glycyrrhizin, a compound extracted from the roots of *Glycyrrhiza glabra*, also has been known to have anti-inflammatory [85] and anti-obesity properties [86]. In HFD-induced obese rats there was a significant reduction in weight gain as well as insulin resistance [86]. The reduction in liver malondialdehyde levels suggested a reduction in obesity-induced oxidative stress [86]. There currently is no literature that assesses the implications of glycyrrhizin on the gut microbiome; however, most literature only reports antiviral properties as opposed to antimicrobial properties [85].

Fecal Microbiota Transplantation

In the case of *C. difficile*, fecal microbiota transplantation is a simple process that takes only one transplantation event to rectify the infection [87,88]. Fecal transplantation via gastroscopy can rarely result in aspiration [88], colonoscopy appearing to be a safer alternative route. Unfortunately, there is very little research on fecal microbiota transplantation and obesity, outside of animal models. Some have argued that the causation of gut microbiota influencing obesity is still debatable [89,90]. However, fecal material from obese animals caused normal weight mice to shift toward obese states [10].

CONCLUSIONS

The structure of the gut microbiome plays a critical role in the development of obesity by being subjected to functional changes. The "obese microbiome" can result in the increase of adipose tissue buildup, which potentially leads to chronic inflammation characteristic of metabolic diseases. Alternatives to antibiotic usage, such as phage therapy, may be used to avoid structural changes in normal microbiota, but may also be modified to make intentional changes away from the obese microbiome. Diet plays a significant role in the development of obesity, but changing cultural diets is impractical. Utilizing natural, diet-borne compounds as supplements or alternatives may prove more manageable.

LIST OF ACRONYMS AND ABBREVIATIONS

ALA α-Lipoic acid
BMI Body Mass Index
cAMP Cyclic adenosine monophosphate
F/B ratio Firmicutes to Bacteroidetes ratio
GALT Gut-associated lymphoid tissue
HFD High-fat diets
HOMA-IR Homeostasis model assessment index
IL Interleukin
NF-kB Nuclear factor kappa-light-chain-enhancer of activated B cells
rDNA Ribosomal DNA
SCFA Short-chain fatty acids
Th T helper
TMAO Trimethylamine-N-oxide
Treg Regulatory T

ACKNOWLEDGMENTS

We would like to thank the Biology department at EWU for supporting the completion of this book chapter.

REFERENCES

[1] Organization WH. Obesity and overweight. October 2017. http://www.who.int/mediacentre/factsheets/fs311/en/2017.

[2] Collaboration NRF. Trends in adult body-mass index in 200 countries from 1975 to 2014: a pooled analysis of 1698 population-based measurement studies with 19·2 million participants. Lancet 2016;387(10026):1377—96.

[3] de Onis M, Blossner M, Borghi E. Global prevalence and trends of overweight and obesity among preschool children. Am J Clin Nutr 2010;92(5):1257—64.

[4] Association AOT. What is obesity. 2017. http://www.americanobesity.org/obesitybasics.htm2017.

[5] Yanovski SZ, Yanovski JA. Long-term drug treatment for obesity: a systematic and clinical review. J Am Med Assoc 2014;311(1):74—86.

[6] Casazza K, Brown A, Astrup A, et al. Weighing the evidence of common beliefs in obesity research. Crit Rev Food Sci Nutr 2015;55(14):2014—53.

[7] Yanovski JA. Pediatric obesity. An introduction. Appetite 2015;93:3—12.

[8] Hollister EB, Riehle K, Luna RA, et al. Structure and function of the healthy pre-adolescent pediatric gut microbiome. Microbiome 2015;3:36.

[9] Ridaura VK, Faith JJ, Rey FE, et al. Gut microbiota from twins discordant for obesity modulate metabolism in mice. Science 2013;341(6150):1241214.

[10] Turnbaugh PJ, Ley RE, Mahowald MA, Magrini V, Mardis ER, Gordon JI. An obesity-associated gut microbiome with increased capacity for energy harvest. Nature 2006;444:1027—31.

[11] Finucane MM, Sharpton TJ, Laurent TJ, Pollard KS. A taxonomic signature of obesity in the microbiome? Getting to the guts of the matter. PLoS One 2014;9(1):e84689.

[12] Turnbaugh PJ, Backhed F, Fulton L, Gordon JI. Diet-induced obesity is linked to marked but reversible alterations in the mouse distal gut microbiome. Cell Host Microbe 2008;3(4):213—23.

[13] Backhed F, Manchester JK, Semenkovich CF, Gordon JI. Mechanisms underlying the resistance to diet-induced obesity in germ-free mice. Proc Natl Acad Sci U S A 2007;104(3):979—84.

[14] Ley RE, Bäckhed F, Turnbaugh P, Lozupone CA, Knight RD, Gordon JI. Obesity alters gut microbial ecology. Proc Natl Acad Sci U S A 2005;102(31):11070—5.

[15] Cani PD, Bibiloni R, Knauf C, et al. Changes in gut microbiota control metabolic endotoxemia-induced inflammation in high-fat diet-induced obesity and diabetes in mice. Diabetes 2008;57(6):1470—81.

[16] Matthews DR, Hosker JP, Rudenski AS, Naylor BA, Treacher DF, Turner RC. Homeostasis model assessment: insulin resistance and beta-cell function from fasting plasma glucose and insulin concentrations in man. Diabetologia 1985;28(7):412−9.

[17] Mahana D, Trent CM, Kurtz ZD, et al. Antibiotic perturbation of the murine gut microbiome enhances the adiposity, insulin resistance, and liver disease associated with high-fat diet. Genome Med 2016;8(1):48.

[18] Turnbaugh PJ, Hamady M, Yatsunenko T, et al. A core gut microbiome in obese and lean twins. Nature 2009;457(7228):480−4.

[19] Walters WA, Xu Z, Knight R. Meta-analyses of human gut microbes associated with obesity and IBD. FEBS Lett 2014;588(22):4223−33.

[20] Ochoa-Reparaz J, Kasper LH. Gut microbiome and the risk factors in central nervous system autoimmunity. FEBS Lett 2014;588(22):4214−22.

[21] Sze MA, Schloss PD. Looking for a signal in the noise: revisiting obesity and the microbiome. mBio 2016;7(4).

[22] Arumugam M, Raes J, Pelletier E, et al. Enterotypes of the human gut microbiome. Nature 2011;473(7346):174−80.

[23] Knights D, Ward TL, McKinlay CE, et al. Rethinking "enterotypes". Cell Host Microbe 2014;16(4):433−7.

[24] Wu GD, Chen J, Hoffmann C, et al. Linking long-term dietary patterns with gut microbial enterotypes. Science 2011;334(6052):105−8.

[25] David LA, Maurice CF, Carmody RN, et al. Diet rapidly and reproducibly alters the human gut microbiome. Nature 2014;505(7484):559−63.

[26] Lee S, Sung J, Lee J, Ko G. Comparison of the gut microbiotas of healthy adult twins living in South Korea and the United States. Appl Environ Microbiol 2011;77(20):7433−7.

[27] Tims S, Derom C, Jonkers DM, et al. Microbiota conservation and BMI signatures in adult monozygotic twins. ISME J 2013;7(4):707−17.

[28] Yatsunenko T, Rey FE, Manary MJ, et al. Human gut microbiome viewed across age and geography. Nature 2012;486(7402):222−7.

[29] Hansen EE, Lozupone CA, Rey FE, et al. Pan-genome of the dominant human gut-associated archaeon, *Methanobrevibacter smithii*, studied in twins. Proc Natl Acad Sci U S A 2011;108(Suppl. 1):4599−606.

[30] Goodrich JK, Waters JL, Poole AC, et al. Human genetics shape the gut microbiome. Cell 2014;159(4):789−99.

[31] Rosa BA, Hallsworth-Pepin K, Martin J, Wollam A, Mitreva M. Genome sequence of *Christensenella minuta* DSM 22607^T. Genome Announc 2017;5(2). pii:e01451−16.

[32] Riccio P, Rossano R. Nutrition facts in multiple sclerosis. ASN Neuro 2015;7(1).

[33] Swank RL, Goodwin J. Review of MS patient survival on a Swank low saturated fat diet. Nutrition 2003;19(2):161−2.

[34] Kleinewietfeld M, Manzel A, Titze J, et al. Sodium chloride drives autoimmune disease by the induction of pathogenic TH17 cells. Nature 2013;496(7446):518−22.

[35] Wu C, Yosef N, Thalhamer T, et al. Induction of pathogenic TH17 cells by inducible salt-sensing kinase SGK1. Nature 2013;496(7446):513−7.

[36] Veldhoen M, Hirota K, Westendorf AM, et al. The aryl hydrocarbon receptor links TH17-cell-mediated autoimmunity to environmental toxins. Nature 2008;453(7191):106−9.

[37] Gupta SC, Tyagi AK, Deshmukh-Taskar P, Hinojosa M, Prasad S, Aggarwal BB. Downregulation of tumor necrosis factor and other proin-flammatory biomarkers by polyphenols. Arch Biochem Biophys 2014;559:91−9.

[38] Wang S, Moustaid-Moussa N, Chen L, et al. Novel insights of dietary polyphenols and obesity. J Nutr Biochem 2014;25(1):1−18.

[39] Liuzzi GM, Latronico T, Brana MT, et al. Structure-dependent inhibition of gelatinases by dietary antioxidants in rat astrocytes and sera of multiple sclerosis patients. Neurochem Res 2011;36(3):518−27.

[40] Schuster I. Cytochromes P450 are essential players in the vitamin D signaling system. Biochim Biophys Acta 2011;1814(1):186−99.

[41] Bavarsad Shahripour R, Harrigan MR, Alexandrov AV. N-acetylcysteine (NAC) in neurological disorders: mechanisms of action and therapeutic opportunities. Brain Behav 2014;4(2):108−22.

[42] Fessenden M. Metabolomics: small molecules, single cells. Nature 2016;540:153−5.

[43] Sonnenburg JL, Backhed F. Diet-microbiota interactions as moderators of human metabolism. Nature 2016;535(7610):56−64.

[44] Devaraj S, Hemarajata P, Versalovic J. The human gut microbiome and body metabolism: implications for obesity and diabetes. Clin Chem 2013;59(4):617−28.

[45] Bäckhed F, Ding H, Wang T, Hooper LV, Koh GY, Nagy A, Semenkovich CF, Gordon JI. The gut microbiota as an environmental factor that regulates fat storage. Proc Natl Acad Sci U S A 2004;101(44):15718−23.

[46] Ilhan ZE, DiBaise JK, Isern NG, et al. Distinctive microbiomes and metabolites linked with weight loss after gastric bypass, but not gastric banding. ISME J 2017;11(9):2047−58.

[47] Turroni F, Bottacini F, Foroni E, et al. Genome analysis of *Bifidobacterium bifidum* PRL2010 reveals metabolic pathways for host-derived glycan foraging. Proc Natl Acad Sci U S A 2010;107(45):19514−9.

[48] Association AH. About metabolic syndrome (AHA). 2016. http://www.heart.org/HEARTORG/Conditions/More/MetabolicSyndrome/About-Metabolic-Syndrome_UCM_301920_Article.jsp#.

[49] Boulange CL, Neves AL, Chilloux J, Nicholson JK, Dumas ME. Impact of the gut microbiota on inflammation, obesity, and metabolic disease. Genome Med 2016;8(1):42.

[50] Schäfer N, Zhonghao Y, Wagener A, et al. Changes in metabolite profiles caused by genetically determined obesity in mice. Metabolomics 2014;10:461−72.

[51] Dumas ME, Rothwell AR, Hoyles L, et al. Microbial-host co-metabolites are prodromal markers predicting phenotypic heterogeneity in behavior, obesity, and impaired glucose tolerance. Cell Rep 2017;20(1):136−48.

[52] Zhao H, Shen J, Djukovic D, et al. Metabolomics-identified metabolites associated with body mass index and prospective weight gain among Mexican American women. Obes Sci Pract 2016;2(3):309−17.

[53] Aminov RI. A brief history of the antibiotic era: lessons learned and challenges for the future. Front Microbiol 2010;1:134.

[54] Spellberg B. The future of antibiotics. Crit Care 2014;18(228):1−7.

[55] Zarandi ER, Mansouri S, Nakhaee N, Sarafzadeh F, Iranmanesh Z, Moradi M. Frequency of antibiotic associated diarrhea caused by *Clostridium difficile* among hospitalized patients in intensive care unit, Kerman, Iran. Gastroenterol Hepatol 2017;10(3):229−34.

[56] Ross CL, Spinler JK, Savidge TC. Structural and functional changes within the gut microbiota and susceptibility to *Clostridium difficile* infection. Anaerobe 2016;41:37−43.

[57] Yassour M, Vatanen T, Siljander H, et al. Natural history of the infant gut microbiome and impact of antibiotic treatment on bacterial strain diversity and stability. Sci Transl Med 2016;8(343):343−81.

[58] Leong KSW, Derraik JGB, Hofman PL, Cutfield WS. Antibiotics, gut microbiome and obesity. Clin Endocrinol 2017;88(2).

[59] Chanishvili N. Phage therapy−history from Twort and d'Herelle through Soviet experience to current approaches. Adv Virus Res 2012;83:3−40.

[60] Bruttin A, Brussow H. Human volunteers receiving *Escherichia coli* phage T4 orally: a safety test of phage therapy. Antimicrob Agents Chemother 2005;49(7):2874−8.

[61] Merabishvili M, Pirnay JP, Verbeken G, et al. Quality-controlled small-scale production of a well-defined bacteriophage cocktail for use in human clinical trials. PLoS One 2009;4(3):e4944.

[62] Gorski A, Wazna E, Dabrowska BW, Dabrowska K, Switala-Jelen K, Miedzybrodzki R. Bacteriophage translocation. FEMS Immunol Med Microbiol 2006;46(3):313−9.

[63] Guarner F, Malagelada J-R. Gut flora in health and disease. Lancet 2003;361(9356):512−9.

[64] Gebbers JO, Laissue JA. Bacterial translocation in the normal human appendix parallels the development of the local immune system. Ann N Y Acad Sci 2004;1029:337−43.

[65] Lichtman SM. Bacterial translocation in humans. J Ped Gastroenterol Nutr 2001;33(1):1−10.

[66] Païsse S, Valle C, Servant F, Courtney M, Burcelin R, Amar J, Lelouvier B. Comprehensive description of blood microbiome from healthy donors assessed by 16S targeted metagenomic sequencing. Transfusion 2016;56(5):1138−47.

[67] Arnold JW, Roach J, Azcarate-Peril MA. Emerging technologies for gut microbiome research. Trends Microbiol 2016;24(11):887−901.

[68] Breitbart M, Hewson I, Felts B, et al. Metagenomic analyses of an uncultured viral community from human feces. J Bacteriol 2003;185(20):6220−3.

[69] Mills S, Shanahan F, Stanton C, Hill C, Coffey A, Ross RP. Movers and shakers: influence of bacteriophages in shaping the mammalian gut microbiota. Gut Microb 2013;4(1):4−16.

[70] Lusiak-Szelachowska M, Weber-Dabrowska B, Jonczyk-Matysiak E, Wojciechowska R, Gorski A. Bacteriophages in the gastrointestinal tract and their implications. Gut Pathog 2017;9:44.

[71] Modi SR, Lee HH, Spina CS, Collins JJ. Antibiotic treatment expands the resistance reservoir and ecological network of the phage metagenome. Nature 2013;499(7457):219−22.

[72] Lin DM, Koskella B, Lin HC. Phage therapy: an alternative to antibiotics in the age of multi-drug resistance. World J Gastrointest Pharmacol Ther 2017;8(3):162−73.

[73] Regeimbal JM, Jacobs AC, Corey BW, et al. Personalized therapeutic cocktail of wild environmental phages rescues mice from *Acinetobacter baumannii* wound infections. Antimicrob Agents Chemother 2016;60(10):5806−16.

[74] Byrne CS, Chambers ES, Morrison DJ, Frost G. The role of short chain fatty acids in appetite regulation and energy homeostasis. Int J Obes 2015;39(9):1331−8.

[75] Cani PD, Amar J, Iglesias MA, et al. Metabolic endotoxemia initiates obesity and insulin resistance. Diabetes 2007;56(7):1761−72.

[76] Stenman LK, Holma R, Korpela R. High-fat-induced intestinal permeability dysfunction associated with altered fecal bile acids. World J Gastroenterol 2012;18(9):923−9.

[77] Dalby MJ, Ross AW, Walker AW, Morgan PJ. Dietary uncoupling of gut microbiota and energy harvesting from obesity and glucose tolerance in mice. Cell Rep 2017;21(6):1521−33.

[78] Heiss CN, Olofsson LE. Gut microbiota-dependent modulation of energy metabolism. J Innate Immun 2017;10(3).

[79] Totsch SK, Quinn TL, Strath LJ, McMeekin LJ, Cowell RM, Gower BA, Sorge RE. The impact of the Standard American Diet in rats: effects on behavior, physiology and recovery from inflammatory injury. Scand J Pain 2017;17:30181−7.

[80] Meng S, Cao J, Feng Q, Peng J, Hu Y. Roles of chlorogenic acid on regulating glucose and lipids metabolism: a review. Evid Based Complement Alternat Med 2013:801457.

[81] Santana-Gálvez J, Cisneros-Zevallos L, Jacobo-Velázquez D. Chlorogenic acid: recent advances on its dual role as a food additive and a nutraceutical against metabolic syndrome. Molecules 2017;22(3):358.

[82] Naveed M, Hejazi V, Abbas M, et al. Chlorogenic acid (CGA): a pharmacological review and call for further research. Biomed Pharmacother 2017;97:67−74.

[83] Yukawa G, Mune M, Otani H, Tone Y, Liang X-M, Iwahashi H, Sakamoto W. Effects of coffee consumption on oxidative susceptibility of low-density lipoproteins and serum lipid levels in humans. Biochemistry (Mosc) 2004;69(1):70−4.

[84] Ayseli MT, Ayseli YI. Flavors of the future: health benefits of flavor precursors and volatile compounds in plant foods. Trends Food Sci Technol 2016;48:69−77.

[85] Sakai-Sugino K, Uematsu J, Kamada M, et al. Glycyrrhizin inhibits human parainfluenza virus type 2 replication by the inhibition of genome RNA, mRNA and protein syntheses. Drug Discov Ther 2017;11(5):246−52.

[86] Abo El-Magd NF, El-Mesery M, El-Karef A, El-Shishtawy MM. Glycyrrhizin ameliorates high fat diet-induced obesity in rats by activating NrF2 pathway. Life Sci 2017;17:30582−9.

[87] Bamba S, Nishida A, Imaeda H, et al. Successful treatment by fecal microbiota transplantation for Japanese patients with refractory *Clostridium difficile* infection: a prospective case series. J Microbiol Immunol Infect 2017;1182(17).

[88] Friedman-Korn T, Livovsky DM, Maharshak N, et al. Fecal transplantation for treatment of *Clostridium difficile* infection in elderly and debilitated patients. Dig Dis Sci 2017;63(1).

[89] Bakker GJ, Nieuwdorp M. Fecal microbiota transplantation: therapeutic potential for a multitude of diseases beyond *Clostridium difficile*. Microbiol Spectr 2017;5(4).

[90] Gérard P. Gut microbiome and obesity. How to prove causality? Ann Am Thorac Soc 2017;14(5):S354−6.

Chapter 24

The Gut Microbiome After Bariatric Surgery

Camila Solar[1], Alex Escalona[2] and Daniel Garrido[1]

[1]Department of Chemical and Bioprocess Engineering, School of Engineering, Pontificia Universidad Catolica de Chile, Vicuñ, Santiago, Chile;
[2]Department of Surgery, Faculty of Medicine, Universidad de Los Andes, Santiago, Chile

INTRODUCTION

According to the World Health Organization (2016), 1.9 billion adults over 18 years are overweight (BMI ≥ 25 kg/m^2), whereas 650 million are obese [1]. Obesity is thus considered as an epidemic worldwide, which is widespread across age, socioeconomic status, and development level of the countries [1]. According to the McKinsey World Institute (2014), obesity is the third social burden on which more money is invested in the world [2].

Bariatric surgery (BS) is considered as an effective treatment for weight loss, which is sustained in the long-term, improving also comorbidities related to obesity such as type 2 diabetes [3]. Currently two procedures are commonly practiced: sleeve gastrectomy (SG) and Roux-en-Y gastric bypass (GB). These surgeries have important differences regarding the procedure, preoperative period, and short- and long-term outcomes. SG is considered a restrictive procedure, consisting in the reduction of the size of the stomach, therefore reducing food intake. In GB, a division is made in the stomach, to create a stomach pouch that connects with the distal jejunum, preventing food from passing through the distal stomach, the duodenum, and the proximal jejunum. GB is characterized both by a restriction in food intake and a reduction in the absorption of nutrients (malabsorptive).

A metaanalysis of 21 studies concluded that at the second year postprocedure, GB patients have a significantly greater weight loss compared to SG [4]. In addition, a better control of type 2 diabetes has been observed in GB, displaying a similar improvement in the rest of the comorbidities. GB is the preferred procedure for severe obesity, accounting for 46.6% of all bariatric surgeries [5].

HORMONE-MEDIATED WEIGHT LOSS IN BARIATRIC SURGERY

BS procedures result in a reduction of hormones that promote appetite and, therefore, food intake. One of these hormones is ghrelin, which is synthesized in the stomach. Ghrelin regulates the activity of the central nervous system in areas associated with reward, which promote appetite [6]. On the other hand, this hormone modulates the metabolism of glucose, through the inhibition of glucose stimulated-insulin release. Its concentration appears diminished in GB patients for up to 2 years [7,8].

Adipokines are signaling molecules synthesized in adipose tissue that regulate several processes through autocrine, paracrine, and endocrine pathways [9]. One of them is leptin, which has a negative effect on adiposity, through the regulation of food intake and energy expenditure. One of its main action sites is the central nervous system, where it promotes satiety [10]. However, obesity is associated with increased levels of leptin, so a resistance to its adiposity-reducing effect has been proposed [11].

Under standard conditions, leptin promotes insulin sensitivity through alterations in muscle metabolism. However, leptin stimulates inflammatory processes associated with obesity and the development of type 2 diabetes [12–14]. Interestingly, leptin levels decrease in GB patients before weight loss and remain low for at least 12 months [15].

Adiponectin is one of the most abundant adipokines in the blood [16,17]. Unlike most adipokines, an increase in BMI is associated with a lower concentration of adiponectin [15]. The hormone is associated with various metabolic processes, such as lipid trafficking and glucose homeostasis [18−20]. It decreases endogenous glucose production through the reduction of key enzymes in gluconeogenesis. It is considered an agent that promotes insulin sensitivity by phosphory-lation in the insulin signaling cascade and the inhibition of triglyceride deposition in the liver and muscle through the promotion of β-oxidation and fatty acid combustion [21,22]. In both GB and SG patients, adiponectin levels increase in parallel with weight loss [23].

Another hormone altered in BS procedures is peptide YY, a polypeptide of 36 amino acids, released by enteroendocrine L cells in the mucosa of the digestive tract. It is expressed especially in the ileum and colon, in response to food. It participates in appetite reduction, promoting gastric emptying and intestinal motility [24]. Its levels are drastically reduced in obese subjects [25], and they appear to be restored after GB [26,27]. This results in appetite reduction, less food intake, and therefore weight loss.

After 35 months post-GB, the fasting plasma peptide YY concentration was similar to controls with similar BMI. However, these subjects displayed a much better postprandial response [28]. A very important intestinal hormone in energy homeostasis is glucagon-like peptide-1 (GLP-1), which is secreted mainly by L cells. GLP-1 is released in response to food intake, promoting the secretion of insulin and suppressing the release of glucagon, so it has a central role in glycemic control [29]. An increase in postprandial GLP-1 levels is observed in both GB and SG patients [30−32].

THE INTESTINAL MICROBIOME

The microbiome releases large quantities of short-chain fatty acids (SCFAs), mainly acetate, propionate, and butyrate, in a proportion of 60:20:20, respectively [33]. SCFAs are metabolized by the epithelial cells of the large intestine, representing 60%−70% of their energy requirement [34]. Butyrate is the preferred source of energy of colonic cells [35], followed by propionate and finally acetate [36]. It is estimated that 95% of the SCFAs are produced and absorbed within the colon [37], whereas the rest is excreted in the feces or directed to the peripheral circulation.

In the liver, acetate and butyrate function as lipogenic substrates, whereas propionate is used for de novo gluconeo-genesis, also suppressing lipogenesis [38,39]. In skeletal muscle, SCFAs induce the expression of genes involved in oxidative and glucose metabolism, storage, and inhibition of glycolysis [39,40]. Butyrate, on the other hand, stimulates fat oxidation [41]. In addition, SCFAs stimulate the secretion of GLP-1 in a mechanism dependent on receptors of free fatty acids. GPR41 and GPR43 are epithelial G protein-coupled receptors that mediate in great part the biological effects of SCFAs. The knockout (KO) mice develops inflammatory diseases such as obesity and colitis [42].

Butyrate has been associated with antiinflammatory effects and a positive role in obesity treatment [43,44]. In obese patients, a higher fecal concentration of SCFAs has been observed, in comparison to healthy controls [45,46]. This could not only reflect a major fermentative activity but it could also represent a decrease in the absorption of SCFA in the intestine, to inhibit energy consumption. A lower abundance of butyrate-producing bacteria in fecal samples from patients with type 2 diabetes is reported [47]. Mice fed a high-fat diet are protected against insulin resistance, inflammation, and increase in energy expenditure, in the presence of butyrate [33,48].

MICROBIOME DYSBIOSIS AND OBESITY

Dysbiosis results from either the loss of keystone bacterial species, reduction of diversity of the gut microbiome, or overgrowth of potentially toxigenic species such as *Clostridium difficile* or *Bacteroides fragilis* [44]. Conventionally raised mice have 40% more body fat than germ-free mice, even when the latter consumed a high-fat diet [48]. Certain protocols indicated a decrease in the Bacteroidetes phylum and a proportional increase in Firmicutes in obese individuals [49]. These observations were also reproduced in mice [50]. Firmicutes could extract energy from carbohydrates more efficiently than Bacteroidetes, an effect mediated by SCFAs, which are known to contribute significantly to the total energy obtained from diet [51]. In addition, butyrate-producing bacteria are significantly reduced, when the host is put on a dietary weight loss treatment [33].

A recent study alternatively identified the *Prevotella/Bacteroides* ratio, both genera of the Bacteroidetes phylum, as a possible marker associated with obesity [52]. Individuals who consumed a diet high in fiber and had a greater ratio gained 3.5 kg more compared with those with a lower ratio on the same diet. It is considered that the *Prevotella* genus is characterized by a higher production of SCFAs compared to other gut microbes [53], and that dietary fiber consumption correlates negatively with the abundance of these microorganisms in Western populations [54].

Individual Markers of Obesity

Another interesting taxon associated with a lean phenotype is *Christensenella minuta*. This microorganism from the Firmicutes phylum is able to reduce weight gain in a murine model of obesity [55].

Akkermansia muciniphila (phylum Verrucomicrobia) is a mucin-degrading bacteria, believed to contribute to intestinal health and glucose homeostasis [56,57]. It could represent 3%−5% of the bacterial community [58], mainly residing in the intestinal mucosa, an interface between the gut microbiome and host tissues [59]. Its abundance is negatively correlated with body mass [60−63]. Mice fed a high-fat diet [64] significantly reduced body weight and adiposity with such a supplement.

An increase in the gene expression of adipocyte differentiation and lipid oxidation markers was observed, suggesting that *A. muciniphila* could regulate adipose tissue metabolism, and therefore fat storage [64]. In addition, fasting diet-induced hyperglycemia was reversed. It is thought that fermentation products derived from *A. muciniphila* mucin utilization could serve as energy source for other gut microbes, in a cross-feeding mechanism [65], which could have a positive effect on host metabolism [66,67].

Yet metaanalyses indicate that the association between obesity, BMI, and gut microbiome alterations is rather weak and of small effect. Obesity is a multifactorial disease, and there are differences in methodologies assessing gut microbiome composition and dietary patterns of the studied subjects [66,67].

GUT MICROBIOME AND BARIATRIC SURGERY

Restrictive and malabsorptive effects of the GB procedure would not be sufficient to explain the weight loss observed in these subjects [68,69].

Animal evidence suggests a direct contribution of the gut microbiome to weight loss in BS [70]. The effects of weight loss and improvement in glucose metabolism are independent of the postoperative diet [70].

Obese mice were subjected to either GB or placebo surgery (SHAM), and the latter were fed ad libitum, or with caloric restriction, weight-matching GB animals. The GB group presented a marked increase in the genus *Escherichia* and *Akkermansia* after surgery compared with sham animals, independent of weight change or diet, and thus related to the surgical procedure. There was a higher proportion of cecum propionate and a decrease in acetate in GB mice. A lower concentration of acetate is associated with a lower rate of lipogenesis in the adipose tissue, which could explain the lower adiposity [71].

Germ-free mice were inoculated with fecal samples from mice undergoing GB (GB-R), sham surgery (SHAM-R), or weight-matched control. Only GB-R mice displayed significant weight loss and reduced adiposity. In addition, GB-R animals had a significant improvement in insulin sensitivity and a decrease in fasting triglyceride levels, compared with SHAM-R.

Microbiome Signature of Gastric Bypass Operation

The anatomical and physiological changes occurring in GB subjects, such as decreased gastric acid production and reduction of the small intestine, promote the growth of facultative anaerobes such as Proteobacteria over obligate anaerobes such as *Clostridia* (phylum Firmicutes). This could also be explained by the increased presence of dissolved oxygen in the gut. By bypassing the upper small intestine, it is believed that certain species commonly residing in the small intestine, such as *Enterobacteriaceae*, relocate to lower segments [72]. In addition, GB has been mostly associated with a decrease in Firmicutes (Table 24.1).

FINGERPRINTING SLEEVE GASTRECTOMY

After SG, certain studies found no significant differences between the pre and postoperative microbiome. We recently correlated changes in morbidly obese subjects, undergoing medical treatment, GB, or SG. As expected, weight loss in the first group was not effective after 6 months, no major changes were observed in their metabolism, and no significant alterations in their gut microbiome were observed. In contrast, subjects in GB and SG groups showed marked weight loss, and several metabolic and anthropometric markers were improved.

In both groups a significant increase in Proteobacteria was found, in keeping with previous studies. Although Bacteroidetes levels were increased in GB, they appeared lower 6 months after SG treatment. This could be explained by the specific physiological changes associated with both surgeries. Finally, we observed that weight loss and BMI reduction in GB positively correlated with changes in Proteobacteria, whereas in SG subjects the change in Bacteroidetes presented several positive correlations, with improvements in metabolic and liver function markers [73]. Diabetes remission could similarly be a relevant variable [74].

TABLE 24 1 Summary of Clinical Studies in Obese Subjects Evaluating the Impact of Obesity Treatments on the Gut Microbiome

Study	MT	GB	SG	Number of Subjects
Medina et al. [73]	No change	Bacteroidetes ↑ Proteobacteria ↑ Firmicutes ↓ Actinobacteria ↑	Proteobacteria ↑ Bacteroidetes ↓	19
Murphy et al. [74]	—	Firmicutes ↑ Actinobacteria ↑ Bacteroidetes ↓	Bacteroidetes ↑	14
Zhang et al. [75]	—	Firmicutes ↓ Gammaproteobacteria ↑	—	9
Damms-Machado et al. [76]	Bacteroidetes ↓ Firmicutes↑	—	Bacteroidetes ↑ Firmicutes ↓	10
Palleja et al. [77]	—	Proteobacteria ↑ *Escherichia coli* ↑ *Klebsiella pneumoniae* ↑ *Akkermansia muciniphila* ↑ *Faecalibacterium prausnitzii* ↓	—	13
Furet et al. [78]	—	*E. coli* ↑ *F. prausnitzii* ↑	—	43
Kong et al. [79]	—	Proteobacteria ↑ Firmicutes ↓ Actinobacteria ↓ Bacteroidetes ↑	—	30

GB, gastric bypass; *MT*, medical treatment; *SG*, sleeve gastrectomy; ↑, significant increase; ↓, significant decrease.

It was shown using shotgun metagenomics that the changes induced in the gut microbiome after GB are stable for up to 9 years, when matching the patients by weight loss. The postsurgery microbiome was characterized by increases in Proteobacteria. Transplanting the microbiome from these subjects to germ-free mice reduced adiposity and altered metabolism in these animals [80]. Metagenomics was also useful to correlate changes in functional categories of the microbiome, with the observed changes.

CONTRIBUTION OF BILE ACIDS

There is evidence that in both SG and GB there is an increase in the concentration of bile acids [81], which could help in decreasing luminal pH. Secondary biliary acids appear to increase with GB, which could contribute to a higher utilization of fats in the gut, therefore contributing to weight loss [82,83].

On the other hand, bile acids are molecules that modulate lipid metabolism, secretion of intestinal peptides and glycemic control [84]. It has been shown that bile acids promote the expression of GLP-1. An increased effect of bile acids on L cells would contribute to an improvement in glycemic control, in patients undergoing BS [85,86]. An increase in the concentration of bile acids, secretion of GLP-1 and PYY, and improvement in HOMA-IR index has been documented after both GB and SG [87]. The elevation was much more progressive and slow (first year postprocedure), than of GLP-1 and PYY (first week). The authors concluded that the bile acids contribute to a long-term metabolic improvement through a GLP-1-independent mechanism.

In addition, bile acids can bind to the farnesoid X receptor (FXR), which is expressed in the liver, adipose tissue, and intestine. It is associated with the regulation of lipid and glucose metabolism. An increase in FXR signaling has been observed in SG and GB, paralleling to the increase in bile acids [88,89]. A study in obese FXR KO mice subjected to SG observed that at 5 weeks they recovered the weight lost in the surgery and were equal to KO mice undergoing sham operation [89]. In addition, surgery was associated with a 20% decrease in plasma glucose for wild-type mice, whereas KO mice had an increase of 24%.

CONCLUSIONS AND FUTURE DIRECTIONS

Bariatric surgeries are very effective for weight loss and reducing the risk of comorbidities. This effect is not only due to mechanical changes such as on gastric volume and emptying, or to hormone changes, but also by changes in the gut microbiome. The gut microbiome interacts directly with host metabolism, modulating bile acids and SCFA production, adding complexity to our understanding of BS. It is possible that integrating advances in hormones, bile acids, and gut microbiome changes after bariatric procedures will allow the design of microbiome-tailored therapeutics. Certain gut microbes, such as *Akkermansia muciniphila* and *Faecalibacterium prausnitzii*, are currently being investigated as next-generation probiotics. New clinical knowledge gained regarding the gut microbiome after BS, along with experimental evidence emerging from simulation of the procedure in vitro, should conduct to more efficient and personalized therapy.

ACKNOWLEDGMENTS

We appreciate the funding of grants Fondecyt de Iniciacion 11130518 and Fondef IDEA ID16i10045.

REFERENCES

[1] World Health Organization. Obesity and overweight. WHO; 2017.

[2] McKinsey Global Institute. How the world could better fight obesity. McKinsey & Company; 2014.

[3] Buchwald H, Oien DM. Metabolic/bariatric surgery worldwide 2011. Obes Surg 2013;23:427−36.

[4] Zhang Y, Ju W, Sun X, et al. Laparoscopic sleeve gastrectomy versus laparoscopic Roux-en-Y gastric bypass for morbid obesity and related comorbidities: a meta-analysis of 21 studies. Obes Surg 2015;25:19−26.

[5] Sjöström L. Review of the key results from the Swedish Obese Subjects (SOS) trial - a prospective controlled intervention study of bariatric surgery. J Intern Med 2013;273:219−34.

[6] Seeley RJ, Chambers AP, Sandoval DA. The role of gut adaptation in the potent effects of multiple bariatric surgeries on obesity and diabetes. Cell Metab 2015;21:369−78.

[7] Tong J, Prigeon RL, Davis HW, et al. Ghrelin suppresses glucose-stimulated insulin secretion and deteriorates glucose tolerance in healthy humans. Diabetes 2010;59:2145−51.

[8] Malin SK, Samat A, Wolski K, et al. Improved acylated ghrelin suppression at 2 years in obese patients with type 2 diabetes: effects of bariatric surgery vs standard medical therapy. Int J Obes 2014;38:364−70.

[9] Lee W-J, Chen C-Y, Chong K, Lee Y-C, Chen S-C, Lee S-D. Changes in postprandial gut hormones after metabolic surgery: a comparison of gastric bypass and sleeve gastrectomy. Surg Obes Relat Dis 2011;7:683−90.

[10] Booth A, Magnuson A, Fouts J, Foster MT. Adipose tissue: an endocrine organ playing a role in metabolic regulation. Horm Mol Biol Clin Investig 2016;26:25−42.

[11] Satoh N, Ogawa Y, Katsuura G, et al. The arcuate nucleus as a primary site of satiety effect of leptin in rats. Neurosci Lett 1997;224: 149−52.

[12] Enriori PJ, Evans AE, Sinnayah P, et al. Diet-induced obesity causes severe but reversible leptin resistance in arcuate melanocortin neurons. Cell Metab 2007;5:181−94.

[13] Gokulakrishnan K, Amutha A, Ranjani H, et al. Relationship of adipokines and proinflammatory cytokines among Asian Indians with obesity and youth onset type 2 diabetes. Endocr Pract 2015;21:1143−51.

[14] Kern PA, Ranganathan S, Li C, Wood L, Ranganathan G. Adipose tissue tumor necrosis factor and interleukin-6 expression in human obesity and insulin resistance. Am J Physiol Metab 2001;280:E745−51.

[15] Spranger J, Kroke A, Möhlig M, et al. Inflammatory cytokines and the risk to develop type 2 diabetes: results of the prospective population-based European Prospective Investigation into Cancer and Nutrition (EPIC)-Potsdam Study. Diabetes 2003;52:812−7.

[16] Rubino F, Gagner M, Gentileschi P, et al. The early effect of the Roux-en-Y gastric bypass on hormones involved in body weight regulation and glucose metabolism. Ann Surg 2004;240:236−42.

[17] Merl V, Peters A, Oltmanns KM, et al. Serum adiponectin concentrations during a 72-hour fast in over- and normal-weight humans. Int J Obes 2005;29:998−1001.

[18] Duncan BB, Schmidt MI, Pankow JS, et al. Adiponectin and the development of type 2 diabetes: the atherosclerosis risk in communities study. Diabetes 2004;53:2473−8.

[19] Hotta K, Funahashi T, Bodkin NL, et al. Circulating concentrations of the adipocyte protein adiponectin are decreased in parallel with reduced insulin sensitivity during the progression to type 2 diabetes in rhesus monkeys. Diabetes 2001;50:1126−33.

[20] Schneider JG, von Eynatten M, Schiekofer S, Nawroth PP, Dugi KA. Low plasma adiponectin levels are associated with increased hepatic lipase activity in vivo. Diabetes Care 2005;28:2181−6.

[21] Wang C, Mao X, Wang L, et al. Adiponectin sensitizes insulin signaling by reducing p70 S6 kinase-mediated serine phosphorylation of IRS-1. J Biol Chem 2007;282:7991−6.

[22] Yamauchi T, Kamon J, Waki H, et al. The fat-derived hormone adiponectin reverses insulin resistance associated with both lipoatrophy and obesity. Nat Med 2001;7:941−6.

[23] Woelnerhanssen B, Peterli R, Steinert RE, Peters T, Borbély Y, Beglinger C. Effects of postbariatric surgery weight loss on adipokines and metabolic parameters: comparison of laparoscopic Roux-en-Y gastric bypass and laparoscopic sleeve gastrectomy—a prospective randomized trial. Surg Obes Relat Dis 2011;7:561–8.

[24] Karra E, Chandarana K, Batterham RL. The role of peptide YY in appetite regulation and obesity. J Physiol 2009;587:19–25.

[25] Chakravartty S, Tassinari D, Salerno A, Giorgakis E, Rubino F. What is the mechanism behind weight loss maintenance with gastric bypass? Curr Obes Rep 2015;4:262–8.

[26] le Roux CW, Aylwin SJB, Batterham RL, et al. Gut hormone profiles following bariatric surgery favor an anorectic state, facilitate weight loss, and improve metabolic parameters. Ann Surg 2006;243:108–14.

[27] Korner J, Inabnet W, Febres G, et al. Prospective study of gut hormone and metabolic changes after adjustable gastric banding and Roux-en-Y gastric bypass. Int J Obes 2009;33:786–95.

[28] Korner J, Bessler M, Cirilo LJ, et al. Effects of Roux-en-Y gastric bypass surgery on fasting and postprandial concentrations of plasma ghrelin, peptide YY, and insulin. J Clin Endocrinol Metab 2005;90:359–65.

[29] Holst JJ. The physiology of glucagon-like peptide 1. Physiol Rev 2007;87:1409–39.

[30] Chambers AP, Smith EP, Begg DP, et al. Regulation of gastric emptying rate and its role in nutrient-induced GLP-1 secretion in rats after vertical sleeve gastrectomy. Am J Physiol Metab 2014;306:E424–32.

[31] Jiménez A, Casamitjana R, Flores L, et al. Long-term effects of sleeve gastrectomy and Roux-en-Y gastric bypass surgery on type 2 diabetes mellitus in morbidly obese subjects. Ann Surg 2012;256:1023–9.

[32] Umeda LM, Silva EA, Carneiro G, Arasaki CH, Geloneze B, Zanella MT. Early improvement in glycemic control after bariatric surgery and its relationships with insulin, GLP-1, and glucagon secretion in type 2 diabetic patients. Obes Surg 2011;21:896–901.

[33] Bergman EN. Energy contributions of volatile fatty acids from the gastrointestinal tract in various species. Physiol Rev 1990;70:567–90.

[34] Brahe LK, Astrup A, Larsen LH. Is butyrate the link between diet, intestinal microbiota and obesity-related metabolic diseases? Obes Rev 2013;14:950–9.

[35] McNeil NI. The contribution of the large intestine to energy supplies in man. Am J Clin Nutr 1984;39:338–42.

[36] Roy CC, Kien CL, Bouthillier L, Levy E. Short-chain fatty acids: ready for prime time? Nutr Clin Pract 2006;21:351–66.

[37] Topping DL, Clifton PM. Short-chain fatty acids and human colonic function: roles of resistant starch and nonstarch polysaccharides. Physiol Rev 2001;81:1031–64.

[38] Demigné C, Morand C, Levrat MA, Besson C, Moundras C, Rémésy C. Effect of propionate on fatty acid and cholesterol synthesis and on acetate metabolism in isolated rat hepatocytes. Br J Nutr 1995;74:209–19.

[39] Yamashita H, Maruta H, Jozuka M, et al. Effects of acetate on lipid metabolism in muscles and adipose tissues of type 2 diabetic Otsuka Long-Evans Tokushima Fatty (OLETF) rats. Biosci Biotechnol Biochem 2009;73:570–6.

[40] Fushimi T, Tayama K, Fukaya M, et al. Acetic acid feeding enhances glycogen repletion in liver and skeletal muscle of rats. J Nutr 2001;131:1973–7.

[41] Gao Z, Yin J, Zhang J, et al. Butyrate improves insulin sensitivity and increases energy expenditure in mice. Diabetes 2009;58:1509–17.

[42] Ang Z, Ding JL. GPR41 and GPR43 in obesity and inflammation - protective or causative? Front Immunol 2016;7. https://doi.org/10.3389/fimmu.2016.00028.

[43] Meijer K, de Vos P, Priebe MG. Butyrate and other short-chain fatty acids as modulators of immunity: what relevance for health? Curr Opin Clin Nutr Metab Care 2010;13:715–21.

[44] Cani PD, Delzenne NM. Interplay between obesity and associated metabolic disorders: new insights into the gut microbiota. Curr Opin Pharmacol 2009;9:737–43.

[45] Patil DP, Dhotre DP, Chavan SG, et al. Molecular analysis of gut microbiota in obesity among Indian individuals. J Biosci 2012;37:647–57.

[46] Schwiertz A, Taras D, Schäfer K, et al. Microbiota and SCFA in lean and overweight healthy subjects. Obesity 2010;18:190–5.

[47] Qin J, Li Y, Cai Z, et al. A metagenome-wide association study of gut microbiota in type 2 diabetes. Nature 2012;490:55–60.

[48] Backhed F, Ding H, Wang T, et al. The gut microbiota as an environmental factor that regulates fat storage. Proc Natl Acad Sci 2004;101:15718–23.

[49] Ley RE, Turnbaugh PJ, Klein S, Gordon JI. Microbial ecology: human gut microbes associated with obesity. Nature 2006;444:1022–3.

[50] Turnbaugh PJ, Ley RE, Mahowald MA, Magrini V, Mardis ER, Gordon JI. An obesity-associated gut microbiome with increased capacity for energy harvest. Nature 2006;444:1027–131.

[51] Chen T, Long W, Zhang C, Liu S, Zhao L, Hamaker BR. Fiber-utilizing capacity varies in Prevotella- versus Bacteroides-dominated gut microbiota. Sci Rep 2017;7:2594.

[52] Wu GD, Chen J, Hoffmann C, et al. Linking long-term dietary patterns with gut microbial enterotypes. Science 2011;334:105–8.

[53] Goodrich JK, Waters JL, Poole AC, et al. Human genetics shape the gut microbiome. Cell 2014;159:789–99.

[54] Png CW, Lindén SK, Gilshenan KS, et al. Mucolytic bacteria with increased prevalence in IBD mucosa augment in vitro utilization of mucin by other bacteria. Am J Gastroenterol 2010;105:2420–8.

[55] Swidsinski A, Dorffel Y, Loening-Baucke V, et al. Acute appendicitis is characterised by local invasion with *Fusobacterium nucleatum/necrophorum*. Gut 2011;60:34–40.

[56] Belzer C, de Vos WM. Microbes inside—from diversity to function: the case of Akkermansia. ISME J 2012;6:1449–58.

[57] Johansson MEV, Larsson JMH, Hansson GC. The two mucus layers of colon are organized by the MUC2 mucin, whereas the outer layer is a legislator of host-microbial interactions. Proc Natl Acad Sci 2011;108:4659–65.

[58] Santacruz A, Collado MC, García-Valdés L, et al. Gut microbiota composition is associated with body weight, weight gain and biochemical parameters in pregnant women. Br J Nutr 2010;104:83−92.

[59] Karlsson CLJ, Önnerfält J, Xu J, Molin G, Ahrné S, Thorngren-Jerneck K. The microbiota of the gut in preschool children with normal and excessive body weight. Obesity 2012;20:2257−61.

[60] Collado MC, Isolauri E, Laitinen K, Salminen S. Distinct composition of gut microbiota during pregnancy in overweight and normal-weight women. Am J Clin Nutr 2008;88:894−9.

[61] Everard A, Lazarevic V, Derrien M, et al. Responses of gut microbiota and glucose and lipid metabolism to prebiotics in genetic obese and diet-induced leptin-resistant mice. Diabetes 2011;60:2775−86.

[62] Everard A, Belzer C, Geurts L, et al. Cross-talk between Akkermansia muciniphila and intestinal epithelium controls diet-induced obesity. Proc Natl Acad Sci U S A 2013;110:9066−71.

[63] Dao MC, Everard A, Aron-Wisnewsky J, et al. Akkermansia muciniphila and improved metabolic health during a dietary intervention in obesity: relationship with gut microbiome richness and ecology. Gut 2016;65:426−36.

[64] Lukovac S, Belzer C, Pellis L, et al. Differential modulation by Akkermansia muciniphila and *Faecalibacterium prausnitzii* of host peripheral lipid metabolism and histone acetylation in mouse gut organoids. mBio 2014;5:e01438−14.

[65] Belzer C, Chia LW, Aalvink S, et al. Microbial metabolic networks at the mucus layer lead to diet-independent butyrate and Vitamin B12 production by intestinal symbionts. mBio 2017;8. https://doi.org/10.1128/mBio.00770-17.

[66] Walters WA, Xu Z, Knight R. Meta-analyses of human gut microbes associated with obesity and IBD. FEBS Lett 2014;588:4223−33.

[67] Finucane MM, Sharpton TJ, Laurent TJ, Pollard KS. A taxonomic signature of obesity in the microbiome? Getting to the guts of the matter. PLoS One 2014;9. https://doi.org/10.1371/journal.pone.0084689.

[68] Pott J, Hornef M. Innate immune signalling at the intestinal epithelium in homeostasis and disease. EMBO Rep 2012;13:684−98.

[69] Hooper LV, Macpherson AJ. Immune adaptations that maintain homeostasis with the intestinal microbiota. Nat Rev Immunol 2010;10: 159−69.

[70] Liu R, Hong J, Xu X, et al. Gut microbiome and serum metabolome alterations in obesity and after weight-loss intervention. Nat Med 2017;23:859−68.

[71] Hong Y-H, Nishimura Y, Hishikawa D, et al. Acetate and propionate short chain fatty acids stimulate adipogenesis via GPCR43. Endocrinology 2005;146:5092−9.

[72] Xu X, Xu P, Ma C, Tang J, Zhang X. Gut microbiota, host health, and polysaccharides. Biotechnol Adv 2013;31:318−37.

[73] Medina DA, Pedreros JP, Turiel D, Quezada N, Pimentel F, Escalona A, Garrido D. Distinct patterns in the gut microbiota after surgical or medical therapy in obese patients. Peer J 2017;5:e3443.

[74] Murphy R, Tsai P, Jüllig M, Liu A, Plank L, Booth M. Differential changes in gut microbiota after gastric bypass and sleeve gastrectomy bariatric surgery vary according to diabetes remission. Obes Surg 2017;27:917−25.

[75] Zhang H, DiBaise JK, Zuccolo A, Kudrna D, Braidotti M, Yu Y, Parameswaran P, Crowell MD, Wing R, Rittmann BE, Krajmalnik-Brown R. Human gut microbiota in obesity and after gastric bypass. Proc Nat Acad Sci USA 2008;106:2365−70.

[76] Damms-Machado A, Mitra S, Schollenberger AE, et al. Effects of surgical and dietary weight loss therapy for obesity on gut microbiota composition and nutrient absorption. BioMed Res Int 2015;2015:1−12.

[77] Palleja A, Kashani A, Allin KH, et al. Roux-en-Y gastric bypass surgery of morbidly obese patients induces swift and persistent changes of the individual gut microbiota. Genome Med 2016;8:67.

[78] Furet J-P, Kong L-C, Tap J, et al. Differential adaptation of human gut microbiota to bariatric surgery-induced weight loss: links with metabolic and low-grade inflammation markers. Diabetes 2010;59:3049−57.

[79] Kong LC, Tap J, Aron-Wisnewsky J, et al. Gut microbiota after gastric bypass in human obesity: increased richness and associations of bacterial genera with adipose tissue genes. Am J Clin Nutr 2013;98:16−24.

[80] Tremaroli V, Karlsson F, Werling M, Stahlmann M, Kovatcheva-Datchary P, Olbers T, Fandriks L, le Roux CW, Nielsen J, Backhed F. Roux-en-Y gastric bypass and vertical banded gastroplasty induce long-term changes on the human gut microbiome contributing to fat mass regulation. Cell Metab 2015;22:228−38.

[81] Kohli R, Kirby M, Setchell KDR, et al. Intestinal adaptation after ileal interposition surgery increases bile acid recycling and protects against obesity-related comorbidities. Am J Physiol Liver Physiol 2010;299:G652−60.

[82] Sayin SI, Wahlström A, Felin J, et al. Gut microbiota regulates bile acid metabolism by reducing the levels of tauro-beta-muricholic acid, a naturally occurring FXR antagonist. Cell Metab 2013;17:225−35.

[83] Werling M, Fändriks L, Björklund P, et al. Long-term results of a randomized clinical trial comparing Roux-en-Y gastric bypass with vertical banded gastroplasty. Br J Surg 2013;100:222−30.

[84] Thomas C, Pellicciari R, Pruzanski M, Auwerx J, Schoonjans K. Targeting bile-acid signalling for metabolic diseases. Nat Rev Drug Discov 2008;7:678−93.

[85] Pournaras DJ, Glicksman C, Vincent RP, et al. The role of bile after Roux-en-Y gastric bypass in promoting weight loss and improving glycaemic control. Endocrinology 2012;153:3613−9.

[86] Simonen M, Dali-Youcef N, Kaminska D, et al. Conjugated bile acids associate with altered rates of glucose and lipid oxidation after Roux-en-Y gastric bypass. Obes Surg 2012;22:1473−80.

[87] Steinert RE, Peterli R, Keller S, et al. Bile acids and gut peptide secretion after bariatric surgery: a 1-year prospective randomized pilot trial. Obesity 2013;21:E660−8.

[88] Myronovych A, Kirby M, Ryan KK, et al. Vertical sleeve gastrectomy reduces hepatic steatosis while increasing serum bile acids in a weight-loss-independent manner. Obesity 2014;22:390—400.

[89] Ryan KK, Tremaroli V, Clemmensen C, et al. FXR is a molecular target for the effects of vertical sleeve gastrectomy. Nature 2014;509:183—8.

FURTHER READING

[1] Hjorth MF, Roager HM, Larsen TM, et al. Pre-treatment microbial Prevotella-to-Bacteroides ratio, determines body fat loss success during a 6-month randomized controlled diet intervention. Int J Obes October 2017;42(3). https://doi.org/10.1038/ijo.2017.220.

Chapter 25

Gut Dysbiosis in Arterial Hypertension: A Candidate Therapeutic Target for Blood Pressure Management

José Luiz de Brito Alves[1], Evandro Leite de Souza[1], Josiane de Campos Cruz[2], Camille de Moura Balarini[2,3], Marciane Magnani[4], Hubert Vidal[5] and Valdir de Andrade Braga[2]

[1]Department of Nutrition, Health Sciences Center, Federal University of Paraíba, João Pessoa, Brazil; [2]Biotechnology Center, Federal University of Paraíba, João Pessoa, Brazil; [3]Department of Physiology and Pathology, Health Sciences Center, Federal University of Paraíba, João Pessoa, Brazil; [4]Department of Food Engineering, Technology Center, Federal University of Paraíba, João Pessoa, Brazil; [5]Univ-Lyon, CarMeN (Cardio, Metabolism, Diabetes and Nutrition) Laboratory, Université Claude Bernard Lyon 1, INSA Lyon, Oullins, France

INTRODUCTION

Arterial hypertension affects more than 1 billion people worldwide [1]. It has been recognized as main risk factor for the development of cardiovascular disease (CD), including coronary heart disease, stroke, and renal failure [2]. For example, 20-mmHg elevation of systolic blood pressure (BP) and 10-mmHg higher diastolic BP have been associated with doubled risk of death from stroke, heart disease, or other vascular diseases [3].

Physical inactivity, obesity, high alcohol consumption, tobacco use, high stress levels, and poor food intake pattern (including a diet rich in calories, sugar, salt, saturated fatty acids, and cholesterol) are lifestyle and environmental factors associated with increased arterial hypertension occurrence [4,5]. Also gut microbiota dysfunction may influence a number of processes that affect the control of BP and predispose to hypertension [6].

GUT DYSBIOSIS AND ARTERIAL HYPERTENSION

The increase in sympathetic nerve discharges alters vascular reactivity and contributes to the development of hypertension in animal models of hypertension [7–9]. Sympathetic outflow can be modulated by the baroreflex sensitivity [10], rhythmicity of the respiratory nervous system [11], and peripheral and central respiratory chemoreceptors [12]. Additionally, gastrointestinal hormones, such as leptin, cholecystokinin, Glucagon-like peptide 1, and ghrelin, also exert effects on sympathetic modulation and BP [13–15].

More recently, a growing body of evidence supports the role of oxidative stress [16], and augmented proinflammatory cytokines [17], in the development of sympathetic overactivity and hypertension. The gut microbiota and its metabolites can exert strong modulatory effects of the immune and inflammatory responses, oxidative stress, sympathetic activity, and, consequently, BP.

The gut microbiota displays an intrinsic capacity to adapt to nutrient availability and changes in environmental conditions. However, exogenous factors (e.g., high dietary fat, salt and sugar, nonsteroidal antiinflammatory drugs, antibiotics, and oxidative stress) can induce gut dysbiosis [18], impairing the gut microbiota composition and function.

Metagenomic Studies

Germ-free rats exhibit lower cardiac output and a markedly decreased microvasculature response to catecholamines [19,20]. An increased ratio of Firmicutes to Bacteroidetes, the two most prevalent bacterial phyla which inhabit the gut, has been used as a measurement of dysbiosis and disease state [18]. On examining experimental model of hypertension, such as the Dahl salt-sensitive rat and the spontaneously hypertensive rat (SHR), important alterations in gut microbiota compared with sham animals were reported. For example, hypertensive rats exhibit (1) enhanced Firmicutes to Bacteroidetes ratio (approximately fivefold); (2) augmented occurrence of bacteria from the phylum Actinobacteria; and (3) lower bacterial load, richness, evenness, and diversity [21–23].

The gut microbiota in patients with prehypertension or hypertension exhibits lower gene richness and microbial diversity, when compared with healthy subjects [23,24]. In addition, hypertensive patients display elevated percentage of *Prevotella*, *Klebsiella*, *Porphyromonas*, and *Actinomyces*. On the other hand, levels of bacteria belonging to genera *Faecalibacterium*, *Roseburia*, and *Bifidobacterium* are reduced [24,25].

The Metabolome and Short-Chain Fatty Acids

Findings from rodent models and clinical trials show that short-chain fatty acids (SCFAs) modulate vasodilatation and induce hypotension [26–28]. An important question to be answered is how SCFA could reduce BP? It has been suggested that SCFA receptors (gpr41, gpr43, and olfactory receptor 78 [Olfr78]), expressed in the kidney, heart, sympathetic ganglia, and blood vessels, when activated by gut microbe-produced SCFAs, contribute to modulate various host physiological pathways, such as inflammatory responses, energy metabolism, sympathetic nerve activity, and BP [28–30].

Transcriptomic Investigation

Interestingly, a single cell transcriptome analysis performed in carotid body (CB) glomus cells of mouse identified abundant G protein-coupled receptor signaling pathway components. Olfr78 was the most abundant G protein-coupled receptor found in the CB [31]. The peripheral chemoreceptors, located in CB glomus cells, are responsible for part of the sympathetic-respiratory modulatory effect. Activation of these chemoreceptors induces strong activation of the cardiorespiratory neuronal network and enhances the sympathetic and respiratory outflow [11]. Future studies should investigate if SCFAs can modulate sympathetic-respiratory activities and BP via CB action. Ongoing studies in our laboratory will contribute to elucidate these key points.

TRIMETHYLAMINE N-OXIDE AND CARDIOVASCULAR FUNCTION

Trimethylamine N-oxide (TMAO), a gut metabolite synthetized from dietary choline and phosphatidylcholine metabolism, has been associated with development of atherosclerosis [32,33]. However, the contribution of TMAO to BP modulation is less clear. By using Sprague–Dawley rats, Ufnal and colleagues demonstrated that TMAO infusion did not augment BP in normotensive condition, but it prolonged the hypertensive effects of angiotensin II in this animal model [34]. A recent study performed in western diet-induced obese mice model demonstrated that long-term treatment with 3,3-dimethyl-1-butanol (an inhibitor of TMAO formation), for 8 weeks, reduced inflammation, improved cardiac function, but had no effect on BP or heart rate [33].

Protein Metabolites

Metabolites of amino acids produced by gut microbiota, such as tryptophan, glutamate, and gamma-aminobutyric acid, can directly influence the central and peripheral nervous systems [35,36]. Many of these metabolites are freely accessible centrally and systemically and have the potential to influence sympathetic activity and BP, at multiple sites in the cardiovascular control centers of the brain, as well as peripherally.

Proinflammatory Environment

Alteration in gut microbiota homeostasis, including a low survival of lactic acid bacteria (LAB) and gut dysbiosis, can help passage of lipopolysaccharide (LPS) from intestinal barriers to systemic circulation. For this, two mechanisms have been proposed: (1) chylomicron-facilitated transport and (2) extracellular leakage through tight junction in the epithelial lining [37]. At systemic level, LPSs are transferred from circulating LPS-biding proteins to CD14 molecules on the surface of leukocytes, stimulating the production of proinflammatory cytokines, such as interleukin (IL) 1 and IL-18 [38]. The relationship between LPS and BP is complex and not entirely clear. A recent study observed that increased LPS levels is a predictive of major adverse cardiovascular events in atrial fibrillation patients [39], but no association has been found with hypertension.

Evidence Relating Probiotic Effects on Blood Pressure

Some authors have suggested that the benefits of probiotic consumption on BP may be related to their antiinflammatory, antioxidant, and gut-modulating properties [40−42]. An alternative mechanism could be the production of bioactive peptides, with angiotensin-converting enzyme (ACE) inhibitory properties, during fermentation processes [43]. ACE inhibition causes decrease in synthesis of angiotensin II and attenuation of vasoconstriction and BP.

In SHRs, treatment with *Lactobacillus (L.) fermentum* CECT5716, or with a mix of *L. coryniformis* CECT5711 and *L. gasseri* CECT5714 (3.3×10^{10} CFU/day, for 5 weeks), significantly reduced systolic arterial pressure, improved endothelial function, reduced vascular oxidative stress, and decreased vascular inflammation in the aorta [44]. Similarly, the supplementation with *L. casei* C1 (1×10^{11} CFU/day, for 8 weeks) attenuated vascular changes and reduced BP in the SHR model [45].

Oral administration for 15 days of a recombinant *L. plantarum* NC8 strain, expressing ACE inhibitory peptide, was effective to reduce BP in SHR [46]. Treatment with VSL#3 (*Streptococcus thermophilus*, *Bifidum (B.) longum*, *B. breve*, *B. infantis*, *L. acidophilus*, *L. plantarum*, *L. casei* and, *L. bulgaricus*) prevented endothelial dysfunction and improved vascular oxidative stress, in rats with portal hypertension [47].

L. plantarum 299v (400 mL/day of a drink containing 5×10^7 CFU/mL, for 6 weeks) reduced SBP, improved metabolic disturbance, and attenuated generation of reactive oxygen species in smokers [48]. In type 2 diabetes, daily administration of 200 mL probiotic soy milk (containing *L. planetarium* A7), for 8 weeks, reduced both systolic and diastolic BPs [49].

A hypocaloric diet (1500 kcal/day) supplemented with 50 g probiotic cheese (containing *L. plantarum* TENSIA DSM 21380, 1.5×10^{11} CFU/g), for 3 weeks, in obese hypertensive patients, effectively decreased BP when compared with a control group, receiving the same diet without probiotic administration [50]. Capsules containing *L. acidophilus* (10^9 CFU/capsule) and *B. bifidum* (10^9 CFU/capsule), administered thrice a day for 6 weeks, significantly reduced systolic BP in patients with hypercholesterolemia, but no alterations were observed in diastolic BP [51].

In contrast, postmenopausal women with metabolic syndrome, receiving fermented milk with *L. plantarum* (80 mL/day, for 14 days) did not improve systolic or diastolic BP [52]. The same occurred with overweight adults, treated with *L. acidophilus* La-05 and *B. animalis* subsp. *lactis* Bb12 (at a dose of 3×10^9 CFU/day), with no significantly altered BP [53].

Evidence Relating Prebiotic Effects on the Blood Pressure

Reports indicate that every 7 g of fibers consumed daily could lower BP and the risk of CD by 9% [54]. Dietary fiber intake in the form of pill supplementation (7 g/day, for 12 weeks) promoted a reduction in diastolic BP of 5 mmHg in adult hypertensive patients [55]. Oat bran fiber (8 g/day, for 12 weeks) reduced 2 mmHg of systolic BP and 1 mmHg of diastolic BP in hypertensive subjects [56]. A diet containing soy protein isolate and supplemented with psyllium fiber (12 g fiber/day, for 8 weeks) caused a reduction in systolic BP of 5.9 mmHg in men and women, using antihypertensive drug therapy [57].

In 18 untreated hypertensives, beta-glucan from whole oat (5.52 g/day, for 6 weeks) reduced 7.5 mmHg of systolic BP and 5.5 mmHg of diastolic BP [58]. Two metaanalysis, including randomized clinical trials, demonstrated that augmenting total intake of dietary fiber there is a significant reduction of systolic and diastolic BPs in hypertensive patients [59,60].

Evidence Relating Synbiotic Effects on Blood Pressure

In SHR rats, the treatment with kefir, a synbiotic matrix containing LAB and yeasts, at a dose of 0.3 mL/100 g body weight, for 60 days (1) attenuated hypertension and endothelial dysfunction; (2) reduced impairment of the cardiac autonomic control of heart rate; and (3) recovered baroreflex sensitivity [61,62]. In hypertensive rats treated with nitro-L-arginine methyl ester (L-NAME), supplementation with fermented blueberries containing *L. plantarum* (2 g/day, for 4 weeks, containing 10^9 CFU) reduced both systolic and diastolic BPs (about 45%) [63]. However, another study showed that adding probiotics to a blueberry-enriched diet did not enhance, and might have actually impaired, the antihypertensive effects of blueberry consumption [64].

FIGURE 25.1 Schematic drawing showing the impact of probiotic, prebiotic, or synbiotic interventions on the inflammation, oxidative stress, and gut microbiota composition. The administration of these gut microbiota modulators may change the microbial composition and its derived-metabolites, contributing to reduce gut dysbiosis, inflammation, and oxidative stress, lastly reducing blood pressure. *SCFA*, short-chain fatty acid; *LPS*, lipopolysaccharide.

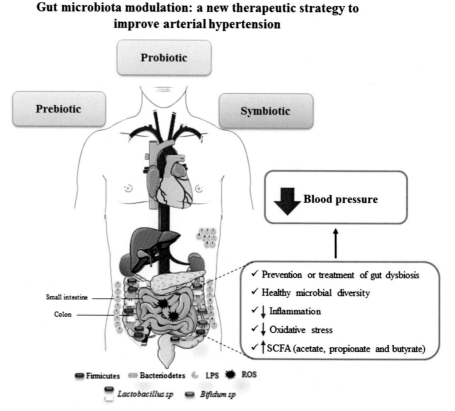

A commercial encapsulated synbiotic containing *L. casei*, *L. rhamnosus*, *S. thermophilus*, *B. breve*, *L. acidophilus*, *B. longum*, *L. bulgaricus*, and prebiotic (fructooligosaccharide) resulted in decreased of systolic BP in 60 nonalcoholic fatty liver disease patients but did not affect diastolic BP [65].

CONCLUSION

Disordered composition of gut microbiota and microbial products, called gut dysbiosis, is linked to increase BP. Further work will be necessary to understand the precise cellular and molecular mechanisms underlying to complex interrelationship between gut microbiota and BP regulation. Current findings demonstrate that administration of gut microbiota modulators (probiotics, prebiotics, or synbiotics) is clinically feasible and could be used as a coadjuvant treatment for hypertensive subjects (Fig. 25.1).

LIST OF ACRONYMS AND ABBREVIATIONS

BP Blood pressure
CD Cardiovascular disease
CO₂ Carbon dioxide
DBP Diastolic blood pressure
GLP-1 Glucagon-like peptide 1
Gpr G protein-coupled receptors
SBP Systolic blood pressure
SFCA Short-chain fatty acids

REFERENCES

[1] Whelton PK, Carey RM, Aronow WS, Casey Jr DE, Collins KJ, Dennison Himmelfarb C, et al. 2017 ACC/AHA/AAPA/ABC/ACPM/AGS/APhA/ ASH/ASPC/NMA/PCNA guideline for the prevention, detection, evaluation, and management of high blood pressure in adults: a report of the American College of Cardiology/American Heart Association Task Force on Clinical Practice Guidelines. Hypertension 2017;71(6).

[2] Hedner T, Kjeldsen SE, Narkiewicz K. State of global health—hypertension burden and control. Blood Pressure 2012;21(Suppl. 1):1—2.

[3] Lewington S, Clarke R, Qizilbash N, Peto R, Collins R. Age-specific relevance of usual blood pressure to vascular mortality: a meta-analysis of individual data for one million adults in 61 prospective studies. Lancet 2002;360(9349):1903—13.

[4] Ezzati M, Riboli E. Behavioral and dietary risk factors for noncommunicable diseases. N Engl J Med 2013;369(10):954—64.

[5] Bjerregaard P. Nutritional transition - where do we go from here? J Hum Nutr Diet 2010;23(Suppl 1):1—2.

[6] de Brito Alves JL, de Sousa VP, Cavalcanti Neto MP, Magnani M, Braga VA, da Costa-Silva JH, et al. New insights on the use of dietary polyphenols or probiotics for the management of arterial hypertension. Front Physiol 2016;7:448.

[7] de Brito Alves JL, Costa-Silva JH. Maternal protein malnutrition induced-hypertension: new evidence about the autonomic and respiratory dysfunctions and epigenetic mechanisms. Clin Exp Pharmacol Physiol 2017;45(5).

[8] Grassi G. Role of the sympathetic nervous system in human hypertension. J Hypertens 1998;16(12 Pt 2):1979—87.

[9] Grassi G, Mark A, Esler M. The sympathetic nervous system alterations in human hypertension. Circ Res 2015;116(6):976—90.

[10] Meguro Y, Miura Y, Kimura S, Noshiro T, Sugawara T, Takahashi M, et al. A sympathetic component of baroreflex function in patients with essential hypertension. Clin Exp Pharmacol Physiol Suppl 1989;15:93—5.

[11] Machado BH, Zoccal DB, Moraes DJA. Neurogenic hypertension and the secrets of respiration. Am J Physiol Regul Integr Comp Physiol 2017;312(6):R864—72.

[12] Prabhakar NR. Sensing hypoxia: physiology, genetics and epigenetics. J Physiol 2013;591(9):2245—57.

[13] Yu M, Moreno C, Hoagland KM, Dahly A, Ditter K, Mistry M, et al. Antihypertensive effect of glucagon-like peptide 1 in Dahl salt-sensitive rats. J Hypertens 2003;21(6):1125—35.

[14] How JM, Fam BC, Verberne AJ, Sartor DM. High-fat diet is associated with blunted splanchnic sympathoinhibitory responses to gastric leptin and cholecystokinin: implications for circulatory control. Am J Physiol Heart Circ Physiol 2011;300(3):H961—7.

[15] Rodriguez A. Novel molecular aspects of ghrelin and leptin in the control of adipobiology and the cardiovascular system. Obesity facts 2014;7(2):82—95.

[16] Cruz JC, Flor AF, Franca-Silva MS, Balarini CM, Braga VA. Reactive oxygen species in the paraventricular nucleus of the hypothalamus alter sympathetic activity during metabolic syndrome. Front Physiol 2015;6:384.

[17] Wenzel U, Turner JE, Krebs C, Kurts C, Harrison DG, Ehmke H. Immune mechanisms in arterial hypertension. J Am Soc Nephrol 2016;27(3):677—86.

[18] Weiss GA, Hennet T. Mechanisms and consequences of intestinal dysbiosis. Cell Mol Life Sci 2017;74(16):2959—77.

[19] Gordon HA, Wostmann BS, Bruckner-Kardoss E. Effects of microbial flora on cardiac output and other elements of blood circulation. Proc Soc Exp Biol Med 1963;114:301—4.

[20] Baez S, Gordon HA. Tone and reactivity of vascular smooth muscle in germfree rat mesentery. J Exp Med 1971;134(4):846—56.

[21] Yang T, Santisteban MM, Rodriguez V, Li E, Ahmari N, Carvajal JM, et al. Gut dysbiosis is linked to hypertension. Hypertension 2015;65(6):1331—40.

[22] Adnan S, Nelson JW, Ajami NJ, Venna VR, Petrosino JF, Bryan Jr RM, et al. Alterations in the gut microbiota can elicit hypertension in rats. Physiol Genom 2017;49(2):96—104.

[23] Marques FZ, Mackay CR, Kaye DM. Beyond gut feelings: how the gut microbiota regulates blood pressure. Nat Rev Cardiol 2018;15(1):20—32.

[24] Li J, Zhao F, Wang Y, Chen J, Tao J, Tian G, et al. Gut microbiota dysbiosis contributes to the development of hypertension. Microbiome 2017;5(1):14.

[25] Marques FZ, Mackay CR, Kaye DM. Beyond gut feelings: how the gut microbiota regulates blood pressure. Nat Rev Cardiol 2017.

[26] Nutting CW, Islam S, Daugirdas JT. Vasorelaxant effects of short chain fatty acid salts in rat caudal artery. Am J Physiol 1991;261(2 Pt. 1):H561—7.

[27] Mortensen FV, Nielsen H, Mulvany MJ, Hessov I. Short chain fatty acids dilate isolated human colonic resistance arteries. Gut 1990;31(12):1391—4.

[28] Miyamoto J, Kasubuchi M, Nakajima A, Irie J, Itoh H, Kimura I. The role of short-chain fatty acid on blood pressure regulation. Curr Opin Nephrol Hypertens 2016;25(5):379—83.

[29] Richards EM, Pepine CJ, Raizada MK, Kim S. The gut, its microbiome, and hypertension. Curr Hypertens Rep 2017;19(4):36.

[30] Ono S, Karaki S, Kuwahara A. Short-chain fatty acids decrease the frequency of spontaneous contractions of longitudinal muscle via enteric nerves in rat distal colon. Jpn J Physiol 2004;54(5):483—93.

[31] Zhou T, Chien MS, Kaleem S, Matsunami H. Single cell transcriptome analysis of mouse carotid body glomus cells. J Physiol 2016;594(15):4225—51.

[32] Li XS, Obeid S, Klingenberg R, Gencer B, Mach F, Raber L, et al. Gut microbiota-dependent trimethylamine N-oxide in acute coronary syndromes: a prognostic marker for incident cardiovascular events beyond traditional risk factors. Eur Heart J 2017;38(11):814—24.

[33] Chen K, Zheng X, Feng M, Li D, Zhang H. Gut microbiota-dependent metabolite trimethylamine N-Oxide contributes to cardiac dysfunction in western diet-induced obese mice. Front Physiol 2017;8:139.

[34] Ufnal M, Jazwiec R, Dadlez M, Drapala A, Sikora M, Skrzypecki J. Trimethylamine-N-oxide: a carnitine-derived metabolite that prolongs the hypertensive effect of angiotensin II in rats. Can J Cardiol 2014;30(12):1700—5.

[35] Morris G, Berk M, Carvalho A, Caso JR, Sanz Y, Walder K, et al. The role of the microbial metabolites including tryptophan catabolites and short chain fatty acids in the pathophysiology of immune-inflammatory and neuroimmune disease. Mol Neurobiol 2017;54(6):4432–51.

[36] O'Mahony SM, Clarke G, Borre YE, Dinan TG, Cryan JF. Serotonin, tryptophan metabolism and the brain-gut-microbiome axis. Behav Brain Res 2015;277:32–48.

[37] Caesar R, Fak F, Backhed F. Effects of gut microbiota on obesity and atherosclerosis via modulation of inflammation and lipid metabolism. J Int Med 2010;268(4):320–8.

[38] Lu YC, Yeh WC, Ohashi PS. LPS/TLR4 signal transduction pathway. Cytokine 2008;42(2):145–51.

[39] Pastori D, Carnevale R, Nocella C, Novo M, Santulli M, Cammisotto V, et al. Gut-derived serum lipopolysaccharide is associated with enhanced risk of major adverse cardiovascular events in atrial fibrillation: effect of adherence to mediterranean diet. J Am Heart Assoc 2017;6(6).

[40] Resta-Lenert S, Barrett KE. Probiotics and commensals reverse TNF-alpha- and IFN-gamma-induced dysfunction in human intestinal epithelial cells. Gastroenterology 2006;130(3):731–46.

[41] Bouhafs L, Moudilou EN, Exbrayat JM, Lahouel M, Idoui T. Protective effects of probiotic Lactobacillus plantarum BJ0021 on liver and kidney oxidative stress and apoptosis induced by endosulfan in pregnant rats. Ren Fail 2015;14:1–9.

[42] Thomas LV, Suzuki K, Zhao J. Probiotics: a proactive approach to health. A symposium report. Br J Nutr 2015;114(Suppl. 1):S1–15.

[43] Thushara RM, Gangadaran S, Solati Z, Moghadasian MH. Cardiovascular benefits of probiotics: a review of experimental and clinical studies. Food Funct 2016;7(2):632–42.

[44] Gomez-Guzman M, Toral M, Romero M, Jimenez R, Galindo P, Sanchez M, et al. Antihypertensive effects of probiotics Lactobacillus strains in spontaneously hypertensive rats. Mol Nutr Food Res 2015;59(11):2326–36.

[45] Yap WB, Ahmad FM, Lim YC, Zainalabidin S. Lactobacillus casei strain C1 attenuates vascular changes in spontaneously hypertensive rats. Korean J Physiol Pharmacol 2016;20(6):621–8.

[46] Yang G, Jiang Y, Yang W, Du F, Yao Y, Shi C, et al. Effective treatment of hypertension by recombinant Lactobacillus plantarum expressing angiotensin converting enzyme inhibitory peptide. Microb Cell Factories 2015;14:202.

[47] Rashid SK, Idris-Khodja N, Auger C, Alhosin M, Boehm N, Oswald-Mammosser M, et al. Probiotics (VSL#3) prevent endothelial dysfunction in rats with portal hypertension: role of the angiotensin system. PLoS One 2014;9(5):e97458.

[48] Naruszewicz M, Johansson ML, Zapolska-Downar D, Bukowska H. Effect of Lactobacillus plantarum 299v on cardiovascular disease risk factors in smokers. Am J Clin Nutr 2002;76(6):1249–55.

[49] Hariri M, Salehi R, Feizi A, Mirlohi M, Kamali S, Ghiasvand R. The effect of probiotic soy milk and soy milk on anthropometric measures and blood pressure in patients with type II diabetes mellitus: a randomized double-blind clinical trial. ARYA Atheroscler 2015;11(Suppl. 1):74–80.

[50] Sharafedtinov KK, Plotnikova OA, Alexeeva RI, Sentsova TB, Songisepp E, Stsepetova J, et al. Hypocaloric diet supplemented with probiotic cheese improves body mass index and blood pressure indices of obese hypertensive patients—a randomized double-blind placebo-controlled pilot study. Nutr J 2013;12:138.

[51] Rerksuppaphol S, Rerksuppaphol L. A randomized double-blind controlled trial of Lactobacillus acidophilus plus Bifidobacterium bifidum versus placebo in patients with hypercholesterolemia. J Clin Diagn Res 2015;9(3):KC01–4.

[52] Barreto FM, Colado Simao AN, Morimoto HK, Batisti Lozovoy MA, Dichi I, Helena da Silva Miglioranza L. Beneficial effects of Lactobacillus plantarum on glycemia and homocysteine levels in postmenopausal women with metabolic syndrome. Nutrition 2014;30(7–8):939–42.

[53] Ivey KL, Hodgson JM, Kerr DA, Thompson PL, Stojceski B, Prince RL. The effect of yoghurt and its probiotics on blood pressure and serum lipid profile; a randomised controlled trial. Nutr Metab Cardiovasc Dis 2015;25(1):46–51.

[54] Threapleton DE, Greenwood DC, Evans CE, Cleghorn CL, Nykjaer C, Woodhead C, et al. Dietary fibre intake and risk of cardiovascular disease: systematic review and meta-analysis. BMJ 2013;347:f6879.

[55] Eliasson K, Ryttig KR, Hylander B, Rossner S. A dietary fibre supplement in the treatment of mild hypertension. A randomized, double-blind, placebo-controlled trial. J Hypertens 1992;10(2):195–9.

[56] He J, Streiffer RH, Muntner P, Krousel-Wood MA, Whelton PK. Effect of dietary fiber intake on blood pressure: a randomized, double-blind, placebo-controlled trial. J Hypertens 2004;22(1):73–80.

[57] Burke V, Hodgson JM, Beilin LJ, Giangiulioi N, Rogers P, Puddey IB. Dietary protein and soluble fiber reduce ambulatory blood pressure in treated hypertensives. Hypertension 2001;38(4):821–6.

[58] Keenan JM, Pins JJ, Frazel C, Moran A, Turnquist L. Oat ingestion reduces systolic and diastolic blood pressure in patients with mild or borderline hypertension: a pilot trial. J Fam Pract 2002;51(4):369.

[59] Whelton SP, Hyre AD, Pedersen B, Yi Y, Whelton PK, He J. Effect of dietary fiber intake on blood pressure: a meta-analysis of randomized, controlled clinical trials. J Hypertens 2005;23(3):475–81.

[60] Aljuraiban GS, Griep LM, Chan Q, Daviglus ML, Stamler J, Van Horn L, et al. Total, insoluble and soluble dietary fibre intake in relation to blood pressure: the INTERMAP Study - CORRIGENDUM. Br J Nutr 2015;114(9):1534.

[61] Klippel BF, Duemke LB, Leal MA, Friques AG, Dantas EM, Dalvi RF, et al. Effects of kefir on the cardiac autonomic tones and baroreflex sensitivity in spontaneously hypertensive rats. Front Physiol 2016;7:211.

[62] Friques AG, Arpini CM, Kalil IC, Gava AL, Leal MA, Porto ML, et al. Chronic administration of the probiotic kefir improves the endothelial function in spontaneously hypertensive rats. J Transl Med 2015;13:390.

[63] Ahren IL, Xu J, Onning G, Olsson C, Ahrne S, Molin G. Antihypertensive activity of blueberries fermented by Lactobacillus plantarum DSM 15313 and effects on the gut microbiota in healthy rats. Clin Nutr 2015;34(4):719–26.

[64] Blanton C, He Z, Gottschall-Pass KT, Sweeney MI. Probiotics blunt the anti-hypertensive effect of blueberry feeding in hypertensive rats without altering hippuric acid production. PLoS One 2015;10(11):e0142036.

[65] Ekhlasi G, Zarrati M, Agah S, Hosseini AF, Hosseini S, Shidfar S, et al. Effects of symbiotic and vitamin E supplementation on blood pressure, nitric oxide and inflammatory factors in non-alcoholic fatty liver disease. EXCLI J 2017;16:278−90.

FURTHER READING

[1] Bezirtzoglou E, Stavropoulou E. Immunology and probiotic impact of the newborn and young children intestinal microflora. Anaerobe 2011;17(6):369−74.

[2] Lederberg J. The microbe's contribution to biology−50 years after. Int Microbiol 2006;9(3):155−6.

[3] Lederberg JMA. "Ome Sweet" Omics−a genealogical treasury of words. Scientist 2001;15:8.

[4] Li J, Jia H, Cai X, Zhong H, Feng Q, Sunagawa S, et al. An integrated catalog of reference genes in the human gut microbiome. Nat Biotechnol 2014;32(8):834−41.

[5] Xu X, Wang Z, Zhang X. The human microbiota associated with overall health. Crit Rev Biotechnol 2015;35(1):129−40.

[6] Le Chatelier E, Nielsen T, Qin J, Prifti E, Hildebrand F, Falony G, et al. Richness of human gut microbiome correlates with metabolic markers. Nature 2013;500(7464):541−6.

[7] Arumugam M, Raes J, Pelletier E, Le Paslier D, Yamada T, Mende DR, et al. Enterotypes of the human gut microbiome. Nature 2011;473(7346):174−80.

[8] Zoetendal EG, Rajilic-Stojanovic M, de Vos WM. High-throughput diversity and functionality analysis of the gastrointestinal tract microbiota. Gut 2008;57(11):1605−15.

[9] LeBlanc JG, Milani C, de Giori GS, Sesma F, van Sinderen D, Ventura M. Bacteria as vitamin suppliers to their host: a gut microbiota perspective. Curr Opin Biotechnol 2013;24(2):160−8.

[10] Bird AR, Brown IL, Topping DL. Starches, resistant starches, the gut microflora and human health. Curr Issues Intest Microbiol 2000;1(1):25−37.

[11] Ramakrishna BS, Roediger WE. Bacterial short chain fatty acids: their role in gastrointestinal disease. Dig Dis 1990;8(6):337−45.

[12] Everard A, Cani PD. Gut microbiota and GLP-1. Rev Endocr Metab Disord 2014;15(3):189−96.

[13] Windey K, De Preter V, Verbeke K. Relevance of protein fermentation to gut health. Mol Nutr Food Res 2012;56(1):184−96.

[14] Koppe L, Pillon NJ, Vella RE, Croze ML, Pelletier CC, Chambert S, et al. p-Cresyl sulfate promotes insulin resistance associated with CKD. J Am Soc Nephrol 2013;24(1):88−99.

[15] Ghazalpour A, Cespedes I, Bennett BJ, Allayee H. Expanding role of gut microbiota in lipid metabolism. Curr Opin Lipidol 2016;27(2):141−7.

[16] Lin Y, Vonk RJ, Slooff MJ, Kuipers F, Smit MJ. Differences in propionate-induced inhibition of cholesterol and triacylglycerol synthesis between human and rat hepatocytes in primary culture. Br J Nutr 1995;74(2):197−207.

[17] Begley M, Hill C, Gahan CG. Bile salt hydrolase activity in probiotics. Appl Environ Microbiol 2006;72(3):1729−38.

[18] Jones ML, Martoni CJ, Prakash S. Cholesterol lowering and inhibition of sterol absorption by Lactobacillus reuteri NCIMB 30242: a randomized controlled trial. Eur J Clin Nutr 2012;66(11):1234−41.

[19] de Aguiar Vallim TQ, Tarling EJ, Edwards PA. Pleiotropic roles of bile acids in metabolism. Cell Metab 2013;17(5):657−69.

[20] Parvez S, Malik KA, Ah Kang S, Kim HY. Probiotics and their fermented food products are beneficial for health. J Appl Microbiol 2006;100(6):1171−85.

[21] Fijan S. Microorganisms with claimed probiotic properties: an overview of recent literature. Int J Environ Res Publ Health 2014;11(5):4745−67.

[22] Gibson GR, Probert HM, Loo JV, Rastall RA, Roberfroid MB. Dietary modulation of the human colonic microbiota: updating the concept of prebiotics. Nutr Res Rev 2004;17(2):259−75.

[23] Gibson GR, Roberfroid MB. Dietary modulation of the human colonic microbiota: introducing the concept of prebiotics. J Nutr 1995;125(6):1401−12.

[24] Yeo SK, Ooi LG, Lim TJ, Liong MT. Antihypertensive properties of plant-based prebiotics. Int J Mol Sci 2009;10(8):3517−30.

[25] Gibson GR, Hutkins R, Sanders ME, Prescott SL, Reimer RA, Salminen SJ, et al. Expert consensus document: the International Scientific Association for Probiotics and Prebiotics (ISAPP) consensus statement on the definition and scope of prebiotics. Nat Rev Gastroenterol Hepatol 2017;14(8):491−502.

[26] Roberfroid M, Gibson GR, Hoyles L, McCartney AL, Rastall R, Rowland I, et al. Prebiotic effects: metabolic and health benefits. Br J Nutr 2010;104(Suppl. 2):S1−63.

[27] Pandey KR, Naik SR, Vakil BV. Probiotics, prebiotics and synbiotics- a review. J Food Sci Technol 2015;52(12):7577−87.

[28] Kao D, Roach B, Silva M, Beck P, Rioux K, Kaplan GG, et al. Effect of oral capsule- vs colonoscopy-delivered fecal microbiota transplantation on recurrent Clostridium difficile infection: a randomized clinical trial. JAMA 2017;318(20):1985−93.

[29] Petrof EO, Gloor GB, Vanner SJ, Weese SJ, Carter D, Daigneault MC, et al. Stool substitute transplant therapy for the eradication of Clostridium difficile infection: 'RePOOPulating' the gut. Microbiome 2013;1(1):3.

[30] Garcia-Lezana T, Raurell I, Bravo M, Torres-Arauz M, Salcedo MT, Santiago A, et al. Restoration of a healthy intestinal microbiota normalizes portal hypertension in a rat model of nonalcoholic steatohepatitis. Hepatology 2017.

[31] Vrieze A, Van Nood E, Holleman F, Salojarvi J, Kootte RS, Bartelsman JF, et al. Transfer of intestinal microbiota from lean donors increases insulin sensitivity in individuals with metabolic syndrome. Gastroenterology 2012;143(4). 913.e7−916 e7.

Chapter 26

The Emerging Role of Microbiome–Gut–Brain Axis in Functional Gastrointestinal Disorders

Karolina Skonieczna-Żydecka[1], Igor Loniewski[1,2], Anastasios Koulaouzidis[3] and Wojciech Marlicz[4]

[1]Department of Biochemistry and Human Nutrition, Pomeranian Medical University, Szczecin, Poland; [2]Sanprobi Sp. z o.o. Sp. k., Szczecin, Poland; [3]Endoscopy Unit, The Royal Infirmary of Edinburgh, Edinburgh, United Kingdom; [4]Department of Gastroenterology, Pomeranian Medical University, Szczecin, Poland

ROME IV CRITERIA AND FUNCTIONAL GASTROINTESTINAL DISORDERS

Functional gastrointestinal disorders (FGIDs) are currently classified as gastrointestinal (GI) symptom–related disorders of gut–brain interaction, related to any combination of motility disturbances, visceral hypersensitivity, altered mucosal and immune function, gut microbiota, and central nervous system (CNS) processing [1]. Functional dyspepsia (FD) is an umbrella term, which refers to patients with postprandial distress syndrome and epigastric pain syndrome [2]. FGID's symptom overlaps are common, which makes the clinical diagnosis often challenging [3]. The microbiome–gut–brain axis has become an attractive target for prophylaxis and treatment of these common disorders (Scheme 26.1).

MICROBIOTA AND INTESTINAL BARRIER

GI tract is a microbiological niche, housing roughly 39×10^9 bacterial cells [4–8]. The evidence is growing that microbiota shapes our immune response and controls host angiogenesis, bone density, and fat mass. The human gut ecosystem was found to be responsible for vitamin, amino acid, steroid hormone and neurotransmitter biosynthesis, xenobiotic metabolism, and modification of the nervous system in terms of structure and function [9]. Many human and animal association studies have shown that particular alterations of the microecological composition of the gut may serve as a bacterial imprint of the disease. Along with biostatistical approaches, predicting functional capacity of the gut microenvironment is of clinical importance to identify microbial-based signature of the disease [10,11].

Microbiota housed in the GI tract acts as a barrier between external-luminal environment and the internal milieu. In fact, the combination of microbiota, the single epithelial cell layer, the gut-associated lymphoid tissue (GALT), and

---

Recurrent abdominal pain on average at least 1 day/week in the last 3 months, associated with two or more of the following criteria:

1. Related to defecation

2. Associated with a change in the frequency of stool

3. Associated with a change in the form (appearance) of stool

*These criteria should be fulfilled for the last 3 months, with symptom onset at least 6 months prior to diagnosis with no alarm symptoms present.

SCHEME 26.1 The ROME IV IBS diagnostic criteria.

Microbiome and Metabolome in Diagnosis, Therapy, and other Strategic Applications. https://doi.org/10.1016/B978-0-12-815249-2.00026-9

circulatory, lymphatic, and nervous system cocreates a physical/functional structure called the intestinal barrier [12]. (Fig 26.1) It seems that the microbiota and the epithelial cell monolayer serve as crucial constituents of this barrier [13].

In the intestine, microbiota was found to regulate epithelial structure, as demonstrated in germ-free studies and probiotic management [14,15]. Eubiosis within the gut ecosystem positively affects mucus and antimicrobial peptide expression, preventing therefore pathogenic infections [16–18]. Epithelial cells directly regulate paracellular passage of antigens and pathogens, which in turn affects the development of immunological activity in the underlying submucosa. Tight junctions (TJ) serve as the most important regulators of intestinal permeability. These are multiprotein complexes built of four transmembrane proteins: claudins, occludins, junctional adhesion molecules, and tricellulins.

They interact with cytosol proteins, i.e., zonula occludens (ZO-1, ZO-2, ZO-3), switched to actin cytoskeleton of enterocytes, to induce myosin light chain phosphorylation, allowing for undisturbed paracellular transport [19]. Several studies found that microbiota reestablishment elevated the expression of crucial TJ proteins [20,21]. Consequently, dysbiosis and subsequent myosin phosphorylation imbalance, as well as TJ proteins endocytosis, significantly increases the permeability of gut barrier and leads to GALT activation [22,23].

Functional Gastrointestinal Disorder Pathophysiology

It is known that FGIDs can be precipitated by an acute GI infection, and mucosal immune activation is proven to play part in their pathogenesis. A few studies confirmed that mucosal immunocytes count, with neutrophil absence, was increased in irritable bowel syndrome (IBS) and FD patients [24,25], though this inflammatory response was of a subclinical nature [7,26,27]. Along with immune activation, gut sensory and motor functions may be altered. It was elegantly shown that mucosal biopsy supernatant of IBS subjects was enriched in histamine, tryptase, and prostaglandins, considered to be due to mast cell degranulation.

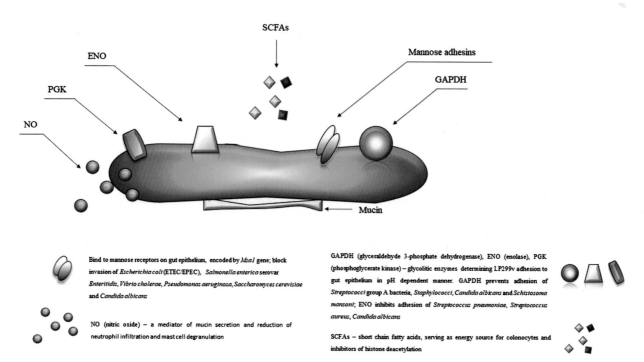

FIGURE 26.1 Microbiota–gut–brain axis in functional gastrointestinal disorders (FGIDs). In patients suffering from FGIDs, microbial imbalance within the gut ecosystem was found. Dysbiosis may induce abnormal fermentation processes (skewed gas volume and composition) and diminish the expression of TJ assembly, elevating gut epithelium permeability. Consequently, luminal antigens (food, bacteria-LPS) enter the lamina propria, leading to GALT activation. Once they upregulate T lymphocytes, an activation of mast cells occurs. Due to mastocyte degranulation, histamine, proteases, and serotonin act with corresponding receptors, resulting in enteric neuron excitation and smooth muscle contraction. Of note, mast cell activation affects spinal, vagal, and pelvic neuronal pathways, implicating gut–brain axis in pathophysiology of the disease. Consequently, patients experience bloating, abdominal pain, and abnormal intestinal transit. Possible targets counteracted by probiotics are marked with grey arrows. The legend is correct but the figure should be swaped with one from below.

In an animal study, Barbara et al. [28] proved that mucosal biopsy supernatants of rodents with IBS significantly increased nerve discharge of rat mesenteric sensory nerves, in comparison to samplings of healthy controls. Further in vitro studies confirmed that proteases or polyunsaturated fatty acids metabolites, present in biopsy supernatants of IBS patients, mediate visceral pain, as these mediators bind to proteinase-activated receptors and transient receptor potential cation channel, respectively [29,30]. It was demonstrated that mucosal IBS supernatants elevated the growth rate of neurites and increased the expression of neuronal growth protein, proving that immune activation typically found in FGIDs may alter the structure of neural network within the mucosa [31]. Recently, it has been demonstrated that visceral hypersensitivity in IBS is linked to intestinal barrier loss of integrity [32].

GUT—BRAIN AXIS

In the late 1940s, Almy and Tulin demonstrated that stress negatively affected gut functions, making the colon irritable and vulnerable to numerous stimuli [33]. Since then, a number of colonic functions were found to be influenced by stressors, including motor, sensory, and secretory ones [34—36]. Additionally, the burden of psychological morbidity among patients with FGIDs is substantial. This group may suffer from anxiety, depression, and cognition decline, more frequently compared to controls [37,38]. In fact, gnotobiotic mice were shown to experience stress in more acute way and develop memory dysfunctions [39,40]. More recently, it has been discovered that independent to any psychological comorbidity, there is a stress-related impairment in visuospatial memory in IBS [41].

Nerves, Neurotransmitters, and Hormones

Intestine and brain communicate through the gut—brain axis [42,43]. The latter encompasses a set of neurons, including the vagus nerve and the plexuses of Auerbach and Meissner, connecting the human CNS with the GI tract and its glands, i.e., pancreas and liver [44]. (Fig 26 1) Communication between the intestine and the brain is of bidirectional nature, with neuronal, endocrine, and immunological factors being involved in this signaling pathway [45].

Several studies have proved that gut microbiota is essential for the proper structure and function of brain regions involved in emotion control and cognitive abilities [46—49]. As far as FGIDs are concerned, gut microbiota is responsible for serotonin production, a key neurotransmitter within CNS and enteric nervous system [50—52]. Peptide YY, produced in the gut, acts as an inhibitor of digestive tract motility and secretion [53,54]. Other signaling molecules synthesized in the GI tract, such as methane and short-chain fatty acids (SCFAs), have a great impact on intestinal function.

Microbial metabolites may affect contractile activity, neuromuscular function, and calcium concentration within the myenteric plexus, all being involved in visceral perception, motility, as well as secretory and motor functions of the GI tract [55—57]. When the gut—brain axis is disrupted, endotoxins released by pathogenic microorganisms (LPS), can affect CNS indirectly through the activation of the immune system or directly via toll-like receptors (TLRs), located on the surface of glial cells [58]. Germ-free studies proved that TLRs were not expressed along proximal—distal axis, but microbiota reestablishment restored the synthesis [59,60]. TLRs were discovered to affect GI motility [61], and the expression of TLRs genes in IBS persons is skewed, in comparison to controls [62,63].

FUNCTIONAL GASTROINTESTINAL DISORDERS AND INTESTINAL BARRIER DISRUPTIONS

It is a great challenge to obtain representative small bowel (SB) samples, as the material is difficult to access, and there are frequent changes in SB microbiota composition, even on a 1-day rhythm [64,65]. Consequently, the SB microbiota is not sufficiently known, which adds uncertainty of the results obtained in experimental studies [66]. As far as SB microbiota function goes, this is even less understood, [65,67]. In 2016, Zhong et al. [68] assessed the duodenal mucosa microbiota composition, in a small cohort of patients with FD, and compared to controls matched for gender, age, and body mass index. The authors found that the *Streptococcus* count in FD was elevated, although not statistically different, while the relative abundance of *Prevotella, Veillonella, Actinomyces, Atopobium,* and *Leptotrichia* significantly decreased in FD patients [68].

Antibiotic Prescriptions

Tan et al. [69] administered 400 mg of rifaximin for 8 weeks to 86 subjects with FD. The treatment resulted in global improvement of symptoms, as well as belching and postprandial fullness/bloating, predominantly in female patients, as fast as 2 weeks. Vanheel et al. [24] evaluated the transepithelial electrical resistance (TEER) and paracellular permeability of duodenal biopsies, collected from 15 FD patients and 15 controls, and found a decrease in TEER and increase in permeability to fluorescein isothiocyanate in the patients. Expression of ZO-1 and occludin, along with phosphorylation rate, was lower in the patients. Furthermore, a low-grade inflammation manifested by increased infiltration of mucosal mast cells and eosinophils was found, providing evidence to duodenal mucosal barrier alteration in FD, as evidenced more recently by Talley et al. [70].

Contribution of Probiotics

Mujagic et al. [21] demonstrated that probiotic strains of *Lactobacillus plantarum*, namely WCFS1, CIP104448, TIFN101, administered orally for a week to 10 healthy individuals, elevated the expression of genes involved in cell to cell adhesion, among them TJ protein synthesis and degradation proving that probiotic therapy may improve duodenal barrier integrity.

Symptom-Related Signatures

A very early study by Balsari et al. [71] showed that the counts of *Bifidobacterium* and *Lactobacillus* were decreased in IBS patients. In 2007, Malinen et al. [72] confirmed that the relative abundance of *Lactobacillus* spp. was lowered in diarrhea-predominant IBS (IBS-D), whereas constipation-predominant IBS (IBS-C) stools were enriched in *Veillonella* spp. The authors showed that *Bifidobacterium catenulatum* and *Clostridium coccoides* were also lowered. Rajilic-Stojanovic et al. [73] discovered that the more symptoms of IBS were experienced by a patient, the more abundant were *Gammaproteobacteria* and less numerous *Bifidobacterium* spp. Furthermore, Jeffery et al. [74] reported that satiety, bloating, and global IBS symptoms were positively correlated with *Cyanobacteria*, whereas the psychosomatic component of IBS was more prevalent in patients enriched for *Proteobacteria* and *Actinomycetales*.

Importantly, Chung et al. [75] identified significant differences between the microbiota composition in stool and mucosal samples of faecal bacterial representatives. Kerckhoffs et al. [76] utilized mucosa brush samples to isolate bacterial DNA and found that intestinal microbiota of IBS patients was poorly enriched in *Bifidobacterium catenulatum*. More recently, Giamarellos-Bourboulis et al. [77], in a study of duodenal aspirates obtained during endoscopy, reported a significant reduction in the diversity of SB microbiota, and the number of species was typical of IBS. *Escherichia/Shigella* genera were found to be increased in IBS patients, and the counts of *Acinetobacter, Citrobacter, Microvirgula, Flavobacterium, Enhydrobacter, Weissella, Leuconostoc,* and *Lactococcus* genera lowered.

Analyzing capsule biopsies from the jejunum of 35 patients with IBS and 16 healthy volunteers Długosz et al. [78] found only five operational taxonomic units, with a trend toward differential expression between IBS and controls. Parkes et al. [79] observed that daily bowel movements were negatively correlated with the mucosa-associated *Bifidobacterium* and *Lactobacillus* counts. Table 26.1.

Intestinal Permeability and Leaky Gut

Martinez et al. [80] examined jejunal fluid from IBS-D patients and found that claudin-2 expression was elevated and occludin phosphorylation diminished in patients. On the other hand, ZO-1 and ZO-3 mRNA expression decreased in IBS-D patients, and cellular protein redistribution to the cytosol occurred. Consequently, the number of dilated junctions in IBS-D was increased, as was the intercellular distance between epithelial cells. Kong et al. [81] discovered low Claudin-1 and Claudin-4 expression in SB mucosa of IBS-D patients, whereas the elevated expression of Claudin-1, -3, and -4 in IBS-C was noted. Zhou et al. [82] identified elevated Claudin-1 concentration in samplings collected from patients with elevated intestinal permeability. Such expression differences were previously found by Bertiaux-Vandaële et al. [83].

Altered Immunoinflammatory Profile

Low-grade immune response has also been described during the course of IBS. Vicario et al. [84] demonstrated a high number of mucosal B lymphocytes and plasma cells in IBS-D patients. Different mucosal-humoral activity, i.e., transcription of Ig genes and percentage of IgG (+) cells, was noted. Earlier, Park et al. [85] and Barbara et al. [28] demonstrated that mast cell counts in terminal ileum, ascending colon, and rectum were increased.

TABLE 26.1 Microbiota Alterations in Functional Gastrointestinal Disorders

Irritable Bowel Syndrome

Age	Sampling	Microbiota analysis	Significant outcome	References
47.3	Feces	16S rRNA sequencing	Lower abundance of *Collinsella aerofaciens, C.* cocleatum, and *Coprococcus eutactus*	a
37	Feces	16S rRNA sequencing	Increased Firmicutes and decreased Bacteroidetes abundance	b
46.5	Feces	16S rRNA sequencing	IBS-D: Decreased *Lactobacillus* spp. abundance; IBS-C: increased *Veillonella* spp. abundance IBS: decreased *Clostridium coccoides* and *Bifidobacterium catenulatum* count	[72]
46	Feces	Culture, PCR-DGGE	Higher abundance of coliforms, increased aerobe:anaerobe ratio	c
47.7	Feces	Culture, 16S rRNA sequencing	Decreased *Lactobacillus* spp. and *Bifidobacterium* spp. and increased count of aerobes; IBS-D: decreased *Clostridium thermosuccinogenes*, increased *Ruminococcus torques* IBS-C: *increased Ruminococcus bromii*-like abundance	d
34.3	rectal biopsy	FISH	Increased total bacterial count, elevated *Bacteroide*s and *Eubacterium rectale—Clostridium coccoides*; IBS-D: decreased B*ifidobacteria*	[79]
42	Feces, duodenum	FISH, real-time PCR	Low abundance of *Lactobacillus* spp., decrease count of *Bifidobacterium catenulatum*	[76]
9.4	Feces	16S rRNA PhyloChip analysis	Increased *Gammaproteobacteria* and *Veillonella*, decreased *Eubacterium* and *Anaerovorax* counts	e
49	Feces	qPCR, microarray analysis	Increased *Firmicutes: Bacteroidetes* ratio, *Dorea, Ruminococcus,* and *Clostridium* spp., decreased *Bacteroidetes, Bifidobacterium, Faecalibacterium* spp. methanogens	[74]
31.9	Feces sigmoidosc.	qPCR	Reduction of aerobic bacteria and increased *Lactobacillu*s spp. in fecal samples	f
41	Feces colonosc.	16S rRNA sequencing	Decreased bacterial diversity in postinfection subject	g
42.2	Feces	qPCR, phylogenetic microarray analysis	Increased *Bacteroidetes* phylum concentration and decreased *Clostridia* count	h
21.7	Feces	qRT-PCR	increased abundance of *Veilonella* spp. and *Lactobacillus* spp.	[97]
28	Feces	culture, FISH	low counts of *Bifidobacterium* spp. and *Lactobacillus* spp., increased *Enterobacteriaceae*	i
39.7	Feces	qPCR	Decreased abundance of *Bifidobacterium* spp. *leptum* group; increased count of *E.coli*	j
47.5	Feces	qPCR	Small diversity of *Bifidobacter* and *Clostridium coccoides*	k
48	Feces	qPCR	High counts of *S.aureus*	l
34	Feces	qPCR	Low *Bifidobacterium* spp. and *Lactobacillus* spp., high *Ruminococcus productus-Clostridium coccoides, Veilonella, Bacteroides thetaiotamicron, Pseudomonas aeruginosa, Bacteroides thetaiotamicron*	m
35	Duod. rectal	16S rRNA sequencing	Small difference between duodenal and rectum; *Faecalibacterium* and *Hyphomicrobium*, clinical data	n

Continued

TABLE 26.1 Microbiota Alterations in Functional Gastrointestinal Disorders—cont'd

Functional Dyspepsia				
42.5	Gastric fluid	T-RFLP	Lower counts of *Prevotella*	o
44	Gastric fluid	16S rRNA sequencing	Low abundance of *Proteobacteria*, high concentration of *Bacteroidetes*	p
nd	Duodenum	16S rRNA sequencing	Low abundance of *Actinomyces, Atopobium, Leptotrichia, Prevotella, Veilonella*	[68]

[a]Kassinen A, Krogius-Kurikka L, Mäkivuokko H, Rinttilä T,Paulin L, Corander J, Malinen E, et al. The fecal microbiota of irritable bowel syndrome patients differs significantly from that of healthy subjects. Gastroenterology 2007;133(1):24—33.

[b]Jeffery IB, O'Toole PW, Ohman L, Claesson MJ, Deane J, Quigley EM, et al. An irritable bowel syndrome subtype defined by species-specific alterations in faecal microbiota. Gut 2012;61(7):997—1006.

[c]Mättö J, Maunuksela L, Kajander K, Palva A, Korpela R, Kassinen A, et al. Composition and temporal stability of gastrointestinal microbiota in irritable bowel syndrome—a longitudinal study in IBS and control subjects. FEMS Immunol Med Microbiol 2005;43:213—22.

[d]Lyra A. Diarrhoea-predominant irritable bowel syndrome distinguishable by 16S rRNA gene phylotype quantification. World J Gastroenterol 2009;15(47):5936.

[e]Saulnier DM, Riehle K, Mistretta TA, Diaz MA, Mandal D, Raza S, et al. Gastrointestinal microbiome signatures of pediatric patients with irritable bowel syndrome. Gastroenterology 2011;141(5):1782—91.

[f]Carroll IM, Chang YH, Park J, Sartor RB, Ringel Y. Luminal and mucosal-associated intestinal microbiota in patients with diarrhea-predominant irritable bowel syndrome. Gut Pathogens 2010;2(1):19.

[g]Sundin J, Rangel I, Fuentes S, Heikamp-de Jong I, Hultgren-Hornquist E, et al. Altered faecal and mucosal microbial composition in postinfectious irritable bowel syndrome patients correlates with mucosal lymphocyte phenotypes and psychological distress. Aliment Pharmacol Ther 2015;41(4):342—51.

[h]Jalanka-Tuovinen J, Salojarvi J, Salonen A, Immonen O, Garsed K, Kelly FM, et al. Faecal microbiota composition and host—microbe cross-talk following gastroenteritis and in postinfectious irritable bowel syndrome. Gut 2014;63(11):1737—45.

[i]Chassard C, Dapoigny M, Scott KP, Crouzet L, Del'homme C, Marquet P, et al. Functional dysbiosis within the gut microbiota of patients with constipated—irritable bowel syndrome. Aliment Pharmacol Ther 2012;35(7):828—38.

[j]Duboc H, Rainteau D, Rajca S, Humbert L, Farabos D, Maubert M, et al. Increase in fecal primary bile acids and dysbiosis in patients with diarrhea-predominant irritable bowel syndrome. Neurogastroenterol Motil 2012;24(6):513—520. e246-517.

[k]Ponnusamy K, Choi JN, Kim J, Lee SY, Lee CH. Microbial community and metabolomic comparison of irritable bowel syndrome faeces. J Med Microbiol 2011;60(Pt 6):817—27.

[l]Rinttila T, Lyra A, Krogius-Kurikka L, Palva A. Real-time PCR analysis of enteric pathogens from fecal samples of irritable bowel syndrome subjects. Gut Pathogens 2011;3(1):6.

[m]Shukla R, Ghoshal U, Dhole TN, Ghoshal UC. Fecal Microbiota in Patients with Irritable Bowel Syndrome Compared with Healthy Controls Using Real-Time Polymerase Chain Reaction: An Evidence of Dysbiosis. Dig Dis Sci 2015;60:2953—62.

[n]Li G, Yang M, Jin Y, Li Y, Qian W, Xiong H, Song J, Hou X. Involvement of shared mucosal-associated microbiota in the duodenum and rectum in diarrhea-predominant irritable bowel syndrome. J Gastroenterol Hepatol November 30, 2017.

[o]Nakae H, Tsuda A, Matsuoka T, Mine T, Koga Y. Gastric microbiota in the functional dyspepsia patients treated with probiotic yogurt. BMJ Open Gastroenterol 2016;3(1):e000109.

[p]Igarashi M, Nakae H, Matsuoka T, Takahashi S, Hisada T, Tomita J, et al. Alteration in the gastric microbiota and its restoration by probiotics in patients with functional dyspepsia. BMJ Open Gastroenterol 2017;4(1):e000144.

MANAGEMENT AND RECOMMENDATIONS

Interventions believed to positively affect gut microbial niche composition include modification of diet, nonabsorbable antibiotics, and probiotic supplementation [86—96].

Dietary Modifications

There are minor differences in eating habits between FGIDs patients and healthy persons, including few meals during a week, with more common adoption of small portions and little snacks [88]. Yet the majority report an association between certain food types and symptoms [89,90]. Fruits and vegetables enriched in high fermenting oligosaccharides, disaccharides, monosaccharides, and polyols, namely FODMAPs [7], are often contraindicated, as such approach was shown to regulate the density of serotonin cells in the colon and normalize fecal level of LPS, both affecting visceral hypersensitivity [91,92].

There are few randomized controlled trials (RCTs) assessing the efficacy of low FODMAP diet in IBS [92—94]. As many as 80% of the patients have seen reduction in symptom severity [95]. Some experience abdominal pain and bloating after wheat product ingestion, although coeliac disease was not clinically proven. This phenomenon is called non-coeliac gluten sensitivity (NCGS) [7].

Complex carbohydrates, including fiber, determine the spectrum of metabolites produced by the microbiota, predominantly SCFAs, which seem beneficial for gut health, by reducing mechanoreceptor hypersensitivity and altering neurotransmitter release, which may reduce luminal pressure [96,97]. However, there is still no coherent conclusion on fiber intake, as the constituent has both positive (increased stool volume, incretin effect, histone deacetylation inhibition) as negative effects (gas production and water retention in the gut) [98]. Carbohydrate fermentation, producing hydrogen gas, may stimulate the growth of *Archaea*, which are well-known sulfide producers, recognized as toxic [7].

Another criticism for the low carbohydrate diet is that it favors protein fermentation, thus protease expression and harmful amine, sulfur, and phenolic compound synthesis [99,100]. Lastly, chylomicron flow stimulated by a low carbohydrate regimen was discovered to facilitate the diffusion of LPS through the gut epithelium, triggering systemic low-grade inflammation [101]. Other dietary therapies to treat FGIDs manifestation are discussed elsewhere [102].

Probiotics

Lactobacillus plantarum 299v (Lp299v) expresses *Msa* gene coding mannose adhesins on the cell surface, determining strong affinity to mannose receptors in epithelium (Scheme 26.2). This linkage prevents the invasion of pathogenic bacteria such as *Escherichia coli*, *Salmonella enterica serovar Enteritidis*, *Vibrio cholerae*, *Pseudomonas aeruginosa*, *Saccharomyces cerevisiae*, and *Candida albicans*. Other surface adhesins, i.e., GAPDH, ENO, and PGK, inhibit other invasions. Generated NO was found to increase mucus secretion and reduce mast cell degranulation and neutrophil infiltration [66] (Fig 26.2). Nevertheless, precise choice of effective probiotic for FGIDs patients is difficult, as the efficacy is strain-specific and may depend on preparation protocol, dosage, and duration of administration [103–105].

Probiotic strains were found not only to restore the microenvironment of the intestine but also to improve biliary salt metabolism and fermentation in the colon, both engaged in clinical problems associated with functional bowel disorders [106,107]. Immune activation may play role in FGIDs onset, and probiotics were shown to have immunoregulatory effects [108,109].

Moayyedi et al. [110] analyzed 19 RCTs, comprising 1650 IBS patients, and found that relative risk of IBS not improved during probiotic prescription. Three years later, it was discovered that strains of *Bifidobacterium longum* and *Lactobacillus acidophilus* improved pain scores, whereas distension scores were found to be diminished by *B. breve*, *B. infantis*, *L. casei*, and *L. plantarum*, and flatulence by *B. breve*, *B. infantis*, *L. casei*, *L. plantarum*, *B. longum*, *L. acidophilus*, *L. bulgaricus*, and *Streptococcus salivarius* ssp. *Thermophilus* [111]. Ford et al. [112] concluded that patients may symptomatically benefit from probiotic formulas. In addition, metanalytic data provided by Dimidi et al. [113] demonstrated that probiotics improve stool number and consistency, both attributable to *B. lactis* strains.

Probiotic Metaanalyses

In the metaanalysis of 2015 by Didari et al. [114], benefits in the relative risk/RR for responders, based on abdominal pain and global symptom score, were around 2.0. In the experience of Tiequn et al. [115], with *Lactobacillus* spp., RR advantage was 7.69. The World Gastroenterology Organization (WGO) recommends a few probiotic formulas for IBS and functional constipation [109]. (Table 26.2).

Functional dyspepsia(FD) diagnostic criteria

1. One or more of the following:

a. Bothersome postprandial fullness
b. Bothersome early satiation
c. Bothersome epigastric pain
d. Bothersome epigastric burning

AND
2. No evidence of structural disease (including upper endoscopy) that is likely to explain the symptoms

* Criteria fulfilled for the last three months, with symptom onset at least six months before diagnosis.

SCHEME 26.2 The ROME IV functional dyspepsia criteria.

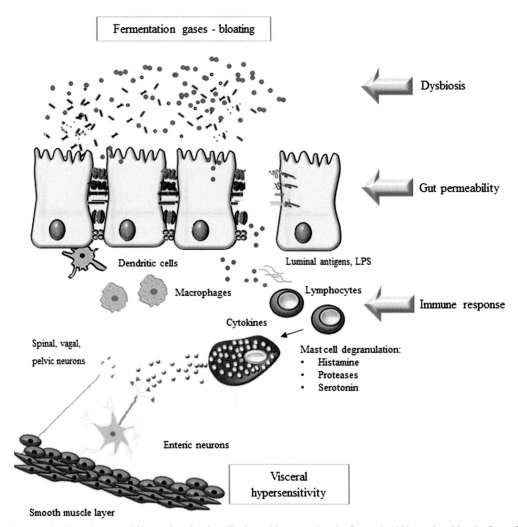

FIGURE 26.2 *Lactobacillus plantarum* 299v modes of action. The legend is correct but the figure should be replaced by the figure from above.

Pharmacological Approach

The most potent microbiota modulators are antibiotics. Although most of the broad spectrum antibiotics might exert negative effect on microbiome and clinical symptoms (for example toxin related *Clostridium difficile* diarrhea), nonabsorbable antimicrobial drugs met scientific and clinical attentions in FGIDs. Currently, safer therapy, whit short-term administration of rifaximin is recommended by the Americal Gastroenterological Association (AGA) for the treatment of patients with IBD-D based on numerous well-conducted clinical studies [116].

Faecal Microbiota Transplantation

Johnsen et al. [117] examined the efficacy of FMT in moderate to severe IBS-D and IBS-C patients. A total of 80 g fresh or frozen faeces were transplanted to 83 participants into cecum by a colonoscope with patients' own faeces used as placebo. After 3 months, 65% of the patients reported a relief of more than 75 points, assessed by IBS severity scoring system. A systematic review comprising data from 48 IBS patients reported that FMT improved the clinical course of the disease in 58% of cases [118]. On the other hand, when focusing on FMT in IBS, there are still a lot of questions that remain to be answered, including route of administration and optimal dosage.

TABLE 26.2 WGO Probiotic Recommendations in 2017

Recommended Probiotics	Dose	References
IBS		
Bifidobacterium bifidum MIMBb75	1×10^9 CFU/daily	q
Lactobacillus plantarum 299v	5×10^7 billion CFU/daily	r
Escherichia coli DSM17252	10^7 CFU/3 times a day	[112]
Lactobacillus rhamnosus NCIMB 30174, *L. plantarum* NCIMB 30173, *L. acidophilus* NCIMB 30175, and *Enterococcus faecium* NCIMB 30176.	10×10^9 CFU	s
Bacillus coagulans and fructooligosaccharides	15×10^7 CFU/3 times a day	t
Lactobacillus animalis subsp. *lactis* BB-12®, *L. acidophilus* LA-5®, *L. delbrueckii* subsp. *bulgaricus* LBY-27, *Streptococcus thermophilus* STY-31	4×10^9 CFU/twice a day	u
Saccharomyces boulardii CNCM I-745	5×10^9 CFU/d or 250 mg twice a day	v
Bifidobacterium infantis 35624	1×10^8 CFU/d	w
Bifidobacterium animalis DN-173 010	1×10^{10} CFU/twice a day	x
Lactobacillus acidophilus SDC 2012, 2013	1×10^{10} CFU/d	y
Lactobacillus rhamnosus GG, *L. rhamnosus* LC705, *Propionibacterium freudenreichii* subsp. *shermanii* JS DSM 7067,*Bifidobacterium animalis* subsp. *lactis* Bb12 DSM 15954	1×10^{10} CFU/d	z
Pediococcus acidilactici CECT 7483, *Lactobacillus plantarum* CECT 7484, *L. plantarum* CECT 7485	$3-6 \times 10^9$ CFU/capsule/d	aa
Bacillus coagulans GBI-30, 6086	2×10^9 CFU/d	bb
Functional constipation		
Bifidobacterium bifidum (KCTC 12199BP), *B. lactis* (KCTC 11904BP), *B. longum* (KCTC 12200BP), *Lactobacillus acidophilus* (KCTC 11906BP), *L. rhamnosus* (KCTC 12202BP), and *Streptococcus thermophilus* (KCTC 11870BP)	2×10^8 CFU/d	cc
Lactobacillus reuteri DSM 17938	1×10^8 CFU/twice a day	dd
Fructooligosaccharide (FOS) and *Lactobacillus paracasei* (Lpc-37), *L. rhamnosus* (HN001), *L. acidophilus* (NCFM) and *Bifidobacterium lactis*	6 g FOS and 10^8-10^9 CFU/d	ee

qGuglielmetti S, Mora D, Gschwender M, Popp K. Randomised clinical trial: Bifidobacterium bifidum MIMBb75 significantly alleviates irritable bowel syndrome and improves quality of life–a double-blind, placebo-controlled study. Aliment Pharmacol Ther 2011;33(10):1123–32.
rDucrotté P, Sawant P, Jayanthi V. Clinical trial: Lactobacillus plantarum 299v (DSM 9843) improves symptoms of irritable bowel syndrome. World J Gastroenterol 2012;18(30):4012–18.
sSisson G, Ayis S, Sherwood RA, Bjarnason I. Randomised clinical trial: A liquid multi-strain probiotic vs. placebo in the irritable bowel syndrome–a 12 week double-blind study. Aliment Pharmacol Ther 2014;40(1):51–62.
tRogha M, Esfahani MZ, Zargarzadeh AH. The efficacy of a synbiotic containing Bacillus Coagulans in treatment of irritable bowel syndrome: a randomized placebo-controlled trial. Gastroenterol Hepatol Bed Bench 2014;7(3):156–63.
uJafari E, Vahedi H, Merat S, Momtahen S, Riahi A. Therapeutic effects, tolerability and safety ofa multi-strain probiotic in Iranian adults with irritable bowel syndrome and bloating. Arch Iran Med 2014;17(7):466–470.
vChoi CH, Jo SY, Park HJ, Chang SK, Byeon J-S, Myung S-J. A randomized, double-blind, placebo-controlled multicenter trial of saccharomyces boulardii in irritable bowel syndrome: effect on quality of life. J Clin Gastroenterol 2011;45(8):679–83.
wMoayyedi P, Ford AC, Talley NJ, Cremonini F, Foxx-Orenstein AE, Brandt LJ, et al.The efficacy of probiotics in the treatment of irritable bowel syndrome: a systematic review. Gut 2010;59(3):325–32.
xAgrawal A, Houghton LA, Morris J, Reilly B, Guyonnet D,Goupil Feuillerat N, et al. Clinical trial: the effects ofa fermented milk product containing Bifidobacterium lactis DN-173 010 on abdominal distension and gastrointestinal transit in irritable bowel syndrome with constipation. Aliment Pharmacol Ther 2009;29(1):104–14.
ySinn DH, Song JH, Kim HJ, Lee JH, Son HJ, Chang DK, et al. Therapeutic effect of Lactobacillus acidophilus-SDC 2012, 2013 in patients with irritable bowel syndrome. Dig Dis Sci 2008;53(10):2714–18.
zKajander K, Myllyluoma E, Rajilić-Stojanović M, Kyrönpalo S, Rasmussen M, Järvenpää S, et al. Clinical trial: multispecies probiotic supplementatio-nalleviates the symptoms of irritable bowel syndrome and stabilizes intestinal microbiota. Aliment Pharmacol Ther 2008;27(1):48–57.
aaLorenzo-Zúñiga V, Llop E, Suárez C, Alvarez B, Abreu L, Espadaler J, et al. I.31, a new combination of probiotics, improves irritable bowel syndrome-related quality of life. World J Gastroenterol 2014;20(26):8709–16.
bbDolin BJ. Effects of a proprietary Bacillus coagulans preparation on symptoms of diarrhea predominant irritable bowel syndrome. Methods Find Exp Clin Pharmacol 2009;31(10):655–9.
ccYeun Y, Lee J. Effect of a double-coated probiotic formulation on functional constipation in the elderly: a randomized, double blind, controlled study Arch Pharm Res 2015;38(7):1345–1350.
ddOjetti V, Ianiro G, Tortora A, D'Angelo G, Di Rienzo TA, Bibbò S, et al. The effect of Lactobacillus reuteri supplementation in adults with chronic functional constipation:a randomized, double-blind, placebo-controlled trial. J Gastrointest Liver Dis JGLD 2014;23(4):387–91.
eeWaitzberg DL, Logullo LC, Bittencourt AF, Torrinhas RS, Shiroma GM, Paulino NP, et al. Effect of synbiotic in constipated adult women - a random-ized, double-blind, placebo-controlled study of clinical response. Clin Nutr Edinb Scotl 2013;32(1):27–33.

LIST OF ACRONYMS AND ABBREVIATIONS

AGA American Gastroenterological Association
CNS Central nervous system
ENS Enteric nervous system
FD Functional dyspepsia
FGIDs Functional gastrointestinal disorders
FMT Fecal microbiota transplantation
FODMAPs Fermenting oligosaccharides, disaccharides, monosaccharides, and polyols
GALT Gut-associated lymphoid tissue
GI Gastrointestinal
IBS— Irritable bowel syndrome
IBS-C Constipation predominant IBS
IBS-D Diarrhea-predominant IBS
LPS Lipopolysaccharide
NCGS Non celiac gluten sensitivity
OUT Operational taxonomic unit
RCT Randomized controlled trial
SCFAs Short-chain fatty acid
TEER Transepithelial electrical resistance
TJ Tight junctions
TLR Toll-like receptor SB 0 small bowel
WGO World Gastroenterology Organization
WHO World Health Organization
ZO Zonula occludens

REFERENCES

[1] Schmulson MJ, Drossman DA. What is new in Rome IV. J Neurogastroenterol Motil 2017;23(2):151. The Korean Society of Neurogastroenterology and Motility.

[2] Stanghellini V, Chan FK, Hasler WL, Malagelada JR, Suzuki H, Tack J, et al. Gastroduodenal disorders. Gastroenterology 2016;150(6):1380—92. Elsevier.

[3] Stanghellini V. Functional dyspepsia and irritable bowel syndrome: beyond Rome IV. Dig Dis 2018;35(Suppl. 1):14—7.

[4] Abbott A. Scientists bust myth that our bodies have more bacteria than human cells. Nat News 2016:8.

[5] Huttenhower C, Gevers D, Knight R, Abubucker S, Badger JH, Chinwalla AT, et al. Structure, function and diversity of te healthy human microbiome. Nature 2012;486(7402):207. Nature Publishing Group.

[6] Caporaso JG, Lauber CL, Costello EK, Berg-Lyons D, Gonzalez A, Stombaugh J, et al. Moving pictures of the human microbiome. Genome Biol 2011;12(5):R50. BioMed Central.

[7] Barbara G, Feinle-Bisset C, Ghoshal UC, Santos J, Vanner SJ, Vergnolle N, et al. The intestinal microenvironment and functional gastrointestinal disorders. Gastroenterology 2016;150(6):1305—18. Elsevier.

[8] Moeller AH, Caro-Quintero A, Mjungu D, Georgiev AV, Lonsdorf EV, Muller MN, et al. Cospeciation of gut microbiota with hominids. Science 2016;353(6297):380—2. American Association for the Advancement of Science.

[9] Dicks L, Geldenhuys J, Mikkelsen L, Brandsborg E, Marcotte H. Our gut microbiota: a long walk to homeostasis. Benef Microbes 2017:1—18. Wageningen Academic Publishers.

[10] Lynch SV, Pedersen O. The human intestinal microbiome in health and disease. NEJM Mass Medical Soc 2016;375(24):2369—79.

[11] Jandhyala SM, Talukdar R, Subramanyam C, Vuyyuru H, Sasikala M, Reddy DN. Role of the normal gut microbiota. World J Gastroenterol 2015;21(29):8787. Baishideng Publishing Group Inc.

[12] Vancamelbeke M, Vermeire S. The intestinal barrier: a fundamental role in health and disease. Expert Rev Gastroenterol Hepatol 2017;11(9):821—34. Taylor & Francis.

[13] Salvo RE, Alonso CC, Pardo CC, Casado BM, Vicario M. The intestinal barrier function and its involvement in digestive disease. Rev Esp Enferm Dig 2015;107(11):686—96.

[14] Abrams GD, Bauer H, Sprinz H. Influence of the normal flora on mucosal morphology and cellular renewal in the ileum. A comparison of germ-free and conventional mice. Lab Invest 1963 Mar;12:355—64.

[15] Swanson PA, Kumar A, Samarin S, Vijay-Kumar M, Kundu K, Murthy N, et al. Enteric commensal bacteria potentiate epithelial restitution via reactive oxygen species-mediated inactivation of focal adhesion kinase phosphatases. Proc Natl Acad Sci USA 2011;108(21):8803—8. National Acad Sciences.

[16] Ghosh D, Porter E, Shen B, Lee SK, Wilk D, Drazba J, et al. Paneth cell trypsin is the processing enzyme for human defensin-5. Nat Immunol 2002;3(6):583. Nature Publishing Group.

[17] Mack D, Ahrné S, Hyde L, Wei S, Hollingsworth M. Extracellular MUC3 mucin secretion follows adherence of Lactobacillus strains to intestinal epithelial cells in vitro. Gut 2003;52(6):827—33. BMJ Publishing Group.

[18] Mattar A, Teitelbaum DH, Drongowski R, Yongyi F, Harmon C, Coran A. Probiotics up-regulate MUC-2 mucin gene expression in a Caco-2 cell-culture model. Pediatr Surg Int 2002;18(7):586—90. Springer.

[19] Lee SH. Intestinal permeability regulation by tight junction: implication on inflammatory bowel diseases. Intest Res 2015;13(1):11—8.

[20] Zyrek AA, Cichon C, Helms S, Enders C, Sonnenborn U, Schmidt MA. Molecular mechanisms underlying the probiotic effects of *Escherichia coli* Nissle 1917 involve ZO-2 and PKCζ redistribution resulting in tight junction and epithelial barrier repair. Cell Microbiol 2007;9(3):804—16. Wiley Online Library.

[21] Mujagic Z, De Vos P, Boekschoten MV, Govers C, Pieters H-JH, De Wit NJ, et al. The effects of Lactobacillus plantarum on small intestinal barrier function and mucosal gene transcription; a randomized double-blind placebo controlled trial. Sci Rep 2017;7:40128. Nature Publishing Group.

[22] Groschwitz KR, Hogan SP. Intestinal barrier function: molecular regulation and disease pathogenesis. J Allergy Clin Immunol 2009;124(1):3—20. Elsevier.

[23] Brown EM, Sadarangani M, Finlay BB. The role of the immune system in governing host-microbe interactions in the intestine. Nat Immunol 2013;14(7):660. Nature Publishing Group.

[24] Vanheel H, Vicario M, Vanuytsel T, Van Oudenhove L, Martinez C, Keita ÅV, et al. Impaired duodenal mucosal integrity and low-grade inflammation in functional dyspepsia. Gut 2014;63(2):262—71. BMJ Publishing Group.

[25] Cremon C, Gargano L, Morselli-Labate AM, Santini D, Cogliandro RF, De Giorgio R, et al. Mucosal immune activation in irritable bowel syndrome: gender-dependence and association with digestive symptoms. TAm J Gastroenterol 2009;104(2):392. Nature Publishing Group.

[26] Simrén M, Barbara G, Flint HJ, Spiegel BM, Spiller RC, Vanner S, et al. Intestinal microbiota in functional bowel disorders: a Rome foundation report. Gut 2012. BMJ Publishing Group.

[27] Barbara G, Cremon C, Carini G, Bellacosa L, Zecchi L, De Giorgio R, et al. The immune system in irritable bowel syndrome. J Neurogastroenterol Motil Korean Soci Neurogastroenterol Motil 2011;17(4):349.

[28] Barbara G, Wang B, Stanghellini V, De Giorgio R, Cremon C, Di Nardo G, et al. Mast cell-dependent excitation of visceral-nociceptive sensory neurons in irritable bowel syndrome. Gastroenterology 2007;132(1):26—37. Elsevier.

[29] Valdez-Morales EE, Overington J, Guerrero-Alba R, Ochoa-Cortes F, Ibeakanma CO, Spreadbury I, et al. Sensitization of peripheral sensory nerves by mediators from colonic biopsies of diarrhea-predominant irritable bowel syndrome patients: a role for PAR2. Am J Gastroenterol 2013;108(10):1634. Nature Publishing Group.

[30] Cenac N, Bautzova T, Le Faouder P, Veldhuis NA, Poole DP, Rolland C, et al. Quantification and potential functions of endogenous agonists of transient receptor potential channels in patients with irritable bowel syndrome. Gastroenterology 2015;149(2):433—44. Elsevier.

[31] Dothel G, Barbaro MR, Boudin H, Vasina V, Cremon C, Gargano L, et al. Nerve fiber outgrowth is increased in the intestinal mucosa of patients with irritable bowel syndrome. Gastroenterology 2015;148(5):1002—11. Elsevier.

[32] Zhou X-Y, Li M, Li X, Long X, Zuo X-L, Hou X-H, et al. Visceral hypersensitive rats share common dysbiosis features with irritable bowel syndrome patients. World J Gastroenterol 2016;22(22):5211. Baishideng Publishing Group Inc.

[33] Almy TP, Tulin M. Alterations in colonic function in man under stress; experimental production of changes simulating the irritable colon. Gastroenterology 1947;8(5):616.

[34] Sarna S, Latimer P, Campbell D, Waterfall W. Effect of stress, meal and neostigmine on rectosigmoid electrical control activity (ECA) in normals and in irritable bowel syndrome patients. Dig Dis Sci 1982;27(7):582—91. Springer.

[35] Soffer E, Scalabrini P, Pope C, Wingate D. Effect of stress on oesophageal motor function in normal subjects and in patients with the irritable bowel syndrome. Gut 1988;29(11):1591—4. BMJ Publishing Group.

[36] Kellow J, Langeluddecke P, Eckersley G, Jones M, Tennant C. Effects of acute psychologic stress on small-intestinal motility in health and the irritable bowel syndrome. Scand J Gastroenterol 1992;27(1):53—8. Taylor & Francis.

[37] Tayama J, Nakaya N, Hamaguchi T, Saigo T, Takeoka A, Sone T, et al. Maladjustment to academic life and employment anxiety in university students with irritable bowel syndrome. PLoS One 2015;10(6):e0129345. Public Library of Science.

[38] Qi R, Liu C, Ke J, Xu Q, Zhong J, Wang F, et al. Intrinsic brain abnormalities in irritable bowel syndrome and effect of anxiety and depression. Brain Imaging Behav 2016;10(4):1127—34. Springer.

[39] Gareau MG, Wine E, Rodrigues DM, Cho JH, Whary MT, Philpott DJ, et al. Bacterial infection causes stress-induced memory dysfunction in mice. Gut 2010. BMJ Publishing Group.

[40] O'Mahony SM, Marchesi JR, Scully P, Codling C, Ceolho A-M, Quigley EM, et al. Early life stress alters behavior, immunity, and microbiota in rats: implications for irritable bowel syndrome and psychiatric illnesses. Biol Psychiatry 2009;65(3):263—7. Elsevier.

[41] Kennedy P, Clarke G, O'Neill A, Groeger J, Quigley E, Shanahan F, et al. Cognitive performance in irritable bowel syndrome: evidence of a stress-related impairment in visuospatial memory. Psychol Med 2014;44(7):1553—66. Cambridge University Press.

[42] Powell N, Walker MM, Talley NJ. The mucosal immune system: master regulator of bidirectional gut-brain communications. Nat Rev Gastroenterol Hepatol 2017;14(3):143—59. Nature Research.

[43] Quigley EM. The gut-brain Axis and the microbiome: clues to pathophysiology and Opportunities for Novel management strategies in irritable bowel syndrome (IBS). J Clin Med 2018;7(1):6. Multidisciplinary Digital Publishing Institute.

[44] Gautron L. Molecular anatomy of the gut-brain axis revealed with transgenic technologies: implications in metabolic research. Front Neurosci 2013;7:134. Frontiers.

[45] Mayer EA, Knight R, Mazmanian SK, Cryan JF, Tillisch K. Gut microbes and the brain: paradigm shift in neuroscience. J Neurosci 2014;34(46):15490—6. Soc Neuroscience.

[46] Cryan JF, Dinan TG. Mind-altering microorganisms: the impact of the gut microbiota on brain and behaviour. Nat Rev Neurosci 2012;13(10):701. Nature Publishing Group.

[47] Foster JA, Neufeld K-AM. Gut-brain axis: how the microbiome influences anxiety and depression. Trends Neurosci 2013;36(5):305—12. Elsevier.

[48] Hoban A, Stilling R, Ryan F, Shanahan F, Dinan T, Claesson M, et al. Regulation of prefrontal cortex myelination by the microbiota. Transl Psychiatry 2016;6(4):e774. Nature Publishing Group.

[49] Luczynski P, Whelan SO, O'sullivan C, Clarke G, Shanahan F, Dinan TG, et al. Adult microbiota-deficient mice have distinct dendritic morphological changes: differential effects in the amygdala and hippocampus. Eur J Neurosci 2016;44(9):2654—66. Wiley Online Library.

[50] Ruddick JP, Evans AK, Nutt DJ, Lightman SL, Rook GA, Lowry CA. Tryptophan metabolism in the central nervous system: medical implications. Expert Rev Mol Med 2006;8(20):1—27. Cambridge University Press.

[51] Mawe GM, Hoffman JM. Serotonin signalling in the gut—functions, dysfunctions and therapeutic targets. Nat Rev Gastroenterol Hepatol 2013;10(8):473. Nature Publishing Group.

[52] Clarke G, Grenham S, Scully P, Fitzgerald P, Moloney R, Shanahan F, et al. The microbiome-gut-brain axis during early life regulates the hippocampal serotonergic system in a sex-dependent manner. Mol Psychiatry 2013;18(6):666—73. Nature Publishing Group.

[53] Holzer P, Reichmann F, Farzi A, Neuropeptide Y. Peptide YY and pancreatic polypeptide in the gut-brain axis. Neuropeptides 2012;46(6):261—74. Elsevier.

[54] Holzer P, Farzi A. Neuropeptides and the microbiota-gut-brain axis. Microbial Endocrinol Microbiota Gut Brain Axis Health Dis 2014:195—219. Springer.

[55] Pimentel M, Lin HC, Enayati P, van den Burg B, Lee H-R, Chen JH, et al. Methane, a gas produced by enteric bacteria, slows intestinal transit and augments small intestinal contractile activity. Am J Physiol Gastrointest Liver Physiol Am Physiol Soc 2006;290(6):G1089—95.

[56] Carbonero F, Benefiel AC, Gaskins HR. Contributions of the microbial hydrogen economy to colonic homeostasis. Nat Rev Gastroenterol Hepatol 2012;9(9). Nature Publishing Group.

[57] Teague B, Asiedu S, Moore P. The smooth muscle relaxant effect of hydrogen sulphide in vitro: evidence for a physiological role to control intestinal contractility. Br J Pharmacol 2002;137(2):139—45. Wiley Online Library.

[58] McCusker RH, Kelley KW. Immune-neural connections: how the immune system's response to infectious agents influences behavior. J Exp Biol 2013;216(1):84—98. The Company of Biologists Ltd.

[59] Wang Y, Devkota S, Musch MW, Jabri B, Nagler C, Antonopoulos DA, et al. Regional mucosa-associated microbiota determine physiological expression of TLR2 and TLR4 in murine colon. PLoS One 2010;5(10):e13607. Public Library of Science.

[60] Rousseaux C, Thuru X, Gelot A, Barnich N, Neut C, Dubuquoy L, et al. Lactobacillus acidophilus modulates intestinal pain and induces opioid and cannabinoid receptors. Nat Med 2007;13(1):35. Nature Publishing Group.

[61] Tattoli I, Petitta C, Scirocco A, Ammoscato F, Cicenia A, Severi C. Microbiota, innate immune system, and gastrointestinal muscle: ongoing studies. J Clin Gastroenterol 2012;46:S6—11. LWW.

[62] Brint EK, MacSharry J, Fanning A, Shanahan F, Quigley EM. Differential expression of toll-like receptors in patients with irritable bowel syndrome. Am J Gastroenterol 2011;106(2):329. Nature Publishing Group.

[63] Belmonte L, Youmba SB, Bertiaux-Vandaele N, Antonietti M, Lecleire S, Zalar A, et al. Role of toll like receptors in irritable bowel syndrome: differential mucosal immune activation according to the disease subtype. PLoS One 2012;7(8):e42777. Public Library of Science.

[64] Booijink CC, El-Aidy S, Rajili'c-Stojanovi'c M, Heilig HG, Troost FJ, Smidt H, et al. High temporal and inter-individual variation detected in the human ileal microbiota. Environ Microbiol 2010;12(12):3213—27. Wiley Online Library.

[65] Zoetendal EG, Raes J, Van Den Bogert B, Arumugam M, Booijink CC, Troost FJ, et al. The human small intestinal microbiota is driven by rapid uptake and conversion of simple carbohydrates. ISME 2012;6(7):1415. Nature Publishing Group.

[66] Marlicz W, Yung D, Skonieczna-Żydecka K, Loniewski I, van Hemert S, Loniewska B, et al. From clinical uncertainties to precision medicine: the emerging role of the gut barrier and microbiome in small bowel functional diseases. Expert Rev Gastroenterol Hepatol 2017. Taylor & Francis.

[67] El Aidy S, Van Den Bogert B, Kleerebezem M. The small intestine microbiota, nutritional modulation and relevance for health. Curr Opin Biotechnol 2015;32:14—20. Elsevier.

[68] Zhong L, Shanahan ER, Raj A, Koloski NA, Fletcher L, Morrison M, et al. Dyspepsia and the microbiome: time to focus on the small intestine. Gut 2016. BMJ Publishing Group.

[69] Tan V, Liu K, Lam F, Hung I, Yuen M, Leung W. Randomised clinical trial: rifaximin versus placebo for the treatment of functional dyspepsia. Alim Pharmacol Ther 2017;45(6):767—76. Wiley Online Library.

[70] Talley NJ. Moving away from focussing on gastric pathophysiology in functional dyspepsia: new insights and therapeutic implications. Am J Gastroenterol 2017;112(1):141. Nature Publishing Group.

[71] Balsari A, Ceccarelli A, Dubini F, Fesce E, Poli G. The fecal microbial population in the irritable bowel syndrome. Microbiologica 1982;5(3):185—94.

[72] Malinen E, Rinttilä T, Kajander K, Mättö J, Kassinen A, Krogius L, et al. Analysis of the fecal microbiota of irritable bowel syndrome patients and healthy controls with real-time PCR. Am J Gastroenterol 2005;100(2):373. Nature Publishing Group.

[73] Rajili'c—Stojanovi'c M, Biagi E, Heilig HG, Kajander K, Kekkonen RA, Tims S, et al. Global and deep molecular analysis of microbiota signatures in fecal samples from patients with irritable bowel syndrome. Gastroenterology 2011;141(5):1792—801. Elsevier.

[74] Jeffery IB, O'toole PW, Öhman L, Claesson MJ, Deane J, Quigley EM, et al. An irritable bowel syndrome subtype defined by species-specific alterations in faecal microbiota. Gut 2012;61(7):997—1006. BMJ Publishing Group.

[75] Chung C-S, Chang P-F, Liao C-H, Lee T-H, Chen Y, Lee Y-C, et al. Differences of microbiota in small bowel and faeces between irritable bowel syndrome patients and healthy subjects. Scand J Gastroenterol 2016;51(4):410—9. Taylor & Francis.

[76] Kerckhoffs AP, Samsom M, van der Rest ME, de Vogel J, Knol J, Ben-Amor K, et al. Lower Bifidobacteria counts in both duodenal mucosa-associated and fecal microbiota in irritable bowel syndrome patients. World J Gastroenterol 2009;15(23):2887. Baishideng Publishing Group Inc.

[77] Giamarellos-Bourboulis E, Tang J, Pyleris E, Pistiki A, Barbatzas C, Brown J, et al. Molecular assessment of differences in the duodenal microbiome in subjects with irritable bowel syndrome. Scand J Gastroenterol 2015;50(9):1076—87. Taylor & Francis.

[78] Dlugosz A, Winckler B, Lundin E, Zakikhany K, Sandström G, Ye W, et al. No difference in small bowel microbiota between patients with irritable bowel syndrome and healthy controls. Sci Rep 2015:8508. Nature Publishing Group.

[79] Parkes G, Rayment N, Hudspith B, Petrovska L, Lomer M, Brostoff J, et al. Distinct microbial populations exist in the mucosa-associated microbiota of sub-groups of irritable bowel syndrome. Neurogastroenterol Motil 2012;24(1):31—9. Wiley Online Library.

[80] Martinez C, Vicario M, Ramos L, Lobo B, Mosquera JL, Alonso C, et al. The jejunum of diarrhea-predominant irritable bowel syndrome shows molecular alterations in the tight junction signaling pathway that are associated with mucosal pathobiology and clinical manifestations. Am J Gastroenterol 2012;107(5):736. Nature Publishing Group.

[81] Kong W, Gong J, Dong L, Xu J. Changes of tight junction claudin-1,-3,-4 protein expression in the intestinal mucosa in patients with irritable bowel syndrome. Nan Fang Yi Ke Da Xue Xue Bao 2007;27(9):1345—7.

[82] Zhou Q, Costinean S, Croce CM, Brasier AR, Merwat S, Larson SA, et al. MicroRNA 29 targets nuclear factor-κB-repressing factor and Claudin 1 to increase intestinal permeability. Gastroenterology 2015;148(1):158—69. Elsevier.

[83] Bertiaux-Vandaële N, Youmba SB, Belmonte L, Lecleire S, Antonietti M, Gourcerol G, et al. The expression and the cellular distribution of the tight junction proteins are altered in irritable bowel syndrome patients with differences according to the disease subtype. Am J Gastroenterol 2011;106(12):2165. Nature Publishing Group.

[84] Vicario M, González-Castro AM, Mart'inez C, Lobo B, Pigrau M, Guilarte M, et al. Increased humoral immunity in the jejunum of diarrhoea-predominant irritable bowel syndrome associated with clinical manifestations. Gut 2014. BMJ Publishing Group.

[85] Park JH, RHEE P-L, Kim HS, Lee JH, KIM Y-H, Kim JJ, et al. Mucosal mast cell counts correlate with visceral hypersensitivity in patients with diarrhea predominant irritable bowel syndrome. J Gastroenterol Hepatol 2006;21(1):71—8. Wiley Online Library.

[86] Drossman DA, Hasler WL. Rome IV—functional GI disorders: disorders of gut-brain interaction. Gastroenterology 2016;150(6):1257—61. Elsevier.

[87] Quigley EM. Gut bacteria in health and disease. Gastroenterol Hepatol 2013;9(9):560. Millenium Medical Publishing.

[88] Pilichiewicz AN, Horowitz M, Holtmann GJ, Talley NJ, Feinle—Bisset C. Relationship between symptoms and dietary patterns in patients with functional dyspepsia. Clin Gastroenterol 2009;7(3):317—22. Elsevier.

[89] Feinle-Bisset C, Azpiroz F. Dietary and lifestyle factors in functional dyspepsia. Nat Rev Gastroenterol Hepatol 2013;10(3):150. Nature Publishing Group.

[90] Böhn L, Störsrud S, Törnblom H, Bengtsson U, Simrén M. Self-reported food-related gastrointestinal symptoms in IBS are common and associated with more severe symptoms and reduced quality of life. Am J Gastroenterol 2013;108(5):634. Nature Publishing Group.

[91] Staudacher HM, Whelan K. The low FODMAP diet: recent advances in understanding its mechanisms and efficacy in IBS. Gut 2017. BMJ Publishing Group.

[92] Staudacher HM, Irving PM, Lomer MC, Whelan K. Mechanisms and efficacy of dietary FODMAP restriction in IBS. Nat Rev Gastroenetrol Hepatol 2014;11(4):256. Nature Publishing Group.

[93] Böhn L, Störsrud S, Liljebo T, Collin L, Lindfors P, Törnblom H, et al. Diet low in FODMAPs reduces symptoms of irritable bowel syndrome as well as traditional dietary advice: a randomized controlled trial. Gastroenterology 2015;149(6):1399—407. Elsevier.

[94] Marsh A, Eslick EM, Eslick GD. Does a diet low in FODMAPs reduce symptoms associated with functional gastrointestinal disorders? A comprehensive systematic review and meta-analysis. Eur J Nutr 2016;55(3):897—906. Springer.

[95] Drossman DA, Morris CB, Schneck S, Hu YJ, Norton NJ, Norton WF, et al. International survey of patients with IBS: symptom features and their severity, health status, treatments, and risk taking to achieve clinical benefit. J Clin Gastroenterol 2009:541. NIH Public Access.

[96] Banasiewicz T, Krokowicz L, Stojcev Z, Kaczmarek B, Kaczmarek E, Maik J, et al. Microencapsulated sodium butyrate reduces the frequency of abdominal pain in patients with irritable bowel syndrome. Colorectal Dis 2013;15(2):204—9. Wiley Online Library.

[97] Tana C, Umesaki Y, Imaoka A, Handa T, Kanazawa M, Fukudo S. Altered profiles of intestinal microbiota and organic acids may be the origin of symptoms in irritable bowel syndrome. Neurogastroenterol Motil 2010;22(5):512. Wiley Online Library.

[98] Harper A, Naghibi MM, Garcha D. The role of bacteria, probiotics and diet in irritable bowel syndrome. Foods Multidiscip 2018;7(2):13. Digital Publishing Institute.

[99] Nyangale EP, Mottram DS, Gibson GR. Gut microbial activity, implications for health and disease: the potential role of metabolite analysis. J Proteome Res 2012;11(12):5573—85. ACS Publications.

[100] Tooth D, Garsed K, Singh G, Marciani L, Lam C, Fordham I, et al. Characterisation of faecal protease activity in irritable bowel syndrome with diarrhoea: origin and effect of gut transit. Gut 2013. BMJ Publishing Group.

[101] Sonnenburg JL, Bäckhed F. Diet-microbiota interactions as moderators of human metabolism. Nature 2016;535(7610):56. Nature Publishing Group.

[102] Tuck C, Vanner S. Dietary therapies for functional bowel symptoms: recent advances, challenges, and future directions. Neurogastroenterol Motil 2017. Wiley Online Library.

[103] Currò D, Ianiro G, Pecere S, Bibbò S, Cammarota G. Probiotics, fibre and herbal medicinal products for functional and inflammatory bowel disorders. Br J Pharmacol 2017;174(11):1426–49. Wiley Online Library.

[104] Williams MD, Ha CY, Ciorba MA. Probiotics as therapy in gastroenterology: a study of physician opinions and recommendations. J Clin Gastroenterol 2010;44(9):631. NIH Public Access.

[105] Principi N, Cozzali R, Farinelli E, Brusaferro A, Esposito S. Gut dysbiosis and irritable bowel syndrome: the potential role of probiotics. J Infect 2017. Elsevier.

[106] Joyce SA, MacSharry J, Casey PG, Kinsella M, Murphy EF, Shanahan F, et al. Regulation of host weight gain and lipid metabolism by bacterial bile acid modification in the gut. Proc Natl Acad Sci USA 2014;111(20):7421–6. National Acad Sciences.

[107] King T, Elia M, Hunter J. Abnormal colonic fermentation in irritable bowel syndrome. Lancet 1998;352(9135):1187–9. Elsevier.

[108] Cammarota G, Ianiro G, Cianci R, Bibbò S, Gasbarrini A, Currò D. The involvement of gut microbiota in inflammatory bowel disease pathogenesis: potential for therapy. Pharmacol Ther 2015;149:191–212. Elsevier.

[109] Guarner F, Sanders ME, Eliakim R, Fedorak R, Gangl A, et al. WGO practice guideline - probiotics and prebiotics. 2017.

[110] Moayyedi P, Ford AC, Talley NJ, Cremonini F, Foxx-orenstein A, Brandt L, et al. The efficacy of probiotics in the therapy of irritable bowel syndrome: a systematic review. Gut 2008. BMJ Publishing Group.

[111] Ortiz-Lucas M, Tobias A, Saz P, Sebastián JJ. Effect of probiotic species on irritable bowel syndrome symptoms: a bring up to date meta-analysis. Rev Esp Enferm Dig 2013;105(1):19–36.

[112] Ford AC, Quigley EM, Lacy BE, Lembo AJ, Saito YA, Schiller LR, et al. Efficacy of prebiotics, probiotics, and synbiotics in irritable bowel syndrome and chronic idiopathic constipation: systematic review and meta-analysis. Am J Gastroenterol 2014;109(10):1547. Nature Publishing Group.

[113] Dimidi E, Christodoulides S, Fragkos KC, Scott SM, Whelan K. The effect of probiotics on functional constipation in adults: a systematic review and meta-analysis of randomized controlled trials-. Am J Clin Nutr 2014;100(4):1075–84. Oxford University Press.

[114] Didari T, Mozaffari S, Nikfar S, Abdollahi M. Effectiveness of probiotics in irritable bowel syndrome: Updated systematic review with meta-analysis. World J Gastroenterol 2015;21(10):3072 [publisher=Baishideng Publishing Group Inc].

[115] Tiequn B, Guanqun C, Shuo Z. Therapeutic effects of Lactobacillus in treating irritable bowel syndrome: a meta-analysis. Intern Med 2015;54(3):243–9. The Japanese Society of Internal Medicine.

[116] Pimentel M, Lembo A, Chey WD, Zakko S, Ringel Y, Yu J, et al. Rifaximin therapy for patients with irritable bowel syndrome without constipation. N Engl J Med 2011;364(1):22–32. Mass Medical Soc.

[117] Johnsen PH, Hilpüsch F, Cavanagh JP, Leikanger IS, Kolstad C, Valle PC, et al. Faecal microbiota transplantation versus placebo for moderate-to-severe irritable bowel syndrome: a double-blind, randomised, placebo-controlled, parallel-group, single-centre trial. Lancet Gastroenterol Hepatol 2018;3(1):17–24.

[118] Halkjær SI, Boolsen AW, Günther S, Christensen AH, Petersen AM. Can fecal microbiota transplantation cure irritable bowel syndrome? World J Gastroenterol 2017;23(22):4112. Baishideng Publishing Group Inc.

Chapter 27

The Microbiome and Metabolome in Nonalcoholic Fatty Liver Disease

Silvia M. Ferolla[1], Cláudia A. Couto[1], Maria de Lourdes A. Ferrari[1], Luciana Costa Faria[1], Murilo Pereira[2] and Teresa C.A. Ferrari[1]

[1]Department of Internal Medicine, Faculty of Medicine, Federal University of Minas Gerais, Belo Horizonte, Brazil; [2]Post graduation in Functional Clinical Nutrition, VP Institute, Cruzeiro do Sul University, São Paulo, Brazil

INTRODUCTION

Nonalcoholic fatty liver disease (NAFLD) is the most prevalent chronic liver disease in the world [1]. It encompasses a spectrum of conditions that ranges from hepatic steatosis to steatosis associated with necroinflammatory lesions (nonalcoholic steatohepatitis/NASH), which may progress to fibrosis and cirrhosis. It is usually associated with metabolic syndrome (MS). The pathogenesis of NAFLD is related to insulin resistance (IR), as a key factor that initiates hepatic fat accumulation and, potentially, NASH [2,3]. The excessive deposition of triglyceride in the hepatocytes leads to a shift from carbohydrates to FFA mitochondrial beta-oxidation and may promote lipid peroxidation, and accumulation of reactive oxygen species (ROS). These compounds produce a variety of cellular stimuli with inflammatory response, hepatocellular injury, and fibrosis [2].

Nowadays, a growing interest has been devoted to gut—liver axis (GLA) dysfunction, i.e., intestinal dysbiosis, bacterial overgrowth, and alteration of mucosa permeability, as relevant in NAFLD progression. It is also a possible alternative therapeutic target, in those patients unable to get benefits deriving from lifestyle modification, healthy diet, and physical activity promotion [4].

Gut—Liver Axis

The term "gut—liver axis" is defined as the strong anatomical and functional interaction, between the gastrointestinal tract and the liver. GLA is characterized by bidirectional traffic. Nutrients and factors derived from gut lumen reach the liver, through the portal circulation. Bile acids produced by hepatocytes are released in the small intestine, through the biliary tract. However, this description is simplistic, as the GLA does not have only a nutritional role. is a complex structure, and the change of two of its components, namely gut barrier and intestinal microbiota, and seems to play an important role in liver lesions and NASH progression.

INTESTINAL ENVIRONMENT

The gut microbiota composition is also influenced by diet, age, body weight, infections, medications, intestinal surgeries, and several liver diseases [5—7]. The gut microbiota is important to the host metabolism by secreting bioactive metabolites. It also participates in the development of the intestinal microvilli defense against pathogens, maintains gut immunity, performs the digestion of complex polysaccharides, synthesizes vitamins, and plays a role in fat storage [7—9].

The gut epithelium is a natural barrier that selects entry of useful substances present in the lumen, such as nutrients, and keeps at bay bacteria and their bioproducts. Tight junctions, specialized intercellular structures, assist this control. Derangement of the homeostasis between bacteria and the host, as occurs in small intestine bacterial overgrowth (SIBO),

Microbiome and Metabolome in Diagnosis, Therapy, and other Strategic Applications. https://doi.org/10.1016/B978-0-12-815249-2.00027-0

may cause disruption of the intercellular tight junctions, and subsequent increase in intestinal permeability, leading to bacterial translocation (BT), i.e., transportation of bacteria and their products from the gut lumen into the blood [9].

The portal vein and the hepatic artery supply blood to the liver. The portal blood contains products of digestion and also microbial products derived from the gut microbiota. Therefore, the liver consists in the first site, of exposure and filtration of bacterial products from the intestine, such as lipopolysaccharides (LPS), lipopeptides, unmethylated DNA, and double-stranded RNA [9].

Endotoxin and Liver Inflammation

The liver contains components of the immune system, such as macrophages, dendritic cells and natural killer T cells, which act as a first-line defense against endotoxin and microorganisms. Toll-like receptors (TLRs), present on the innate immune cells, consist of a family of type I transmembrane proteins, which recognize pathogen-associated molecular patterns (PAMPs), and damage associated molecular patterns (DAMPs) present on endogenous ligands. They initiate an adaptive immune response signaling cascade, leading to activation of proinflammatory genes, such as TNF-α IL-6, IL-8, and IL-12 genes.

LPS, the active component of endotoxin, is the most studied PAMP. The liver is exposed to these PAMPs due to BT [40]. LPS has affinity to LPS-binding protein, which in turn, binds to CD14 and activates TLR4 in Kupffer cells, triggering an essential inflammatory cascade [7]. The synthesis of proinflammatory cytokines leads to prolonged inflammation and hepatocyte damage [10].

Intestinal Inflammasomes

The TLRs, together with other sensors of PAMPs and DAMPs, are the inflammasomes, which are formed by a molecular macrocomplex that includes the enzyme caspase-1, whose activation causes the release of bioactive IL-1β and/or IL-18 [9]. Recent evidence suggests that the inflammasome is involved in NAFLD/NASH progression, via modulation of the gut microbiota [11]. Genetic inflammasome deficiency, associated with dysbiosis, determines increased concentration of bacterial products in the portal blood, which may exacerbate steatosis and increase tumor necrosis factor (TNF-α) expression [11].

Another mechanism that supports the role of the gut microbiota in the pathogenesis of NAFLD is the fact that certain types of bacteria inhibit gut epithelial expression of fasting-induced adipocyte factor (FIAF), a suppressor of lipoprotein lipase (LPL). FIAF is produced not only by the gut, but also by liver and adipose tissue, being an essential regulator of peripheral fat storage. By suppressing FIAF, the microbiota increases LPL activity in adipose tissue, enhancing the delivery of adipocyte-derived triglycerides, which determine storage of triacylglycerols in the liver [4]. Additionally, the microbiota is related to IR [12]. The transfer of gut microbiota from lean human donors to recipients with MS, via duodenal tube, resulted in increased insulin sensitivity within 6 weeks [13].

Bile Acids and Farnesoid X Receptor

Bile acids are ligands of the farnesoid X receptor (FXR), which is expressed in the liver and gut [14]. The activation of FXR reduces circulating bile acids (feedback mechanism), and participates in the control of the gut-microbiota composition, and in the regulation of lipids and glucose homeostasis in the gut-liver axis. It is known to have a crucial role in hepatic de-novo lipogenesis, VLDL export, and plasma triglyceride turnover [4]. A recent human study showed low FXR protein expression in patients with NASH versus simple NAFLD, suggesting a protective role of FXR in liver disease progression [15].

Gut Microbiome in NAFLD Patients

Obese individuals present predominance of Firmicutes and relative low proportion of Bacteroidetes [14]. In animal models, this has been associated with a propensity to NAFLD features like fasting hyperglycemia, insulin resistance, liver steatosis, and high expression of genes involved in de novo lipogenesis [16]. In humans with NASH, the microbiota composition also seems to have a lower proportion of Bacteroidetes, independently of hight fat diet and BMI.

The low prevalence of Bacteroidetes might facilitate the growth of other phyla that are more efficient in harvesting energy from the diet [17]. Indeed, obese subjects present more H_2-producing Prevotellaceae, and H_2-utilizing methanogenic Archaea. The coexistence of H_2-producing bacteria with H_2-utilizing Archaea suggests that H_2 transfer might increase energy uptake by the human large intestine in the obese person [18].

Response to High-Fat Diet

In two mice fed with high-fat diet (HFD) and with similar body weight gain, one mouse (called responder) developed hyperglycemia and presented high serum concentrations of proinflammatory cytokines, whereas the other (called nonresponder), did not. The gut microbiota was transplanted from both the responder, and the nonresponder, to germ-free mice. These animals were fed with the same HFD, and also developed comparable obesity. The responder stool-receiving mice presented fasting hyperglycemia, hyperinsulinemia, hepatic steatosis and high expression of genes involved in de novo lipogenesis. The nonresponder-receiver animals did not develop metabolic abnormalities. Gut microbiota associated with the NAFLD-prone and NAFLD-resistant phenotypes were, respectively, Firmicutes and Bacteroidetes [16].

Mouzaki et al. investigated the NASH human gut microbiota [19] and found a lower proportion of *Bacteroides/Prevotella* ratio (herein referred to as Bacteroidetes), when compared to simple steatosis or healthy controls (living liver donors), independently of BMI and high fatty diet. NASH subjects also presented increased number of *Clostridium coccoides* in their stool, in comparison to the individuals with simple steatosis.

In the experience of Zhu et al. [20] in NASH increased abundance of alcohol-producing *Escherichia* occurred, as well as high blood−ethanol levels, leading to increased oxidative stress, and liver inflammation due to alcohol metabolism. The gut microbiota also produced LPS, which contributes to liver damage, disrupts normal hepatocyte function, leads to mitochondrial oxidative stress, and reduces the clearance of toxins by the hepatocytes.

SIBO AND INCREASED INTESTINAL PERMEABILITY IN NAFLD PATIENTS

Evidences suggest that NAFLD subjects present a high prevalence of small intestine bacterial overgrowth/SIBO [21−26] and increased gut permeability [24,27,28]. Most controlled trials demonstrated that the prevalence of SIBO in NAFLD, ranges from 50% to 77.8% [21,22,24,25]. SIBO in NASH individuals is also associated with the rise in hepatic TLR4 expression and release of interleukin (IL)-8 [25].

SIBO has been independently associated with the severity of liver steatosis in histology studies [23,24]. As liver biopsy is not always available, markers of hepatic damage, such as plasminogen activator inhibitor 1 (PAI-1), have been used [27,29]. Some studies have showed that high serum concentrations of PAI-1 are associated with hepatic steatosis, fibrosis [30,31], and elevated serum endotoxin concentrations [27,29].

SIBO may increase gut permeability, leading to endotoxemia, and oxidative stress in the liver. Miele et al. [24] showed high prevalence of SIBO, increased intestinal permeability (urinary excretion of 51Cr-ethylene diamine tetra acetate/51Cr-EDTA test), and disruption of the intercellular tight junctions of the gut (immunohistochemical analysis of zona occludens-1/ZO-1 expression in duodenal biopsy specimens), in subjects with NAFLD. Indeed, increased gut permeability was associated with severity of liver steatosis. Other authors, using the lactulose−mannitol test, also identified increased intestinal permeability in NAFLD patients [27,28].

Studies using plasma endotoxin levels and TLR4 expression (endotoxin receptor), as markers of intestinal permeability, found similar results [7,25,29,32]. According to them, NAFLD subjects present higher serum concentrations of the endotoxin core antibodies EndoCAb IgG (marker of endotoxin exposure) [32], increased endotoxin levels [27,29], and high expression of TLR4 in liver [29] and on CD14$^+$ cells [25].

USE OF PROBIOTICS AND SYNBIOTICS IN NAFLD PATIENTS

Several interventional studies [28,33−45] and meta-analyses [46−48], on the use of oral probiotics and synbiotics to modify gut microbiota in NAFLD subjects, have shown improvement of inflammatory markers, oxidative stress parameters and liver biochemistry. However, it is important to emphasize that the studies differ regarding probiotic doses, strains of bacteria and duration of treatment, which hamper the establishment of the best intervention.

CONCLUSIONS

The evidences discussed here support the notion that gut microbiota changes induce an immune imbalance, leading to a state of metabolic alterations, hepatic fat accumulation, chronic inflammation, and, even NASH. More clinical studies are necessary, to better understand the cell-specific recognition and intracellular signaling events, involved in recognizing gut-derived microbes, and to set up an optimal balance in the GLA, in order to prevent and treat NAFLD.

REFERENCES

[1] Chalasani N, Younossi Z, Lavine JE, Diehl AM, Brunt EM, Cusi K, Charlton M, Sanyal AJ. The diagnosis and management of non-alcoholic fatty liver disease: practice guideline by the American association for the study of liver diseases, American College of Gastroenterology, and the American Gastroenterological association. Hepatology 2012;55:2005−23.

[2] Day C. Pathogenesis of steatohepatitis. Best Pract Res Clin Gastroenterol 2002;16:663−78.

[3] Chitturi S, Abeygunasekera S, Farrell GC, Holmes-Walker J, Hui JM, Fung C, Karim R, Lin R, Samarasinghe D, Liddle C, et al. NASH and insulin resistance: insulin hypersecretion and specific association with the insulin resistance syndrome. Hepatology 2002;35:373−9.

[4] Poeta M, Pierri L, Vajro P. Gut-liver axis derangement in non-alcoholic fatty liver disease. Children 2017;2:E66.

[5] Neish AS. Microbes in gastrointestinal health and disease. Gastroenterology 2009;136:65−80.

[6] Bäckhed F, Ding H, Wang T, Hooper LV, Koh GY, Nagy A, Semenkovich CF, Gordon JI. The gut microbiota as an environmental factor that regulates fat storage. Proc Natl Acad Sci USA 2004;101:15718−23.

[7] Duseja A, Chawla YK. Obesity and NAFLD: the role of bacteria and microbiota. Clin Liver Dis 2014;18:59−71.

[8] Zoetendal EG, Vaughan EE, de Vos WM. A microbial world within us. Mol Microbiol 2006;59:1639−50.

[9] Turnbaugh PJ, Ley RE, Mahowald MA, Magrini V, Mardis ER, Gordon JI. An obesity-associated gut microbiome with increased capacity for energy harvest. Nature 2006;444:1027−31.

[10] Szabo G, Mandrekar P, Dolganiuc A. Innate immune response and hepatic inflammation. Semin Liver Dis 2007;27:339−50.

[11] Takeuchi O, Akira S. Pattern recognition receptors and inflammation. Cell 2010;140:805−20.

[12] Xie G, Zhong W, Li H, Li Q, Qiu Y, Zheng X, Chen H, Zhao X, Zhang S, Zhou Z, Zeisel SH, Jia W. Alteration of bile acid metabolism in the rat induced by chronic ethanol consumption. FASEB J 2013;27:3583−93.

[13] Zhang H, DiBaise JK, Zuccolo A, Kudrna D, Braidotti M, Yu Y, Parameswaran P, Crowell MD, Wing R, Rittmann BE, Krajmalnik-Brown R. Human gut microbiota in obesity and after gastric bypass. Proc Natl Acad Sci USA 2009;106:2365−70.

[14] Roh YS, Seki E. Toll-like receptors in alcoholic liver disease, non-alcoholic steatohepatitis and carcinogenesis. J Gastroenterol Hepatol 2013;28:38−42.

[15] Aguilar Olivos NE, CarilloCórdova D, Oria-Hernandez J. The nuclear receptor FXR, but not LXR, up-regulates bile acid transporter expression in Non-alcoholic fatty liver disease. Ann Hepatol 2015;14(4):487−93.

[16] Ley RE, Turnbaugh PJ, Klein S, Gordon JI. Microbial ecology: human gut microbes associated with obesity. Nature 2006;444:1022−3.

[17] Le Roy T, Llopis M, Lepage P, Bruneau A, Rabot S, Bevilacqua C, Martin P, Philippe C, Walker F, Bado A, Perlemuter G, Cassard-Doulcier AM, Gérard P. Intestinal microbiota determines development of non-alcoholic fatty liver disease in mice. Gut 2013;62:1787−94.

[18] Rychlik JL, May T. The effect of a methanogen, *Methanobrevibacter smithii*, on the growth rate, organic acid production, and specific ATP activity of three predominant ruminal cellulolytic bacteria. Curr Microbiol 2000;40:176−80.

[19] Mouzaki M, Comelli EM, Arendt BM, Bonengel J, Fung SK, Fischer ID, McGilvray S, Allard JP. Intestinal microbiota in patients with nonalcoholic fatty liver disease. Hepatology 2013;58:120−7.

[20] Zhu L, Baker SS, Gill C, Liu W, Alkhouri R, Baker RD, Gill SR. Characterization of gut microbiomes in nonalcoholic steatohepatitis (NASH) patients: a connection between endogenous alcohol and NASH. Hepatology 2013;57:601−9.

[21] Wigg AJ, Roberts-Thomson IC, Dymock RB, McCarthy PJ, Grose RH, Cummins AG. The role of small intestinal bacterial overgrowth, intestinal permeability, endotoxaemia, and tumour necrosis factor alpha in the pathogenesis of non-alcoholic steatohepatitis. Gut 2001;48:206−11.

[22] Sajjad A, Mottershead M, Syn WK, Jones R, Smith S, Nwokolo CU. Ciprofloxacin suppresses bacterial overgrowth, increases fasting insulin but does not correct low acylated ghrelin concentration in non-alcoholic steatohepatitis. Aliment Pharmacol Ther 2005;22:291−9.

[23] Sabaté JM, Jouët P, Harnois F, Mechler C, Msika S, Grossin M, Coffin B. High prevalence of small intestinal bacterial overgrowth in patients with morbid obesity: a contributor to severe hepatic steatosis. Obes Surg 2008;18:371−7.

[24] Miele L, Valenza V, la Torre G, Montalto M, Cammarota G, Ricci R, Mascianà R, Forgione A, Gabrieli ML, Perotti G, Vecchio FM, Rapaccini G, Gasbarrini G, Day CP, Grieco A. Increased intestinal permeability and tight junction alterations in nonalcoholic fatty liver disease. Hepatology 2009;49:1877−87.

[25] Shanab AA, Scully P, Crosbie O, Buckley M, O'Mahony L, Shanahan F, Gazareen S, Murphy E, Quigley EM. Small intestinal bacterial overgrowth in nonalcoholic steatohepatitis: association with toll-like receptor 4 expression and plasma levels of interleukin 8. Dig Dis Sci 2011;56:1524−34.

[26] Kapil S, Duseja A, Sharma BK, Singla B, Chakraborti A, Das A, Ray P, Dhiman RK, Chawla Y. Small intestinal bacterial over growth and toll-like receptor signaling in patients with non-alcoholic fatty liver disease. J Gastroenterol Hepatol 2016;31:213−21.

[27] Volynets V, Küper MA, Strahl S, Maier IB, Spruss A, Wagnerberger S, Königsrainer A, Bischoff SC, Bergheim I. Nutrition, intestinal permeability, and blood ethanol levels are altered in patients with nonalcoholic fatty liver disease (NAFLD). Dig Dis Sci 2012;57:1932−41.

[28] Ferolla SM, Couto CA, Costa-Silva L, Armiliato GN, Pereira CA, Martins FS, Ferrari Mde L, Vilela EG, Torres HO, Cunha AS, Ferrari TC. Beneficial effect of synbiotic supplementation on hepatic steatosis and anthropometric parameters, but not on gut permeability in a population with nonalcoholic steatohepatitis. Nutrients 2016;28:E397.

[29] Alessi MC, Bastelica D, Mavri A, Morange P, Berthet B, Grino M, Juhan-Vague I. Plasma PAI-1 levels are more strongly related to liver steatosis than to adipose tissue accumulation. Arterioscler Thromb Vasc Biol 2003;23:1262−8.

[30] Bergheim I, Guo L, Davis MA, Lambert JC, Beier JI, Duveau I, Luyendyk JP, Roth RA, Arteel GE. Metformin prevents alcohol-induced liver injury in the mouse: critical role of plasminogen activator inhibitor-1. Gastroenterology 2006;130:2099−112.

[31] Bergheim I, Guo L, Davis MA, Duveau I, Arteel GE. Critical role of plasminogen activator inhibitor-1 in cholestatic liver injury and fibrosis. J Pharmacol Exp Ther 2006;316:592−600.

[32] Thuy S, Ladurner R, Volynets V, Wagner S, Strahl S, Königsrainer A, Maier KP, Bischoff SC, Bergheim I. Nonalcoholic fatty liver disease in humans is associated with increased plasma endotoxin and plasminogen activator inhibitor 1 concentrations and with fructose intake. J Nutr 2008;138:1452−5.

[33] Loguercio C, de Simone T, Federico A, Terracciano F, Tuccillo C, di Chicco M, Cartenì M. Gut-liver axis: a new point of attack to treat chronic liver damage? Am J Gastroenterol 2002;97:2144−6.

[34] Loguercio C, Federico A, Tuccillo C, Terracciano F, D'Auria MV, de Simone C, del Vecchio Blanco C. Beneficial effects of a probiotic VSL#3 on parameters of liver dysfunction in chronic liver diseases. J Clin Gastroenterol 2005;39:540−3.

[35] Vajro P, Mandato C, Licenziati MR, Franzese A, Vitale DF, Lenta S, Caropreso M, Vallone G, Meli R. Effects of Lactobacillus rhamnosus strain GG in pediatric obesity-related liver disease. J Pediatr Gastroenterol Nutr 2011;52:740−3.

[36] Aller R, de Luis DA, Izaola O, Conde R, Gonzalez Sagrado M, Primo D, de la Fuente B, Gonzalez J. Effect of a probiotic on liver aminotransferases in nonalcoholic fatty liver disease patients: a double blind randomized clinical trial. Eur Rev Med Pharmacol Sci 2011;15:1090−5.

[37] Wong VW, Won GL, Chim AM, Chu WC, Yeung DK, Li KC, Chan HL. Treatment of nonalcoholic steatohepatitis with probiotics. A proof-of-concept study. Ann Hepatol 2013;12:256−62.

[38] Eslamparast T, Poustchi H, Zamani F, Sharafkhah M, Malekzadeh R, Hekmatdoost A. Synbiotic supplementation in nonalcoholic fatty liver disease: a randomized, double-blind, placebo-controlled pilot study. Am J ClinNutr 2014;99:535−42.

[39] Alisi A, Bedogni G, Baviera G, Giorgio V, Porro E, Paris C, Giammaria P, Reali L, Anania F, Nobili V. Randomised clinical trial: the beneficial effects of VSL#3 in obese children with non-alcoholic steatohepatitis. Aliment Pharmacol Ther 2014;39:1276−85.

[40] Famouri F, Shariat Z, Hashemipour M, Keikha M, Kelishadi R. Effects of probiotics on nonalcoholic fatty liver disease in obese children and adolescents. J Pediatr Gastroenterol Nutr 2017;64:413−7.

[41] Mykhal'chyshyn HP, Bodnar PM, Kobyliak NM. Effect of probiotics on proinflammatory cytokines level in patients with type 2 diabetes and nonalcoholic fatty liver disease. Likars' ka Sprava 2013;2:56−62.

[42] Nabavi S, Rafraf M, Somi MH, Homayouni-Rad A, Asghari-Jafarabadi M. Effects of probiotic yogurt consumption on metabolic factors in individuals with nonalcoholic fatty liver disease. J Dairy Sci 2014;97:7386−93.

[43] Sepideh A, Karim P, Hossein A, Leila R, Hamdollah M, Mohammad EG, Mojtaba S, Mohammad S, Ghader G, SeyedMoayed A. Effects of multistrain probiotic supplementation on glycemic and inflammatory indices in patients with nonalcoholic fatty liver disease: a double-blind randomized clinical trial. J Am Coll Nutr 2016;35:500−5.

[44] Shavakhi A, Minakari M, Firouzian H, Assali R, Hekmatdoost A, Ferns G. Effect of a probiotic and metformin on liver aminotransferases in non-alcoholic steatohepatitis: a double blind randomized clinical trial. Int J Prev Med 2013;4:531−7.

[45] Mofidi F, Poustchi H, Yari Z, Nourinayyer B, Merat S, Sharafkhah M, Malekzadeh R, Hekmatdoost A. Synbiotic supplementation in lean patients with non-alcoholic fatty liver disease: a pilot, randomised, double-blind, placebo-controlled, clinical trial. Br J Nutr 2017;117:662−8.

[46] Ma YY, Li L, Yu CH, Shen Z, Chen LH, Li YM. Effects of probiotics on nonalcoholic fatty liver disease: a meta-analysis. World J Gastroenterol 2013;19:6911−8.

[47] Khalesi S, Johnson DW, Campbell K, Williams S, Fenning A, Saluja S, Irwin C. Effect of probiotics and synbiotics consumption on serum concentrations of liver function test enzymes: a systematic review and meta-analysis. Eur J Nutr 2017. https://doi.org/10.1007/s00394-017-1568-y [Epub ahead of print].

[48] S Lavekar A, V Raje D, Manohar T, A Lavekar A. Role of probiotics in the treatment of nonalcoholic fatty liver disease: a meta-analysis. Euroasian J Hepato-Gastroenterol 2017;7:130−7.

Chapter 28

The Microbiome and Metabolome in Alcoholic Liver Disease

Kalpesh G. Patel and Nikolaos T. Pyrsopoulos

Division of Gastroenterology and Hepatology Rutgers, The State University of New Jersey, Newark, NJ, United States

INTRODUCTION

Alcoholic liver disease is a leading cause of liver cirrhosis and liver-related death worldwide [1,2]. It is a constellation of several types of liver pathologies, including fatty liver, steatohepatitis, alcoholic hepatitis, fibrosis, cirrhosis, and hepatocellular carcinoma [3,4]. In the liver-gut axis, the former receives blood supply from the intestine which includes nutrients and gut-derived microbial products, among others [5]. The microbiota maintains a symbiotic relationship within the intestine, and is responsible for assistance in digestion, vitamin synthesis, and resistance to colonization of the intestine by pathogens [6].

The microbiota represents approximately 500−1000 highly prevalent species, and there are about 10−100 trillion microorganisms present in each gram of feces [6]. The gut microbiota evokes innate and adaptive immune responses to protect intestinal homeostasis [6]. In normal homeostasis, there is a delicate balance between gut barrier function, gut permeability, and equilibrium between commensal and pathogenic organisms, which ultimately prevents bacterial gut translocation if leaky gut exists [5]. In patients with evidence of alcoholic liver disease and in individuals who consume alcohol, there is increased intestinal permeability and load of bacterial products in the portal blood flow [5]. Overall, alcohol leads to intestinal dysbiosis and triggers inflammation in alcoholic liver disease, by increasing bacterial translocation of bacterial products into the liver [5].

GUT-LIVER AXIS

There is a close functional and anatomical relationship between the liver and the gut, which is known as the gut-liver axis [3,7]. The liver receives a majority of its blood supply, approximately two-thirds to three-fourths, from the portal vein, and is the first organ to filter nutrients, toxins [8] and gut-derived microbial products [5] absorbed by the intestines [8]. It is imperative for the microbiota, mucosal barrier, the liver, and immune system to maintain complex interactions to prevent translocation of harmful microbial products from the gut [5] and to ensure homeostasis [7] (Fig. 28.1).

Communication Between the Liver and Intestine

Bile acids are produced by the liver from cholesterol, and aid in the absorption of dietary fats and vitamins [9]. They are secreted by the liver as glycine or taurine conjugates via the bile duct into the small intestine, and are reabsorbed in the terminal ileum and returned to the liver [10]. They act via the farnesoid X receptor (FXR) and the G-protein coupled receptor TGR5 [10]. The liver and small intestine communicate with each other by means of enterohepatic circulation of bile acids [10]. Bile acids have a bacteriostatic effect, as they bind to the FXR in the intestine [11], which increases the production of antimicrobial proteins angiogenin 1 and RNase family member 4 [10]. Overall, bile acids inhibit bacterial proliferation, prevent bacterial overgrowth, and promote epithelial cell integrity [10], all of which can be altered by alcohol use [12].

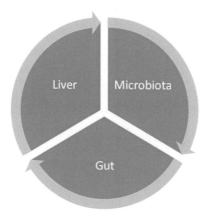

FIGURE 28.1 Relationship between the liver, microbiota, and gut.

The Intestinal Microbiota

The intestinal microbiota plays an important role in digestion, vitamin synthesis, resistance to colonization by pathogenic microorganisms [6], and providing energy substrates for cells [8]. It is home to greater than 500 species of bacteria, some of which are fixed to the intestine, and others which only pass through the intestine [3]. The majority of bacteria in the intestinal lumen are Gram-negative and anaerobes, and are estimated to be about 100—1000 times more numerous than aerobic bacteria [3]. The major phyla are Firmicutes, Bacteroidetes, Actinobacteria, Proteobacteria, and Verrucomicrobia [13] (Fig. 28.2). The most frequently encountered bacteria are *Bacteroides, Porphyromonas, Bifidobacterium, Lactobacillus, Clostridium*, and *Escherichia coli* [3]. Autochthonous bacteria are classified as beneficial taxa. Besides digestion of complex carbohydrates and fermentation of simple sugars, they produce short-chain fatty acids and help maintain the intestinal barrier [13,14].

Gut Barrier Function

A small portion of the intestinal contents reaches the portal circulation, secondary to the limiting effects of the specialized intestinal epithelium [7]. This barrier regulates water and nutrient transport, while protecting from gut-derived pathogens. The intestinal barrier consists of a mucous lining, a layer of specialized epithelial cells, and intercellular junctions. The mucous lining, which is made up predominantly of mucin, prevents large particles and intact bacteria from coming into contact with the epithelium. The specialized epithelium forms a single continuous layer of cells, which prevents the passage of most hydrophilic substances [7]. Claudins, occludins, and zonula occludens are proteins which formulate tight junctions and adherens junctions and play an important role in paracellular transport, securing the space between the epithelial cells [7].

Gut immune cells in the walls of the intestine also contribute to the integrity of the intestinal mucosa [2]. Substances such as alcohol can cause direct injury to intestinal epithelial tight junctions, which can lead to microbial product translocation and endotoxemia [7]. Short-chain fatty acids are vital in maintaining the integrity of the mucosal barrier [8,15]. Butyrate is an essential substrate in the maintenance of mucosal integrity and expression of tight junction proteins [15]. *Faecalibacterium prausnitzii* is a well-known member of the Firmicutes phylum which produces butyrate [15]. On another note, zinc deficiency is commonly seen in alcoholic liver disease. Animal models have shown zinc deficiency impairs the epithelial mucosal barrier [16].

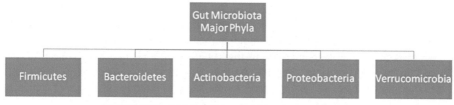

FIGURE 28.2 Division of gut microbiota into major phyla.

ALCOHOL AND THE INTESTINAL MICROBIOME

Qualitative and quantitative changes in the intestinal microbiota occur in individuals with moderate alcohol consumption, alcoholics, and those who have alcoholic cirrhosis. Chronic dysbiosis ratio (CDR) is a term used to indicate the changes in "good" and "bad" bacteria occurring in the intestine of a cirrhotic patient [11]. The CDR is the ratio of autochthonous to nonautochthonous taxa in cirrhosis [14]. The lower the CDR, more is the endotoxemia and more advanced the cirrhosis [14]. Alcohol has direct effects on functions of the intestine, and there are indirect effects of alcohol and/or its metabolites via the blood [2].

Alterations of Gut Microbiome and Metabolome

After exposure to excessive amounts of alcohol, there is higher abundance of *Enterobacteriaceae* [8], a family of the phylum Proteobacteria, and other potentially pathogenic bacteria, with lower abundance of Bacteroides [5] and Firmicutes [3]. Proteobacteria are Gram-negative bacteria which include pathogenic species such as *Salmonella* [3,9], *Campylobacter* [9], *Helicobacter*, *Vibrio*, and *Escherichia* [3]. High levels of Proteobacteria cause an increase in acetaldehyde [8], a product of alcohol metabolism which increases intestinal permeability, by disrupting integrity and expression of tight junction molecules, zonula occludens-1, and claudin-1 [7].

Bacterial translocation occurs when bacteria are able to adhere to the epithelium, compromise the intestinal wall, and reach the lymph nodes [7]. Exposure to bacterial endotoxins by the immune system causes an upregulation of inflammatory mediators [14]. Metagenomic analysis indicates decreased diversity of the intestinal microbiome in liver cirrhosis, which is associated with lack of protective species, an abundance of pathogenic species, and reduced integrity of the gut mucosa barrier [5,17]. Ultimately, intestinal dysbiosis triggers local intestinal inflammatory response [11] and high levels of endotoxins in the blood [5].

Alcohol and Liver Disease

Endogenous ethanol is also produced by bacterial flora from the fermentation of carbohydrates [3]. This process is amplified by the presence of gut dysmotility as seen in obesity, diabetes, but also chronic alcohol use [3]. The intestinal oxidation of alcohol produces higher concentrations of acetaldehyde responsible for alteration in intestinal permeability and microbiota homeostasis [3]. Acetaldehyde is also responsible for mitochondrial dysfunction, and has direct hepatocyte damaging effect by means of affinity to DNA-binding proteins [3].

Leaky Gut

Alcohol-compromised integrity of the intestinal barrier, or leaky gut, activates transcription of nuclear factor kappaB (NF-κB) gene, and overexpression of nitric oxide (NO) production [3]. NO produced via constitutive NOS (cNOS) contributes to the intestinal barrier integrity [3]. On the other hand, NO produced by inducible NOS/iNOS is seen in inflammation, and may play a role in impairing the mucosal barrier. Endothelial cells, hepatocytes, macrophages, neutrophils all express iNOS [3]. Increased levels of iNOS contribute to the decreased stability of tubulin, which damages the microtubule cytoskeleton leading to impaired barrier function. Additionally, increased NO leads to oxidative stress in hepatocytes [3].

Gut-Derived Bacterial Products and Liver Injury

Gut-derived bacterial products transported by the portal vein, as a result of leaky gut, cause injury in alcoholic liver disease by their structural components and pathogen-associated molecular patterns (PAMPs), which activate the immune system in the liver [5]. PAMPs activate the innate immune system by attaching to toll-like receptors (TLRs), leading to a cascade of inflammatory processes [5]. A well-known PAMP is lipopolysaccharide (LPS), an activator of TLR4 [5], and a component of the wall of Gram-negative bacteria [3].

Normally, a small quantity of endotoxin is absorbed from the intestine, and is neutralized in the liver by Kupffer cells. In chronic alcohol consumption, there is a larger amount of endotoxin absorbed, as the dysbiotic bowel flora produces more LPS, which leads to elevated levels of LPS-binding protein, activation of the inflammatory cascade, and progression of alcoholic liver disease [3]. LPS causes hepatic and extrahepatic macrophages to produce inflammatory cytokines, such as IL-1β, TNF- α, IL-6, and IL-8 [3]. Endotoxins in the circulation and inflammatory cytokines lead to increased intestinal permeability, leading to even more endotoxins into the circulation, causing a detrimental cycle [3].

Immune Dysfunction

Alcohol alters both the innate and adaptive immune system by suppression of natural killer cell activity, and antibody-dependent cell-mediated cytotoxicity. The innate immune system uses TLR to identify PAMPs. TLRs are found on every type of liver cell [3]. TLR1 is found on hepatocytes, and TLR 2,3,4 are found in stellate cells, bile duct epithelium, and Kupffer cells. The bile duct epithelium also expresses TLR5 as well [3]. Bacterial LPS binds to cluster of differentiation 14 (CD-14) to create a complex, which is facilitated by LPS binding protein [3,18].

This LPS-LBP complex is recognized by Kupffer cells via TLR, and activates transcription factor NF-κβ and activator protein. This causes an increase in inflammatory cytokine production such as interferon gamma, TNF-alpha, IL-6, IL-1, chemokines, and reactive oxygen species. The LPS/TLR-4 interaction induces fibrogenesis, by causing hepatic stellate cells to overproduce chemokines, and the release of transforming growth factor beta (TGF-β) by Kupffer cells. Ultimately, oxidative stress and proinflammatory cytokines produce hepatocellular injury and damage [3].

Other Factors Contributing to Alcoholic Liver Disease

Environmental factors, genetics, intestinal dysmotility, increased gastric pH, and altered bile flow may contribute to alcohol-induced dysbiosis [11]. Dietary habits and exposure to medications or antibiotics are the strongest usual causes affecting the composition of the intestinal microbiome [11]. The combination of alcohol and obesity is shown in experimental animals and humans to worsen liver disease [11]. Fatty liver develops in a majority of patients with chronic alcohol abuse, while fibrosis and cirrhosis occur in 40%−60% of alcoholics [11].

Genetic factors, for example, female gender, can enhance susceptibility to alcohol-induced liver disease [11]. Additionally, polymorphisms in cytochrome and alcohol dehydrogenase-3 genes are risk factors for developing liver disease in alcoholics [11]. However, it is not clear if genetic variations contribute to liver disease by affecting the intestinal microbiome [11]. Alcohol-precipitated intestinal dysmotility can lead to a proliferation of intestinal bacteria. Patients with cirrhosis are found to have small intestinal bacterial overgrowth and prolonged gastric transit times [11].

Alcohol does not affect gastric acid release; however, alcoholics have decreased acid production, likely secondary to altered gastric histology, and higher rates of atrophic gastritis. This may favor small intestinal bacterial overgrowth [11]. Chronic alcohol use leads to higher bile acid levels in the stools, but in patients with cirrhosis, the total stool bile acid secretion decreases [11], as they have decreased bile flow [10]. Bile acids have a bacteriostatic effect and inhibit intestinal bacterial overgrowth [10]. Therefore, having a decreased gut concentration of bile acids leads to decreased protective effects of bile acids in the intestine.

IN VITRO AND IN VIVO MODELS

Mice given alcohol for 3 weeks displayed increased Bacteroidetes in the cecum, compared to a predominance of Firmicutes in control mice [4]. In addition, alcohol-fed mice also had an increase in *Bacteroides* and *Akkermansia* with reduced *Lactobacillus* [4]. In rats given alcohol (8 g/kg per day) for 10 weeks, ileal and colonic dysbiosis followed [12,19]. Intestinal oxidative stress, increased intestinal permeability, higher levels of endotoxin in the blood, and steatohepatitis have equally been demonstrated in rats [12].

Human studies have shown that chronic alcohol use leads to bacterial overgrowth and dysbiosis [12]. Alcohol altered the mucosa-associated microbiota composition in sigmoid biopsies [12,20]. Higher Proteobacteria and lower Bacteroidetes were features of alcoholics, with and without alcoholic liver disease [12,20]. High levels of endotoxins are observed, suggesting that changes in the microbiota may lead to alcohol provoked effects on the liver and small intestine [12].

METABOLOME, TRANSCRIPTOME, AND PROTEOME

In plasma samples from 896 Japanese men, 19 metabolites were associated with alcohol intake [21−23]. Threonine, guanidinosuccinate, and glutamine were related to alcohol-induced liver injury [23]. In advanced alcoholic cirrhosis, glutamate, acetate, and N-acetyl glycoproteins in the serum correlated with hepatocellular carcinoma, whereas lipids and glutamine correlated with cirrhosis [24]. Alcoholic hepatitis featured higher levels of betaine, and lower levels of creatinine, phenylalanine, citrulline, tyrosine, homocitrulline, octanoylcarnitine, and symmetric dimethyl-larginine [22]. Betaine and citrulline were able to accurately differentiate between alcoholic hepatitis and alcohol decompensation [22].

THERAPEUTIC PROTOCOLS (PROBIOTICS, PREBIOTICS, SYNBIOTICS, FECAL TRANSPLANT)

Probiotics

Probiotics such as *Lactobacillus reuteri* may produce substances which have antimicrobial properties, controlling growth, binding, and invasion of harmful bacteria. They may improve intestinal barrier function by supporting epithelial cell survival and growth, as well as help to suppress the release of proinflammatory cytokines such as TNF-alpha, stimulating protective cytokines such as IL-10 and TGF-β. Lastly, certain probiotics can induce microopioid and cannabinoid receptor expression, providing an analgesic effect on abdominal pain [4].

Attenuated endotoxemia and improved liver function in cirrhosis, along with alcoholic-induced liver injury, are reported [4]. These patients were given *Lactobacillus plantarum* and *Bifidobacterium bifidum* [4,25]. *Lactobacillus casei* Shirota repaired neutrophil dysfunction in patients with compensated liver cirrhosis [4,26,27]. In addition, a study involving treatment with *Lactobacillus subtilis* and *Streptococcus faecium* in those with alcoholic hepatitis had a reduction of gut-derived LPS [4,28].

Mouse studies with *Lactobacillus rhamnosus* and *Lactobacillus acidophilus* for 4 weeks, indicated lower TLR4 levels compared to the control group [4,29]. As mentioned, TLR4 is responsible for the inflammatory cascade that leads to the pathogenesis of alcoholic liver disease [4,29]. Additionally, the study demonstrated a decrease of proinflammatory cytokines like IL-1B and TNF-alpha [4,29].

Prebiotics

A well-known nonabsorbable sugar, lactulose, has been used in the treatment and prevention of hepatic encephalopathy in cirrhotic patients [4,8]. The exact mechanism to how lactulose produces these effects is unknown. Stool acidification and the purging effect could ultimately alter the stool flora and promote ammonia excretion [8]. Human studies of prebiotics on alcoholic liver disease have been limited, secondary to difficulty with restricting dietary behavior [8].

Synbiotics

A study in 2004 evaluated 20 patients with cirrhosis and minimal hepatic encephalopathy. Increase in nonurease producing *Lactobacillus* occurred 14 days following treatment, which led to a reduction of hepatic encephalopathy, ammonia levels, and endotoxemia [15,30].

Fecal Microbiota Transplant

The premise of fecal transplant is to reestablish commensal microorganisms in the gut, and have them produce antimicrobial products such as bacteriocins [4]. A small safety study evaluated the changes in the microbiota composition and urinary metabolome of fecal transplant versus no treatment in patients with recurrent hepatic encephalopathy. They found fecal transplant had a favorable influence on cognitive function [15,31]. A pilot study involving eight patients with alcoholic hepatitis who underwent fecal microbiota transplant had favorable microbiota changes after 1 year, and improved mortality [13,32].

Antibiotics

Various antibiotics are well studied in the selective decontamination of the intestine [33]. Norfloxacin given over long periods of time has demonstrated to decrease aerobic Gram-negative bacilli without affecting other microorganisms, and reducing recurrent spontaneous bacterial peritonitis [33]. However, with the increase in resistance, and a higher chance for methicillin-resistant *Staphylococcus aureus*, its use has been less favored, except in patients who are classified as high-risk cirrhotics [33].

Rifaximin has in-vitro activity against Gram-positive and Gram-negative aerobic and anaerobic bacteria, with low risk for causing bacterial resistance. A lower risk of hepatic encephalopathy recurrence and related hospitalizations can be achieved [33]. Furthermore, short-term treatment with 4 weeks of rifaximin decreased hepatic venous pressure gradient and improved endotoxemia [33,34]. Rifaximin has also found to be beneficial in those with decompensated cirrhosis, as it decreases risks of cirrhotic complications such as variceal bleeding, spontaneous bacterial peritonitis, hepatorenal syndrome, and hepatic encephalopathy [33].

OTHER STRATEGICAL INTERVENTIONS

As patients with alcoholic liver disease are deficient in nutrients and vitamins, dietary supplementation has received interest in alcoholic liver disease [2]. As discussed earlier, zinc deficiency amplifies the effects of alcohol on the liver and is associated with impaired mucosal integrity [2]. In mice who were fed alcohol, administration of zinc has shown to restore alcohol-induced gut dysfunction [2]. Further studies in humans regarding the effects of zinc on the microbiome are warranted. Additional studies could focus on anti-LPS antibodies, TLR4 inhibition [2], and FXR agonists [16] as potential targets in the alcoholic liver disease and its interaction with the gut-liver axis.

LIST OF ACRONYMS AND ABBREVIATIONS

CD cluster of differentiation
CDR chronic dysbiosis ratio
cNOS constitutive nitric acid synthases
eNOS endothelial nitric acid synthases
FXR farnesoid X receptor
IL interleukin
iNOS inducible nitric acid synthases
LPB LPS-binding protein
LPS lipopolysaccharide
NF-κB nuclear factor kappaB
nNOS neuronal nitric acid synthases
NO nitric oxide
NOS nitric acid synthases
PAMP pathogen-associated molecular patterns
TGF-β transforming growth factor-beta
TLR toll-like receptor
TNF tumor necrosis factor

REFERENCES

[1] Chen P, Schnabl B. Host-microbiome interactions in alcoholic liver disease. Gut Liver 2014;8(3):237−41.

[2] Szabo G. Gut-liver axis in alcoholic liver disease. Gastroenterology 2015;148(1):30−6.

[3] Malaguarnera G, Giordano M, Nunnari G, Bertino G, Malaguarnera M. Gut microbiota in alcoholic liver disease: pathogenetic role and therapeutic perspectives. World J Gastroenterol 2014;20(44):16639−48.

[4] Sung H, Kim SW, Hong M, Suk KT. Microbiota-based treatments in alcoholic liver disease. World J Gastroenterol 2016;22(29):6673−82.

[5] Szabo G, Petrasek J. Gut-liver axis and sterile signals in the development of alcoholic liver disease. Alcohol Alcohol 2017;52(4):414−24.

[6] Betrapally NS, Gillevet PM, Bajaj JS. Changes in the intestinal microbiome and alcoholic and nonalcoholic liver diseases: causes or effects? [Internet] Gastroenterology 2016;150(8):1745−55. Available from: https://doi.org/10.1053/j.gastro.2016.02.073.

[7] Arab JP, Martin-Mateos RM, Shah VH. Gut−liver axis, cirrhosis and portal hypertension: the chicken and the egg. Hepatol Int 2017:1−10.

[8] Haque TR, Barritt AS. Intestinal microbiota in liver disease [Internet] Best Pract Res Clin Gastroenterol 2016;30(1):133−42. Available from: http://linkinghub.elsevier.com/retrieve/pii/S1521691816000081.

[9] Bluemel S, Williams B, Knight R, Schnabl B. Precision medicine in alcoholic and nonalcoholic fatty liver disease via modulating the gut microbiota [Internet] Am J Physiol Gastrointest Liver Physiol 2016;311(6):G1018−36. Available from: http://ajpgi.physiology.org/lookup/doi/10.1152/ajpgi.00245.2016.

[10] Schnabl B, Brenner DA. Interactions between the intestinal microbiome and liver diseases [Internet] Gastroenterology 2014;146(6):1513−24. Available from: https://doi.org/10.1053/j.gastro.2014.01.020.

[11] Hartmann P, Seebauer CT, Schnabl B. Alcoholic liver disease: the gut microbiome and liver cross talk. Alcohol Clin Exp Res 2015;39(5):763−75.

[12] Engen PA, Green SJ, Voigt RM, Forsyth CB, Keshavarzian A. The gastrointestinal microbiome: alcohol effects on the composition of intestinal microbiota [Internet] Alcohol Res 2015;37(2):223−36. Available from: http://www.ncbi.nlm.nih.gov/pubmed/26695747%5Cnhttp://www.pubmedcentral.nih.gov/articlerender.fcgi?artid=PMC4590619.

[13] Acharya C, Sahingur SE, Bajaj JS. Microbiota, cirrhosis, and the emerging oral-gut-liver axis [Internet] JCI Insight 2017;2(19):1−11. Available from: http://www.ncbi.nlm.nih.gov/pubmed/28978799%0Ahttps://insight.jci.org/articles/view/94416.

[14] Acharya C, Bajaj JS. Gut microbiota and complications of liver disease [Internet] Gastroenterol Clin N Am March 2017;46(1):155−69. Available from: https://doi.org/10.1016/j.gtc.2016.09.013.

[15] Woodhouse CA, Patel VC, Singanayagam A, Shawcross DL. Review article: the gut microbiome as a therapeutic target in the pathogenesis and treatment of chronic liver disease [Internet] Aliment Pharmacol Ther May 2017:1−11. Available from: http://doi.wiley.com/10.1111/apt.14397.

[16] Dunn W, Shah VH. Pathogenesis of alcoholic liver disease [Internet] Clin Liver Dis 2016;20(3):445−56. Available from: https://doi.org/10.1016/j.cld.2016.02.004.

[17] Qin N, Yang F, Li A, Prifti E, Chen Y, Shao L, et al. Alterations of the human gut microbiome in liver cirrhosis [Internet] Nature 2014;513(7516):59−64. Available from: http://www.nature.com/doifinder/10.1038/nature13568.

[18] Fung P, Pyrsopoulos N. Emerging concepts in alcoholic hepatitis. World J Hepatol 2017;9(12):567−85.

[19] Mutlu E, Keshavarzian A, Engen P, Forsyth CB, Sikaroodi M, Gillevet P. Intestinal dysbiosis: a possible mechanism of alcohol-induced endotoxemia and alcoholic steatohepatitis in rats [Internet] Alcohol Clin Exp Res October 2009;33(10):1836−46. Available from: http://www.ncbi.nlm.nih.gov/pubmed/19645728.

[20] Mutlu EA, Gillevet PM, Rangwala H, Sikaroodi M, Naqvi A, Engen PA, et al. Colonic microbiome is altered in alcoholism [Internet] AJP Gastrointest Liver Physiol 2012;302(9):G966−78. Available from: http://ajpgi.physiology.org/cgi/doi/10.1152/ajpgi.00380.2011.

[21] Yu M, Zhu Y, Cong Q, Wu C. Metabonomics research progress on liver diseases. Can J Gastroenterol Hepatol 2017;2017.

[22] Ascha M, Wang Z, Ascha MS, Dweik R, Zein NN, Grove D, et al. Metabolomics studies identify novel diagnostic and prognostic indicators in patients with alcoholic hepatitis. World J Hepatol 2016;8(10):499−508.

[23] Harada S, Takebayashi T, Kurihara A, Akiyama M, Suzuki A, Hatakeyama Y, et al. Metabolomic profiling reveals novel biomarkers of alcohol intake and alcohol-induced liver injury in community-dwelling men. Environ Health Prev Med 2016;21(1):18−26.

[24] Nahon P, Amathieu R, Triba MN, Bouchemal N, Nault JC, Ziol M, et al. Identification of serum proton NMR metabolomic fingerprints associated with hepatocellular carcinoma in patients with alcoholic cirrhosis. Clin Cancer Res 2012;18(24):6714−22.

[25] Kirpich IA, Solovieva NV, Leikhter SN, Shidakova NA, Lebedeva OV, Sidorov PI, et al. Probiotics restore bowel flora and improve liver enzymes in human alcohol-induced liver injury: a pilot study [Internet] Alcohol 2008;42(8):675−82. Available from: https://doi.org/10.1016/j.alcohol.2008.08.006.

[26] Stadlbauer V, Mookerjee RP, Hodges S, Wright GAK, Davies NA, Jalan R. Effect of probiotic treatment on deranged neutrophil function and cytokine responses in patients with compensated alcoholic cirrhosis. J Hepatol 2008;48(6):945−51.

[27] Mookerjee RP, Stadlbauer V, Lidder S, Wright GAK, Hodges SJ, Davies NA, et al. Neutrophil dysfunction in alcoholic hepatitis superimposed on cirrhosis is reversible and predicts the outcome. Hepatology 2007;46(3):831−40.

[28] Han SH, Suk KT, Kim DJ, Kim MY, Baik SK, Kim YD, et al. Effects of probiotics (cultured Lactobacillus subtilis/Streptococcus faecium) in the treatment of alcoholic hepatitis: randomized-controlled multicenter study. Eur J Gastroenterol Hepatol 2015;27(11):1300−6.

[29] Bang CS, Hong SH, Suk KT, Kim JB, Han SH, Sung H, et al. Effects of Korean red ginseng (Panax ginseng), urushiol (Rhus vernicifera Stokes), and probiotics (Lactobacillus rhamnosus R0011 and Lactobacillus acidophilus R0052) on the gut-liver axis of alcoholic liver disease [Internet] J Ginseng Res 2014;38(3):167−72. Available from: https://doi.org/10.1016/j.jgr.2014.04.002.

[30] Liu J-P, Zou W-L, Chen S-J, Wei H-Y, Yin Y-N, Zou Y-Y, et al. Effects of different diets on intestinal microbiota and nonalcoholic fatty liver disease development [Internet] World J Gastroenterol 2016;22(32):7353. Available from: http://www.wjgnet.com/1007-9327/full/v22/i32/7353.htm.

[31] Bajaj JS, Kassam Z, Fagan A, Gavis EA, Liu E, Cox IJ, et al. Fecal microbiota transplant from a rational stool donor improves hepatic encephalopathy: a randomized clinical trial. Hepatology 2017;66(6):1727−38.

[32] Philips CA, Pande A, Shasthry SM, Jamwal KD, Khillan V, Chandel SS, et al. Healthy donor fecal microbiota transplantation in steroid-ineligible severe alcoholic hepatitis: a pilot study [Internet] Clin Gastroenterol Hepatol 2017;15(4):600−2. Available from: https://doi.org/10.1016/j.cgh.2016.10.029.

[33] Fukui H. Gut microbiome-based therapeutics in liver cirrhosis: basic consideration for the next step [Internet] J Clin Transl Hepatol 2017;5:249−60. Available from: http://www.xiahepublishing.com/ArticleFullText.aspx?sid=2&jid=1&id=10.14218%2FJCTH.2017.00008.

[34] Vlachogiannakos J, Saveriadis AS, Viazis N, Theodoropoulos I, Foudoulis K, Manolakopoulos S, et al. Intestinal decontamination improves liver haemodynamics in patients with alcohol-related decompensated cirrhosis. Aliment Pharmacol Ther 2009;29(9):992−9.

Chapter 29

The Microbiome, Metabolome, and Proteome in Preterm Neonatal Sepsis

Andrew Nelson[1] and Christopher J. Stewart[2,3]

[1]Northumbria University, Faculty of Health and Life Sciences, Newcastle upon Tyne, United Kingdom; [2]Newcastle University, Institute of Cellular Medicine, Newcastle upon Tyne, United Kingdom; [3]Baylor College of Medicine, The Alkek Center for Metagenomics and Microbiome Research, Houston TX, United States

OVERVIEW

Neonatal sepsis is an infection of the bloodstream by bacteria and fungi, and remains a significant cause of morbidity and mortality, in both the developed and developing world. These infections lead to increased financial burden due to a greater number of bed days, as well as associated treatment costs (e.g., administration of antibiotics). It is estimated that neonatal sepsis causes 3.6 million deaths per year, with 90% attributed to low-income countries [1]. The most important factor in development of neonatal sepsis is the level of physiological development, which is measured as gestation age (i.e., degree of prematurity) or the weight of the infant at birth. In the neonatal intensive care unit (NICU), approximately 25% of very low birthweight infants, 35% extremely low birthweight infants, and 50% of infants >750 g will develop suspected sepsis [2]. Infants who survive these infections are more likely to suffer from morbidity later in life, and preterm infants who develop sepsis are at a greater long-term risk of neurodevelopmental impairment.

There are two definitions of neonatal sepsis based on the time to diagnosis following birth (Table 29.1 [3]). Early onset neonatal sepsis (EOS) is defined as an infection acquired within the 72 h postpartum, whereas late onset sepsis (LOS) occurs after 72 h of life. The aetiology of EOS is most commonly associated with vertical transmission of organisms from mother to infant during the birthing process or through intraamniotic infections. Risk factors for EOS include preterm birth, immunological immaturity, and maternal infection such as chorioamnionitis. EOS occurs in approximately 1 out of every 1000 live births in the United States, but the incidence increases with decreasing birthweight. Unsurprisingly, the rate of mortality is 51% higher in infants delivered at 22 weeks gestation (54%) compared to infants delivered at term (3%) [4].

The most prevalent organisms associated with EOS are *Escherichia coli* and Group B Streptococcus (GBS), also known as *Streptococcus agalactiae*. Rates of *E. coli* EOS are greater in preterm deliveries, whereas GBS is common in term cases. LOS occurs after day three of life, and is commonly a complication associated with preterm birth, with infants <32 weeks being most at risk. Other risk factors include exposure to intravenous treatments and failure to thrive. The most common organism isolated from LOS patients is coagulase negative *Staphylococcus* (CoNS) although other bacterial species, as well as fungi, have been isolated from blood culture.

Although the risk of sepsis has been shown to be the same for vaginal and cesarean deliveries, the risk of CoNS sepsis is increased in infants who were delivered by cesarean section [5]. In preterm infants, the organism isolated in diagnostic blood culture is usually dominant in the gut microbiota [6] and the same strain has been isolated in both stool and blood within the same infant [7]. Thus, translocation from an immature "leaky" gut represents the likely route of infection in preterm infants. Other possible routes of transmission include skin, for example, through insertion of catheters or during injury to the skin barrier.

Microbiome and Metabolome in Diagnosis, Therapy, and other Strategic Applications. https://doi.org/10.1016/B978-0-12-815249-2.00029-4

TABLE 29.1 Neonatal Sepsis: Summary Table

Types of Neonatal Sepsis	Early Onset Sepsis	Late Onset Sepsis
Definition	Neonatal infection during the 1st 7 d of life or during the 1st 72 h of life in case of VLBW infants.	Neonatal infection after 7 d from delivery.
Epidemiology	Incidence of 1–2 per 1000 live newborns.	Prevalence of ~25%–30% in VLBW infants; incidence of ~6%–10% in late preterm newborns (gestational age, 34–37 weeks).
Mortality	3% among term newborns, & ~16% in VLBW infants.	36% in VLBW babies aged between 8 d & 14 d & 52% in those aged between 15 d & 28 d.
Physiopathology	Vertical transmission from mother: infection contracted from bacteria colonizing the maternal perineum, maternal hematogenous transmission or chorioamnionitis.	Infection is acquired after the delivery; preterm & VLBW infants are most frequently involved.
Predisposing factors	Maternal factors: premature birth (<37 week), premature or prolonged time (>18 h) of membranes rupture, maternal peripartum infection, a low socioeconomic status, maternal age <20 y & > 35 y, cesarean section, black ethnicity, obstetric practices, having previously had an infant with GBS infection. Neonatal factors: alterations of the innate immune response, defects of immunoregulatory genes, prematurity, birth weight, newborn jaundice, male sex, neonatal Apgar scoring, wet lung, fetal distress, anemia, intraventricular hemorrhage, hypothermia & metabolic disorders.	Neonatal factors: the risk is inversely related to gestational age & birth weight; other risk factors are maternal intake of corticosteroids, antenatal administration of corticosteroids in babies, prolonged hospitalization, mechanical ventilation, invasive procedures & devices implantation.
Causative microorganisms	Streptococcus agalactiae & Escherichia coli are the agents most commonly found, but all microorganisms may be responsible.	About 70% of the 1st episodes of LOS are caused by Gram-positive bacteria; CoNS are the most common pathogens. Gram-negative organisms are responsible for 18% of cases. The remaining 12% are caused by fungal organisms.
Clinical manifestations	Clinical manifestations are common & unspecific: fever, cyanosis, apnea, tachycardia, bradycardia, hypotension, jaundice, drowsiness, irritability, lethargy, convulsions, anorexia, regurgitation, abdominal distension, vomiting, diarrhea, skin lesions, involvement of cardiovascular system, septic shock, urinary tract infections, osteomyelitis & deep infections.	
Diagnosis	Serum inflammatory biomarkers (acute-phase reactants, inflammatory cytokines, alterations in blood tests); Identification of causative agent through molecular genetics techniques (amplification of target DNA/RNA fragments); Microbiological exams on biological samples (blood, urine, cerebrospinal fluid).	
Prevention	Universal GBS screening of all pregnant women at 35–37 weeks of gestation & in case of positive test, intrapartum antibiotic prophylaxis at least 4 h before the delivery.	Reduce, as far as possible, the sources of contamination (ensuring a sterile environment in NICUs, minimizing the invasive procedures).
Therapy	Empiric therapy as 1st line: ampicillin & an aminoglycoside are recommended. Then, target antibiotic therapy on the base of culture exams.	Empiric therapy as 1st line: vancomycin & an aminoglycoside are recommended. Then, target antibiotic therapy on the base of results of culture exams.

NICUs, neonatal intensive care units; VLBW, very low birth weight.
Adapted from Cortese F, Scicchitano P, Gesualdo M, Filaninno A, De Giorgi E, Schettini F, et al. Early and late infections in newborns: where do we stand? A review. Pediatr Neonatol 2016;57:265–73. https://doi.org/10.1016/J.PEDNEO.2015.09.007 with permission from Elsevier.

MICROBIOME

Role of the Microbiome in Early Onset Sepsis

The major route of infection in cases of EOS is via ascending infections or colonization of the infant during the birthing process. Therefore, understanding the mechanisms and agents of ascending infections may help to further our understanding of the disease, and offer early intervention. Examination of the microbiota of cord blood and amniotic fluid from 36 preterm deliveries has given insights into the transmission of bacteria from mother to infant [8]. In 6 EOS cases, the infant developed a positive blood culture, a further 16 EOS cases were blood culture negative, and 14 infants did not develop sepsis and were included as controls. In the confirmed EOS group, there were a greater number of polymicrobial infections using molecular techniques than by culture. Of the 6 confirmed EOS cases, 3 cases were in accordance, in detecting the same causative organism by both molecular and culture techniques. This study highlights the utility of molecular techniques for bacterial identification in cases of sepsis, but further work is required to ensure that these organisms are in fact aetiological agents rather than contaminants during collection and sample processing.

In a separate study, 87 preterm infants (mean gestational age of 29 weeks) underwent screening by amplicon sequencing to match bacteria in cord blood to cases of EOS [9]. Of the 87 preterm infants, 8 were diagnosed with confirmed EOS. In 7 out of 8 confirmed EOS cases, a genus level match in cord blood was observed. In the eighth case the blood culture grew *Gardnerella vaginalis* and *E. coli*, and these were confirmed by polymerase chain reaction (PCR) and Sanger sequencing. However, *E. coli* was not detected in the cord blood sample and sequences attributed to *Gardnerella* accounted for less than 1% of the total reads for that patient. The authors also found that the alpha diversity was lower in bacterial communities from patients with confirmed EOS. These results could potentially be used to predict EOS aetiology from maternal tissues and inform treatment strategy earlier than blood culture from the neonate.

Using whole genome sequencing it was found that a patient with suspected sepsis had *Enterobacter cloacae* identified in the gut microbiome. Furthermore, whole genome sequencing (WGS) of this community was able to identify a fosA2 gene, which is a fosfomycin resistance gene, present in *E. cloacae*. While no clinical microbiology was carried out to confirm the result, this research highlights how metagenomics could be used for surveillance, and to subsequently guide clinical management of high-risk patients in the clinic [10].

Blood culture is an unreliable source of viable organisms, due to poor recovery from low sample volumes, and prophylactic antibiotic treatment rendering organisms inert. In EOS, where the infecting organism is usually present in the birth canal, these samples offer a potential reservoir of viable organisms that can be readily accessed to inform treatment (e.g., placental tissues). The placental membrane microbiota in 40 healthy or chorioamnionitis tissues has been investigated [11]. In cases where the infant was delivered preterm, the authors found that patients with severe chorioamnionitis had a reduced alpha diversity. Furthermore, preterm membranes with severe chorioamnionitis had a greater abundance of *Ureaplasma parvum*, GBS, and *Fusobacterium nucleatum*, all of which have been previously identified as causative agents of EOS. Although the authors found *E. coli* to be universally part of the placental microbiome, there was no link between *E. coli* presence and sepsis caused by *E. coli*, suggesting that the organism may be nonviable.

Another study examined 1097 women in Malawi, and found that 68.1% of fetal membranes and 46.8% of placental tissues contained bacterial DNA [12]. Women with severe chorioamnionitis had lower alpha diversity in fetal membranes and placental tissue. Fusobacteria and Tenericutes (to which *Ureaplasma* belong) had a higher abundance in women with severe chorioamnionitis. While these studies potentially offer insights into intraamniotic infection, it is worthwhile noting that in many cases healthy control placental tissues are used as part of these studies. A body of evidence exists, which asserts that the placental tissues of healthy women are in fact sterile, which casts doubt when the same organisms are found in the diseased samples from the same studies [13].

Role of the Microbiome in Late Onset Sepsis

Several studies have purported dysbiosis in the preterm gut, although a true definition of a healthy preterm gut microbiota remains elusive, and is likely to vary from one infant to the next. An early example of this found that preterm infants who went on to develop LOS had a gut bacterial community that had a lower alpha diversity [14]. Interestingly, they demonstrate that the path from meconium to a healthy community is associated with an increased alpha diversity, whereas infants who developed LOS did not show any increase in diversity in the gut microbiota. While this may be due to increased use of antibiotics in infants who develop LOS, there may also be an underlying mechanism where a pathogen gets a foothold and becomes dominant, preventing a healthy community from assembling by filling a niche or altering the environment of the gut.

Furthermore, they found that patients who developed disease had a gut microbiota with a greater abundance of *Staphylococcus* compared to controls. Similarly, a study of 38 preterm infants with a mean gestation age of 27 weeks who developed LOS had a greater presence of *Staphylococcus* [15]. A longitudinal study of 22 LOS cases found a lower abundance of obligate anaerobes (i.e., higher abundance of facultative anaerobes such as *Staphylococcus*), in cases 1 week prior to and at time of LOS diagnosis [7]. More specifically, infants with a higher abundance of *Staphylococcus* in the first week of life were more likely to go on to develop LOS. However, a longitudinal study of 28 preterm infants (mean postmenstrual age of 27 weeks), 10 of whom developed LOS found increased relative abundance of Proteobacteria in LOS cases compared to controls [16].

A recent study examined 66 infants from two different NICUs (median gestational age of 25 weeks) [17]. They found that 3 out of 17 patients had discordance between the whole genome sequencing/WGS and blood culture. In the first case, blood culture identified *S. aureus* whereas WGS identified *S. epidermidis*. In the second case, blood culture identified GBS (i.e., *S. agalactiae*), whereas WGS identified *S. salivarius*. In the third case, blood culture found *Klebsiella pneumoniae*, whereas WGS identified *Klebsiella oxytoca* and an unclassified *Klebsiella*. Stewart et al. [6] performed a longitudinal study of the preterm gut microbiota with 7 LOS patients matched to 28 controls (gestational age between 25 and 28 weeks). Six of the 7 LOS patients had an abundant genus in the gut that matched with the genus of the blood culture, suggesting the gut as the primary site of translation in preterm infants.

A separate longitudinal prospective study of 11 LOS cases using genetic markers and WGS found that in 7 of the cases the organism identified from the patient stool was the same as that which was identified by blood culture [18]. In 2 cases where *E. coli* was identified as the pathogen by blood culture, the same isolate had been present in the stool since birth. In the other 5 cases the isolate was only identified closer to the time of infection. In 4 cases of LOS, the isolate that caused infection was also identified from the stool of a control infant who was in close proximity to the case infant at the time of infection. This demonstrates the potential of species to crosscontaminate infants in environments, such as NICUs, which has important connotations for clinical management of infants in hospitals. Indeed, most associated strains detected in the developing preterm infant gut can be detected in the hospital room, highlighting that a component of preterm infant gut colonization involves microbial exchange between room and occupant [19,20].

The causative organism is most often an opportunistic pathogen, and is therefore a usual constituent of the endogenous infant microbiota (i.e., a commensal organism). Indeed, even beneficial organisms deliberately administered in probiotics have been reported to translocate from the gut into the bloodstream in preterm infants, further highlighting the inherently leaky gut (and not the specific organism) as the central factor in sepsis pathobiology [21]. The ability to identify the aetiological agents of LOS in these studies is met with several difficulties. Firstly, in many cases blood culture for microorganisms is unsuccessful due to antimicrobial prophylaxis. Secondly, while the gut remains one potential avenue of infection in LOS other routes exist such as infection at the site of IV insertion or access at sites of damage to mucous membranes caused by intubation.

In 106 infants that were delivered preterm [22], during the first week of life aberrant microbiota in the gut occurred in those exposed to intraamniotic infection. Organisms that have been previously identified as sepsis-causing entities, such as *Ureaplasma* and Fusobacterium, as well as organisms associated with bacterial vaginosis, were in higher abundance. Several of these findings persisted into the second and third week postpartum, suggesting that these organisms are not sensitive to the commonly prescribed antibiotics, or that the infant is being repeatedly exposed to a reservoir of these organisms. Furthermore, this study implicated chorioamnionitis and a dysbiotic gut microbiota in fatal LOS cases. Notably, despite major differences in clinical practices (e.g., increased antibiotic administration), infants diagnosed with sepsis can recover the lack of diversity and possess a gut microbiota similar to healthy term infants following discharge from the NICU [23].

Potential of Specific Organisms to Prevent Bacterial Translocation in Preterm Infants

A lack of organisms thought to promote gut health, such as *Lactobacilli* and *Bifidobacteria*, is seen in preterm populations. Several studies have identified that infants who develop, or are born, with communities that contain these organisms do not develop LOS. Stewart et al. [6] identified six preterm gut community types (PGCT). It was apparent that none of the infants who developed LOS had a bacterial community that belonged to PGCT 6 prior to diagnosis. PGCT 6 was characterized by high alpha diversity and high abundance of *Bifidobacteria*. Mai et al. [16] also found that *Bifidobacteria* were present in control bacterial communities at a higher level than infants who developed sepsis. *Bifidobacteria* could either protect the patient from sepsis caused by LOS, or at the very least be a biomarker for a healthy preterm gut. Preterm control infants had a higher abundance of *Lactobacilli* in two different NICUs [17]. Results such as these have guided probiotic development for preterm infants, with *Lactobacilli* and *Bifidobacteria* being the main organisms currently implemented.

While probiotics are rarely shown to have detrimental effects, the results from these studies are conflicting, possibly due to differences in the selection of probiotic species and the tested population. Of major justification for the use of probiotics to prevent sepsis is a recent trial in a population of 4556 rural Indian newborns. The study showed a significant reduction in sepsis and death in infants given oral *Lactobacillus plantarum* (plus fructooligosaccharide) [24].

The exact pre- and probiotic combination likely depends on a multitude of factors including host genetics, environment, existing gut microbiota community, age, and degree of prematurity, amongst other factors, which leaves this an avenue that is well worth further exploration.

METABOLOMIC AND PROTEOMIC BIOMARKERS OF SEPSIS

In most cases when an infant is delivered preterm (<32 weeks gestation), clinicians will prescribe a course of antibiotics, especially when the infant has been exposed to intraamniotic infection, which could be a possible reason for failure to culture sepsis causing organisms. Furthermore, in cases where the organism can be cultured, it still takes between 24 and 48 h for a positive result with confirmation of the organism. If antimicrobial resistance profiles are sought, this would add a further day. Many authors have looked to identify individual or panels of biomarkers to aid clinicians in the diagnosis of sepsis with varying degrees of success.

Urine Metabolome

Urine samples from a group of 25 preterm infants have been used to identify biomarkers of EOS and LOS [25]. Nuclear magnetic resonance spectroscopy/NMR identified several metabolites that were increased in infants with sepsis including acetate, glycine, glucose, acetone, lactate and lysine, whereas control samples were characterized by higher levels of citrate and creatinine. Using gas chromatography mass spectrometry/GCMS septic samples had increased levels of maltose, glucose, and lactose. However, control samples were characterized by ribitol, ribonic acid, pseudouridine, 2,3,4-trihydroxybutyric acid, 2-ketogluconic acid, 3,4-dihydroxybutanoic acid, and 3,4,5-trihydroxypentanoic acid.

Levels of the enzyme pyruvate dehydrogenase are lower in adult patients with sepsis [26]. Pyruvate dehydrogenase is responsible for the conversion of pyruvate to acetyl-coA, which then enters the tricarboxylic acid cycle (Krebs cycle). When pyruvate dehydrogenase is not active, pyruvate is converted to lactate via anaerobic metabolism which produces less ATP. This may induce a starvation like state, which would introduce ketosis and the production of ketone bodies such as acetone.

Metabolomic urine analysis in 32 preterm infants successfully discriminated LOS infants from controls at time of diagnosis and 3 days prior to diagnosis, although 3 days prior to diagnosis was less pronounced. Ten days prior to diagnosis, case samples were indistinguishable from controls [27]. Glucose and acetone were elevated in the urine of sepsis infants, which is in line with the findings of the previous study [25]. However, the authors found no individual metabolites that were significantly different between cases and controls at 3 and 10 days before diagnosis. Using liquid chromatography mass spectrometry (LCMS) the authors were again able to discriminate between urine from case and control infants, at diagnosis and 3 days prior to diagnosis. While 17 metabolites were significantly different at time of diagnosis, there were also 10 that remained significant 3 days prior to diagnosis, unlike NMR profiling. When using NMR and LCMS approaches combined, glucose, taurine, inosine, and pyruvic acid were identified as significant metabolites.

Biomarkers of fungal sepsis have been identified using metabolomics [28]. This was a study of 13 healthy infants with only one case infant included in the study. Glucose and amino acid levels were increased in the urine of the infants with sepsis. More work is needed to identify whether these biomarkers are specific for sepsis or are merely a sign of ill-health.

Stool Metabolome

Stewart et al. [6] examined the stool metabolome of 14 preterm infants, 4 of which were diagnosed with LOS. Diagnosis with a sensitivity of between 75% and 89% was achieved, when 5 or 25 metabolites were used as the biomarker panel, respectively. Seven metabolites were more abundant in control infants prior to and at the time of diagnosis with no metabolites found to be increased in infants with LOS. The 2 most significant metabolites were putatively identified as sucrose and raffinose, which both belong to the galactose metabolism pathway. Upon successful treatment of LOS, these metabolites increased in line with levels of the controls.

Also 32 LOS infants matched to 40 controls, with a mean GA of 27 weeks were investigated [29]. The authors used GCMS analysis of stool from 1, 2, or 3 days prior to LOS diagnosis, which could be successfully discriminated from control samples. However, the accuracy of the test was too low for diagnostic purposes.

Serum Proteome and Metabolome

A recent study of the serum proteome and metabolome in 19 preterm infants was done using LCMS [30]. Four proteins were identified that were associated with LOS infants; Haptoglobin; Isoform Xk of plasma membrane calcium-transporting ATPase 4; Transthyretin, and U5 small nuclear ribonucleoprotein. However, there was a poor predictive value for these tests, probably due to the number of conserved proteins and metabolites between diseased and control infants, as well as some heterogeneity in the diseased infants.

Recently, a protein biomarker panel for the detection of EOS was developed, using cord blood from three infants with confirmed EOS versus 3 controls, using two-dimensional difference in gel electrophoresis/2D-DiGE [31]. The authors found that proteins related to transport and immunity were most prevalent pathways in the biomarkers, and identified haptoglobin and haptoglobin-related protein as most increased in cord blood of EOS. An ELISA assay, coupled with IL-6 levels and other hematology data, was more accurate than clinical indicators of postnatal injury. Interestingly, both of these studies identified levels of haptoglobin to be increased in infants with sepsis.

CONCLUSION

The use of "omics" technologies has begun to expand our knowledge of the mechanisms through which neonatal sepsis develops, and has offered potential markers to allow for earlier diagnosis. However, the specificities of the diagnosis would need to be conclusive, with differences observed between sepsis and other diseases (e.g., NEC) being evident. Similarly, studies of the microbiota are interesting, and have shown that dysbiosis occurs prior to sepsis, and that members of the gut microbiota can translocate to cause sepsis. The identification of bacterial strains from one infant on the NICU infecting another is an interesting finding that warrants further investigation to see how these organisms cause disease in susceptible infants.

REFERENCES

[1] Mwaniki MK, Atieno M, Lawn JE, Newton C. Long-term neurodevelopmental outcomes after intrauterine and neonatal insults: a systematic review. Lancet 2012;379:445−52. https://doi.org/10.1016/S0140-6736(11)61577-8.

[2] Stoll BJ, Hansen NI, Adams-Chapman I, Fanaroff AA, Hintz SR, Vohr B, et al. Neurodevelopmental and growth impairment among extremely low-birth-weight infants with neonatal infection. J Am Med Assoc 2004;292:2357. https://doi.org/10.1001/jama.292.19.2357.

[3] Cortese F, Scicchitano P, Gesualdo M, Filaninno A, De Giorgi E, Schettini F, et al. Early and late infections in newborns: where do we stand? A review. Pediatr Neonatol 2016;57:265−73. https://doi.org/10.1016/J.PEDNEO.2015.09.007.

[4] Shah BA, Padbury JF. Neonatal sepsis. Virulence 2014;5:170−8. https://doi.org/10.4161/viru.26906.

[5] Olivier F, Bertelle V, Shah PS, Drolet C, Piedboeuf B. Association between birth route and late-onset sepsis in very preterm neonates. J Perinatol 2016;36:1083−7. https://doi.org/10.1038/jp.2016.146.

[6] Stewart CJ, Embleton ND, Marrs ECL, Smith DP, Fofanova T, Nelson A, et al. Longitudinal development of the gut microbiome and metabolome in preterm neonates with late onset sepsis and healthy controls. Microbiome BioMed Central 2017;5:75. https://doi.org/10.1186/s40168-017-0295-1.

[7] Shaw AG, Sim K, Randell P, Cox MJ, McClure ZE, Li M-S, et al. Late-onset bloodstream infection and perturbed maturation of the gastrointestinal microbiota in premature infants. PLoS One 2015;10:e0132923. https://doi.org/10.1371/journal.pone.0132923.

[8] Wang X, Buhimschi CS, Temoin S, Bhandari V, Han YW, Buhimschi IA. Comparative microbial analysis of paired amniotic fluid and cord blood from pregnancies complicated by preterm birth and early-onset neonatal sepsis. PLoS One 2013;8:e56131. https://doi.org/10.1371/journal.pone.0056131.

[9] Mithal L, Malczynski M, Qi C, Green S, Seed PC, Mestan KK. Umbilical cord blood diagnostics for early onset sepsis in premature infants: detection of bacterial DNA and systemic inflammatory response. bioRxiv 2017:200337. https://doi.org/10.1101/200337.

[10] Leggett RM, Alcon-Giner C, Heavens D, Caim S, Brook TC, Kujawska M, et al. Rapid MinION metagenomic profiling of the preterm infant gut microbiota to aid in pathogen diagnostics. bioRxiv 2017:180406. https://doi.org/10.1101/180406.

[11] Prince AL, Ma J, Kannan PS, Alvarez M, Gisslen T, Harris RA, et al. The placental membrane microbiome is altered among subjects with spontaneous preterm birth with and without chorioamnionitis. Am J Obstet Gynecol 2016;214:627.e1−627.e16. https://doi.org/10.1016/j.ajog.2016.01.193.

[12] Doyle RM, Harris K, Kamiza S, Harjunmaa U, Ashorn U, Nkhoma M, et al. Bacterial communities found in placental tissues are associated with severe chorioamnionitis and adverse birth outcomes. In: Terry J, editor. PLoS one, vol. 12; 2017. e0180167. https://doi.org/10.1371/journal.pone.0180167.

[13] Perez-Muñoz ME, Arrieta M-C, Ramer-Tait AE, Walter J. A critical assessment of the "sterile womb" and "in utero colonization" hypotheses: implications for research on the pioneer infant microbiome. Microbiome 2017;5:48. https://doi.org/10.1186/s40168-017-0268-4.

[14] Madan JC, Salari RC, Saxena D, Davidson L, O'Toole GA, Moore JH, et al. Gut microbial colonisation in premature neonates predicts neonatal sepsis. Arch Dis Child Fetal Neonatal Ed 2012;97:F456−62. https://doi.org/10.1136/fetalneonatal-2011-301373.

[15] Stewart C, Marrs E, Magorrian S, Nelson A, Lanyon C, Perry J, et al. The preterm gut microbiota: changes associated with necrotizing enterocolitis and infection. Acta Paediatr 2012;101:1121−7. https://doi.org/10.1111/j.1651-2227.2012.02801.x.

[16] Mai V, Torrazza RM, Ukhanova M, Wang X, Sun Y, Li N, et al. Distortions in development of intestinal microbiota associated with late onset sepsis in preterm infants. PLoS One 2013;8:e52876. https://doi.org/10.1371/journal.pone.0052876.

[17] Taft DH, Ambalavanan N, Schibler KR, Yu Z, Newburg DS, Deshmukh H, et al. Center variation in intestinal microbiota prior to late-onset sepsis in preterm infants. PLoS One 2015;10:e0130604. https://doi.org/10.1371/journal.pone.0130604.

[18] Carl MA, Ndao IM, Springman AC, Manning SD, Johnson JR, Johnston BD, et al. Sepsis from the gut: the enteric habitat of bacteria that cause late-onset neonatal bloodstream infections. Clin Infect Dis 2014;58:1211−8. https://doi.org/10.1093/cid/ciu084.

[19] Brooks B, Firek BA, Miller CS, Sharon I, Thomas BC, Baker R, et al. Microbes in the neonatal intensive care unit resemble those found in the gut of premature infants. Microbiome 2014;2:1. https://doi.org/10.1186/2049-2618-2-1.

[20] Brooks B, Olm MR, Firek BA, Baker R, Thomas BC, Morowitz MJ, et al. Strain-resolved analysis of hospital rooms and infants reveals overlap between the human and room microbiome. Nat Commun 2017;8:1814. https://doi.org/10.1038/s41467-017-02018-w.

[21] Jenke A, Ruf E-M, Hoppe T, Heldmann M, Wirth S. Bifidobacterium septicaemia in an extremely low-birthweight infant under probiotic therapy. Arch Dis Child Fetal Neonatal Ed 2012;97:F217−8. https://doi.org/10.1136/archdischild-2011-300838.

[22] Puri K, Taft DH, Ambalavanan N, Schibler KR, Morrow AL, Kallapur SG. Association of chorioamnionitis with aberrant neonatal gut colonization and adverse clinical outcomes. PLoS One 2016;11:e0162734. https://doi.org/10.1371/journal.pone.0162734.

[23] Stewart CJ, Skeath T, Nelson A, Fernstad SJ, Marrs ECL, Perry JD, et al. Preterm gut microbiota and metabolome following discharge from intensive care. Sci Rep 2015;5:17141. https://doi.org/10.1038/srep17141.

[24] Panigrahi P, Parida S, Nanda NC, Satpathy R, Pradhan L, Chandel DS, et al. A randomized synbiotic trial to prevent sepsis among infants in rural India. Nature 2017;548:407−12. https://doi.org/10.1038/nature23480.

[25] Fanos V, Caboni P, Corsello G, Stronati M, Gazzolo D, Noto A, et al. Urinary 1H-NMR and GC-MS metabolomics predicts early and late onset neonatal sepsis. Early Hum Dev 2014;90:S78−83. https://doi.org/10.1016/S0378-3782(14)70024-6.

[26] Nuzzo E, Liu X, Berg K, Andersen L, Doninno M. Pyruvate dehydrogenase levels are low in sepsis. Crit Care 2015;19:P33. https://doi.org/10.1186/cc14113.

[27] Sarafidis K, Chatziioannou AC, Thomaidou A, Gika H, Mikros E, Benaki D, et al. Urine metabolomics in neonates with late-onset sepsis in a case-control study. Sci Rep 2017;7:45506. https://doi.org/10.1038/srep45506.

[28] Dessì A, Liori B, Caboni P, Corsello G, Giuffrè M, Noto A, et al. Monitoring neonatal fungal infection with metabolomics. J Matern Neonatal Med 2014;27:34−8. https://doi.org/10.3109/14767058.2014.954787.

[29] Berkhout DJC, Niemarkt HJ, Buijck M, van Weissenbruch MM, Brinkman P, Benninga MA, et al. Detection of sepsis in preterm infants by fecal volatile organic compounds analysis. J Pediatr Gastroenterol Nutr 2017;65:e47−52. https://doi.org/10.1097/MPG.0000000000001471.

[30] Stewart CJ, Nelson A, Treumann A, Skeath T, Cummings SP, Embleton ND, et al. Metabolomic and proteomic analysis of serum from preterm infants with necrotising enterocolitis and late-onset sepsis. Pediatr Res 2016;79:425−31. https://doi.org/10.1038/pr.2015.235.

[31] Buhimschi CS, Bhandari V, Dulay AT, Nayeri UA, Abdel-Razeq SS, Pettker CM, et al. Proteomics mapping of cord blood identifies haptoglobin "Switch-On" pattern as biomarker of early-onset neonatal sepsis in preterm newborns. PLoS One 2011;6:e26111. https://doi.org/10.1371/journal.pone.0026111.

Chapter 30

Gut Microbiome in the Elderly Hospitalized Patient: A Marker of Disease and Prognosis?

Andrea Ticinesi[1,2,3], Christian Milani[1,4], Fulvio Lauretani[2,3], Antonio Nouvenne[1,2,3], Claudio Tana[2,3], Marco Ventura[1,4] and Tiziana Meschi[1,2,3]

[1]*Microbiome Research Hub, University of Parma, Italy;* [2]*Department of Medicine and Surgery, University of Parma, Parma, Italy;* [3]*Dipartimento Medico-Geriatrico-Riabilitativo, Azienda Ospedaliero-Universitaria di Parma, Parma, Italy;* [4]*Laboratory of Probiogenomics, Department of Chemistry, Life Sciences and Environmental Sustainability, University of Parma, Parma, Italy*

AGING AND GUT MICROBIOTA: A CLINICAL PERSPECTIVE

In the last decade, a large body of scientific literature has investigated the associations between human gut microbiota composition and diseases [1]. However, only few studies have been specifically focused on the possible role of gut bacteria in modulating the aging process [2,3].

The gut microbiome acts as a fundamental transducer of metabolic signals from diet to the host, representing a crossroad between the environment and human pathophysiology [4,5]. Thus, its active involvement in modulating aging is biologically plausible, and the putative underlying mechanisms have been recently reviewed by Vaiserman and colleagues [6]. In this context, gut microbiota alterations, due to changes in dietary and lifestyle habits or to the presence of acute or chronic diseases, could influence the biology of aging, and their detection could have significant prognostic implications.

The gut microbiota, however, faces important age-related modifications even when healthy active aging is present. The studies by Biagi and colleagues [7,8] showed that centenarians have a significantly lower species richness in their gut microbiota than young adults or old individuals aged between 70 and 80 years. Moreover, some taxa with purported health-promoting, antiinflammatory and proanabolic activities on the host, such as *Faecalibacterium prausnitzii*, have a significantly decreased relative abundance in fecal samples from centenarians [7,8]. A core network of health-associated bacterial groups, including *Akkermansia*, *Bifidobacterium,* and Christensenellaceae, is anyway maintained in the fecal microbiota of supercentenarians, allowing to consider these specific taxa as longevity biomarkers [8].

Population-based studies have shown a dramatic increase of gut microbiota interindividual variability with aging [9,10]. These differences, which cannot generally be observed before the age of 70 years, depend on a large number of environmental factors and their interaction with the host [9,10]. The most-frequent age-related medical conditions, such as chronic diseases, multimorbidity, polypharmacy, malnutrition, and mobility limitations [11,12], are all listed as major determinants of gut microbiota composition in the largest data set of human fecal microbiota analyses published to date [10].

In line with these concepts, Claesson and colleagues identified significant differences in the gut microbiota composition between community-dwelling and institutionalized older individuals [11]. These differences concerned species richness, decreased in institutionalized subjects, and the representation of key taxa, shifting from *Ruminococcus* and *Prevotella* dominance to *Oscillibacter* dominance in long-stay residents [13]. The overall microbiota composition was also significantly associated with parameters of inflammation, nutrition, cognitive, and functional performance [13].

In a study conducted in a larger sample of 371 Irish older subjects, the same research group also showed that the presence of frailty or disability was associated with a reduced resilience of the gut microbiota [14]. Namely, the microbiome composition of institutionalized older subjects, compared with community-dwellers, showed instability over time and reduced capacity of recovering the steady state after a stressful event, such as a course of antibiotic treatment [14]. Frailty, operationalized according to the Rockwood model as accumulation of functional deficits [15], was also associated with a lower alpha diversity, underexpression of *F. prausnitzii*, and overexpression of *Eubacterium dolichum* and *Eggerthella lenta* [16]. Similar results were also found in a group of 85 adult and older community-dwellers from the United States, where the gut microbiota composition resulted much more correlated with biological, rather than chronological, age [17].

Several studies performed on smaller sample sizes, recently reviewed by Salazar and colleagues [18], confirmed that aging may be associated with relative reduction of abundance of *Faecalibacterium*, *Clostridium* cluster XIVa, *Bifidobacterium*, and *Akkermansia*, but the low statistical power of these studies prevent the generalizability of the results [18].

In spite of these intriguing findings, the current literature state of art does not support the assumption that alterations in gut microbiota composition and functionality drive the biologic trajectory of aging. In fact, the detected differences in the microbiota composition between frail and robust subjects could also represent a consequence, and not a cause, of the condition. Gut microbiota biodiversity and composition are in fact significantly influenced by exercise and diet [18,19].

Older individuals with increasing mobility limitations over time may experience progressive changes in the gut microbiota composition simply due to the reduced level of exercise. Intervention studies have in fact shown that a change from sedentary to healthy active lifestyle is associated with beneficial shifts in gut microbiota composition [5,20], and the inverse phenomenon, although not yet demonstrated, is plausible.

Similarly, the frailty-related alterations observed in the gut microbiota of elderly community-dwellers and nursing-home residents could also reflect changes in their dietary pattern [13,18]. Malnutrition has a high prevalence in older individuals and is often underrecognized [21,22]. Its causes are generally multifactorial and involve, among others, age-related alterations of gastrointestinal physiology, endocrine modifications, oral diseases affecting chewing and swallowing, impaired cognitive functions, depression, and polypharmacy [22]. These conditions lead to the so-called "anorexia of aging," with a progressive loss of the sense of hunger and reduced nutrient intake [22]. As studies carried out on low-income countries have shown, undernutrition is associated with a dramatically dysbiotic gut microbiota, with very low species richness and imbalances between symbionts and pathobionts [23]. Conversely, the restoration of a normal microbiome composition could improve the nutritional status [24], in a complex interplay between nutrients and gut bacteria [5], which has been hypothesized also in the physiopathology of eating disorders [25]. However, the association between malnutrition and microbiota has not been comprehensively studied in older individuals.

In this scenario, the comprehension of the role of gut microbiota in aging physiopathology and its clinical correlates is only at the beginning. Most studies have concentrated on healthy older subjects or clinically stable patients affected by chronic illness. However, little is known about the gut microbiota modifications in acute diseases and their relevance in influencing the disease course and prognosis.

GUT MICROBIOTA DURING HOSPITALIZATION: THE INTENSIVE CARE UNIT

Very few studies have evaluated the characteristics of gut microbiota composition when an acute disease, requiring hospital admission, is present. Most of these studies have been performed in an intensive care unit (ICU) setting, especially in adult patients experiencing severe sepsis/septic shock or adult respiratory distress syndrome (ARDS) [26,27].

In the ICU, the severity of the acute disease, the intensity of care, and the invasiveness of medical or surgical procedures have a dramatic effect on the gut microbiota [26,27]. Drugs, such as proton pump inhibitors (PPIs) or antibiotics, endotracheal intubation, intravascular catheterization, and positioning of gastrointestinal devices like nasogastric or gastrostomy tubes, all contribute to disrupt the natural barriers to environmental bacteria [26].

Antibiotic treatment alone is able to promote significant rearrangements of the gut microbiota, indiscriminately killing symbionts and promoting the selection of resistant strains with elevated pathogenic potential [28]. The most damaged bacterial populations are those who can metabolize dietary fibers into short-chain fatty acids (SCFA), such as *Faecalibacterium*, *Succinivibrio*, and *Butyricimonas*. SCFA are the main energy sources for enterocytes, and a shortage of these metabolites in the gut lumen generally results in increased mucosa permeability, leading to bacterial translocation to the bloodstream [29].

ICU admission represents a stress factor for the patient also independently of the delivered treatment and of severity of the acute disease. The duration of stress is in fact significantly associated with modifications of the gut microbiota composition leading to dysbiosis [30]. The complex interplay between acute disease, nutrition, systemic, and local immunity should also be considered for explaining the disruption of gut microbiota [31].

Several studies performed in animal models, recently reviewed by Kitsios and colleagues [26], have supported these physiopathologic concepts. To date, only five studies, summarized in Table 30.1, characterized the fecal microbiota composition of human adults admitted to ICU for critical illness, by using a next-generation sequencing approach [32−36].

During ICU stay, microbiota composition rapidly changes toward an extremely low diversity and richness, with blooming of pathogens such as *Candida* spp, Enterobacteriaceae, *Enterococcus*, and *Staphylococcus* [32]. In adults undergoing invasive mechanical ventilation, an extreme dysbiosis with relative overrepresentation of Bacteroidetes in all samples and dynamic changes over time reflecting the disease course was observed [33]. These concepts were confirmed in the largest study performed to date in an ICU setting, where dysbiosis was detected in both admission and discharge fecal samples from 115 patients [34]. The taxa experiencing the most significant depletion in their average relative abundance were those with purported antiinflammatory and health-promoting activity (i.e., SCFA producers such as *Faecalibacterium*). Conversely, a significant overrepresentation of pathogens, including *Staphylococcus, Enterobacter, Salmonella, Citrobacter,* and *Serratia*, was shown [34].

Other smaller studies have confirmed these findings (Table 30.1), also underlining the importance of the underrepresentation of Clostridiales, particularly of *Clostridium* clusters IC and XIVa, associated with ICU stay [35,36].

In summary, the acute critical illness and the ICU stay deeply shape the gut microbiota composition toward communities with a relevant pathogenic potential. Some authors have used the term of "pathobiome" to define these particular characteristics of gut microbiota in patients with septic shock [37].

From a clinical point of view, this means a dramatic risk of adverse outcomes, mainly due to secondary infections, including the *Clostridium difficile* enterocolitis [38]. Interestingly, Ojima and colleagues found that a Bacteroidetes to Firmicutes ratio >10 at ICU admission was associated in most cases with death, although the small sample size of their study limited the power and generalizability of the finding [33]. Gut microbiota composition, particularly dysbiosis measured in terms of low species richness, may represent an important marker of disease severity and poor prognosis. Conversely, gut microbiota manipulation through probiotics represents a promising potential therapeutic target in the ICU [39].

GUT MICROBIOTA DURING HOSPITALIZATION: THE MEDICAL AND GERIATRIC UNIT

For the older patient, hospital admission represents itself a very big stressor superimposing in many cases to a preexisting condition of frailty or disability [40]. From a gut microbiota perspective, this means that many disruptors simultaneously act on a microbial ecosystem, which is already undermined by the presence of frailty. Thus, even if in medical wards the intensity of care is generally lower than in the ICU, the effects of hospital admission on the gut microbiota may be very similar [41,42].

Unfortunately, very few studies have addressed this specific topic (Table 30.2). The first one [43] was performed in the early 2000s and used quantitative polymerase chain reaction (qPCR) and an early version of 16S rRNA microbial profiling techniques to detect the abundance of the most common bacterial taxa known at the moment as components of the gut microbiota. Fecal samples from 59 hospitalized elderly showed significant depletion of *Faecalibacterium, Desulfovibrio, Clostridium,* and bifidobacteria, as compared to 35 healthy active community-dwelling age-matched volunteers. However, the average relative abundance of these bacterial taxa was not different between the two groups, since samples from hospitalized patients showed a total 16S rRNA copy number lower than volunteers. Overexpression of enterobacteria was also seen in hospitalized patients [43].

More recently, metagenomics has been used to analyze the fecal microbiota of patients with cirrhosis [44]. Profound depletion of taxa with purported health-promoting activity, including Ruminococcaceae, Lachnospiraceae, and *Clostridium* cluster XIV, and expansion of pathogenic taxa, such as Enterobacteriaceae and Enterococcaceae, have been detected [44].

Moreover, in a large cohort of adult and elderly patients with acute myeloid leukemia hospitalized for chemotherapy regimens, an overexpression of pathogenic taxa in the fecal microbiota was also demonstrated. This finding was associated with a high interindividual variability and temporal instability of the overall microbiota composition [45].

Similar characteristics were also found by Braun and colleagues in a large data set of hospitalized Israeli patients, covering a wide age range, from children to oldest old [46]. Interestingly, they showed an extreme interindividual variability, with overexpression of pathobionts, such as *Escherichia, Citrobacter* and *Acinetobacter*, belonging to the Proteobacteria phylum, but they also found a normal alpha diversity in all the analyzed samples [46].

TABLE 30.1 Summary of the Main Studies Profiling the Characteristics of the Gut Microbiota Composition in Adult Patients Hospitalized in Intensive Care Units (ICU)

First Author, Journal, Year [Ref]	Country	Study Design	Sample Size	Setting/Health Status	Methodology of Microbiota Analysis	Main Findings
Zaborin, mBio, 2014 [32]	USA	Prospective observational	14	Patients admitted to ICU and with prolonged stay, irrespective of diagnosis	16S rRNA microbial profiling on feces	ICU stay associated with extreme dysbiosis and blooming of commensals that may display pathogenic properties (Enterobacteriaceae, Enterococcus, Staphylococcus, Candida)
Ojima, Dig Dis Sci, 2016 [33]	Japan	Prospective observational	12	Patients admitted to ICU with respiratory failure, undergoing invasive mechanical ventilation and enteral nutrition	16S rRNA microbial profiling on rectal swabs	Extreme level of dysbiosis with dynamic changes during stay High representation of Bacteroidetes in the sickest patients Bacteroidetes/Firmicutes ratio predicted mortality
McDonald, mSphere, 2016 [34]	USA Canada	Prospective observational	115	Patients admitted to 4 ICUs, irrespective of diagnosis	16S rRNA microbial profiling on feces	Fecal samples of ICU patients showed extreme dysbiosis, compared to a public database of healthy community-dwellers ICU stay was associated with depletion of Faecalibacterium and overexpression of pathobionts Enterobacter, Staphylococcus, and Enterobacteriaceae
Yeh, Shock, 2016 [36]	USA	Prospective observational	32	Patients admitted to ICU after trauma or acute urgent surgery	16S rRNA microbial profiling on feces	The fecal microbiota composition of ICU patients clustered separately from a group of healthy community-dwellers Overexpression of Enterococcus and depletion of Faecalibacterium, Blautia, and Ruminococcus Similar microbiota composition at different body sites
Buelow, Microbiome, 2017 [35]	Netherlands	Prospective observational	10	Patients acutely admitted to ICU irrespective of diagnosis and free of antibiotic therapy in the previous 6 months	16S rRNA microbial profiling and resistome analysis on feces	Lower microbial diversity in ICU Decreased abundance of butyrate-producing Clostridia Increased abundance of Bacteroidetes and enterococci Increased expression of antibiotic resistance genes

TABLE 30.2 Summary of the Main Studies Profiling the Characteristics of the Gut Microbiota Composition in Adult and Older Patients Hospitalized in Medical Wards

First Author, Journal, Year [Ref]	Country	Study Design	Sample Size	Setting/Health Status	Methodology of Microbiota Analysis	Main Findings
Bartosch, Appl Environ Microbiol, 2004 [43]	United Kingdom	Cross-sectional	94	Elderly (≥65) hospitalized patients receiving (N = 21) and not receiving antibiotics (N = 38), healthy community-dwellers (N = 35)	Culturomics on fecal samples (quantitative polymerase chain reaction)	Marked reduction of the *Bacteroides-Prevotella* group, bifidobacteria, *Desulfovibrio* spp, *Clostridium clostridioforme* and *Faecalibacterium prausnitzii* in the hospitalized groups Antibiotic treatment resulted in suppression of certain bacterial communities
Bajaj, J Hepatol, 2014 [44]	USA	Prospective observational	244	Patients with cirrhosis (N = 219) with various degrees of compensation (44 inpatients), healthy controls (N = 25)	Pyrosequencing and ribosomal data taxa analysis	Inpatients with cirrhosis showed a significant overexpression of pathogens (Enterococcaceae, Enterobacteriaceae, Staphylococcaceae) and changes over time Dysbiosis was more pronounced in those with decompensated disease requiring hospitalization
Milani, Sci Rep, 2016 [41]	Italy	Cross-sectional	84	Elderly inpatients with *Clostridium difficile* infection (N = 25); elderly inpatients with acute extraintestinal disease with (N = 29) and without (N = 30) antibiotic therapy	16S rRNA microbial profiling and shotgun metagenomics	Dysbiosis and interindividual variability more pronounced in those with *Clostridium difficile* infection Poor effect of antibiotic treatment on overall microbiota composition Depletion of *Bacteroides, Alistipes, Lachnospira,* and *Barnesiella* in those with *Clostridium difficile* infection
Vincent, Microbiome, 2016 [47]	Canada	Prospective cohort	98	Elderly patients admitted to a general medical ward and expected to undergo antibiotic therapy during stay	16S rRNA microbial profiling	The use of laxatives was associated with depletion of *Clostridium* and *Eubacterium* and risk of developing *Clostridium difficile* infection The development of *Clostridium difficile* infection was associated with significant gut microbiota dysbiosis
Braun, Sci Rep. 2017 [46]	Israel	Cross-sectional	196	Hospitalized patients of all ages compared with 881 community-dwellers	16S rRNA microbial profiling	Hospitalization was associated with increased species richness in the young and adult age, but not in older subjects Hospitalization was associated with overexpression of Proteobacteria

Continued

TABLE 30.2 Summary of the Main Studies Profiling the Characteristics of the Gut Microbiota Composition in Adult and Older Patients Hospitalized in Medical Wards—cont'd

First Author, Journal, Year [Ref]	Country	Study Design	Sample Size	Setting/Health Status	Methodology of Microbiota Analysis	Main Findings
Galloway-Pena, Genome Med, 2017 [45]	USA	Prospective observational	59	Patients hospitalized with acute myeloid leukemia undergoing induction chemotherapy	16S rRNA microbial profiling	Elevated temporal instability of gut microbiota composition Higher alpha diversity correlated with increased risk of infection
Ticinesi, Sci Rep, 2017 [42]	Italy	Cross-sectional	76	Elderly inpatients with acute extraintestinal disease	16S rRNA microbial profiling	Alpha diversity was inversely correlated with the number of medications and predicted 2-year mortality

In a group of elderly hospitalized patients aged in average 83 ± 8 years, our research group has confirmed the presence of an extreme interindividual variability of fecal microbiota composition [41]. We also showed that those subjects who developed *Clostridium difficile* infection (CDI) had a reduced species richness and a distinct fecal microbiota composition, with depletion of some supposedly beneficial taxa, such as *Alistipes* and *Butyricimonas*, and overexpression of pathobionts, including *Klebsiella* and other Enterobacteriaceae [41]. Interestingly, a post hoc analysis of the same dataset revealed that, in older hospitalized patients without CDI, fecal microbiota dysbiosis was independently associated with polypharmacy, and particularly with chronic treatment with PPIs, antidepressants and antipsychotics, but not with frailty or multimorbidity [42]. In fact, the level of fecal microbiota dysbiosis is associated with the risk of *C. difficile* colonization and infection [47]. Moreover, the drugs administered during hospital stay seem to specifically reduce the relative abundance of species, such as *Clostridium* cluster XIV and *Eubacterium*, involved in protection against *C. difficile* colonization [47].

In summary, these data support the concept that hospitalization may cause a disruption in the gut microbiota, also through the effect of medical treatments (Table 30.2).

THE CLINICAL RELEVANCE OF GUT MICROBIOTA ALTERATIONS IN HOSPITAL

The disruption in gut microbiota composition, observed in hospitalized patients and particularly in those over 70 years old, may have relevant clinical correlates.

The prognostic impact of gut microbiota on the clinical course and patient outcome has been studied particularly in the context of cirrhosis. The current pathophysiological models of this chronic illness underline the role of gut microbiota as both modulator of hepatocyte damage and possible source of systemic infection when the gut mucosal barrier is altered. In decompensated cirrhosis, the dysbiotic gut microbiota promotes increased intestinal permeability, leading to systemic inflammation and bloodstream infection [48]. In this context, the overrepresentation of pathobionts is significantly associated with the clinical course of the disease. Thus, the highest levels of dysbiosis belong to patients with encephalopathy or spontaneous bacterial peritonitis requiring hospitalization [44]. Similarly, in patients with compensated stable cirrhosis, specific gut microbiota alterations, such as severe underrepresentation of Bacteroidaceae, Lachnospiraceae, Ruminococcaceae, and Clostridiales XIV and overrepresentation of Enterococcaceae and Enterobacteriaceae, can predict the 90-day risk of hospitalization [49].

In patients undergoing inhospital induction chemotherapy for acute myeloid leukemia, reduced fecal microbiota species richness and increased temporal instability of its composition were significantly associated to clinically relevant outcomes, including the 90-day risk of infection after discharge [45]. Moreover, in a small cohort of Italian elderly patients hospitalized for acute extraintestinal diseases, fecal microbiota Chao1 alpha diversity index, representing the number of independent taxa detected by 16S rRNA microbial profiling, below the threshold of 1105 within the first 48 h of hospital stay was significantly associated with a higher mortality risk after 2 years of follow-up [42]. However, the same parameter was not able to predict the risk of infection-related hospitalization [42].

The prognostic relevance of gut dysbiosis can also be indirectly confirmed by epidemiologic studies, showing that clinical events, known to induce dysbiosis, are significantly associated with patient-centered outcomes, such as sepsis and hospitalization. In 10,996 participants to the Health and Retirement Study-Medicare in the United States, the 90-day risk of hospital admission for sepsis was significantly increased after an index hospitalization [50]. The magnitude of the risk was greater when the index hospitalization was for an infectious disease, other than sepsis, and maximum when the index hospitalization was for CDI [50]. Interestingly, this risk trend could inversely reflect the alpha diversity index of fecal microbiota, as suggested by the findings in similar categories of older patients [41].

GUT MICROBIOTA MANIPULATIONS IN HOSPITALIZED OLDER INDIVIDUALS

Gut microbiota manipulation during hospital stay could thus represent an intriguing target of care for improving patient health. A recent systematic review with metaregression analysis has demonstrated that the administration of probiotics to hospitalized patients receiving antibiotic treatment can reduce the risk for developing CDI during the hospital stay [51]. The preventive efficacy of probiotic administration is enhanced when it is started within 2 days of the initiation of antibiotic therapy [51]. Another systematic review has also shown that the benefits of probiotic administration on the development of CDI are relevant only for those subjects who have a high baseline risk of developing the infection [52]. Probiotic administration may also influence the clinical course of CDI [53] and of other infections by multidrug resistant bacteria, such as carbapenem-resistant *Klebsiella pneumoniae* [54,55].

However, the currently available probiotics are only able to introduce a single microbial taxon, or, at most, few taxa in the gut microbiota. In many cases, these taxa are not targeted on the mechanisms of disease development. For example, recent experimental evidence has shown that the metabolic activity of a limited number of species, and particularly of *Clostridium scindens*, is deeply involved in protection against the development of CDI, through synthesis of bile acids [56]. However, at the current state of art, there is no possibility of administering *C. scindens* as probiotic. Thus, even if next-generation probiotics specifically targeting the molecular mechanisms of microbiome-mediated resistance to *C. difficile* colonization are being studied [57], there is still a substantial gap between microbiome science and therapeutic possibilities.

One way proven effective to reconstitute a normal gut microbiota composition is fecal microbiota transplantation. However, this procedure is complex and expensive, and its efficacy has been clearly demonstrated only in the treatment of recurrent CDI, especially after failure of first- and second-line antibiotic treatments [58]. In other clinical settings, particularly in older individuals, the indications to perform this procedure are controversial, and cost-effectiveness is not proved [58].

The role of microbiota manipulations in older hospitalized patients is understudied and underestimated in the scientific literature to date. The outcomes of the existing studies have been mainly centered on gastrointestinal disorders, such as prevention of CDI [59] and constipation [60], or on microbiological outcomes, i.e., changes of the gut microbial composition after probiotic administration [61].

However, several studies support the concept that gut microbiota composition is able to modulate the host metabolism and inflammation, influencing the physiopathology of aging and the clinical course of disease [5]. The influence of gut microbiota composition and its dynamic modifications in aging on patient-centered outcomes, such as hospital readmissions, frailty, disability, and mortality, is a territory that clinical science has yet to explore.

FUTURE PERSPECTIVES: GUT MICROBIOTA AND TRAJECTORIES OF AGING

In older patients, hospitalization implies a transient condition of clinical vulnerability, especially in the 30 days after discharge [62]. This condition, sometimes called "the posthospital syndrome," represents the physiological period of recovery after acute illness but may also reflect the period of recovery from hospitalization-induced gut microbiota dysbiosis. Interestingly, the main determinants of posthospitalization functional decline trajectory, i.e., disease severity, impaired mobility, continence care, hospital length of stay, and malnutrition [63,64], are also factors that can enhance gut microbiota dysbiosis.

Epidemiologic data also underline the concept that, in older patients approaching the end of life, hospital admission is the strongest determinant of the severity of disability after discharge, and hospitalization itself represents a risk factor for earlier mortality [65]. At the current state of art, no studies have explored the association between gut microbiota dysbiosis and aging trajectories. However, our understanding of the deep symbiosis between gut microbiota and the host physiology allows to hypothesize such an association.

The gut microbiota composition, in terms of species richness and presence and abundance of key taxa, could thus represent a promising and intriguing marker of aging and disease status in older patients. More clinical and translational studies are however needed to confirm these assumptions.

Metagenomics analyses of the fecal microbiota are currently being performed mainly for research purposes, and their diagnostic applications implemented only in few centers. However, the speed we are improving our knowledge of its specific alterations in various conditions and diseases makes a clinical translation of these analyses nearer and nearer. In the foreseeable future, microbiome analyses could be routinely used for diagnostic and prognostic purposes to establish personalized targets of care and guide clinical management [66].

In the hospital setting, microbiome analysis could thus help to identify those subjects at higher risk for complications, including but not limited to CDI, and for a complicated postdischarge course. Aging represents the ideal territory of clinical application of microbiome science for the complexity of the geriatric patient, which is sometimes difficult to manage with traditional tools.

REFERENCES

[1] Nayak RR, Turnbaugh PJ. Mirror, mirror on the wall: which microbiomes will help heal them all. Microbiome 2016;14:72.

[2] Zapata HJ, Quagliarello VJ. The microbiota and microbiome in aging: potential implications in health and age-related diseases. J Am Geriatr Soc 2015;63(4):776—81.

[3] Bischoff SC. Microbiota and aging. Curr Opin Clin Nutr Metab Care 2016;19:26—30.

[4] Shanahan F, van Sinderen D, O'Toole PW, Stanton C. Feeding the microbiota: transducer of nutrient signals for the host. Gut 2017;66:1709—17.

[5] Ticinesi A, Lauretani F, Milani C, Nouvenne A, Tana C, Del Rio D, Maggio M, Ventura M, Meschi T. Aging gut microbiota at the cross-road between nutrition, physical frailty, and sarcopenia: is there a gut-muscle axis? Nutrients 2017;9(12):E1303.

[6] Vaiserman AM, Koliada AK, Marotta F. Gut microbiota: a player in aging and a target for anti-aging intervention. Ageing Res Rev 2017;35:36—45.

[7] Biagi E, Nylund L, Candela M, Ostan R, Bucci L, Pini E, Nikkila J, Monti D, Satokari R, Franceschi C, Brigidi P, de Vos W. Through ageing, and beyond: gut microbiota and inflammatory status in seniors and centenarians. PLoS One 2010;5(5):e10667.

[8] Biagi E, Franceschi C, Rampelli S, Severgnini M, Ostan R, Turroni S, Consolandi C, Quercia S, Scurti M, Monti D, Capri M, Brigidi P, Candela M. Gut microbiota and extreme longevity. Curr Biol 2016;26(11):1480—5.

[9] Claesson MJ, Cusack S, O'Sullivan O, Greene-Diniz R, de Weerd H, Flannery E, Marchesi JR, Falush D, Dinan T, Fitzgerald G, Stanton C, van Sinderen D, O'Connor M, Harnedy N, O'Connor K, Henry C, O'Mahony D, Fitzgerald AP, Shanahan F, Twomey C, Hill C, Ross RP, O'Toole PW. Composition, variability, and temporal stability of the intestinal microbiota of the elderly. Proc Natl Acad Sci USA 2011;108(S1):4586—91.

[10] Zhernakova A, Kurilshikov A, Bonder MJ, Tigchelaar EF, Schirmer M, Vatanen T, Mujagic Z, Vich Vila A, Falony G, Vieira-Silva S, Wang J, Imhann F, Brandsma E, Jankipersadsing SA, Joossens M, Cenit MC, Deelen P, Swertz MA, LifeLines cohort study, Weersma RK, Feskens EJM, Netea MG, Gevers D, Jonkers D, Franke L, Aulchenko YS, Huttenhower C, Raes J, Hofker MH, Xavier RJ, Wijmenga C, Fu J. Population-based metagenomics analysis reveals markers for gut microbiome composition and diversity. Science 2016;352(6285):565—9.

[11] Mannucci PM, Nobili A. REPOSI Investigators. Multimorbidity and polypharmacy in the elderly: lessons from REPOSI. Intern Emerg Med 2014;9(7):723—34.

[12] Cederholm T, Nouvenne A, Ticinesi A, Maggio M, Lauretani F, Ceda GP, Borghi L, Meschi T. The role of malnutrition in older persons with mobility limitations. Curr Pharm Des 2014;20(19):3173—7.

[13] Claesson MJ, Jeffery IB, Conde S, Power SE, O'Connor EM, Cusack S, Harris HBM, Coakley M, Lakshminarayanan B, O'Sullivan O, Fitzgerald GF, Deane J, O'Connor M, Harnedy N, O'Connor K, O'Mahony D, van Sinderen D, Wallace M, Brennan L, Stanton C, Marchesi JR, Fitzgerald AP, Shanahan F, Hill C, Ross RP, O'Toole PW. Gut microbiota composition correlates with diet and health in the elderly. Nature 2012;488:178—84.

[14] Jeffery IB, Lynch DB, O'Toole PW. Composition and temporal stability of the gut microbiota in older persons. ISME J 2016;10:170—82.

[15] Clegg A, Young J, Iliffe S, Rikkert MO, Rockwood K. Frailty in elderly people. Lancet 2013;381(9868):752—62.

[16] Jackson MA, Jeffery IB, Beaumont M, Bell JT, Clark AG, Ley RE, O'Toole PW, Spector TD, Steves CJ. Signatures of early frailty in the gut microbiota. Genome Med 2016;8:8.

[17] Maffei VJ, Kim S, Blanchard IV E, Luo M, Jazwinski SM, Taylor CM, Welsh DA. Biological aging and the human gut microbiota. J Gerontol A Biol Sci Med Sci 2017;72(11):1474—82.

[18] Salazar N, Valdes-Varela L, Gonzalez S, Gueimonde M, De los Reyes-Gavilan CG. Nutrition and the gut microbiome in the elderly. Gut Microb 2017;8(2):82—97.

[19] Clarke SF, Murphy EF, O'Sullivan O, Lucey AJ, Humphreys M, Hogan A, Hayes P, O'Reilly M, Jeffery IB, Wood-Martin R, Kerins DM, Quigley E, Ross RP, O'Toole PW, Molloy MG, Falvey E, Shanahan F, Cotter PD. Exercise and associated dietary extremes impact on gut microbial diversity. Gut 2014;63(12):1913—20.

[20] Allen JM, Mailing LJ, Niemiro GM, Moore R, Cook MD, White BA, Holscher HD, Woods JA. Exercise alters gut microbiota composition and function in lean and obese humans. Med Sci Sports Exerc 2018;50(4):747—57.

[21] Rojer AG, Kruizenga HN, Trappenburg MC, Reijnierse EM, Sipila S, Narici MV, Hogrel JY, Butler-Browne G, McPhee JS, Paasuke M, Meskers CG, Maier AB, de van der Schueren MA. Prevalence of malnutrition according to the new ESPEN definition in four diverse populations. Clin Nutr 2016;35(6):758—62.

[22] Landi F, Calvani R, Tosato M, Martone AM, Ortolani E, Savera G, Sisto A, Marzetti E. Anorexia of aging: risk factors, consequences, and potential treatments. Nutrients 2016;8(2):69.

[23] Alou MT, Million M, Traore SI, Mouelhi I, Khelaifia S, Bachar D, Caputo A, Delerce J, Brah S, Alhousseini D, Sokhna C, Robert C, Diallo BA, Diallo A, Parola P, Golden M, Lagier JC, Raoult D. Gut bacteria missing in severe acute malnutrition, can we identify potential probiotics by culturomics? Front Microbiol 2017;8:899.

[24] Blanton LV, Barratt MJ, Charbonneau MR, Ahmed T, Gordon JI. Childhood undernutrition, the gut microbiota, and microbiota-directed therapeutics. Science 2016;352(6293):aad9359.

[25] Lam YY, Maguire S, Palacios T, Caterson ID. Are the gut bacteria telling us to eat or not to eat? Reviewing the role of gut microbiota in the etiology, disease progression and treatment of eating disorders. Nutrients 2017;9:602.

[26] Kitsios GD, Morowitz MJ, Dickson RP, Huffnagle GB, McVerry BJ, Morris A. Dysbiosis in the intensive care unit: microbiome science coming to the bedside. J Crit Care 2017;38:84−91.

[27] Haak BW, Levi M, Wiersinga WJ. Microbiota-targeted therapies on the intensive care unit. Curr Opin Crit Care 2017;23:167−74.

[28] Modi SR, Collins JJ, Relman DA. Antibiotics and the gut microbiota. J Clin Invest 2014;124:4212−8.

[29] McKenney PT, Pamer EG. From hype to hope: the gut microbiota in enteric infectious diseases. Cell 2015;163:1326−32.

[30] Watanabe Y, Arase S, Nagaoka N, Kawai M, Matsumoto S. Chronic psychological stress disrupted the composition of the murine colonic microbiota and accelerated a murine model of inflammatory bowel disease. PLoS One 2016;11(3):e0150559.

[31] Weiss GA, Hennet T. Mechanisms and consequences of intestinal dysbiosis. Cell Mol Life Sci 2017;74:2959−77.

[32] Zaborin A, Smith D, Garfield K, Quensen J, Shakhsheer B, Kade M, Tirrell M, Tiedje J, Gilbert JA, Zaborina O, Alverdy JC. Membership and behavior of ultra-low-diversity pathogen communities present in the gut of humans during prolonged critical illness. mBio 2014;5(5):e01361−14.

[33] Ojima M, Motooka D, Shimizu K, Gotoh K, Shintani A, Yoshiya K, Nakamura S, Ogura H, Iida T, Shimazu T. Metagenomic analysis reveals dynamic changes of whole gut microbiota in the acute phase of intensive care unit patients. Dig Dis Sci 2016;61:1628−34.

[34] McDonald D, Ackermann G, Khailova L, Baird C, Heyland D, Kozar R, Lamieux M, Derenski K, King J, Vis-Kampen C, Knight R, Wischmeyer PE. Extreme dysbiosis of the microbiome in critical illness. mSphere 2016;1(4):e00199−16.

[35] Buelow E, Bello Gonzalez TDJ, Fuentes S, de Steenhuijsen Piters WAA, Lahti L, Bayjanov JR, Majoor EAM, Braat JC, van Mourik MSM, Oostdijk EAN, Willems RJL, Bonten MJM, van Passel MWJ, Smidt H, van Schaik W. Comparative gut microbiota and resistome profiling of intensive care patients receiving selective digestive tract decontamination and healthy subjects. Microbiome 2017;5:88.

[36] Yeh A, Rogers MB, Firek B, Neal MD, Zuckerbraun BS, Morowitz MJ. Dysbiosis across multiple body sites in critically ill adult surgical patients. Shock 2016;46(6):649−54.

[37] Krezalek MA, DeFazio J, Zaborina O, Zaborin A, Alverdy JC. The shift of an intestinal "microbiome" to a "pathobiome" governs the course and outcome of sepsis following surgical injury. Shock 2016;45(5):475−82.

[38] Akrami K, Sweeney DA. The microbiome of the critically ill patient. Curr Opin Crit Care 2018;24:49−54.

[39] Morrow LE, Wischmeyer P. Blurred lines. Dysbiosis and probiotics in the ICU. Chest 2017;151(2):492−9.

[40] Romero-Ortuno R, Wallis S, Biram R, Keevil V. Clinical frailty adds to acute illness severity in predicting mortality in hospitalized older adults: an observational study. Eur J Intern Med 2016;35:24−34.

[41] Milani C, Ticinesi A, Gerritsen J, Nouvenne A, Lugli GA, Mancabelli L, Turroni F, Duranti S, Mangifesta M, Viappiani A, Ferrario C, Maggio M, Lauretani F, De Vos W, van Sinderen D, Meschi T, Ventura M. Gut microbiota composition and *Clostridium difficile* infection in hospitalized elderly individuals: a metagenomic study. Sci Rep 2016;6:25945.

[42] Ticinesi A, Milani C, Lauretani F, Nouvenne A, Mancabelli L, Lugli GA, Turroni F, Duranti S, Mangifesta M, Viappiani A, Ferrario C, Maggio M, Ventura M, Meschi T. Gut microbiota composition is associated with polypharmacy in elderly hospitalized patients. Sci Rep 2017;7:11102.

[43] Bartosch S, Fite A, Macfarlane GT, McMurdo MET. Characterization of bacterial communities in feces from healthy elderly volunteers and hospitalized elderly patients by using real-time PCR and effects of antibiotic treatment on the fecal microbiota. Appl Environ Microbiol 2004;70(6):3575−81.

[44] Bajaj JS, Heuman DM, Hylemon PB, Sanyal AJ, White MB, Monteith P, Noble NA, Unser AB, Daita K, Fisher AR, Sikaroodi M, Gillevet PM. Altered profile of human gut microbiome is associated with cirrhosis and its complications. J Hepatol 2014;60:940−7.

[45] Galloway-Pena JR, Smith DP, Sahasrabjohane P, Wadsworth WD, Fellman BM, Ajami NJ, Shpall EJ, Daver N, Guindani M, Petrosino JM, Kontoyiannis DP, Shelburne SA. Characterization of oral and gut microbiome temporal variability in hospitalized cancer patients. Genome Med 2017;9:21.

[46] Braun T, Di Segni A, BenShoshan M, Asaf R, Squires JE, Barhom SF, Saar EG, Cesarkas K, Smollan G, Weiss B, Amit S, Keller N, Haberman Y. Fecal microbiota characterization of hospitalized patients with suspected infectious diarrhea shows significant dysbiosis. Sci Rep 2017;7:1088.

[47] Vincent C, Miller MA, Edens TJ, Mehrotra S, Dewar K, Manges AR. Bloom and blust: intestinal microbiota dynamics in response to hospital exposures and *Clostridium difficile* colonization or infection. Microbiome 2016;4:12.

[48] Tilg H, Cani PD, Mayer EA. Gut microbiome and liver diseases. Gut 2016;65:2035−44.

[49] Bajaj JS, Betrapally NS, Hylemon PB, Thacker LR, Daita K, Kang DJ, White MB, Unser AB, Fagan A, Gavis EA, Sikaroodi M, Dalmet S, Heuman DM, Gillevet PM. Gut microbiota alterations can predict hospitalizations in cirrhosis independent of diabetes mellitus. Sci Rep 2015;5:18559.

[50] Prescott HC, Dickson RP, Rogers MAM, Langa KM, Iwashyna TJ. Hospitalization type and subsequent severe sepsis. Am J Respir Crit Care Med 2015;192(5):581−8.

[51] Shen NT, Maw A, Tmanova LL, Pino A, Ancy K, Crawford CV, Simon MS, Evans AT. Timely use of probiotics in hospitalized adults prevents *Clostridium difficile* infection: a systematic review with meta-regression analysis. Gastroenterology 2017;152:1889−900.

[52] Goldenberg JZ, Yap C, Lytvyn L, Lo CK, Beardsley J, Mertz D, Johnston BC. Probiotics for the prevention of Clostridium difficile-associated diarrhea in adults and children. Cochrane Database Syst Rev 2017;12:CD006095.

[53] Barker AK, Duster M, Valentine S, Hess T, Archbald-Pannone L, Guerrant R, Safdar N. A randomized controlled trial of probiotics for *Clostridium difficile* infection in adults (PICO). J Antimicrob Chemother 2017;72:3177−80.

[54] Nouvenne A, Ticinesi A, Meschi T. Carbapenemase-producing *Klebsiella pneumoniae* in elderly frail patients admitted to medical wards. Ital J Med 2015;9:112−5.

[55] Nouvenne A, Ticinesi A, Lauretani F, Maggio M, Lippi G, Guida L, Morelli I, Ridolo E, Borghi L, Meschi T. Comorbidities and disease severity as risk factors for carbapenemase-resistant *Klebsiella pneumoniae* colonization: report of an experience in an internal medicine unit. PLoS One 2014;9(10):e110001.

[56] Buffie CG, Bucci V, Stein RR, McKenney PT, Ling L, Gobourne A, No D, Liu H, Kinnebrew M, Viale A, Littmann E, van den Brink MRM, Jenq RR, Taur Y, Sander C, Cross JR, Toussaint NC, Xavier JB, Pamer EG. Precision microbiome reconstitution restores bile acid mediated resistance to *Clostridium difficile*. Nature 2015;517:205−8.

[57] Spinler JK, Auchtung J, Brown A, Boonma P, Oezguen N, Ross CL, Luna RA, Runge J, Versalovic J, Peniche A, Dann SM, Britton RA, Haag A, Savidge TC. Next-generation probiotics targeting *Clostridium difficile* through precursor-directed antimicrobial biosynthesis. Infect Immun 2017;85(10):e00303−17.

[58] Kelly CR, Kahn S, Kashyap P, Layne L, Rubin D, Atreja A, Moore T, Wu G. Update on fecal microbiota transplantation 2015: indications, methodologies, mechanisms, and outlook. Gastroenterology 2015;149:223−37.

[59] Spinler JK, Ross CL, Savidge TC. Probiotics as adjunctive therapy for preventing *Clostridium difficile* infection − what are we waiting for? Anaerobe 2016;41:51−7.

[60] Zaharoni H, Rimon E, Vardi H, Friger M, Bolotin A, Shahar DR. Probiotics improve bowel movements in hospitalized elderly patients − the PROAGE study. J Nutr Health Aging 2011;15(3):215−20.

[61] Lahtinen SJ, Forssten S, Aakko J, Granlund L, Rautonen N, Salminen S, Viitanen M, Ouwehand AC. Probiotic cheese containing Lactobacillus rhamnosus HN001 and Lactobacillus acidophilus NCFM® modifies subpopulations of fecal lactobacilli and *Clostridium difficile* in the elderly. Age 2012;34:133−43.

[62] Krumholz HM. Post-hospital syndrome − an acquired, transient condition of generalized risk. N Engl J Med 2013;368(2):100−2.

[63] Zisberg A, Shadmi E, Gur-Yaish N, Tonkikh O, Sinoff G. Hospital-associated functional decline: the role of hospitalization processes beyond individual risk factors. J Am Geriatr Soc 2015;63:55−62.

[64] Fimognari FL, Pierantozzi A, De Alfieri W, Salani B, Zuccaro SM, Arone A, Palleschi G, Palleschi L. The severity of acute illness and functional trajectories in hospitalized older medical patients. J Gerontol A Biol Sci Med Sci 2017;72(1):102−8.

[65] Gill TM, Gahbauer EA, Han L, Allore HG. The role of intervening hospital admissions on trajectories of disability in the last year of life: prospective cohort study of older people. BMJ 2015;350:h2361.

[66] Kashyap PC, Chia N, Nelson H, Segal E, Elinav E. Microbiome at the frontier of personalized medicine. Mayo Clin Proc 2017;92(12):1855−64.

Chapter 31

The Lung Microbiome, Metabolome, and Breath Volatolome in the Diagnosis of Pulmonary Disease

Samuel M. Gonçalves[1,2], Cláudio Duarte-Oliveira[1,2], Cristina Cunha[1,2] and Agostinho Carvalho[1,2]

[1]Life and Health Sciences Research Institute (ICVS), School of Medicine, University of Minho, Braga, Portugal; [2]ICVS/3B's - PT Government Associate Laboratory, Braga/Guimarães, Portugal

INTRODUCTION

The increased success of novel therapeutic approaches involving extensive immunosuppression has driven a remarkable expansion in immunocompromised populations at high risk of opportunistic pulmonary infections [1−3]. Because of the underlying immune dysfunction, these patients are not able to efficiently eliminate opportunistic fungi, and are extremely susceptible to life-threatening, invasive infections [4]. Among these, invasive pulmonary aspergillosis (IPA) caused primarily by the fungus *Aspergillus fumigatus* is typically diagnosed among recipients of solid organ and stem-cell transplantation and patients undergoing cancer chemotherapy [5].

In addition to these, IPA has been increasingly reported in patients under intensive care, namely those suffering from chronic obstructive pulmonary disease (COPD) or following influenza infection [6,7]. Beside invasive disease, patients with asthma or cystic fibrosis (CF) are prone to develop fungal-induced allergic airway diseases, the most severe form being allergic bronchopulmonary aspergillosis (ABPA) [8], whereas chronic pulmonary aspergillosis (CPA) is instead a typical feature of patients with preexisting lung cavities caused by tuberculosis or COPD [9−11].

THE LUNG MICROBIOTA

An emerging body of evidence has highlighted the significant role of the pulmonary microbiota in inflammation, metabolism, and other physiological processes that regulate the antifungal immune responses, thereby conditioning host susceptibility to fungal disease [12]. The lung microbiota comprises an ecological community of commensal, symbiotic, and pathogenic organisms from six predominant phyla: Firmicutes, Bacteroidetes, Proteobacteria, Actinobacteria, Fusobacteria, and Cyanobacteria [13].

Historically, medical texts allude to a sterile lung environment, and this dogma has persisted in contemporary medicine mainly because of limitations in standard microbial culture techniques to mimic the lung habitat. The development of cutting-edge molecular biology approaches for the analysis of bacterial communities has revealed that the lung is not sterile [14], and that the airways harbor an abundant and cross-interacting microbiota, with Bacteroidetes and Firmicutes being the predominant phyla [15].

The existence of a resident lung microbiota remains however to be clearly demonstrated, with the currently most accepted view, involving the immigration and persistence of bacterial species from surrounding microenvironments, such as the oral cavity [16−19]. The mucociliary clearance which is often impaired in many chronic inflammatory airway diseases [20] and the establishment of a proinflammatory microenvironment [21] may also represent major permissive factors for the adaptation and establishment of bacterial species colonizing the lower airways.

THE INTERACTION BETWEEN THE LUNG MICROBIOTA AND THE IMMUNE SYSTEM

The immune system is composed of a complex network of innate and adaptive components endowed with the capacity to adapt and respond to pathogens, and to react to tissue damage. In this context, the complex bacterial communities that inhabit the lungs play a pivotal role in maintaining tissue and organ immune homeostasis [13]. The permanent interaction of the microbiota with the innate immune system and its various microbe-sensing pattern recognition receptors (PRRs) has evolved into elaborate regulatory mechanisms, through which both systems maintain a symbiotic relationship [22]. This means that the function of the immune system is promoted and calibrated by its interaction with the microbiota, allowing the establishment of protective responses to pathogens, the control of overgrowth of indigenous pathobionts, and the tolerance to antigens derived from commensals [23].

THE INDIGENOUS COMMUNITY AND INVASIVE PATHOGENS

In the respiratory tract, an important role of the upper airway commensals in shaping and modulating host immune responses has been reported [24]. For example, upper airway commensals can trigger effective antiviral immune responses against influenza A virus in the lower airways through the production of M2 macrophage-induced antiinflammatory cytokines and inhibitory ligands [25]. Moreover, commensal neomycin-sensitive bacteria regulate the initiation of adaptive immunity after influenza A infection, by generating virus-specific CD4(+) and CD8(+) T cells through inflammasome activation and dendritic cell migration from the site of infection to the lymph nodes [26].

Likewise, opportunistic fungi such as *Aspergillus* spp. and *Candida* spp., members of the fungal mycobiome (thought to represent as little as 0.1% of the total microbiome [27]), are believed to expand and potentially contribute to disease, upon disturbances in the environment or when the host is immunocompromised. One such example is illustrated by mice lacking dectin-1, the innate immune receptor for β-glucans, present in the fungal cell wall; the mice display increased susceptibility to colitis, as the result of the expansion of the *Candida* and *Trichosporon* genera and impaired immune responses to these commensal fungi [28]. Given that β-glucans in the cell wall of *A. fumigatus* are also recognized by dectin-1 [29], this brings up the interesting possibility that the functional disruption of this receptor could also impact the pulmonary microbiota and the local immune response to the fungus. In fact, genetic variants influencing dectin-1 function have been associated with the development of IPA in hematological patients [30,31] and recipients of stem-cell transplantation [32]. Future investigation of bacterial and fungal communities in dectin-1-deficient patients might provide further insights into the genetic regulation of microbiota composition and function.

THE DYNAMICS OF LUNG MICROBIOTA PROFILES IN RESPIRATORY FUNGAL DISEASE

The composition of the lung microbiota suffers dynamic changes during life, and is determined at least by the balance between three factors: the microbial immigration into the airways; their elimination by mechanisms such as cough, mucociliary clearance, and host immune responses; and the regional growth conditions including nutrient availability and oxygen tension, local microbial competition, and activation of inflammatory cells [19]. The homeostasis in the composition of the normal flora in the respiratory tract is therefore essential to prevent the expansion of species with pathogenic potential, which may ultimately contribute to the development, progression, and exacerbation of various inflammatory disorders of the lung, including asthma, COPD, and CF [33]. These alterations in the lung microenvironment may ultimately be permissive to fungal colonization and establishment, providing an explanation for their typical association with respiratory fungal diseases such as IPA.

Chronic Obstructive Pulmonary Disease

COPD is a multifactorial disease that involves airway inflammation, mucociliary dysfunction, and lung structural changes. For these reasons, COPD is acknowledged as an important risk factor for IPA in critically ill patients [34,35]. Despite the lack of coherence between the many studies investigating the pulmonary microbiota in COPD mostly due to sampling variability and heterogeneous therapeutic regimens, recent studies agree on the fact that *Haemophilus* spp., *Streptococcus* spp., *Prevotella* spp., and *Staphylococcus* spp. are the most common bacteria found in COPD airways [36–39]. Of note, the large FUNGI-COPD study reported that *Aspergillus* spp. were frequently isolated from the sputum of COPD patients during acute exacerbations, and that the concomitant isolation of *Pseudomonas aeruginosa* was a contributing factor to fungal colonization [40].

Cystic Fibrosis

Chronic airway colonization by *Aspergillus* spp. can be found in up to 60% of CF patients [41]. Recent studies targeting the composition of the airway microbiota identified *P. aeruginosa* and *Streptococcus aureus* as the predominant bacterial pathogens, whereas *Candida* and *Aspergillus* were the most commonly observed fungal genera [42–44]. The complex polymicrobial community present in the lungs of CF patients, together with the pulmonary dysfunction due to mucus formation, oxygen tension, and inflammatory cell recruitment might favor the formation of bacterial and fungal biofilms and influence pathogenicity [45].

P. aeruginosa and *A. fumigatus* Coinfection

The interaction between *P. aeruginosa* and *A. fumigatus* is among the most important microbial interactions taking place in the CF lung. Many studies have proposed an antagonistic relationship between these two organisms, triggered by direct contact and release of small molecules, affecting quorum-sensing networks (e.g., pyocyanin and siderophore pyoverdine), influencing the ability of *A. fumigatus* to germinate and develop biofilms [46–48]. However, cocolonized CF patients have also been reported to display poor clinical outcomes compared to patients persistently colonized with *P. aeruginosa* [49]. In support of this, antibiotic treatment targeting *P. aeruginosa* decreased the detection of *Aspergillus* spp. in sputum samples, emphasizing the clinical significance of cocolonization by these microorganisms [50].

Both *A. fumigatus* and *P. aeruginosa* are bound by the soluble pattern recognition receptor pentraxin 3 (PTX3), which in turn favors their recognition and clearance by the innate immune system [51]. Genetic variants influencing PTX3 expression in the lung have been associated with both *P. aeruginosa* colonization in CF patients [52] and IPA in stem-cell and lung transplant recipients [53–55] and COPD [56]. PTX3 may therefore be proposed to act as a master regulator of microbiota homeostasis in the lung, and that under specific circumstances such as the CF lung microenvironment, commensals may be able to subvert PTX3 expression to control conflicting species.

Asthma

Asthma is frequently caused or exacerbated by sensitization to fungal allergens, with *Aspergillus* spp. being among the growing list of allergens that can aggravate asthmatic responses [57]. The airway hyperresponsiveness and bronchial constriction, in response to fungal sensitization, are associated with poor control of asthma [58], increasing the likelihood for more severe disease and complications such as bronchiectasis and CPA [59]. Significant pulmonary pathology is associated with *Aspergillus*-induced allergic and asthmatic lung disease [57,60], and different studies have reported that fungal-sensitized asthmatic patients have poorer lung function, mainly associated with IgE sensitization, neutrophilic airway inflammation, bronchiectasis, and airflow obstruction [61,62].

The association between early childhood antibiotic exposure and subsequent development of asthma and allergies observed in multiple studies [39] suggests that disruption of the normal gut microbiota may contribute to the pathogenesis of this disease. In fact, severe asthma patients were found to have a specific pulmonary bacterial community, when compared to both healthy individuals and patients with mild-to-moderate asthma [63]. Besides the bacterial dysbiosis, several fungal species were found to be more abundant in induced sputum of asthmatic patients [64], and the reduction in their pulmonary capacity was also significantly correlated with the presence of *Aspergillus* and *Penicillium* genera [65].

THE MICROBIOTA-METABOLOME CROSSTALK IN RESPIRATORY FUNGAL DISEASES

The pivotal role of microbial metabolites in the regulation of host immune responses is illustrated by the recognition of short-chain fatty acids (SCFAs), products of the intestinal microbiota, by the G-coupled protein receptor GPR43, expressed on innate immune cells [66–68]. Mice lacking GPR43 were shown to be unable to resolve intestinal inflammation during experimental colitis [69], and more importantly, displayed also a more inflammatory allergic response during experimental asthma [16], illustrating the systemic regulation of immune responses in the lung by bacterial metabolites. In this context, mice fed with fermentable fibers presented marked alterations in the gut microbiota composition, with the resulting increased levels of circulating SCFAs conferring them protection against house dust mite–induced allergic pulmonary inflammation [70].

Another metabolic pathway whereby tryptophan metabolites produced by *Lactobacillus reuteri* in the gut microbiota regulate mucosal immunity has also been identified [71]. The tryptophan derivative produced indole-3-aldehyde acts as an aryl hydrocarbon receptor ligand that regulates the commensal colonization by *Candida albicans* and the development of recurrent vulvovaginal candidiasis [72]. Whether tryptophan metabolites also display distal effects on the lung remains to be elucidated.

In any case, *Lactobacillus* spp. have been found to dampen allergic airway responses by inducing the expansion of T regulatory cells (Tregs) [73,74], and the tryptophan catabolism was found to regulate the T helper (Th)1/Treg balance and lung homeostasis in mouse models of CF [75].

In the CF lung, *P. aeruginosa* can produce metabolites, with important consequences in its interaction with *A. fumigatus* [47]. Recent molecular networking-based metabolomics revealed that the chemical makeup of the CF sputa comprises xenobiotics, specialized metabolites from *P. aeruginosa*, and host sphingolipids [76]. Importantly, because the microbial metabolites in the sputa from patients did not match those produced by laboratory cultures, it suggests that the metabolism of *P. aeruginosa* in vivo critically relies on signals provided by the chemical nature of the CF lung environment. As for the host sphingolipids, these contain the inflammatory mediator ceramide and may therefore have a potential role in perpetuating inflammation in CF [77].

METABOLOMIC PROFILING IN FUNGAL DIAGNOSTICS

The usefulness of metabolite profiling as a biomarker for lung disease development and severity has been recently highlighted (Table 31.1). For example, CF patients presenting pulmonary exacerbation showed distinct sputum metabolic profiles when compared to clinically stable patients or non-CF individuals [78]. Similarly, metabolic profiling of serum and urine allowed the discrimination between patients with COPD and healthy individuals [79]. In addition to these profiles, the volatile fraction of the metabolome composed by volatile organic compounds (VOCs) is also critical when considering the development of novel diagnostic approaches.

The recent availability of sophisticated analytical techniques such as gas chromatography mass spectrometry (GC-MS) and novel gas sensor devices (e-Nose) has provided the necessary tools to identify disease-specific VOCs in clinical matrices [80,81]. For example, higher alveolar concentrations of pentane, an alkane that belongs to the broader class of hydrocarbons, and lower levels of dimethyl sulfide allowed the accurate discrimination of CF patients from healthy individuals [82]. In COPD, decane and decane,6-ethyl-2-methyl were found to be present in higher levels [83], and more recently hexanal, an end metabolite of the lipid peroxidation of cell membrane phospholipids, was also significantly increased [84]. Although VOC analysis has been primarily evaluated in the screening of diseased versus healthy states, its ability to "measure" host immune responses suggests that it may be exploited in strategies aimed at the early detection of infection in patients with underlying lung diseases (Table 31.2).

Fungal Breath Fingerprinting (VOCs)

Fungal VOCs can also be detected in the breath of patients with lung infection [85] and used in diagnostic approaches. Initial studies have revealed that *Aspergillus* spp. and *Fusarium* spp. can produce 2-pentylfuran and that this metabolite is present in breath samples from patients with bronchiectasis and CF, or in those undergoing immune suppression [86,87]. More recently, the monoterpenes—camphene, α-pinene and limonene—and the sesquiterpenes—α-trans-bergamotene and β-transbergamotene—were found to be consistently emitted in a wide range of clinical and type fungal strains on a variety of growth substrates [88].

Other pathogenic *Aspergillus* species (e.g., *Aspergillus terreus*, *Aspergillus calidoustus*, *Aspergillus flavus*, and *Aspergillus niger*) also displayed unique VOC signatures thereby suggesting that the species-specific discrimination of IPA in the breath may allow guiding selection of antifungal therapy. In a clinical study, the sesquiterpene metabolites—α-trans-bergamotene and β-transbergamotene—in combination with a β-vatirenene-like sesquiterpene and trans-geranylacetone were able to identify patients with IPA with 94% sensitivity and 93% specificity [88].

While there was some overlap in sesquiterpene metabolites between in vitro cultures and patient breath samples, additional sesquiterpene metabolites or sesquiterpene derivatives were only identified in vivo, suggesting that the mold metabolism is also shaped by the host lung microbiome and immune response. It is therefore critically important to derive biomarker signatures that identify infections in vivo, with the pathogen metabolizing in the nutritional and immunological milieu of the host.

THE CLINICAL APPLICATION OF THE MICROBIOTA—METABOLOME CROSSTALK

The host—fungus interaction is currently being exploited to project more efficient and reliable fungal diagnostics [89], and efforts are being devoted to the implementation of clinical models, aimed at the prediction of infection in high-risk

TABLE 31.1 Examples of Volatile Organic Compounds Detected in Pulmonary Diseases

Disease	Metabolic Profiling	VOC	Reference
Lung cancer	PTR-MS	Isoprene ↓ Acetone ↓ Methanol ↓	[94]
Lung cancer	GC-MS and SPME	1-Methyl-4-(1-methylethyl)benzene ↓ Toluene ↑ Dodecane ↓ 3,3-Dimethyl pentane ↑ 2,3,4-Trimethyl hexane ↑	[95]
Lung cancer	GC-MS	Cyclohexane ↑ Xylene ↑	[96]
Lung cancer	GC-MS	CHN↑ Methanol↑ CH_3CN↑ Isoprene↑ 1-Propanol↑	[97]
CF	GC-MS	Pentane ↑ Dimethyl sulfide ↓	[82]
CF	GC-MS	C5–C16 hydrocarbons ↑ N-Methyl-2-methylpropylamine ↑	[98]
COPD	e-Nose and GC-MS	Decane ↑ Decane, 6-ethyl-2-methyl- ↑	[83]
COPD	GC-MS	hexanal ↑	[84]
AECOPD	TD-GC-MS	2-Pentanone ↑ Cyclohexanone ↑ 4-Heptanone ↑ n-Butane ↓	[99]
Asthma	e-Nose for smellprints and GC-MS for VOC	4-Methyloctane ↑ 2,4-Dimethylheptane ↑ Isopropanol ↑ Toluene ↑ Isoprene ↑ Alkane ↑ Acetic acid ↑ Acetone ↑ 2,6,11-Trimethyl dodecane ↑ 3,7-Dimethyl undecane ↑ 2,3-Dimethyl heptane ↑	[100]

AECOPD, acute exacerbation of COPD; *CF*, cystic fibrosis; *COPD*, chronic obstructive pulmonary disease; *GC-MS*, gas chromatography-MS; *PTR-MS*, proton transfer reaction-mass spectrometry; *SPME*, solid phase microextraction; *TD-GC-MS*, thermal desorption GC-MS; *VOC*, volatile organic compound.

TABLE 31.2 Examples of Volatile Organic Compounds Used to Discriminate Fungal Colonization and Infection in Patients With Different Underlying Pulmonary Diseases

Underlying Condition	Fungal Disease	Metabolic Profiling	VOC	Reference
CF	*Aspergillus* spp. colonization	GC-MS and SPME	2-Pentylfuran ↑	[87]
Bronchiectasis, CF, and immune suppression	*Aspergillus* spp. colonization	GC-MS	2-Pentylfuran ↑	[86]
Cancer and transplantation	IPA	GC-MS	Camphene ↑ α- and β-pinene ↑ Limonene↑ α-and β-trans-bergamotene ↑	[88]

CF, cystic fibrosis; *GC-MS*, gas chromatography-mass spectrometry; *IPA*, invasive pulmonary aspergillosis; *SPME*, solid phase microextraction; *VOC*, volatile organic compound.

patients [90]. For example, a specific set of alveolar cytokines was found to be differentially expressed in patients with IPA, allowing their accurate discrimination from uninfected, matched individuals [91].

Pulmonary complications and mortality of patients undergoing allogeneic hematopoietic stem-cell transplantation were found to be predicted by the expansion of gammaproteobacteria in the gut [92]. This study represents a critical example of how microbiota-derived information may be used in the management of patients at risk of fungal disease. Along the same line, the manipulation of the microbiota toward its homeostatic state is nowadays regarded as one important therapeutic option in patients suffering from infection, in some cases being more effective than antibiotherapy [93]. Likewise, the targeted manipulation of the lung microbiota using probiotics, applied either directly via inhalational route or indirectly via the intestinal route, may be a promising therapeutic option.

CONCLUDING REMARKS

Increasing evidence supports a role for gut and lung dysbiosis in deregulated immune responses and inflammation, leading to the disruption of the balance between fungal colonization and overt disease. Although the study of the lung microbiota remains an active field of research, discrepancies between studies in terms of sampling procedures, sequencing approaches, and heterogeneity of patients have precluded the definitive clinical application of microbiota detection and manipulation in the diagnosis and treatment of respiratory fungal diseases.

However, the bacterial populations that have been consistently involved in several clinical conditions associated with susceptibility to fungal disease could serve as a starting point for the clinical studies required to bridge fundamental and clinical research. Ultimately, by linking deep clinical phenotyping to molecular and genetic signatures of the host—fungus—microbiota interaction, novel approaches of personalized prognosis, diagnosis, and treatment of respiratory fungal diseases can be devised.

ACKNOWLEDGMENTS

This work was supported by a Mérieux Research Grant 2016 from Institut Mérieux, the Northern Portugal Regional Operational Programme (NORTE 2020), under the Portugal 2020 Partnership Agreement, through the European Regional Development Fund (FEDER) (NORTE-01-0145-FEDER-000013), and by Fundação para a Ciência e Tecnologia (FCT) (IF/00735/2014 to A.C., and SFRH/BPD/96176/2013 to C.C.).

REFERENCES

[1] Kontoyiannis DP, Marr KA, Park BJ, et al. Prospective surveillance for invasive fungal infections in hematopoietic stem cell transplant recipients, 2001—2006: overview of the Transplant-Associated Infection Surveillance Network (TRANSNET) Database. Clin Infect Dis 2010;50(8):1091—100.

[2] Pagano L, Caira M, Candoni A, et al. Invasive aspergillosis in patients with acute myeloid leukemia: a SEIFEM-2008 registry study. Haematologica 2010;95(4):644—50.

[3] Pappas PG, Alexander BD, Andes DR, et al. Invasive fungal infections among organ transplant recipients: results of the Transplant-Associated Infection Surveillance Network (TRANSNET). Clin Infect Dis 2010;50(8):1101—11.

[4] Brown GD, Denning DW, Gow NA, Levitz SM, Netea MG, White TC. Hidden killers: human fungal infections. Sci Transl Med 2012;4(165):165rv13.

[5] Pfaller MA, Diekema DJ. Epidemiology of invasive mycoses in North America. Crit Rev Microbiol 2010;36(1):1—53.

[6] Taccone FS, Van den Abeele AM, Bulpa P, et al. Epidemiology of invasive aspergillosis in critically ill patients: clinical presentation, underlying conditions, and outcomes. Crit Care 2015;19:7.

[7] Wauters J, Baar I, Meersseman P, et al. Invasive pulmonary aspergillosis is a frequent complication of critically ill H1N1 patients: a retrospective study. Intensive Care Medicine 2012;38(11):1761—8.

[8] King J, Brunel SF, Warris A. Aspergillus infections in cystic fibrosis. J Infect 2016;72(Suppl):S50—5.

[9] Denning DW, Cadranel J, Beigelman-Aubry C, et al. Chronic pulmonary aspergillosis: rationale and clinical guidelines for diagnosis and management. Eur Respir J 2016;47(1):45—68.

[10] Alanio A, Bretagne S. Challenges in microbiological diagnosis of invasive Aspergillus infections. F1000Res 2017;6.

[11] Perlin DS, Rautemaa-Richardson R, Alastruey-Izquierdo A. The global problem of antifungal resistance: prevalence, mechanisms, and management. Lancet Infect Dis 2017;17(12):e383—92.

[12] Goncalves SM, Lagrou K, Duarte-Oliveira C, Maertens JA, Cunha C, Carvalho A. The microbiome-metabolome crosstalk in the pathogenesis of respiratory fungal diseases. Virulence 2016:1—12.

[13] O'Dwyer DN, Dickson RP, Moore BB. The lung microbiome, immunity, and the pathogenesis of chronic lung disease. J Immunol 2016;196(12):4839—47.

[14] Dickson RP, Erb-Downward JR, Martinez FJ, Huffnagle GB. The microbiome and the respiratory tract. Annu Rev Physiol 2016;78:481—504.

[15] Marsland BJ, Gollwitzer ES. Host-microorganism interactions in lung diseases. Nat Rev Immunol 2014;14(12):827—35.

[16] Beck JM, Young VB, Huffnagle GB. The microbiome of the lung. Transl Res J Lab Clin Med 2012;160(4):258–66.

[17] Charlson ES, Bittinger K, Chen J, et al. Assessing bacterial populations in the lung by replicate analysis of samples from the upper and lower respiratory tracts. PLoS One 2012;7(9):e42786.

[18] Charlson ES, Bittinger K, Haas AR, et al. Topographical continuity of bacterial populations in the healthy human respiratory tract. Am J Respir Crit Care Med 2011;184(8):957–63.

[19] Dickson RP, Huffnagle GB. The lung microbiome: new principles for respiratory bacteriology in health and disease. PLoS Pathogens 2015;11(7):e1004923.

[20] Chilvers MA, Rutman A, O'Callaghan C. Functional analysis of cilia and ciliated epithelial ultrastructure in healthy children and young adults. Thorax 2003;58(4):333–8.

[21] Segal LN, Clemente JC, Tsay JC, et al. Enrichment of the lung microbiome with oral taxa is associated with lung inflammation of a Th17 phenotype. Nat Microbiol 2016;1:16031.

[22] Thaiss CA, Zmora N, Levy M, Elinav E. The microbiome and innate immunity. Nature 2016;535(7610):65–74.

[23] Belkaid Y, Hand TW. Role of the microbiota in immunity and inflammation. Cell 2014;157(1):121–41.

[24] Shukla SD, Budden KF, Neal R, Hansbro PM. Microbiome effects on immunity, health and disease in the lung. Clin Transl Immunol 2017;6(3):e133.

[25] Wang J, Li F, Sun R, et al. Bacterial colonization dampens influenza-mediated acute lung injury via induction of M2 alveolar macrophages. Nat Commun 2013;4:2106.

[26] Ichinohe T, Pang IK, Kumamoto Y, et al. Microbiota regulates immune defense against respiratory tract influenza A virus infection. Proc Natl Acad Sci USA 2011;108(13):5354–9.

[27] Huffnagle GB, Noverr MC. The emerging world of the fungal microbiome. Trends Microbiol 2013;21(7):334–41.

[28] Iliev ID, Funari VA, Taylor KD, et al. Interactions between commensal fungi and the C-type lectin receptor Dectin-1 influence colitis. Science 2012;336(6086):1314–7.

[29] Steele C, Rapaka RR, Metz A, et al. The beta-glucan receptor dectin-1 recognizes specific morphologies of Aspergillus fumigatus. PLoS Pathogens 2005;1(4):e42.

[30] Chai LY, de Boer MG, van der Velden WJ, et al. The Y238X stop codon polymorphism in the human beta-glucan receptor dectin-1 and susceptibility to invasive aspergillosis. J Infect Dis 2011;203(5):736–43.

[31] Sainz J, Lupianez CB, Segura-Catena J, et al. Dectin-1 and DC-SIGN polymorphisms associated with invasive pulmonary Aspergillosis infection. PLoS One 2012;7(2):e32273.

[32] Cunha C, Di Ianni M, Bozza S, et al. Dectin-1 Y238X polymorphism associates with susceptibility to invasive aspergillosis in hematopoietic transplantation through impairment of both recipient- and donor-dependent mechanisms of antifungal immunity. Blood 2010;116(24):5394–402.

[33] Kolwijck E, van de Veerdonk FL. The potential impact of the pulmonary microbiome on immunopathogenesis of Aspergillus-related lung disease. Eur J Immunol 2014;44(11):3156–65.

[34] Delsuc C, Cottereau A, Frealle E, et al. Putative invasive pulmonary aspergillosis in critically ill patients with chronic obstructive pulmonary disease: a matched cohort study. Crit Care 2015;19:421.

[35] Meersseman W, Lagrou K, Maertens J, Van Wijngaerden E. Invasive aspergillosis in the intensive care unit. Clin Infect Dis 2007;45(2):205–16.

[36] Cameron SJ, Lewis KE, Huws SA, et al. Metagenomic sequencing of the chronic obstructive pulmonary disease upper bronchial tract microbiome reveals functional changes associated with disease severity. PLoS One 2016;11(2):e0149095.

[37] Galiana A, Aguirre E, Rodriguez JC, et al. Sputum microbiota in moderate versus severe patients with COPD. Eur Respir J 2014;43(6):1787–90.

[38] Wang Z, Bafadhel M, Haldar K, et al. Lung microbiome dynamics in COPD exacerbations. Eur Respir J 2016;47(4):1082–92.

[39] Dickson RP, Erb-Downward JR, Huffnagle GB. The role of the bacterial microbiome in lung disease. Expet Rev Respir Med 2013;7(3):245–57.

[40] Huerta A, Soler N, Esperatti M, et al. Importance of Aspergillus spp. isolation in Acute exacerbations of severe COPD: prevalence, factors and follow-up: the FUNGI-COPD study. Respir Res 2014;15:17.

[41] Warren TA, Yau Y, Ratjen F, Tullis E, Waters V. Serum galactomannan in cystic fibrosis patients colonized with Aspergillus species. Med Mycol 2012;50(6):658–60.

[42] Hogan DA, Gladfelter AS. Editorial overview: host-microbe interactions: fungi: heterogeneity in fungal cells, populations, and communities. Curr Opin Microbiol 2015;26:vii–ix.

[43] Kim SH, Clark ST, Surendra A, et al. Global analysis of the fungal microbiome in cystic fibrosis patients reveals loss of function of the transcriptional repressor Nrg1 as a mechanism of pathogen adaptation. PLoS Pathogens 2015;11(11):e1005308.

[44] Willger SD, Grim SL, Dolben EL, et al. Characterization and quantification of the fungal microbiome in serial samples from individuals with cystic fibrosis. Microbiome 2014;2:40.

[45] Boisvert AA, Cheng MP, Sheppard DC, Nguyen D. Microbial biofilms in pulmonary and critical care diseases. Ann Am Thorac Soc 2016;13(9):1615–23.

[46] Mowat E, Paterson S, Fothergill JL, et al. *Pseudomonas aeruginosa* population diversity and turnover in cystic fibrosis chronic infections. Am J Respir Crit Care Med 2011;183(12):1674–9.

[47] Mowat E, Rajendran R, Williams C, et al. *Pseudomonas aeruginosa* and their small diffusible extracellular molecules inhibit Aspergillus fumigatus biofilm formation. FEMS Microbiology Letters 2010;313(2):96–102.

[48] Sass G, Nazik H, Penner J, et al. Studies of *Pseudomonas aeruginosa* mutants indicate pyoverdine as the central factor in inhibition of Aspergillus fumigatus biofilm. J Bacteriol 2018;200(1).

[49] Baxter CG, Rautemaa R, Jones AM, et al. Intravenous antibiotics reduce the presence of Aspergillus in adult cystic fibrosis sputum. Thorax 2013;68(7):652–7.

[50] Reece E, Segurado R, Jackson A, McClean S, Renwick J, Greally P. Co-colonisation with Aspergillus fumigatus and *Pseudomonas aeruginosa* is associated with poorer health in cystic fibrosis patients: an Irish registry analysis. BMC Pulm Med 2017;17(1):70.

[51] Garlanda C, Jaillon S, Doni A, Bottazzi B, Mantovani A. PTX3, a humoral pattern recognition molecule at the interface between microbe and matrix recognition. Curr Opin Immunol 2016;38:39–44.

[52] Chiarini M, Sabelli C, Melotti P, et al. PTX3 genetic variations affect the risk of *Pseudomonas aeruginosa* airway colonization in cystic fibrosis patients. Gene Immun 2010;11(8):665–70.

[53] Cunha C, Aversa F, Lacerda JF, et al. Genetic PTX3 deficiency and aspergillosis in stem-cell transplantation. N Engl J Med 2014;370(5):421–32.

[54] Cunha C, Monteiro AA, Oliveira-Coelho A, et al. PTX3-Based genetic testing for risk of aspergillosis after lung transplant. Clin Infect Dis 2015;61(12):1893–4.

[55] Wojtowicz A, Lecompte TD, Bibert S, et al. PTX3 polymorphisms and invasive mold infections after solid organ transplant. Clin Infect Dis 2015;61(4):619–22.

[56] He Q, Li H, Rui Y, et al. Pentraxin 3 gene polymorphisms and pulmonary aspergillosis in chronic obstructive pulmonary disease patients. Clin Infect Dis 2018;66(2):261–7.

[57] Hogaboam CM, Carpenter KJ, Schuh JM, Buckland KF. Aspergillus and asthma—any link? Med Mycol 2005;43(Suppl. 1):S197–202.

[58] Ghosh S, Hoselton SA, Dorsam GP, Schuh JM. Eosinophils in fungus-associated allergic pulmonary disease. Front Pharmacol 2013;4:8.

[59] Denning DW, Pashley C, Hartl D, et al. Fungal allergy in asthma-state of the art and research needs. Clin Transl Allergy 2014;4:14.

[60] Chaudhary N, Marr KA. Impact of Aspergillus fumigatus in allergic airway diseases. Clin Transl Allergy 2011;1(1):4.

[61] Fairs A, Agbetile J, Hargadon B, et al. IgE sensitization to Aspergillus fumigatus is associated with reduced lung function in asthma. Am J Respir Crit Care Med 2010;182(11):1362–8.

[62] Menzies D, Holmes L, McCumesky G, Prys-Picard C, Niven R. Aspergillus sensitization is associated with airflow limitation and bronchiectasis in severe asthma. Allergy 2011;66(5):679–85.

[63] Huang YJ, Nariya S, Harris JM, et al. The airway microbiome in patients with severe asthma: associations with disease features and severity. J Allergy Clin Immunol 2015;136(4):874–84.

[64] van Woerden HC, Gregory C, Brown R, Marchesi JR, Hoogendoorn B, Matthews IP. Differences in fungi present in induced sputum samples from asthma patients and non-atopic controls: a community based case control study. BMC Infect Dis 2013;13:69.

[65] Agbetile J, Fairs A, Desai D, et al. Isolation of filamentous fungi from sputum in asthma is associated with reduced post-bronchodilator FEV1. Clin Exp Allergy J Br Soc Allergy Clin Immunol 2012;42(5):782–91.

[66] Hooper LV, Littman DR, Macpherson AJ. Interactions between the microbiota and the immune system. Science 2012;336(6086):1268–73.

[67] Blumberg R, Powrie F. Microbiota, disease, and back to health: a metastable journey. Sci Transl Med 2012;4(137):137rv7.

[68] McKenzie CI, Mackay CR, Macia L. GPR43-A prototypic metabolite sensor linking metabolic and inflammatory diseases. Trends Endocrinol Metab 2015;26(10):511–2.

[69] Maslowski KM, Vieira AT, Ng A, et al. Regulation of inflammatory responses by gut microbiota and chemoattractant receptor GPR43. Nature 2009;461(7268):1282–6.

[70] Trompette A, Gollwitzer ES, Yadava K, et al. Gut microbiota metabolism of dietary fiber influences allergic airway disease and hematopoiesis. Nat Med 2014;20(2):159–66.

[71] Zelante T, Iannitti RG, Cunha C, et al. Tryptophan catabolites from microbiota engage aryl hydrocarbon receptor and balance mucosal reactivity via interleukin-22. Immunity 2013;39(2):372–85.

[72] De Luca A, Carvalho A, Cunha C, et al. IL-22 and Ido1 affect immunity and tolerance to murine and human vaginal candidiasis. PLoS Pathogens 2013;9(7):e1003486.

[73] Jang SO, Kim HJ, Kim YJ, et al. Asthma prevention by *Lactobacillus rhamnosus* in a mouse model is associated with CD4(+)CD25(+)Foxp3(+) T cells. Allergy Asthma Immunol Res 2012;4(3):150–6.

[74] Karimi K, Inman MD, Bienenstock J, Forsythe P. Lactobacillus reuteri-induced regulatory T cells protect against an allergic airway response in mice. Am J Respir Crit Care Med 2009;179(3):186–93.

[75] Iannitti RG, Carvalho A, Cunha C, et al. Th17/Treg imbalance in murine cystic fibrosis is linked to indoleamine 2,3-dioxygenase deficiency but corrected by kynurenines. Am J Respir Crit Care Med 2013;187(6):609–20.

[76] Quinn RA, Phelan VV, Whiteson KL, et al. Microbial, host and xenobiotic diversity in the cystic fibrosis sputum metabolome. ISME J 2016;10(6):1483–98.

[77] Ghidoni R, Caretti A, Signorelli P. Role of sphingolipids in the pathobiology of lung inflammation. Mediat Inflamm 2015;2015:487508.

[78] Twomey KB, Alston M, An SQ, et al. Microbiota and metabolite profiling reveal specific alterations in bacterial community structure and environment in the cystic fibrosis airway during exacerbation. PLoS One 2013;8(12):e82432.

[79] Wang L, Tang Y, Liu S, et al. Metabonomic profiling of serum and urine by (1)H NMR-based spectroscopy discriminates patients with chronic obstructive pulmonary disease and healthy individuals. PLoS One 2013;8(6):e65675.

[80] Sethi S, Nanda R, Chakraborty T. Clinical application of volatile organic compound analysis for detecting infectious diseases. Clin Microbiol Rev 2013;26(3):462–75.

[81] Shirasu M, Touhara K. The scent of disease: volatile organic compounds of the human body related to disease and disorder. J Biochem 2011;150(3):257–66.

[82] Barker M, Hengst M, Schmid J, et al. Volatile organic compounds in the exhaled breath of young patients with cystic fibrosis. Eur Respir J 2006;27(5):929−36.

[83] Cazzola M, Segreti A, Capuano R, et al. Analysis of exhaled breath fingerprints and volatile organic compounds in COPD. COPD Res Pract 2015;1(1).

[84] Jareno-Esteban JJ, Munoz-Lucas MA, Gomez-Martin O, et al. Study of 5 volatile organic compounds in exhaled breath in chronic obstructive pulmonary disease. Arch Bronconeumol 2017;53(5):251−6.

[85] Heddergott C, Calvo AM, Latge JP. The volatome of Aspergillus fumigatus. Eukaryot Cell 2014;13(8):1014−25.

[86] Chambers ST, Syhre M, Murdoch DR, McCartin F, Epton MJ. Detection of 2-pentylfuran in the breath of patients with *Aspergillus fumigatus*. Med Mycol 2009;47(5):468−76.

[87] Syhre M, Scotter JM, Chambers ST. Investigation into the production of 2-Pentylfuran by Aspergillus fumigatus and other respiratory pathogens in vitro and human breath samples. Med Mycol 2008;46(3):209−15.

[88] Koo S, Thomas HR, Daniels SD, et al. A breath fungal secondary metabolite signature to diagnose invasive aspergillosis. Clin Infect Dis 2014;59(12):1733−40.

[89] Oliveira-Coelho A, Rodrigues F, Campos Jr A, Lacerda JF, Carvalho A, Cunha C. Paving the way for predictive diagnostics and personalized treatment of invasive aspergillosis. Front Microbiol 2015;6:411.

[90] Stanzani M, Lewis RE, Fiacchini M, et al. A risk prediction score for invasive mold disease in patients with hematological malignancies. PLoS One 2013;8(9):e75531.

[91] Gonçalves SM, Lagrou K, Rodrigues CS, et al. Evaluation of bronchoalveolar lavage fluid cytokines as biomarkers for invasive pulmonary aspergillosis in at-risk patients. Front Microbiol 2017;8:2362.

[92] Harris B, Morjaria SM, Littmann ER, et al. Gut microbiota predict pulmonary infiltrates after allogeneic hematopoietic cell transplantation. Am J Respir Crit Care Med 2016;194(4):450−63.

[93] van Nood E, Vrieze A, Nieuwdorp M, et al. Duodenal infusion of donor feces for recurrent *Clostridium difficile*. N Engl J Med 2013;368(5):407−15.

[94] Bajtarevic A, Ager C, Pienz M, et al. Noninvasive detection of lung cancer by analysis of exhaled breath. BMC Canc 2009;9:348.

[95] Peng G, Hakim M, Broza YY, et al. Detection of lung, breast, colorectal, and prostate cancers from exhaled breath using a single array of nanosensors. Br J Cancer 2010;103(4):542−51.

[96] Oguma T, Nagaoka T, Kurahashi M, et al. Clinical contributions of exhaled volatile organic compounds in the diagnosis of lung cancer. PLoS One 2017;12(4):e0174802.

[97] Sakumura Y, Koyama Y, Tokutake H, et al. Diagnosis by volatile organic compounds in exhaled breath from lung cancer patients using support vector machine algorithm. Sensors 2017;17(2).

[98] Robroeks CM, van Berkel JJ, Dallinga JW, et al. Metabolomics of volatile organic compounds in cystic fibrosis patients and controls. Pediatr Res 2010;68(1):75−80.

[99] Pizzini A, Filipiak W, Wille J, et al. Analysis of volatile organic compounds in the breath of patients with stable or acute exacerbation of chronic obstructive pulmonary disease. J Breath Res 2018;12(3):036002.

[100] Dragonieri S, Schot R, Mertens BJ, et al. An electronic nose in the discrimination of patients with asthma and controls. J Allergy Clin Immunol 2007;120(4):856−62.

Chapter 32

The Oral, Genital and Gut Microbiome in HIV Infection

P. Pérez-Matute[1], M. Íñiguez[1], M.J. Villanueva-Millán[1] and J.A. Oteo[1,2]

[1]Infectious Diseases, Microbiota and Metabolism Unit, Infectious Diseases Department, Center for Biomedical Research of La Rioja (CIBIR), Logroño, Spain; [2]Infectious Diseases Department, Hospital San Pedro, Logroño, Spain

INTRODUCTION

The human body is an excellent culture vessel, providing nutrients and a hospitable environment that support the growth of a wide variety of microorganisms. Collectively, these microbial species constitute what is known as human microbiota, or microbial flora. In *stricto sensu*, it is composed of millions of microorganisms belonging not only to bacteria, but also to viruses, fungi, protozoa, and archaea. However, the term *microbiota* is commonly used to describe the bacterial population only. Therefore, the populations of viruses and fungi are described as *virome* and *mycobiome*, respectively [1]. The term "microbiome" is used to describe the whole genetic information of a microbial complex. In this chapter, we will focus on bacteria, with a limited description of the virome in HIV infection.

Human microbiota plays important roles in health, as described elsewhere [2,3]. It is considered as a metabolic "organ" very well-tuned to our physiology. The prompt development of next-generation sequencing technologies [4,5], and the increasing number of articles published in this field demonstrates that the microbiota, its evolutive dynamics, and the influence on the host through its protective, trophic, and metabolic actions, has a key role in health. However, when disrupted, the same indigenous microbes can cause disease. Because of that, the study of microbiota opens unique opportunities for the identification of new markers of the pathophysiological state of each individual [1,6].

HIV INFECTION AND MICROBIOTA

Alterations of Microbiota at the Genital and Rectal Sites in HIV-Infection

The mucosal immune system of the female reproductive tract is one of the first lines of defense against HIV-infection and other pathogens [7]. Lying superficial to the epithelial cells of the vaginal tract is the microbiota, which exists in a symbiotic relationship with the female host. This vaginal microbiota living on the mucus layer helps to inactivate the HIV by secreting H_2O_2 or by decreasing the pH of the environment, among other mechanisms [8]. In fact, lactic acid and short-chain fatty acids (SCFAs) produced by the vaginal microbiota have reportedly antimicrobial and immune modulatory activities, indicating their potential as biomarkers of disease and/or disease susceptibility [9]. Microbiota present in female genital tract is also associated with vaginal antiretroviral drug concentrations in HIV-infected women on antiretroviral therapy (ART) [10].

The current concept of a "healthy" vaginal microbiota includes a low-diversity, *Lactobacillus* rich environment. Four species of *Lactobacilli* (Lactobacillus *crispatus*, Lactobacilli. *gasseri*, Lactobacilli. *iners*, and *Lactobacilli. jensenii*) are known to be the most common dominant species in the vaginal microbiome of Caucasian, Asian, Black, and Hispanic women [11]. However, alterations of diet, inflammation, menstrual cycle, usage of hormonal contraceptives, and infection by other viruses such as HPV (human papillomavirus) or HSV (herpes simplex virus) will affect the composition and activity of the vaginal microbiota, which may enhance the chance of HIV infection [12].

Bacterial vaginosis is a symptomatic clinical condition characterized by a significant reduction of *Lactobacillus* (*Lactobacillus iners* and *Lactobacillus crispatus*) populations, and overgrowth of anaerobes such as *Prevotella bivia* and *Lachnospiraceae* [13]. Although bacterial vaginosis has been consistently linked to increased susceptibility to HIV infection [13−15], it is becoming increasingly clear that bacterial diversity, even in the absence of bacterial vaginosis, might also confer greater susceptibility to disease [16−18].

Bacterial diversity is also a key target that has been investigated in three studies focused on rectal mucosa microbiota, although with disparate results (reviewed by Dubourg et al.) [19]. Yu et al. [20] also reported reduced alpha diversity in anal microbiota from HIV-positive men, compared with negative subjects, partially attributable to the use of antibiotics. A study carried out in stomach fluid also observed a reduced bacterial diversity in HIV-infected people [21]. Concerning specific alterations in microbiota, the study of Mutlu et al. [22] demonstrated that HIV patients under ART presented decreased microbial diversity in the right colon and terminal ileum compared to uninfected controls with the loss of commensals, as well as a gain of some pathogenic bacterial taxa.

Higher proportions of Lactobacillales in distal gut of HIV-infected men were also found to be associated with lower markers of bacterial translocation (BT) (see Section 2.5 of the chapter), higher CD4 + T lymphocytes, and lower viral loads, before ART was started. During ART, higher proportions of gut Lactobacillales were associated with higher CD4 percentage in plasma, less BT, less systemic immune activation, less gut T lymphocyte proliferation, and higher CD4 in the gut [23], corroborating the role of Lactobacillales in HIV infection in distal gut similar to what was observed in the female genital tract. Finally, among species belonging to the *Bacteroides* genus, *Bacteroides fragilis*, which has been found to be significantly reduced in HIV mucosal samples [24], as well in colon biopsies [25], may also play a key role in the persistence of the adaptive immune system [26].

Alterations of Microbiota in Blood, Semen, and Brain in HIV Infection

Human plasma in healthy individuals has been assumed to be sterile, and although recent studies suggest that pleomorphic bacteria exist in the blood of healthy humans [27], a very recent study has demonstrated that the bacteria-like vesicles and refringent particles observed represent nonliving membrane vesicles and protein aggregates derived from blood [28]. This might not be the case in immunocompromised hosts. The amount of bacterial DNA was lower in healthy controls than in HIV-infected patients without receiving ART, as observed in a previous study [29]. In addition, huge amounts of bacterial DNA were identified in HIV-infected patients, being Pseudomonadales the predominant component. Furthermore, the bacterial elements found in blood were very similar than those living in the gastrointestinal tract, suggesting BT [29].

Following the previous idea that highlights the role of *Lactobacillus* in susceptibility to HIV infection in genital tract, the study by Merlini et al. [30] also suggests that the lack of *Lactobacilli* in PBMCs (peripheral blood mononuclear cells) could be related to the null or partial immune response of patients against ART.

HIV could also be transmitted through semen and, to our knowledge, only one group has analyzed the semen microbiome and its relationship with local immunology and viral load in HIV infection [32]. The authors state that semen microbiome has an important role in HIV sexual transmission. Thus, this is an interesting area that deserves further investigation.

Another organ that has been considered to be sterile is the brain. However, the study by Branton et al. [31] showed that Proteobacteria predominate in primate and human brains, regardless of underlying immune status. Unfortunately, there are not a lot of publications in this field.

Alterations of Microbiota in Oral Cavity and Airways in HIV Infection

HIV infection is associated with alterations in the respiratory microbiota [33,34]. Most of the studies focused on lung microbiota derived from the *Lung HIV Microbiome Project* (LHMP) [35]. This project was driven by the recognition that pulmonary complications continued to be a major cause of morbidity in HIV-infected individuals, even in the era of ART [36]. To date, evidence suggests that lung microbiota in healthy HIV-infected individuals with preserved CD4 counts is similar to uninfected individuals. However, in individuals with more advanced disease, there is an altered alveolar microbiota, characterized by a loss of richness and diversity within individuals, but an increase in beta diversity differences between individuals.

These differences decline with ART, but even after effective therapy, the alveolar microbiota in some HIV-infected individuals contain increased amounts of signature bacteria, some of which have been previously associated with chronic lung inflammation [34]. Other studies have failed to find out such differences in lung microbiota [37]. A very recent study has suggested that microbial communities and their interactions with the host may have functional metabolic impact in the lung [38].

HIV infection has also been associated with a variety of oral manifestations, and oral microbiota has been suggested to be involved in such HIV-related complications [39,40]. Thus, Dang et al. [41] observed a shift of microbial composition in the lingual region, which was related to the viral load in early stage HIV patients. ART therapy also has significant effects on salivary microbial colonization [39]. In addition, the site of collection of oral microbiota could be determinant, as plaque and saliva showed a distinct microbial composition, and, only in saliva, minor but significant differences were observed, when comparing non-HIV infected patients with HIV-infected patients [42].

Bacterial Translocation and Alterations of Gut Microbiota in HIV Infection

The gastrointestinal tract is a major site of HIV replication, independent of the way of acquisition of the virus [43]. Gut-associated lymphoid tissue (GALT) is where the majority of CD4 + T lymphocytes are housed [44,45]. These lymphocytes, in contrast to what happens in peripheral blood mononuclear cells, are memory T lymphocytes in a continuous state of activation, which makes them very susceptible to HIV infection [46]. In fact, HIV infection is mainly considered as a gut disease, as it significantly depletes CD4 T cells from mucosal sites, particularly from GALT [47].

The first stage in HIV infection is precisely the preferential depletion of CD4 + cells that express the coreceptors CCR5 and CXR4 that reside in the intestine, although it subsequently affects all CD4 + T cells in the body. After ART, a significant restoration of CD4 + cells in the peripheral blood is observed, which contrasts with the partial and slow restoration observed in the cells of the gastrointestinal tract [48]. The reason for this poor restoration at the intestinal level is still unknown. The infection of macrophages present in the intestinal mucosa also favors the loss of the CD4 + cells, due to transmission of the virus "cell by cell," and also to cytokine secretion that attracts and recruits activated T lymphocytes, which subsequently are infected, enhancing intestinal infection (reviewed by Nasi et al. [49]). HIV infection also depletes mucosal Th17 cells that play a key role in the antimicrobial defense [50,51].

Depletion of gastrointestinal CD4+ T lymphocytes during HIV infection is followed by alterations of lymphoid tissue architecture, as well as integrity and function of the mucosal barrier. Loss of immune protection of the intestinal mucosa allows translocation of microbial products (such as lipopolysaccharides/LPS) into the lamina propria of the gastrointestinal tract and, eventually, into the systemic circulation. This is known as bacterial translocation, which was first described in HIV-infected patients in 2006 [52]. This BT has been shown to persist throughout the course of the infection, and contributes to the immune activation that is observed in the chronic phases of the infection.

These bacterial products (LPS) induce a significant increase in proinflammatory cytokine production via toll-like receptors (TLR), contributing to immune activation and inflammation. There are other cell wall receptors that also recognize LPS, and induce an inflammatory response such as the MSR (macrophage scavenger receptor), K+ channels, and CD11/CD18 receptors. In all cases, the binding causes the activation of kinases, and promotes the mobilization of NF-Kβ, which, in turn, activates the genes responsible for the inflammatory response such as TNFα, IL-1, or IL-6 (Fig. 32.1).

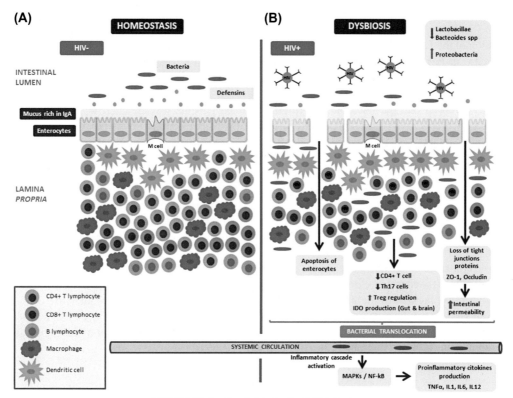

FIGURE 32.1 Gut dysbiosis after HIV infection.

The degree of BT is linked to the severity of HIV progression, independently of viremia, and several studies have demonstrated that increased BT and proinflammatory cytokines are partially responsible for HIV-related comorbidities [53,54], which implies a significant increase in morbidity and mortality.

Gut Microbiota Composition: Bacterial Diversity/Richness

Most of the studies carried out in HIV-infected people have been focused on gut microbiota. Thus, lots of studies will be cited and discussed in the following section. HIV infection has been mostly associated with reduced bacterial diversity [22,55,56], although others have observed a significant diversity increase in HIV + (naïve) patients compared to HIV + (cART) [57]. Our results show a significant and clear collapse in α-diversity in HIV + (naive) compared with the controls (non-HIV-infected subjects) [58] as occurs in other pathologies (Tables 32.1 and 32.2).

Microbial Composition

Depletion in Clostridia class and *Bacteroides* spp., an overgrowth of *Enterobacteriaceae*, and an increase in the genus *Prevotella* have been widely observed (summarized in Dubourg et al. [19]).

More specifically, it has been demonstrated that fecal microbiota of naïve HIV patients had a higher abundance of *Prevotellaceae* (*Prevotella*), *Erysipelotrichaceae* (*Catenibacterium* and *Bulleidia*), *Veillonellaceae* (*Dialister* and *Mitsuokella*), *Clostridium cluster XIII*, and the genus *Desulfovibrio*, compared with a non-HIV population. In contrast, the control population had higher amounts of *Bacteroidaceae* (*Bacteroides*), *Rikenellaceae* (*Alistipes*), and *Porphyromonadaceae* (*Parabacteroides*) [57]. Similarly, a pilot study observed, through real-time PCR, an increase in the proportion of Enterobacteriales and Bacteroidales in naïve HIV patients, and demonstrated important associations between this different bacterial composition and systemic immunity parameters in the HIV patients [59].

More recent studies have shown a greater abundance of Proteobacteria (which promote inflammation), and a lower abundance of Bacteroidia (known to limit inflammation) in naïve HIV patients, which has been associated with increased T cell activation, a lower secretion of lymphocytes IL-17/IL-22, and a greater presence of inflammatory markers [24]. A lower abundance of several *Bacteroides* species influences and impairs the functions of the invariant natural killer T cells, which are known to limit BT and chronic pathologic immune activation in HIV-1 infection [60]. Finally, among the Bacteroidales, the *Rikenellaceae* family was reported to be decreased during HIV infection, and the genus *Alistipes* seems the most affected, although these data have not been interpreted yet [19].

The study by Dillon et al. [25] also found a greater presence of Proteobacteria and, on the contrary, a lower abundance of Firmicutes in naïve HIV patients. Among the Firmicutes phylum, particular consideration had been given during the last years to *Faecalibacterium prausnitzii*, an obligate anaerobe belonging to *Lachnospiraceae*, with known antiinflammatory properties. Depletion of the *Lachnospiraceae* family has been reported during HIV infection [22,25,61]. Our group observed a significant decrease in *F. prausnitzii* in HIV-infected patients under ART, specifically in those under a combination of nucleoside reverse transcriptase inhibitors + protease inhibitors/NRTIs + PIs, suggesting loss of protection and persistent inflammation [58].

Mutlu et al. [22] showed that gut microbiota from HIV-infected patients was enriched with a number of potentially pathogenic bacteria such as *Prevotella*, as previously mentioned, and, in contrast, has poor content of the commensal *Bacteroides*, similar to what was observed by McHardy et al., [55]. In the same line, Gori et al. [62] reported the predominance of opportunistic pathogens in fecal flora of HIV-infected subjects. *Candida albicans* and *Pseudomonas aeruginosa* were overrepresented in the early stage of the infection. On the other hand, the abundance of protective bacteria such as *Bifidobacteria* and *Lactobacilli* was decreased [62].

The Firmicutes/Bacteroidetes ratio has also been increased in HIV patients (naïve and under ART) [63]. Furthermore, HIV + elite controllers (those that have the capacity to control viremia in the absence of ART) have enriched genera such as *Succinivibrio*, *Sutterella*, *Rhizobium*, *Delftia*, *Anaerofilum*, and *Oscillospira*, but depleted in *Blautia* and *Anaerostipes* [64].

However, some contradictory results have arisen, especially with respect to the *Prevotella* genus. While some studies have observed a significant increase in chronic HIV infection [65], others have failed to find such increase [24,56]. This may be due, among other factors, to the different methodology used (PCR vs. massive sequencing of the 16s and the region chosen from 16s), to the characteristics of the population (recent infection vs. chronic vs. naïve or low ART), etc. Other factors such as sexual orientation, the way of HIV acquisition/infection, or addiction to drugs can also contribute to the different composition of gut microbiota in HIV patients [20,66,67].

TABLE 32.1 Metagenomic Studies in HIV-Infection and Impact on Microbial Diversity/Richness

Author	Sample	Control	HIV + (naïve)	HIV + (cART)	α-Diversity/Richness	Is ART Able to Restore Gut Microbiota Diversity?
Gastrointestinal Tract						
Lozupone et al. [57]	Stool	13	3 recent HIV-infected + 11 chronic HIV-infected untreated	6 (short-term) + 8 (long-term)	Untreated HIV-positive patients had significantly higher alpha diversity compared to ART-treated and HIV-negative individuals.	Partially
McHardy et al. [55]	Rectal mucosal	20	20	20	HIV infection resulted in a slight reduction of alpha diversity in untreated HIV-positive patient subjects that was reversed to near equivalence with negative individuals in ART-treated subjects.	Partially
Vujkovic-Cvijin et al. [24]	Recto sigmoid biopsy	9	6 HIV-infected untreated and 1 HIV-infected long-term nonprogressor	18	No significant differences between HIV-positive and negative individuals.	No
Dillon et al. [25]	Rectal swab, rectal aspirate, and colon biopsies	14	18	–	No significant difference was found.	–
Mutlu et al. [22]	Stool and colon and terminal ileum biopsies	22	–	21	HIV-positive patients had less alpha diversity than negative individuals.	–
Nowak et al. [56]	Stool	9	28 (19 were analyzed after ART introduction) + 3 elite controllers	19	Decreased alpha diversity in untreated HIV-positive patients compared to negative individuals.	Partially
Ling et al. [63]	Stool	16	35	32	No significant difference was found.	Partially
Villanueva-Millan et al. [58]	Stool analysis	21	5	45	HIV infection decreased alpha diversity.	Partially
Oral Cavity and Airway						
Li et al. [39]	Saliva	10	10	10 (same naïve patients after 6 months on HAART)	Decreased oral biodiversity in HIV-infected persons relative to uninfected controls.	Partially
Kistler et al. [42]	Saliva Plaque	37	–	37	Reduced bacterial diversity in the saliva of HIV-positive patients compared to HIV-negative individuals.	–

Continued

TABLE 32.1 Metagenomic Studies in HIV-Infection and Impact on Microbial Diversity/Richness—cont'd

Author	Sample	Control	HIV + (naïve)	HIV + (cART)	α-Diversity/Richness	Is ART Able to Restore Gut Microbiota Diversity?
Dang et al. [41]	Oral swab of the dorsal tongue surface	9	6	6	Chronic HIV infection leads to substantial disruptions in the community structure of the lingual microbiota although no significant differences in the number of bacterial species was found between treated or untreated HIV infected groups and healthy controls.	–
Beck et al. [37]	BAL Oral washes	86 67	18 14	38 30	No significant differences were found among groups.	No
Cribbs et al. [38]	BAL	20		39	No significant differences were found between the lung microbiota of HIV-infected and HIV-uninfected groups.	–
Semen						
Liu et al. [32]	Semen	22	27	27 (same naïve patients after 6 months on ART)	HIV infection was associated with decreased semen microbiome diversity.	Yes

TABLE 32.2 Other Metagenomic Studies Focused on Microbial Diversity

Author	Sample	Control	HIV+	α-Diversity
Gastrointestinal Tract				
Yu et al. [20]	2 rectal swab (1−5 weeks)	32 samples 1 41 samples 2	41	No significant differences were found when examining sample 1. Reduced alpha diversity in sample 2.
Noguera-Julian et al. [66]	Stool	57	296	Decreased alpha diversity in HIV-positive patients compared to negative individuals.
von Ronsenvinge et al. [21]	Stomach fluid	21	4	Reduced bacterial diversity in HIV-infected people.
Brain				
Branton et al. [31]	Brain tissue	6	4	No significant differences were found among groups.
Genital Site				
Gosmann et al. [18]	Recto-cervical swab	205	31	HIV acquisition is increased in women with high-diversity, low Lactobacillus abundance.

Influence of ART

After ART, intestinal damage as well as BT are partially, but not completely, improved [68]. In fact, it has been observed that both the so-called "partial responders," and those designated as "non-responders" (do not reach 250 CD4+), have similar plasma levels of LPS and sCD14 after 12 months on ART and also without ART [30]. Several bacteria in long-term HIV individuals have been shown to return to levels comparable to HIV uninfected subjects, such as *Peptococcus*, *Catenibacterium*, and *Desulfovibrio* spp., whereas *Bacteroides* and *Prevotella* remain more similar to untreated patients [57], although again, the results regarding *Prevotella* are not conclusive [56].

To date, only three research groups have evaluated the effects of distinct ARV combinations, rather than ART as a whole (reviewed by Pinto-Cardoso et al. [69]). The study of Nowak et al. [56] addressed the role of nonnucleoside reverse transcriptase inhibitors (NNRTIs) versus protease inhibitors (PIs) on gut microbiota composition, although no differences were observed. However, both Pinto-Cardoso et al. [58] and Villanueva-Millan et al. [69] concluded that the least favorable ART combination was PI as it increased BT and endothelial damage, and reduced bacterial diversity, and this was associated with increased inflammation. Furthermore, Villanueva-Millán et al. demonstrated that INSTI-based regimen (integrase strand transfer inhibitors), and specially, Raltegravir-based regimen, is associated with levels of systemic inflammation, sCD14 plasma levels, and microbial diversity similar to uninfected controls, suggesting a healthier gut, and potentially fewer HIV-related complications.

One possible explanation for the differential effects of ARV drugs on gut microbiota and markers of BT and inflammation could be the differential penetration of ARV drugs on the gastrointestinal tissue and their pharmacokinetics. In fact, Raltegravir has been shown to penetrate faster in the gastrointestinal tract [70]. Furthermore, the superior capacity of ART with INSTIs to restore gut microbiota diversity may be due to the fact that INSTIs induce a greater reduction in proviral DNA, which could lead to rapid immunologic reconstitution.

Our group has also carried out a pilot study in an animal model of obesity where the effects of Maraviroc (MVC), a CCR5 antagonist, were analyzed. This antiretroviral was able to induce several changes in the abundance of the main bacterial orders present in the gastrointestinal tract. In addition, MVC was able to reverse in many cases the effects caused by a high fat diet, which seems to emphasize the direct effects of this antagonist of CCR5 on gut microbiota, and also its potential usage in some pathologies that appear frequently in HIV infection, such as obesity [71]. However, its effects on HIV-infected individuals have not been tested yet.

HIV Infection and Virome

The virome is composed by viruses that infect eukaryotic cells and bacteriophages that infect others members of the microbiota. Handley et al. [72] used WMG (whole metagenomics sequencing) to define the enteric virome of rhesus monkeys during simian immunodeficiency virus (SIV) infection. While no changes in bacterial microbiome composition were detected, pathogenic infection was associated with expansion of the enteric virome that disrupted the intestinal epithelial lining and, in turn, activated the immune system, contributing to the progression of SIV infection to AIDS. Adenovirus, parvovirus, calicivirus, picornavirus, and polyomavirus, all of them inductors of diseases, were detected in this study. Similarly, enteric virome dysbiosis was found in wild chimpanzees infected with SIV [73].

In agreement with these studies, an enteric expansion of adenovirus, independent of ART therapy, has been described in HIV-infected subjects with low CD4 T cell counts [74], suggesting that alterations in the enteric virome are associated with immunodeficiency in progressive HIV infection in humans. Progression of AIDS has also been associated to changes in plasma virome in HIV-infected patients from the United States and Uganda [75]. Viral sequences of HIV, hepatitis G virus (GBV-C), hepatitis B virus (HBV), hepatitis C virus (HCV), anellovirus, and human endogenous retrovirus (HERV) were detected in the plasma of these patients. This study also found that anellovirus levels were higher in patients with low CD4 T cell counts, suggesting that AIDS was associated with reduced immunological control of anellovirus replication.

HERV virus, mainly HERV-H and HERV-W, which have been reported to be upregulated in autoimmune disorders such as multiple sclerosis [76] were also described to be significantly elevated in Ugandan low CD4 T cell patients. Another study reported the presence of a large proportion of bacteriophages and endogenous retroviruses in naïve HIV/AIDS patients from China [29]. However, no anellovirus presence was reported by these authors. In addition, this study only evaluated DNA viral sequences; thus, RNA viruses like HIV itself could not be detected [29].

Anelloviruses were detected in the respiratory tract of naïve HIV-infected patients. In addition, a small percentage of bacteriophages, predominantly phages of *Enterobacteria, Salmonella, Pseudomonas, Streptococcus,* and *Yersinia* were also detectable [77]. In preliminary data obtained with shotgun sequencing on DNA preps, the presence of lymphocryptovirus in the bronchoalveolar lavage (BAL) of HIV-infected individuals, which was associated with higher concentrations of the inflammatory chemokines CXCL10 and CXCL9, was demonstrated. In another study performed with shotgun sequencing on DNA and RNA preps, bacteriophages—Herpesviridae, Retroviridae, and Parvoviridae—were found in lung and blood compartments of HIV-treated patients.

Anelloviruses were also found in blood compartments, but unlike the study of Young et al. [77] no annelloviruses were detected in the lung of HIV patients, maybe attributable to a healthier state of the subjects. This investigation also pointed to the fact that several DNA virus were found in RNA preparations, indicating an active replication of these viruses.

HIV INFECTION, METABOLIC PATHWAYS, AND MICROBE-ASSOCIATED METABOLITES

Understanding key bacterial functional pathways that are altered in HIV infection will provide a better understanding of the relationship between the host microbiota and the immune system, thereby potentially providing novel treatments to improve the health of HIV-infected subjects [78]. In this context, the reduced inflammation and immune recovery observed in HIV-infected patients under optimal ART response is associated with an active fraction of gut microbiota (at both protein synthesis and metabolic activity levels), which differs from the responses observed in untreated or nonresponder patients [79]. The role of bacterial metabolites is also pivotal to understand HIV-infection pathogenesis, and in order to find out potential targets to treat this disease.

Metabolic Pathways and Microbe-Associated Metabolites in HIV

An augmented rate of tryptophan catabolism through the kynurenine pathway has been shown in HIV infection and was associated with disease progression [24,80–83]. In fact, increased levels of tryptophan catabolites, particularly 3-hydroxyanthranilic acid, are directly involved in the biased balance of Th17 to Treg cells, which further contributes to immune suppression, BT, and systemic inflammation [80].

Impairment in the capacity to produce proline, phenylalanine, and lysine has also been observed in HIV-infected patients [82]. The study of Vazquez-Castellanos et al. [65] also demonstrated an altered metabolism of gut microbiota characterized by enrichment of genes involved in pathogenic processes such as LPS biosynthesis, BT and inflammation, with depletion of genes encoding factors contributing to amino acid metabolism and energy processes. Furthermore, the metabolic changes that exist in HIV infection seem to be disease-dependent, and different from those observed in other pathologies [82].

In comparison with progressors, elite controllers have a distinct microbiota metabolic profile that favors fatty acid metabolism, peroxisome proliferator-activated receptor signaling, and lipid biosynthesis protein pathways, combined with a decrease in carbohydrate metabolism and secondary bile acid synthesis [64].

Concerning rectal microbiota, the study of McHardy et al. [61] demonstrated enrichment in genes and pathways involved in glutathione metabolism, selenocompound metabolism, folate biosynthesis, and siderophore biosynthetic genes. In contrast, genes involved in amino acid production, fructose/mannose metabolism, and CoA biosynthesis were depleted [61].

Reductions in mucosal butyrate, because of diminished colonic butyrate-producing bacteria, may exacerbate gut T-cell activation and HIV replication, thereby contributing to HIV-associated mucosal pathogenesis [84]. In addition, patients with higher Firmicutes to Bacteroidetes ratio had higher trimethylamine-N-oxide (TMAO) production. This microbiota-dependent metabolite and cardiovascular disease marker lacks an important role in the pathogenesis of coronary heart disease in HIV subjects. However, levels of TMA, the precursor of TMAO, are positively associated with a number of coronary plaque features [85].

DIAGNOSTIC IMPLICATIONS

Changes in gut viroma have been associated to rapidly progressive disease [72,73]. Similarly, the presence in feces (a noninvasive sample) of specific bacteria or genera such as *Prevotella* that are clearly associated with BT, persistent inflammation, and immune activation can give clues about possible future complications that could be early treated. Even more importantly, these biomarkers could be used as indicators of effectiveness of treatment and progression of the disease.

HIV/AIDS patients without ART are very susceptible to infections [86] which can cause severe complications [87]. Thus, the plasma microbiome could lead to early diagnosis, when bacteremia or viremia are still asymptomatic, and prophylaxis can still be performed [88]. The best sample to analyze biomarkers should also deserve attention, with preference for noninvasive material such as saliva and feces. Stool samples have low amounts of human DNA, which makes them optimal for virome studies in which host contamination could be a challenge.

THERAPEUTICAL APPROACHES

Investigation and restoration of gut microbes and their metabolites represent novel therapeutic targets to prevent or reverse HIV-induced BT, chronic immunoactivation, and, lastly, HIV-associated alterations [89—91].

Probiotics

The use of probiotics, such as *Lactobacillus*, *Bifidobacterium*, and *Enterococcus* that can benefit the intestinal and immune system may be inexpensive and clinically important to reduce HIV-related morbidity and mortality [92]. The mechanisms by which probiotics may interfere with HIV have been brilliantly summarized by D'Angelo et al. [93].

Effects on CD4, Microbial Translocation, and Inflammation

Probiotics improve CD4 counts in HIV-infected patients [94—99]. Most of these studies tested the effects of yoghurt containing *Lactobacillus rhamnosus* and *L. reuteri* [96,97,100], capsules with *Bacillus coagulans GBI-30* [99] and *L. casei* [101], among others. Treatment with *Saccharomyces boulardii* significantly decreased plasma BT and inflammation parameters in HIV-treated patients, half of whom had an immunodiscordant response to ART [102]. The same authors demonstrated that this probiotic affects gut microbiota composition, with lower concentrations of some gut species, such as *Clostridiaceae*, after 12 weeks of administration [103].

Treatment for 48 weeks with a mix of *Streptococcus salivarius* ssp. *Termophilus*, *Bifidobacteria—represented by Bifidobacteria breve*, *Bifidobacteria Infantis*, and *Bifidobacteria longum*; *L. acidophilus*, *Lactobacillus plantarum*, *Lactobacillus casei*, *Lactobacillus delbrueckii* ssp.; *Bulgaricus*; and *Streptococcus faecium* was also able to reduce immune activation of CD4 T lymphocytes settling at values comparable to controls [104]. Treatment for 6 months with Vivomixx or Visbiome was also very effective in reducing T cell activation, and in increasing Th17 frequencies. In fact these probiotics were able to recover GALT integrity in HIV-infected patients under ART [105].

When probiotics are administered in yoghurt, the effects can be more related to the micronutrients present in lactic beverages [106], although other studies have failed to confirm these suggestions. Length of the treatment (ranging from 2 to 25 weeks), doses, and end points (inflammatory cytokines, etc.) could affect the results. Thus, more standardized clinical trials are needed.

Effects on Diarrhea

Overall data are inconclusive for a role of bacterially based probiotics in the management of HIV-associated diarrhea as outcomes varied in terms of frequency of stools, consistency, duration, hospital stay, and use of concomitant antimicrobials and other medications. However, most of the studies showed beneficial actions [100,107].

Effects on Bacterial Vaginosis

One randomized clinical trial has demonstrated that supplementation of *L. rhamnosus GR-1* and *L. reuteri RC-14* did not enhance the cure of bacterial vaginosis among women living with HIV, but may prevent the condition [108]. Obviously, more studies are needed.

Other Effects

Six months of probiotic supplementation (mix of bacteria) improved some neurocognitive functions (short and long memory, abstract reasoning) [109]. A promising role of probiotics in combating periodontal disease, a comorbidity that could act as a trigger for chronic inflammation with risk of systemic diseases such as diabetes, hyperlipidemia, and chronic kidney diseases, has also been suggested [110,111].

Side Effects

Case reports in HIV-infected people documented probiotic-related sepsis [112,113]. *L. acidophilus* was the agent of bacteremia in a patient with AIDS [114]. However, the observed rate of side effects was far lower than with conventional medications, and participants taking probiotics were not more likely to experience adverse events than controls [94,115].

Prebiotics

Prebiotics can have an impact on markers of translocation, inflammation, and immune activation in HIV-infection [116,117]. Gori et al. [116] suggested that in nonsymptomatic naïve HIV-1-infected adults, gut microbiota can at least be partially restored by a unique prebiotic mixture consisting of scGOS/lcFOS/pAOS, with stimulation of *Bifidobacteria* growth and reduction in fecal pathogenic load. In addition, a significant reduction in sCD14 and LPS levels, CD4 + T-cell activation (CD25), and improved NK cell cytolytic activity were observed. Six weeks with the mixture of scGOS/lcFOS/glutamine attenuated HIV-associated dysbiosis, which was most apparent in naïve individuals. Reduction in markers of BT and T-cell activation, improvement of thymic output, and changes in butyrate production were also observed [118].

Only mild side effects, such as some flatulence and abdominal distension, have been reported [116].

Symbiosis Between Probiotics and Prebiotics

Synbiotic therapy augmented the levels of probiotic species in the gut during chronic HIV-1 infection, with absence of changes in BT and markers of systemic immune activation [117,119,120].

Fecal Bacteriotherapy or Fecal Transplantation (FMT)

Fecal Transplantation (FMT), which is generating many patent applications (Table 32.3), has been highly successful for *Clostridium difficile* infection in immunocompromised patients [121]. However, the greatest microbiota shift in SIV-infected rhesus macaques was after antibiotic treatment. The bacterial composition 2 weeks post-FMT resembled the pre-FMT structure, with differences in the abundances of minor bacterial populations. Increases in the number of peripheral Th17 and Th22 cells and reduced CD4 + T cell activation in gastrointestinal tissue were also observed. In this study, FMT was well tolerated, with no negative clinical side effects [122].

In the study of Vujkovic-Cvjin et al. [123] shift toward the donor microbiota were observed, but were less potent in comparison to those observed in recurrent *C. difficile* infection. In addition, markers of HIV-associated inflammation remained stable [123].

TABLE 32.3 Patent Applications Related to Fecal Transplantation

Patent Number	Title	Assignee/Inventor	Publication Date
EP 2338989 A1	Method for preparing fecal sample, fecal sample preparation solution, and feces collection kit	Olympus Corporation	August 25, 2009
WO2011033310 A1	Encapsulated intestinal flora extracted from feces	William John Martin	September 17, 2010
WO 2012122478 1	Compositions and methods for transplantation of colon microbiota	Regents Of The University Of Minnesota	March 9, 2012
WO 2013053836 1	Composition comprising anaerobically cultivated human intestinal microbiota	Quantum Pharmaceuticals Sa	October 11, 2012
WO 2013090825A1	Device for the collection, refinement, and administration of gastrointestinal microflora	Pureflora, Inc.	December 14, 2012
WO 2014078911 A1	Compositions for the restoration of a fecal microbiota and methods for making and using them	Borody Thomas J	May 30, 2014
US 20140238154 A1	Fecal microbiome transplant material preparation method and apparatus	Christopher J. Stevens	August 28, 2014
WO 2014152484A1	Freeze dried fecal microbiota for use in fecal microbial transplantation	Regents Of The University Of Minnesota	September 25, 2014
US 20150093360 A1	Compositions and methods	Seres Health, Inc.	April 2, 2015
US20150238544	Microbiota restoration therapy (mrt), compositions and methods of manufacture	Rebiotix, Inc.	August 27, 2015
WO 2016183577 A1	Compositions for fecal floral transplantation and methods for making and using them and devices for delivering them	Crestovo Llc	November 17, 2016
WO 2017075098 A1	Compositions and methods for fecal microbiota-related therapy	Crestovo Llc	May 4, 2017

Other Interventions

Sevelamer is a phosphate-binding drug that decreased BT while also decreasing inflammation and immune activation in an SIV model [124]. However, a study in naïve HIV-infected patients evidenced no changes in BT, inflammation, or immune activation [125]. IL-7 has also been investigated. After IL-7 administration, patients showed increased CD4+ and CD8+ T cells, as well as an increase in gut-homing lymphocytes. Patients also displayed a decrease in sCD14, indicating an improvement in the gut barrier integrity [126]. However, effects on gut microbiota composition have not been evaluated yet.

CONCLUSIONS

More functional studies of microbiota are warranted, exploring the changes in immune response and metabolic functions combining genomics, metabolomics, and proteomics without overlooking microbiology, bioinformatics, immunology, and cell biology.

It is important to remember that the intestinal microbiota is not constituted by bacteria only, but also by viruses. As the virome could play a crucial role in the regulation of intestinal immunity and homeostasis, this field deserves special attention, and opens the door to future work.

LIST OF ACRONYMS AND ABBREVIATIONS

ART Antiretroviral therapy
BT Bacterial translocation
CD Cluster of differentiation
GALT Gut-associated lymphoid tissue
GBV Hepatitis G virus
HBV Hepatitis B virus
HCV Hepatitis C virus
HSV Herpes simplex virus
HERV Human endogenous retrovirus
HIV Human immunodeficiency virus
HPV Human papilloma virus
INSTIs Integrase strand transfer inhibitors
IL-1 Interleukin-1
IL-6 Interleukin-6
IL-17 Interleukin-17
IL-22 Interleukin-22
LPS Lipopolysaccharide
MSR Macrophage scavenger receptor
MVC Maraviroc
NRTIs Nucleoside reverse transcriptase inhibitors
NNRTIs Nonnucleoside reverse transcriptase inhibitors
NF-KB Nuclear factor-kappaB
PIs Protease inhibitors
SCFAs Short-chain fatty acids
SIV Simian immunodeficiency virus
sCD14 Soluble CD14
TLR Toll-like receptor
TNFα Tumor necrosis factor alpha
WMG Whole metagenomic sequencing

REFERENCES

[1] Zhang Y, Lun CY, Tsui SK. Metagenomics: a new way to illustrate the crosstalk between infectious diseases and host microbiome. Int J Mol Sci 2015;16(11):26263–79.

[2] Fujimura KE, Slusher NA, Cabana MD, Lynch SV. Role of the gut microbiota in defining human health. Expert Rev Anti Infect Ther 2010;8(4):435–54.

[3] Villanueva-Millan MJ, Perez-Matute P, Oteo JA. Gut microbiota: a key player in health and disease. A review focused on obesity. J Physiol Biochem 2015;71(3):509–25.

[4] Vandeputte D, Kathagen G, D'Hoe K, Vieira-Silva S, Valles-Colomer M, Sabino J, et al. Quantitative microbiome profiling links gut community variation to microbial load. Nature 2017;551(7681):507–11.

[5] Yarza P, Yilmaz P, Pruesse E, Glockner FO, Ludwig W, Schleifer KH, et al. Uniting the classification of cultured and uncultured bacteria and archaea using 16S rRNA gene sequences. Nat Rev Microbiol 2014;12(9):635–45.

[6] Waldman AJ, Balskus EP. The human microbiota, infectious disease, and global health: challenges and opportunities. ACS Infect Dis 2018;4(1):14–26.

[7] Vitali D, Wessels JM, Kaushic C. Role of sex hormones and the vaginal microbiome in susceptibility and mucosal immunity to HIV-1 in the female genital tract. AIDS Res Ther 2017;14(1):39.

[8] Haase AT. Perils at mucosal front lines for HIV and SIV and their hosts. Nat Rev Immunol 2005;5(10):783–92.

[9] Aldunate M, Srbinovski D, Hearps AC, Latham CF, Ramsland PA, Gugasyan R, et al. Antimicrobial and immune modulatory effects of lactic acid and short chain fatty acids produced by vaginal microbiota associated with eubiosis and bacterial vaginosis. Front Physiol 2015;6:164.

[10] Donahue CR, Sheth AN, Read TD, Frisch MB, Mehta CC, Martin A, et al. The female genital tract microbiome is associated with vaginal antiretroviral drug concentrations in human immunodeficiency virus-infected women on antiretroviral therapy. J Infect Dis 2017;216(8):10.

[11] Huang B, Fettweis JM, Brooks JP, Jefferson KK, Buck GA. The changing landscape of the vaginal microbiome. Clin Lab Med 2014;34(4):747–61.

[12] Murphy K, Irvin SC, Herold BC. Research gaps in defining the biological link between HIV risk and hormonal contraception. Am J Reprod Immunol 2014;72(2):228–35.

[13] Hummelen R, Fernandes AD, Macklaim JM, Dickson RJ, Changalucha J, Gloor GB, et al. Deep sequencing of the vaginal microbiota of women with HIV. PLoS One 2010;5(8):e12078.

[14] Schellenberg JJ, Card CM, Ball TB, Mungai JN, Irungu E, Kimani J, et al. Bacterial vaginosis, HIV serostatus and T-cell subset distribution in a cohort of East African commercial sex workers: retrospective analysis. AIDS 2012;26(3):387–93.

[15] Atashili J, Poole C, Ndumbe PM, Adimora AA, Smith JS. Bacterial vaginosis and HIV acquisition: a meta-analysis of published studies. AIDS 2008;22(12):1493–501.

[16] Jespers V, Kyongo J, Joseph S, Hardy L, Cools P, Crucitti T, et al. A longitudinal analysis of the vaginal microbiota and vaginal immune mediators in women from sub-Saharan Africa. Sci Rep 2017;7(1):11974.

[17] Anahtar MN, Byrne EH, Doherty KE, Bowman BA, Yamamoto HS, Soumillon M, et al. Cervicovaginal bacteria are a major modulator of host inflammatory responses in the female genital tract. Immunity 2015;42(5):965–76.

[18] Gosmann C, Anahtar MN, Handley SA, Farcasanu M, Abu-Ali G, Bowman BA, et al. Lactobacillus-deficient cervicovaginal bacterial communities are associated with increased HIV acquisition in Young South African women. Immunity 2017;46(1):29–37.

[19] Dubourg G, Surenaud M, Levy Y, Hue S, Raoult D. Microbiome of HIV-infected people. Microb Pathog 2017;106:85–93.

[20] Yu G, Fadrosh D, Ma B, Ravel J, Goedert JJ. Anal microbiota profiles in HIV-positive and HIV-negative MSM. AIDS 2014;28(5):753–60.

[21] von Rosenvinge EC, Song Y, White JR, Maddox C, Blanchard T, Fricke WF. Immune status, antibiotic medication and pH are associated with changes in the stomach fluid microbiota. ISME J 2013;7(7):1354–66.

[22] Mutlu EA, Keshavarzian A, Losurdo J, Swanson G, Siewe B, Forsyth C, et al. A compositional look at the human gastrointestinal microbiome and immune activation parameters in HIV infected subjects. PLoS Pathog 2014;10(2):e1003829.

[23] Perez-Santiago J, Gianella S, Massanella M, Spina CA, Karris MY, Var SR, et al. Gut Lactobacillales are associated with higher CD4 and less microbial translocation during HIV infection. AIDS 2013;27(12):1921–31.

[24] Vujkovic-Cvijin I, Dunham RM, Iwai S, Maher MC, Albright RG, Broadhurst MJ, et al. Dysbiosis of the gut microbiota is associated with HIV disease progression and tryptophan catabolism. Sci Transl Med 2013;5(193):193ra91.

[25] Dillon SM, Lee EJ, Kotter CV, Austin GL, Dong Z, Hecht DK, et al. An altered intestinal mucosal microbiome in HIV-1 infection is associated with mucosal and systemic immune activation and endotoxemia. Mucosal Immunol 2014;7(4):983–94.

[26] Round JL, Mazmanian SK. Inducible Foxp3+ regulatory T-cell development by a commensal bacterium of the intestinal microbiota. Proc Natl Acad Sci USA 2010;107(27):12204–9.

[27] Nelson KE, Weinstock GM, Highlander SK, Worley KC, Creasy HH, Wortman JR, et al. A catalog of reference genomes from the human microbiome. Science 2010;328(5981):994–9.

[28] Martel J, Wu CY, Huang PR, Cheng WY, Young JD. Pleomorphic bacteria-like structures in human blood represent non-living membrane vesicles and protein particles. Sci Rep 2017;7(1):10650.

[29] Li SK, Leung RK, Guo HX, Wei JF, Wang JH, Kwong KT, et al. Detection and identification of plasma bacterial and viral elements in HIV/AIDS patients in comparison to healthy adults. Clin Microbiol Infect 2012;18(11):1126–33.

[30] Merlini E, Bai F, Bellistri GM, Tincati C, d'Arminio Monforte A, Marchetti G. Evidence for polymicrobic flora translocating in peripheral blood of HIV-infected patients with poor immune response to antiretroviral therapy. PLoS One 2011;6(4):e18580.

[31] Branton WG, Ellestad KK, Maingat F, Wheatley BM, Rud E, Warren RL, et al. Brain microbial populations in HIV/AIDS: alpha-proteobacteria predominate independent of host immune status. PLoS One 2013;8(1):e54673.

[32] Liu CM, Osborne BJ, Hungate BA, Shahabi K, Huibner S, Lester R, et al. The semen microbiome and its relationship with local immunology and viral load in HIV infection. PLoS Pathog 2014;10(7):e1004262.

[33] Lawani MB, Morris A. The respiratory microbiome of HIV-infected individuals. Expert Rev Anti Infect Ther 2016;14(8):719–29.

[34] Twigg 3rd HL, Weinstock GM, Knox KS. Lung microbiome in human immunodeficiency virus infection. Transl Res 2017;179:97–107.

[35] Cui L, Morris A, Huang L, Beck JM, Twigg 3rd HL, von Mutius E, et al. The microbiome and the lung. Ann Am Thorac Soc 2014;11(Suppl. 4):S227–32.

[36] Grubb JR, Moorman AC, Baker RK, Masur H. The changing spectrum of pulmonary disease in patients with HIV infection on antiretroviral therapy. AIDS 2006;20(8):1095−107.

[37] Beck JM, Schloss PD, Venkataraman A, Twigg 3rd H, Jablonski KA, Bushman FD, et al. Multicenter comparison of lung and oral microbiomes of HIV-infected and HIV-uninfected individuals. Am J Respir Crit Care Med 2015;192(11):1335−44.

[38] Cribbs SK, Uppal K, Li S, Jones DP, Huang L, Tipton L, et al. Correlation of the lung microbiota with metabolic profiles in bronchoalveolar lavage fluid in HIV infection. Microbiome 2016;4:3.

[39] Li Y, Saxena D, Chen Z, Liu G, Abrams WR, Phelan JA, et al. HIV infection and microbial diversity in saliva. J Clin Microbiol 2014;52(5):1400−11.

[40] Moyes DL, Saxena D, John MD, Malamud D. The gut and oral microbiome in HIV disease: a workshop report. Oral Dis 2016;22(Suppl. 1):166−70.

[41] Dang AT, Cotton S, Sankaran-Walters S, Li CS, Lee CY, Dandekar S, et al. Evidence of an increased pathogenic footprint in the lingual microbiome of untreated HIV infected patients. BMC Microbiol 2012;12:153.

[42] Kistler JO, Arirachakaran P, Poovorawan Y, Dahlen G, Wade WG. The oral microbiome in human immunodeficiency virus (HIV)-positive individuals. J Med Microbiol 2015;64(9):1094−101.

[43] Brenchley JM, Douek DC. HIV infection and the gastrointestinal immune system. Mucosal Immunol 2008;1(1):23−30.

[44] Guadalupe M, Reay E, Sankaran S, Prindiville T, Flamm J, McNeil A, et al. Severe CD4+ T-cell depletion in gut lymphoid tissue during primary human immunodeficiency virus type 1 infection and substantial delay in restoration following highly active antiretroviral therapy. J Virol 2003;77(21):11708−17.

[45] Smit-McBride Z, Mattapallil JJ, McChesney M, Ferrick D, Dandekar S. Gastrointestinal T lymphocytes retain high potential for cytokine responses but have severe CD4(+) T-cell depletion at all stages of simian immunodeficiency virus infection compared to peripheral lymphocytes. J Virol 1998;72(8):6646−56.

[46] Lapenta C, Boirivant M, Marini M, Santini SM, Logozzi M, Viora M, et al. Human intestinal lamina propria lymphocytes are naturally permissive to HIV-1 infection. Eur J Immunol 1999;29(4):1202−8.

[47] Brenchley JM, Schacker TW, Ruff LE, Price DA, Taylor JH, Beilman GJ, et al. CD4+ T cell depletion during all stages of HIV disease occurs predominantly in the gastrointestinal tract. J Exp Med 2004;200(6):749−59.

[48] Mehandru S, Poles MA, Tenner-Racz K, Horowitz A, Hurley A, Hogan C, et al. Primary HIV-1 infection is associated with preferential depletion of CD4+ T lymphocytes from effector sites in the gastrointestinal tract. J Exp Med 2004;200(6):761−70.

[49] Nasi M, Pinti M, Mussini C, Cossarizza A. Persistent inflammation in HIV infection: established concepts, new perspectives. Immunol Lett 2014;161(2):184−8.

[50] Brenchley JM, Paiardini M, Knox KS, Asher AI, Cervasi B, Asher TE, et al. Differential Th17 CD4 T-cell depletion in pathogenic and nonpathogenic lentiviral infections. Blood 2008;112(7):2826−35.

[51] Elhed A, Unutmaz D. Th17 cells and HIV infection. Curr Opin HIV AIDS 2010;5(2):146−50.

[52] Brenchley JM, Price DA, Schacker TW, Asher TE, Silvestri G, Rao S, et al. Microbial translocation is a cause of systemic immune activation in chronic HIV infection. Nat Med 2006;12(12):1365−71.

[53] Hsu DC, Sereti I, Ananworanich J. Serious Non-AIDS events: immunopathogenesis and interventional strategies. AIDS Res Ther 2013;10(1):29.

[54] Lozupone CA, Stombaugh JI, Gordon JI, Jansson JK, Knight R. Diversity, stability and resilience of the human gut microbiota. Nature 2012;489(7415):220−30.

[55] McHardy IH, Li X, Tong M, Ruegger P, Jacobs J, Borneman J, et al. HIV Infection is associated with compositional and functional shifts in the rectal mucosal microbiota. Microbiome 2013;1(1):26.

[56] Nowak P, Troseid M, Avershina E, Barqasho B, Neogi U, Holm K, et al. Gut microbiota diversity predicts immune status in HIV-1 infection. AIDS 2015;29(18):2409−18.

[57] Lozupone CA, Li M, Campbell TB, Flores SC, Linderman D, Gebert MJ, et al. Alterations in the gut microbiota associated with HIV-1 infection. Cell Host Microbe 2013;14(3):329−39.

[58] Villanueva-Millan MJ, Perez-Matute P, Recio-Fernandez E, Lezana Rosales JM, Oteo JA. Differential effects of antiretrovirals on microbial translocation and gut microbiota composition of HIV-infected patients. J Int AIDS Soc 2017;20(1):21526.

[59] Ellis CL, Ma ZM, Mann SK, Li CS, Wu J, Knight TH, et al. Molecular characterization of stool microbiota in HIV-infected subjects by panbacterial and order-level 16S ribosomal DNA (rDNA) quantification and correlations with immune activation. J Acquir Immune Defic Syndr 2011;57(5):363−70.

[60] Paquin-Proulx D, Ching C, Vujkovic-Cvijin I, Fadrosh D, Loh L, Huang Y, et al. Bacteroides are associated with GALT iNKT cell function and reduction of microbial translocation in HIV-1 infection. Mucosal Immunol 2017;10(1):69−78.

[61] McHardy IH, Goudarzi M, Tong M, Ruegger PM, Schwager E, Weger JR, et al. Integrative analysis of the microbiome and metabolome of the human intestinal mucosal surface reveals exquisite inter-relationships. Microbiome 2013;1(1):17.

[62] Gori A, Tincati C, Rizzardini G, Torti C, Quirino T, Haarman M, et al. Early impairment of gut function and gut flora supporting a role for alteration of gastrointestinal mucosa in human immunodeficiency virus pathogenesis. J Clin Microbiol 2008;46(2):757−8.

[63] Ling Z, Jin C, Xie T, Cheng Y, Li L, Wu N. Alterations in the fecal microbiota of patients with HIV-1 infection: an observational Study in A Chinese population. Sci Rep 2016;6:30673.

[64] Vesterbacka J, Rivera J, Noyan K, Parera M, Neogi U, Calle M, et al. Richer gut microbiota with distinct metabolic profile in HIV infected Elite Controllers. Sci Rep 2017;7(1):6269.

[65] Vazquez-Castellanos JF, Serrano-Villar S, Latorre A, Artacho A, Ferrus ML, Madrid N, et al. Altered metabolism of gut microbiota contributes to chronic immune activation in HIV-infected individuals. Mucosal Immunol 2015;8(4):760–72.

[66] Noguera-Julian M, Rocafort M, Guillen Y, Rivera J, Casadella M, Nowak P, et al. Gut microbiota linked to sexual preference and HIV infection. EBioMedicine 2016;5:135–46.

[67] Volpe GE, Ward H, Mwamburi M, Dinh D, Bhalchandra S, Wanke C, et al. Associations of cocaine use and HIV infection with the intestinal microbiota, microbial translocation, and inflammation. J Stud Alcohol Drugs 2014;75(2):347–57.

[68] Costiniuk CT, Angel JB. Human immunodeficiency virus and the gastrointestinal immune system: does highly active antiretroviral therapy restore gut immunity? Mucosal Immunol 2012;5(6):596–604.

[69] Pinto-Cardoso S, Lozupone C, Briceno O, Alva-Hernandez S, Tellez N, Adriana A, et al. Fecal Bacterial Communities in treated HIV infected individuals on two antiretroviral regimens. Sci Rep 2017;7:43741.

[70] Patterson KB, Prince HA, Stevens T, Shaheen NJ, Dellon ES, Madanick RD, et al. Differential penetration of raltegravir throughout gastrointestinal tissue: implications for eradication and cure. AIDS 2013;27(9):1413–9.

[71] Perez-Matute P, Perez-Martinez L, Aguilera-Lizarraga J, Blanco JR, Oteo JA. Maraviroc modifies gut microbiota composition in a mouse model of obesity: a plausible therapeutic option to prevent metabolic disorders in HIV-infected patients. Rev Esp Quimioter 2015;28(4):200–6.

[72] Handley SA, Thackray LB, Zhao G, Presti R, Miller AD, Droit L, et al. Pathogenic simian immunodeficiency virus infection is associated with expansion of the enteric virome. Cell 2012;151(2):253–66.

[73] Barbian HJ, Li Y, Ramirez M, Klase Z, Lipende I, Mjungu D, Moeller AH, Wilson ML, Pusey AE, Lonsdorf EV, Bushman FD, Hahn BH, et al. Destabilization of the gut microbiome marks the end-stage of simian immunodeficiency virus infection in wild chimpanzees. Am J Primatol 2018 Jan;80(1).

[74] Monaco CL, Gootenberg DB, Zhao G, Handley SA, Ghebremichael MS, Lim ES, et al. Altered virome and bacterial microbiome in human immunodeficiency virus-associated acquired immunodeficiency syndrome. Cell Host Microbe 2016;19(3):311–22.

[75] Li L, Deng X, Linsuwanon P, Bangsberg D, Bwana MB, Hunt P, et al. AIDS alters the commensal plasma virome. J Virol 2013;87(19):10912–5.

[76] Christensen T. Association of human endogenous retroviruses with multiple sclerosis and possible interactions with herpes viruses. Rev Med Virol 2005;15(3):179–211.

[77] Young JC, Chehoud C, Bittinger K, Bailey A, Diamond JM, Cantu E, et al. Viral metagenomics reveal blooms of anelloviruses in the respiratory tract of lung transplant recipients. Am J Transplant 2015;15(1):200–9.

[78] Dillon SM, Frank DN, Wilson CC. The gut microbiome and HIV-1 pathogenesis: a two-way street. AIDS 2016;30(18):2737–51.

[79] Serrano-Villar S, Rojo D, Martinez-Martinez M, Deusch S, Vazquez-Castellanos JF, Bargiela R, et al. Gut bacteria metabolism impacts immune recovery in HIV-infected individuals. EBioMedicine 2016;8:203–16.

[80] Favre D, Mold J, Hunt PW, Kanwar B, Loke P, Seu L, et al. Tryptophan catabolism by indoleamine 2,3-dioxygenase 1 alters the balance of TH17 to regulatory T cells in HIV disease. Sci Transl Med 2010;2(32):32ra6.

[81] Jenabian MA, Patel M, Kema I, Kanagaratham C, Radzioch D, Thebault P, et al. Distinct tryptophan catabolism and Th17/Treg balance in HIV progressors and elite controllers. PLoS One 2013;8(10):e78146.

[82] Serrano-Villar S, Rojo D, Martinez-Martinez M, Deusch S, Vazquez-Castellanos JF, Sainz T, et al. HIV infection results in metabolic alterations in the gut microbiota different from those induced by other diseases. Sci Rep 2016;6:26192.

[83] Vyboh K, Jenabian MA, Mehraj V, Routy JP. HIV and the gut microbiota, partners in crime: breaking the vicious cycle to unearth new therapeutic targets. J Immunol Res 2015;2015:614127.

[84] Dillon SM, Kibbie J, Lee EJ, Guo K, Santiago ML, Austin GL, et al. Low abundance of colonic butyrate-producing bacteria in HIV infection is associated with microbial translocation and immune activation. AIDS 2017;31(4):511–21.

[85] Srinivasa S, Fitch KV, Lo J, Kadar H, Knight R, Wong K, et al. Plaque burden in HIV-infected patients is associated with serum intestinal microbiota-generated trimethylamine. AIDS 2015;29(4):443–52.

[86] Mootsikapun P. Bacteremia in adult patients with acquired immunodeficiency syndrome in the northeast of Thailand. Int J Infect Dis 2007;11(3):226–31.

[87] Palella Jr FJ, Baker RK, Moorman AC, Chmiel JS, Wood KC, Brooks JT, et al. Mortality in the highly active antiretroviral therapy era: changing causes of death and disease in the HIV outpatient study. J Acquir Immune Defic Syndr 2006;43(1):27–34.

[88] Lepage P, Leclerc MC, Joossens M, Mondot S, Blottiere HM, Raes J, et al. A metagenomic insight into our gut's microbiome. Gut 2013;62(1):146–58.

[89] Assimakopoulos SF, Dimitropoulou D, Marangos M, Gogos CA. Intestinal barrier dysfunction in HIV infection: pathophysiology, clinical implications and potential therapies. Infection 2014;42(6):951–9.

[90] Cani PD, Knauf C. How gut microbes talk to organs: the role of endocrine and nervous routes. Mol Metab 2016;5(9):743–52.

[91] Rajasuriar R, Khoury G, Kamarulzaman A, French MA, Cameron PU, Lewin SR. Persistent immune activation in chronic HIV infection: do any interventions work? AIDS 2013;27(8):1199–208.

[92] Hummelen R, Vos AP, van't Land B, van Norren K, Reid G. Altered host-microbe interaction in HIV: a target for intervention with pro- and prebiotics. Int Rev Immunol 2010;29(5):485–513.

[93] D'Angelo C, Reale M, Costantini E. Microbiota and probiotics in health and HIV infection. Nutrients 2017;9(6).

[94] Carter GM, Esmaeili A, Shah H, Indyk D, Johnson M, Andreae M, et al. Probiotics in human immunodeficiency virus infection: a systematic review and evidence synthesis of benefits and risks. Open Forum Infect Dis 2016;3(4):ofw164.

[95] Cunningham-Rundles S, Ahrne S, Johann-Liang R, Abuav R, Dunn-Navarra AM, Grassey C, et al. Effect of probiotic bacteria on microbial host defense, growth, and immune function in human immunodeficiency virus type-1 infection. Nutrients 2011;3(12):1042−70.

[96] Hummelen R, Changalucha J, Butamanya NL, Koyama TE, Cook A, Habbema JD, et al. Effect of 25 weeks probiotic supplementation on immune function of HIV patients. Gut Microbes 2011;2(2):80−5.

[97] Irvine SL, Hummelen R, Hekmat S, Looman CW, Habbema JD, Reid G. Probiotic yogurt consumption is associated with an increase of CD4 count among people living with HIV/AIDS. J Clin Gastroenterol 2010;44(9):e201−5.

[98] Miller H, Ferris R, Phelps BR. The effect of probiotics on CD4 counts among people living with HIV: a systematic review. Benef Microbes 2016;7(3):345−51.

[99] Yang OO, Kelesidis T, Cordova R, Khanlou H. Immunomodulation of antiretroviral drug-suppressed chronic HIV-1 infection in an oral probiotic double-blind placebo-controlled trial. AIDS Res Hum Retroviruses 2014;30(10):988−95.

[100] Anukam KC, Osazuwa EO, Osadolor HB, Bruce AW, Reid G. Yogurt containing probiotic Lactobacillus rhamnosus GR-1 and L. reuteri RC-14 helps resolve moderate diarrhea and increases CD4 count in HIV/AIDS patients. J Clin Gastroenterol 2008;42(3):239−43.

[101] Falasca K, Vecchiet J, Ucciferri C, Di Nicola M, D'Angelo C, Reale M. Effect of probiotic supplement on cytokine levels in HIV-infected individuals: a preliminary study. Nutrients 2015;7(10):8335−47.

[102] Villar-Garcia J, Hernandez JJ, Guerri-Fernandez R, Gonzalez A, Lerma E, Guelar A, et al. Effect of probiotics (Saccharomyces boulardii) on microbial translocation and inflammation in HIV-treated patients: a double-blind, randomized, placebo-controlled trial. J Acquir Immune Defic Syndr 2015;68(3):256−63.

[103] Villar-Garcia J, Guerri-Fernandez R, Moya A, Gonzalez A, Hernandez JJ, Lerma E, et al. Impact of probiotic Saccharomyces boulardii on the gut microbiome composition in HIV-treated patients: a double-blind, randomised, placebo-controlled trial. PLoS One 2017;12(4):e0173802.

[104] d'Ettorre G, Ceccarelli G, Giustini N, Serafino S, Calantone N, De Girolamo G, et al. Probiotics reduce inflammation in antiretroviral treated, HIV-infected individuals: results of the "Probio-HIV" clinical trial. PLoS One 2015;10(9):e0137200.

[105] d'Ettorre G, Rossi G, Scagnolari C, Andreotti M, Giustini N, Serafino S, et al. Probiotic supplementation promotes a reduction in T-cell activation, an increase in Th17 frequencies, and a recovery of intestinal epithelium integrity and mitochondrial morphology in ART-treated HIV-1-positive patients. Immun Inflamm Dis 2017;5(3):244−60.

[106] Hemsworth JC, Hekmat S, Reid G. Micronutrient supplemented probiotic yogurt for HIV-infected adults taking HAART in London, Canada. Gut Microbes 2012;3(5):414−9.

[107] Guinane S. The effectiveness of probiotics for managing diarrhoea in people with HIV infection: a critically appraised topic. HIV Med 2013;14(3):187−90.

[108] Hummelen R, Changalucha J, Butamanya NL, Cook A, Habbema JD, Reid G. Lactobacillus rhamnosus GR-1 and L. reuteri RC-14 to prevent or cure bacterial vaginosis among women with HIV. Int J Gynaecol Obstet 2010;111(3):245−8.

[109] Ceccarelli G, Fratino M, Selvaggi C, Giustini N, Serafino S, Schietroma I, et al. A pilot study on the effects of probiotic supplementation on neuropsychological performance and microRNA-29a-c levels in antiretroviral-treated HIV-1-infected patients. Brain Behav 2017;7(8):e00756.

[110] Gupta G. Probiotics and periodontal health. J Med Life 2011;4(4):387−94.

[111] Staab B, Eick S, Knofler G, Jentsch H. The influence of a probiotic milk drink on the development of gingivitis: a pilot study. J Clin Periodontol 2009;36(10):850−6.

[112] Doron S, Snydman DR. Risk and safety of probiotics. Clin Infect Dis 2015;60(Suppl. 2):S129−34.

[113] Wilson NL, Moneyham LD, Alexandrov AW. A systematic review of probiotics as a potential intervention to restore gut health in HIV infection. J Assoc Nurses AIDS Care 2013;24(2):98−111.

[114] Haghighat L, Crum-Cianflone NF. The potential risks of probiotics among HIV-infected persons: bacteraemia due to Lactobacillus acidophilus and review of the literature. Int J STD AIDS 2016;27(13):1223−30.

[115] Hempel S, Newberry S, Ruelaz A, Wang Z, Miles JN, Suttorp MJ, et al. Safety of probiotics used to reduce risk and prevent or treat disease. Evid Rep Technol Assess 2011;(200):1−645.

[116] Gori A, Rizzardini G, Van't Land B, Amor KB, van Schaik J, Torti C, et al. Specific prebiotics modulate gut microbiota and immune activation in HAART-naive HIV-infected adults: results of the "COPA" pilot randomized trial. Mucosal Immunol 2011;4(5):554−63.

[117] Klatt NR, Canary LA, Sun X, Vinton CL, Funderburg NT, Morcock DR, et al. Probiotic/prebiotic supplementation of antiretrovirals improves gastrointestinal immunity in SIV-infected macaques. J Clin Investig 2013;123(2):903−7.

[118] Serrano-Villar S, Vazquez-Castellanos JF, Vallejo A, Latorre A, Sainz T, Ferrando-Martinez S, et al. The effects of prebiotics on microbial dysbiosis, butyrate production and immunity in HIV-infected subjects. Mucosal Immunol 2017;10(5):1279−93.

[119] Gonzalez-Hernandez LA, Jave-Suarez LF, Fafutis-Morris M, Montes-Salcedo KE, Valle-Gutierrez LG, Campos-Loza AE, et al. Synbiotic therapy decreases microbial translocation and inflammation and improves immunological status in HIV-infected patients: a double-blind randomized controlled pilot trial. Nutr J 2012;11:90.

[120] Schunter M, Chu H, Hayes TL, McConnell D, Crawford SS, Luciw PA, et al. Randomized pilot trial of a synbiotic dietary supplement in chronic HIV-1 infection. BMC Complement Altern Med 2012;12:84.

[121] Kelly CR, Ihunnah C, Fischer M, Khoruts A, Surawicz C, Afzali A, et al. Fecal microbiota transplant for treatment of Clostridium difficile infection in immunocompromised patients. Am J Gastroenterol 2014;109(7):1065−71.

[122] Hensley-McBain T, Zevin AS, Manuzak J, Smith E, Gile J, Miller C, et al. Effects of fecal microbial transplantation on microbiome and immunity in simian immunodeficiency virus-infected macaques. J Virol 2016;90(10):4981−9.

[123] Vujkovic-Cvijin I, Rutishauser RL, Pao M, Hunt PW, Lynch SV, McCune JM, et al. Limited engraftment of donor microbiome via one-time fecal microbial transplantation in treated HIV-infected individuals. Gut Microbes 2017;8(5):440−50.

[124] Kristoff J, Haret-Richter G, Ma D, Ribeiro RM, Xu C, Cornell E, et al. Early microbial translocation blockade reduces SIV-mediated inflammation and viral replication. J Clin Investig 2014;124(6):2802−6.

[125] Sandler NG, Zhang X, Bosch RJ, Funderburg NT, Choi AI, Robinson JK, et al. Sevelamer does not decrease lipopolysaccharide or soluble CD14 levels but decreases soluble tissue factor, low-density lipoprotein (LDL) cholesterol, and oxidized LDL cholesterol levels in individuals with untreated HIV infection. J Infect Dis 2014;210(10):1549−54.

[126] Sereti I, Estes JD, Thompson WL, Morcock DR, Fischl MA, Croughs T, et al. Decreases in colonic and systemic inflammation in chronic HIV infection after IL-7 administration. PLoS Pathogens 2014;10(1):e1003890.

Chapter 33

The Gut Microbiome in Autoimmune Diseases

Gislane Lellis Vilela de Oliveira[1,2]

[1]Microbiome Study Group, School of Health Sciences Paulo Prata (FACISB), Barretos, Brazil; [2]Microbiology Department, São Paulo State University (UNESP), Institute of Biosciences, Humanities and Exact Sciences (IBILCE), Sao Jose do Rio Preto, Sao Paulo, Brazil

GUT MICROBIOME AND AUTOIMMUNE DISEASES

Dysbiosis observed in autoimmune diseases is associated with decreased bacteria function and diversity, impaired epithelial barrier, bacterial translocation, inflammation, and decreased regulatory T cells (Tregs) in the gut [1,2]. The hypotheses proposed to link dysbiosis with autoimmune diseases include molecular mimicry, perpetuation of autoimmunity by inflammatory milieu amplified by altered gut microbiota, altered expression of toll-like receptors (TLRs) in antigen presenting cells, T helper 17 (Th17)/Tregs imbalance, and the posttranslational modification of luminal proteins (PTMP) promoted by enzymes from dysbiotic microbiota, which modify substrates in a different way than that performed under homeostatic conditions. The defective PTMP might generate neoepitopes that become immunogenic, induce systemic autoimmunity, and might trigger autoimmune diseases [3,4]. Furthermore, studies showed that commensal microbes and the gut microbiota metabolites could regulate immune cells and cytokines via epigenetic modifications suggesting their possible role in autoimmunity [4].

MULTIPLE SCLEROSIS

Studies have shown that the gut-brain axis can trigger multiple sclerosis (MS), and these investigations implied dysbiosis as one of the possible causes of extraintestinal disease development [5]. The enteric nervous system regulates the motility and homeostasis in the gut, affecting the gut microbiota composition [6]. On the other hand, gut microbes might influence the enteric nervous system, and the vagus nerve links the gut and the central nervous system (CNS). Commensal bacteria produce neurotransmitters, precursors, and metabolites that directly influence the brain. Furthermore, fermenting bacteria produce short-chain fatty acids (SCFAs), which can translocate into the central nervous system (CNS) and inhibit deacetylases, promoting epigenetic changes [7].

The induction of experimental autoimmune encephalomyelitis (EAE) in germ-free mice, resulted in the reduction of IFN-γ and IL-17A in the CNS, accompanied by an increase in the number of Treg cells in the gut [8]. The colonization of the germ-free mice with segmented filamentous bacteria promotes an increase in the number of Th17 cells in the lamina propria and CNS, worsening the disease score [9]. The colonization of the same mice with *Bacteroides fragilis* and polysaccharide A, which induces Foxp3 Treg cells, ameliorates symptoms in EAE mice, by inducing IL-10 secretion and suppressing IL-17 [10].

In humanized transgenic mice expressing the MS-associated HLA-DR2a gene and T cell receptors specific for myelin basic protein/DR2a derived from an MS patient, it was demonstrated that intestinal dysbiosis triggers the development of EAE during adolescence and early young adulthood. Furthermore, dysbiosis induces the expression of complement C3, production of the anaphylatoxin C3a, decreased Foxp3 gene expression, and anergy-related E3 ubiquitin ligase genes. Consequently, intestinal dysbiosis can induce the development of encephalitogenic T cells and EAE in young adulthood [11].

Microbiome and Metabolome in Diagnosis, Therapy, and other Strategic Applications. https://doi.org/10.1016/B978-0-12-815249-2.00033-6

Studies in relapsing-remitting MS patients (RRMS) showed decreased Firmicutes, Bacteroidetes, Proteobacteria [12,13], and *Clostridia* clusters XIVa and IV [14], with increased *Pseudomonas, Mycoplasma, Haemophilus, Blautia,* and *Dorei* genera [15]. Moreover, after vitamin D supplementation in RRMS patients, researchers reported decreased relative abundance of *Faecalibacterium*, and increased *Akkermansia* and *Coprococcus* [16].

In untreated MS patients, studies reported increased *Methanobrevibacter smithii* and *Akkermansia*, and reduction in Firmicutes and *Butyricimonas* [13,16]. *Methanobrevibacter* is involved in inflammatory conditions by recruiting macrophages and activating dendritic cells [17]. *Akkermansia* species have immunoregulatory effects by converting mucin into SCFAs; however, they could play a role in degrading the mucous layer and promoting inflammation [18,19]. *Butyricimonas* species are butyrate-producing bacteria and have immunomodulatory properties by inducing Treg cells in the gut [20]. In patients treated with IFN-β or glatiramer acetate there was an increase in *Prevotella* [21], which is associated with high-fiber ingestion and has regulatory roles via butyrate generation [22].

Less relative abundance of α1,3-galactose in MS patients was reported as well, and it is hypothesized that decreased circulating anti-α1,3-galactose IgG, and lower content of galactosyl transferase A1 in the gut microbiota could serve as a precipitating environmental factor for the disease [23]. The relationship between immunity in the gastrointestinal mucosa and commensal bacteria seems important for host homeostasis, including CNS demyelinating diseases [24].

PROBIOTICS IN MULTIPLE SCLEROSIS

In EAE animals, *Lactobacillus* species, *Pediococcus acidilactici, Bifidobacterium bifidum, Bifidobacterium animalis,* and *Bacteroides fragilis* decreased disease score and CNS inflammation through the induction of Treg cells in the gastrointestinal mucosa and in the mesenteric lymph nodes; IL-4, IL-10, and TGF-β secretion; and decreased expansion of Th1 and Th17 inflammatory subsets [25–29].

In MS patients, improved Expanded Disability Status Score (EDSS) and a decrease in inflammatory markers followed prescription of *Lactobacillus acidophilus, Lactobacillus casei, Lactobacillus fermentum,* and *B. bifidum* [30]. Likewise, administration of *L. acidophilus, L. casei, B. bifidum,* and *L. fermentum* induced a decrease in gene expression of IL-8 and TNF genes in peripheral blood mononuclear cells [31].

A pilot study with RRMS patients who had consumed a high-vegetable/low-protein diet versus a Western diet, demonstrated increased *Lachnospiraceae* family members, decreased T CD4 lymphocytes producing IL-17 and expressing PD-1, and EDSS in those with high-vegetable/low-protein diet. Because *Lachnospiraceae* induce Treg cell differentiation and TGF-β and IL-10 secretion, this effect could be associated with an increase of these bacteria [32].

Type 1 Diabetes/T1D

It is postulated that an impaired tolerance process in infancy with associated dysbiosis leads to a susceptibility to develop autoimmune diseases, and may result in activation of autoreactive T cells and autoantibodies generation. Dysbiosis could also predispose children with genetic susceptibility and positive autoantibodies to develop clinical disease [33].

The composition of the gut microbiota modulates innate and adaptive immunity, and enhances the disease in MyD88$^{-/-}$ nonobese diabetic/NOD mice. Protection against diabetes in these mice is averted by antibiotic administration and in germ-free conditions, suggesting that commensal microbes are important to decrease disease susceptibility [34].

An altered balance among Th1, Th17, and Treg cell differentiation in the gut was detected in germ-free NOD mice, and it is associated with insulitis and pancreas inflammation [35]. Interestingly, the segmented filamentous bacteria, which are associated with the exacerbation of Th17 responses in other autoimmune diseases are involved in protection against diabetes in NOD mice [36]. These filamentous bacteria are present in animals and in human infants; however, they are not present in human adults and seem to play a role in immune modulation. NOD mice treated with a conventional diet have an impaired tolerance to gut microbiota, increased barrier permeability and peritoneum macrophages, and decreased Firmicutes members [37].

In biobreeding diabetes-prone rats, increased *Bacteroides, Ruminococcus,* and *Eubacterium* genera were observed, and there was a higher abundance of *Bifidobacterium* and *Lactobacillus* in diabetes-resistant rats [38]. The augmented Bacteroidetes members could promote increased intestinal permeability, and precede the clinical onset of T1D in animal models and in prediabetic and diabetic patients [39,40].

In T1D children, there is decreased gut microbiota diversity, diminished abundance of mucin-degrading and butyrate-producing microbes, and reduced Firmicutes/Bacteroidetes ratio, along with *Lactobacillus, Bifidobacterium,* and *Prevotella* species [41]. Smaller numbers of lactate-producing bacteria, such as *Bifidobacterium longum*, and increased *Clostridium, Bacteroides,* and *Veillonella* species are equally observed [42].

The composition of the gut microbiota is altered in prediabetic children with genetic susceptibility and autoantibodies against β-cells [43]. Augmented *Bacteroides dorei* and *Bacteroides vulgatus* species in seroconverted T1D patients are registered 8 months prior to β-cell autoimmunity, suggesting that early dysbiosis may predict T1D in genetically predisposed subjects [44]. Additionally, children with β-cell autoantibodies exhibited increased Bacteroidetes members and decreased lactate and butyrate-producing bacteria [45]. These studies support the hypothesis that there is a gut microbiome signature associated with T1D development in seropositive children [46].

Even though studies have shown that intestinal dysbiosis can affect gut permeability via bacterial metabolites and play a role in T1D development, the real role of intestinal microbiota in the development of autoimmunity to β-cells and in tissue damage in humans is not well elucidated. Additional studies are needed to find the specific microbial ligands that signal through immune cells in the gut, and might be involved in the autoreactivity to β-cells [47].

PROBIOTICS IN T1D

Early oral administration of *Lactobacillaceae*-enriched probiotic prevented diabetes in NOD mice by abrogating insulitis and β-cell destruction. This protection was associated with increased IL-10 secretion in Peyer patches, spleen, and pancreas. The protective effect was transferable to irradiated mice receiving diabetogenic cells and splenocytes from treated mice [48]. Likewise, the administration of *Lactobacillaceae*-enriched probiotic alone or in combination with retinoic acid protected NOD mice from diabetes by suppressing inflammasome activation, IL-1β expression, and by inducing the immunomodulatory indoleamine 2,3-dioxygenase and IL-33 secretion. In addition, treated NOD mice showed modulation of the gut immunity by induction of CD103 dendritic cell differentiation and suppression of Th1 and Th17 subsets in the gut mucosa [49].

Fecal transplantation between NOD and resistant NOD mice (NOR) and oral antibiotic and probiotic treatment of NOD mice modulate pancreatic function. The intestinal microbiota from NOD mice housed more pathobionts and fewer beneficial microbes. Fecal transplantation of NOD microbes induced insulitis in NOR hosts, suggesting that the NOD microbiome is diabetogenic. Moreover, antibiotic exposure accelerated diabetes onset in NOD mice, accompanied by increased Th1 and Th17 cells in the mucosal-associated lymphoid tissues.

Somewhat conflicting experimental results are announced as well. Diabetes susceptibility correlated with reduced fecal SCFAs [50]. However, *Lactobacillaceae*-enriched probiotic in NOD mice poorly colonized the intestine and were insufficient to overcome the effects of a diabetogenic microbiome [50].

The administration of probiotic *Clostridium butyricum* induces Treg cells in the pancreas, and consequently inhibits the diabetes onset in NOD mice. Insulitis was suppressed, disease onset was delayed, and glucose metabolism improved. These beneficial effects could involve the migration of intestinal Treg cells to the pancreatic lymph nodes, and alterations in the Th1/Th2/Th17 balance, favoring an antiinflammatory milieu in the gut and pancreas. Additionally, probiotic supplementation increased the Firmicutes/Bacteroidetes ratio, *Clostridium* species, and butyrate-producing bacteria in the gut [51].

Probiotic supplementation has been hypothesized to affect innate and adaptive immune responses to environmental antigens by supporting healthy gut microbiota, and could therefore be used to prevent the onset of T1D-associated islet autoimmunity and treat the established disease [52]. In humans, early probiotic administration could decrease the risk of islet autoimmune reactions in children, with high genetic risk to develop the disease [53].

Lactobacillus and *Bifidobacterium* are important genera in the colon microbiota of humans. Beneficial effects of *Lactobacillus rhamnosus* and *Bifidobacterium lactis* on β-cell function would create a rationale for its use in patients with newly diagnosed T1D [54]. Bacteriotherapy is a potential tool to modulate the immune system in the prevention of islet cell autoimmunity [54,55].

RHEUMATOID ARTHRITIS

Dysbiosis might trigger joint inflammation [56]. Mechanisms in rheumatoid arthritis (RA) include activation of antigen presenting cells through TLR or Nod-like receptors, production of citrullinated peptides by enzymatic action, molecular mimicry, impaired gut permeability, and interference with host immune system by triggering T cell differentiation and activation of Th17-mediated mucosal inflammation [57].

Studies in collagen-induced arthritic (CIA) mice showed that the administration of antibiotics exacerbates the disease, and increases the levels of IL-6, IFN-γ, and IL-17 inflammatory cytokines [58]. An abundance of *Desulfovibrio*, *Prevotella*, *Parabacteroides*, *Odoribacter*, *Acetatifactor*, *Blautia*, *Coprococcus*, and *Ruminococcus* genera is registered in arthritic mice in addition to increased serum levels of IL-17 and CD4 Th17 cells in the spleen [59].

Increased intestinal permeability and a Th17 profile in AR susceptible mice were also observed, suggesting that genetic background influences the individual's microbiota profile [60]. Furthermore, Th17 cells induced by segmented filamentous bacteria in the lamina propria may induce autoantibodies involved in RA development in animal models [61].

In humans, recent studies have shown that Gram-negative *Prevotella* members dominated the gut microbiota of newly diagnosed RA patients, especially *Prevotella copri* [62]. The transplantation of the gut microbiota from RA patients to germ-free arthritis-prone SKG mice, induced increased Th17 cells in the gut and severe arthritis. Additionally, the coculture of SKG dendritic cells with *P. copri* increased IL-17 secretion in response to RA autoantigens, suggesting that RA gut microbiota may induce autoreactive cells in the gut and joint inflammation [63].

In RA patients, less species richness (alpha-diversity) is positively correlated with increased rheumatoid factor levels and disease progression [64]. In untreated recently diagnosed RA patients, *Lactobacillus salivarius*, *Lactobacillus iners*, and *Lactobacillus ruminis* were prominent [65]. In contrast, methotrexate-treated patients exhibited more abundant *Eggerthella*, *Actinomyces*, *Turicibacter*, *Streptococcus*, and *Collinsella* with positive correlation with IL-17 inflammatory cytokine. Rheumatoid factor, C-reactive protein, disease progression, and methotrexate correlated with beta-diversity in patients, suggesting that these clinical data might play a role in microbiota modulation [64,66].

PROBIOTICS IN RHEUMATOID ARTHRITIS

The spore-forming probiotic strain *Bacillus coagulans* has antiinflammatory and immunomodulatory effects on RA animal models and humans. The treatment with this probiotic and the prebiotic inulin significantly inhibits serum amyloid A protein in arthritic rats and promotes a significant decrease in the release of TNF, a proinflammatory cytokine [67].

The oral administration of *Lactobacillus helveticus* was associated with decreased joint swelling and improved body weight and bovine type II collagen (CII)−specific antibodies in the CIA mouse model. In addition, the intraperitoneal inoculation of *L. helveticus* decreased the arthritis incidence, IL-6 serum concentrations, and joint damage. Immune cells including B cells, germinal center B cells, and CD4 T cells in the draining lymph nodes also diminished following inoculation. These findings suggest the ability of this probiotic to downregulate the immune response, and the subsequent production of CII-specific antibodies and IL-6 thereby suppressing CIA symptoms [68]. *L. helveticus* inhibits the proliferation of lymphocytes through the suppression of the JNK signaling pathway [69].

In RA patients, the oral administration of *L. rhamnosus* and *L. reuteri* for 3 months promoted an improvement in the Health Assessment Questionnaire score with functional improvement [70]. Another trial revealed improvement in disease activity score, increased levels of serum IL-10, and lower levels of proinflammatory TNF-alpha, IL-6, and IL-12 cytokines with oral *L. casei* [71]. Likewise, *L. acidophilus*, *L. casei*, and *B. bifidum*, for 8 weeks, improved Disease Activity Score of 28 joints (DAS-28). In addition, there was a significant decrease in serum insulin levels, homeostatic model assessment-B cell function (HOMA-B), and serum high-sensitivity C-reactive protein concentrations in RA patients [72].

SYSTEMIC LUPUS ERYTHEMATOSUS

In female systemic lupus erythematosus (SLE)-prone mice, *Lactobacillus* species are less abundant, with expansion of *Lachnospiraceae* members. Early disease onset and severe symptoms correlated with *Lachnospiraceae* abundance. *Clostridiaceae* and *Lachnospiraceae* were more prominent at specific time points during disease progression [73]. Interestingly, caloric restriction in NZB/WF1 mice promoted changes in the gut microbiota and suppressed disease progression [74].

Reduced Firmicutes/Bacteroidetes ratio is also observed in SLE patients. Silico analysis suggested higher oxidative phosphorylation and glycan metabolism pathways induced by intestinal microbiota [75]. Less Firmicutes, elevated Bacteroidetes, and prevalence of *Rhodococcus*, *Eggerthella*, *Klebsiella*, *Prevotella*, *Eubacterium*, *Flavonifractor*, and *Incertae sedis* genera could be the microbiota fingerprint in SLE patients [76].

Fecal samples of affected patients induced Th17 differentiation, and the supplementation with Treg-inducing bacteria significantly decreased the Th17/Th1 balance, supporting the use of these strains as therapeutic probiotics for autoimmune diseases [77].

PROBIOTICS IN SLE

The administration of retinoic acid restored *Lactobacillus* species in lupus-like animal model and improved symptoms, suggesting the use of these species as a probiotic [78]. Some *Lactobacillus* species have immunomodulatory activities in the gut mucosa and inhibit neutrophil extracellular trap formation, improving antioxidant status and expression of adhesion molecules in the gut barrier [79,80].

The oral administration of *Lactobacillus paracasei* and *L. reuteri* to NZB/W F1 mice was followed by significant decrease in IL-6 and TNF serum concentrations stimulating antioxidant activity in liver and serum. Differentiation of CD4 + CD25 + FoxP3+ Treg cells in NZB/W F1 mice was also noticed with *L. reuteri* suggesting that this strain could be used as adjuvant treatment of SLE patients [81]. Experimentally, *L. paracasei* and *L. reuteri* ameliorate hepatic apoptosis, matrix metalloproteinase-9 activity, C-reactive protein and inducible nitric oxide synthase expression, along with reduction of hepatic IL-1β, IL-6, and TNF proteins, by suppressing the mitogen-activated protein kinase and NF-κB signaling pathways [82].

Although some studies in SLE animal models showed promising results using probiotic supplementation, currently, there are no clinical trials.

CONCLUSIONS

Dysbiosis have been associated with autoimmune disease pathogenesis. The gastrointestinal mucosa could represent a trigger site of autoimmunity, by neo-antigen generation under dysbiotic conditions. If this hypothesis is validated, new studies could collaborate in the discovery of immunomodulatory probiotics, predictive bacterial biomarkers of disease onset and disease progression, and new therapeutic approaches for use in clinical settings.

REFERENCES

[1] Wu HJ, Wu E. The role of gut microbiota in immune homeostasis and autoimmunity. Gut Microbes 2012;3:4–14.

[2] Rosser EC, Mauri C. A clinical update on the significance of the gut microbiota in systemic autoimmunity. J Autoimmune 2016;74:85–93.

[3] Lerner A, Aminov R, Matthias T. Dysbiosis may trigger autoimmune diseases via inappropriate post-translational modification of host proteins. Front Microbiol 2016;7:84.

[4] Chen B, Sun L, Zhang X. Integration of microbiome and epigenome to decipher the pathogenesis of autoimmune diseases. J Autoimmune 2017;83:31–42.

[5] Berer K, Mues M, Koutrolos M, Rasbi ZA, Boziki M, Johner C, Wekerle H, Krishnamoorthy G. Commensal microbiota and myelin autoantigen cooperate to trigger autoimmune demyelination. Nature 2011;479:538–41.

[6] Rolig AS, Mittge EK, Ganz J, Troll JV, Melancon E, Wiles TJ, Alligood K, Stephens WZ, Eisen JS, Guillemin K. The enteric nervous system promotes intestinal health by constraining microbiota composition. PLoS Biol 2017;15:e2000689.

[7] Van den Hoogen WJ, Laman JD, Hart BA. Modulation of multiple sclerosis and its animal model experimental autoimmune encephalomyelitis by food and gut microbiota. Front Immunol 2017;8:1081.

[8] Lee YK, Menezes JS, Umesaki Y, Mazmanian SK. Proinflammatory T-cell responses to gut microbiota promote experimental autoimmune encephalomyelitis. Proc Natl Acad Sci USA 2011;108:4615–22.

[9] Lee YK, Menezes JS, Umesaki Y, Mazmanian SK. Proinflammatory T-cell responses to gut microbiota promote experimental autoimmune encephalomyelitis. Proc Natl Acad Sci USA 2011;108:4615–22.

[10] Takata K, Kinoshita M, Okuno T, Moriya M, Kohda T, Honorat JA, Sugimoto T, Kumanogoh A, Kayama H, Takeda K, Sakoda S, Nakatsuji Y. The lactic acid bacterium *Pediococcus acidilactici* suppresses autoimmune encephalomyelitis by inducing IL-10-producing regulatory T cells. PLoS One 2011;6:e27644.

[11] Yadava SK, Boppanaa S, Ito N, Mindur JE, Mathay MT, Patel A, Dhib-Jalbut S, Ito K. Gut dysbiosis breaks immunological tolerance toward the central nervous system during young adulthood. Proc Natl Acad Sci USA 2017;114:E9318–27.

[12] Bhargava P, Mowey EM. Gut microbiome and multiple sclerosis. Curr Neurol Neurosci Rep 2014;14:492.

[13] Jhangi S, Gandhi R, Glanz B, Cook S, Nejad P, Ward D, Li N, Gerber G, Bry L, Weiner H. Increased Archaea species and changes with therapy in gut microbiome of multiple sclerosis subjects. Neurology 2014;82:S24.001.

[14] Miyake S, Kim S, Suda W, Oshima K, Nakamura M, Matsuoka T, Chihara N, Tomita A, Sato W, Kim SW, Morita H, Hattori M, Yamamura T. Dysbiosis in the gut microbiota of patients with multiple sclerosis, with a striking depletion of species belonging to Clostridia XIVa and IV clusters. PLoS One 2015;10:e0137429.

[15] Chen J, Chia N, Kalari KR, Yao JZ, Novotna M, Soldan MM, Luckey DH, Marietta EV, Jeraldo PR, Chen X, Weinshenker BG, Rodriguez M, Kantarci OH, Nelson H, Murray JA, Mangalam AK. Multiple sclerosis patients have a distinct gut microbiota compared to healthy controls. Sci Rep 2016;6:28484.

[16] Cantarel BL, Waubant E, Chehoud C, Kuczynski J, DeSantis TZ, Warrington J, Venkatesan A, Fraser CM, Mowry EM. Gut microbiota in multiple sclerosis: possible influence of immunomodulators. J Investig Med 2015;63:729–34.

[17] Bang C, Weidenbach K, Gutsmann T, Heine H, Schmitz RA. The intestinal Archaea *Methanosphaera stadtmanae* and *Methanobrevibacter smithii* activate human dendritic cells. PLoS One 2014;9:e99411.

[18] Derrien M, Van Baarlen P, Hooiveld G, Norin E, Müller M, de Vos WM. Modulation of mucosal immune response, tolerance, and proliferation in mice colonized by the mucin-degrader Akkermansia muciniphila. Front Microbiol 2011;2:166.

[19] Ganesh BP, Klopfleisch R, Loh G, Blaut M. Commensal *Akkermansia muciniphila* exacerbates gut inflammation in *Salmonella Typhimurium*-infected gnotobiotic mice. PLoS One 2013;8:e74963.

[20] Furusawa Y, Obata Y, Fukuda S, Endo TA, Nakato G, Takahashi D, Nakanishi Y, Uetake C, Kato K, Kato T, Takahashi M, Fukuda NN, Murakami S, Miyauchi E, Hino S, Atarashi K, Onawa S, Fujimura Y, Lockett T, Clarke JM, Topping DL, Tomita M, Hori S, Ohara O, Morita T, Koseki H, Kikuchi J, Honda K, Hase K, Ohno H. Commensal microbe-derived butyrate induces the differentiation of colonic regulatory T cells. Nature 2013;504:446–50.

[21] Jangi S, Gandhi R, Cox LM, Li N, Von Glehn F, Yan R, Patel B, Mazzola MA, Liu S, Glanz BL, Cook S, Tankou S, Stuart F, Melo K, Nejad P, Smith K, Topçuolu BD, Holden J, Kivisäkk P, Chitnis T, De Jager PL, Quintana FJ, Gerber GK, Bry L, Weiner HL. Alterations of the human microbiome in multiple sclerosis. Nat Commun 2016;7:12015.

[22] Wu GD, Chen J, Hoffmann C, Bittinger K, Chen YY, Keilbaugh SA, Bewtra M, Knights D, Walters WA, Knight R, Sinha R, Gilroy E, Gupta K, Baldassano R, Nessel L, Li H, Bushman FD, Lewis JD. Linking long-term dietary patterns with gut microbial enterotypes. Science 2011;334:105–8.

[23] Montassier E, Berthelot L, Soulillou JP. Are the decrease in circulating anti-α1,3-Gal IgG and the lower content of galactosyl transferase A1 in the microbiota of patients with multiple sclerosis a novel environmental risk factor for the disease? Mol Immunol 2018;93:162–5.

[24] Colpitts SL, Kasper LH. Influence of the gut microbiome on autoimmunity in the central nervous system. J Immunol 2017;198:596–604.

[25] Takata K, Kinoshita M, Okuno T, Moriya M, Kohda T, Honorat JA, Sugimoto T, Kumanogoh A, Kayama H, Takeda K, Sakoda S, Nakatsuji Y. The lactic acid bacterium Pediococcus acidilactici suppresses autoimmune encephalomyelitis by inducing IL-10-producing regulatory T cells. PLoS One 2011;6:e27644.

[26] Ochoa-Repáraz J, Mielcarz DW, Wang Y, Begum-Haque S, Dasgupta S, Kasper DL, Kasper LH. A polysaccharide from the human commensal Bacteroides fragilis protects against CNS demyelinating disease. Mucosal Immunol 2010;3:487–95.

[27] Lavasani S, Dzhambazov B, Nouri M, Fåk F, Buske S, Molin G, Thorlacius H, Alenfall J, Jeppsson B, Weström B. A novel probiotic mixture exerts a therapeutic effect on experimental autoimmune encephalomyelitis mediated by IL-10 producing regulatory T cells. PLoS One 2010;5:e9009.

[28] Kwon HK, Kim GC, Kim Y, Hwang W, Jash A, Sahoo A, Kim JE, Nam JH, Im SH. Amelioration of experimental autoimmune encephalomyelitis by probiotic mixture is mediated by a shift in T helper cell immune response. Clin Immunol 2013;146:217–27.

[29] Salehipour Z, Haghmorad D, Sankian M, Rastin M, Nosratabadi R, Soltan Dallal MM, Tabasi N, Khazaee M, Nasiraii LR, Mahmoudi M. Bifidobacterium animalis in combination with human origin of Lactobacillus plantarum ameliorate neuroinflammation in experimental model of multiple sclerosis by altering CD4+ T cell subset balance. Biomed Pharmacother 2017;95:1535–48.

[30] Kouchaki E, Tamtaii OR, Salami M, Bahmani F, Daneshvar Kakhaki R, Akbari E, Tajabadi-Ebrahimi M, Jafari P, Azemi Z. Clinical and metabolic response to probiotic supplementation in patients with multiple sclerosis: a randomized, double-blind, placebo-controlled trial. Clin Nutr 2016. S0261-5614:30214–X.

[31] Tamtaji OR, Kouchaki E, Salami M, Aghadavod E, Akbari E, Tajabadi-Ebrahimi M, Asemi Z. The effects of probiotic supplementation on gene expression related to inflammation, insulin, and lipids in patients with multiple sclerosis: a randomized, double-blinded, placebo-controlled trial. J Am Coll Nutr 2017;36:660–5.

[32] Saresella M, Mendozzi L, Rossi V, Mazzali F, Piancone F, LaRosa F, Marventano I, Caputo D, Felis GE, Clerici M. Immunological and clinical effect of diet modulation of the gut microbiome in multiple sclerosis patients: a pilot study. Front Immunol 2017;8:1391.

[33] Knip M, Honkanen J. Modulation of Type 1 diabetes risk by the intestinal microbiome. Curr Diab Rep 2017;17:105.

[34] Wen L, Ley RE, Volchkov PY, Stranges PB, Avanesyan L, Stonebraker AC, Hu C, Wong FS, Szot GL, Bluestone JA, Gordon JI, Chervonsky AV. Innate immunity and intestinal microbiota in the development of Type 1 diabetes. Nature 2008;455:1109–13.

[35] Alam C, Bittoun E, Bhagwat D, Valkonen S, Saari A, Jaakkola U, Eerola E, Huovinen P, Hänninen A. Effects of a germ-free environment on gut immune regulation and diabetes progression in non-obese diabetic (NOD) mice. Diabetologia 2011;54:1398–406.

[36] Kriegel MA, Sefik E, Hill JA, Wu HJ, Benoist C, Mathis D. Naturally transmitted segmented filamentous bacteria segregate with diabetes protection in nonobese diabetic mice. Proc Natl Acad Sci USA 2011;108:11548–53.

[37] Emani R, Alam C, Pekkala S, Zafar S, Emani MR, Hänninen A. Peritoneal cavity is a route for gut-derived microbial signals to promote autoimmunity in non-obese diabetic mice. Scand J Immunol 2015;81:102–9.

[38] Roesch LF, Lorca GL, Casella G, Giongo A, Naranjo A, Pionzio AM, Li N, Mai V, Wasserfall CH, Schatz D, Atkinson MA, Neu J, Triplett EW. Culture-independent identification of gut bacteria correlated with the onset of diabetes in a rat model. ISME J 2009;3:536–48.

[39] Bosei E, Molteni L, Radaelli MG, Folini L, Fermo I, Bazzigaluppi E, Piemonti L, Pastore MR, Paroni R. Increased intestinal permeability precedes clinical onset of type 1 diabetes. Diabetologia 2006;49:2824–7.

[40] Vaarala O. Human intestinal microbiota and type 1 diabetes. Curr Diab Rep 2013;13:601–7.

[41] Giongo A, Gano KA, Crabb DB, Mukherjee N, Novelo LL, Casella G, Drew JC, Ilonen J, Knip M, Huöty H, Veijola R, Simell T, Simell O, Neu J, Wasserfall CH, Schatz D, Atkinson MA, Triplett EW. Toward defining the autoimmune microbiome for type 1 diabetes. ISME J 2011;5:82–91.

[42] Murri M, Leiva I, Gomez-Zumaquero JM, Tinahones FJ, Cardona F, Soriguer F, Queipo-Ortuño MI. Gut microbiota in children with type 1 diabetes differs from that in healthy children: a case-control study. BMC Med 2013;21:11–46.

[43] De Goffau MC, Luopajärvi K, Knip M, Ilonen J, Ruohtula T, Härkönen T, Orivuori L, Hakala S, Welling GW, Harmsen HJ, Vaarala O. Fecal microbiota composition differs between children with β-cell autoimmunity and those without. Diabetes 2013;62:1238–44.

[44] Davis-Richardson AG, Ardissone AN, Dias R, Simell V, Leonard MT, Kemppainen KM, Drew JC, Schatz D, Atkinson MA, Kolaczkowski B, Ilonen J, Knip M, Toppari J, Nurminen N, Hyöty H, Veijola R, Simell T, Mykkänen J, Simell O, Triplett EW. Bacteroides dorei dominates gut microbiome prior to autoimmunity in Finnish children at high risk for type 1 diabetes. Front Microbiol 2014;10(5):678.

[45] Li X, Atkinson MA. The role for gut permeability in the pathogenesis of type 1 diabetes - a solid or leaky concept? Pediatr Diabetes 2015;16:485–92.

[46] Meiía-León ME, Petrosino JF, Aiami NJ, Domínguez-Bello MG, De La Barca AM. Fecal microbiota imbalance in Mexican children with type 1 diabetes. Sci Rep 2014;4:3814.

[47] Paun A, Yau C, Danska JS. The influence of the microbiome on type 1 diabetes. J Immunol 2017;198:590−5.

[48] Calcinaro F, Dionisi S, Marinaro M, Candeloro P, Bonato V, Marzotti S, Corneli RB, Ferretti E, Gulino A, Grasso F, De Simone C, Di Mario U, Falorni A, Boirivant M, Dotta F. Oral probiotic administration induces interleukin-10 production and prevents spontaneous autoimmune diabetes in the non-obese diabetic mouse. Diabetologia 2005;48:1565−75.

[49] Dolpady J, Sorini C, Di Pietro C, Cosorich I, Ferrarese R, Saita D, Clementi M, Canducci F, Falcone M. Oral probiotic VSL#3 prevents autoimmune diabetes by modulating microbiota and promoting indoleamine 2,3-dioxygenase-enriched tolerogenic intestinal environment. J Diabetes Res 2016;2016:7569431.

[50] Brown K, Godovannyi A, Ma C, Zhang Y, Ahmadi-Vand Z, Dai C, Gorzelak MA, Chan Y, Chan JM, Lochner A, Dutz JP, Vallance BA, Gibson DL. Prolonged antibiotic treatment induces a diabetogenic intestinal microbiome that accelerates diabetes in NOD mice. ISME J 2016;10:321−32.

[51] Jia L, Shan K, Pan LL, Feng N, Lv Z, Sun Y, Li J, Wu C, Zhang H, Chen W, Diana J, Sun J, Chen YQ. Clostridium butyricum CGMCC0313.1 protects against autoimmune diabetes by modulating intestinal immune homeostasis and inducing pancreatic regulatory T cells. Front Immunol 2017;8:1345.

[52] De Oliveira GLV, Leite AZ, Higuchi BS, Gonzaga MI, Mariano VS. Intestinal dysbiosis and probiotic applications in autoimmune diseases. Immunology 2017;152:1−12.

[53] Uusitalo U, Liu X, Yang J, Aronsson CA, Hummel S, Butterworth M, Lernmark Å, Rewers M, Hagopian W, She JX, Simell O, Toppari J, Ziegler AG, Akolkar B, Krischer J, Norris JM, Virtanen SM, TEDDY Study Group. Association of early exposure of probiotics and islet auto-immunity in the TEDDY study. JAMA Pediatr 2016;170:20−8.

[54] Groele L, Szajewska H, Szypowska A. Effects of Lactobacillus rhamnosus GG and Bifidobacterium lactis Bb12 on beta-cell function in children with newly diagnosed type 1 diabetes: protocol of a randomized controlled trial. BMJ Open 2017;7:e017178.

[55] Papizadeh M, Rohani M, Nahrevanian H, Javadi A, Pourshafie MR. Probiotic characters of *Bifidobacterium* and *Lactobacillus* are a result of the ongoing gene acquisition and genome minimization evolutionary trends. Microb Pathog 2017;111:118−31.

[56] Brusca SB, Abramson SB, Scher JU. Microbiome and mucosal inflammation as extra-articular triggers for rheumatoid arthritis and autoimmunity. Curr Opin Rheumatol 2014;26:101−7.

[57] Horta-Baas G, Romero-Figueroa MS, Montiel-Jarquín AJ, Pizano-Zárate ML, García-Mena J, Ramírez-Durán N. Intestinal dysbiosis and rheu-matoid arthritis: a link between gut microbiota and the pathogenesis of rheumatoid arthritis. J Immunol Res 2017;2017:4835189.

[58] Dorożyńska I, Majewska-Szczepanik M, Marcińska K, Szczepanik M. Partial depletion of natural gut flora by antibiotic aggravates collagen induced arthritis (CIA) in mice. Pharmacol Rep 2014;66:250−5.

[59] Liu X, Zeng B, Zhang J, Li W, Mou F, Wang H, Zou Q, Zhong B, Wu L, Wei H, Fang Y. Role of the gut microbiome in modulating arthritis progression in mice. Sci Rep 2016;6:30594.

[60] Gomez A, Luckey D, Yeoman CJ, Marietta EV, Berg Miller ME, Murray JA, White BA, Taneja V. Loss of sex and age driven differences in the gut microbiome characterize arthritis-susceptible 0401 mice but not arthritis-resistant 0402 mice. PLoS One 2012;7:e36095.

[61] Wu HJ, Ivanov II, Darce J, Hattori K, Shima T, Umesaki Y, Littman DR, Benoist C, Mathis D. Gut-residing segmented filamentous bacteria drive autoimmune arthritis via T helper 17 cells. Immunity 2010;32:815−27.

[62] Scher JU, Sczesnak A, Longman RS, Segata N, Ubeda C, Bielski C, Rostron T, Cerundolo V, Pamer EG, Abramson SB, Huttenhower C, Littman DR. Expansion of intestinal Prevotella copri correlates with enhanced susceptibility to arthritis. Elife 2013;2:e01202.

[63] Maeda Y, Kurakawa T, Umemoto E, Motooka D, Ito Y, Gotoh K, Hirota K, Matsushita M, Furuta Y, Narazaki M, Sakaguchi N, Kayama H, Nakamura S, Iida T, Saeki Y, Kumanogoh A, Sakaguchi S, Takeda K. Dysbiosis contributes to arthritis development via activation of autoreactive T cells in the intestine. Arthritis Rheumatol 2016;68:2646−61.

[64] Chen J, Wright K, Davis JM, Jeraldo O, Marietta EV, Murray J, Nelson H, Matteson EL, Taneja V. An expansion of rare lineage intestinal microbes characterizes rheumatoid arthritis. Genome Med 2016;8:43.

[65] Liu X, Zou Q, Zeng B, Fang Y, Wei H. Analysis of fecal *Lactobacillus* community structure in patients with early rheumatoid arthritis. Curr Microbiol 2013;67:170−6.

[66] Di Paola M, Cavalieri D, Albanese D, Sordo M, Pindo M, Donati C, Pagnini I, Giani T, Simonini G, Paladini A, Lionetti P, De Filippo C, Cimaz R. Alteration of fecal microbiota profiles in juvenile idiopathic arthritis. Associations with HLA-B27 allele and disease status. Front Microbiol 2016;7:979.

[67] Abhari K, Shekarforoush SS, Hosseinzadeh S, Nazifi S, Sajedianfard J, Eskandari MH. The effects of orally administered *Bacillus coagulans* and inulin on prevention and progression of rheumatoid arthritis in rats. Food Nutr Res 2016;60:30876.

[68] Yamashita M, Matsumoto K, Endo T, Ukibe K, Hosoya T, Matsubara Y, Nakagawa H, Sakai F, Miyazaki T. Preventive effect of *Lactobacillus helveticus* SBT2171 on collagen-induced arthritis in mice. Front Microbiol 2017;8:1159.

[69] Hosoya T, Sakai F, Yamashita M, Shiozaki T, Endo T, Ukibe K, Uenishi H, Kadooka Y, Moriya T, Nakagawa H, Nakayama Y, Miyazaki T. Lactobacillus helveticus SBT2171 inhibits lymphocyte proliferation by regulation of the JNK signaling pathway. PLoS One 2014;9:e108360.

[70] Pineda Mde L, Thompson SF, Summers K, De Leon F, Pope J, Reid G. A randomized, double-blinded, placebo-controlled pilot study of probiotics in active rheumatoid arthritis. Med Sci Monit 2011;17:CR347−54.

[71] Vaghef-Mehrabany E, Alipour B, Homayouni-Rad A, Sharif SK, Asghari-Jafarabadi M, Zavvari ZS. Probiotic supplementation improves in-flammatory status in patients with rheumatoid arthritis. Nutrition 2014;30:430−5.

[72] Zamani B, Golkar HR, Farshbaf S, Emadi-Baygi M, Tajabadi-Ebrahimi M, Jafari P, Akhavan R, Taghizadeh M, Memarzadeh MR, Asemi Z. Clinical and metabolic response to probiotic supplementation in patients with rheumatoid arthritis: a randomized, double-blind, placebo-controlled trial. Int J Rheum Dis 2016;19:869−79.

[73] Zhang H, Liao X, Sparks JB, Luo XM. Dynamics of gut microbiota in autoimmune lupus. Appl Environ Microbiol 2014;80:7551−60.

[74] Hsieh CC, Lin BF. Dietary factors regulate cytokines in murine models of systemic lupus erythematosus. Autoimmun Rev 2011;11:22−7.

[75] Hevia A, Milani C, López P, Cuervo A, Arboleya S, Duranti S, Turroni F, González S, Suárez A, Gueimonde M, Ventura M, Sánchez B, Margolles A. Intestinal dysbiosis associated with systemic lupus erythematosus. mBio 2014;5:e01548−14.

[76] He Z, Shao T, Li H, Xie Z, Wen C. Alterations of the gut microbiome in Chinese patients with systemic lupus erythematosus. Gut Pathog 2016;8:64.

[77] López P, de Paz B, Rodrígues-Carrio J, Hevia A, Sánchez B, Margolles A, Suárez A. Th17 responses and natural IgM antibodies are related to gut microbiota composition in systemic lupus erythematosus patients. Sci Rep 2016;6:24072.

[78] Zhang H, Liao X, Sparks JB, Luo XM. Dynamics of gut microbiota in autoimmune lupus. Appl Environ Microbiol 2014;80:7551−60.

[79] Vong L, Lorentz RJ, Assa A, Glogauer M, Sherman PM. Probiotic *Lactobacillus rhamnosus* inhibits the formation of neutrophil extracellular traps. J Immunol 2014;192:1870−7.

[80] Mu Q, Zhang H, Luo XM. SLE: another autoimmune disorder influenced by microbes and diet? Front Immunol 2015;6:608.

[81] Tzang BS, Liu CH, Hsu KC, Chen YH, Huang CY, Hsu TC. Effects of oral Lactobacillus administration on antioxidant activities and CD4+CD25+forkhead box P3 (FoxP3)+ T cells in NZB/W F1 mice. Br J Nutr 2017;118:333−42.

[82] Hsu TC, Huang CY, Liu CH, Hsu KC, Chen YH, Tzang BS. *Lactobacillus paracasei* GMNL-32, Lactobacillus reuteri GMNL-89 and L. reuteri GMNL-263 ameliorate hepatic injuries in lupus-prone mice. Br J Nutr 2017;117:1066−74.

Chapter 34

The Gut Microbiome and Metabolome in Multiple Sclerosis

Shailendra Giri[1] and Ashutosh Mangalam[2]

[1]Department of Neurology, Henry Ford Health System, Detroit, MI, United States; [2]Department of Pathology, University of Iowa Carver College of Medicine, Iowa City, IA, United States

INTRODUCTION

The critical role of the gut microbiota in multiple sclerosis (MS) is explained by studies demonstrating the presence of gut microbial dysbiosis in MS patients (reviewed in Ref. [1]), suppression of disease in germ-free mice on fecal transfer from healthy controls (HCs), and exacerbation of disease on fecal transfer from MS patients.

In this chapter we will discuss the following: (1) the composition of gut microbiota in MS patients, (2) the factors regulating the composition of the gut microbiota, (3) the regulation of host metabolism by the gut microbiota, (4) the potential for their use as diagnostic/prognostic markers, and (5) the usage of gut bacteria as a drug (brug).

GUT MICROBIOTA

Changes in the gut microbiota, and subsequent changes in its metabolic network, disrupt homeostasis within the host and can lead to both intestinal and systemic disorders such as MS, rheumatoid arthritis, and other autoimmune diseases. The gut microbiota consists of a variety of organisms, including bacteria, viruses, archaea, and fungi, yet the majority of studies have focused only on bacteria and archaea [1–4].

MS AND THE GUT MICROBIOTA

In the last few years our group and several others have shown that the gut microbiota profile in MS patients is distinct from that of HCs, with both depletion and enrichment of certain bacteria observed, suggesting that MS patients exhibit gut dysbiosis. The majority of bacteria ($\sim$90%) in the adult human gut belong to either the Firmicutes or Bacteroidetes phyla, with the remaining bacteria belonging to the Actinobacteria, Proteobacteria, or other phyla [5].

Despite the large amount of heterogeneity in the human microbiome, a small number of specific gut bacteria are depleted or enriched in patients with MS, as summarized in Table 34.1. Compared with HCs, MS patients had a decrease in the abundance of *Prevotella, Sutterella, Adlercreutzia, Collinsella, Clostridium, Faecalibacterium,* and *Parabacteroides* (Table 34.1) [1,4]. Using more stringent cutoff criteria to select gut bacteria, reported to be lower in at least two independent studies of adult MS subjects (>10 subjects), three gut bacteria *Prevotella, Parabacteroides,* and *Adlercreutzia,* showed lower abundance in MS.

Our group and several others have shown lower levels of *Prevotella* in MS patients compared with HCs (Table 34.1) [1,4], and duodenal biopsies from MS patients with active disease showed a lower abundance of *Prevotella* compared with HCs or patients in remission [6]. In the same study, levels of *Prevotella* were also negatively correlated with a T helper 17 (Th17) cells response, which is associated with increased disease severity in MS. In two of the three studies analyzing the adult MS population in the USA, a lower abundance of *Parabacteroides* in MS patients was observed, relative to HCs [7,8].

Cekanaviciute et al. [8] specifically reported a lower abundance of *Parabacteroides distasonis* in MS patients, suggesting that *P. distasonis* may provide protection from MS. Additionally, germ-free (GF) mice colonized with fecal material from HCs had higher levels of *Adlercreutzia,* compared with those colonized with fecal materials from MS

334 BLOCK | IV Diagnostic and Therapeutic Applications

TABLE 34.1 Comparison of Adult MS Microbiome Studies

MS Microbiome Study # Samples Tissue (Country) (Reference)	Bacteria Genera Showing Lower Abundance in MS Patients Versus HC	Bacteria Genera Showing Lower Abundance in MS Patients After Treatment
RRMS (n = 31) HC (n = 36) Fecal (USA) Chen et al. [7]	*Prevotella, Parabacteroides*, Adlercreutzia, Collinsella Lactobacillus, Coprobacillus	
RRMS (n = 60) HC (n = 43) Fecal (USA) Jangi et al. [16]	Butyricimonas *Prevotella Parabacteroides* (see text)	*Prevotella* Sutterella
RRMS (n = 20) HC (n = 40) Fecal (Japan) Miyake et al. [29]	Bacteroides, Faecalibacterium *Prevotella*, Anaerostipes Clostridium, Sutterella	
RRMS (n = 30) HC (n = 14) Fecal (UK) Castillo et al. [30]		*Prevotella*
Treatment-naïve MS (n = 64) Fecal (USA) Cekanaviciute et al. [8]	*Parabacteroides*	
Monozygotic twins (n = 32) Fecal (Germany) Berer et al. [9]	Lower levels of *Adlercreutzia* in GF mice colonized with fecal samples from MS patients versus GF mice colonized with fecal samples from HCs	
RRMS (n = 19) HC (n = 17) Mucosa (Italy) Cosorich et al. [7]	*Prevotella*	

MS, multiple sclerosis; HCs, healthy controls; GF, germ-free. Gut bacteria common between different MS microbiome studies have been bolded. ECTRIMS 2016.

patients [9]. Thus these studies indicate that the loss of *Prevotella, Parabacteroides, and Adlercreutzia* plays a significant role in MS pathogenesis.

Certain bacteria are also present in greater abundance in MS patients compared with HCs, including *Akkermansia, Acinetobacter, Dorea, Blautia, Pseudomonas,* and Archaea-*Methanobrevibacter*. Interestingly *Acinetobacter* species and *Pseudomonas aeruginosa*, both of which are more abundant in MS patients, are common multidrug-resistant bacteria, responsible for hospital-acquired infections [10]. Thus microbiome studies indicate that gut dysbiosis occurs in MS patients, and current research is consistent with the depletion of bacteria, rather than enrichment, as a potential causative factor for disease.

GUT MICROBIOTA AND MODULATION OF THE IMMUNE RESPONSE

Because MS is an inflammatory disease and the majority of MS drugs work by blocking proinflammatory pathways, it is reasonable to believe that the gut microbiota could influence the disease, by modulating the host immune response. The importance of gut bacteria in maintaining immune homeostasis, in part by keeping a balance between antiinflammatory regulatory T cells (Tregs) and proinflammatory Th1/Th17 cells, is now well established [11,12]. GF mice have a normal Th1 response, but have significantly reduced Tregs and Th17 cells, indicating that gut bacteria are essential for the generation of these populations.

MS is an inflammatory disease in which both Th1 and Th17 cells are pathogenic; thus gut bacteria that are more abundant in MS patients can be considered proinflammatory/pathobionts, whereas less abundant gut bacteria can be considered antiinflammatory/symbionts. Indeed, an MS microbiome study suggests that MS patients are depleted of bacteria that induce immunoregulatory cells and have a greater abundance of bacteria that induce a proinflammatory response [4].

Bacterial Fingerprint

In this context *Prevotella, Parabacteroides, Clostridium,* and *Adlercreutzia* can be considered as antiinflammatory symbionts, while *Akkermansia, Methanobrevibacter, Pseudomonas,* and *Acinetobacter* can be classified as proinflammatory pathobionts [1]. This is supported by the fact that *Prevotella* is negatively associated with Th17 cells in MS patients and that *Prevotella histicola*, the bacteria belonging to the *Prevotella* genus, can induce CD4$^+$FoxP3$^+$ Tregs and suppress disease in an animal model of MS [13].

Additionally, *P. histicola* can suppress disease in an animal model of arthritis [14]. Culture of human peripheral blood mononuclear cells (PBMCs) with *P. distasonis* (reported as lower in MS patients), promotes the generation of CD4$^+$CD25$^+$IL10$^+$ T cells [8]. Furthermore, a mixture of rationally isolated 17 *Clostridium* strains from human microbiota promoted the generation of CD4$^+$FoxP3$^+$ Tregs and production of the antiinflammatory cytokine IL-10, by generating a Transforming growth factor (TGF)β-rich environment [15].

Short-Chain Fatty Acids

This could be due to the ability of *Clostridium* strains to metabolize fiber and complex plant carbohydrates into short-chain fatty acids (SCFAs) that might play a crucial role in promoting the generation of the Treg population. Interestingly only MS patients from Japan showed a lower abundance of *Clostridium* in their gut microbiota, whereas MS microbiome studies from the USA and Europe did not observe a difference in the abundance of *Clostridium* [1]. This raises the possibility that geographical differences in gut microbial communities exist. *Prevotella* has emerged as a common gut microbe, with lower abundance in MS patients across geographical regions (USA, Europe, and Asia) [1,4].

Conversely *Akkermansia* and *Acinetobacter calcoaceticus* are linked with induction of a proinflammatory response. The presence of *Methanobrevibacter* (Archae) and *Akkermansia* are associated with increased expression of innate and adaptive immune response genes [16], and *A. calcoaceticus* induces a Th1/Th17 response and reduces Treg frequencies [8].

THE GUT MICROBIOTA AND METABOLIC PATHWAYS

The gut microbiota affects host physiology either directly, through production of chemicals such as lipopolysaccharide, peptidoglycan, or polysaccharide A (PSA), or indirectly through the metabolism of dietary components (fibers, phytoestrogens, tryptophan) or host-derived compounds (bile acids) [1] (Fig. 34.1). The generation of SCFAs by gut bacteria, from indigestible starch and sugar, is among the most studied gut microbiota-induced pathways. With help of specific gut microbiota, SCFAs produced in the colon are essential for both the health of colonic epithelial cells and the induction of regulatory CD4$^+$FoxP3$^+$ Tregs.

Regulatory T cells

A lack of Tregs and/or abnormal Tregs, with decreased suppressive abilities, is linked to MS development and relapse. MS patients lack some bacterial species capable of producing SCFAs, such as *Prevotella, Parabacteroides, Faecalibacterium, Clostridium,* and *Lactobacillus*, which could play a crucial role in MS pathogenesis. Bacteria from both Firmicutes and Bacteroidetes phyla can produce SCFAs, and thus can help in maintaining immune homeostasis.

Phytoestrogens

Gut bacteria are also involved in the metabolism of phytoestrogens, and a number of phytoestrogen-metabolizing bacteria (*Parabacteroides, Lactobacillus, Prevotella, and Adlercreutzia*) are less abundant in patients with MS [1]. The metabolism of phytoestrogen compounds, such as plant lignans present in flaxseeds and isoflavones present in soybeans/soy-based products, leads to the generation of enterolactone and S-equol. Owing to structural similarities with human estrogens, enterolactone and S-equol can modulate the host immune response and provide protection from development of MS [17].

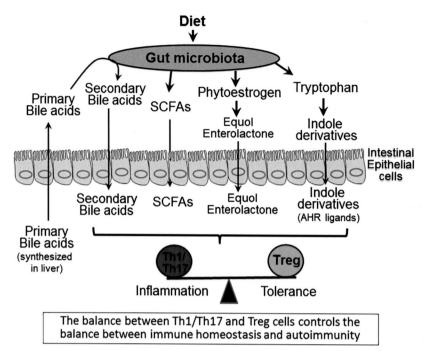

FIGURE 34.1 Gut bacteria help in the metabolism of multiple pathways, which can modulate host immune response. Gut bacteria metabolize a number of diet or host compounds, such as phytoestrogen, secondary bile acids, and tryptophan, to produce metabolites such as SCFAs, equol/enterolactone, secondary bile acids, and indole derivatives, respectively. These small metabolites help in maintaining immune homeostasis, by keeping the balance between proinflammatory Th1/Th17 cells and antiinflammatory Tregs. SCFAs, short-chain fatty acids; Tregs, regulatory T cells.

Animal studies suggest that phytoestrogen, in particular the isoflavone genistein, can suppress experimental autoimmune encephalomyelitis (EAE) [18] and reduce proinflammatory cytokine levels. Furthermore, S-equol has a direct protective effect on neuronal and glial cells, in the central nervous system (CNS). Owing to their health-promoting capabilities, phytoestrogens are being tested for use as potential therapeutics for multiples diseases, including cancer, cardiovascular diseases, and type 2 diabetes.

Tryptophan and Indole Metabolites

Tryptophan, an essential amino acid present in a number of plant-based foods, can be metabolized by gut bacteria into indole and its derivatives and modulates the host immune system by activating aryl hydrocarbon receptor (AhR) or pregnane X receptor [19,20]. Mice lacking the AhR develop an abnormal immune system and gut dysbiosis, supporting a role for AhR/tryptophan in the regulation of immune response. Additionally, AhR ligands induced by the gut microbiota attenuate an inflammatory immune response.

MS patients show lower levels of AhR agonists, as well as *Lactobacillus*, the latter being a known metabolizer of tryptophan. Consistent with this, development of clinical symptoms in mouse model of MS (EAE) is associated with the loss of *Lactobacillus,* and treatment with the brug, *P. histicola,* suppresses disease and restores *Lactobacillus* levels to those observed before disease onset [13]. Thus the metabolism of tryptophan by *Lactobacillus* promotes signaling through the AhR, resulting in modulation of the immune system.

The metabolism of tryptophan is also affected by high salt intake. A study in mice identified that high salt intake depleted the gut microbiota, mainly *Lactobacillus murinus*, and exacerbated EAE [21]. Supplementation of *L. murinus* ameliorated the high salt—induced exacerbation of EAE, by attenuating Th17 cells. Because the *Lactobacillus* species is known to metabolize tryptophan into indole metabolites that are protective, depletion of *L murinus* is associated with a significant reduction in fecal indole-3-lactic acid (ILA) and indole-3-acetic acid (IAA), in high salt-induced EAE.

GF mice monocolonized with an *L. murinus* isolate exhibited higher levels of fecal ILA and IAA, indicating the capability of *L. murinus* to produce these indoles. Notably ILA significantly reduced the differentiation of Th17 cells, suggesting that the disease-suppressive mechanism of *L. murinus* may be through indole-mediated suppression of Th17 cell differentiation.

These findings are well corroborated with an exploratory pilot study in human male volunteers, where participants were subjected to high salt intake for 2 weeks. Nine of 10 *Lactobacillus* populations that were initially present in high salt intake participants could no longer be detected, and participants had a significantly higher frequency of Th17 cells in the blood.

Bile Acid Metabolism

Gut bacteria also play an important role in the metabolism of primary and secondary bile acids. Bile acids are produced by the liver and secreted into the intestine, where they may be biotransformed into secondary bile acids, by enzymes produced by gut bacteria, such as *Parabacteroides*. Primary bile acids are proinflammatory, whereas secondary bile acids are thought to be antiinflammatory. Currently there is no report of altered secondary bile acid metabolites in MS patients.

Patients with inflammatory bowel disease have higher levels of primary, proinflammatory bile acids and lower levels of secondary, antiinflammatory bile acids. In addition, reduced levels of secondary bile acids, including deoxycholate, taurodeoxycholate, and alpha-mucholic acid, were found in the plasma of EAE-induced mice [22], suggesting that decreased levels of secondary, antiinflammatory bile acids could be a contributing factor to the severity of EAE disease.

Bile acids are ligands for the nuclear hormone receptor, Farnesoid X receptor (FXR), which regulates bile acid synthesis, transport, and cholesterol metabolism. It was reported that a synthetic bile acid agonist, obeticholic acid, ameliorated clinical disease in established EAE [23], suggesting that bile acids act as agonists of FXR and consequently function as negative regulators of neuroinflammation.

Human Studies With Bile Acids

A phase 1/2 clinical trial of bile acid (tauroursodeoxycholic acid) supplementation in progressive MS patients (NCT03423121) is currently underway to test its safety and tolerability as a therapeutic option in MS patients. The findings from this study will assess the impact of bile acid on a patient's immune system and gut microbiome.

THE INTERFACE BETWEEN THE GUT MICROBIOTA AND PHARMACEUTICAL AGENTS

The contribution of the gut microbiota to MS pathogenesis was first confirmed through antibiotic-depletion studies, performed in an animal model of the disease. EAE is a widely used animal model of MS and simulates a number of features of the disease. Depletion of gut flora with broad-spectrum antibiotics resulted in milder EAE, suggesting that the gut microbiota plays a crucial role in maintaining peripheral immune homeostasis [24].

We have also observed that induction of disease leads to loss of *Prevotella* and *Sutterella* in EAE-challenged mice compared with naïve mice [13]. In humans, MS patients receiving disease-modifying therapies, such as Copaxone or interferon beta, have higher levels of *Prevotella* and *Sutterella* compared with untreated patients [16].

Furthermore, the gut microbiota is crucial for the beneficial effects of interferon-beta therapy in MS patients [25]. Interferon-beta treatment suppresses inflammation in the CNS. However, depletion of gut bacteria through the use of antibiotics disrupts this effect. The antiinflammatory effects of beta-interferons are mediated, in part, through the AhR. As described previously the gut microbiota plays a major role in production of AhR agonists through the metabolism of tryptophan. However, more work is needed to decipher the bidirectional interactions between drugs and the gut microbiota.

DIAGNOSTIC IMPLICATIONS

In MS patients, enrichment of certain bacteria such as *Akkermansia and Acinetobacter* and depletion of others such as *Prevotella, Parabacteroides, Lactobacillus*, and *Adlercreutzia* suggests that a microbial signature panel can be developed for use as a potential diagnostic marker for MS patients. However, studies to date have reported more consistent results, regarding bacteria that are depleted in MS patients, relative to bacteria that are enriched. Thus disease-linked bacteria need to be confirmed in large cohorts of patients, followed by independent validation in a fresh cohort. Another possible avenue to explore is to assemble a panel of existing bacteria, that are known to be in lower or higher abundance in MS patients, and validate this in a larger cohort.

Diagnostic and Therapeutic Potential of Gut Metabolites

Metabolites derived from bacteria could also be isolated and administered therapeutically to MS patients. SCFA production is dependent on the gut microbiota, and a correlation between higher levels of SCFAs and disease protection has been observed in certain animal studies. Furthermore, phytoestrogen-metabolizing bacteria (*Prevotella*, *Parabacteroides*, *Lactobacillus*, and *Adlercreutzia*) are lower in abundance in MS patients. S-equol and enterolactone are the final metabolites of this pathway and can modulate immune response in an animal model of MS. Thus investigation into their use as potential biomarkers for diagnosis and prognosis is a promising area of continued study.

USE OF GUT BACTERIA AS POTENTIAL THERAPEUTIC AGENTS

There is great interest in developing gut bacteria–based therapies for MS, termed brug, for bacteria-based drugs. Lower levels of *Prevotella* in MS patients and the ability of *P. histicola* to suppress disease in an animal model of MS highlight an important immunomodulatory role of *Prevotella* in MS. *P. histicola* suppressed disease through induction of Tregs and dendritic cells (DCs) [13]. Additionally, *P. histicola* can suppress disease in animal model of rheumatoid arthritis [14].

The therapeutic potential of other bacteria has also been explored. The *Bacteroid fragilis*–derived molecule, PSA, can suppress EAE and promote the generation of Tregs from naïve CD4 T cells, derived from human PBMCs. Studies are currently underway to test the safety and efficacy of *P histicola* and PSA for human use. Multistrain (a mixture of more than one strain of the same bacteria genera) or multispecies (a mixture of more than one bacterial species) probiotics have been tested in an EAE animal model with promising results [26].

In mice, treatment with multistrain *Lactobacillus* suppresses EAE because of induction of Tregs and suppression of the inflammatory response. Similarly a mixture of three bacterial species including *Lactobacillus*, *Bifidobacteria*, and *Streptococcus thermophilus* has also shown to be beneficial in an EAE model. While promising, the efficacy of these treatments needs to be tested in MS patients.

Fecal Microbiota Transplantation/FMT

Thus far, Fecal microbiota transplantation (FMT) is only approved in cases of *Clostridium difficile*–induced diarrhea, and the direct translation of these results to other diseases is difficult to discern, for a number of reasons [28]. Primarily patients with *C. difficile*–induced diarrhea lack most of the gut microbes except *C. difficile*, thus colonization of these patients is relatively easy. In contrast the gut microbiota is intact in patients with MS, albeit with increased pathobionts. Therefore methods need to be standardized to ensure that FMT results in an increase in symbionts, rather than pathobionts. Currently there is a clinical trial underway to test the safety and efficacy of FMT in MS patients. Continued testing of the safety and efficacy of this technology is required before this can be offered as a treatment option to MS patients.

DIET AND THE GUT MICROBIOTA

In MS the importance of a healthy diet and lifestyle is highlighted by the fact that obesity is a risk factor for both disease development and disease severity. In addition, in countries closer to Polar regions (i.e., North America, Europe, and Australia) there is a well-established association between lower levels of Vitamin D and MS. Besides vitamin D, lower levels of vitamin A and E are also linked with MS development. Currently there are a number of on-going clinical trials to assess the therapeutic effect of different diets on MS, including the Paleo diet versus Swank diet, fasting mimicking diet, and gluten-free diets.

A fiber-rich diet has been linked with healthy state because of increased production of SCFAs. In EAE, long-chain fatty acid intake decreases SCFAs in the gut and exacerbates disease by expanding pathogenic Th1 and/or Th17 populations in the small intestine. Treatment with SCFAs ameliorates EAE by inducing Tregs. Fatty acids can be beneficial regardless of the chain length, especially if they are unsaturated. In particular, polyunsaturated fatty acids (PUFAs) are known for their antiinflammatory effects, in a number of human diseases, and there is considerable interest in the potential antiinflammatory effect of PUFAs on MS.

There is a modest association between consumption of low levels of unsaturated fat and an increased incidence of MS. Among the specific types of PUFAs, only alpha-linolenic acid (ALA) was inversely associated with MS risk. ALA is an essential omega-3 fatty acid found in plants, similar to omega-3 fatty acids in fish oil. Similar to omega-3 fatty acids, ALA is metabolized into eicosapentaenoic acid (EPA) and docosahexaenoic acid (DHA). Notably MS patients and mice with EAE disease have lower levels of omega-3 fatty acids and their downstream metabolite, resolvin D1, and supplementation of resolvin D1 in mice ameliorated EAE disease [27].

Additionally, EPA- or DHA-enriched diets are protective in EAE models, although their impact on the gut microbiota has not been studied. Studies in healthy adults showed changes in the gut microbiota after omega-3 supplementation. In particular a decrease in *Faecalibacterium* often associated with increases in Bacteroidetes and *Lachnospiraceae* family has been reported. Omega-3 supplementation could exert its positive effects by reverting the gut microbiota dysbiosis in MS patients and by increasing antiinflammatory compound–producing bacteria, such as those related to SCFAs.

CONCLUSION

Prospective studies with large patient cohorts are required to establish a causal relationship between the gut microbiota and MS pathogenesis. Future studies are also needed to analyze the gut microbiome at multiple time points during disease progression to identify bacteria linked with disease incidence and relapse. Identifying such bacteria will help utilize the gut microbiota and their metabolites as diagnostic and prognostic markers of MS. Modulation of the gut microbiota through introduction of a single bacterium or a mixture of bacteria or through FMT could provide novel treatment options for MS patients. Additionally, owing to the major impact of diet on the gut microbiota, dietary manipulation may provide an alternative treatment option.

REFERENCES

[1] Freedman SN, Shahi SK, Mangalam AK. The "gut feeling": breaking down the role of gut microbiome in multiple sclerosis. Neurotherapeutics 2018;15(1):109–25. https://doi.org/10.1007/s13311-017-0588-x. PubMed PMID:29204955; PubMed Central PMCID:PMCPMC5794701.

[2] Rescigno M. The intestinal epithelial barrier in the control of homeostasis and immunity. Trends Immunol 2011;32(6):256–64. https://doi.org/10.1016/j.it.2011.04.003. PubMed PMID:21565554.

[3] Blacher E, Levy M, Tatirovsky E, Elinav E. Microbiome-modulated metabolites at the interface of host immunity. J Immunol 2017;198(2):572–80. https://doi.org/10.4049/jimmunol.1601247. PubMed PMID:28069752.

[4] Shahi SK, Freedman SN, Mangalam AK. Gut microbiome in multiple sclerosis: the players involved and the roles they play. Gut Microb 2017:1–9. https://doi.org/10.1080/19490976.2017.1349041. PubMed PMID:28696139.

[5] Human Microbiome Project C. Structure, function and diversity of the healthy human microbiome. Nature 2012;486(7402):207–14. https://doi.org/10.1038/nature11234. PubMed PMID:22699609; PubMed Central PMCID:PMCPMC3564958.

[6] Cosorich I, Dalla-Costa G, Sorini C, Ferrarese R, Messina MJ, Dolpady J, et al. High frequency of intestinal TH17 cells correlates with microbiota alterations and disease activity in multiple sclerosis. Sci Adv 2017;3(7):e1700492. https://doi.org/10.1126/sciadv.1700492. PubMed PMID:28706993; PubMed Central PMCID:PMCPMC5507635.

[7] Chen J, Chia N, Kalari KR, Yao JZ, Novotna M, Soldan MM, et al. Multiple sclerosis patients have a distinct gut microbiota compared to healthy controls. Sci Rep 2016;6:28484. https://doi.org/10.1038/srep28484. PubMed PMID:27346372; PubMed Central PMCID:PMCPMC4921909.

[8] Cekanaviciute E, Yoo BB, Runia TF, Debelius JW, Singh S, Nelson CA, et al. Gut bacteria from multiple sclerosis patients modulate human T cells and exacerbate symptoms in mouse models. Proc Natl Acad Sci U S A 2017;114(40):10713–8. https://doi.org/10.1073/pnas.1711235114. PubMed PMID:28893978; PubMed Central PMCID: PMCPMC5635915.

[9] Berer K, Gerdes LA, Cekanaviciute E, Jia X, Xiao L, Xia Z, et al. Gut microbiota from multiple sclerosis patients enables spontaneous autoimmune encephalomyelitis in mice. Proc Natl Acad Sci U S A 2017;114(40):10719–24. https://doi.org/10.1073/pnas.1711233114. PubMed PMID:28893994; PubMed Central PMCID:PMCPMC5635914.

[10] Wieland K, Chhatwal P, Vonberg RP. Nosocomial outbreaks caused by *Acinetobacter baumannii* and *Pseudomonas aeruginosa*: results of a systematic review. Am J Infect Control 2018. https://doi.org/10.1016/j.ajic.2017.12.014. PubMed PMID:29398072.

[11] Belkaid Y, Hand TW. Role of the microbiota in immunity and inflammation. Cell 2014;157(1):121–41. https://doi.org/10.1016/j.cell.2014.03.011. PubMed PMID:24679531; PubMed Central PMCID:PMCPMC4056765.

[12] Tanoue T, Atarashi K, Honda K. Development and maintenance of intestinal regulatory T cells. Nat Rev Immunol 2016;16(5):295–309. https://doi.org/10.1038/nri.2016.36. PubMed PMID:27087661.

[13] Mangalam A, Shahi SK, Luckey D, Karau M, Marietta E, Luo N, et al. Human gut-derived commensal bacteria suppress CNS inflammatory and demyelinating disease. Cell Rep 2017;20(6):1269–77. https://doi.org/10.1016/j.celrep.2017.07.031. PubMed PMID:28793252.

[14] Marietta EV, Murray JA, Luckey DH, Jeraldo PR, Lamba A, Patel R, et al. Suppression of inflammatory arthritis by human gut-derived *Prevotella histicola* in humanized mice. Arthritis Rheumatol 2016;68(12):2878–88. https://doi.org/10.1002/art.39785. PubMed PMID:27337150; PubMed Central PMCID:PMCPMC5125894.

[15] Atarashi K, Tanoue T, Oshima K, Suda W, Nagano Y, Nishikawa H, et al. Treg induction by a rationally selected mixture of Clostridia strains from the human microbiota. Nature 2013;500(7461):232–6. https://doi.org/10.1038/nature12331. PubMed PMID:23842501.

[16] Jangi S, Gandhi R, Cox LM, Li N, von Glehn F, Yan R, et al. Alterations of the human gut microbiome in multiple sclerosis. Nat Commun 2016;7:12015. https://doi.org/10.1038/ncomms12015. PubMed PMID:27352007; PubMed Central PMCID:PMCPMC4931233.

[17] Sirotkin AV, Harrath AH. Phytoestrogens and their effects. Eur J Pharmacol 2014;741:230–6. https://doi.org/10.1016/j.ejphar.2014.07.057. PubMed PMID:25160742.

[18] De Paula ML, Rodrigues DH, Teixeira HC, Barsante MM, Souza MA, Ferreira AP. Genistein down-modulates pro-inflammatory cytokines and reverses clinical signs of experimental autoimmune encephalomyelitis. Int Immunopharm 2008;8(9):1291–7. https://doi.org/10.1016/j.intimp.2008.05.002. PubMed PMID:18602076.

[19] Zhang LS, Davies SS. Microbial metabolism of dietary components to bioactive metabolites: opportunities for new therapeutic interventions. Genome Med 2016;8(1):46. https://doi.org/10.1186/s13073-016-0296-x. PubMed PMID:27102537; PubMed Central PMCID:PMCPMC4840492.

[20] Rooks MG, Garrett WS. Gut microbiota, metabolites and host immunity. Nat Rev Immunol 2016;16(6):341—52. https://doi.org/10.1038/nri.2016.42. PubMed PMID:27231050.

[21] Wilck N, Matus MG, Kearney SM, Olesen SW, Forslund K, Bartolomaeus H, et al. Salt-responsive gut commensal modulates TH17 axis and disease. Nature 2017;551(7682):585—9. https://doi.org/10.1038/nature24628. PubMed PMID:29143823.

[22] Mangalam A, Poisson L, Nemutlu E, Datta I, Denic A, Dzeja P, et al. Profile of circulatory metabolites in a relapsing-remitting animal model of multiple sclerosis using global metabolomics. J Clin Cell Immunol 2013;4. https://doi.org/10.4172/2155-9899.1000150. PubMed PMID:24273690; PubMed Central PMCID:PMCPMC3837296.

[23] Ho PP, Steinman L. Obeticholic acid, a synthetic bile acid agonist of the farnesoid X receptor, attenuates experimental autoimmune encephalomyelitis. Proc Natl Acad Sci U S A 2016;113(6):1600—5. https://doi.org/10.1073/pnas.1524890113. PubMed PMID:26811456; PubMed Central PMCID:PMCPMC4760777.

[24] Ochoa-Reparaz J, Mielcarz DW, Ditrio LE, Burroughs AR, Foureau DM, Haque-Begum S, et al. Role of gut commensal microflora in the development of experimental autoimmune encephalomyelitis. J Immunol 2009;183(10):6041—50. https://doi.org/10.4049/jimmunol.0900747. PubMed PMID:19841183.

[25] Rothhammer V, Mascanfroni ID, Bunse L, Takenaka MC, Kenison JE, Mayo L, et al. Type I interferons and microbial metabolites of tryptophan modulate astrocyte activity and central nervous system inflammation via the aryl hydrocarbon receptor. Nat Med 2016;22(6):586—97. https://doi.org/10.1038/nm. 4106. PubMed PMID:27158906; PubMed Central PMCID:PMCPMC4899206.

[26] Calvo-Barreiro L, Eixarch H, Montalban X, Espejo C. Combined therapies to treat complex diseases: the role of the gut microbiota in multiple sclerosis. Autoimmun Rev 2018;17(2):165—74. https://doi.org/10.1016/j.autrev.2017.11.019. PubMed PMID:29191793.

[27] Poisson LM, Suhail H, Singh J, Datta I, Denic A, Labuzek K, et al. Untargeted plasma metabolomics identifies endogenous metabolite with drug-like properties in chronic animal model of multiple sclerosis. J Biol Chem 2015;290(52):30697—712. https://doi.org/10.1074/jbc.M115.679068. PubMed PMID:26546682; PubMed Central PMCID:PMCPMC4692201.

[28] Choi HH, Cho YS. Fecal microbiota transplantation: current applications, effectiveness, and future perspectives. Clin Endosc 2016;49(3):257—65. https://doi.org/10.5946/ce.2015.117. PubMed PMID:26956193; PubMed Central PMCID:PMCPMC4895930.

[29] Miyake S, Kim S, Suda W, Oshima K, Nakamura M, Matsuoka T, Chihara N, Tomita A, Sato W, Kim SW, Morita H, Hattori M, Yamamura T. Dysbiosis in the Gut Microbiota of Patients with Multiple Sclerosis, with a Striking Depletion of Species Belonging to Clostridia XIVa and IV Clusters. PLoS One 2015 Sep 14;10(9):e0137429. https://doi.org/10.1371/journal.pone.0137429. eCollection 2015. PubMed PMID: 26367776; PubMed Central PMCID: PMC4569432.

[30] Castillo-Álvarez F, Pérez-Matute P, Oteo JA, Marzo-Sola ME. The influence of interferon β-1b on gut microbiota composition in patients with multiple sclerosis. Neurologia 2018 Jun 9. S0213-4853(18)30158-0. https://doi.org/10.1016/j.nrl.2018.04.006. [Epub ahead of print] English, Spanish. PubMed PMID: 29895466.

Chapter 35

Connections Between Gut Microbiota and Bone Health

P. D'Amelio and F. Sassi

Department of Medical Sciences, Gerontology and Bone Metabolic Disease Section, University of Torino, Torino, Italy

INTRODUCTION

In physiological conditions the gut microbiota/GM relationship with the host is complex and encompasses various forms of relationship such as parasitic, commensal, and mutualistic. Immune homeostasis disruption can be the mechanism of several chronic noncommunicable human diseases (NCDs) such as allergy, asthma, some autoimmune, cardiovascular, and metabolic diseases, and neurodegenerative disorders. These disorders are often characterized by a low grade of inflammation. Although inflammation and the pathways to disease are multifactorial, the altered gut colonization patterns, associated with decreasing microbial diversity, are increasingly implicated in the physiologic, immunologic, and metabolic deregulation seen in many NCDs [1,2].

MOLECULES AND PATHWAYS

Products of microbial metabolism signal the host and influence metabolism. Butyrate is particularly important as an energy substrate for cellular metabolism in the colonic epithelium, whereas acetate and propionate are taken up by the liver and used as substrates for lipogenesis and gluconeogenesis. Short-chain fatty acids/SCFAs also affect proliferation, differentiation, and modulation of gene expression in mammalian colonic epithelial cells [3,4]. Acting as a potent histone deacetylase inhibitor, butyrate may regulate the mammalian transcriptome.

SCFAs have antiinflammatory effects on intestinal mucosa, thus protecting the bowel from the development of inflammatory bowel disease [5–7]. SCFAs signal several nonenteral cell types through G-protein–coupled receptors, and the signal between GM and the immune system is fundamental to regulate the homeostasis and to maintain the balance between immune tolerance to commensal bacteria and immunity to pathogens.

Depending on the cytokine milieu, interaction between SCFAs and their receptors influences T cell differentiation not only toward T regulatory cells (Tregs) but also toward effector T cells. Furthermore, butyrate and propionate also modulate antigen presentation, inhibiting the development of dendritic cells by gene expression regulation [8–11] and by interaction with SCFAs receptors [12,13].

Indole derivatives and polyamines, from digested food, similarly have important immunomodulatory function. These metabolites are derived from dietary tryptophan and arginine. Indole derivatives favor the integrity of the enteral mucosa and the barrier defense toward pathogens by stimulating the production of antimicrobial peptides, mucins, and proliferation of intestinal goblet cells. Polyamines such as putrescine, spermidine, and spermine fulfill important roles in gene expression and proliferation. They enhance the development and maintenance of the intestinal mucosa and resident immune cells [2].

The primary bile acids are synthesized in the human liver from cholesterol and are important for ensuring that cholesterol, dietary fats, and fat-soluble vitamins from the small intestine are soluble and absorbable. Primary bile acids are conjugated to glycine in humans and are taken up in the distal ileum for transport to the liver. However, bacteria in this part of the ileum deconjugate these bile acids, which then escape intestinal uptake and can be further metabolized by the GM into secondary bile acids.

Microbiome and Metabolome in Diagnosis, Therapy, and other Strategic Applications. https://doi.org/10.1016/B978-0-12-815249-2.00035-X

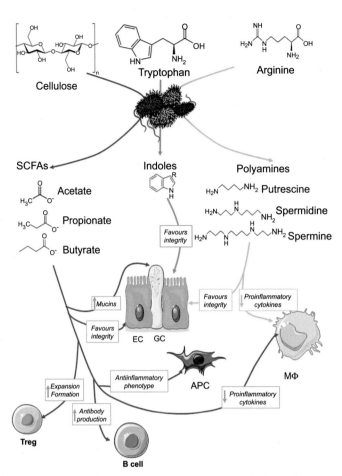

FIGURE 35.1 Gut microbiota metabolites and their effects on bone cells. *SCFAs,* short-chain fatty acids; *EC,* enteral cells; *GC,* goblet cells; *APC,* antigen presenting cells; *MØ,* Macrophages; *Treg,* T regulatory cells.

Bile acids are taken up from the gut and circulate throughout the body, acting as signaling molecules and binding to cellular receptors in peripheral organs. An immunomodulating role has also been postulated for metabolized bile acids. However, physiological role of these metabolites in health and disease is still an open question [14].

Polysaccharide A is a bacterial product that influences T cell fate through its interaction with the toll-like receptor 2. It favors immune tolerance by inhibiting T helper-17 differentiation and favoring Tregs activity [15]. GM influences T cell differentiation through the production of bacterial metabolites at the intestinal mucosa level through cognate bacterial antigens [16]. Various models are revealing roles for the microbiota in skeletal biology [17,18]. Several molecular pathways are summarized in Fig. 35.1.

GM, IMMUNE SYSTEM, AND BONE LOSS

Osteoporosis increases dramatically the risk of fractures: major osteoporotic fractures are a social and economic burden. In developed countries the lifetime risk for osteoporotic fractures at the wrist, hip, or spine is 30%–40%, very close to that for coronary heart disease. The number of new fractures in 2010 in the EU was estimated at 3.5 million, comprising approximately 620,000 hip fractures; 520,000 vertebral fractures; 560,000 forearm fractures; and 1,800,000 other fractures [19]. Osteoporotic fractures impair patients' quality of life and increase mortality: 20% of elderly patients suffering from femoral fractures will die within a year, and 50% of the survivors will lose independence.

The most frequent cause of bone loss is postmenopausal osteoporosis (PMO) that is driven by estrogen deficiency at menopause. In PMO there is an imbalance in bone turnover, with increased bone resorption and reduced bone formation. It has been demonstrated both in experimental models and in humans that estrogen deficiency affects bone cell number and activation as well as bone turnover, partially through its effect on immune system [20]. During estrogen deficiency, T cells increase their production of proinflammatory and proosteoclastogenic cytokines, such as TNF alpha and Receptor

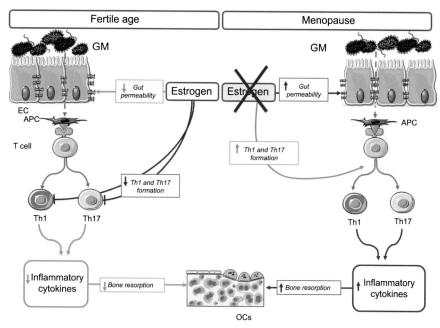

FIGURE 35.2 Relationships between immune system, estrogen deficiency-bone loss, and gut microbiota: enteral barrier integrity, cytokine production, immune and bone cells are involved. *GM,* gut microbiota; *EC,* enteral cells; *APC,* antigen presenting cells; *Treg,* T regulatory cells; *Th1,* T helper-1 cells; *Th17,* T helper-17 cells; *OCs,* osteoclasts.

Activator of Nuclear factor Kappa-B Ligand (RANKL) [21]; however, the reasons for this increased activity in osteoporotic women are unknown.

Germ-free (GF) mice have increased bone mass. An acute effect of GF colonization (conventionalization) can be reduction of bone mass due to increased bone resorption, although this is not a universal phenomenon [18]. Long-term colonization resulted in a net skeletal growth in young animals [22].

Studies in mice treated with broad-spectrum antibiotics, to alter GM, bring to different conclusions regarding the effect on bone density. These discrepancies are possibly due to differences in animal age, sex, and protocols applied for antibiotic treatment [23–26]. However, the majority of the reports suggest that antibiotic-treated mice have increased bone density [26] and also better bone mechanical properties than conventionally raised mice.

It is possible to induce a bone loss condition in mice pharmacologically with the GnRH agonists leuprolide and to investigate the role of GM in bone loss induced by sex steroid deficiency. These studies demonstrated that GM plays an important role in sex steroid deficiency–induced osteoporosis: GF mice are protected against osteoporosis and the increase in bone turnover induced by sex steroid deprivation because of the lack of increase in TNF, RANKL, and IL-17. Furthermore, sex steroid depletion augments inflammation in the intestine by increasing gut permeability to bacterial antigens by decreasing the expression of modulators of intestinal barrier integrity [27].

In humans, scarce data support results obtained in mice. The relationship between GM, immune system, and bone in PMO is summarized in Fig. 35.2.

GM AND BONE HEALTH BEYOND IMMUNE SYSTEM

GM composition and manipulation may affect bone health by influencing calcium absorption and the production of gut-derived serotonin [28]. *Lactobacillus reuteri* in healthy subjects increases the level of serum 25OH vitamin D that influences calcium absorption and benefits bone health. The mechanism through which this probiotic influences vitamin D level is not clear. This may be due to a modification in the gut environment that specifically favors vitamin D absorption or to indirect effect on increased hepatic 25-hydroxylase activity or 7-dehydrocholesterol concentration, due to reduced absorption of dietary and biliary cholesterol [2,29].

The relation between GM and vitamin D may also be inverse as it has been proposed that decreased vitamin D intake is associated with different GM profiles [30,31]. Another possible mechanism through which GM benefits bone health is the increase in calcium absorption. It is well known that maintaining a positive calcium balance is important in achieving a good peak of bone mass that protects from the development of osteoporosis in older age. Dietary intake of fibers influences

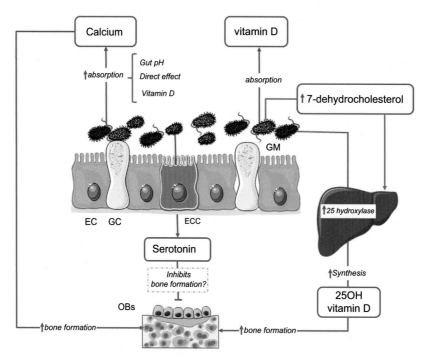

FIGURE 35.3 The link between gut microbiota and bone turnover beyond immune system. *GM,* gut microbiota; *EC,* enteral cells; *ECC,* enterochromaffin cells; *GC,* goblet cells; *OBs,* osteoblasts.

calcium absorption: after being fermented by GM, fibers improve calcium absorption by reduction of gut pH, thus reducing the formation of calcium phosphates and increasing the calcium absorption, and by increasing the production of SCFAs such as butyrate. The effect of SCFAs may be more complex than the effect on gut pH, and they could increase calcium transport through signaling pathway modulation [32,33].

Another possible mechanism through which GM influences bone health is mediated by its effect on the production of gut serotonin (5HT). In the recent past a dual effect of serotonin in the regulation of bone mass has been described, depending on the site of production of this molecule [34]. A role as a bone mass regulator was proposed to gut-derived 5HT (g5HT), which is influenced by GM. Enterochromaffin cells of the duodenum are responsible for the synthesis of g5HT, which is partially modulated by GM, as SCFAs increase the synthesis of g5HT [35,36]. It has been shown that 5HT interacts with bone cells and, in particular, decreases osteoblast proliferation. These observations suggest that regulation of g5HT by GM may be a potential therapeutic strategy to improve bone health.

In ovariectomy-induced bone loss, pharmacological inhibition of g5HT synthesis results in the prevention of osteoporosis, mediated by increased bone formation [37]. However, data on the effect of 5HT on bone health are quite controversial.

The relationship between GM and bone beyond immune system is summarized in Fig. 35.3.

GM MANIPULATION AND BONE HEALTH

Prebiotic supplementation in animal models favors the proliferation of *Bifidobacteria* and increases SCFA production. Prebiotics elevated calcium absorption and bone density in animal models [38,39]. Also in humans, different prebiotics in adolescent girls improved calcium absorption and bone density [40,41]. More abundant presence of Bacteroidetes and Firmicutes known to ferment starch and fiber was simultaneously documented [42,43]. Firmicutes are positively correlated with calcium absorption, suggesting that the role of GM in calcium absorption is complex due to different species [42].

Prebiotic fiber may influence bone metabolism both by the change in the composition of GM-favoring microbes, with higher antiinflammatory potential, and by increasing SCFAs production, thus increasing calcium absorption. It has also been suggested that prebiotics could have a direct effect on immune system modulation and an antipathogen effect regardless to their effect on GM [44]. However, until now, in human studies on prebiotics and bone health, only calcium absorption, markers of bone metabolism, and bone density have been investigated, whereas immune phenotype and inflammation have not been investigated. Table 35.1 summarizes the results obtained in humans with different prebiotics.

TABLE 35.1 Summary of the Results Obtained in Humans With Different Prebiotics

Subjects Recruited	Number of Subjects	Prebiotic Used	Dose and Time	Outcomes	References
Adolescents	50 males 50 females	Oligofructanes and long-chain inulin	8 g/day 12 months	↑ Ca absorption ↑ Bone mineral density	[41]
Preadolescent	31 females	Galactooligosaccharides	5–10 g/day 3 weeks	↑ Ca absorption	[40]
Adolescents	15 males 9 females	Soluble corn fiber	12 g/day 3 weeks	↑ Ca absorption ↑ Bacteroides	[43]
Adolescent	28 females	Soluble corn fiber	10–20 g/day 4 weeks	↑ Ca absorption and bone markers ↑ Bacteroidetes and Firmicutes	[42]

Probiotics are known to increase bone mass and reduce sex steroid–associated bone loss [45–47]. Prescription of *Lactobacillus* spp. demonstrated high antiinflammatory and bone-protective effect. Also, yogurt, which contains different probiotics, but is also a source of calcium and proteins that are fundamental for bone health, has been recommended [48,49].

The use of probiotics has been proposed also as an adjuvant treatment in focal bone loss such as alveolar erosion in periodontitis. The ability of different lactobacilli strains in reducing osteoclast number, alveolar erosions, and tooth movement in rat and mice has been demonstrated [50–52]. In humans a recent metaanalysis concludes that the current evidence suggests a possible use of probiotics as an adjuvant therapy in gingivitis and periodontitis [53]. In a geriatric population the administration of *Lactobacillus helveticus* increases serum calcium [54]. Also, in osteopenic women the administration of a multispecies probiotic (6 different species) increases markers of bone formation, decreases TNF alpha level, but has no effect on bone density during a 6-month period [55].

FUTURE PERSPECTIVES

The great amount of information about the GM will allow to detect potential targets for therapeutical strategies and to manipulate the GM through diet (use of functional food), prebiotics and probiotics, and administration of antibiotics directed to specific microbial species. As we know more about the mechanisms by which the GM contributes to pathologies, we could set up an "intelligent modulation" of the GM and to precisely determine its effects on the intestinal community and the host, thus having the possibility to be extremely beneficial for human health [56].

LIST OF ACRONYMS AND ABBREVIATIONS

FOS Fructooligosaccharides
GF Germ free
GM Gut microbiota
GOS Galactooligosaccharides
NCDs Noncommunicable human diseases
PMO Postmenopausal osteoporosis
SCF Soluble corn fiber
SCFAs Short-chain fatty acids
Tregs T regulatory cells
Th T helper cells
5HT/g5HT Serotonin/gut-derived serotonin

REFERENCES

[1] Peterson CT, Sharma V, Elmén L, Peterson SN. Immune homeostasis, dysbiosis and therapeutic modulation of the gut microbiota. Clin Exp Immunol 2015;179:363–77. https://doi.org/10.1111/cei.12474.

[2] D'Amelio P, Sassi F. Gut microbiota, immune system, and bone. Calcif Tissue Int 2017. https://doi.org/10.1007/s00223-017-0331-y [Epub ahead of print].

[3] Davie JR. Inhibition of histone deacetylase activity by butyrate. J Nutr 2003;133:2485S−93S.

[4] Tremaroli V, Bäckhed F. Functional interactions between the gut microbiota and host metabolism. Nature 2012;489:242−9. https://doi.org/10.1038/nature11552.

[5] Ferreira CM, Vieira AT, Vinolo MA, Oliveira FA, Curi R, Martins Fdos S. The central role of the gut microbiota in chronic inflammatory diseases. J Immunol Res 2014;2014:689492. https://doi.org/10.1155/2014/689492.

[6] Furusawa Y, Obata Y, Fukuda S, Endo TA, Nakato G, Takahashi D, Nakanishi Y, Uetake C, Kato K, Kato T, Takahashi M, Fukuda NN, Murakami S, Miyauchi E, Hino S, Atarashi K, Onawa S, Fujimura Y, Lockett T, Clarke JM, Topping DL, Tomita M, Hori S, Ohara O, Morita T, Koseki H, Kikuchi J, Honda K, Hase K, Ohno H. Commensal microbe-derived butyrate induces the differentiation of colonic regulatory T cells. Nature 2013;504:446−50. https://doi.org/10.1038/nature12721.

[7] Smith PM, Howitt MR, Panikov N, Michaud M, Gallini CA, Bohlooly-Y M, Glickman JN, Garrett WS. The microbial metabolites, short-chain fatty acids, regulate colonic Treg cell homeostasis. Science 2013;341:569−73. https://doi.org/10.1126/science.1241165.

[8] Liu L, Li L, Min J, Wang J, Wu H, Zeng Y, Chen S, Chu Z. Butyrate interferes with the differentiation and function of human monocyte-derived dendritic cells. Cell Immunol 2012;277:6673. https://doi.org/10.1016/j.cellimm.2012.05.011.

[9] Millard AL, Mertes PM, Ittelet D, Villard F, Jeannesson P, Bernard J. Butyrate affects differentiation, maturation and function of human monocyte-derived dendritic cells and macrophages. Clin Exp Immunol 2002;130:245−55.

[10] Singh N, Thangaraju M, Prasad PD, Martin PM, Lambert NA, Boettger T, Offermanns S, Ganapathy V. Blockade of dendritic cell development by bacterial fermentation products butyrate and propionate through a transporter (Slc5a8)-dependent inhibition of histone deacetylases. J Biol Chem 2010;285:27601−8. https://doi.org/10.1074/jbc.M110.102947.

[11] Wang B, Morinobu A, Horiuchi M, Liu J, Kumagai S. Butyrate inhibits functional differentiation of human monocytederived dendritic cells. Cell Immunol 2008;253:54−8. https://doi.org/10.1016/j.cellimm.2008.04.016.

[12] Singh N, Gurav A, Sivaprakasam S, Brady E, Padia R, Shi H, Thangaraju M, Prasad PD, Manicassamy S, Munn DH, Lee JR, Offermanns S, Ganapathy V. Activation of Gpr109a, receptor for niacin and the commensal metabolite butyrate, suppresses colonic inflammation and carcinogenesis. Immunity 2014;40:128−39. https://doi.org/10.1016/j.immuni.2013.12.007.

[13] Trompette A, Gollwitzer ES, Yadava K, Sichelstiel AK, Sprenger N, Ngom-Bru C, Blanchard C, Junt T, Nicod LP, Harris NL, Marsland BJ. Gut microbiota metabolism of dietary fiber influences allergic airway disease and hematopoiesis. Nat Med 2014;20:159−66. https://doi.org/10.1038/nm.3444.

[14] Postler TS, Ghosh S. Understanding the holobiont: how microbial metabolites affect human health and shape the immune system. Cell Metab 2017;26:110−30. https://doi.org/10.1016/j.cmet.2017.05.008.

[15] Round JL, Lee SM, Li J, Tran G, Jabri B, Chatila TA, Mazmanian SK. The Toll-like receptor 2 pathway establishes colonization by a commensal of the human microbiota. Science 2011;332:974−7. https://doi.org/10.1126/science.1206095.

[16] Longman RS, Yang Y, Diehl GE, Kim SV, Littman DR. Microbiota: host interactions in mucosal homeostasis and systemic autoimmunity. Cold Spring Harb Symp Quant Biol 2013;78:193−201. https://doi.org/10.1101/sqb.2013.78.020081.

[17] Sjogren K, Engdahl C, Henning P, Lerner UH, Tremaroli V, Lagerquist MK, Bäckhed F, Ohlsson C. The gut microbiota regulates bone mass in mice. J Bone Miner Res 2012;27:1357−67.

[18] Quach D, Collins F, Parameswaran N, McCabe L, Britton RA. Microbiota reconstitution does not cause bone loss in germ-free mice. mSphere January 3, 2018;3(1).

[19] Hernlund E, Svedbom A, Ivergård M, Compston J, Cooper C, Stenmark J, McCloskey EV, Jönsson B, Kanis JA. Osteoporosis in the European union: medical management, epidemiology and economic burden. A report prepared in collaboration with the international osteoporosis foundation (IOF) and the European federation of pharmaceutical industry associations (EFPIA). Arch Osteoporos 2013;8:136. https://doi.org/10.1007/s11657-013-0136-1.

[20] Mori G, D'Amelio P, Faccio R, Brunetti G. Bone-immune cell crosstalk: bone diseases. J Immunol Res 2015;2015:108451. https://doi.org/10.1155/2015/108451.

[21] D'Amelio P, Grimaldi A, Di Bella S, Brianza SZ, Cristofaro MA, Tamone C, Giribaldi G, Ulliers D, Pescarmona GP, Isaia G. Estrogen deficiency increases osteoclastogenesis up-regulating T cells activity: a key mechanism in osteoporosis. Bone 2008;43:92−100. https://doi.org/10.1016/j.bone.2008.02.017.

[22] Yan J, Herzog JW, Tsang K, Brennan CA, Bower MA, Garrett WS, Sartor BR, Aliprantis AO, Charles JF. Gut microbiota induce IGF-1 and promote bone formation and growth. Proc Natl Acad Sci USA 2016;113:E7554−63. https://doi.org/10.1073/pnas.1607235113.

[23] Cox LM, Yamanishi S, Sohn J, Alekseyenko AV, Leung JM, Cho I, Kim SG, Li H, Gao Z, Mahana D, Zárate Rodriguez JG, Rogers AB, Robine N, Loke P, Blaser MJ. Altering the intestinal microbiota during a critical developmental window has lasting metabolic consequences. Cell 2014;158:705−21. https://doi.org/10.1016/j.cell.2014.05.052.

[24] Nobel YR, Cox LM, Kirigin FF, Bokulich NA, Yamanishi S, Teitler I, Chung J, Sohn J, Barber CM, Goldfarb DS, Raju K, Abubucker S, Zhou Y, Ruiz VE, Li H, Mitreva M, Alekseyenko AV, Weinstock GM, Sodergren E, Blaser MJ. Metabolic and metagenomic outcomes from early-life pulsed antibiotic treatment. Nat Commun 2015;6:7486. https://doi.org/10.1038/ncomms8486.

[25] Pytlik M, Folwarczna J, Janiec W. Effects of doxycycline on mechanical properties of bones in rats with ovariectomy induced osteopenia. Calcif Tissue Int 2004;75:225−30. https://doi.org/10.1007/s00223-004-0097-x.

[26] Guss JD, Horsfield MW, Fontenele FF, Sandoval TN, Luna M, Apoorva F, Lima SF, Bicalho RC, Singh A, Ley RE, van der Meulen MC, Goldring SR, Hernandez CJ. Alterations to the gut microbiome impair bone strength and tissue material properties. J Bone Miner Res 2017;32:1343−53. https://doi.org/10.1002/jbmr.3114.

[27] Li JY, Chassaing B, Tyagi AM, Vaccaro C, Luo T, Adams J, Darby TM, Weitzmann MN, Mulle JG, Gewirtz AT, Jones RM, Pacifici R. Sex steroid deficiency-associated bone loss is microbiota dependent and prevented by probiotics. J Clin Investig 2016;126:2049−63.

[28] Wang J, Wang Y, Gao W, Wang B, Zhao H, Zeng Y, Ji Y, Hao D. Diversity analysis of gut microbiota in osteoporosis and osteopenia patients. Peer J 2017;5:e3450. https://doi.org/10.7717/peerj.3450.

[29] Jones ML, Martoni CJ, Prakash S. Oral supplementation with probiotic L. reuteri NCIMB 30242 increases mean circulating 25-hydroxyvitamin D: a post hoc analysis of a randomized controlled trial. J Clin Endocrinol Metab 2013;98:2944−51. https://doi.org/10.1210/jc.2012-4262.

[30] Yoon SS, Sun J. Probiotics, nuclear receptor signaling, and anti-inflammatory pathways. Gastroenterol Res Pract 2011;2011:971938. https://doi.org/10.1155/2011/971938.

[31] Ly NP, Litonjua A, Gold DR, Celedòn JC. Gut microbiota, probiotics, and vitamin D: interrelated exposures influencing allergy, asthma, and obesity? J Allergy Clin Immunol 2011;127:1087−94. https://doi.org/10.1016/j.jaci.2011.02.015.

[32] Weaver CM. Diet, gut microbiome, and bone health. Curr Osteoporos Rep 2015;13:125−30.

[33] Wallace TC, Marzorati M, Spence L, Weaver CM, Williamson PS. New frontiers in fibers: innovative and emerging research on the gut microbiome and bone health. J Am Coll Nutr 2017;36:218−22. https://doi.org/10.1080/07315724.2016.1257961.

[34] D'Amelio P, Panico A, Spertino E, Isaia GC. Energy metabolism and the skeleton: reciprocal interplay. World J Orthop 2012;3:190−8. https://doi.org/10.5312/wjo.v3.i11.190.

[35] Reigstad CS, Salmonson CE, Rainey 3rd JF, Szurszewski JH, Linden DR, Sonnenburg JL, Farrugia G, Kashyap PC. Gut microbes promote colonic serotonin production through an effect of short-chain fatty acids on enterochromaffin cells. FASEB J 2015;29:1395−403. https://doi.org/10.1096/fj.14-259598.

[36] Yano JM, Yu K, Donaldson GP, Shastri GG, Ann P, Ma L, Nagler CR, Ismagilov RF, Mazmanian SK, Hsiao EY. Indigenous bacteria from the gut microbiota regulate host serotonin biosynthesis. Cell 2015;161:264−76. https://doi.org/10.1016/j.cell.2015.02.047.

[37] Yadav VK, Balaji S, Suresh PS, Liu XS, Lu X, Li Z, Guo XE, Mann JJ, Balapure AK, Gershon MD, Medhamurthy R, Vidal M, Karsenty G, Ducy P. Pharmacological inhibition of gut-derived serotonin synthesis is a potential bone anabolic treatment for osteoporosis. Nat Med 2010;16:308−12. https://doi.org/10.1038/nm.2098.

[38] Weaver CM, Martin BR, Nakatsu CH, Armstrong AP, Clavijo A, McCabe LD, McCabe GP, Duignan S, Schoterman MH, van den Heuvel EG. Galactooligosaccharides improve mineral absorption and bone properties in growing rats through gut fermentation. J Agric Food Chem 2011;59:6501−10. https://doi.org/10.1021/jf2009777.

[39] Scholz-Ahrens KE, Schaafsma G, van den Heuvel EG, Schrezenmeir J. Effects of prebiotics on mineral metabolism. Am J Clin Nutr 2001;73:459S−64S.

[40] Whisner CM, Martin BR, Schoterman MH, Nakatsu CH, McCabe LD, McCabe GP, Wastney ME, van den Heuvel EG, Weaver CM. Galacto-oligosaccharides increase calcium absorption and gut Bifidobacteria in young girls: a double-blind cross-over trial. Br J Nutr 2013;110:1292−303. https://doi.org/10.1017/S000711451300055X.

[41] Abrams SA, Griffin IJ, Hawthorne KM, Liang L, Gunn SK, Darlington G, Ellis KJ. A combination of prebiotic short- and long-chain inulin-type fructans enhances calcium absorption and bone mineralization in young adolescents. Am J Clin Nutr 2005;82:471−6.

[42] Whisner CM, Martin BR, Nakatsu CH, Story JA, MacDonald-Clarke CJ, McCabe LD, McCabe GP, Weaver CM. Soluble corn fiber increases calcium absorption associated with shifts in the gut microbiome: a randomized dose-response trial in free-living pubertal females. J Nutr 2016;146:1298−306. https://doi.org/10.3945/jn.115.227256.

[43] Whisner CM, Martin BR, Nakatsu CH, McCabe GP, McCabe LD, Peacock M, Weaver CM. Soluble maize fibre affects short-term calcium absorption in adolescent boys and girls: a randomised controlled trial using dual stable isotopic tracers. Br J Nutr 2014;112:446−56. https://doi.org/10.1017/S0007114514000981.

[44] Bindels LB, Delzenne NM, Cani PD, Walter J. Towards a more comprehensive concept for prebiotics. Nat Rev Gastroenterol Hepatol 2015;12:303−10. https://doi.org/10.1038/nrgastro.2015.47.

[45] Ohlsson C, Engdahl C, Fåk F, Andersson A, Windahl SH, Farman HH, Movérare-Skrtic S, Islander U, Sjögren K. Probiotics protect mice from ovariectomy-induced cortical bone loss. PLoS One 2014;9:e92368. https://doi.org/10.1371/journal.pone.0092368.

[46] Britton RA, Irwin R, Quach D, Schaefer L, Zhang J, Lee T, Parameswaran N, McCabe LR. Probiotic L. reuteri treatment prevents bone loss in a menopausal ovariectomized mouse model. J Cell Physiol 2014;229:1822−30. https://doi.org/10.1002/jcp.24636.

[47] Parvaneh K, Ebrahimi M, Sabran MR, Karimi G, Hwei AN, Abdul-Majeed S, Ahmad Z, Ibrahim Z, Jamaluddin R. Probiotics (Bifidobacterium longum) increase bone mass density and upregulate SPARC and BMP-2 genes in rats with bone loss resulting from ovariectomy. BioMed Res Int 2015;2015:897639. https://doi.org/10.1155/2015/897639.

[48] Rozenberg S, Body JJ, Bruyère O, Bergmann P, Brandi ML, Cooper C, Devogelaer JP, Gielen E, Goemaere S, Kaufman JM, Rizzoli R, Reginster JY. Effects of dairy products consumption on health: benefits and beliefs−a commentary from the Belgian bone club and the European society for clinical and economic aspects of osteoporosis, osteoarthritis and musculoskeletal diseases. Calcif Tissue Int 2016;98:1−17. https://doi.org/10.1007/s00223-015-0062-x.

[49] Laird E, Molloy AM, McNulty H, Ward M, McCarroll K, Hoey L, Hughes CF, Cunningham C, Strain JJ, Casey MC. Greater yogurt consumption is associated with increased bone mineral density and physical function in older adults. Osteoporos Int 2017;28:2409−19. https://doi.org/10.1007/s00198-017-4049-5.

[50] Pazzini CA, Pereira LJ, da Silva TA, Montalvany-Antonucci CC, Macari S, Marques LS, de Paiva SM. Probiotic consumption decreases the number of osteoclasts during orthodontic movement in mice. Arch Oral Biol 2017;79:30−4. https://doi.org/10.1016/j.archoralbio.2017.02.017.

[51] Ricoldi MST, Furlaneto FAC, Oliveira LFF, Teixeira GC, Pischiotini JP, Moreira ALG, Ervolino E, de Oliveira MN, Bogsan CSB, Salvador SL, Messora MR. Effects of the probiotic *Bifidobacterium animalis* subsp. lactis on the non-surgical treatment of periodontitis. A histomorphometric, microtomographic and immunohistochemical study in rats. PLoS One 2017;12:e0179946. https://doi.org/10.1371/journal.pone.0179946.

[52] Kobayashi R, Kobayashi T, Sakai F, Hosoya T, Yamamoto M, Kurita-Ochiai T. Oral administration of *Lactobacillus gasseri* SBT2055 is effective in preventing Porphyromonas gingivalis-accelerated periodontal disease. Sci Rep 2017;7:545. https://doi.org/10.1038/s41598-017-00623-9.

[53] Gruner D, Paris S, Schwendicke F. Probiotics for managing caries and periodontitis: systematic review and metaanalysis. J Dent 2016;48:16—25. https://doi.org/10.1016/j.jdent.2016.03.002.

[54] Gohel MK, Prajapati JB, Mudgal SV, Pandya HV, Singh US, Trivedi SS, Phatak AG, Patel RM. Effect of probiotic dietary intervention on calcium and haematological parameters in geriatrics. J Clin Diagn Res 2016;10:5—9. https://doi.org/10.7860/JCDR/2016/18877.7627.

[55] Jafarnejad S, Djafarian K, Fazeli MR, Yekaninejad MS, Rostamian A, Keshavarz SA. Effects of a multispecies probiotic supplement on bone health in osteopenic postmenopausal women: a randomized, double-blind, controlled trial. J Am Coll Nutr 2017;19:1—10. https://doi.org/10.1080/07315724.2017.1318724.

[56] Pacifici R. Bone remodeling and the microbiome. Cold Spring Harb Perspect Med 2017;8. pii: a031203. https://doi.org/10.1101/cshperspect.a031203.

Chapter 36

The Gut Microbiome in Chronic Kidney Disease

Natália Alvarenga Borges[1,2] and Denise Mafra[1,3]

[1]Graduate Program in Cardiovascular Sciences, Fluminense Federal University (UFF), Niterói, Brazil; [2]Unidade de Pesquisa Clínica, Niterói, Brazil; [3]Graduate Program in Medical Sciences, Fluminense Federal University (UFF), Niterói, Brazil

INTRODUCTION

Several factors can contribute to imbalance of the gut microbiota, such as enteric infection, diet, xenobiotics, antibiotics, obesity, smoking, and host genetics, leading to compositional and functional alterations, which can characterize a dysbiosis state. Studies have suggested that the gut microbial environment also plays a critical role in chronic kidney disease (CKD) [1−3].

Wang et al. [4] were the first to analyze gut microbiota by real-time polymerase chain reaction in CKD patients on peritoneal dialysis treatment, detecting less *Bifidobacterium* and *Lactobacillus* species. Vaziri et al. [5], by microarray analysis, showed significant differences in the abundance of 190 microbial operational taxonomic units (OTUs) between CKD patients and healthy individuals. OTUs from *Brachybacterium*, *Catenibacterium*, *Enterobacteriaceae*, *Halomonadaceae*, *Moraxellaceae*, *Nesterenkonia*, *Polyangiaceae*, *Pseudomonadaceae*, and *Thiothrix* families were markedly increased in CKD patients.

To isolate the effect of uremia from interindividual variations, comorbid conditions, and dietary and medicinal interventions, rats were studied for 8 weeks after five out of six nephrectomies. A significant difference in the abundance of 175 bacterial OTUs, most notably, a decrease in the *Lactobacillaceae* and *Prevotellaceae* families, was confirmed. Recently Jiang et al. [6] reported that total bacteria were reduced in CKD patients compared with healthy individuals.

The gut microbiota imbalance may be associated with increased intestinal permeability and production and influx of uremic toxins, leading to inflammation and oxidative stress, which are related to increased mortality risk in these patients [7−10].

CKD ALTERING THE GUT ECOSYSTEM

CKD patients are constantly exposed to various factors that compromise the intestinal homeostasis, such as malnutrition, edema, stress (physical, psychological, or pharmacological), constipation, and uremia [9]. Dietary restriction of potassium-rich fruits and vegetables is advised to prevent hyperkalemia, and these foods are also the main sources of dietary fibers, which have fermentability properties in the colon, acting as prebiotics [9,11].

The low fiber intake together with the fluid restriction, which many hemodialysis (HD) patients are exposed to, contributes to intestinal constipation. In turn, constipation prolongs the contact of pathogens and toxins with the gastrointestinal mucus layer [12].

However, the major cause of the alterations in gut ecosystem of CKD patients seems to be uremia. In CKD the high levels of urea in the blood leads to its massive influx into the gastrointestinal tract, resulting in profound changes in the gut biochemical milieu and also in the structure and function of the gut barrier. Biochemical shifts can modulate the transcriptional machinery and, consequently, the enzymatic activity of gut bacteria, with an impact on the generated metabolites [7,13].

High intraluminal urea provides greater urea hydrolysis by microbial urease, leading to formation of ammonia $[CO(NH_2)2 + H_2O \rightarrow CO_2 + 2NH_3]$ which, in turn, is converted to ammonium hydroxide $[NH_3 + H_2O \rightarrow NH_4OH]$. This product is harmful to the gut barrier, damaging tight junction proteins. The disruption of the epithelial barrier facilitates the paracellular translocation of endotoxins, noxious microbes or their fragments, and other waste products from the gut to systemic circulation. This process known as endotoxemia has the potential to activate innate immunity, resulting in systemic inflammation, which plays a central role in the pathogenesis of CVD, progression of CKD, and numerous CKD-associated complications [7,13,14].

DISTURBED GUT ECOSYSTEM AS A CATALYZER FOR CKD METABOLIC DISORDERS

Healthy or unbalanced gut microbiota can influence the host immune system via two types of signal: microbial cell components and metabolites. Once dysbiosis has been established, both gut-associated lymphoid tissue and systemic immune cells are substantially affected by these signals, contributing to perpetuate the inflammation [3].

Increased exposure to lipopolysaccharides (LPS) and uremic toxins produced by gut microbiota occurs. LPS initiates proinflammatory responses by binding to Toll-like receptors (TLRs) in epithelial cells, monocytes, and macrophages, initiating the signaling cascade that leads to the transcription of proinflammatory genes, and promoting endothelial damage. TLR-4 is the specific receptor for bacterial LPS which will activate, after several reactions, the nuclear factor-κB and protein activator-1. These factors result in the production of proinflammatory cytokines, such as tumor necrosis factor alfa (TNF-α), interleukin-6 (IL-6), interleukin-1 (IL-1), and reactive oxygen species, which can cause endothelial damage and dysfunction and promote atherosclerosis [8].

Uremic toxins are bacterial metabolites that exert cytotoxicity and promote inflammation, tissue injury, and dysfunction. Indoxyl sulfate (IS), indole 3-acetic acid (IAA), p-cresyl sulfate (p-CS), and trimethylamine N-oxide (TMAO) are products of bacterial fermentation of diet components [1,15,16].

Certain bacteria degrade the amino acid tyrosine originating p-CS or convert tryptophan to indole, and this compound is metabolized to IAA, directly in the gut, and to IS, in the liver. Both toxins derived from tryptophan are ligands of aryl hydrocarbon receptor (AhR), a ligand-activated transcription factor involved in atherogenesis, vascular inflammation, and oxidative stress.

TMAO is a metabolite of trimethylamine-containing compounds, such as L-carnitine, choline, and betaine. Initial catabolism of these substrates by intestinal microbiota forms trimethylamine, which is efficiently absorbed and metabolized to TMAO in the liver. It has been suggested that uremic toxins are a potential "missing link" between CKD and CVD because traditional risk factors are insufficient to explain the high incidence of CVD in CKD patients [17]. In fact, recent studies have reported that these toxins induce proinflammatory responses and are reliable markers of CVD and mortality in CKD patients [18–21].

In CKD patients there is an accumulation of these toxins due to loss of kidney function, and moreover, they are not efficiently removed by dialysis treatment [15,16].

THERAPEUTIC APPROACHES

Several novel therapeutic approaches are being considered (Fig. 36.1).

Oral Adsorbents

AST-120 (Kureha Corporation, Tokyo, Japan) is an orally administered carbon preparation, which can help reduce the absorption of urea, urea-derived ammonia, and uremic toxins. AST-120 was approved in Japan in 1991 for delaying the initiation of dialysis and attenuating uremia symptoms in CKD patients. It has also been approved in Korea and the Philippines [7,22].

In nephrectomized rats, administration of AST-120 has reduced IS and p-CS serum and urine levels [5,23] and, in another experimental study, partially restored epithelial tight junction proteins, with reduction in plasma endotoxin, and markers of oxidative stress and inflammation [24].

However, clinical data are not consistent. Small clinical trials suggest potential benefits of AST-120 in inhibiting the synthesis of IS by blocking the gastrointestinal absorption of its precursor indole, but a recent trial concluded that there was no benefit [22].

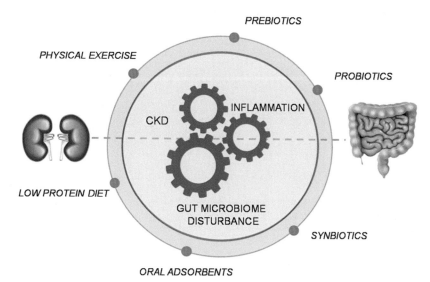

FIGURE 36.1 Gut microbiome disturbance acting as a catalyzer for inflammation-associated disorders in CKD and the therapeutic strategies to modulate microbiome and metabolome in CKD patients.

Prebiotics

Prebiotics have been implicated in the rise of SCFA-forming bacteria and attenuation of urease-possessing bacterial expansion, thereby reducing generation of ammonia and ammonium hydroxide, which are harmful to the gut barrier in CKD patients. SCFA production rise may also reduce intestinal pH, decreasing formation of proinflammatory and prooxidant uremic toxins from colonic bacteria [11,14].

Colonic transit time is a factor involved in the regulation of bacterial protein fermentation and consequently in uremic toxin production. Prebiotics help to prevent constipation. Finally, prebiotics may increase fecal excretion of ammonia, thereby attenuating nitrogen accumulation in CKD patients [11,26,27].

Data from an analysis of 14,543 participants in the National Health and Nutrition Examination Survey III showed that high dietary fiber intake was associated with lower risk of inflammation and mortality, and the associations were stronger in magnitude in CKD individuals than in those without CKD [28].

Because CKD diet is poor in fiber, some studies have investigated the effects of supplementation with these prebiotic agents, as shown in Table 36.1.

Probiotics

Probiotics may be an adjuvant therapy for CKD given the role of the gut microbiota alterations in uremic toxicity and systemic inflammation [13,30–32]. However, results are contradictory about their effectiveness (Table 36.1).

Uremia-induced changes in the biochemical/biophysical gut environment, along with other factors such as diet and medicines, create an unfavorable milieu for symbiotic microbiota. Furthermore, many bacterial genes are expressed only under specific conditions, as revealed by transcriptomics or proteomics. Attempts to restore the microbiome by introducing probiotics, without simultaneously improving the gut biochemical milieu, could be useless [1,7,13].

Synbiotics

Few studies have been published about this therapy in CKD patients (Table 36.1).

Low-Protein Diet

Uremic toxins are produced by amino acid fermentation in the gut [33], and dietary intake can change the gut microbiota profile [33–35]. A low-protein diet/LPD (0.6 g/kg per day) has been recommended for nondialysis CKD patients to reduce

TABLE 36.1 Studies Involving Prebiotics, Probiotics, and Synbiotics in Chronic Kidney Disease

References	Study Design, Follow-Up	Intervention	Results
Prebiotics			
Younes et al. [26]	RCT, crossover; 9 nondialysis CKD patients, 5 weeks	40 g/day of fermentable carbohydrate: 25 g from bread and 15 g from powder (4.5 g inulin + 10.5 g resistant starch)	↑ fecal nitrogen excretion ↓ urinary nitrogen excretion ↓ plasma urea concentration
Meijers et al. [27]	Noncontrolled trial; 20 HD patients; 1 month	20 g/day of oligofructose-enriched inulin	↓ serum p-CS No significant change in IS serum concentrations
Sirich et al. [45]	RCT; 40 HD patients; 6 weeks	15 g/day of resistant starch	↓ plasma IS ↓ serum p-Cresol and TMAO
Poesen et al. [21]	RCT, crossover; 39 non–dialysis-dependent CKD patients; 1 month	20 g/day of arabinoxylan oligosaccharides	No significant change in IS and p-CS serum concentrations
Probiotics			
Taki et al. [46]	Noncontrolled trial; 27 HD patients; 2 months	6×10^9 CFU/day of *Bifidobacterium longum*	↓ serum IS ↓ serum homocysteine
Ranganathan et al. [30]	RCT crossover, multicenter trial, 46 non–dialysis-dependent CKD patients; 6 months	9×10^{10} CFU/day of a probiotic mix: *S. thermophilus, L. acidophilus*, and *B. longum*	↓ serum urea levels Overall improvement in the quality of life No significant change in creatinine or uric acid serum concentrations
Miranda et al. [32]	RCT; 15 non–dialysis-dependent CKD patients; 2 months	1.6×10^{10} CFU/day of *Lactobacillus casei shirota*	↓ serum urea
Natarajan et al. [31]	RCT, crossover; 22 HD patients; 2 months	1.8×10^{14} CFU/day of a probiotic mix: *S. thermophilus, L. acidophilus*, and *B. longum*	No significant change in uremic toxins or inflammatory markers.
Borges et al. [13]	RCT; 16 HD patients; 3 months	9×10^{13} CFU/day of a probiotic mix: *S. thermophilus, L. acidophilus*, and *B. longum*	↑ plasma IS, potassium, urea No significant change in p-CS, IAA, and inflammatory markers (CRP and IL-6)

Synbiotics

Study	Design	Intervention	Outcomes
Nakabayashi et al. [47]	Noncontrolled trial; 9 HD patients; 2 weeks	Three times a day: 1×10^8 CFU of a probiotic mix: *L. casei shirota* and *B. breve* Yakult 4 g of galacto-oligosaccharides	↓ p-cresol Improvement in bowel habit No significant change in IS
Guida et al. [48]	RCT; 18 non—dialysis-dependent CKD patients; 1 month	Three times a day: 5×10^9 *L. plantarum*, 2×10^9 *L. casei* subsp. *rhamnosus*, 2×10^9 *L. gasseri*, 1×10^9 *B. infantis*, 1×10^9 *B. longum*, 1×10^9 *L. acidophilus*, 1×10^9 *L. salivarius*, 1×10^9 *L. sporogenes*, and 5×10^9 *S. thermophilus* 2.2 g of inulin and 1.3 g of resistant starch	↓ plasma p-Cresol
Viramontes-Homer et al. [49]	RCT; 22 HD patients; 2 months	1.1×10^7 CFU/day of a probiotic mix: *Lactobacillus acidophilus* and *Bifidobacterium lactis* 2.31 g/day of inulin (plus omega-3 and vitamins)	↓ severity of gastrointestinal symptoms No significant change in CRP, IL-6, or TNF-α.
Rossi et al. [50]	RCT, crossover; 31 non—dialysis-dependent CKD patients; 6 weeks	9×10^{13} CFU/day of a probiotic mix: nine different strains across the *Lactobacillus*, *Bifidobacteria*, and *Streptococcus* genera 15 g of a prebiotic mix: inulin, fructooligosaccharides, and galacto-oligosaccharides	↓ serum p-CS No significant change in serum IS Favorable change in microbiome: ↑ *Bifidobacterium* and ↓ *Ruminococcaceae*

CFU, colony forming unit; *CKD*, chronic kidney disease; *CRP*, C-reactive protein; *HD*, hemodialysis; *IAA*, indole 3-acetic acid; *IL-6*, Interleukin-6; *IS*, indoxyl sulfate; *p-CS*, p-cresyl sulfate; *RCT*, randomized clinical trial; *TMAO*, trimethylamine N-oxide; *TNF-α*, tumor necrosis factor alfa.

uremic symptoms [36]. In addition, LPD may reduce uremic toxin generation, as proposed by Mafra et al. [33] and confirmed by Black et al. [37], after 6 months of nutritional intervention. However, it is important to note that LPD is impractical for HD patients, who require 1.2 g of protein/kg per day.

Physical Exercise

Exercise has been associated with improvement of physical capacity and quality of life, antiinflammatory effects, and cardiovascular benefits in CKD patients [38–40]. Therefore some guidelines recommend regular physical exercise for CKD patients (The National Kidney Foundation Kidney Disease Outcome Quality Initiative [K/DOQI]; The Kidney Disease: Improving Global Outcomes [KDIGO]; and The American College of Sports Medicine) [41–43]. The main hypotheses are as follows:

1. Reestablishing gut barrier integrity: The increase in vagal tone by exercise may decrease intestinal permeability and foster an antiinflammatory milieu at the intestinal–luminal interface.
2. Increasing commensal bacteria: High noradrenaline plasma levels stimulate the growth of nonpathogenic commensal bacteria.
3. Increasing innate immunity: TLR signaling pathway and inflammatory cytokine production were decreased, and immune tolerance was improved.
4. Reducing colonic transit time: contact of harmful substances (such as ammonium hydroxide and uremic toxins) with the gut barrier was decreased, diminishing the arrival of these substances in the circulatory system [44].

Some studies in non-CKD patients or in experimental animal models have observed changes in composition and metabolites from the gut microbiota. However, the effects of physical exercise on gut microbiota of CKD are still being investigated.

CONCLUSION

The interaction between microbiome and host genome seems to be the target for future approaches in CKD. Available studies are still limited by poor methodologic quality, limited sample size, heterogeneity of the study populations, different experimental designs, different doses and treatment periods, limited control for dietary intake, and lack of exploration of effects on the gut microbiome, among others.

LIST OF ABBREVIATIONS

AhR Aryl hydrocarbon receptor
AP-1 Protein activator-1
CKD Chronic kidney disease
CVD Cardiovascular disease
GALT Gut-associated lymphoid tissue
IAA Indole 3-acetic acid
IL-1 Interleukin-1
IL-6 Interleukin-6
IS Indoxyl sulfate
K/DOQI Kidney Disease Outcome Quality Initiative
KDIGO Kidney disease improving global outcomes
LPS Lipopolysaccharides
NF-κB Nuclear factor-κB
OTUs Operational taxonomic units
PCR Polymerase chain reaction
p-CS p-cresyl sulfate
ROS Reactive oxygen species
RS Resistant starch
SCFA Short-chain fatty acids
TLR Toll-like receptor
TMA Trimethylamine
TMAO Trimethylamine N-oxide
TNF-α Tumor necrosis factor alfa

REFERENCES

[1] Lozupone CA, Stombaugh JI, Gordon JI, Jansson JK, Knight R. Diversity, stability and resilience of the human gut microbiota. Nature 2012;489(7415):220–30.

[2] Barros AF, Borges NA, Ferreira DC, Carmo FL, Rosado AS, Fouque D, Mafra D. Is there interaction between gut microbial profile and cardiovascular risk in chronic kidney disease patients? Future Microbiol 2015;10(4):517–26.

[3] Levy M, Kolodziejczyk AA, Thaiss CA, Elinav E. Dysbiosis and the immune system. Nat Rev Immunol 2017;17(4):219–32.

[4] Wang IK, Lai HC, Yu CJ, Liang CC, Chang CT, Kuo HL, Yang YF, Lin CC, Lin HH, Liu YL, Chang YC, Wu YY, Chen CH, Li CY, Chuang FR, Huang CC, Lin CH, Lin HC. Real-time PCR analysis of the intestinal microbiotas in peritoneal dialysis patients. Appl Environ Microbiol 2012;78:1107–12.

[5] Vaziri ND, Wong J, Pahl M, Piceno YM, Yuan J, DeSantis TZ, Ni Z, Nguyen TH, Andersen GL. Chronic kidney disease alters intestinal microbial flora. Kidney Int 2013;83:308–15.

[6] Jiang S, Xie S, Lv D, Wang P, He H, Zhang T, Zhou Y, Lin Q, Zhou H, Jiang J, Nie J, Hou F, Chen Y. Alteration of the gut microbiota in Chinese population with chronic kidney disease. Sci Rep 2017;7:2870.

[7] Vaziri ND, Zhao YY, Pahl MV. Altered intestinal microbial flora and impaired epithelial barrier structure and function in CKD: the nature, mechanisms, consequences and potential treatment. Nephrol Dial Transplant 2015;31:737–46.

[8] Mafra D, Lobo JC, Barros AF, Koppe L, Vaziri ND, Fouque D. Role of altered intestinal microbiota in systemic inflammation and cardiovascular disease in chronic kidney disease. Future Microbiol 2014;9(3):399–410.

[9] Lau WL, Kalantar-Zadeh K, Vaziri ND. The gut as a source of inflammation in chronic kidney disease. Nephron 2015;130:92–8.

[10] Mafra D, Fouque D. Gut microbiota and inflammation in chronic kidney disease patients. Clin Kidney J 2015;8:332–4.

[11] Moraes C, Borges NA, Mafra D. Resistant starch for modulation of gut microbiota: promising adjuvant therapy for chronic kidney disease patients? Eur J Nutr 2016;55(5):1813–21.

[12] Bermon S, Petriz B, Kajėnienė A, Prestes J, Castell L, Franco OL. The microbiota: an exercise immunology perspective. Exerc Immunol Rev 2015;21:70–9.

[13] Borges NA, Carmo FL, Stockler-Pinto MB, de Brito JS, Dolenga CJ, Ferreira DC, Nakao LS, Rosado A, Fouque D, Mafra D. Probiotic supplementation in chronic kidney disease: a double-blind, randomized, placebocontrolled trial. J Ren Nutr 2018;28(1):28–36.

[14] Vaziri ND, Liu S-M, Lau WL, Khazaeli M, Nazertehrani S, Farzaneh SH, Kieffer DA, Adams SH, Martin RJ. High amylose resistant starch diet ameliorates oxidative stress, inflammation, and progression of chronic kidney disease. PLoS One 2014;9(12):e114881.

[15] Wong J, Piceno YM, Desantis TZ, Pahl M, Andersen GL, Vaziri ND. Expansion of urease and uricase-containing, indole- and p- cresol-forming and contraction of short- chain fatty acid-producing intestinal microbiota in ESRD. Am J Nephrol 2014;39:230–7.

[16] Borges NA, Barros AF, Nakao LS, Dolenga CJ, Fouque D, Mafra D. Protein-bound uremic toxins from gut microbiota and inflammatory markers in chronic kidney disease. J Ren Nutr 2016;26:396–400.

[17] Mafra D, Borges NA, Cardozo LFMF, Anjos JS, Black AP, Moraes C, Bergman P, Bengt P, Stenvinkel P. Red meat intake in chronic kidney disease patients: two sides of the coin. Nutrition 2018;46:26–32.

[18] Dalton TP, Puga A, Shertzer HG. Induction of cellular oxidative stress by aryl hydrocarbon receptor activation. Chem Biol Interact 2002;141:77–95.

[19] Sallee M, Dou L, Cerini C, Poitevin S, Brunet P, Burtey S. The aryl hydrocarbon receptor-activating effect of uremic toxins from tryptophan metabolism: a new concept to understand cardiovascular complications of chronic kidney disease. Toxins 2014;6:934–49.

[20] Deltombe O, Biesen WV, Glorieux G, Massy Z, Dhondt A, Eloot S. Exploring protein binding of uremic toxins in patients with different stages of chronic kidney disease and during hemodialysis. Toxins 2015;7:3933–46.

[21] Poesen R, Windey K, Neven E, Kuypers D, De Preter V, Augustijns P, D'Haese P, Evenepoel P, Verbeke K, Meijers B. The influence of CKD on colonic microbial metabolism. J Am Soc Nephrol 2016;27:1389–99.

[22] Schulman G, Berl T, Beck GJ, Remuzzi G, Ritz E, Arita K, Kato A, Shimizu M. Randomized placebo-controlled EPPIC trials of AST-120 in CKD. J Am Soc Nephrol 2015;26:1732–46.

[23] Kikuchi K, Itoh Y, Tateoka R, Ezawa A, Murakami K, Niwa T. Metabolomic search for uremic toxins as indicators of the effect of an oral sorbent AST-120 by liquid chromatography/tandem mass spectrometry. J Chromatogr B 2010;878:2997–3003.

[24] Vaziri ND, Yuan J, Khazaeli M, Masuda Y, Ichii H, Liu S. Oral activated charcoal adsorbent (AST-120) ameliorates CKD-induced intestinal epithelial barrier disruption and systemic inflammation. Am J Nephrol 2013;37:518–25.

[25] Gibson GR, Roberfroid MB. Dietary modulation of the human colonic microbiota: introducing the concept of prebiotics. J Nutr 1995;125:1401–12.

[26] Younes H, Egret N, Hadj-Abdelkader M, Rémésy C, Demigné C, Gueret C, Deteix P, Alphonse JC. Fermentable carbohydrate supplementation alters nitrogen excretion in chronic renal failure. J Ren Nutr Off J Counc Ren Nutr Natl Kidney Found 2006;16:67–74.

[27] Meijers BKI, De Preter V, Verbeke K, Vanrenterghem Y, Evenepoel P. p-Cresyl sulfate serum concentrations in haemodialysis patients are reduced by the prebiotic oligofructose-enriched inulin. Nephrol Dial Transpl Off Publ Eur Dial Transpl Assoc Eur Ren Assoc 2010;25:219–24.

[28] Krishnamurthy VMR, Wei G, Baird BC, Murtaugh M, Chonchol MB, Raphael KL, Greene T, Beddhu S. High dietary fiber intake is associated with decreased inflammation and allcause mortality in patients with chronic kidney disease. Kidney Int 2012;81:300–6.

[29] Joint FAO/WHO. Working group report on drafting guidelines for the evaluation of probiotics in food. London, Ontario, Canada. April 30 and May 1, 2002. Available at: http://www.fda.gov/ohrms/dockets/dockets/95s0316/95s-0316-rpt0282-tab-03-ref-19-joint-faowho-vol219.pdf.

[30] Ranganathan N, Ranganathan P, Friedman EA, Joseph A, Delano B, Goldfarb DS, Tam P, Rao AV, Anteyi E, Musso CG. Pilot study of probiotic dietary supplementation for promoting healthy kidney function in patients with chronic kidney disease. Adv Ther 2010;27:634–47.

[31] Natarajan R, Pechenyak B, Vyas U, Ranganathan P, Weinberg A, Liang P, Mallappallil MC, Norin AJ, Friedman EA, Saggi SJ. Randomized controlled trial of strain-specific probiotic formulation (Renadyl) in dialysis patients. BioMed Res Int 2014;2014:568571.

[32] Miranda Alatriste PV, Urbina Arronte R, Gomez Espinosa CO, Espinosa Cuevas Mde L. Effect of probiotics on human blood urea levels in patients with chronic renal failure. Nutr Hosp 2014;29:582−90.

[33] Mafra D, Barros AF, Fouque D. Dietary protein metabolism by gut microbiota and its consequences for chronic kidney disease patients. Future Microbiol 2013;8:1317−23.

[34] Wu GD, Chen J, Hoffmann C, Bittinger K, Chen Y, Keilbaugh AS, Bewtra M, Knights D, Walters WA, Knight R, Sinha R, Gilroy E, Gupta K, Baldassano R, Nessel L, Li H, Bushman FD, Lewis JD. Linking long-term dietary patterns with gut microbial enterotypes. Science 2011;334:105.

[35] Patel KP, Luo FJ, Plummer NS, Hostetter TH, Meyer TW. The production of p-cresol sulfate and indoxyl sulfate in vegetarians versus omnivores. Clin J Am Soc Nephrol 2012;7(6):982−8.

[36] Di Iorio B, De Santo NG, Anastasio P, Perna A, Pollastro MR, Di Micco L, Cirillo M. The Giordano-Giovannetti diet. J Nephol 2013;26(22):S143−52.

[37] Black AP, Saraiva J, Cardozo L, Carmo FL, Dolenga CJ, Nakao LS, Ferreira DC, Rosado A, Carraro Eduardo JC, Mafra D. Does low-protein diet influence the uremic toxin serum levels from the gut microbiota in Nondialysis chronic kidney disease patients? J Ren Nutr 2018;28(3):208−14.

[38] Painter P, Roshanravan B. The association of physical activity and physical function with clinical outcomes in adults with chronic kidney disease. Curr Opin Nephrol Hypertens 2013;22(6):615−23.

[39] Petersen AM, Pedersen BK. The anti-inflammatory effect of exercise. J Appl Physiol 2005;98:1154−62.

[40] Gleeson M, McFarlin BK, Flynn MG. Exercise and toll-like receptors. Exerc Immunol Rev 2006;12:34−53.

[41] K/DOQI Workgroup. K/DOQI clinical practice guidelines for cardiovascular disease in dialysis patients. Am J Kidney Dis 2005;45:S1−153.

[42] KDIGO. Clinical practice guideline for the evaluation and management of chronic kidney disease. J Int Soc Nephrol 2012;2013(3):1−163.

[43] American College of Sports Medicine. ACSM's guidelines for exercise testing and prescription. 9th ed. (PA, USA): Wolters Kluwer/Lippincott Williams & Wilkins; 2013. p. 305−8.

[44] Esgalhado M, Borges NA, Mafra D. Could physical exercise help modulate the gut microbiota in chronic kidney disease? Future Microbiol 2016;11:699−707.

[45] Sirich TL, Plummer NS, Gardner CD, Hostetter TH, Meyer TW. Effect of increasing dietary fiber on plasma levels of colon-derived solutes in hemodialysis patients. Clin J Am Soc Nephrol 2014;9:1603−10.

[46] Taki K, Takayama F, Niwa T. Beneficial effects of Bifidobacteria in a gastroresistant seamless capsule on hyperhomocysteinemia in hemodialysis patients. J Ren Nutr 2005;15(1):77−80.

[47] Nakabayashi I, Nakamura M, Kawakami K, Ohta T, Kato I, Uchida K, Yoshida M. Effects of synbiotic treatment on serum level of p-cresol in hemodialysis patients: a preliminary study. Nephrol Dial Transplant 2011;26(3):1094−8.

[48] Guida B, Germano R, Trio R, Russo D, Memoli B, Grumetto L, Barbato F, Cataldi M. Effect of short-term synbiotic treatment on plasma p-cresol levels in patients with chronic renal failure: a randomized clinical trial. Nutr Metab Cardiovasc Dis 2014;24:1043−9.

[49] Viramontes-Hörner D, Márquez-Sandoval F, Martín-del-Campo F, Vizmanos-Lamotte B, Sandoval-Rodríguez A, Armendáriz-Borunda J, García-Bejarano H, Renoirte-López K, García-García G. Effect of a symbiotic gel (Lactobacillus acidophilus + Bifidobacterium lactis + inulin) on presence and severity of gastrointestinal symptoms in hemodialysis patients. J Ren Nutr 2015;25(3):284−91.

[50] Rossi M, Johnson DW, Morrison M, Pascoe EM, Coombes JS, Forbes JM, Szeto CC, McWhinney BC, Ungerer JP, Campbell KL. Synbiotics easing renal failure by improving gut microbiology (SYNERGY): a randomized trial. Clin J Am Soc Nephrol 2016;11:223−31.

Chapter 37

Dysbiosis in Benign and Malignant Diseases of the Exocrine Pancreas

Robert Memba[1,2], Sinead N. Duggan[1], Rosa Jorba[2] and Kevin C. Conlon[1]

[1]Professorial Surgical Unit, Department of Surgery Trinity College Dublin, Tallaght Hospital, Dublin, Ireland; [2]Hepatobiliary and Pancreatic Surgery Unit, Department of Surgery, Joan XXIII University Hospital, Tarragona, Spain

INTRODUCTION

The pancreas has two basic functional compartments: the exocrine pancreas (which secretes enzymes for digestion of fat, carbohydrate, and protein), and the endocrine pancreas, which maintains glucose homeostasis. The exocrine pancreas represents 98%–99% of total pancreatic mass, consisting of a highly branched, tubular, epithelial tree-like network. This portion is comprised mainly of acinar, centroacinar, and ductal cells [1]. The most common benign conditions affecting the exocrine pancreas are acute pancreatitis (AP) and chronic pancreatitis (CP). AP is an inflammation of the pancreatic gland frequently caused by structural blockage such as gallstones or gland toxicity due to excess alcohol consumption [2]. CP is an inflammatory disease of the pancreas characterized by irreversible morphological changes, typically causing chronic pain, exocrine dysfunction, and endocrine dysfunction [3]. The necrosis-fibrosis hypothesis suggests that the initial damage is caused by an initial acute inflammatory process, progressing to chronic irreversible damage as a result of repeated acute attacks.

Although alcohol consumption has long been considered the dominant risk factor in the development of CP, only 3% of the alcoholic population develop this disease [4,5]. Therefore, additional triggers or initiating factors may play a significant role. The most common malignancy of exocrine pancreas is pancreatic ductal cancer (PDC). PDC is one of the most aggressive cancers. For the majority of patients it remains a lethal disease, and is the fourth leading cause of cancer death in Europe. There is no effective screening diagnostic test, and most patients have advanced disease at presentation; therefore, the lack of biomarkers for early detection contributes to poor outcomes.

This is exacerbated by other factors such as its aggressive biology and its resistance to conventional and targeted therapeutic agents. Thus, identification of modifiable risk factors, and an effective primary prevention strategy are critical to reducing the aggressiveness of this malignancy. Some of the known risk factors are smoking, obesity, diet, genetics, and CP [6,7]. In fact, occupational and environmental exposures may act synergistically with inherited acquired genetic polymorphisms, resulting in earlier occurrence of PDC [8].

THE MICROBIOME AND "OMICS" TOOLS

The ecological community of commensal, symbiotic, and pathogenic microorganisms present in our body is known as the microbiota. The microbiome is the entire genome sequence of this microbial community. Microbiota is predominantly formed by bacteria but also comprises fungi, yeast, viruses, and archaea [9]. The increasing number of genomic-based molecular techniques such as transcriptome, metabolome, and metagenome analyses, combined with the use of various in vivo models such as germ-free mice, has expanded our current knowledge on the microbiome [10].

CLINICAL IMPLICATIONS OF DYSBIOSIS

Fluctuations in the composition of gut microbiota are associated with the development of several disorders, while microbial stability is associated with health [7,11]. Microbiota imbalance (also known as dysbiosis or dysbacteriosis) has been linked

Microbiome and Metabolome in Diagnosis, Therapy, and other Strategic Applications. https://doi.org/10.1016/B978-0-12-815249-2.00037-3

to dysregulation of immune effector cells and activation of inflammatory cytokines playing a role in several inflammatory-mediated diseases. For example, multiple sclerosis, a chronic demyelinating disease triggered by an autoimmune mechanism, has been linked with dysbiosis [12]. Similarly, patients with systemic lupus erythematosus, also an autoimmune disease, are reported to have subgingival dysbiosis [13]. Regarding Alzheimer disease, a lower level of *Bifidobacterium* has been found [14].

In the same way, increases in *Prevotella* have been linked to ankylosing spondylitis [15]. Finally, detection of a fecal bacteria subset including *Fusobacterium nucleatum, Bacteroides clarus,* and *Roseburia intestinalis* have been successfully identified in patients with colorectal cancer, which gives hope to the development of a novel diagnostic biomarker [16]. Furthermore, dysbiosis has been linked to other GI disorders such as celiac disease [17], irritable bowel syndrome (IBS) [18], and inflammatory bowel disease (IBD) [19]. Some studies have also suggested a link between dysbiosis and obesity [20]. However, data for benign or malignant pancreatic diseases remain scarce.

SMALL BOWEL BACTERIAL OVERGROWTH IN CHRONIC PANCREATITIS

Chronic pancreatitis generally results in gastrointestinal (GI) tract microbial abnormalities. In fact, small intestinal bacterial overgrowth (SIBO) occurs in up to 14.7% of patients with CP who have not had GI surgery [21]. SIBO is a condition in which the increased bacterial load in the small bowel results in excessive fermentation and inflammation. It is associated with chronic intestinal symptoms, such as abdominal discomfort, bloating, diarrhea, and malabsorption [22]. The clinical manifestations and signs of SIBO can sometimes be attributed to CP itself; therefore all CP patients should be screened for SIBO, irrespective of symptoms, to differentiate those related with unresponsive steatorrhea, in order to avoid an unnecessary increased dosage of pancreatic enzyme replacement therapy (PERT). SIBO could be associated with an imbalance of colonic microbiota in CP patients; however, this has not yet been investigated [23].

RELEVANT MICROBIAL POPULATIONS

To date, only a few studies have assessed specific microbial abnormalities linked to pancreatic diseases (summarized in Table 37.1) [23].

Acute Pancreatitis

Regarding AP, there has only been one study published on dysbiosis to date. Tan et al. [24] found that *Enterobacteriaceae* and *Enterococcus* populations are higher in patients with AP. There were no differences between mild and severe AP patients. In this study, they also found *Bifidobacterium* was lower in AP patients when compared to healthy participants.

Chronic Pancreatitis

Jandhyala et al. [25] compared bacterial DNA from fecal samples using 16S mRNA in CP patients. Patients with CP and pancreatic endocrine insufficiency (known as type 3c diabetes, T3cDM) had higher levels of Bacteroidetes and lower levels of *Faecalibacterium* compared to those with CP but without T3cDM. Patients with T3cDM and pancreatic exocrine insufficiency (PEI) had lower amounts of *Bifidobacterium* compared to those without PEI. Another publication by Gorovits et al. [26] found lower levels of *Bifidobacterium* and *Lactobacillus* and higher levels of *Enterobacter, Proteus, Klebsiella,* and *Morganella* in patients with CP when compared to normal literature reference ranges.

Farrel et al. [27] assessed microbial composition in salivary samples, extracting bacterial DNA by using Human Oral Microbe Identification Microarray and quantitative real-time polymerase chain reaction (PCR) in both CP and PDC patients. The CP group had lower levels of *Granulicatella adiacens* compared to those with PDC. Savitskaya et al. [28] also compared CP to normal literature reference ranges finding lower levels of *Lactobacillus* and higher levels of *Escherichia coli, Enterococcus faecalis,* and *Enterococcus faecium* in CP patients. However, there was no difference in the amount of *Bifidobacterium* in CP patients.

Pancreatic Ductal Cancer

Michaud et al. [29] determined antibodies against different oral bacteria. There was significantly higher antibody levels of periodontal pathogen *Porphyromonas gingivalis,* found in PDC patients. The highest concentration of *P. gingivalis* was associated with a twofold increase in PDC risk. Moreover, higher levels of antibodies to commensal oral bacteria were

TABLE 37.1 Studies Published on the Levels of Specific Bacteria in the GI Tract, of Patients With Pancreatic Disease, Compared to Controls Where Applicable

Disease	Author, year	Results	Study Type
CP	Jandhyala 2017	Bacteroidetes higher in CP with T3cDM versus CP no T3cDM. Faecalibacterium lower in CP with T3cDM versus CP no T3cDM. Bifidobacterium lower in CP/T3c DM patients with PEI versus without PEI.	Controlled
	Gorovits 2013	Bifidobacterium and Lactobacillus lower in CP. Enterobacter, Proteus, Klebsiella and Morganella higher in CP. versus healthy participants.	Observational
	Farrell 2012	G. adiacens higher in PDC versus CP patients, S. mitis lower in PDC patients	Controlled
	Savitskaya 2002	Lactobacillus lower in CP. Bifidobacterium no significant differences. E. coli, E. faecalis and E. faecium higher in CP.	Observational
AP	Tan 2015	Enterobacteriaceae and Enterococcus populations higher in AP groups versus healthy participants. Bifidobacterium lower in AP versus healthy participants.	Controlled
PDC	Michaud 2016	High antibody levels of P. gingivalis more common in PDC patients than controls; the highest concentration of P. gingivalis was associated with a twofold increase in PDC risk.	Prospective cohort
	Fan 2016	Higher P. gingivalis and Aggregatibacter in PDC. Lower Leptotrichia and Fusobacteria in PDC.	Nested case-control
	Torres 2015	Leptotrichia higher and P. gingivalis lower in PDC. Bacteroides higher (not significant) in PDC. N. elongata and Aggregatibacter lower (not significant) in PDC. No difference in S. mitis and G. adiacens.	Controlled
	Half 2015	Bacteroides and Verrucomicrobia increased twofold in PDC. Sutterella, Veillonella, Bacteroides, Odoribacter, and Akkermansia also higher in PDC. Firmicutes and Actinobacteria lower in PDC.	Pilot
	Lin 2013	Bacteroides higher in PDC and pancreatitis. Corynebacterium and Aggregatibacter lower in PDC.	Pilot
	Farrell 2012	Species within 6 genera different between PDC and healthy participants (Streptococcus, Prevotella, Campylobacter, Granulicatella, Atopobium, Neisseria). N. elongata and S. mitis lower in PDC patients versus healthy participants. G. adiacens higher in PDC versus CP patients, S. mitis lower in PDC patients versus healthy participants.	Controlled

AP, acute pancreatitis; CP, chronic pancreatitis; PDC, pancreatic ductal cancer.

linked to a lower risk of PDC. *P. gingivalis* is the main microorganism involved in periodontal disease, a very common oral disorder affecting around 90% of general population [29,30]. Another publication by Fan et al. [31] mirrored this association. Conversely, other publications by Torres [32], Half [33], and Lin [34] found a link between an increase of Bacteroides and the risk of developing PDC. Two studies addressed the oral cavity [32,34] and one investigated fecal samples [33]. As mentioned earlier, Farrell [27] also analyzed salivary microbiota in PDC, reporting lower levels of *Neisseria elongata* and *Streptococcus mitis* in PDC patients, while *Granulicatella adiacens* was higher.

H. PYLORI AND PDC

Several groups have analyzed *Helicobacter pylori* as risk factor for the development of PDC. However, reports are inconsistent, with several metaanalyses finding an association between *H. pylori* and PDC [35,36] and others reporting no association [37,38].

CRUCIAL MOLECULES AND PATHWAYS

In the only study on AP, *Bifidobacterium* was found to be lower in patients [24]. It is well-known that repeated AP episodes commonly precede CP; therefore, this finding would be compatible with the hypothesis of a causal association between microbiota imbalance and CP rather than a reactive response in CP patients [7,11]. There is insufficient evidence to suggest a microbial link in the pathogenesis of AP or CP. However, the common finding of a decrease in *Bifidobacteria* and other microbial abnormalities in both disorders might entail similar mechanisms to link dysbiosis to pancreatitis.

As mentioned before, CP patients with PEI often develop SIBO. In both AP and CP, there is a disruption to the gut barrier, and a systemic inflammatory response secondary to cytokine release, which increases bacterial translocation. Both abnormal motility and bacterial translocation are also factors linked to bacterial overgrowth (SIBO) that occur in CP patients [39–41]. It is reasonable to posit that CP patients with overgrowth of bacteria in the small bowel may also have colonic microbiota imbalance, as malabsorption secondary to PEI has a significant impact on the nutrient absorption, and therefore on the availability of nutrients for intestinal microorganisms.

EXCESSIVE PROINFLAMMATORY ENTEROBACTERIA

A relative abundance of Enterobacteriaceae could promote a systemic inflammatory response, thus contributing to the development of CP. Research on IBD and cystic fibrosis (CF), have shown that these diseases are also associated with high levels of Enterobacteriaceae [42]. There has been significant interest in the role of *Escherichia coli*, particularly in patients with ileal Crohn disease (CD), and biopsies of patients with this disorder have revealed invasion of the mucosa by *E. coli* [19,43].

The involvement of Enterobacteriaceae in the development of pancreatic diseases is likely to be immune-mediated. The commensal bacteria are environmental factors capable of inducing autoimmunity. Yanagisawa et al. [44] showed that *E. coli* was able to induce AIP-like pathological alterations in the pancreas of normal mice. Furthermore, this study also showed higher antibody titres against *E. coli* in patients with AIP compared to disease-free people. Therefore, the findings of Tan et al. [24], were consistent with these data in CP and in other diseases. Microbial-mediated procarcinogenic mechanisms of *P. gingivalis* are secondary to two particularities: Firstly these bacteria show both in vitro and in vivo ability to evade the host immune homeostasis, increasing systemic inflammation. Secondly, they also increase nitrosamine exposure.

CHRONIC INFLAMMATION AND CARCINOGENESIS

Commensal bacteria hypothetically inhibit the growth of pathogenic ones. It is thought that periodontal disease is linked to carcinogenesis due to an abnormal inflammatory response, rather than by having a direct mutagenic effect [29,45,46]. *S. mitis* has been shown to have a protective role against carcinogenic pathogens, which may allow for the overgrowth of *G. adiacens* [9,20,30,46]. The association between CP and PDC suggests that inflammation may be involved in the initiation and/or promotion of the mutagenesis process [8,27].

ALCOHOLIC DYSBIOSIS

Alcohol is a significant etiological factor for CP. Alcohol has been shown to alter jejunal microflora [47], since almost 50% of alcoholics with documented recent ethanol abuse displayed an increase in total number of bacteria, most of which originated from the fecal flora. There is substantial evidence that alcohol increases gut permeability to large molecules of endotoxin size. Likewise, acetaldehyde as a result of alcohol metabolism by colonic bacteria has the capacity to disrupt epithelial junctions, suggesting that the increased serum endotoxin concentrations observed in alcoholics may also be of colonic origin [48,49]. Since all alcoholics may be expected to have bacterial translocation, the fact that only a minority develop overt pancreatitis indicates that genetic polymorphism plays a significant role [50]. Therefore, further investigations in this field might help to prevent development of CP, other than advising alcohol abstinence.

IN VITRO AND IN VIVO MODELS

There are no interventional human studies in CP, investigating the potential to modify gut microbiota to improve outcomes. Only a few animal studies have been reported. In a mouse model, Ren et al. [51] investigated the administration of seleno-lentinan, a prebiotic-like preparation. They reported that this selenium-based product increased the proportion of beneficial bacteria, and suggested that seleno-lentinan may prevent CP development by elevating antioxidant status and modulating gut microbiota. Another study by Hu et al. [52] showed a lower diversity and richness in the gut microbiota of CP mice, with a relative lack of Firmicutes and higher levels of Bacteroidetes.

CLINICAL IMPLICATIONS

Gut microbiota is normally measured by oral, bowel, or fecal samples using DNA-based analysis, as well as by specific antibodies against known pathogens [7,53]. Microbial abnormalities may be explored as early biomarkers for PDC, and therefore may potentially have a significant impact on the long-term survival of this lethal disease. Therefore, based on the current evidence, specific bacterial profiling has been suggested as a potential biomarker for PDC [9]. For instance, the identified salivary biomarkers [27,45] have high specificity and sensitivity for the detection of PDC.

On the therapeutic front, proving a causal association between microbiota imbalance and pancreatic diseases could potentially lead to the development of therapeutic or prevention tools in diseases such as CP or PDC. The consistent findings of low levels of "beneficial" *bifidobacteria* in pancreatitis (as with other conditions) raise compelling questions about potential therapeutic interventions [54]. Fecal microbiota transplantation (FMT) has already been shown to be effective in treating recurrent or resistant *Clostridium difficile* infection [55] and is showing promise as a potential treatment for IBD [19]. Probiotic preparations containing either *Lactobacillus* alone or in combination with *Bifidobacterium* can be effective in the prophylaxis of severe necrotizing enterocolitis in preterm infants, as well as in the prevention of *C. difficile*-associated diarrhea [56].

Probiotic studies in animal models and also in humans have presented conflicting results for the outcome of pancreatitis [5,54]. With only four studies investigating dysbiosis in CP, there is, as yet, insufficient evidence. There are several mechanisms that explain the health beneficial effects of *bifidobacteria*: they reduce transit time (relieving diarrhea and malabsorption) and produce short chain fatty acids (SCFAs) and lactic acid. SCFAs reduce luminal pH, and this in itself, inhibits pathogenic microorganisms, and increases the absorption of some nutrients. SCFAs also have a trophic effect on the intestinal epithelium.

Bifidobacteria are saccharolytic and therefore play an important role in carbohydrate fermentation in the colon. *Bifidobacteria* produce lactate that may be transformed into butyrate, which reduces the rate of transformed cell growth, thus leading to cell reversion from neoplastic to nonneoplastic. In addition, to fermentation products, in vitro, *bifidobacteria* are able to synthesize B vitamins. *Bifidobacteria* also stabilize the intestinal mucosa, normalizing intestinal permeability and improving gut immunology, leading to the prevention of the overgrowth of pathogenic microorganisms [20,56–60].

Regarding diet and microbiota, a "western-style" diet (typically high in saturated fat and carbohydrate and low in fibre) is positively correlated to *Bacteroides* enterotype, while an Asian diet (typically lower in fat) may promote a predominance of beneficial *Bifidobacteria*, and a greater microbial diversity [7,61,62].

ANTIBIOTICS VERSUS PROBIOTICS

Since bacteria or bacterial products appear to play a primary role in the initiation, progression, and rate of complications of alcoholic pancreatitis, it appears logical to target gut bacteria either within the lumen via bacterial decontamination with nonabsorbable antibiotics, or once translocation has occurred, via systemic administration of antibiotics [5]. However, results in the absence of positive bacterial culture are still debated.

In turn, the administration of prophylactic probiotics in patients during the acute phase of severe AP increased mortality in a randomized controlled trial [63]. This multicenter study from the Netherlands (The Propatria study) compared the use of a multispecies probiotic preparation (delivered twice per day via a jejunal feeding tube) to a placebo. The primary outcome was infectious complications and this did not differ between the groups. However, nine patients in the probiotic arm developed bowel ischemia, of which eight died. Overall, 24 patients in the probiotic arm died, compared to nine in the placebo group. The authors of this controversial study concluded that probiotics should not be administered during the acute phase of severe AP. The administration of prophylactic antibiotics after the resolution of AP, in order to prevent dysbiosis and potential evolution to CP, has not been yet investigated.

CONCLUSIONS

Current evidence suggests a possible link between specific microbiota abnormalities and PDC. Specifically, salivary abundance of *P. gingivalis* and *G. adiacens* appear to be risk factors, while *S. mitis* appears to be a protecting factor in PDC. Regarding pancreatitis, the available data suggest a plausible association between dysbiosis and CP, including lower levels of *Bifidobacterium* and a higher level of Enterobacteriaceae in the gut. Further research is needed to confirm the potential role of the microbiota in pancreatic disorders, which may open a new avenue for biomarkers and targeted therapies.

REFERENCES

[1] Pictet RL, Clark WR, Williams RH, Rutter WJ. An ultrastructural analysis of the developing embryonic pancreas. Dev Biol 1972;29(4):436−67.

[2] Yadav D, Lowenfels AB. Trends in the epidemiology of the first attack of acute pancreatitis: a systematic review. Pancreas 2006;33(4):323−30.

[3] Sarner M, Cotton PB. Classification of pancreatitis. Gut 1984;25(7):756−9.

[4] Yadav D, Hawes RH, Brand RE, Anderson MA, Money ME, Banks PA, et al. Alcohol consumption, cigarette smoking, and the risk of recurrent acute and chronic pancreatitis. Arch Intern Med 2009;169(11):1035−45.

[5] Vonlaufen A, Spahr L, Apte MV, Frossard J-L. Alcoholic pancreatitis: a tale of spirits and bacteria. World J Gastrointest Pathophysiol 2014;5(2):82−90.

[6] Becker AE, Hernandez YG, Frucht H, Lucas AL. Pancreatic ductal adenocarcinoma: risk factors, screening, and early detection. World J Gastroenterol 2014;20(32):11182−98.

[7] Power SE, O'Toole PW, Stanton C, Ross RP, Fitzgerald GF. Intestinal microbiota, diet and health. Br J Nutr 2014;111(03):387−402.

[8] Yeo TP, Hruban RH, Brody J, Brune K, Fitzgerald S, Yeo CJ. Assessment of "gene-environment" interaction in cases of familial and sporadic pancreatic cancer. J Gastrointest Surg 2009;13(8):1487−94.

[9] Leal-Lopes C, Velloso FJ, Campopiano JC, Sogayar MC, Correa RG. Roles of commensal microbiota in pancreas homeostasis and pancreatic pathologies. J Diabetes Res 2015:2015.

[10] Redinbo MR. The microbiota, chemical symbiosis, and human disease. J Mol Biol 2014;426(23):3877−91.

[11] Milani C, Ferrario C, Turroni F, Duranti S, Mangifesta M, van Sinderen D, et al. The human gut microbiota and its interactive connections to diet. J Hum Nutr Diet 2016;29(5):539−46.

[12] Yadav SK, Boppana S, Ito N, Mindur JE, Mathay MT, Patel A, et al. Gut dysbiosis breaks immunological tolerance toward the central nervous system during young adulthood. Proc Natl Acad Sci USA 2017;114(44):E9318−27.

[13] Correa JD, Calderaro DC, Ferreira GA, Mendonca SM, Fernandes GR, Xiao E, et al. Subgingival microbiota dysbiosis in systemic lupus erythematosus: association with periodontal status. Microbiome 2017;5(1):34.

[14] Vogt NM, Kerby RL, Dill-McFarland KA, Harding SJ, Merluzzi AP, Johnson SC, et al. Gut microbiome alterations in Alzheimer's disease. Sci Rep 2017;7(1):13537.

[15] Wen C, Zheng Z, Shao T, Liu L, Xie Z, Le Chatelier E, et al. Quantitative metagenomics reveals unique gut microbiome biomarkers in ankylosing spondylitis. Genome Biol 2017;18(1):142.

[16] Liang Q, Chiu J, Chen Y, Huang Y, Higashimori A, Fang J, et al. Fecal bacteria act as novel biomarkers for noninvasive diagnosis of colorectal cancer. Clin Cancer Res 2017;23(8):2061−70.

[17] Verdu EF, Galipeau HJ, Jabri B. Novel players in coeliac disease pathogenesis: role of the gut microbiota. Nat Rev Gastroenterol Hepatol 2015;12(9):497−506.

[18] Distrutti E, Monaldi L, Ricci P, Fiorucci S. Gut microbiota role in irritable bowel syndrome: new therapeutic strategies. World J Gastroenterol 2016;22(7):2219−41.

[19] Sheehan D, Moran C, Shanahan F. The microbiota in inflammatory bowel disease. J Gastroenterol 2015;50(5):495−507.

[20] Ramakrishna BS. Role of the gut microbiota in human nutrition and metabolism. J Gastroenterol Hepatol 2013;28(S4):9−17.

[21] Capurso G, Signoretti M, Archibugi L, Stigliano S, Delle Fave G. Systematic review and meta-analysis: small intestinal bacterial overgrowth in chronic pancreatitis. United Eur Gastroenterol J 2016. https://doi.org/10.1177/2050640616630117.

[22] Kim DB, Paik C-N, Sung HJ, Chung WC, Lee K-M, Yang J-M, et al. Breath hydrogen and methane are associated with intestinal symptoms in patients with chronic pancreatitis. Pancreatology 2015;15(5):514−8.

[23] Memba R, Duggan SN, Ni Chonchubhair HM, Griffin OM, Bashir Y, O'Connor DB, et al. The potential role of gut microbiota in pancreatic disease: a systematic review. Pancreatology 2017;17(6):867−74.

[24] Tan C, Ling Z, Huang Y, Cao Y, Liu Q, Cai T, et al. Dysbiosis of intestinal microbiota associated with inflammation involved in the progression of acute pancreatitis. Pancreas 2015;44(6):868−75.

[25] Jandhyala SM, Madhulika A, Deepika G, Rao GV, Reddy DN, Subramanyam C, et al. Altered intestinal microbiota in patients with chronic pancreatitis: implications in diabetes and metabolic abnormalities. Sci Rep 2017;7:43640.

[26] Gorovits ES, Tokareva EV, Khlynova OV, Zhelobov VG, El'kin VD. Complex evaluation of intestine microbiocenosis condition in patients with chronic pancreatitis. Zh Mikrobiol Epidemiol Immunobiol 2013;(4):73−6.

[27] Farrell JJ, Zhang L, Zhou H, Chia D, Elashoff D, Akin D, et al. Variations of oral microbiota are associated with pancreatic diseases including pancreatic cancer. Gut 2011. https://doi.org/10.1136/gutjnl-2011−300784.

[28] Savitskaya KMY, Vorobyev A, Zagalskaya N. Evaluation of the colonic content microenvironment in patients with chronic pancreatitis. Vestn Ross Akad Med Nauk 2002;4:20−3.

[29] Michaud DS, Izard J, Wilhelm-Benartzi CS, You DH, Grote VA, Tjonneland A, et al. Plasma antibodies to oral bacteria and risk of pancreatic cancer in a large European prospective cohort study. Gut 2013;62(12):1764−70. https://doi.org/10.1136/gutjnl-2012−303006.

[30] Chang JS, Tsai C-R, Chen L-T, Shan Y-S. Investigating the association between periodontal disease and risk of pancreatic cancer. Pancreas 2016;45(1):134−41.

[31] Fan X, Alekseyenko AV, Wu J, Peters BA, Jacobs EJ, Gapstur SM, et al. Human oral microbiome and prospective risk for pancreatic cancer: a population-based nested case-control study. Gut 2018;67(1):120−7.

[32] Torres PJ, Fletcher EM, Gibbons SM, Bouvet M, Doran KS, Kelley ST. Characterization of the salivary microbiome in patients with pancreatic cancer. PeerJ 2015;3:e1373.

[33] Half E, Keren N, Dorfman T, Reshef L, Lachter I, Kluger Y. Specific changes in fecal microbiota may differentiate pancreatic cancer patients from healthy individuals. Ann Oncol 2015;26:iv48.

[34] Lin IH, Wu J, Cohen SM, Chen C, Bryk D, Marr M. Pilot study of oral microbiome and risk of pancreatic cancer. Cancer Res 2013;73(8).

[35] Schulte A, Pandeya N, Fawcett J, Fritschi L, Risch HA, Webb PM, et al. Association between *Helicobacter pylori* and pancreatic cancer risk: a meta-analysis. Cancer Causes Control 2015;26(7):1027−35.

[36] Trikudanathan G, Philip A, Dasanu CA, Baker WL. Association between *Helicobacter pylori* infection and pancreatic cancer. A cumulative meta-analysis. JOP 2011;12(1):26−31.

[37] Wang Y, Zhang FC, Wang YJ. *Helicobacter pylori* and pancreatic cancer risk: a meta- analysis based on 2,049 cases and 2,861 controls. Asian Pac J Cancer Prev 2014;15(11):4449−54.

[38] Xiao M, Wang Y, Gao Y. Association between *Helicobacter pylori* Infection and Pancreatic cancer development: a meta-analysis. PLoS One 2013;8(9):1−11.

[39] Layer P, von der Ohe MR, Holst JJ, Jansen JB, Grandt D, Holtmann G, et al. Altered postprandial motility in chronic pancreatitis: role of malabsorption. Gastroenterology 1997;112(5):1624−34.

[40] Balzan S, de Almeida Quadros C, de Cleva R, Zilberstein B, Cecconello I. Bacterial translocation: overview of mechanisms and clinical impact. J Gastroenterol Hepatol 2007;22(4):464−71.

[41] Dukowicz AC, Lacy BE, Levine GM. Small intestinal bacterial overgrowth: a comprehensive review. Gastroenterol Hepatol 2007;3(2):112−22.

[42] Li L, Somerset S. The clinical significance of the gut microbiota in cystic fibrosis and the potential for dietary therapies. Clin Nutr 2014;33(4):571−80.

[43] Morgan XC, Tickle TL, Sokol H, Gevers D, Devaney KL, Ward DV, et al. Dysfunction of the intestinal microbiome in inflammatory bowel disease and treatment. Genome Biol 2012;13(9):1.

[44] Yanagisawa N, Haruta I, Shimizu K, Furukawa T, Higuchi T, Shibata N, et al. Identification of commensal flora-associated antigen as a pathogenetic factor of autoimmune pancreatitis. Pancreatology 2014;14(2):100−6.

[45] Zhang L, Farrell JJ, Zhou H, Elashoff D, Akin D, Park NH, et al. Salivary transcriptomic biomarkers for detection of resectable pancreatic cancer. Gastroenterology 2010;138(3):949−57. e1−1e7.

[46] Michaud DS, Izard J. Microbiota, oral microbiome, and pancreatic cancer. Cancer J 2014;20(3):203.

[47] Bode JC, Bode C, Heidelbach R, Durr HK, Martini GA. Jejunal microflora in patients with chronic alcohol abuse. Hepato Gastroenterol 1984;31(1):30−4.

[48] Parlesak A, Schafer C, Schutz T, Bode JC, Bode C. Increased intestinal permeability to macromolecules and endotoxemia in patients with chronic alcohol abuse in different stages of alcohol-induced liver disease. J Hepatol 2000;32(5):742−7.

[49] Jokelainen K, Roine RP, Vaananen H, Farkkila M, Salaspuro M. In vitro acetaldehyde formation by human colonic bacteria. Gut 1994;35(9):1271−4.

[50] Miyasaka K, Ohta M, Takano S, Hayashi H, Higuchi S, Maruyama K, et al. Carboxylester lipase gene polymorphism as a risk of alcohol-induced pancreatitis. Pancreas 2005;30(4):e87−91.

[51] Ren GYM, Li K, Hu Y, Wang Y, Xu X, Qu J. Seleno-lentinan prevents chronic pancreatitis development and modulates gut microbiota in mice. J Funct Foods 2016;22:177−88.

[52] Hu Y, Teng C, Yu S, Wang X, Liang J, Bai X, et al. Inonotus obliquus polysaccharide regulates gut microbiota of chronic pancreatitis in mice. AMB Express 2017;7(1):39.

[53] Claesson MJ, Cusack S, O'Sullivan O, Greene-Diniz R, de Weerd H, Flannery E, et al. Composition, variability, and temporal stability of the intestinal microbiota of the elderly. Proc Natl Acad Sci USA 2011;108(Suppl. 1):4586−91.

[54] Gou S, Yang Z, Liu T, Wu H, Wang C. Use of probiotics in the treatment of severe acute pancreatitis: a systematic review and meta-analysis of randomized controlled trials. Crit Care 2014;18(2):1.

[55] Kelly CR, Khoruts A, Staley C, Sadowsky MJ, Abd M, Alani M, et al. Effect of fecal microbiota transplantation on recurrence in multiply recurrent *Clostridium difficile* infection: a randomized trial. Ann Intern Med 2016;165(9):609−16.

[56] AlFaleh K, Anabrees J, Bassler D, Al-Kharfi T. Cochrane Review: probiotics for prevention of necrotizing enterocolitis in preterm infants. Evid Base Child Health A Cochrane Rev J 2012;7(6):1807−54.

[57] Picard C, Fioramonti J, Francois A, Robinson T, Neant F, Matuchansky C. Review article: bifidobacteria as probiotic agents—physiological effects and clinical benefits. Aliment Pharmacol Ther 2005;22(6):495−512.

[58] Ríos-Covián D, Ruas-Madiedo P, Margolles A, Gueimonde M, de los Reyes-Gavilán CG, Salazar N. Intestinal short chain fatty acids and their link with diet and human health. Front Microbiol 2016;7.

[59] O'Callaghan A, van Sinderen D. Bifidobacteria and their role as members of the human gut microbiota. Front Microbiol 2016;7:925.

[60] Ventura M, Van Sinderen D, Fitzgerald GF, Zink T. Insights into the taxonomy, genetics and physiology of bifidobacteria. Antonie Van Leeuwenhoek 2004;86(3):205−23.

[61] Graf D, Di Cagno R, Fak F, Flint HJ, Nyman M, Saarela M, et al. Contribution of diet to the composition of the human gut microbiota. Microb Ecol Health Dis 2015;26:26164.

[62] Simpson HL, Campbell BJ. Review article: dietary fibre-microbiota interactions. Aliment Pharmacol Ther 2015;42(2):158−79.

[63] Besselink MG, van Santvoort HC, Buskens E, Boermeester MA, van Goor H, Timmerman HM, et al. Probiotic prophylaxis in predicted severe acute pancreatitis: a randomised, double-blind, placebo-controlled trial. Lancet 2008;371(9613):651−9.

Chapter 38

Importance of the Microbiome and the Metabolome in Cancer

Liliane Martins dos Santos, Ana Clara Matoso Montuori de Andrade and Mateus Eustáquio Moura Lopes
Department of Biochemistry and Immunology, Federal University of Minas Gerais, Belo Horizonte, Brazil

Recent studies show that imbalances in the host—microbiota relationship, also called dysbiosis, can affect the formation and development of tumors and influence the outcome of treatments. The common denominator in the interplay between microbiota, genetics, diet, and drugs are the metabolites. Microbiota generate a plethora of metabolic products from ingested food and host metabolic products, some of which may become carcinogenic under certain physiologic circumstances. Thus, understanding the origin, production, and function of metabolites will lead to a better understanding of host and bacterial activities, contributing to efficacy of therapeutic protocols for cancers.

MICROBES ASSOCIATED WITH CARCINOGENESIS

Colorectal Cancer

Colorectal cancer (CRC) is the third most commonly occurring cancer in men, and the second in women. CRC encompasses tumors that affect a more distal part of the large intestine (the colon) and rectum, being one of the most accessible types of cancer in the world. Risk factors for colorectal cancer are still not well understood; however, it is known that a genetic component, where damaged DNA and genetic instability begin a malignant transformation, is associated with the appearance of CRC. In view of the constant exposure of the colon mucosa to microbes, it is possible that the microorganisms and their metabolites are involved in the development of tumors.

For carcinogenesis of the colon, it is becoming increasingly evident that a large and complex bacterial population of the large intestine plays an important role. Studies have observed higher numbers of some groups of bacteria (such as Fusobacteria, *Porphyromonadaceae*, Coriobacteridae, and *Staphylococcaceae*) while others are diminished during CRC (such as *Bifidobacterium*, *Lactobacillus*, *Ruminococcus,* and *Treponema*).

Several bacteria normally present in the intestinal microbiota have already been implicated in the development of CRC, including *Streptococcus gallolyticus*, *Bacteroides*, and *Escherichia coli*. Some bacterial species, including *Lactobacillus acidophilus* and *Bifidobacterium longum*, are underrepresented and are able to inhibit the development of tumors in rats after induction by a potent carcinogen, 1,2-dimethylhydrazine [1].

As a commensal microorganism, *E. coli* harmoniously coexists with its mammalian host, contributing to the maintenance of intestinal homeostasis, and rarely causing disease. However, some virulent strains of *E. coli* that have acquired islands of pathogenicity may colonize the human gastrointestinal tract and induce disease. These microorganisms are able to encode a genotoxin called colibactin, which can induce single strand breaks in DNA, leading to an increased rate of mutations in infected cells [2].

The bacterium *Enterococcus faecalis* normally present in the colon is able to produce reactive oxygen species (ROS), which have the potential to promote damage to the genetic material of epithelial cells, by breaking the double-stranded DNA or promoting chromosomal instability. The presence of this microbe, associated with a dysregulation of the homeostasis between microbiota and host can lead to the development of colon cancer [3].

Microbiome and Metabolome in Diagnosis, Therapy, and other Strategic Applications. https://doi.org/10.1016/B978-0-12-815249-2.00038-5

Gastric Cancer

Gastric cancer is the fourth most common malignancy, and the second most responsible for cancer-related deaths. Despite the extremely acidic environment, the presence of gastric mucus generates a pH gradient, which allows the existence of a gastric microbiota. Stomach cancer is commonly associated with infection by the bacterium *Helicobacter pylori,* which induces chronic gastritis and peptic ulcers. *H. pylori* is considered a component of the normal gastric microbiota in humans. Several studies have demonstrated the increased risk of developing gastric cancer in stomach samples from individuals infected with *H. pylori* [4].

Oral Cancer

Fusobacterium nucleatum and *Porphyromonas gingivalis* are known to synergistically promote oral cancer progression. It has been proposed that *F. nucleatum* may act as a bridging organism, allowing for other oral microbes to bind via compatible adhesins (*Porphyromonas* spp., *Peptostreptococcus* spp., and *Parvimonas* spp.). The oral microbes form a biofilm community that alters epithelial tight junctions and promotes infiltration and inflammation, leading to transformation of cells and oncogenesis. As tumor tissues grow, peptides and proteins that are metabolized by the oral microbes are released. A persistent state of inflammation continues to sustain the biofilm and to promote tumorigenesis in the colon. This reveals the contribution of an entire microbial tumor community, and not just a single microbe in the carcinogenic process [5].

MICROBIAL METABOLOME AND CANCER

The microbial community has a dichotomous role in cancer. Recent data have demonstrated the role of short-chain fatty acids (SCFAs), such as propionate and butyrate in suppressing inflammation and tumors, while other microbial metabolites such as secondary bile acids and ROS promote tumorigenesis. This dichotomy depends on the specific microorganism and their metabolic products.

Detrimental Metabolites

Protein-derived metabolites are increased, while SCFA levels decrease throughout the development of cancer. Undigested dietary components that reach the large intestine are fermented by the anaerobic microbial community for the production of various metabolites, which may participate in epigenetic processes. Hydrogen sulfide (H_2S), secondary bile acids, polyamines, and ROS have the potential to directly cause DNA damage. In addition, they are capable of promoting inflammation, thus contributing to carcinogenesis.

Some nitrogenous products derived from protein dietary intake such as NOCs and *N*-nitroso compounds are widely used as food additives in the processing of meat products, and can induce DNA alkylation, promoting mutation in cells and carcinogenesis. This metabolite has been associated with CRC in humans [6].

Other products of protein fermentation—polyamines including putrescine, spermidine, and spermine—are produced from arginine in host tissues, but also occur in gut bacteria. Polyamines are related to cancer and cause damage to cellular protein and DNA, besides activating oxidative pathways [7]. In colon cancer, biofilms are associated with the disease and linked to cancer location. More specifically, there is a direct correlation between biofilm formation in colon cancers and the upregulation of N1, N12-diacetylspermine, a polyamine metabolite. Therefore, it seems that bacterial polyamine metabolites act synergistically to promote biofilm formation and cellular proliferation, creating conditions conducive to oncogenic transformation in epithelial cells [8].

Hydrogen sulfide (H_2S) and secondary bile acids can promote direct DNA damage or drive an inflammatory environment, which is protumorigenic. H_2S is produced in the gut via the reduction of diet-derived sulfate and the metabolism of other compounds, including sulfur amino acids and taurine. H_2S inhibits butyrate oxidation, which promotes cancer and is toxic to colonocytes. Sulfate-reducing bacteria (SRB) are detectable in low numbers in most individuals, and are able to use lactate as a cosubstrate for growth and formation of sulfide [9].

Secondary bile acids are produced in the large intestine from primary bile acids, generated in the liver from cholesterol, and are then conjugated to glycine or taurine and excreted into the duodenum to facilitate fat digestion. In the large intestine, microbial bile salt hydrolases cleave glycine and taurine residues from the primary bile acids, which converts them into several different secondary bile acids, by dehydrogenation and dehydroxylation reactions. The major secondary bile acids, deoxycholic acid and lithocholic acid, are able to induce mitosis and an inflammatory response in the stellate cells of the liver, increasing the chances of developing liver cancer [10].

They can cause changes in the microbiota composition, decrease SCFA production, and promote oxidative stress, DNA damage, and also resistance to apoptosis [11]. Therefore, secondary bile acids are considered procarcinogenic metabolites and possible therapeutic targets.

Beneficial Metabolites

Undigested dietary compounds that reach the large intestine are fermented by the anaerobic microbial community to produce a wide range of metabolites. The most important fermentation products in healthy adults are gases and the SCFAs—acetate, propionate, and butyrate.

SCFAs are related to protection against intestinal tumorigenesis. Butyrate inhibits the activity of histone deacetylases (enzymes that promote the hyperacetylation of histones), thus stimulating apoptosis of CRC cells, and activation of the G-protein-coupled receptors (GPCRs) producing an immunosuppressive response [12]. Propionate, another SCFA, can reduce cancer cell proliferation in the liver. Propionate also has the ability to inhibit tumor growth using an adenosine cyclic monophosphate (cAMP)-inflammatory mediator dependent pathway and GPR43, a free fatty acid receptor, to which the propionate binds [13]. The antiinflammatory effects of SCFAs are not only important because of their influence on host cells, but it is also possible that they contribute to homeostasis of the gut microbiota reducing cancer risk.

Ursodeoxycholic acid is a beneficial secondary bile acid, produced by *Ruminococcus gnavus*, which inhibits oxidative stress and protects from cytotoxic activity of deoxycholic acid [14]. Regarding the earliest stages of cancer, the role of microbiota in inhibiting tumor initiation and progression is not restricted to direct effects on tumorigenesis. For example, higher estrogen levels are a major risk factor for endometrial and breast cancers, whereas the gut microbiota is able to partially metabolize this hormone [15].

EXPERIMENTAL MODELS TO EVALUATE THE MICROBIAL METABOLOME INFLUENCE IN CANCER

The metabolome is a result of both host and microbial activities. Microbial species have distinct enzymatic capabilities; therefore, it is often possible to separate the role of metabolites derived from individual bacterial species from effects caused by whole communities of microorganisms.

Cell culture models can be used for in vitro evaluation of the direct effects of microbial metabolites on tumor cells. For in vivo studies, animal models were extensively used for many decades. These models permit the evaluation of the effects of individual microorganisms or metabolites in the host organism.

In vitro models are based on tumor cell culture, patient-derived or immortalized cells from human cancers. Cell culture permits highly controlled conditions, accuracy, and reproducibility. On the other hand, they are inefficient in expressing the complex organic conditions and systemic factors that influence cancer [16,17].

Mice models are also important, because anticancer therapies require preclinical tests in animals, before human trials. Immunocompromised mice, athymic nude or SCID, are examples of mice lineages with heterozygous mutation, which develop spontaneous or induced tumors [16]. Gnotobiotic, germ-free, and antibiotic treated mice are valuable throughout the carcinogenesis process by compounds such as dextran sulfate sodium and azoxymethane [17,18].

Butyrate also presents tumor-suppressive functions in vivo. When gnotobiotic animals were colonized with a microbial strain of bacteria that produces butyrate, and then given high-fiber diet, they were protected against CRC. Tumor cells metabolize less butyrate, and this compound accumulates in tumors, affecting apoptosis and cell proliferation [18]. Germ-free mice can also be monoassociated or polyassociated with a small number of microbes. This technique makes it possible to assign function to a specific microbe and sometimes a specific microbial gene or metabolite [19]. Also, treatment of animals with an antibiotic cocktail is extensively used to evaluate how the microbiome and their metabolome can influence cancer.

MICROBIOME ALTERATION AND METABOLOME: INTERFACE OF ANTIBIOTICS AND OTHER DRUGS IN CANCER

The gut microbiota has a main role in the metabolism of xenobiotics. The microbes can influence the metabolism and are also influenced by it during antibiotic treatment and cancer therapy. The indiscriminate use of antibiotics and dysbiosis can be conducive to the initiation of tumor development or progression of an existing tumor. Depletion of the intestinal microbiota using an oral antibiotic cocktail can increase the severity of lung cancer, indicating that the balance of the intestinal microbiota is essential for tumor control at certain sites [17]. Antibiotic-associated perturbations of the gut microbiota support the notion that bacteria are key drivers of deoxycholic acid production in mice [10] and dysbiosis is related to cancer predisposition and development in humans [20].

Cisplatin, oxaliplatin, and cyclophosphamide are used in cancer therapy. These compounds are cytotoxic, inhibit proliferation, and induce apoptosis in cancer cells [16,21]. The effectiveness of the treatment with these drugs is dependent on the microbiome homeostasis. For example, germ-free mice and antibiotic-treated mice suffer from reduced efficacy of this treatment [22,23].

DIAGNOSTIC IMPLICATIONS

Gut microbiota associated with CRC differs from that of healthy individuals, and fecal microbial markers have the potential to provide alternatives, for the noninvasive diagnosis of CRC [24]. There are limitations, as they do not necessarily perform as well in the detection of polyps, adenomas, and other early-stage lesions as they do in the detection of more advanced disease [24]. This fact inhibits their use since, to fulfill the promise of providing a noninvasive and relatively inexpensive means for CRC screening, marker-based diagnostics need to be able to accurately detect both early- and late-stage disease, as well as differentiate between stages [25].

THERAPEUTICAL PROTOCOLS

Probiotics

Probiotic effects can include regulation of the immune system, suppression of microbiota growth involved in the production of carcinogens and mutagens, protection of DNA from oxidative damage, and modification in carcinogen metabolism. Studies in vitro and in vivo, including animal models and human studies, have shown the potential of probiotics in reducing invasion and metastasis in cancer cells, reprogramming anticancer immune response, influencing anticancer treatments such as alkylating or immune checkpoint blockade agents, and optimizing the immune response against solid cancer [26].

Lactobacillus plantarum enriched with selenium nanoparticles provides an efficient immune response toward breast cancer by increasing cell-mediated innate host defense systems such as macrophages, NK cells, and neutrophils, with higher production of cytokines such as TNF-alpha, IFN-γ, and IL-2 [27]. Another strain, *Lactobacillus casei* BL23, showed the potential to modulate mice immune responses, and to significantly protect against 1,2-dimethylhydrazine induced colorectal cancer. The animals treated with *L. casei* BL23 showed decreased number of multiple plaque lesions in the gut, reduction of histological damages, decreased levels of proinflammatory cytokines like MCP-1 and TNF-alpha, and increased amounts of the antiinflammatory ones, such as IL-10.

Moreover, mice treated with this probiotic showed lower levels of regulatory T (Treg) cells and a slight increase of Th17 cells suggesting that *L. casei* BL23 triggers a Th17/Treg mixed-type immune response [28].

A study of subcutaneous B16.SIY melanoma growth compared genetically similar mice that harbor distinct microbiota, namely Jackson Laboratory (JAX mice) versus Taconic Farms (TAC mice). Growth was more aggressive in TAC mice; however, the difference disappeared upon cohousing the animals or when fecal material from JAX mice was transferred into the gut of TAC mice.

When the JAX fecal material was administered in combination with monoclonal antibodies (mAb) and anti-programmed death-ligand 1 (PD-L1) to TAC mice, there was a control of the tumor size and circulating tumor antigen—specific T cell response improvement. The major component of JAX microbiota that promotes these beneficial antitumor effects is *Bifidobacterium*. *Bifidobacterium* modifies dendritic cell activity, which can lead to improved tumor-specific CD8$^+$ T cell function [28].

Enterococcus hirae and *Barnesiella intestinihominis* are both involved in response to cyclophosphamide therapy. *E. hirae* translocates from the small intestine to secondary lymphoid organs and increases the intratumoral CD8/Treg ratio. *B. intestinihominis* accumulates in the colon, and promotes the infiltration of IFN-y-producing γδ-T cells in cancer lesions, and both promote specific-memory Th1 cell immune responses which selectively predicted longer progression-free survival in advanced lung and ovarian cancer patients treated with cyclophosphamide [29].

DIET, NUTRITION, EXERCISE, OTHER LIFESTYLE REPERCUSSIONS, AND PREVENTION
Diet and Exercise

Nutrition can affect cancer development by inducing alterations in the microbiome, such as changes in carcinogenic or tumor-suppressive metabolites, and microbe-associated molecular patterns generated in the gut. High food intake of fat or greater consumption of red meat or processed meat compared to fruits and vegetables is associated with growth of bacteria that may contribute to a more hostile intestinal environment, and is generally considered a risk factor in the development of colorectal adenomas.

This effect appears to be modulated by potent N-nitroso compounds (discussed above), which can be generated by the bacterial metabolism of salts, used as food additives in the processing of meat products [6]. High-fat and high-energy diets also facilitate the absorption of bacterial lipopolysaccharide (LPS) from intestinal bacteria, resulting in increased intestinal permeability and decreased antimicrobial peptides in the protective mucin layer [30].

Fat-rich diets are related to increased incidence of CRC, leading to an increase in bile secretion and fecal bile acid in CRC patients. Folic acid or folate is a vitamin involved in many metabolic pathways such as nucleotide synthesis and synthesis of other vitamins and some amino acids. The efficiency of DNA replication, repair, and methylation are affected by the availability of folate. Rapidly proliferating cells such as leukocytes, erythrocytes, and enterocytes require large amounts of folate. Folate is widely distributed in the biological world, and intestinal bacteria are important sources of this vitamin [31].

Higher ingestion of some nutrients like sugar and fat favors the excessive circulation of these nutrients in the blood, increases the levels of trophic hormones, and favors the generation of a subclinical inflammatory status, with increased circulating proinflammatory cytokines. All these changes meet the metabolic needs of cancer cells, since they produce a vicious cycle that leads to inhibition of autophagy. The excess of these types of nutrients can modify the composition and the metabolism of the microbiota, compromising immunological functions. Such alterations impair the immunological response against tumor and the response to anticancer drugs [30].

Exercise can determine changes in the qualitative and quantitative gut microbial composition, with possible benefits for the host. Both diet and exercise can increase microbial biodiversity, and stimulate the development of bacteria able to determine adaptive changes in host metabolism. These changes could help reduce the risk of colon cancer, diverticulosis, and inflammatory bowel diseases, and have protective effects on the pathogenesis of several conditions [32].

Obesity

Obesity contributes to 15%−20% of cancer cases worldwide. Weight gain and the metabolic consequences of obesity occur in the context of alterations in the gut microbiota, with decreased diversity of microbes and intestinal inflammation [10].

A high-fat diet coupled with dysbiosis of the microbiota generated by treatment with antibiotics or in an obese person, coupled with genetic predisposition to tumors, can amplify tumor growth and its spread. This same diet can also induce a procarcinogenic environment through production of bile salts by members of the microbiome. Mainly bacteria of the Clostridia class produce secondary bile salts to aid in the digestion of food. Obese people present a higher prevalence of this bacterial class in their intestinal microbiota [10].

High-fat and high-energy diets facilitate absorption of LPS from intestinal Gram-negative bacteria. LPS can interact with soluble CD14 and LPS binding protein (LBP) activating toll-like receptor (TLR)-4 signaling and promoting colon cancer cell invasion into the extracellular matrix. This pathway has been suggested as a potential target for prevention of CRC [33].

Alcohol and Tobacco

Ethanol itself is not considered a carcinogen; however, its immediate oxidation generates acetaldehyde, which is highly toxic and carcinogenic, having effects ranging from DNA damage to degradation of the vitamin folate. Microorganisms from the oral microbiota may contribute to the production of acetaldehyde from ethanol, which suggests an important role for the microbiota in the tumorigenesis process. Excessive alcohol consumption causes dysbiosis, and destroys the gut permeability barrier, increasing the risk of CRC [31].

Tobacco use is associated with cancer of the lung, larynx, mouth, esophagus, throat, bladder, kidney, liver, stomach, pancreas, colon, rectum, and cervix, as well as acute myeloid leukemia. People who consume tobacco products or who are

TABLE 38.1 Major Metabolites Formed From Dietary or Environmental Compounds, Which Have Pro- And/Or Anti-carcinogenic Properties

Dietary/Environmental Compound	Microbial Metabolite	Effect on Host Organism
Fibres	SCFAs[a]	Apoptosis of tumor cells immunosuppressive responses
Cholesterol	Secondary bile acids[b]	Production from cytotoxic activity of other acids Genotoxicity, dysbiosis, inflammation
Protein	Polyamines	Genotoxicity, activation of oxidative stress Bacterial biofilm formation in colon
Protein	Hydrogen Sulfide	Inhibition of butyrate oxidation Genotoxicity
Protein	NOCs	Genotoxicity
Tobacco	————————	Dysbiosis
Ethanol	Acetaldehyde	Genotoxicity, vitamin degradation

[a]Mainly anticarcinogenic responses.
[b]Both pro- and anticarcinogenic responses.

regularly around environmental tobacco smoke (also called secondhand smoke) have an increased risk of cancer because of many chemicals that cause DNA damage.

There are some associations between the use of tobacco and decreased diversity of both oral and respiratory tract microbiota. These studies observed increases in periodontal and respiratory bacterial pathogens, and decreases in bacteria capable of interfering with pathogen growth, suggesting that this substance may cause dysbiosis, which is associated to increased risk of cancer [34] (Table 38.1).

LIST OF ACRONYMS AND ABBREVIATIONS

AOM Azoxymethane
cAMP Cyclic adenosine monophosphate
CRC Colorectal cancer
DC Dendritic cell
DSS Dextran sulfate sodium
GPCR G protein-coupled receptor
H₂S Hydrogen sulfide
IBD Inflammatory bowel disease
IFN-γ Interferon-gamma
IL-10 Interleukin-10
IL-2 Interleukin-2
LBP Lipopolysaccharide binding protein
LPS Lipopolysaccharide
MCP-1 Monocyte chemoattractant protein-1
NK Natural killer
PD-L1 Programmed death-ligand-1
RNS Reactive nitrogen species
ROS Reactive oxygen species
SCFA Short-chain fatty acid
SRB Sulfate-reducing bacteria
TLR Toll-like receptor
TNF Tumor necrosis factor

REFERENCES

[1] Louis P, Hold GL, Flint HJ. The gut microbiota, bacterial metabolites and colorectal cancer. Nat Rev Microbiol 2014;12(10):661−72.

[2] Arthur JC, Perez-Chanona E, Muhlbauer M, et al. Intestinal inflammation targets cancer-inducing activity of the microbiota. Science 2012;338(6103):120−3.

[3] Huycke MM, Gaskins HR. Commensal bacteria, redox stress, and colorectal cancer: mechanisms and models. Exp Biol Med 2004;229(7):586−97.

[4] Lee CW, Rickman B, Rogers AB, Ge Z, Wang TC, Fox JG. *Helicobacter pylori* eradication prevents progression of gastric cancer in hyper-gastrinemic INS-GAS mice. Cancer Res 2008;68(9):3540−8.

[5] Flynn KJ, Baxter NT, Schloss PD. Metabolic and community synergy of oral bacteria in colorectal cancer. mSphere 2016;1(3).

[6] Nistal E, Fernandez-Fernandez N, Vivas S, Olcoz JL. Factors determining colorectal cancer: the role of the intestinal microbiota. Front Oncol 2015;5:220.

[7] Di Martino ML, Campilongo R, Casalino M, Micheli G, Colonna B, Prosseda G. Polyamines: emerging players in bacteria-host interactions. Int J Med Microbiol 2013;303(8):484−91.

[8] Johnson CH, Dejea CM, Edler D, et al. Metabolism links bacterial biofilms and colon carcinogenesis. Cell Metab 2015;21(6):891−7.

[9] Carbonero F, Benefiel AC, Gaskins HR. Contributions of the microbial hydrogen economy to colonic homeostasis. Nat Rev Gastroenterol Hepatol 2012;9(9):504−18.

[10] Yoshimoto S, Loo TM, Atarashi K, et al. Obesity-induced gut microbial metabolite promotes liver cancer through senescence secretome. Nature 2013;499(7456):97−101.

[11] Barrasa JI, Olmo N, Lizarbe MA, Turnay J. Bile acids in the colon, from healthy to cytotoxic molecules. Toxicol in Vitro 2013;27(2):964−77. An International Journal Published in Association with BIBRA.

[12] Singh N, Gurav A, Sivaprakasam S, et al. Activation of Gpr109a, receptor for niacin and the commensal metabolite butyrate, suppresses colonic inflammation and carcinogenesis. Immunity 2014;40(1):128−39.

[13] Bindels LB, Porporato P, Dewulf EM, et al. Gut microbiota-derived propionate reduces cancer cell proliferation in the liver. Br J Cancer 2012;107(8):1337−44.

[14] Lee JY, Arai H, Nakamura Y, Fukiya S, Wada M, Yokota A. Contribution of the 7beta-hydroxysteroid dehydrogenase from *Ruminococcus gnavus* N53 to ursodeoxycholic acid formation in the human colon. J Lipid Res 2013;54(11):3062−9.

[15] Goedert JJ, Jones G, Hua X, et al. Investigation of the association between the fecal microbiota and breast cancer in postmenopausal women: a population-based case-control pilot study. J Natl Cancer Inst 2015;107(8).

[16] Cekanova M, Rathore K. Animal models and therapeutic molecular targets of cancer: utility and limitations. Drug Des Dev Therapy 2014;8:1911−21.

[17] Gui QF, Lu HF, Zhang CX, Xu ZR, Yang YH. Well-balanced commensal microbiota contributes to anti-cancer response in a lung cancer mouse model. Genet Mol Res 2015;14(2):5642−51.

[18] Donohoe DR, Holley D, Collins LB, et al. A gnotobiotic mouse model demonstrates that dietary fiber protects against colorectal tumorigenesis in a microbiota- and butyrate-dependent manner. Cancer Discov 2014;4(12):1387−97.

[19] Wong SH, Zhao L, Zhang X, et al. Gavage of fecal samples from patients with colorectal cancer promotes intestinal carcinogenesis in germ-free and conventional mice. Gastroenterology 2017;153(6):e61621−33.

[20] Boursi B, Mamtani R, Haynes K, Yang YX. Recurrent antibiotic exposure may promote cancer formation−Another step in understanding the role of the human microbiota? Eur J Cancer 2015;51(17):2655−64.

[21] Zitvogel L, Galluzzi L, Viaud S, et al. Cancer and the gut microbiota: an unexpected link. Sci Transl Med 2015;7(271):271ps1.

[22] Dzutsev A, Goldszmid RS, Viaud S, Zitvogel L, Trinchieri G. The role of the microbiota in inflammation, carcinogenesis, and cancer therapy. Eur J Immunol 2015;45(1):17−31.

[23] Viaud S, Saccheri F, Mignot G, et al. The intestinal microbiota modulates the anticancer immune effects of cyclophosphamide. Science 2013;342(6161):971−6.

[24] Zackular JP, Rogers MA, Ruffin MT, Schloss PD. The human gut microbiome as a screening tool for colorectal cancer. Cancer Prev Res (Phila) 2014;7(11):1112−21.

[25] DeSantis TZ, Shah MS, Cope JL, Hollister EB. Microbial markers in the diagnosis of colorectal cancer: the promise, reality and challenge. Future Microbiol 2017;12:1341−4.

[26] Motevaseli E, Dianatpour A, Ghafouri-Fard S. The role of probiotics in cancer treatment: emphasis on their in vivo and in vitro anti-metastatic effects. Int J Mol Cell Med 2017;6(2):66−76.

[27] Yazdi MH, Mahdavi M, Kheradmand E, Shahverdi AR. The preventive oral supplementation of a selenium nanoparticle-enriched probiotic increases the immune response and lifespan of 4T1 breast cancer bearing mice. Arzneimittelforschung 2012;62(11):525−31.

[28] Lenoir M, Del Carmen S, Cortes-Perez NG, et al. Lactobacillus casei BL23 regulates Treg and Th17 T-cell populations and reduces DMH-associated colorectal cancer. J Gastroenterol 2016;51(9):862−73.

[29] Daillere R, Vetizou M, Waldschmitt N, et al. Enterococcus hirae and Barnesiella intestinihominis facilitate cyclophosphamide-induced therapeutic immunomodulatory effects. Immunity 2016;45(4):931−43.

[30] Zitvogel L, Pietrocola F, Kroemer G. Nutrition, inflammation and cancer. Nat Immunol 2017;18(8):843—50.

[31] Seitz HK, Stickel F. Molecular mechanisms of alcohol-mediated carcinogenesis. Nat Rev Cancer 2007;7(8):599—612.

[32] Monda V, Villano I, Messina A, et al. Exercise modifies the gut microbiota with positive health effects. Oxid Med Cell Longev 2017;2017:3831972.

[33] Djuric Z. Obesity-associated cancer risk: the role of intestinal microbiota in the etiology of the host proinflammatory state. Transl Res 2017;179:155—67.

[34] Thomas AM, Gleber-Netto FO, Fernandes GR, et al. Alcohol and tobacco consumption affects bacterial richness in oral cavity mucosa biofilms. BMC Microbiol 2014;14:250.

Chapter 39

The Microbiome in Graft Versus Host Disease

Mathilde Payen[1] and Clotilde Rousseau[1,2]

[1]EA4065, Ecosystème intestinal, probiotiques, antibiotiques, Faculté de Pharmacie, Université Paris Descartes. Paris, France; [2]Laboratoire de microbiologie, Hôpital Saint-Louis, Paris, France

INTRODUCTION

Allo-Hematopoietic Stem Cell Transplantation (allo-HSCT) is a potentially curative treatment for a range of hematopoietic disorders such as leukemia, lymphoma, myeloma, or medullar aplasia. More than 25,000 allo-HSCTs are annually performed around the world, and this number is expected to rise [1]. Before transplantation, patients undergo full or partial myeloablative conditioning regimen (chemotherapy, with or without total body irradiation), which destroys patient hematopoietic cells. Donor stem cells are transferred to the patient in order to reconstitute a new healthy hematopoietic system.

Even if allo-HSCT is considered to be the most effective treatment available to date for these hematopoietic diseases, it is associated with a variety of complications including infections and graft-versus-host disease (GVHD). Currently, GVHD remains a major source of morbidity and mortality following allo-HSCT, and understanding the mechanisms of GVHD has been highlighted as a key research priority. The gastrointestinal tract is particularly impacted by allo-HSCT, and a relationship between the microbiota and GVHD has long been suspected and is recently being extensively investigated by researchers.

GRAFT-VERSUS-HOST DISEASE

About 40% of recipients of HLA-identical grafts develop systemic acute GVHD (aGVHD) requiring treatment with high dose of steroids. aGVHD occurs within 100 days of allo-HSCT, and is characterized by significant inflammatory lesions in the skin, digestive tract, or liver, and frequently by multisite involvement. Several risk factors for aGVHD have been identified: HLA mismatch (the most important factor), high intensity of conditioning regimen, and type of aGVHD prophylaxis [2]. However, the exact etiological cause of aGVHD is still unknown.

Diagnosis

Few tools are available for aGVHD diagnosis that is almost exclusively based on clinical signs. Two classifications are commonly used for diagnosis: the Glucksberg classification [3] (Table 39.1) and the International Bone Marrow Transplant Registry grading system. Regarding Glucksberg grade, doctors first assign organ damage stage, according to skin, liver, and gut symptoms. Then, in the light of all the organ stages, a grade is assigned to the patient (Table 39.2).

The most commonly affected site is the skin (80% of the patients), followed by the digestive tract and gut; aGVHD being observed in 50%—70% of the patients [1]. Gastrointestinal tract involvement usually presents as diarrhea, but can also include vomiting, anorexia, abdominal pain, or a combination thereof, when severe. Diarrhea in aGVHD is secretory, and usually voluminous. Bleeding which has poor prognosis happens as a result of mucosal ulceration, but patchy involvement of mucosa generally leads to a normal appearance on endoscopy. Diagnosis can also be confirmed by histological features. However, no biological marker can be used to help diagnosis of aGVHD.

Microbiome and Metabolome in Diagnosis, Therapy, and other Strategic Applications. https://doi.org/10.1016/B978-0-12-815249-2.00039-7

TABLE 39.1 Determination of aGVHD Clinical Stage for Each Targeted Organ (Skin, Liver, and Gut) According to the Glucksberg Classification

	Targeted Organ		
Stage	Skin	Gut	Liver
1	Maculopapular rash <25% of body surface	Diarrhea 500–999 mL/day	Bilirubin 34–50 μmol/L
2	Maculopapular rash 25%–50% of body surface	Diarrhea 1000–1499 mL/day	Bilirubin 51–102 μmol/L
3	Generalized erythroderma	Diarrhea>1500 mL/day	Bilirubin 103–225 μmol/L
4	Generalized erythroderma with desquamation and bullous formation	Severe abdominal pain, with or without ileus	Bilirubin >225 μmol/L

TABLE 39.2 Determination of aGVHD Overall Grade According to the Glucksberg Classification, Depending on Targeted Organ (Skin, Liver, and Gut), and patient's Clinical Stage

	Degree of General and Organ Involvement			
Grade	Skin	Gut	Liver	Clinical Performance
I	Stage 1–2	no	no	No decrease
II	Stage 1–3	Gut stage 1 **and/or** liver stage 1		Mild decrease
III	Stage 2–3	Gut stage 2–3 **and/or** liver stage 2–4		Marked decrease
IV	Stage 2–4	Gut stage 2-4 **and/or** liver stage 2–4		Extreme decrease

Pathophysiology

aGVHD is the result of systemic inflammation that takes place in several successive steps [1]:

- Aggressive conditioning regimen causes injuries in recipient tissues, and induces inflammation. In the gut, proinflammatory cytokines such as TNFα and IL1 are produced, and digestive inflammation allows bacteria to pass the intestinal barrier.
- Antigen-presenting cells, macrophages, are activated by danger signals, and induce immune activation in a Th1-way.
- CD8+ T cells and activated macrophages induce direct damage to target cells.

Autoimmune attack results in destruction of tissues in organs such as lungs, liver, gut, or skin.

In addition, several elements suggest that the intestinal microbiota may be involved in the pathophysiology of aGVHD, as in some other inflammatory bowel diseases.

Prevention and Treatment

aGVHD can be prevented by immunosuppressive therapy, such as Cyclosporine or Tacrolimus [4]. Antibiotics are still used as a prophylactic treatment, with the aim of digestive decontamination, even if their effectiveness is much discussed, and recently a deleterious role is even advanced by some investigations.

aGVHD treatment is based on corticosteroids [5]. Around 50% of the patients do not respond to steroids. Immunosuppressive therapy, like Cyclosporine or Tacrolimus, have been tried but without real success. Survival rate in patients with steroid resistance is poor, with 70% mortality after 2 years [6]. Global aGVHD mortality in allo-HSCT recipients is estimated at 25%. Fecal calprotectin and α1-antitrypsin, two inflammatory markers, are predictive for responses to treatment at the time of aGVHD diagnosis [7].

GUT MICROBIOTA AND aGVHD

The allo-HSCT procedure has a strong impact on the intestinal microbiota. The conditioning regimen induces a shift in gut microbiota composition, with a decrease in Firmicutes and an increase in Proteobacteria [8]. In addition, other parameters may modify the intestinal microbiota in this context; some patients receive antibiotics for prevention of aGVHD and most of all as a treatment of infection or nonspecific fever. Thus, the cause-and-effect relationship between microbiota alterations and aGVHD is difficult to establish.

Antibiotics and Decontamination

Gut microbiota and aGVHD relationship is assumed since the 1970s when a study showed that germ-free mice developed less gut GVHD than conventional mice [9]. Another study observed that mice treated by high-dose antibiotics before HSCT developed less GVHD than those free of antibiotics [10]. Thereafter, clinical studies tried to decontaminate the gastrointestinal tract of the patients with high dose of antibiotics. Some showed a protective effect [11,12]; however, results are still controversial [13—15]. A recent study compared aGVHD incidence and survival in 500 allo-HSCT recipients receiving antibiotic prophylaxis or not [14]. They showed that gut decontamination with antibiotics increased the rate of grade II—IV aGVHD and reduced the median overall survival.

Dysbiosis and aGVHD

In 2012, Jenq et al. showed that inflammation due to aGVHD was associated with gut microbiota shift in mice and in 5 humans [16]. First, a loss of microbial diversity was observed. In animals, the proportions of *Lactobacillus* and *Enterobacteriaceae* increased whereas the relative abundance of *Clostridium* decreased. In humans, similar shifts were observed, with an increase in the proportions of *Lactobacillus* and a decrease in the abundance of *Clostridium*. However, the proportions of the *Enterobacteriaceae* showed no significant increase in aGVHD patients.

Another study by Holler et al. also showed a drastic decrease of Firmicutes abundance (*Clostridium* and *Eubacterium rectale*) during aGVHD in humans [17]. In addition, they observed that an increase in the relative abundance of the genus *Enterococcus* after allo-HSCT could be a risk factor for aGVHD development. The increase in the *Enterococcus* abundance might be explained by both concomitant loss of bacterial diversity and of several *Clostridium* species more than by actual *Enterococcus* expansion. In this study, antibiotics appeared to be a risk factor for increase of *Enterococcus* proportions and as a consequence for dysbiosis, questioning again the relevance of pretransplant digestive decontamination.

In a pediatric population, Biagi et al. conducted a longitudinal study from before allo-HSCT to at least 51 days after transplant [18]. The authors associated aGVHD with a gut microbiota signature, characterized by increase in the abundance of *Enterococcus* and Clostridiales, and decrease in the abundance of *Faecalibacterium*, *Ruminococcus*, and Bacteroidetes. In addition, they noted that among bacterial sequences obtained from the microbiota of aGVHD patients, there were more unassigned OTUs (unidentified species) than in patients free of aGVHD during all the follow-up.

Short-Chain Fatty Acids and aGVHD

Acetate, propionate, and butyrate have been shown to improve inflammatory conditions such as inflammatory bowel diseases (IBD) [19]. In these diseases, the decrease in short-chain fatty acid (SCFA) levels might be explained by the loss of *Faecalibacterium* and *Ruminococcus* species, which are known to be SCFA producers.

Mathewson et al. showed that following allo-HSCT, the amount of intraepithelial butyrate decreased [20]. Butyrate is an important energy source for intestinal epithelial cells, and participates in maintaining the integrity of the intestinal barrier. One can hypothesize that butyrate could have a protective role against GVHD through two major functions: (1) expression of antiapoptotic genes, inducing better cell survival, (2) expression of tight junction proteins, allowing better integrity of the gut epithelial barrier. In this study, administration of butyrate-producing bacteria belonging to the Clostridia class decreased the severity and mortality of murine aGVHD. Thus, in animal model, butyrate appears to be protective against GVHD.

In children, Biagi et al. showed that SCFAs in the gut decreased by 76% after allo-HSCT, in both aGVHD and non-aGVHD patients [18]. However, patients that did not develop aGVHD had higher levels of SCFAs before transplantation.

Antimicrobial Peptides and aGVHD

Gut microbiota is by several mechanisms regulated by the host immunity. Paneth cells control bacterial development, thanks to antimicrobial peptide secretion. Among these molecules, α and β defensins are part of innate immunity. Defensins may be defective, for example, during the IBDs, for which they serve as a diagnostic biomarker [21]. In mice, a study showed that the Paneth cells are destroyed during aGVHD leading to a reduction in α-defensin levels [22].

A recent study analyzed the expression of Paneth cell antimicrobial peptides (AMPs), human defensins 5 and 6 (HD5, HD 6), and the lectin regenerating islet-derived 3 α (Reg3α) in intestinal biopsies as well as Reg3α serum levels, in relation to gut aGVHD in 200 patients [23]. aGVHD stage 2–4 was associated with reduced expression of AMPs in the small intestine in comparison to stage 0–1 disease. In contrast, in the large intestine stage 2–4 gut aGVHD correlated with higher expression of HD5, HD6, and Reg3α compared to mild or no acute GVHD.

MICROBIOTA AND POSTTRANSPLANT MORTALITY

Taur et al. analyzed the bacterial diversity of the microbiota of 80 allo-HSCT recipients, and identified three groups of patients: low, intermediate, and high microbial diversity [24]. They showed that low bacterial diversity was strongly correlated with a higher mortality at 3 years post HSCT. Only 36% of the patients with a low diversity microbiota were alive at 3 years, compared to 67% with high diversity.

High abundance of the genus *Blautia*, 12 days after allo-HSCT, was also correlated with a reduced risk of death from aGVHD, and of relapse [25]. Bacteria of the genus *Blautia* belong to the Clostridia class and are known to be butyrate-producers.

Use of antibiotics during engraftment has also been linked to a worse survival after allo-HSCT [26]. The administration of imipenem and piperacillin-tazobactam, two major beta-lactam antibiotics frequently used in probabilistic treatment of neutropenic fever, has been correlated to an increase in aGVHD-related mortality, 5 years post HSCT.

FECAL MICROBIOTA TRANSPLANTATION: A NEW TREATMENT?

Recent studies seem to show the deleterious effect of antibiotics on allo-HSCT patients and risk of aGVHD development [14,26]. By eradicating health-promoting bacteria (such as *Faecalibacterium* spp.), they enable the growth of other bacteria including *Enterococcus*, thus precipitating dysbiosis.

Fecal microbiota transplantation (FMT) is used to restore healthy microbiota in *Clostridium difficile* infection. As bacterial dysbiosis is also observed during aGVHD and may be implicated in aGVHD pathogenesis, some advanced the hypothesis that FMT can be used to cure aGVHD.

The first FMT series for gut aGVHD was described in 2016 [27]. This study enrolled four patients with steroid-resistant or steroid-dependent GVHD, and showed promising results. FMT was safely performed in these immunocompromised patients. Three were cured (loss of clinical signs and corticosteroid decrease), and one was classified as incomplete response. Another study with only three corticoid resistant patients showed complete remission of gut aGVHD after 1 to 6 FMTs [28]. Three other studies are currently going on (Table 39.3).

One trial will enroll 18 patients after allo-HSCT. FMT will be performed 3 weeks after engraftment. Healthy donors will provide the fecal transplant. Feasibility of FMT will be measured by survival and aGVHD incidence after 2 years. In another trial, FMT will be attempted in patients with steroid-resistant or dependent aGVHD. This pilot study will enroll four patients and will evaluate safety and feasibility of FMT. Serious adverse events are not expected within 28 days after FMT. Gut aGVHD response will also be followed. The aim of the third trial is to evaluate the safety and efficacy of FMT for gut aGVHD. The study will enroll 20 patients.

TABLE 39.3 Clinical Trials on Fecal Microbiota Transplantation as a Preventive or Curative Treatment for Acute GVHD After allo-HSCT

Trial Number	Country	Study Name	Transplant	Purpose
NCT02733744	USA	Allo-transplantation after allo-HSCT.	Autologous transplant	Preventive
NCT03214289	Israel	Fecal microbiota transplantation for steroid-resistant and steroid-dependent gut acute graft versus host disease.	Autologous transplant	Curative
NCT03148743	China	Fecal microbiota transplantation in gut aGVHD treated.	Not specified	Curative

Several points must be kept in mind regarding this promising therapy. First, in the short term, infectious risks in immunocompromised patients such as allo-HSCT recipients are high, and require careful administration of fecal bacteria. Second, the long-term consequences of FMT should also be explored.

CONCLUSIONS

During aGVHD, the abundance of Firmicutes and Clostridiales seems to decrease allowing other bacteria to grow. *Enterococcus* proportions have been demonstrated, in several studies, to increase in aGVHD patients. Still, we should be careful about our conclusions. First, observations have been made on relative proportions of bacteria, and absolute amounts may have an impact. Second, many confounding factors must be considered including antibiotics used, allo-HSCT management, diet.

Investigations carried out in mice have provided important elements for understanding pathophysiology, in particular the probable role of SCFAs. These aspects are now to be explored in humans. Predictive markers of GVHD occurrence and clinical outcome are also highly investigated.

FMTs that would restore healthy microbiota in patients with disrupted gut ecology appear as a potential new treatment. FMT seems worth exploring even if we should be very careful with patients as immunocompromised as these transplant recipients.

LIST OF ACRONYMS AND ABBREVIATIONS

aGVHD acute Graft-versus-host disease
Allo-HSCT allo-Hematopoietic stem cell transplant
AMP Paneth cell antimicrobial peptide
FMT Fecal microbiota transplantation
IBD Inflammatory bowel disease
SCFA Short-chain fatty acid

REFERENCES

[1] Ferrara JL, Levine JE, Reddy P, Holler E. Graft-versus-host disease. Lancet May 2, 2009;373(9674):1550−61.

[2] Flowers ME, Inamoto Y, Carpenter PA, Lee SJ, Kiem HP, Petersdorf EW, et al. Comparative analysis of risk factors for acute graft-versus-host disease and for chronic graft-versus-host disease according to National Institutes of Health consensus criteria. Blood March 17, 2011;117(11):3214−9.

[3] Rowlings PA, Przepiorka D, Klein JP, Gale RP, Passweg JR, Henslee-Downey PJ, et al. IBMTR Severity Index for grading acute graft-versus-host disease: retrospective comparison with Glucksberg grade. Br J Haematol June 1997;97(4):855−64.

[4] Nash RA, Antin JH, Karanes C, Fay JW, Avalos BR, Yeager AM, et al. Phase 3 study comparing methotrexate and tacrolimus with methotrexate and cyclosporine for prophylaxis of acute graft-versus-host disease after marrow transplantation from unrelated donors. Blood September 15, 2000;96(6):2062−8.

[5] Deeg HJ. How I treat refractory acute GVHD. Blood May 15, 2007;109(10):4119−26.

[6] Xhaard A, Rocha V, Bueno B, de Latour RP, Lenglet J, Petropoulou A, et al. Steroid-refractory acute GVHD: lack of long-term improved survival using new generation anticytokine treatment. Biol Blood Marrow Transplant March 2012;18(3):406−13.

[7] Rodriguez-Otero P, Porcher R, Peffault de LR, Contreras M, Bouhnik Y, Xhaard A, et al. Fecal calprotectin and alpha-1 antitrypsin predict severity and response to corticosteroids in gastrointestinal graft-versus-host disease. Blood June 14, 2012;119(24):5909−17.

[8] Montassier E, Batard E, Massart S, Gastinne T, Carton T, Caillon J, et al. 16S rRNA gene pyrosequencing reveals shift in patient faecal microbiota during high-dose chemotherapy as conditioning regimen for bone marrow transplantation. Microb Ecol April 2014;67(3):690−9.

[9] van Bekkum DW, Roodenburg J, Heidt PJ, van der Waaij D. Mitigation of secondary disease of allogeneic mouse radiation chimeras by modification of the intestinal microflora. J Natl Cancer Inst February 1974;52(2):401−4.

[10] Heidt PJ, Vossen JM. Experimental and clinical gnotobiotics: influence of the microflora on graft-versus-host disease after allogeneic bone marrow transplantation. J Med 1992;23(3−4):161−73.

[11] Storb R, Prentice RL, Buckner CD, Clift RA, Appelbaum F, Deeg J, et al. Graft-versus-host disease and survival in patients with aplastic anemia treated by marrow grafts from HLA-identical siblings. Beneficial effect of a protective environment. N Engl J Med February 10, 1983;308(6):302−7.

[12] Vossen JM, Heidt PJ, van den Berg H, Gerritsen EJ, Hermans J, Dooren LJ. Prevention of infection and graft-versus-host disease by suppression of intestinal microflora in children treated with allogeneic bone marrow transplantation. Eur J Clin Microbiol Infect Dis January 1990;9(1):14−23.

[13] Passweg JR, Rowlings PA, Atkinson KA, Barrett AJ, Gale RP, Gratwohl A, et al. Influence of protective isolation on outcome of allogeneic bone marrow transplantation for leukemia. Bone Marrow Transplant June 1998;21(12):1231−8.

[14] Routy B, Letendre C, Enot D, Chenard-Poirier M, Mehraj V, Seguin NC, et al. The influence of gut-decontamination prophylactic antibiotics on acute graft-versus-host disease and survival following allogeneic hematopoietic stem cell transplantation. OncoImmunology 2017;6(1):e1258506.

[15] Russell JA, Chaudhry A, Booth K, Brown C, Woodman RC, Valentine K, et al. Early outcomes after allogeneic stem cell transplantation for leukemia and myelodysplasia without protective isolation: a 10-year experience. Biol Blood Marrow Transplant 2000;6(2):109—14.

[16] Jenq RR, Ubeda C, Taur Y, Menezes CC, Khanin R, Dudakov JA, et al. Regulation of intestinal inflammation by microbiota following allogeneic bone marrow transplantation. J Exp Med May 7, 2012;209(5):903—11.

[17] Holler E, Butzhammer P, Schmid K, Hundsrucker C, Koestler J, Peter K, et al. Metagenomic analysis of the stool microbiome in patients receiving allogeneic stem cell transplantation: loss of diversity is associated with use of systemic antibiotics and more pronounced in gastrointestinal graft-versus-host disease. Biol Blood Marrow Transplant May 2014;20(5):640—5.

[18] Biagi E, Zama D, Nastasi C, Consolandi C, Fiori J, Rampelli S, et al. Gut microbiota trajectory in pediatric patients undergoing hematopoietic SCT. Bone Marrow Transplant July 2015;50(7):992—8.

[19] Sun M, Wu W, Liu Z, Cong Y. Microbiota metabolite short chain fatty acids, GPCR, and inflammatory bowel diseases. J Gastroenterol January 2017;52(1):1—8.

[20] Mathewson ND, Jenq R, Mathew AV, Koenigsknecht M, Hanash A, Toubai T, et al. Gut microbiome-derived metabolites modulate intestinal epithelial cell damage and mitigate graft-versus-host disease. Nat Immunol May 2016;17(5):505—13.

[21] Lewis JD. The utility of biomarkers in the diagnosis and therapy of inflammatory bowel disease. Gastroenterology May 2011;140(6):1817—26.

[22] Eriguchi Y, Takashima S, Oka H, Shimoji S, Nakamura K, Uryu H, et al. Graft-versus-host disease disrupts intestinal microbial ecology by inhibiting Paneth cell production of alpha-defensins. Blood July 5, 2012;120(1):223—31.

[23] Weber D, Frauenschlager K, Ghimire S, Peter K, Panzer I, Hiergeist A, et al. The association between acute graft-versus-host disease and anti-microbial peptide expression in the gastrointestinal tract after allogeneic stem cell transplantation. PLoS One 2017;12(9):e0185265.

[24] Taur Y, Jenq RR, Perales MA, Littmann ER, Morjaria S, Ling L, et al. The effects of intestinal tract bacterial diversity on mortality following allogeneic hematopoietic stem cell transplantation. Blood August 14, 2014;124(7):1174—82.

[25] Jenq RR, Taur Y, Devlin SM, Ponce DM, Goldberg JD, Ahr KF, et al. Intestinal Blautia is associated with reduced death from graft-versus-host disease. Biol Blood Marrow Transplant August 2015;21(8):1373—83.

[26] Shono Y, Docampo MD, Peled JU, Perobelli SM, Velardi E, Tsai JJ, et al. Increased GVHD-related mortality with broad-spectrum antibiotic use after allogeneic hematopoietic stem cell transplantation in human patients and mice. Sci Transl Med May 18, 2016;8(339):339ra71.

[27] Kakihana K, Fujioka Y, Suda W, Najima Y, Kuwata G, Sasajima S, et al. Fecal microbiota transplantation for patients with steroid-resistant acute graft-versus-host disease of the gut. Blood October 20, 2016;128(16):2083—8.

[28] Spindelboeck W, Schulz E, Uhl B, Kashofer K, Aigelsreiter A, Zinke-Cerwenka W, et al. Repeated fecal microbiota transplantations attenuate diarrhea and lead to sustained changes in the fecal microbiota in acute, refractory gastrointestinal graft-versus-host-disease. Haematologica May 2017;102(5):e210—3.

Chapter 40

Impact of the Gut Microbiome on Behavior and Emotions

Ingrid Rivera-Iñiguez[1,2], Sonia Roman[1,2], Claudia Ojeda-Granados[1,2] and Panduro Arturo[1,2]
[1]Department of Molecular Biology in Medicine, Civil Hospital of Guadalajara, Guadalajara, Jalisco, Mexico; [2]Health Sciences Center, University of Guadalajara, Guadalajara, Jalisco, Mexico

INTRODUCTION

Behavior and emotions have allowed humans to adapt to the environment, by detecting or reacting toward harmful or dangerous situations. However, chronic incidences of negative emotions or behavioral disorders may interact with genetic and environmental factors to detonate serious diseases. Patients with gastrointestinal disorders are a good example of recurrence of feelings of depression, anxiety, and stressful personalities. Therefore, it has been suspected that behavior and emotions could have a more gastrointestinal origin.

The gastrointestinal tract contains a wide variety of microorganisms that sustain essential and beneficial interactions with the host [1]. Interestingly, gut microbiota communicates with the central nervous system, synthesizes neurotransmitters, and influences neurogenesis. As a result, microbiota modulates human behavior and emotion [2]. Understanding the communication pathways between the gut microbiota and central nervous system via distinct mediators, and the mechanisms of such pathways, could be a promising strategy in the treatment of behavioral and gastrointestinal disorders.

BRAIN STRUCTURES INVOLVED IN BEHAVIOR AND EMOTIONS

Behavior and emotions are regulated in the brain reward circuit, localized in the mesolimbic region [3]. In this region, different brain structures are interconnected by neurons and neurotransmitters to regulate memory, learning, emotions, cognition, and behavior. The amygdala serves as a warning system that detects stimuli inputs related to emotions and sends this information to the reticular activating system [4]. Furthermore, the amygdala regulates the endocrine response to stress, by interacting with the hypothalamic—pituitary—adrenal (HPA) axis [5].

Physiological alterations of the amygdala have been associated with behavioral disturbances in germ-free mice [6]. In humans, an impaired function of the amygdala may interfere with social behavior, as observed in patients with schizophrenia and autism spectrum disorders (ASD) [7]. Furthermore, the amygdala and the prefrontal cortex (PFC) evaluates the stimuli and establishes a memory. This region also controls motor functions that give feedback information to the hippocampal septum. This information is known as working memory [8,9].

The PFC can also suppress negative responses, such as anxiety and fear [9,10]. The orbital frontal cortex (OFC) connects to the nucleus accumbens (NAC), hypothalamus, insula, dorsal striatum area, and medial frontal cortex, and regulates the emotional value to the memorized information [11].

GUT—BRAIN AXIS

The brain maintains a bidirectional communication with the gastrointestinal system by neural, immune, and endocrine pathways that ensure a healthy behavior, as seen in Fig. 40.1. The brain interacts with the gut via sympathetic and parasympathetic fibers, vagus nerve, adrenergic nerve, and portal vein [12]. In fact, the vagus nerve is considered as the principal route of communication because it connects close to 100 million of enteric neurons to the brain.

Microbiome and Metabolome in Diagnosis, Therapy, and other Strategic Applications. https://doi.org/10.1016/B978-0-12-815249-2.00040-3

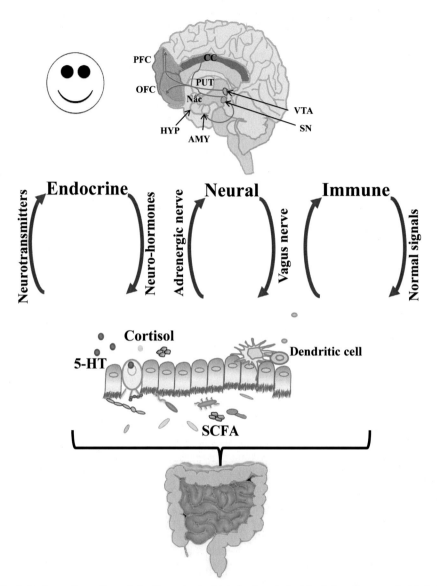

FIGURE 40.1 Healthy behavior and emotions. A healthy gut microbiota is characterized by being diverse and rich in beneficial bacteria that contributes to a normal communication with the brain. This includes a normal neurotransmitter signaling, cytokine profile, and endocrine signals. *5-HT*, Serotonin; *AMY*, Amygdala; *CC*, cingulate cortex; *HYP*, Hypothalamus; *Nac*, Nucleus accumbens; *OFC*, Orbitofrontal cortex; *PFC*, prefrontal cortex; *PUT*, Putamen; *SCFA*, Short-chain fatty acids; *SN*, Substantia nigra; *VTA*, Ventral tegmental área.

Efferent signals from the brain (sympathetic and parasympathetic) control different gastrointestinal functions, like motility and the secretion of hormones and enzymes [13]. Bacteria in the gut produce neurotransmitters, like dopamine and serotonin, from the metabolism of tryptophan and neuropeptides, including P substance, calcitonin, neuropeptide Y (NPY), peptide YY, corticotrophin-releasing factor, pancreatic polypeptide, vasoactive intestinal polypeptide, glucagon-like peptide 1 (GLP-1), and somatostatin, among others [14,15](Fig. 40.1).

Fig. 40.2 highlights how neuroendocrine signals affect mood and behavior. Hormones mediate the secretion of neurotransmitters by gut microbiota [16]. In the presence of physiological or psychological stress, the HPA axis activates the release of corticosterone by the hypothalamus, adrenocorticotropic hormone by the pituitary gland, and cortisol by the adrenal gland [13]. The release of systemic cortisol induces a proinflammatory response, which is capable of modifying gut microbiota composition [17,18]. Also, stress hormones such as adrenaline and noradrenaline promote overgrowth of pathogenic bacteria such as *Escherichia coli* (*E. coli* O157), *Yersinia enterocolitica*, and *Pseudomonas aeruginosa* [16].

Beneficial and pathogenic bacteria interact with the immune system, by secreting antimicrobial peptides (defensins, mucin, and angiogenin 4), mucus secretion, and releasing immunoglobulin A [19]. Bacteria associated with antigen

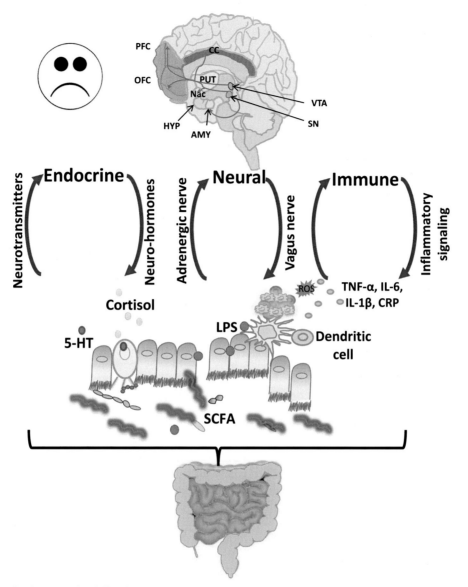

FIGURE 40.2 Behavioral and emotional disorders. Gut microbiota dysbiosis is characterized by low diversity and overgrowth of pathogenic bacteria that contributes to an abnormal communication with the brain. This conditions includes impaired neurotransmitter signaling, a proinflammatory profile (TNFα, INFγ, IL-6, IL-1), and altered endocrine signals. 5-*HT*, Serotonin; *CC*, cingulate cortex; *HYP*, Hypothalamus; *LPS*, Lipopolysaccharides; *MY*, Amygdala; *Nac*, Nucleus accumbens; *OFC*, Orbitofrontal cortex; *PFC*, prefrontal cortex; *PUT*, Putamen; *SCFA*, Short-chain fatty acids; *SN*, Substantia nigra; *VTA*, Ventral tegmental area.

recognition of Pathogen-Associated Molecular Patterns (PAMPS), by toll-like receptors (TLRs), lead to the subsequent activation of the immune response against pathogenic bacteria [20]. Environmental factors such as stress and diet can jeopardize the integrity of the intestinal mucosa [21].

This scenario facilitates that lipopolysaccharides and pathogenic bacteria translocate into the circulation resulting in metabolic endotoxemia, which is associated with chronic disease progression [22,23]. At the same time, the fermentation of nondigestible carbohydrates by gut bacteria generates short-chain fatty acids (SCFA), such as butyrate, acetate, and propionate that modulate the immune response and can also interact with the brain [24] (Fig. 40.2).

THE LINK BETWEEN MICROBIOTA, BEHAVIOR, AND EMOTIONS

Initially, the hypothesis that the brain could communicate with the gut was formulated by observing patients with hepatic encephalopathy, who experienced mood changes and cognitive impairments. Later on, the gut microbiota-brain communication was confirmed in animal models. Studies in germ-free mice have revealed that neurogenesis depends mostly on gut microbiota composition [25]. NIH Swiss mice exhibit a normal behavior, whereas germ-free mice have an anxiety-like behavior [26].

Interestingly, when feces from NIH Swiss mice were transferred to germ-free BALB/c mice, this anxiety-like behavior was reduced, and levels of brain-derived neurotrophic factor (BDNF) increased [27]. Furthermore, the interactions between gut microbiota and brain activity vary according to sex, as shown by male/female differences in BDNF expression in germ-free mice [28].

Also, gut bacteria produce neurotransmitters such as serotonin from tryptophan metabolism. This finding was verified by Liu and colleagues, who administered *Lactobacillus plantarum* PS128 and found increased levels of neurotransmitters dopamine and serotonin in the brain, which were associated with behavior [29].

Gastrointestinal Disorders and Emotions

The relevance of the gut—brain axis is generally accepted in irritable bowel syndrome (IBS), colitis, and other experimental mood disorders [30]. Intrauterine stress in animal models causes gut microbial dysbiosis and this, in turn, affects both gastrointestinal and central nervous system development [27]. Furthermore, IBS patients tend to have more anxiety, depression, and neurotic type personality [31]. In fact, neuroticism is a personality trait that is related to IBS, colitis, Crohn's disease, and dyspepsia [32].

Negative emotions exacerbate symptoms like altered motility, abdominal pain, and hyperreactivity among IBS patients [33]. It is proposed that these negative feelings influence the gastrointestinal system, through signals that arise from the central nervous, endocrine, and immune systems. In fact, stress alters cortisol levels via HPA axis dysregulation, and this is associated with IBS reactivity episodes, especially in women [34].

Cortisol also causes gut dysbiosis and exacerbates the inflammation profiles related to different gastrointestinal diseases, including Crohn's disease and colitis [35]. Moreover, high levels of anxiety and depression induce the overexpression of the heat shock protein 70 (HSP70) in polymorphonuclear cells of patients with severe ulcerative colitis, destroying the intestinal mucosa [36]. In addition, patients with colitis have a poor microbiota diversity, including reductions in *Akkermansia*, which promotes protection against toxins [37].

Patients with inflammatory bowel disease also present reduced abundance of Firmicutes and a higher presence of Proteobacteria, Tenericutes, and *Escherichia coli* [38]. Patients with Crohn's disease have a decreased number of Firmicutes and increased number of Bacteroidetes, when compared with healthy controls [39]. Patients with depression or anxiety are at higher risk of chronic diseases such as cardiovascular disease, diabetes, cerebral vascular disease, and obesity [40—43].

Psychiatric and Social Behavior Disorders

The etiology of mental disorders such as schizophrenia, attention deficit hyperactivity disorder (ADHD), and autism spectrum disorders (ASD) is complex. However, autoimmunity is highlighted as a common key player [44]. In this sense, the gut immune system could detect beneficial bacteria as pathogenic and trigger an autoimmune environment by facilitating the activity of Th17 cells and suppressing the role of the regulatory T cells (Tregs) [45]. In Table 40.1, the gut microbiota composition in different behavioral and emotional disorders is depicted. As noted before, early exposure to adverse events triggers changes in the gut microbiota. Patients with schizophrenia present immunity activation for food protein antigens and pathogens and have a detrimental bacteria profile when compared to healthy control [46].

Gut Microbiota and Emotional Disorders

Neurodevelopmental programming in the offspring depends largely on gut microbiota composition, which could be affected by stressful events [47]. Intrauterine stress in animal models causes a gut dysbiosis and this, in turn, affects central nervous system development [25]. The HPA axis is affected by microbiota alterations and early exposure to stress. Changes occurring in the microbiota composition influence anxiety-type behavior in stress-induced animal models [48].

Gram negative-derived lipopolysaccharides generate a systemic low-grade inflammatory profile, during stress and depression [49]. Also, neuroinflammation provoked by an imbalance in gut microbiota seems to affect the host's behavior [50]. Oxidative stress can also be facilitated and plausibly promote neurodegeneration. Therefore, the impact of gut microbiota dysbiosis on cognition and learning can contribute to Alzheimer's and Parkinson's disease [51].

TABLE 40.1 Gut Microbiota Composition in Behavioral and Emotional Disorders

Disease	Experimental Model (Reference)	Measurements	Bacteria Composition
Irritable bowel syndrome (IBS) with anxiety or depression or low quality of life	37 patients with IBS [80]	Fecal microbiota by pyrosequencing	High Firmicutes/Bacteroidetes ration in IBS patients
Major depressive disorder (MDD)	46 MDD patients and 30 healthy controls [81]	Hamilton's depression scale (HAMDS) Montgomery-Asberg Depression Rating Scale (MADRS), fecal gut microbiota by pyrosequencing, cytokine, and BDNF levels	Increments in fecal bacteria α-diversity in MDD patients related to high levels of Bacteroidetes, Proteobacteria, and Actinobacteria and reductions in Firmicutes Reductions in *Faecalibacterium* were negatively correlated with symptom severity
Depression	34 depressive patients and 33 healthy controls [82] Fecal microbiota transplanted (FMT) adult male Sprague–Dawley rats from depressed patients [82]	Fecal microbiota by next generation sequencing, SCFA concentration, salivary cortisol, kynurenine/tryptophan determination, cytokine levels, lipopolysaccharide-binding protein Rats: Behavioral test, corticosterone, CRP, cytokine, LPS, SCFA levels	Reductions in phylogenetic diversity in depressed group from phylum Prevotellaceae and increases in *Thermoanaerobacteriaceae* FMT rats: presented anhedonia-like behaviors, anxiety-like behaviors, increased plasma kynurenine and kynurenine/tryptophan ratio, reductions in microbiota diversity and increments in acetate levels

In toddlers, a higher phylogenetic diversity was observed among subjects with more social active temperaments, than those with fear temperaments [52]. Patients with anorexia nervosa had lower bacteria diversity when compared to healthy controls, in the inpatient environment, and these results were related to high levels of depression and anxiety [53].

Eating Behavior

It has been proposed that the gut–brain axis interacts with the hunger-reward system, which is modulated by both genes and environmental factors that influence emotionally driven food decisions [54]. Several human genes that encode brain and gut receptors of the hunger-satiety circuit, and also in the brain reward system, express polymorphic alleles. Some of these alleles increase the risk for altered eating behaviors and excess weight, thus indirectly altering the gut microbiota [54–56].

Metabolic syndrome is related to impaired synthesis of neurotransmitters and deficiency in melatonin and tryptophan [57], which contribute to damage in the hippocampus [58]. The hypothesis that microbiota dysbiosis may affect eating behavior was proposed by Dr. Joe Alcok in 2014. He proposed that cravings and eating without control be elicited by microbes through the neuroendocrine system, to optimize their functioning [59]. Later on, Perry and colleagues documented that gut microbiota dysbiosis generates an increased production of acetate, which in turn could elicit overeating behaviors, by affecting the secretion of ghrelin, insulin, and glucose [60].

Postabsorptive signals seem to influence dopamine levels, which is a major neurotransmitter that is implicated in rewarding addictive behaviors [61]. An important single nucleotide polymorphism is the *DRD2/ANKK1* A1 polymorphism, caused by an amino acid substitution (Glu713Lys), within the 11th ankyrin repeat of the *ANKK1* gene. Carriers of the *DRD2/ANKK1*A1 allele are at risk of alcoholism and other addictive behaviors. Recently in a Mexican-Mestizo population, the highest prevalence of the A1 allele known to date worldwide was reported that coincides with the cultural history of alcohol consumption [62].

PSYCHOBIOTICS: FROM PREBIOTICS AND PROBIOTICS TO DIETARY INTERVENTIONS

Table 40.2 summarizes the different strategies aimed to modify the gut microbiota, for the treatment of behavioral and emotional disorders. Animal and clinical studies have demonstrated that probiotic supplementation in autism, personality

TABLE 40.2 Modulation of Gut Microbiota in the Treatment of Behavioral and Emotional Disorders

Component	Experimental Model (Reference)	Measurements	Key Outcomes
Probiotics			
L. Rhamnosus JB-1	Corticosterone-induced stress and anxiety [64]	Anxiety, depression	Reductions in anxiety and stress increases in hippocampal $GABA_{B1b}$ receptor expression and cortical $GABA_{A\alpha2}$ receptor expression
Fermented milk product (*Bifidobacterium animalis, Streptococcus thermophiles, Lactobacillus bulgaricus, Lactococcus lactis*)	Healthy women [65] Twice/day for 4 weeks	MRI: Brain activity and response to face emotions	Reductions in task response in affective viscerosensory and somatosensory networks. Improvements in brain connectivity
Lactobacillus helveticus R0052	$IL-10^{-/-}$ mice and wild-type 129/SvEv mice [83] 21 days of Western diet	Barnes maze: Memory and exploratory and anxiety-like behavior IFNΥ, IL-1β, IL-2, Il-4, IL-5, keratinocyte-derived chemokine, IL-10, IL-12, TNFα and corticosterone levels Fecal microbiota composition	Reductions in weight increases, inflammatory markers and corticosterone. Increased microbiota diversity associated with improvements in memory and reductions in anxiety-like behavior
*L. helveticus*NS8	21 days of stress [84]	Cognition and behavior tests	Reductions in corticosterone levels and anxiety and increases in 5-HT, IL-10, and cognition
Bifidobacterium bifidum W23, *Bifidobacterium lactis* W52, *Lactobacillus acidophilus* W37, *Lactobacillus brevis* W63, *Lactobacillus casei* W56, *Lactobacillus salivarius* W24, *Lactococcus lactis*	Triple-blind randomized control study [85] Healthy volunteers Four weeks	Dysphoria assessed with Leiden index of depression sensitivity scale (LEIDS-r)	Reductions in cognitive reaction to sadness
VSL3 Streptococcus salivarius subsp., *thermophilis, Bifidobacterium (B. breve, B. infanti, B. longum), Lactobacillus acidophilus, L. planarum, L. casei, and L. delbrueekisubsp. Bulgaricus*	Inflammatory murine model of liver inflammation [86] C57BL/6 mice	Body weight, sickness social behavior, gut permeability, fecal microbiota, intravital microscopy cerebral vasculature, cerebral monocyte infiltration	Reductions in TNF-α, intestinal permeability, monocyte recruitment and improvements in social behavior
Probiotic yogurt (*Lactobacillus acidophilus* LA5 and *Bifidobacterium lactis*BB12) + placebo capsule or Probiotic capsule (*Actobacilluscasei, L. Acidophilus, L. Rhamnosus, L. Bulgaricus, Bifidobacterium breve, B. Longum, S. thermophilus*) + conventional yogurt (*Streptococcus thermophilus* and *Lactobacillus bulgaricus* starters) or conventional yogurt + placebo capsule	Double-blind randomized study [87] 75 healthy petrochemical workers (male and female)	Health questionnaire (GHQ-28) Depression anxiety and stress scale (DASS), in kynurenine, tryptophan, neuropeptide Y, cortisol and ACTH levels	Improvements in health, depression, and anxiety with probiotic yogurt and probiotic capsule
Lactobacillus acidophilus, Lactobacillus casei, and Bifidobacterium bifidum	Double-blind randomized study [88] 40 patients with major depressive disorder (MDD) 8-week intervention	Depression with Beck depression inventory (BDI), fasting plasma glucose, insulin metabolism markers, lipids, c-reactive protein, oxidative stress markers, antioxidant capacity, and glutathione GSH levels	Reductions in depression scores and serum metabolic, inflammatory and oxidative stress improvements

Continued

TABLE 40.2 Modulation of Gut Microbiota in the Treatment of Behavioral and Emotional Disorders—cont'd

Component	Experimental Model (Reference)	Measurements	Key Outcomes
B. longum NCC3001 *subspecies longum* strain (*BL*)	Randomized, double-blind, placebo-controlled, pilot study [89] 44 patients with Inflammatory bowel syndrome with anxiety and depression	HAD scale: Anxiety and/or Depression, State-Trait Anxiety Inventory, (STAI), IBS global adequate relief, IBS symptoms, somatization, quality of life, changes in brain activation patterns (functional Magnetic Resonance Imaging, fMRI), serum inflammatory markers, neurotransmitters and BDNF, and fecal microbiota	Reduction in depression scores, in responses to negative emotional stimuli in amygdala and fronto-limbic regions. Improvements in quality of life scores
Prebiotics			
Fructooligosaccharides or Oligosaccharides	Adult male Sprague—Dawley Rats [90] Five weeks	Plasma glucose, PPY, GLP-1 levels, cortex and hippocampus BDNF and NMDARs protein levels and expression, Fecal microbiota	Increments in BDNF hippocampal regions, NR1 expression in hippocampal regions and frontal cortex, and NR2 in hippocampal regions, increments in PYY levels Growth of *Lactobacilli* and *Bifidobacteria*
Fructooligosaccharides (FOS) or Bimuno-galactooligosaccharides (B-GOS)	Double-blind randomized study [91] Healthy volunteers (22 males, 23 females) Three-week supplementation	Levels of salivary cortisol Computer-based Emotional Test Battery IQ test Personality, stress response, mood questionnaires, Beck depression inventory, and state-trait anxiety inventory	B-GOS reduced salivary cortisol levels and attentional vigilance to negative versus positive words
Resistant dextrin (Nutriose06) or maltodextrin	Females with type 2 Diabetes Mellitus [92]	General health questionnaire (GHQ) Depression, anxiety, and stress scale (DASS) White blood cell count, CD4, CD8, IFNγ, IL12, IL4, IL10, cortisol, tryptophan, Adrenocorticotropic hormone, kynurenine, and plasma lipopolysaccharide (LPS)	Reductions in cortisol, kynurenine, IFNγ, IL-12, IL-10, IL-4, LPS, CD8, GHQ, DASS
Synbiotics			
Lactobacillus rhamnosus CGMCC1.3724 (LPR) + oligofructose and inulin + magnesium or Placebo	Randomized, double-blind, placebo-controlled trial [93] Overweight and obese	Visual analogue scale for appetite Satiety quotient Three-factor eating questionnaire State-trait food cravings questionnaire trait Beck depression inventory Body esteem scale Binge eating scale Perceived stress scale State-trait anxiety inventory Energy intake, physical activity	Reductions in weight associated with improvements in food behavior and sad feelings in women. Improvements in food behavior and cognition in males.

disorders, depression, and cognitive impairment are a promising treatment strategy [15,63]. Bravo and colleagues achieved to reduce anxiety and stress, by giving *L. rhamnosus* JB-1 to mice, showing improvements in gamma-aminobutyric acid (GABA) receptor activity in brain structures related to emotion and behaviors [64].

In fact, the use of probiotics ameliorates brain connectivity, as shown in the supplementation of a fermented product containing *Bifidobacterium animalis, Streptococcus thermophilus, Lactobacillus bulgaricus,* and *Lactococcus lactis* to healthy women, finding signs of healthy behavior and emotions [64]. Prebiotics are another alternative, potentially enhancing SCFA such as butyrate, which activates gluconeogenesis, improves insulin sensitivity and energy efficiency and therefore is implicated in appetite and satiety. However, one of the most used prebiotics is resistant starch, with controversial changes in gut microbiota and behavior [66,67].

Healthy Diets and Supplements

A westernized diet is deleterious for gut microbiota and behaviors [50]. In contrast, good sources of fats such as ω-3-rich oils can modify gut microbiota and improve behavior [69]. Other dietary strategies include phytochemicals, which are rich in crucial nutrients for mental health such as magnesium, potassium, zinc, vitamin c, folic acid, and vitamin B12. By adding more fiber through the consumption of fruits and vegetables, depressive symptoms could be reduced [70].

Ethnic and Geographical Considerations

Gut bacteria in the Japanese population, specifically *Bacteroides plebeius*, have a unique gene signature that codifies carbohydrate-active enzymes (CAZymes) porphyranases and agarases. These genes were acquired from the marine Bacteroidetes, *Zobellia galactanivorans*, which naturally degrade seaweeds [71]. The Hazda hunter-gatherers of Tanzania have a distinctive gut microbiota composition, enriched in *Prevotella, Treponema,* and unclassified Bacteroidetes that gives them the ability to digest fibrous vegetable foods and benefit from their nutrients [72].

Another example is the gut microbiome of the Inuit in the Canadian Arctic, with significantly reduced genetic diversity within the *Prevotella,* a genus related to their traditional diet, low in fiber and rich in animal fats and protein [73]. Furthermore, it has been documented that population groups living in communal lifestyle and consuming their seasonally traditional diets have a healthier microbiome and less microbiome-associated diseases [74].

In agreement with the genetic adaptations to traditional diets, and bacteria coevolving with humans, recommendations of the dietary pattern should include regionally available prebiotic foods, to promote a beneficial gut bacteria [75]. For example, dietary strategies based on the food components of the Mediterranean diet have proven beneficial for mental health [76].

Genetic adaptations to traditional and prehispanic foods of the Mexican populations seem to have influenced a healthier and fitter status [77,78]. Mexican prehispanic foods that have shown significant prebiotic and probiotic activities, include a healthy combination of maize-derived products and legumes, prickly pear, amaranth, chia seeds, fiber, and low-degree alcohol fermented beverages [77−79]. In other cultures, traditional foods such as soy, fermented products, turmeric, cocoa, and green tea that are rich in polyphenols, phytochemicals, and nondigestible fibers that could improve altered gut microbiota and general health.

CONCLUSION

Personalized-medicine and genome-based nutrition strategies are necessary to balance the host's genes, emotions, and diet, to prevent and alleviate gut microbiota-related diseases.

LIST OF ACRONYMS AND ABBREVIATIONS

ADHD Attention deficit hyperactivity disorder
ASD Autism spectrum disorders
BDNF Brain-derived neurotrophic factor
CNS Central nervous system
GABA Gamma-aminobutyric acid
GLP-1 Glucagon-like peptide 1
HPA Hypothalamic pituitary adrenal
HSP70 Heat shock protein
IBS Irritable bowel syndrome

NAC Nucleus accumbens
NCCDs Noncommunicable chronic diseases
NPY Neuropeptide Y
OFC Orbital frontal cortex
PAMPS Pathogen-associated molecular patterns
PFC Prefrontal cortex
PYY Peptide YY
SCFA Short-chain fatty acids
TLRs Toll-like receptors
Tregs Regulatory T cells

REFERENCES

[1] Sommer F, Backhed F. The gut microbiota-masters of host development and physiology. Nat Rev Microbiol 2013;11:227–38.

[2] Cani PD. Crosstalk between the gut microbiota and the endocannabinoid system: impact on the gut barrier function and the adipose tissue. Clin Microbiol Infect 2012;18(Suppl. 4):50–3.

[3] Olds J, Milner P. Positive reinforcement produced by electrical stimulation of septal area and other regions of rat brain. J Comp Physiol Psychol 1954;47:419–27.

[4] Schoenbaum G, Roesch M. Orbitofrontal cortex, associative learning, and expectancies. Neuron 2005;47:633–6.

[5] McEwen BS, Nasca C, Gray JD. Stress effects on neuronal structure: hippocampus, amygdala, and prefrontal cortex. Neuropsychopharmacology 2016;41:3–23.

[6] Price JL, Drevets WC. Neurocircuitry of mood disorders. Neuropsychopharmacology 2010;35:192–216.

[7] Rutishauser U, Mamelak AN, Adolphs R. The primate amygdala in social perception - insights from electrophysiological recordings and stimulation. Trends Neurosci 2015;38:295–306.

[8] Arana FS, Parkinson JA, Hinton E, Holland AJ, Owen AM, Roberts AC. Dissociable contributions of the human amygdala and orbitofrontal cortex to incentive motivation and goal selection. J Neurosci 2003;23:9632–8.

[9] Riley MR, Constantinidis C. Role of prefrontal persistent activity in working memory. Front Syst Neurosci 2015;9:181.

[10] Richard JM, Berridge KC. Prefrontal cortex modulates desire and dread generated by nucleus accumbens glutamate disruption. Biol Psychiatry 2013;73:360–70.

[11] Sandoval-Salazar C, Ramirez-Emiliano J, Trejo-Bahena A, Oviedo-Solis CI, Solis-Ortiz MS. A high-fat diet decreases GABA concentration in the frontal cortex and hippocampus of rats. Biol Res 2016;49:15.

[12] Minokoshi Y, Alquier T, Furukawa N, Kim YB, Lee A, Xue B, Mu J, Foufelle F, Ferré P, Birnbaum MJ, Stuck BJ, Kahn BB. AMP-kinase regulates food intake by responding to hormonal and nutrient signals in the hypothalamus. Nature 2004;428:569–74.

[13] Grenham S, Clarke G, Cryan JF, Dinan TG. Brain-gut-microbe communication in health and disease. Front Physiol 2011;2:94.

[14] Bercik P, Denou E, Collins J, Jackson W, Lu J, Jury J, Deng Y, Blennerhassett P, Macri J, McCoy KD, Verdu EF, Collins SM. The intestinal microbiota affect central levels of brain-derived neurotropic factor and behavior in mice. Gastroenterology 2011;141:599–609. 609.e1-3.

[15] Desbonnet L, Garrett L, Clarke G, Bienenstock J, Dinan TG. The probiotic Bifidobacteria infantis: an assessment of potential antidepressant properties in the rat. J Psychiatr Res 2008;43:164–74.

[16] Holzer P, Farzi A. Neuropeptides and the microbiota-gut-brain axis. Adv Exp Med Biol 2014;817:195–219.

[17] Ando T, Brown RF, Berg RD, Dunn AJ. Bacterial translocation can increase plasma corticosterone and brain catecholamine and indoleamine metabolism. Am J Physiol Regul Integr Comp Physiol 2000;279:R2164–72.

[18] Dunn AJ, Ando T, Brown RF, Berg RD. HPA axis activation and neurochemical responses to bacterial translocation from the gastrointestinal tract. Ann NY Acad Sci 2003;992:21–9.

[19] Ostaff MJ, Stange EF, Wehkamp J. Antimicrobial peptides and gut microbiota in homeostasis and pathology. EMBO Mol Med 2013;5:1465–83.

[20] Kalliomaki MA, Walker WA. Physiologic and pathologic interactions of bacteria with gastrointestinal epithelium. Gastroenterol Clin North Am 2005;34:383–99. vii.

[21] Galley JD, Nelson MC, Yu Z, Dowd SE, Walter J, Kumar PS, Lyte M, Bailey MT. Exposure to a social stressor disrupts the community structure of the colonic mucosa-associated microbiota. BMC Microbiol 2014;14:189.

[22] Abraham C, Medzhitov R. Interactions between the host innate immune system and microbes in inflammatory bowel disease. Gastroenterology 2011;140:1729–37.

[23] Sharma V, Garg S, Aggarwal S. Probiotics and liver disease. Perm J 2013;17:62–7.

[24] Cani PD, Everard A, Duparc T. Gut microbiota, enteroendocrine functions and metabolism. Curr Opin Pharmacol 2013;13:935–40.

[25] Selkrig J, Wong P, Zhang X, Pettersson S. Metabolic tinkering by the gut microbiome: implications for brain development and function. Gut Microb 2014;5:369–80.

[26] Crumeyrolle-Arias M, Jaglin M, Bruneau A, Vancassel S, Cardona A, Daugé V, Naudon L, Rabot S. Absence of the gut microbiota enhances anxiety-like behavior and neuroendocrine response to acute stress in rats. Psychoneuroendocrinology 2014;42:207–17.

[27] Neufeld K-AM, Kang N, Bienenstock J, Foster JA. Effects of intestinal microbiota on anxiety-like behavior. Commun Integr Biol 2011;4:492–4.

[28] Clarke G, Grenham S, Scully P, Fitzgerald P, Moloney RD, Shanahan F, Dinan TG, Cryan JF. The microbiome-gut-brain axis during early life regulates the hippocampal serotonergic system in a sex-dependent manner. Mol Psychiatry 2013;18:666–73.

[29] Liu YW, Liu W-H, Wu CC, Juan YC, Wu YC, Tsai HP, Wang S, Tsai YC. Psychotropic effects of Lactobacillus plantarum PS128 in early life-stressed and naive adult mice. Brain Res 2016;1631:1–12.

[30] Muscatello MRA, Bruno A, Scimeca G, Pandolfo G, Zoccali RA. Role of negative affects in pathophysiology and clinical expression of irritable bowel syndrome. World J Gastroenterol 2014;20:7570–86.

[31] Tayama J, Nakaya N, Hamaguchi T, Tomile T, Shinozaki M, Saigo T, Shirabe S, Fukudo S. Effects of personality traits on the manifestations of irritable bowel syndrome. Biopsychosoc Med 2012;6:20.

[32] Lackner JM, Gudleski GD, Ma C-X, Dewanwala A, Naliboff B. Fear of GI symptoms has an important impact on quality of life in patients with moderate-to-severe IBS. Am J Gastroenterol 2014;109:1815–23.

[33] Zhu L, Huang D, Shi L, Liang L, Xu T, Chang M, Chen W, Wu D, Zhang F, Fang X. Intestinal symptoms and psychological factors jointly affect quality of life of patients with irritable bowel syndrome with diarrhea. Health Qual Life Outcomes 2015;13:49.

[34] Suarez-Hitz KA, Otto B, Bidlingmaier M, Schwizer W, Fried M, Ehlert U. Altered psychobiological responsiveness in women with irritable bowel syndrome. Psychosom Med 2012;74:221–31.

[35] Strober W. Impact of the gut microbiome on mucosal inflammation. Trends Immunol 2013;34:423–30.

[36] Vlachos II, Barbatis C, Tsopanomichalou M, Abou-Assabeh L, Goumas K, Ginieri-Coccossis M, Economou M, Papadimitriou GN, Patsouris E, Nicolopoulou-Stamati P. Correlation between depression, anxiety, and polymorphonuclear cells' resilience in ulcerative colitis: the mediating role of heat shock protein 70. BMC Gastroenterol 2014;14:77.

[37] Vigsnaes LK, Brynskov J, Steenholdt C, Wilcks A, Licht TR. Gram-negative bacteria account for main differences between faecal microbiota from patients with ulcerative colitis and healthy controls. Benef Microbes 2012;3:287–97.

[38] Tong M, Li X, Wegener Parfrey L, Roth B, Ippoliti A, Wei B, Borneman J, McGovern DP, Frank DN, Li E, Horvarth S, Knight R, Braun J. A modular organization of the human intestinal mucosal microbiota and its association with inflammatory bowel disease. PLoS One 2013;8:e80702.

[39] Erickson AR, Cantarel BL, Lamendella R, Darzi Y, Mongodin EF, Pan C, Shah M, Halfvarson J, Tsyk C, Henrissat B, Raes J, Verberkmoes NC, Fraser CM, Hettich RL, Jansson JK. Integrated metagenomics/metaproteomics reveals human host-microbiota signatures of Crohn's disease. PLoS One 2012;7:e49138.

[40] Kurth B-M, Ellert U. Perceived or true obesity: which causes more suffering in adolescents? Findings of the German health interview and examination survey for children and adolescents (KiGGS). Dtsch Arzteblatt Int 2008;105:406–12.

[41] Penninx BWJH, Milaneschi Y, Lamers F, Vogelzangs N. Understanding the somatic consequences of depression: biological mechanisms and the role of depression symptom profile. BMC Med 2013;11:129.

[42] Bjorntorp P. Do stress reactions cause abdominal obesity and comorbidities? Obes Rev 2001;2:73–86.

[43] Weber-Hamann B, Werner M, Hentschel F, Bindeballe N, Lederbogen F, Deuschle M, Heuser I. Metabolic changes in elderly patients with major depression: evidence for increased accumulation of visceral fat at follow-up. Psychoneuroendocrinology 2006;31:347–54.

[44] Benros ME, Eaton WW, Mortensen PB. The epidemiologic evidence linking autoimmune diseases and psychosis. Biol Psychiatry 2014;75:300–6.

[45] Severance EG, Tveiten D, Lindstrom LH, Yolken RH, Reichelt KL. The gut microbiota and the emergence of autoimmunity: relevance to major psychiatric disorders. Curr Pharm Des 2016;22:6076–86.

[46] Castro-Nallar E, Bendall ML, Perez-Losada M, Sabuncyan S, Severance EG, Dickerson FB, Schroeder JR, Yolken RH, Crandall KA. Composition, taxonomy and functional diversity of the oropharynx microbiome in individuals with schizophrenia and controls. PeerJ 2015;3:e1140.

[47] Sudo N, Chida Y, Aiba Y, Sonoda J, Oyama N, Yu XN, Kubo C, Koga Y. Postnatal microbial colonization programs the hypothalamic-pituitary-adrenal system for stress response in mice. J Physiol 2004;558:263–75.

[48] Bangsgaard Bendtsen KM, Krych L, Sorensen DB, Pang W, Nielsen DS, Josefsen K, Hansen LH, Sorensen SJ, Hansen AK. Gut microbiota composition is correlated to grid floor induced stress and behavior in the BALB/c mouse. PLoS One 2012;7:e46231.

[49] De Punder K, Pruimboom L. Stress induces endotoxemia and low-grade inflammation by increasing barrier permeability. Front Immunol 2015;6:223.

[50] Pyndt Jorgensen B, Hansen JT, Krych L, Larsen C, Klein AB, Nielsen DS, Josefsen K, Hansen AK, Sorensen DB. A possible link between food and mood: dietary impact on gut microbiota and behavior in BALB/c mice. PLoS One 2014;9:e103398.

[51] Mulak A, Bonaz B. Brain-gut-microbiota axis in Parkinson's disease. World J Gastroenterol 2015;21:10609–20.

[52] Christian LM, Galley JD, Hade EM, Schoppe-Sullivan S, Kamp Dush C, Bailey MT. Gut microbiome composition is associated with temperament during early childhood. Brain Behav Immun 2015;45:118–27.

[53] Kleiman SC, Watson HJ, Bulik-Sullivan EC, Huh EY, Tarantino LM, Bulik CM, Carroll IM. The intestinal microbiota in acute anorexia nervosa and during renourishment: relationship to depression, anxiety, and eating disorder psychopathology. Psychosom Med 2015;77:969–81.

[54] Panduro A, Rivera-Iniguez I, Sepulveda-Villegas M, Roman S. Genes, emotions and gut microbiota: the next frontier for the gastroenterologist. World J Gastroenterol 2017;23:3030–42.

[55] Foley DL, Morley KI, Madden PAF, Heath AC, Whitfield JB, Martin NG. Major depression and the metabolic syndrome. Twin Res Hum Genet 2010;13:347–58.

[56] Wiltink J, Michal M, Wild PS, Schneider A, Konig J, Blettner M, Munzel T, Schulz A, Weber M, Fottner C, Pfeiffer N, Lackner K, Beutel ME. Associations between depression and diabetes in the community: do symptom dimensions matter? Results from the Gutenberg Health Study. PLoS One 2014;9:e105499.

[57] Elovainio M, Hurme M, Jokela M, Pulkki-Raback L, Kivimaki M, Hintsanen M, Hintsa T, Lehtimaki T, Vikari J, Raitakari OT, Keltikangas-Jarvinen L. Indoleamine 2,3-dioxygenase activation and depressive symptoms: results from the Young Finns Study. Psychosom Med 2012;74:675–81.

[58] Schiepers OJG, Wichers MC, Maes M. Cytokines and major depression. Prog Neuro-Psychopharmacol Biol Psychiatry 2005;29:201−17.

[59] Alcock J, Maley CC, Aktipis CA. Is eating behavior manipulated by the gastrointestinal microbiota? Evolutionary pressures and potential mechanisms. BioEssays News Rev Mol Cell Dev Biol 2014;36:940−9.

[60] Perry RJ, Peng L, Barry NA, Cline GW, Zhang D, Cardone RL, Petersen KF, Kibbey RG, Goodman AL, Shulman GI. Acetate mediates a microbiome-brain-beta-cell axis to promote metabolic syndrome. Nature 2016;534:213−7.

[61] Sustkova-Fiserova M, Jerabek P, Havlickova T, Kacer P, Krsiak M. Ghrelin receptor antagonism of morphine-induced accumbens dopamine release and behavioral stimulation in rats. Psychopharmacol (Berl) 2014;231:2899−908.

[62] Panduro A, Ramos-Lopez OM, Campollo O, Zepeda-Carrillo EA, Gonzalez-Aldaco K, Torres-Valadez R, Roman S. High frequency of the DRD2/ANKK1 A1 allele in Mexican Native Amerindians and Mestizos and its associations with alcohol consumption. Drug Alcohol Depend 2017;172:66−72.

[63] Distrutti E, O'Reilly J-A, McDonald C, Cipriani S, Renga B, Lynch MA, Fiorucci S. Modulation of intestinal microbiota by the probiotic VSL#3 resets brain gene expression and ameliorates the age-related deficit in LTP. PLoS One 2014;9:e106503.

[64] Bravo JA, Forsythe P, Chew MV, Escaravage E, Savignac HM, Dinan TG, Bienenstock J, Cryan JF. Ingestion of Lactobacillus strain regulates emotional behavior and central GABA receptor expression in a mouse via the vagus nerve. Proc Natl Acad Sci USA 2011;108:16050−5.

[65] Tillisch K, Labus J, Kilpatrick L, Jiang Z, Stains J, Ebrat B, Guyonnet D, Legrain-Raspaud S, Trotin B, Naliboff B, Mayer EA. Consumption of fermented milk product with probiotic modulates brain activity. Gastroenterology 2013;144:1394−401. 1401.e1-4.

[66] Smith PM, Howitt MR, Panikov N, Michaud M, Gallini CA, Bohlooly-Y M, Glickman JN, Garret WS. The microbial metabolites, short-chain fatty acids, regulate colonic Treg cell homeostasis. Science 2013;341:569−73.

[67] Lyte M, Chapel A, Lyte JM, Ai Y, Proctor A, Jane JL, Philips GJ. Resistant starch alters the microbiota-gut brain axis: implications for dietary modulation of behavior. PLoS One 2016;11:e0146406.

[68] Xu Z, Knight R. Dietary effects on human gut microbiome diversity. Br J Nutr 2015;113(Suppl: S1−5).

[69] Robertson RC, Seira Oriach C, Murphy K, Moloney GM, Cryan JF, Dinan TG, Paul Ross R, Stanton C. Omega-3 polyunsaturated fatty acids critically regulate behaviour and gut microbiota development in adolescence and adulthood. Brain Behav Immun 2017;59:21−37.

[70] Miki T, Eguchi M, Kurotani K, Kochi T, Kuwahara K, Ito R, Tsuruoka H, Akter S, Kashino I, Kabe I, Kawakami N, Mizoue T. Dietary fiber intake and depressive symptoms in Japanese employees: the Furukawa Nutrition and Health Study. Nutrition 2016;32:584−9.

[71] Hehemann J-H, Correc G, Barbeyron T, Helbert W, Czjzek M, Michel G. Transfer of carbohydrate-active enzymes from marine bacteria to Japanese gut microbiota. Nature 2010;464:908−12.

[72] Schnorr SL, Candela M, Rampelli S, Centanni M, Consolandi C, Basaglia G, Turroni S, Biagi E, Peano C, Severgnini M, Fiori J, Gotti R, De Bellis G, Luiselli D, Brigidi P, Mabulla A, Marlowe F, Henry AG, Crittenden AN. Gut microbiome of the Hadza hunter-gatherers. Nat Commun 2014;5:3654.

[73] Girard C, Tromas N, Amyot M, Shapiro BJ. Gut microbiome of the canadian arctic inuit. mSphere 2017;2.

[74] Smits SA, Leach J, Sonnenburg ED, Gonzalez CG, Lichtman JS, Reid G, Knight R, Manjurano A, Changalucha J, Elias JE, Dominguez-Bello MG, Sonnenburg JL. Seasonal cycling in the gut microbiome of the Hadza hunter-gatherers of Tanzania. Science 2017;357:802−6.

[75] Kang JX. Gut microbiota and personalized nutrition. J Nutr Nutr 2013;6:I−II.

[76] Logan AC, Jacka FN. Nutritional psychiatry research: an emerging discipline and its intersection with global urbanization, environmental challenges and the evolutionary mismatch. J Physiol Anthropol 2014;33:22.

[77] Roman S, Ojeda-Granados C, Ramos-Lopez O, Panduro A. Genome-based nutrition: an intervention strategy for the prevention and treatment of obesity and nonalcoholic steatohepatitis. World J Gastroenterol 2015;21:3449−61.

[78] Ojeda-Granados C, Panduro A, Gonzalez-Aldaco K, Sepulveda-Villegas M, Rivera-Iniguez I, Roman S. Tailoring nutritional advice for Mexicans based on prevalence profiles of diet-related adaptive gene polymorphisms. J Pers Med 2017;7.

[79] Avila-Nava A, Noriega LG, Tovar AR, Granados O, Perez-Cruz C, Pedraza-Chaverri J, Torres N. Food combination based on a pre-hispanic Mexican diet decreases metabolic and cognitive abnormalities and gut microbiota dysbiosis caused by a sucrose-enriched high-fat diet in rats. Mol Nutr Food Res 2017;61.

[80] Jeffery IB, O'Toole PW, Ohman L, Claesson MJ, Deane J, Quigley EM, Simrén M. An irritable bowel syndrome subtype defined by species-specific alterations in faecal microbiota. Gut 2012;61:997−1006.

[81] Jiang H, Ling Z, Zhang Y, Mao H, Ma Z, Yin Y, Wang W, Tang W, Tan Z, Shi J, Li L, Ruan B. Altered fecal microbiota composition in patients with major depressive disorder. Brain Behav Immun 2015;48:186−94.

[82] Kelly JR, Borre Y, O' Brien C, Patterson E, EL Aidy S, Deane J, Kennedy PJ, Beers S, Scott K, Moloney G, Hoban AE, Scott L, Fitzgerald P, Ross P, Stanton C, Clarke G, Cryan JF, Dinan TG. Transferring the blues: depression-associated gut microbiota induces neurobehavioural changes in the rat. J Psychiatr Res 2016;82:109−18.

[83] Ohland CL, Kish L, Bell H, Thiesen S, Hotte N, Pankiv E, Madsen KL. Effects of Lactobacillus helveticus on murine behavior are dependent on diet and genotype and correlate with alterations in the gut microbiome. Psychoneuroendocrinology 2013;38:1738−47.

[84] Liang S, Wang T, Hu X, Luo J, Li W, Wu X, Duan Y, Jin F. Administration of Lactobacillus helveticus NS8 improves behavioral, cognitive, and biochemical aberrations caused by chronic restraint stress. Neuroscience 2015;310:561−77.

[85] Steenbergen L, Sellaro R, van Hemert S, Bosch JA, Colzato LS. A randomized controlled trial to test the effect of multispecies probiotics on cognitive reactivity to sad mood. Brain Behav Immun 2015;48:258−64.

[86] D'Mello C, Ronaghan N, Zaheer R, Dicay M, Le T, MacNaughton WK, Surrette MG, Swain MG. Probiotics improve inflammation-associated sickness behavior by altering communication between the peripheral immune system and the brain. J Neurosci 2015;35:10821−30.

[87] Mohammadi AA, Jazayeri S, Khosravi-Darani K, Solati Z, Mohammadpour N, Asemi Z, Adab Z, Djalali M, Tehrani-Doost M, Hosseini M, Eghtesadi S. The effects of probiotics on mental health and hypothalamic-pituitary-adrenal axis: a randomized, double-blind, placebo-controlled trial in petrochemical workers. Nutr Neurosci 2016;19:387–95.

[88] Akkasheh G, Kashani-Poor Z, Tajabadi-Ebrahimi M, Jafari P, Akbari H, Taghizadeh M, Memarzadeh MR, Asemi Z, Esmaillzadeh A. Clinical and metabolic response to probiotic administration in patients with major depressive disorder: a randomized, double-blind, placebo-controlled trial. Nutrition 2016;32:315–20.

[89] Pinto-Sanchez MI, Hall GB, Ghajar K, NArdelli A, Bolino C, Lau JT, Martin FP, Cominetti O, Welsh C, Rieder A, Traynor J, Gregory C, De Palma G, Pigrau M, Ford AC, Macri J, Berger B, Bergonzelli G, Surette MG, Collins SM, Moayyedi P, Bercik P. Probiotic Bifidobacterium longum NCC3001 reduces depression scores and alters brain activity: a pilot study in patients with irritable bowel syndrome. Gastroenterology 2017;153:e8448–59.

[90] Savignac HM, Corona G, Mills H, Chen L, Spencer JP, Tzortzis G, Burnet PW. Prebiotic feeding elevates central brain derived neurotrophic factor. Neurochem Int 2013;63:756–64.

[91] Schmidt K, Cowen PJ, Harmer CJ, Tzortzis G, Errington S, Burnet PWJ. Prebiotic intake reduces the waking cortisol response and alters emotional bias in healthy volunteers. Psychopharmacol (Berl) 2015;232:1793–801.

[92] Farhangi MA, Javid AZ, Sarmadi B, Karimi P, Dehghan P. A randomized controlled trial on the efficacy of resistant dextrin, as functional food, in women with type 2 diabetes: targeting the hypothalamic-pituitary-adrenal axis and immune system. Clin Nutr 2017. published online June 10.

[93] Sanchez M, Darimont C, Panahi S, Drapeau V, Marette A, Taylor VH, Doré J, Tremblay A. Effects of a diet-based weight-reducing program with probiotic supplementation on satiety efficiency, eating behavior traits, and psychosocial behaviors in obese individuals. Nutrients 2017;9.

FURTHER READING

[1] Roman S, Panduro A. Genomic medicine in gastroenterology: a new approach or a new specialty? World J Gastroenterol 2015;21(27):8227–37.

Block V

Applications for Foods, Drugs, and Xenobiotics

Chapter 41

The Gut Microbiome in Vegetarians

Ana Carolina F. Moraes[1], Bianca de Almeida-Pittito[2] and Sandra Roberta G. Ferreira[1]

[1]Department of Epidemiology, School of Public Health, University of Sao Paulo, São Paulo, Brazil; [2]Department of Preventive Medicine, Federal University of Sao Paulo, São Paulo, Brazil

INTRODUCTION

Consistent evidence supports the role for nutrients and dietary patterns in the genesis of noncommunicable chronic diseases such as obesity, type 2 diabetes, hypertension, and cardiovascular disease [1,2]. One of the most recognized protective patterns has been the Mediterranean-style diet [3—5] which includes a lot of vegetables, nuts, and plant-derived foods, with emphasis on olive oil. Also, the vegetarian diet has been associated with a beneficial cardiometabolic profile and lower rates of cardiovascular events [6—8]. Typically, plant-based diets preclude meat, poultry, or fish but may contain egg or dairy products. Therefore the term strict vegetarian or vegan has been attributed to consumers of only plant foods (vegetables, fruits, grains, beans, and nuts) and no animal products, while the lacto-ovo vegetarians can eat egg or dairy products [6,8].

In contrast to the increased cardiovascular risk associated with high-saturated fat low-fiber Western diets [9], a cardioprotective profile has been demonstrated in vegetarian populations [7,8]. The high amount of energy of the Western dietary pattern contributes to body fat accumulation known to trigger inflammation and insulin resistance, involved in the genesis of cardiometabolic diseases [10,11]. On the other hand, vegetarian individuals, such as the Seventh-day Adventists, were shown to have lower body mass index and a more favorable metabolic profile [2,12], which may have a beneficial impact in morbidity and mortality rates. Actually, the Adventist Health Study 2 (AHS-2) showed reduced all-cause death and cardiovascular mortality in vegetarians [13].

GUT MICROBIOME-MODULATED EFFECTS OF DIETARY COMPONENTS

Carbohydrates

Carbohydrates usually represent the most prominent macronutrient of human diet particularly of vegetarians [14,15]. Simple carbs are rapidly absorbed, while the complex carbohydrates—such as dietary fiber—are not enzymatically degraded in the small intestine but are fermented by the microbiome in the colon [16,17].

A marked characteristic of vegetarian diet is its high fiber content [15]. The fermentation of dietary fiber, especially by genera *Bacteroides, Bifidobacterium, Ruminococcus, Eubacterium,* and *Lactobacillus* [18,19], shapes metabolic activities through the production of short-chain fatty acids (SCFAs)—mainly acetate, propionate, and butyrate. These provide energy for colonocytes and alter the intestinal pH, impacting on the growth of pathogens [16,18]. Despite some controversies, SCFAs have been associated with metabolic benefits [19,20]. It was reported that acetate could act as a substrate for hepatic lipogenesis favoring fat deposition [21], a potentially harmful feature. In contrast, butyrate could favor the intestinal barrier function and other desirable effects on glucose metabolism [16,20,22].

Butyrate is mainly produced by *Faecalibacterium prausnitzii, Eubacterium rectale, Eubacterium hallii, Ruminococcus bromii* [20], and *Roseburia* [23], which have been considered markers of health [12,24]. This SCFA contributes to upregulation of tight-junction protein expression (claudin-1, zonula occludens-1, and occludin), enhancing the intestinal barrier function, thus preventing inflammatory diseases related to lipopolysaccharide (LPS) translocation [20,22,25]. An antiinflammatory effect is also described via inhibition of nuclear factor κB in colonocytes [20,26].

Microbiome and Metabolome in Diagnosis, Therapy, and other Strategic Applications. https://doi.org/10.1016/B978-0-12-815249-2.00041-5

SCFAs have been related to the control of satiety and food intake [26]. Their binding to G protein-coupled receptors 41 and 43, also referred as free fatty acids receptors 2 and 3 (FFAR2 and FFAR3), in enteroendocrine L-cells, stimulates the release of the satietogenic hormones glucagon-like peptide 1 (GLP-1) and peptide YY (PYY) [25,27]. GLP-1 stimulates β-cell insulin secretion and favors satiety due to the reduction of gastric emptying [28]. GLP-2 is co-secreted with GLP-1 and also enhances intestinal epithelial proliferation and permeability [29].

PYY regulates intestinal motility, affecting nutrient absorption [26,30] and acts centrally, inhibiting orexigenic neurons at the arcuate nucleus, thus reducing food consumption [31]. FFAR2 is also expressed in immune cells, influencing the inflammatory response [32]. Considering that obesity and type 2 diabetes mellitus are low-grade inflammation conditions, it is possible that the modulation of the immune system by SCFAs may contribute to the benefits observed in vegetarian individuals [20,33] and those following fiber-rich interventions in cardiometabolic diseases.

The potential of prebiotic supplementation to increase bacteria associated to better cardiometabolic profile has been demonstrated, including abundance of *Akkermansia muciniphila* [34]. This mucin-degrading bacterium has been related to improvement in glucose metabolism, with direct correlation to GLP-1 and GLP-2 production [35]. In humans, our group confirmed this relationship and proposed that *A. muciniphila* could be mediating the action of interferon-gamma on glucose metabolism [36]. Acetate production has also been directly correlated to the *A. muciniphila* abundance [37]. However, the effects of this SCFA in metabolism are still controversial [21,38,39].

Fatty Acids

In strict vegetarians, the fat content of the diet is mainly attributed to unsaturated fatty acids [14]. Also lacto-ovo vegetarians have a lower intake of saturated fat in comparison to omnivores [40]. Diverse proportions of fatty acids in diet can modify the gut microbiome composition. In fish oil—fed mice, higher abundance of *Akkermansia, Lactobacillus,* and *Bifidobacterium* was observed in comparison with lard-fed ones [41]. An increase in *Akkermansiaceae* family abundance was found in humans, supplemented with 4 g of mixed docosahexaenoicacid/eicosapentaenoicacid for 8 weeks [42]. Whether vegetarians with a high unsaturated fat intake would have increased proportions of these bacteria was not reported.

Chronic exposure to a high-fat diet has been associated with increased gram-negative bacteria abundance and intestinal permeability, apparently dependent of the reduced expression of tight-junction proteins [43,44]. LPS present in the outer membrane of these bacteria is recognized by toll-like receptors in immune cells and others, leading to an activation of inflammatory pathways that deteriorates insulin signaling [43]. The increase of LPS in circulation (metabolic endotoxemia) leads to insulin resistance, which represents the main pathophysiological mechanism of the metabolic syndrome.

In animals, the ability of a high-fat diet to increase acetate production resulted in activation of parasympathetic pathways that stimulates ghrelin secretion as well as β cell insulin secretion [21]. The combination of high ghrelin and insulin concentration was followed by increased appetite and food intake, favoring the development and/or exacerbation of obesity and comorbidities.

Proteins

An adequate vegetarian diet should include a good selection of plant products such as vegetables, grains, nuts, and seeds which are able to provide the same protein quality as diets that include meat [6]. A beneficial effect of vegetarian diet is linked to this high intake of proteins, also rich in fiber, magnesium, potassium, and folate [45]. Commonly, this contains soy and soy products, which seem to modify the gut microbiome composition at phyla level [46].

In omnivores, a major source of proteins is meat consumption that has been unfavorably associated with cardiometabolic endpoints [2]. This association is in part dependent of the effects of saturated fat content on cholesterol levels. The roles of its proteins, micronutrients, and other components are not been fully understood. A review [47] found that consumption of processed meats, but not red meats, is associated with higher incidence of cardiovascular disease and type 2 diabetes.

Phosphatidylcholine and L-carnitine originated from animal sources (eggs, milk, liver, red meat, poultry, shell fish, and fish) or dietary supplementation are metabolized by the microbiome resulting in the formation of trimethylamine (TMA) that is converted to trimethylamine N-oxide (TMAO) in the liver [48]. There is some evidence that high TMAO levels are predictive of increased cardiovascular risk [48–50], since this compound was related to the suppression of the reverse transport of cholesterol and formation of the foamy cell in arterial wall. This suggests that reduced ingestion of L-carnitine and choline by vegetarians could contribute to cardiovascular benefits [49]. Higher concentrations of TMAO were reported in omnivores. A possible mechanism involves a reduced capacity to synthesize TMA via gut microbial metabolism in vegetarians. Interestingly, even with a supplementation of L-carnitine or consumption of meat, long-term vegetarians (>1 year) have a reduced capacity to synthesize TMAO [49].

Phytochemicals

The role of phytochemicals in modulating the gut microbiome, particularly polyphenols, has been a current research interest [51,52]. These compounds are present in a vast range of fruits and vegetables as well as in cocoa, tea, and wine [53,54]. Health benefits attributed to plant-based diets could be at least in part due to their consumption [55] that is able to favorably modify the microbiome composition. In turn, the increase of beneficial bacteria and inhibition of potentially pathogenic ones is important to polyphenol metabolization into simpler phenolic compounds in order to be absorbed [52,56].

Cranberry extract, a potent polyphenol, when administered to mice, improved insulin tolerance, accompanied by elevation of *A. muciniphila* proportion [57]. Similar results were found using grapes. *A. muciniphila* showed to be the strongest biomarker of polyphenol supplementation and attenuated metabolic endotoxemia [56].

RELATIONSHIP OF DIETARY PATTERNS, GUT MICROBIOME, AND CHRONIC DISEASES

Mediterranean diet and dietary approaches to stop hypertension have been reported as the best regimens to reduce cardiovascular mortality. Some evidence has suggested that the benefits are mediated by a gut microbiome composition with lower inflammatory potential [58,59]. Many characteristics of these diets are adopted by vegetarians, since they stimulate the consumption of plant-based foods and include higher amount of whole grain foods, fruit, vegetables, and sources of unsaturated fat (olive oil and nuts) [5].

The association of plant-based diet with the occurrence of type 2 diabetes was examined in three cohort studies: the Nurses' Health Study (1984–2012), Nurses' Health Study 2 (1991–2011), and the Health Professionals Follow-Up Study (1986–2010). Plant-based diets, especially those including high-quality plant foods, were associated with substantially lower risk of developing diabetes, even adjusted for body mass index [60].

Results from the AHS-2 indicated that a vegetarian diet conferred a lower risk for cancer than other dietary habits [61]. Vegetarian habits (nonvegetarian, semi-vegetarian, pesco-vegetarian, lacto-ovo vegetarian, and vegan) were inversely associated with cardiovascular mortality, noncardiovascular mortality, noncancer mortality, renal mortality, and endocrine mortality, even after controlling for demographic and lifestyle confounders. The study found a 12% risk reduction of all-cause mortality in all vegetarians [13].

Vegetarians from North American and South Asian populations had lower probability of overweight/obesity, compared to nonvegetarians [62]. A stronger inverse association with central obesity was observed in Americans, compared with South Asian vegetarians. However, vegetarians in the American sample represented only 2.4% of the population, in contrast to 33% among South Asians.

Many beneficial effects associated with consumption of whole grains, vegetables, and fruits are at least in part mediated by the end products of microbial metabolism, including SCFAs [49,63–65].

Fiber-enriched diets such as the vegetarian ones are associated with the increased abundance of the genus *Prevotella* from the phylum Bacteroidetes [65–67]. Comparison of European children and vegetarian African children showed less Enterobacteriaceae (*Shigella* and *Escherichia coli*) and higher Bacteroidetes/Firmicutes ratio in the latter [65].

Short-term dietary intervention is able to alter gut microbiome composition. An elevation of *Alistipes, Bilophila,* and *Bacteroides* and a decrease in butyrate-producing bacteria (*Roseburia, E. rectale* and *R. bromii*) was noticed in an animal-based diet, indicating that the microbiome rapidly responds to dietary changes [68]. This statement was reinforced by findings in individuals with cardiometabolic disorders, in whom a strict vegetarian diet was used for 1 month [63]. An increase in the abundance of bacteria that use plant-derived polysaccharides as energy source, such as Lachnospiraceae and Ruminococcacea, in addition to a decrease of Enterobacteriaceae was observed, in parallel with improvement of the metabolic risk profile.

Our group deepened the investigation on the role of low-grade inflammation and insulin resistance for gut microbiome-mediated cardiometabolic risk profile in 268 Brazilians Adventists undergoing distinct dietary patterns [12]. As previously described, a lower risk of cardiometabolic diseases was observed in the strict vegetarians, especially when compared to the omnivores [2,13,60–62]. In agreement with the AHS-2 results [13,69], prevalence rates of obesity, hypertension, and prediabetes were lower among strict vegetarians. They also had less Firmicutes and more Bacteroidetes.

Moreover, among the Firmicutes, there was a predominance of butyrate-producing *Roseburia* and *Faecalibacterium*, which were previously associated with beneficial phenotypes. Our finding of lower LPS concentration in the strict vegetarian group is coherent with these effects. Lower inflammatory markers levels (C-reactive protein and tumor necrosis factor-alpha/interleukin 10 ratio) detected in the same group possibly could be due to antiinflammatory action of butyrate, inhibiting nuclear factor κB in colonic cells [26].

In the Bacteroidetes phylum, a higher proportion of *Prevotella* was confirmed, while a higher proportion of *Faecalibacterium* (Firmicutes phylum) was observed in lacto-ovo vegetarians. Dairy products and eggs might be substrates for these bacteria [70], which is in consonance with a previous report [66].

BACTERIAL ENTEROTYPES

Three enterotypes of bacterial patterns have been traditionally reported: *Bacteroides* (enterotype 1), *Prevotella* (enterotype 2), and *Ruminococcus* (enterotype 3). The *Prevotella* was strongly associated with a carbohydrate rich diet, whereas the *Bacteroides* enterotype with protein and animal fat, such as the Western diet [71]. Interestingly, this study described only two clusters, the *Bacteroides* enterotype fused with the less distinct *Ruminococcus* enterotype [71].

The presence of all three enterotypes in Brazilian Adventists, with distinct dietary habits (strict vegetarians, lacto-ovo vegetarians, and omnivores), was evaluated [72]. As expected, the frequency of strict vegetarians was greater in the *Prevotella* enterotype; however frequencies of other groups did not differ among the clusters. Whether the absence of animal food—derived saturated fatty acids could account for this result requires investigation.

Few studies have examined the association of enterotypes with cardiometabolic risk factors [73,74]. The finding of lower low-density lipoprotein cholesterol in *Prevotella* compared to *Bacteroides* enterotype suggested that certain bacteria might be important drivers of effects on lipid metabolism [72]. The fiber-rich diet consumed by vegetarians with the *Prevotella* enterotype could trigger beneficial effects at intestinal and systemic levels. This is in agreement with the favorable cardiometabolic risk profile described in Adventists, consumers of fruits and vegetables [72].

The association between diet and the gut microbiome composition has also been explored in the context of the pathogenesis of inflammatory bowel diseases (IBDs), since the bacterial composition is decisive for triggering intestinal inflammation [75]. The abundance of certain commensal bacteria is reduced in patients with IBD, especially those belonging to *Clostridium cluster XIVa* and *Clostridium cluster IV*, such as *Ruminococcus, Roseburia, Eubacterium,* and *Faecalibacterium,* known by their ability to produce butyrate [76]. Taking into consideration the antiinflammatory role of butyrate via the nuclear factor κB inhibition, this pathway is supposed to be activated in IBD [26]. Therefore, the importance of fiber-enriched diet was raised for individuals with IBD due to the expected modulation of the gut microbiome composition by inducing an increase in the butyrate-producing bacteria. It was reported that an intervention with a lacto-ovo vegetarian with fish once a week and meat once every 2 weeks was capable to prevent relapse in Crohn's disease [77]. This is in agreement with a systematic review that concluded that fiber-enriched diet with fruits and vegetables was associated with decreased risk while a high-fat and meat diet was associated with increased risk of IBD [78].

In summary, there is plenty of evidence from literature showing that consumption of vegetarian diets compared to omnivorous diet results in distinct gut microbial composition, particularly with increased beneficial bacteria abundances, which in turn influence processes involved in the pathogenesis of several chronic diseases. The recognition that metabolism of most dietary components depends on the gut microbiome has opened a window of opportunity to improve these diseases' prognosis using dietary strategies addressed to modify and sustain a favorable gut microbiome profile. Vegetarianism seems to represent a promising strategy to induce a beneficial profile at least for individuals affected by unhealthy diet-dependent diseases such as the cardiometabolic ones.

LIST OF ABBREVIATIONS

AHS-2 Adventist Health Study 2
CD Crohn's disease
CRP C-reactive protein
DASH dietary approaches to stop hypertension
DHA docosahexaenoicacid
EPA eicosapentaenoicacid
FFAR2 free fatty acids receptors 2
FFAR3 free fatty acids receptors 3
GLP-1 glucagon-like peptide 1
GLP-2 glucagon-like peptide 2
IBDs inflammatory bowel diseases
IL-10 interleukin 10
LDL-c low-density lipoprotein cholesterol
LPS lipopolysaccharides

PYY peptide YY
SCFA short-chain fatty acids
TMA trimethylamine
TMAO trimethylamine N-oxide
TNF-α tumor necrosis factor-alpha

GLOSSARY

Cardiometabolic disease Concerning both heart disease and metabolic disorders, such as diabetes.
DASH diet Dietary recommendations that promote reduction in or prevention of high blood pressure. Recommendations include increasing intake of fruits and vegetables and high-fiber, low-fat foods and reducing the intake of dietary sodium and high-fat foods.
Dietary pattern Quantities, proportions, variety, or combinations of different foods and beverages in diets, and the frequency with which they are habitually consumed.
Eating habits Behavioral responses or sequences associated with eating including modes of feeding, rhythmic patterns of eating, and time intervals.
Mediterranean diet A diet typical of the Mediterranean region characterized by a pattern high in fruits and vegetables, edible grain and bread, potatoes, poultry, beans, nuts, olive oil, and fish while low in red meat and dairy and moderate in alcohol consumption.
Plant-based diet Diet that emphasize legumes, whole grains, vegetables, fruits, nuts, and seeds and discourage most or all animal products.
Vegetarian diet The dietary practice of completely avoiding meat or fish products in the diet, consuming vegetables, cereals, and nuts. Some vegetarian diets called lacto-ovo also include milk and egg products.
Western diet A pattern of food consumption adopted mainly by the people of North America and Western Europe. It is mainly characterized by high intake of meat, processed grains, dietary sugars, dairy products, and dietary fats.

REFERENCES

[1] Le LT, Sabaté J. Beyond meatless, the health effects of vegan diets: findings from the adventist cohorts. Nutrients 2014;6(6):2131−47.
[2] Sabaté J, Wien M. A perspective on vegetarian dietary patterns and risk of metabolic syndrome. Br J Nutr 2015;113(S2):S136−43.
[3] Salas-Salvadó J, Bulló M, Babio N, Martínez-González MA, Ibarrola-Jurado N, Basora J, Estruch R, Covas MI, Corella D, Aós F, Ruiz-Gutiérrez V, Ros E. PREDIMED StudyInvestigators. Reduction in the incidence of type 2 diabetes with the Mediterranean diet. Diabetes Care 2011;34(1):14−9.
[4] Estruch R, Ros E, Salas-Salvadó J, Covas MI, Corella D, Arós F, Gómez-Gracia E, Ruiz-Gutiérrez V, Fiol M, Lapetra J, Lamuela-Raventos RM, Serra-Majem L, Pintó X, Basora J, Muñoz MA, Sorlí JV, Martínez JA, Martínez-González MA. PREDIMED StudyInvestigators. Primary prevention of cardiovascular disease with a Mediterranean diet. N Engl J Med 2013;368(14):1279−90.
[5] Esposito K, Maiorino MI, Bellastella G, Chiodini P, Panagiotakos D, Giugliano D. A journey into a Mediterranean diet and type 2 diabetes: a systematic review with meta-analyses. BMJ Open 2015;5(8):e008222.
[6] Cullum-Dugan D, Pawlak R. Position of the academy of nutrition and dietetics: vegetarian diets. J Acad Nutr Diet 2015;115(5):801−10.
[7] Appleby PN, Key TJ. The long-term health of vegetarians and vegans. Proc Nutr Soc 2016;75(3):287−93.
[8] Kahleova H, Levin S, Barnard N. Cardio-metabolic benefits of plant-based diets. Nutrients 2017;9(8):1−13.
[9] Grandl G, Wolfrum C. Hemostasis, endothelial stress, inflammation, and the metabolic syndrome. Semin Immunopathol December 5, 2017. https://doi.org/10.1007/s00281-017-0666-5.
[10] Calder PC, Ahluwalia N, Albers R, Bosco N, Bourdet-Sicard R, Haller D, Holgate ST, Jönsson LS, Latulippe ME, Marcos A, Moreines J, M'Rini C, Müller M, Pawelec G, van Neerven RJ, Watzl B, Zhao J. A consideration of biomarkers to be used for evaluation of inflammation in human nutritional studies. Br J Nutr 2013;109(S1):S1−34.
[11] Jung UJ, Choi MS. Obesity and its metabolic complications: the role of adipokines and the relationship between obesity, inflammation, insulin resistance, dyslipidemia and nonalcoholic fatty liver disease. Int J Mol Sci 2014;15(4):6184−223.
[12] Franco-De-Moraes AC, De Almeida-Pititto B, Da Rocha Fernandes G, Gomes EP, Da Costa Pereira A, Ferreira SRG. Worse inflammatory profile in omnivores than in vegetarians associates with the gut microbiota composition. Diabetol Metab Syndr 2017;9(1):1−8.
[13] Orlich MJ, Singh PN, Sabaté J, Jaceldo-Siegl K, Fan J, Knutsen S, Beeson WL, Fraser GE. Vegetarian dietary patterns and mortality in adventist health study 2. JAMA Intern Med 2013;173(13):1230−8.
[14] Clarys P, Deliens T, Huybrechts I, Deriemaeker P, Vanaelst B, De Keyzer W, Hebbelinck M, Mullie P. Comparison of nutritional quality of the vegan, vegetarian, semi-vegetarian, pesco-vegetarian and omnivorous diet. Nutrients 2014;6(3):1318−32.
[15] Glick-Bauer M, Yeh M-C. The health advantage of a vegan diet: exploring the gut microbiota connection. Nutrients 2014;6(11):4822−38.
[16] Chassard C, Lacroix C. Carbohydrates and the human gut microbiota. Curr Opin Clin Nutr Metab Care 2013;16(4):453−60.
[17] Singh RK,Chang H-W, Yan D, Lee KM, Ucmak D, Wong K, Abrouk M, Farahnik B, Nakamura M, Zhu TH, Bhutani T, Liao W. Influence of diet on the gut microbiome and implications for human health. J Transl Med 2017;15:73.
[18] Wong JM, de Souza R, Kendall CW, Emam A, Jenkins DJ. Colonic health: fermentation and short chain fatty acids. J Clin Gastroenterol 2006;40(3):235−43.
[19] Roberfroid M, Gibson GR, Hoyles L, McCartney AL, Rastall R, Rowland I, Wolvers D, Watzl B, Szajewska H, Stahl B, Guarner F, Respondek F, Whelan K, Coxam V, Davicco MJ, Léotoing L, Wittrant Y, Delzenne NM, Cani PD, Neyrinck AM, Meheust A. Prebiotic effects: metabolic and health benefits. Br J Nutr 2010;104(Suppl. 2):S1−63.

[20] Morrison DJ, Preston T. Formation of short chain fatty acids by the gut microbiota and their impact on human metabolism. Gut Microb 2016;7(3):189−200.

[21] Perry RJ, Peng L, Barry NA, Cline GW, Zhang D, Cardone RL, Petersen KF, Kibbey RG, Goodman AL, Shulman GI. Acetate mediates a microbiome-brain-β-cell axis to promote metabolic syndrome. Nature 2016;534(7606):213−7.

[22] Yan H, Ajuwon KM. Butyrate modifies intestinal barrier function in IPEC-J2 cells through a selective upregulation of tight junction proteins and activation of the Akt signaling pathway. PLoS One 2017;12(6):1−20.

[23] Van den Abbeele P, Belzer C, Goossens M, Kleerebezem M, De Vos WM, Thas O, De Weirdt R, Kerckhof FM, Van de Wiele T. Butyrate-producing Clostridium cluster XIVa species specifically colonize mucins in an in vitro gut model. ISME J 2013;7(5):949−61.

[24] Tamanai-Shacoori Z, Smida I, Bousarghin L, Loreal O, Meuric V, Fong SB, Bonnaure-Mallet M, Jolivet-Gougeon A. Roseburia spp.: a marker of health? Future Microbiol 2017;12(2):157−70.

[25] Hansen TH, Gøbel RJ, Hansen T, Pedersen O. The gut microbiome in cardio-metabolic health. Genome Med 2015;7(1):33.

[26] Canani RB, Costanzo MD, Leone L, Pedata M, Meli R, Calignano A. Poten- tial beneficial effects of butyrate in intestinal and extraintestinaldiseases. World J Gastroenterol 2011;17(12):1519−28.

[27] Moreno-Indias I, Cardona F, Tinahones FJ, Queipo-Ortuño MI. Impact of the gut microbiota on the development of obesity and type 2 diabetes mellitus. Front Microbiol 2014;5:190.

[28] Cani PD, Lecourt E, Dewulf EM, Sohet FM, Pachikian BD, Naslain D, De Backer F, Neyrinck AM, Delzenne NM. Gut microbiota fermentation of prebiotics increases satietogenic and incretin gut peptide production with consequences for appetite sensation and glucose response after a meal. Am J Clin Nutr 2009;90(5):1236−43.

[29] Cani PD, Possemiers S, Van de Wiele T, Guiot Y, Everard A, Rottier O, Geurts L, Naslain D, Neyrinck A, Lambert DM, Muccioli GG, Delzenne NM. Changes in gut microbiota control inflammation in obese mice through a mechanism involving GLP-2-driven improvement of gut permeability. Gut 2009;58(8):1091−103.

[30] Liszt K, Zwielehner J, Handschur M, Hippe B, Thaler R, Haslberger AG. Characterization of bacteria, clostridia and Bacteroides in faeces of vegetarians using qPCR and PCR-DGGE fingerprinting. Ann Nutr Metab 2009;54(4):253−7.

[31] Holzer P, Reichmann F, Farzi A, Neuropeptide Y. Peptide YY and pancreatic polypeptide in the gut-brain axis. Neuropeptides 2012;46(6):261−74.

[32] Miyamoto J, Hasegawa S, Kasubuchi M, Ichimura A, Nakajima A, Kimura I. Nutritional signaling via free fatty acid receptors. Int J Mol Sci 2016;17(4):450.

[33] Zimmer J1, Lange B, Frick JS, Sauer H, Zimmermann K, Schwiertz A, Rusch K, Klosterhalfen S, Enck P. A vegan or vegetarian diet substantially alters the human colonic faecal microbiota. Eur J Clin Nutr 2012;66(1):53−60.

[34] Everard A, Belzer C, Geurts L, Ouwerkerk JP, Druart C, Bindels LB, Guiot Y, Derrien M, Muccioli GG, Delzenne NM, de Vos WM, Cani PD. Cross-talk between *Akkermansia muciniphila* and intestinal epithelium controls diet-induced obesity. Proc Natl Acad Sci USA 2013;110(22):9066−71.

[35] Everard A, Lazarevic V, Derrien M, Girard M, Muccioli GG, Neyrinck AM, Possemiers S, Van Holle A, François P, de Vos WM, Delzenne NM, Schrenzel J, Cani PD. Responses of gut microbiota and glucose and lipid metabolism to prebiotics in genetic obese and diet-induced leptin-resistant mice. Diabetes 2011;60(11):2775−86.

[36] Greer RL, Dong X, Moraes AC, Zielke RA, Fernandes GR, Peremyslova E, Vasquez-Perez S, Schoenborn AA, Gomes EP, Pereira AC, Ferreira SR, Yao M, Fuss IJ, Strober W, Sikora AE, Taylor GA, Gulati AS, Morgun A, Shulzhenko N. Akkermansia muciniphila mediates negative effects of IFNγ on glucose metabolism. Nat Commun 2016;7:1−13.

[37] Dao MC, Everard A, Aron-Wisnewsky J, Sokolovska N, Prifti E, Verger EO, Kayser BD, Levenez F, Chilloux J, Hoyles L, MICRO-Obes Consortium, Dumas ME, Rizkalla SW, Doré J, Cani PD, Clément K. Akkermansia muciniphila and improved metabolic health during a dietary intervention in obesity: relationship with gut microbiome richness and ecology. Gut 2016;65(3):426−36.

[38] Ríos-Covián D, Ruas-Madiedo P, Margolles A, Gueimonde M, De los Reyes-Gavilán CG, Salazar N. Intestinal short chain fatty acids and their link with diet and human health. Front Microbiol 2016;7:185.

[39] Frost G, Sleeth ML, Sahuri-Arisoylu M, Lizarbe B, Cerdan S, Brody L, Anastasovska J, Ghourab S, Hankir M, Zhang S, Carling D, Swann JR, Gibson G, Viardot A, Morrison D, Louise Thomas E, Bell JD. The short-chain fatty acid acetate reduces appetite via a central homeostatic mechanism. Nat Commun 2014;5:3611.

[40] Rizzo NS, Jaceldo-Siegl K, Sabate J, Fraser GE. Nutrient profiles of vegetarian and nonvegetarian dietary patterns. J Acad Nutr Diet 2013;113(12):1610−9.

[41] Caesar R, Tremaroli V, Kovatcheva-Datchary P, Cani PD, Bäckhed F. Crosstalk between gut microbiota and dietary lipids aggravates WAT inflammation through TLR signaling. Cell Metab 2015;22(4):658−68.

[42] Costantini L, Molinari R, Farinon B, Merendino N. Impact of Omega-3 fatty acids on the gut microbiota. Int J Mol Sci 2017;18(12):2645.

[43] Cani PD, Amar J, Iglesias MA, Poggi M, Knauf C, Bastelica D, Neyrinck AM, Fava F, Tuohy KM, Chabo C, Waget A, Delmée E, Cousin B, Sulpice T, Chamontin B, Ferrières J, Tanti JF, Gibson GR, Casteilla L, Delzenne NM, Alessi MC, Burcelin R. Metabolic endotoxemia initiates obesity and insulin resistance. Diabetes 2007;56:1761−72.

[44] Cani PD, Everard A. Talking microbes: when gut bacteria interact with diet and host organs. Mol Nutr Food Res 2016;60(1):58−66.

[45] Craig WJ. Health effects of vegan diets. Am J Clin Nutr 2009;89(5):1627S−33S.

[46] Huang H, Krishnan HB, Pham Q, Yu LL, Wang TTY. Soy and gut microbiota: interaction and implication for human health. J Agric Food Chem 2016;64(46):8695−709.

[47] Micha R, Wallace SK, Mozaffarian D. Red and processed meat consumption and risk of incident coronary heart disease, stroke, and diabetes: a systematic review and meta-analysis. Circulation 2010;121(21):2271−83.

[48] Wang Z, Klipfell E, Bennett BJ, Koeth R, Levison BS, Dugar B, et al. Gut flora metabolism of phophatidylcholine promotes cardivovascular disease. Nature 2011;472(7341):57−63.

[49] Koeth RA, Wang Z, Levison BS, Buffa JA, Org E, Sheehy BT, Britt EB, Fu X, Wu Y, Li L, Smith JD, DiDonato JA, Chen J, Li H, Wu GD, Lewis JD, Warrier M, Brown JM, Krauss RM, Tang WH, Bushman FD, Lusis AJ, Hazen SL. Intestinal microbiota metabolism of L-carnitine, a nutrient in red meat, promotes atherosclerosis. Nat Med 2013;19(5):576−85.

[50] Yin J, Liao SX, He Y, Wang S, Xia GH, Liu FT, Zhu JJ, You C, Chen Q, Zhou L, Pan SY, Zhou HW. Dysbiosis of gut microbiota with reduced trimethylamine-N-oxide level in patients with large-artery atherosclerotic stroke or transient ischemic attack. J Am Heart Assoc 2015;4(11):e002699.

[51] Neyrinck AM, Van Hée VF, Bindels LB, De Backer F, Cani PD, Delzenne NM. Polyphenol-rich extract of pomegranate peel alleviates tissue inflammation and hypercholesterolaemia in high-fat diet-induced obese mice: potential implication of the gut microbiota. Br J Nutr 2013;109(5):802−9.

[52] Braune A, Blaut M. Bacterial species involved in the conversion of dietary flavonoids in the human gut. Gut Microb 2016;7(3):216−34.

[53] Dueñas M, Muñoz-González I, Cueva C, Jiménez-Girón A, Sánchez-Patán F, Santos-Buelga C, Moreno-Arribas MV, Bartolomé B. A survey of modulation of gut microbiota by dietary polyphenols. BioMed Res Int 2015;2015. 850902.

[54] Sheflin AM, Melby CL, Carbonero F, Weir TL. Linking dietary patterns with gut microbial composition and function. Gut Microb 2017;8(2):113−29.

[55] Meyer KA, Bennet BJ. Diet and gut microbial function in metabolic and cardiovascular disease risk. Curr Diab Rep 2016;16(10):93.

[56] Roopchand DE, Carmody RN, Kuhn P, Moskal K, Rojas-Silva P, Turnbaugh PJ, Raskin I. Dietary polyphenols promote growth of the gut bacterium Akkermansia muciniphila and attenuate high-fat diet-induced metabolic syndrome. Diabetes 2015;64(8):2847−58.

[57] Anhê FF, Roy D, Pilon G, Dudonné S, Matamoros S, Varin TV, Garofalo C, Moine Q, Desjardins Y, Levy E, Marette A. Apolyphenol-rich cranberry extract protects from diet-induced obesity, insulin resistance and intestinal inflammation in association with increased *Akkermansia spp.* population in the gut microbiota of mice. Gut 2015;64(6):872−83.

[58] De Filippis F, Pellegrini N, Vannini L, Jeffery IB, La Storia A, Laghi L, Serrazanetti DI, Di Cagno R, Ferrocino I, Lazzi C, Turroni S, Cocolin L, Brigidi P, Neviani E, Gobbetti M, O'Toole PW, Ercolini D. High-level adherence to a Mediterranean diet beneficially impacts the gut microbiota and associated metabolome. Gut 2016;65(11):1812−21.

[59] Haro C, Montes-Borrego M, Rangel-Zúñiga OA, Alcalá-Díaz JF, Gómez-Delgado F, Pérez-Martínez P, Delgado-Lista J, Quintana-Navarro GM, Tinahones FJ, Landa BB, López-Miranda J, Camargo A, Pérez-Jiménez F. Two healthy diets modulate gut microbial community improving insulin sensitivity in a human obese population. J Clin Endocrinol Metab 2016;101(1):233−42.

[60] Satija A, Bhupathiraju SN, Rimm EB, Spiegelman D, Chiuve SE, Borgi L, Willett WC, Manson JE, Sun Q, Hu FB. Plant-based dietary patterns and incidence of type 2 diabetes in US men and women: results from three prospective cohort studies. PLoS Med 2016;13(6):e1002039.

[61] Tantamango-Bartley Y, Jaceldo-Siegl K, Fan J, Fraser G. Vegetarian diets and the incidence of cancer in a low-risk population. Cancer Epidemiol Biomark Prev 2013;22(2):286−94.

[62] Jaacks LM, Kapoor D, Singh K, Narayan KM, Ali MK, Kadir MM, Mohan V, Tandon N, Prabhakaran D. Vegetarianism and cardiometabolic disease risk factors: differences between South Asian and US adults. Nutrition 2016;32(9):975−84.

[63] Kim MS, Hwang SS, Park EJ, Bae JW. Strict vegetarian diet improves the risk factors associated with metabolic diseases by modulating gut microbiota and reducing intestinal inflammation. Environ Microbiol Rep 2013;5(5):765−75.

[64] Romano KA, Vivas EI, Amador-Noguez D, Rey FE. Intestinal microbiota composition modulates choline bioavailability from diet and accumulation of the proatherogenic metabolite trimethylamine-N-oxide. mBio 2015;6(2):e02481.

[65] De Filippo C, Cavalieri D, Di Paola M, Ramazzotti M, Poullet JB, Massart S, Collini S, Pieraccini G, Lionetti P. Impact of diet in shaping gut microbiota revealed by a comparative study in children from Europe and rural Africa. Proc Natl Acad Sci USA 2010;107(33):14691−6.

[66] Ferrocino I, Di Cagno R, De Angelis M, Turroni S, Vannini L, Bancalari E, Rantsiou K, Cardinali G, Neviani E, Cocolin L. Fecal microbiota in healthy subjects following omnivore, vegetarian and vegan diets: culturable populations and rRNA DGGE profiling. PLoS One 2015;10(6):e0128669.

[67] Ruengsomwong S, Korenori Y, Sakamoto N, Wannissorn B, Nakayama J, Nitisinprasert S. Senior Thai fecal microbiota comparison between vegetarians and non-vegetarians using PCR-DGGE and real-time PCR. J Microbiol Biotechnol 2014;24(8):1026−33.

[68] David LA, Maurice CF, Carmody RN, Gootenberg DB, Button JE, Wolfe BE, Ling AV, Devlin AS, Varma Y, Fischbach MA, Biddinger SB, Dutton RJ, Turnbaugh PJ. Diet rapidly and reproducibly alters the human gut microbiome. Nature 2014;505(7484):559−63.

[69] Pettersen BJ, Anousheh R, Fan J, Jaceldo-Siegl K, Fraser GE. Vegetarian diets and blood pressure among white subjects: results from the Adventist Health Study-2 (AHS-2). Public Health Nutr 2012;15(10):1909−16.

[70] Khan MT, Duncan SH, Stams AJM, van Dijl JM, Flint HJ, Harmsen HJM. The gut anaerobe Faecalibacteriumprausnitzii uses an extracellular electron shuttle to grow at oxic−anoxic interphases. ISME J 2012;6(8):1578−85.

[71] Wu GD, Chen J, Hoffmann C, Bittinger K, Chen YY, Keilbaugh SA, Bewtra M, Knights D, Walters WA, Knight R, Sinha R, Gilroy E, Gupta K, Baldassano R, Nessel L, Li H, Bushman FD, Lewis JD. Linking long-term dietary patterns with gut microbial enterotypes. Science 2011;334(6052):105−8.

[72] de Moraes AC, Fernandes GR, da Silva IT, Almeida-Pititto B, Gomes EP, Pereira AD, Ferreira SR. Enterotype may drive the dietary-associated cardiometabolic risk factors. Front Cell Infect Microbiol 2017;7:47.

[73] Zupancic ML, Cantarel BL, Liu Z, Drabek EF, Ryan KA, Cirimotich S, Jones C, Knight R, Walters WA, Knights D, Mongodin EF, Horenstein RB, Mitchell BD, Steinle N, Snitker S, Shuldiner AR, Fraser CM. Analysis of the gut microbiota in the old order Amish and its relation to the metabolic syndrome. PLoS One 2012;7(8):e43052.

[74] Lim MY, Rho M, Song YM, Lee K, Sung J, Ko G. Stability of gut enterotypes in Korean monozygotic twins and their association with biomarkers and diet. Sci Rep 2014;4:7348.

[75] Albenberg LG, Lewis JD, Wu GD. Food and the gut microbiota in IBD: a critical connection. Curr Opin Gastroenterol 2012;28(4). https://doi.org/10.1097/MOG.0b013e328354586f.

[76] Nagao-Kitamoto H, Kamada N. Host-microbial cross-talk in inflammatory bowel disease. Immune Netw 2017;17(1):1—12.

[77] Chiba M, Abe T, Tsuda H, Sugawara T, Tsuda S, Tozawa H, Fujiwara K, Imai H. Lifestyle-related disease in Crohn's disease: relapse prevention by a semi-vegetarian diet. World J Gastroenterol 2010;16(20):2484—95.

[78] Hou JK, Abraham B, El-Serag H. Dietary intake and risk of developing inflammatory bowel disease: a systematic review of the literature. Am J Gastroenterol 2011;106(4):563—73.

FURTHER READING

[1] Holscher HD. Dietary fiber and prebiotics and the gastrointestinal microbiota. Gut Microb 2017;8(2):172—84.

[2] Kong LC, Holmes BA, Cotillard A, Habi-Rachedi F, Brazeilles R, Gougis S, Gausserès N, Cani PD, Fellahi S, Bastard JP, Kennedy SP, Doré J, Ehrlich SD, Zucker JD, Rizkalla SW, Clément K. Dietary patterns differently associate with inflammation and gut microbiota in overweight and obese subjects. PLoS One 2014;9(10):e109434.

[3] Bäckhed F, Ding H, Wang T, Hooper LV, Koh GY, Nagy A, Semenkovich CF, Gordon JI. The gut microbiota as an environmental factor that regulates fat storage. Proc Natl Acad Sci USA 2004;101(44):15718—23.

[4] Ley RE, Backed F, Turnbaugh P, Lozupone CA, Knight RD, Gordon JI. Obesity alters gut microbial ecology. Proc Natl Acad Sci USA 2005;102(31):11070—5.

[5] Woting A, Blaut M. The intestinal microbiota in metabolic disease. Nutrients 2016;8(4):202.

[6] Benjamin EJ, Blaha MJ, Chiuve SE, Cushman M, Das SR, Deo R, de Ferranti SD, Floyd J, Fornage M, Gillespie C, Isasi CR, Jiménez MC, Jordan LC, Judd SE, Lackland D, Lichtman JH, Lisabeth L, Liu S, Longenecker CT, Mackey RH, Matsushita K, Mozaffarian D, Mussolino ME, Nasir K, Neumar RW, Palaniappan L, Pandey DK, Thiagarajan RR, Reeves MJ, Ritchey M, Rodriguez CJ, Roth GA, Rosamond WD, Sasson C, Towfighi A, Tsao CW, Turner MB, Virani SS, Voeks JH, Willey JZ, Wilkins JT, Wu JHY, Alger HM, Wong SS, Muntner P. Heart disease and stroke statistics—2017 update. Circulation 2017;135(10):e146—603.

[7] Leon BM, Maddox TM. Diabetes and cardiovascular disease: epidemiology, biological mechanisms, treatment recommendations and future research. World J Diabetes 2015;6(13):1246—58.

[8] Arumugam M, Raes J, Pelletier E, Le Paslier D, Yamada T, Mende DR, Fernandes GR, Tap J, Bruls T, Batto JM, Bertalan M, Borruel N, Casellas F, Fernandez L, Gautier L, Hansen T, Hattori M, Hayashi T, Kleerebezem M, Kurokawa K, Leclerc M, Levenez F, Manichanh C, Nielsen HB, Nielsen T, Pons N, Poulain J, Qin J, Sicheritz-Ponten T, Tims S, Torrents D, Ugarte E, Zoetendal EG, Wang J, Guarner F, Pedersen O, de Vos WM, Brunak S, Doré J, MetaHIT Consortium, Antolín M, Artiguenave F, Blottiere HM, Almeida M, Brechot C, Cara C, Chervaux C, Cultrone A, Delorme C, Denariaz G, Dervyn R, Foerstner KU, Friss C, van de Guchte M, Guedon E, Haimet F, Huber W, van Hylckama-Vlieg J, Jamet A, Juste C, Kaci G, Knol J, Lakhdari O, Layec S, Le Roux K, Maguin E, Mérieux A, MeloMinardi R, M'rini C, Muller J, Oozeer R, Parkhill J, Renault P, Rescigno M, Sanchez N, Sunagawa S, Torrejon A, Turner K, Vandemeulebrouck G, Varela E, Winogradsky Y, Zeller G, Weissenbach J, Ehrlich SD, Bork P. Enterotypes of the human gut microbiome. Nature 2011;473(7346):174—80.

Chapter 42

Metformin: A Candidate Drug to Control the Epidemic of Diabetes and Obesity by Way of Gut Microbiome Modification

Kunal Maniar[1], Vandana Singh[2], Deepak Kumar[2], Amal Moideen[1], Rajasri Bhattacharyya[2] and Dibyajyoti Banerjee[2]

[1]*Department of Pharmacology, Postgraduate Institute of Medical Education and Research, Chandigarh, India;* [2]*Department of Experimental Medicine and Biotechnology, Postgraduate Institute of Medical Education and Research, Chandigarh, India*

MULTIFACTORIAL PATHOGENESIS OF OBESITY

From a biological point of view, obesity is a unique inflammatory state, differing from the classical modalities. It is low grade, chronic in nature, mediated primarily by metabolic cells (e.g., adipocytes, hepatocytes), which result in a distinctive immune response, and is associated with a low basal metabolic rate [1].

The genetic predisposition to obesity is currently modeled on two different theories. (1) "Thrifty gene" hypothesis: frequent famines and natural disasters faced by our ancestors and selected genes for efficient storage and acquisition of nutrients from food [2]. (2)" Random drift" hypothesis: disappearance of predators, or a stable supply, coupled with consistent random genetic mutations, causing a drift, relaxed the upper limits of body weight [3]. Monogenic forms of obesity, though severe, are rare [4].

Adipose tissue in lean individuals represents a balanced state of immune cells and proinflammatory and antiin-flammatory mediators, all of which are essential in maintaining the metabolic homeostasis, insulin signaling, and endocrine functions, and preserve standard adipose storage [1]. Important mediators representing the antiinflammatory milieu are cytokines (IL-4, IL-5, IL-10, IL-13, IL-25, and IL-33), immune cells (eosinophils, regulatory T cells, M2 macrophages, and T helper type 2 cells), and proteins (adiponectin and secreted frizzled-related protein 5).

Adipose tissue in obese individuals is characterized not only by an increase in the number and size of adipocytes but also by infiltration of mononuclear cells, mitochondrial dysfunction, cell death, and rarefaction of vascular and neural structures [5]. There is a shift in the composition of immune cells (T helper 1 cells, CD8+ T lymphocytes, M1 macrophages), proinflammatory cytokines (IL-6, IL-1β, TNF-α, IFN-Υ), and hormones (leptin) [1].

Inflammation, Obesity, and Insulin Resistance

The most important implication of the proinflammatory state in the adipose tissue is on insulin signaling pathways, leading to the development of insulin resistance commonly found in obesity. Proinflammatory cytokines (TNF-α, IL-6) and nutrients (e.g., saturated fatty acids) activate numerous protein kinases, such as inhibitor of κ kinase, c-Jun N-terminal kinase (JNK), and mechanistic target of rapamycin (mTOR), which trigger the insulin receptor substrate inhibition, and modulate the expression of transcriptional proteins such as NF-κB, and AP-1 (activator protein 1), to increase the proinflammatory cytokine expression [6−8]. All this results in the formation of a positive feedback loop in the adipose tissue, which maintains the chronic inflammatory characteristic of obesity [9,10].

Another significant effect of proinflammatory cytokines is the downregulation of peroxisome proliferator-activated receptor gamma (PPAR-Υ), a nuclear receptor which is highly expressed in adipose tissue, and acts beneficially at the levels of inflammation (reduction of proinflammatory mediator secretion) and insulin resistance [11]. The fact that immune cells play

Microbiome and Metabolome in Diagnosis, Therapy, and other Strategic Applications. https://doi.org/10.1016/B978-0-12-815249-2.00042-7

an important role in the pathogenesis of obesity is supported by numerous animal studies, in which researchers deleted different immune cell lineages using different techniques and studied the effect on obesity and insulin signaling [12–14].

Insulin Signaling in Hepatocytes

Also in the liver, there is disruption of insulin signaling due to nutritional overload or proinflammatory cytokines. The NF-κB pathway appears to be crucial for the same phenomenon, as well as for regulation of gluconeogenesis [15,16]. The liver, in the obese state, differs from adipose tissue in one crucial aspect, as different liver cells undergo inflammation themselves at the onset of obesity (macrophage infiltration is absent at onset) [1]. It is known that inflammation leads to hepatic lipid accumulation, but the mechanistic insight of the process is lacking.

Other Organs and Tissues

The skeletal muscle in obesity exhibits features of insulin resistance, due to the release of several proinflammatory cytokines (IL-6, TNF-α, retinol binding protein), from peripheral tissues such as adipose tissue and liver [6,17]. Whether muscle tissue releases inflammatory mediators, by itself, is debatable [18].

In the pancreas, insulin resistance is observed in parallel with infiltration of macrophages [19]. While IL-1β—NF-κB, IFN-Υ, and JNK are considered necessary for β-cell apoptosis; their role in the context of obesity is yet not known [1].

The gastrointestinal tract is the source of hormones such as ghrelin, glucagon-like peptide 1 ((GLP-1), peptide YY (PYY)), neurotransmitters (serotonin (5-HT3), cholecystokinin (CCK)), and several other yet unexplored factors that act locally, peripherally and centrally, to influence the state of obesity [20].

Influence of brain in controlling appetite and energy balance should not be overlooked [21]. The hypothalamus is an important regulator of appetite. The neuropeptide Y and Agouti-related protein staining cells of the arcuate nucleus (ARC) are involved in appetite stimulation, whereas the proopiomelanocortin and cocaine- and amphetamine-regulated transcript staining cells of ARC are involved in appetite suppression. These groups of cells influence each other, as well as are influenced by numerous proteins of the body.

GUT MICROBIOME: CAUSE OR EFFECT OF OBESITY?

Early research identified that obese animals and humans had a reduced Bacteroidetes-to-Firmicutes ratio (BFR) and a reduced diversity [22,23]. Transferring the microbiome from either transgenic obese or high-fat diet (HFD) induced animals, to their matched germ-free (GF) counterparts, showed that the GF controls mirrored their donors [22,24]. In a recent study, cohousing of lean and obese mice, enabling microbiota exchange via natural coprophagy, it was seen that the obese mice acquired the phenotypic signature of lean mice [25]. A reduced diversity of the microbiome is associated with an overweight/obese phenotype of individuals [26], and it is documented that fecal microbial transplantation (FMT) from lean to obese humans can reduce some markers of insulin resistance [27,28].

Divergent Results

Yet BFR as a possible taxonomic identity of the obese phenotype has been criticized, and it was shown that using the subjects of the Human Microbiome Project no significant associations were found between BFR and obesity. No alternative taxonomic signature was ascertained, diversity of microbiome and obesity was not correlated, and when the authors reanalyzed the data from early studies, they noted that the BFR variation was greater interstudy rather than intrastudy [29].

Interestingly, a recently published genome-wide association study of a large sample of twins showed that the obese phenotype was associated with a reduced microbial diversity, and an increase in Firmicutes phylum was correlated with obesity [30]. The authors also could partially replicate the findings from other population cohorts. One explanation could be that the obese microbiome is a consequence of several host (diet, genetics, age, hormones) and environmental factors (stress, exercise, jet lag, social status), which could act in a unidirectional or bidirectional manner [30,31].

Dietary Influences

Mice on HFD had a reduction in Bacteroidetes and an increase in Firmicutes and Proteobacteria phyla, independently of obesity [32]. A prospective analysis of fecal samples of a single child, from birth upto 2.5 years, documented the microbiome with changes in diet [33]. Firmicutes phylum dominated during the breastfeeding period, whereas the introduction of solid foods was associated with an increase in Bacteroidetes and Proteobacteria.

Children consuming high-fiber diet showed a higher number of bacteria belonging to Bacteroidetes, *Prevotella*, and *Xylanibacter*, whereas those on a westernized HFD and high-carb diet showed a dominance of Firmicutes [34]. On switching the diets of lean individuals to a calorie-rich diet, stool energy loss was significantly reduced, indicating that the microbiome can change energy harvest based on nutrient availability and adding evidence that the microbiome of obese individuals has an increased energy harvesting capacity [35].

Host Genes and Microbiome Composition

Animal models and human studies have shown that the microbiome can be heritably colonized [30,36,37]. Specific genes (FHIT, ELAV4, TDRG1), were found to be significantly associated with obesity, although exact biological links remain to be determined [30]. The obese microbiome and circadian rhythm appear to share a bidirectional relationship, akin to that between diet and microbiome. HFD-induced changes in gut microbiome (increased number of Firmicutes, Actinobacteria, and Mollicutes) were heritable and associated with a change in hepatic circadian clock, and this was predominantly mediated by PPAR-Υ [38]. FMT from a jet-lagged individual into GF mice resulted in the development of insulin resistance and obesity [39].

The Gut Metabolome in Obesity: Role of Short Chain Fatty Acids

The species specificity of short-chain fatty acids (SCFA) increases in the order of acetate, propionate, and butyrate. Butyrate production is predominantly limited to *Ruminococcus* spp., *Faecalibacterium* spp., and *Eubacteria* spp. [40]. *Akkermansia* spp. produces butyrate and propionate, whereas acetate is generated by multiple families of bacteria. In obese individuals, there is an increase in nonbutyrate-producing bacteria such as *E. coli*. Butyrate-producing taxa were depleted in obese and type 2 diabetes mellitus (T2DM) subjects [41,42].

Contradictory evidence, showing an increase in butyrate-producing taxa (*Faecalibacterium* spp., *Ruminococcus* spp.) in obese subjects, also exists [43,44]. Several authors suggest that fecal SCFA concentrations are significantly higher in obese individuals, despite similar SCFA absorption capacities of obese and lean subjects, consistent with an increase in SCFA production, including butyrate [45,46]. This seems counterintuitive, given that SCFA administration prevents the development of obesity, and despite low SCFA levels, diet-induced obesity does not occur in GF murine animals [47,48].

SCFAs are absorbed in exchange for bicarbonate, thus affecting the luminal pH of the gut, and subsequently the growth of microbiome. For instance, butyrate-producing taxa decrease from the proximal to distal colon [49]. Butyrate is primarily utilized as fuel by colonocytes and shows poor absorption. SCFAs primarily exert their action by G-protein coupled receptors (GPCR). Signaling is involved in gut membrane integrity, glucose and lipid metabolism, appetite regulation, and immune function [40].

Host Microbiome Cross Talk

Two important pathways are highlighted in the literature, mediated principally via Adenosine monophosphate-activated protein kinase (AMPK), and fasting-induced adipose factor (Fiaf)-LPL is a physiological lipoprotein lipase (LPL) antagonist. It was found that GF Fiaf-deficient mice were prone to HFD-induced obesity, and GF mice had increased AMPK expression [50].

Immunoinflammatory Activation

Release of local and peripheral cytokines, in response to the invasion of bacterial components such as LPS (aka metabolic endotoxemia) [51,52], and toll like receptor responses, cannot be overlooked in obesity pathogenesis [53], as occurs in other local and systemic inflammatory diseases.

MICROBIOME MODULATION AND CONTROL OF OBESITY

Microbiome modulation to control obesity is to some extent achieved by (1) FMT, (2) bariatric surgery, (3) prebiotics and probiotics, (4) diet. In FMT, animal and human studies have shown promise [27,54]; however, the stability of the transplanted microbiome remains to be ascertained in long-term studies. One such study, in which participants were followed up over a period of 3 months, showed that transplanted microbial communities were resilient. However, the clinical outcome was not [55,56].

Bariatric interventions are relevant to control obesity and T2DM [57], and the gut microbiome changes composition and richness postbariatric interventions. The changes vary depending on the procedure [58]. There is an increase in particular butyrate-producing taxa such as *Akkermansia muciniphila* and *Faecalibacterium prausnitzii* [59,60]. Secondary bile acid generation, acting via Farnesoid X and GPCR, should be kept in mind [56].

Prebiotic interventions are associated with an increase in SCFA-producing bacteria, and release of anorexigenic hormones such as GLP-1 [61]. A trial exploring the use of *Akkermansia muciniphila* as a probiotic for metabolic syndrome is underway [62−64]. In turn, the MICRO-Obes study showed that successful dieting was associated with improvement of bacterial richness [65].

METFORMIN AND GUT MICROBIOME

Metformin is a dimethylguanide, part of the biguanide group, some of which are already used as antimicrobials. It is widely prescribed as an antidiabetic, and our group recently proposed that it may also show antibiotic action [66]. The microbiome of metformin users exhibits an increase in *Subdoligranulum* genus, *Akkermansia* genus, *Escherichia* spp. and a decrease in *Intestinibacter* spp. [42]. In treatment-naïve newly diagnosed diabetic individuals, metformin was associated with an increase in *Escherichia*, Firmicutes, and butyrate-producing *Akkermansia* spp. and *Bifidobacterium* spp. [67−70].

Obesity is commonly associated with hyperinsulinemia, and our group has reported that supraphysiological concentrations of insulin may lead to proliferation of nonbutyrate producing taxa, such as *E. coli* and *S. aureus* [71]. *E. coli* seems to be more abundant in obesity, and also pregnancy-associated weight gain is correlated with this species [72,73]. In vitro, metformin inhibits *E. coli* growth [74], whereas another study has reported no change, albeit in different experimental conditions [67]. We recently hypothesized that metformin may exert its action on obese prediabetic individuals, by inhibiting nonbutyrate producing taxa and consequently causing an increase in butyrate-producing taxa in gut microbiome [75].

Metformin inhibits isolated bacterial complex 1 [76]; however, it may act as an antibiotic by inhibiting dihydrofolate reductase complex [77], akin to antifolate actions of drugs such as methotrexate, mirroring our hypothesis [66].

Preclinical studies investigating metformin's effect on bacteria are in mM concentration (typically more than 10 mM), whereas blood levels usually range in μM concentrations. This situation appears to be similar to that faced by researchers investigating the anticancer effect of metformin [78]. We conjectured that high concentrations of metformin might be very much achievable in the gut, where it could predominantly act [79]. When researchers administered a special delayed release formulation to T2DM subjects, which reduced gut absorption, the antihyperglycemic effect was maintained, consistent with primary mediation via the gut [80].

METFORMIN IN OBESITY AND DIABETES

Metformin treatment increased glucose uptake in subcutaneous and viscerally derived adipocytes, in the presence and absence of insulin [81]. The stage of differentiation of adipocytes influences metformin's action. Metformin increases lipolysis via uncoupling proteins and inhibits adipogenesis of differentiating adipocytes [82]. When adipocytes from lean individuals were subjected to several folds lower concentration of metformin, for a lesser duration, no effect was observed [83].

Indirectly, metformin has various actions. It inhibits hepatic gluconeogenesis, glycogenesis, as well as carbohydrate and bile salt uptake [84,85]. By inhibiting NF-κB and ERK pathways, metformin administration is associated with a decreased production and reduced action of proinflammatory mediators such as TNF-α [86,87]. Metformin treatment was associated with an increase in Tregs [88] and M2 macrophages [89], mediated via FGF21, and improved mobilization, respectively. Metformin treatment is also associated with increased adiponectin, GLP-1, PYY levels in obese and T2DM patients [90−92].

Metformin restores insulin sensitivity and is a known AMPK activator, leading to mTOR inhibition, thus mitigating insulin's proobesogenic activity [93]. Metformin is known to induce satiety in animals and humans. The mechanisms of inducing satiety are debatable [94,95].

Metformin−Microbiome Antiobesity Link

We have proposed that metformin and butyrate may act synergistically to control obesity [66]. SCFA levels (including butyrate) are increased in obese subjects. It may be possible that obesity is also a state of butyrate receptor resistance to insulin resistance. In such a state, plasma butyrate levels may be insufficient, which may send some signals to gut microbiome to increase butyrate production.

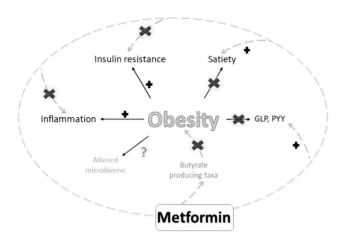

FIGURE 42.1 Obesity is a state of insulin resistance and inflammation and is associated with decreased hormone levels (such as GLP, PYY), abolishment of satiety, and an altered gut microbiome. The question mark is whether the altered microbiome causes obesity or is an effect of obesity. Metformin increases butyrate-producing gut microbes, and we hypothesize that this effect is the reason for its antiobesity effects. *GLP*, glucagon-like peptide; *PYY*, peptide YY.

The BIGPRO trial showed a trend toward weight reduction in metformin users [96]. Evidence that metformin causes weight loss remains poor, and the diabetes prevention program outcome study (DPPOS) remains the landmark study favoring the antiobesity effect of metformin [66].

CONCLUSIONS

Obesity and T2DM show a strong positive correlation. Patients with T2DM are often obese, and in general, the epidemiology of T2DM, especially in developed nations, follows a predictable trend of increasing weight, obesity, metabolic syndrome, and T2DM. Insulin injections are the most common cause of weight gain, and T1DM often purposely forgo their injections to reduce weight (diabulimia). It is also known that poorly controlled T2DM patients have increased energy expenditure, protein turnover, and weight loss.

The cause of death of most obese and diabetic subjects, irrespective of status of control, is those with a fundamental basis of inflammation and obesity. Thus, it may be plausible that diabetes mellitus is body's way of controlling obesity, however, at a cost that increases with time. Antiinflammatory medications have shown promise in trials; however, such drugs do not usually cover the entire specter of inflammation. In such context it is noteworthy that metformin shows controlling effect on T2DM, obesity, and inflammation (Fig. 42.1).

There are reasons to believe that all these positive effects on human health are majorly derived by gut microbiome modulation because metformin is known to be concentrated at the intestine. We feel that focused studies in these directions will further enrich our knowledge in this subject in days to come.

DISCLAIMER

DB acknowledges financial assistance from Postgraduate Institute of Medical Education and Research (PGIMER), Chandigarh, India.

REFERENCES

[1] Gregor MF, Hotamisligil GS. Inflammatory mechanisms in obesity. Annu Rev Immunol 2011;29:415–45.

[2] Speakman JR. Thrifty genes for obesity and the metabolic syndrome—time to call off the search? Diab Vasc Dis Res 2006;3:7–11.

[3] Speakman JR. Thrifty genes for obesity, an attractive but flawed idea, and an alternative perspective: the 'drifty gene' hypothesis. Int J Obes 2008;32:1611–7.

[4] Heymsfield SB, Wadden TA. Mechanisms, pathophysiology, and management of obesity. N Engl J Med 2017;376:254–66.

[5] Coppack SW. Adipose tissue changes in obesity. Biochem Soc Trans 2005;33:1049–52.

[6] Cai D, Yuan M, Frantz DF, et al. Local and systemic insulin resistance resulting from hepatic activation of IKK-beta and NF-kappaB. Nat Med 2005;11:183–90.

[7] Hirosumi J, Tuncman G, Chang L, et al. A central role for JNK in obesity and insulin resistance. Nature 2002;420:333–6.

[8] Boura-Halfon S, Zick Y. Phosphorylation of IRS proteins, insulin action, and insulin resistance. Am J Physiol Endocrinol Metab 2009;296:E581–91.

[9] Uysal KT, Wiesbrock SM, Marino MW, Hotamisligil GS. Protection from obesity-induced insulin resistance in mice lacking TNF-alpha function. Nature 1997;389:610–4.

[10] Chiang S-H, Bazuine M, Lumeng CN, et al. The protein kinase IKKepsilon regulates energy balance in obese mice. Cell 2009;138:961−75.

[11] Sharma AM, Staels B. Review: peroxisome proliferator-activated receptor gamma and adipose tissue−understanding obesity-related changes in regulation of lipid and glucose metabolism. J Clin Endocrinol Metab 2007;92:386−95.

[12] Feuerer M, Herrero L, Cipolletta D, et al. Lean, but not obese, fat is enriched for a unique population of regulatory T cells that affect metabolic parameters. Nat Med 2009;15:930−9.

[13] Nishimura S, Manabe I, Nagasaki M, et al. CD8+ effector T cells contribute to macrophage recruitment and adipose tissue inflammation in obesity. Nat Med 2009;15:914−20.

[14] Ohmura K, Ishimori N, Ohmura Y, et al. Natural killer T cells are involved in adipose tissues inflammation and glucose intolerance in diet-induced obese mice. Arterioscler Thromb Vasc Biol 2010;30:193−9.

[15] Videla LA, Tapia G, Rodrigo R, et al. Liver NF-kappaB and AP-1 DNA binding in obese patients. Obesity (Silver Spring Md) 2009;17:973−9.

[16] Carlsen H, Haugen F, Zadelaar S, et al. Diet-induced obesity increases NF-kappaB signaling in reporter mice. Genes Nutr 2009;4:215−22.

[17] Yang Q, Graham TE, Mody N, et al. Serum retinol binding protein 4 contributes to insulin resistance in obesity and type 2 diabetes. Nature 2005;436:356−62.

[18] Wu H, Ballantyne CM. Skeletal muscle inflammation and insulin resistance in obesity. J Clin Invest 2017;127:43−54.

[19] Apovian CM, Bigornia S, Mott M, et al. Adipose macrophage infiltration is associated with insulin resistance and vascular endothelial dysfunction in obese subjects. Arterioscler Thromb Vasc Biol 2008;28:1654−9.

[20] Adamska E, Ostrowska L, Górska M, Kretowski A. The role of gastrointestinal hormones in the pathogenesis of obesity and type 2 diabetes. Przegląd Gastroenterol 2014;9:69−76.

[21] Berthoud H-R, Morrison C. The brain, appetite, and obesity. Annu Rev Psychol 2008;59:55−92.

[22] Turnbaugh PJ, Ley RE, Mahowald MA, Magrini V, Mardis ER, Gordon JI. An obesity-associated gut microbiome with increased capacity for energy harvest. Nature 2006;444:1027−31.

[23] Turnbaugh PJ, Hamady M, Yatsunenko T, et al. A core gut microbiome in obese and lean twins. Nature 2009;457:480−4.

[24] Turnbaugh PJ, Bäckhed F, Fulton L, Gordon JI. Diet-induced obesity is linked to marked but reversible alterations in the mouse distal gut microbiome. Cell Host Microbe 2008;3:213−23.

[25] Griffin NW, Ahern PP, Cheng J, et al. Prior dietary practices and connections to a human gut microbial metacommunity alter responses to diet interventions. Cell Host Microbe 2017;21:84−96.

[26] Khan MJ, Gerasimidis K, Edwards CA, Shaikh MG. Role of gut microbiota in the aetiology of obesity: proposed mechanisms and review of the literature. J Obes 2016;2016, 7353642.

[27] Vrieze A, Van Nood E, Holleman F, et al. Transfer of intestinal microbiota from lean donors increases insulin sensitivity in individuals with metabolic syndrome. Gastroenterology 2012;143:913−6. e7.

[28] Harley ITW, Karp CL. Obesity and the gut microbiome: striving for causality. Mol Metab 2012;1:21−31.

[29] Finucane MM, Sharpton TJ, Laurent TJ, Pollard KS. A taxonomic signature of obesity in the microbiome? Getting to the guts of the matter. PLoS One 2014;9:e84689.

[30] Beaumont M, Goodrich JK, Jackson MA, et al. Heritable components of the human fecal microbiome are associated with visceral fat. Genome Biol 2016;17:189.

[31] Maruvada P, Leone V, Kaplan LM, Chang EB. The human microbiome and obesity: moving beyond associations. Cell Host Microbe 2017;22:589−99.

[32] Hildebrandt MA, Hoffmann C, Sherrill-Mix SA, et al. High-fat diet determines the composition of the murine gut microbiome independently of obesity. Gastroenterology 2009;137:1716−1724.e1-2.

[33] Koenig JE, Spor A, Scalfone N, et al. Succession of microbial consortia in the developing infant gut microbiome. Proc Natl Acad Sci U S A 2011;108(Suppl 1):4578−85.

[34] De Filippo C, Cavalieri D, Di Paola M, et al. Impact of diet in shaping gut microbiota revealed by a comparative study in children from Europe and rural Africa. Proc Natl Acad Sci U S A 2010;107:14691−6.

[35] Jumpertz R, Le DS, Turnbaugh PJ, et al. Energy-balance studies reveal associations between gut microbes, caloric load, and nutrient absorption in humans. Am J Clin Nutr 2011;94:58−65.

[36] Turnbaugh PJ, Ridaura VK, Faith JJ, Rey FE, Knight R, Gordon JI. The effect of diet on the human gut microbiome: a metagenomic analysis in humanized gnotobiotic mice. Sci Transl Med 2009;1:6ra14.

[37] Goodrich JK, Davenport ER, Beaumont M, et al. Genetic determinants of the gut microbiome in UK twins. Cell Host Microbe 2016;19:731−43.

[38] Murakami M, Tognini P, Liu Y, Eckel-Mahan KL, Baldi P, Sassone-Corsi P. Gut microbiota directs PPARγ-driven reprogramming of the liver circadian clock by nutritional challenge. EMBO Rep 2016;17:1292−303.

[39] Thaiss CA, Levy M, Korem T, et al. Microbiota diurnal rhythmicity programs host transcriptome oscillations. Cell 2016;167:1495−1510.e12.

[40] Morrison DJ, Preston T. Formation of short chain fatty acids by the gut microbiota and their impact on human metabolism. Gut Microb 2016;7:189−200.

[41] Shen J, Obin MS, Zhao L. The gut microbiota, obesity and insulin resistance. Mol Aspects Med 2013;34:39−58.

[42] Forslund K, Hildebrand F, Nielsen T, et al. Disentangling type 2 diabetes and metformin treatment signatures in the human gut microbiota. Nature 2015;528:262−6.

[43] Balamurugan R, George G, Kabeerdoss J, Hepsiba J, Chandragunasekaran AMS, Ramakrishna BS. Quantitative differences in intestinal Faecalibacterium prausnitzii in obese Indian children. Br J Nutr 2010;103:335−8.

[44] Kasai C, Sugimoto K, Moritani I, et al. Comparison of the gut microbiota composition between obese and non-obese individuals in a Japanese population, as analyzed by terminal restriction fragment length polymorphism and next-generation sequencing. BMC Gastroenterol 2015;15:100.

[45] Fernandes J, Su W, Rahat-Rozenbloom S, Wolever TMS, Comelli EM. Adiposity, gut microbiota and faecal short chain fatty acids are linked in adult humans. Nutr Diabetes 2014;4:e121.

[46] Rahat-Rozenbloom S, Fernandes J, Gloor GB, Wolever TMS. Evidence for greater production of colonic short-chain fatty acids in overweight than lean humans. Int J Obes 2014;38:1525–31.

[47] Høverstad T, Midtvedt T. Short-chain fatty acids in germfree mice and rats. J Nutr 1986;116:1772–6.

[48] Lu Y, Fan C, Li P, Lu Y, Chang X, Qi K. Short chain fatty acids prevent high-fat-diet-induced obesity in mice by regulating G protein-coupled receptors and gut microbiota. Sci Rep 2016;6:37589.

[49] den Besten G, van Eunen K, Groen AK, Venema K, Reijngoud D-J, Bakker BM. The role of short-chain fatty acids in the interplay between diet, gut microbiota, and host energy metabolism. J Lipid Res 2013;54:2325–40.

[50] Parekh PJ, Balart LA, Johnson DA. The influence of the gut microbiome on obesity, metabolic syndrome and gastrointestinal disease. Clin Transl Gastroenterol 2015;6:e91.

[51] Marchiando AM, Shen L, Graham WV, et al. Caveolin-1-dependent occludin endocytosis is required for TNF-induced tight junction regulation in vivo. J Cell Biol 2010;189:111–26.

[52] Cani PD, Amar J, Iglesias MA, et al. Metabolic endotoxemia initiates obesity and insulin resistance. Diabetes 2007;56:1761–72.

[53] Yiu JHC, Dorweiler B, Woo CW. Interaction between gut microbiota and toll-like receptor: from immunity to metabolism. J Mol Med (Berl Ger) 2017;95:13–20.

[54] Bäckhed F, Manchester JK, Semenkovich CF, Gordon JI. Mechanisms underlying the resistance to diet-induced obesity in germ-free mice. Proc Natl Acad Sci U S A 2007;104:979–84.

[55] Li SS, Zhu A, Benes V, et al. Durable coexistence of donor and recipient strains after fecal microbiota transplantation. Science 2016;352:586–9.

[56] Dao MC, Clément K. Gut microbiota and obesity: concepts relevant to clinical care. Eur J Intern Med 2017. https://doi.org/10.1016/j.ejim.2017.10.005. published online Oct 27.

[57] Tremaroli V, Karlsson F, Werling M, et al. Roux-en-Y gastric bypass and vertical banded gastroplasty induce long-term changes on the human gut microbiome contributing to fat mass regulation. Cell Metab 2015;22:228–38.

[58] Aron-Wisnewsky J, Doré J, Clement K. The importance of the gut microbiota after bariatric surgery. Nat Rev Gastroenterol Hepatol 2012;9:590–8.

[59] Graessler J, Qin Y, Zhong H, et al. Metagenomic sequencing of the human gut microbiome before and after bariatric surgery in obese patients with type 2 diabetes: correlation with inflammatory and metabolic parameters. Pharmacogenomics J 2013;13:514–22.

[60] Damms-Machado A, Mitra S, Schollenberger AE, et al. Effects of surgical and dietary weight loss therapy for obesity on gut microbiota composition and nutrient absorption. BioMed Res Int 2015;2015, 806248.

[61] John GK, Mullin GE. The gut microbiome and obesity. Curr Oncol Rep 2016;18:45.

[62] Cani PD, Van Hul M. Novel opportunities for next-generation probiotics targeting metabolic syndrome. Curr Opin Biotechnol 2015;32:21–7.

[63] Evaluation of the effects associated with the administration of Akkermansia Muciniphila on parameters of metabolic syndrome—Full Text View—ClinicalTrials.gov. https://clinicaltrials.gov/ct2/show/NCT02637115.

[64] Park S, Bae J-H. Probiotics for weight loss: a systematic review and meta-analysis. Nutr Res 2015;35:566–75.

[65] Cotillard A, Kennedy SP, Kong LC, et al. Dietary intervention impact on gut microbial gene richness. Nature 2013;500:585–8.

[66] Maniar K, Moideen A, Mittal A, Patil A, Chakrabarti A, Banerjee D. A story of metformin-butyrate synergism to control various pathological conditions as a consequence of gut microbiome modification: genesis of a wonder drug? Pharmacol Res 2017;117:103–28.

[67] Wu H, Esteve E, Tremaroli V, et al. Metformin alters the gut microbiome of individuals with treatment-naive type 2 diabetes, contributing to the therapeutic effects of the drug. Nat Med 2017. https://doi.org/10.1038/nm.4345. published online May 22.

[68] Lee H, Ko G. Effect of metformin on metabolic improvement and gut microbiota. Appl Environ Microbiol 2014;80:5935–43.

[69] Rios-Covian D, Gueimonde M, Duncan SH, Flint HJ, de los Reyes-Gavilan CG. Enhanced butyrate formation by cross-feeding between Faecalibacterium prausnitzii and Bifidobacterium adolescentis. FEMS Microbiol Lett 2015;362. https://doi.org/10.1093/femsle/fnv176.

[70] de la Cuesta-Zuluaga J, Mueller NT, Corrales-Agudelo V, et al. Metformin is associated with higher relative abundance of mucin-degrading Akkermansia muciniphila and several short-chain fatty acid–producing microbiota in the gut. Diabetes Care 2017;40:54–62.

[71] Chakraborty S, Mittal A, Banerjee D. Human recombinant insulin in supraphysiological concentration support bacterial growth in glucose independent manner. Int J Drug Dev Res 2016;8:035–40.

[72] Santacruz A, Collado MC, García-Valdés L, et al. Gut microbiota composition is associated with body weight, weight gain and biochemical parameters in pregnant women. Br J Nutr 2010;104:83–92.

[73] Gao X, Jia R, Xie L, Kuang L, Feng L, Wan C. Obesity in school-aged children and its correlation with gut E.coli and Bifidobacteria: a case-control study. BMC Pediatr 2015;15:64.

[74] Cabreiro F, Au C, Leung K-Y, et al. Metformin retards aging in C. elegans by altering microbial folate and methionine metabolism. Cell 2013;153:228–39.

[75] Maniar K, Moideen A, Bhattacharyya R, Banerjee D. Metformin exerts anti-obesity effect via gut microbiome modulation in prediabetics: a hypothesis. Med Hypotheses 2017;104:117–20.

[76] Bridges HR, Jones AJY, Pollak MN, Hirst J. Effects of metformin and other biguanides on oxidative phosphorylation in mitochondria. Biochem J 2014;462:475–87.

[77] Gabel SA, Duff MR, Pedersen LC, et al. A structural basis for biguanide activity. Biochemistry (Mosc) 2017;56:4786–98.

[78] Chandel NS, Avizonis D, Reczek CR, et al. Are metformin doses used in murine cancer models clinically relevant? Cell Metab 2016;23:569–70.

[79] Maniar K, Moideen A, Bhattacharyya R, Banerjee D. Whether 25mM of metformin is achievable in human gut from a therapeutic dose of metformin? Med Hypotheses 2017;108:51.

[80] Buse JB, DeFronzo RA, Rosenstock J, et al. The primary glucose-lowering effect of metformin resides in the gut, not the circulation: results from short-term pharmacokinetic and 12-week dose-ranging studies. Diabetes Care 2016;39:198–205.

[81] Fischer M, Timper K, Radimerski T, et al. Metformin induces glucose uptake in human preadipocyte-derived adipocytes from various fat depots. Diabetes Obes Metab 2010;12:356–9.

[82] Anedda A, Rial E, González-Barroso MM. Metformin induces oxidative stress in white adipocytes and raises uncoupling protein 2 levels. J Endocrinol 2008;199:33–40.

[83] Pedersen O, Nielsen O, Bak J, Richelsen B, Beck-Nielsen H, Sørensen N. The effects of metformin on adipocyte insulin action and metabolic control in obese subjects with type 2 diabetes. Diabet Med J Br Diabet Assoc 1989;6:249–56.

[84] Bailey CJ, Turner RC. Metformin. N Engl J Med 1996;334:574–9.

[85] Scarpello JH, Hodgson E, Howlett HC. Effect of metformin on bile salt circulation and intestinal motility in type 2 diabetes mellitus. Diabet Med J Br Diabet Assoc 1998;15:651–6.

[86] Hyun B, Shin S, Lee A, et al. Metformin down-regulates TNF-α secretion via suppression of scavenger receptors in macrophages. Immune Netw 2013;13:123–32.

[87] Ren T, He J, Jiang H, et al. Metformin reduces lipolysis in primary rat adipocytes stimulated by tumor necrosis factor-alpha or isoproterenol. J Mol Endocrinol 2006;37:175–83.

[88] Kim EK, Lee SH, Jhun JY, et al. Metformin prevents fatty liver and improves balance of White/Brown adipose in an obesity mouse model by inducing FGF21. Mediators Inflamm 2016;2016, 5813030.

[89] Jing Y, Wu F, Li D, Yang L, Li Q, Li R. Metformin improves obesity-associated inflammation by altering macrophages polarization. Mol Cell Endocrinol 2018;461:256–64.

[90] Mannucci E, Ognibene A, Cremasco F, et al. Effect of metformin on glucagon-like peptide 1 (GLP-1) and leptin levels in obese nondiabetic subjects. Diabetes Care 2001;24:489–94.

[91] Adamia N, Virsaladze D, Charkviani N, Skhirtladze M, Khutsishvili M. Effect of metformin therapy on plasma adiponectin and leptin levels in obese and insulin resistant postmenopausal females with type 2 diabetes. Georgian Med News 2007:52–5.

[92] DeFronzo RA, Buse JB, Kim T, et al. Once-daily delayed-release metformin lowers plasma glucose and enhances fasting and postprandial GLP-1 and PYY: results from two randomised trials. Diabetologia 2016;59:1645–54.

[93] Viollet B, Guigas B, Sanz Garcia N, Leclerc J, Foretz M, Andreelli F. Cellular and molecular mechanisms of metformin: an overview. Clin Sci Lond Engl 2012;122:253–70.

[94] Lv W-S, Wen J-P, Li L, et al. The effect of metformin on food intake and its potential role in hypothalamic regulation in obese diabetic rats. Brain Res 2012;1444:11–9.

[95] Lee CK, Choi YJ, Park SY, Kim JY, Won KC, Kim YW. Intracerebroventricular injection of metformin induces anorexia in rats. Diabetes Metab J 2012;36:293–9.

[96] Fontbonne A, Charles MA, Juhan-Vague I, et al. The effect of metformin on the metabolic abnormalities associated with upper-body fat distribution. BIGPRO Study Group. Diabetes Care 1996;19:920–6.

FURTHER READING

[1] Caballero B. The global epidemic of obesity: an overview. Epidemiol Rev 2007;29:1–5.

[2] Bhurosy T, Jeewon R. Overweight and obesity epidemic in developing countries: a problem with diet, physical activity, or socioeconomic status? ScientificWorldJournal 2014;2014, 964236.

[3] Mitchell NS, Catenacci VA, Wyatt HR, Hill JO. Obesity: overview of an epidemic. Psychiatr Clin North Am 2011;34:717–32.

[4] Alexandratos N, Bruinsma J. World agriculture towards 2030/2050: the 2012 revision. ESA Working paper. Rome: FAO; 2012.

[5] Drewnowski A. The real contribution of added sugars and fats to obesity. Epidemiol Rev 2007;29:160–71.

[6] McLaren L. Socioeconomic status and obesity. Epidemiol Rev 2007;29:29–48.

[7] Sturmberg JP. Obesity—a multifaceted approach: one problem—different models—different insights and solutions. In: Sturmberg JP, editor. Health system redesign: how to make health care person-centered, equitable, and sustainable. Cham: Springer International Publishing; 2018. p. 213–32.

[8] Papas MA, Alberg AJ, Ewing R, Helzlsouer KJ, Gary TL, Klassen AC. The built environment and obesity. Epidemiol Rev 2007;29:129–43.

[9] Taylor MM. The obesity epidemic: individual accountability and the social determinants of health. In: Taylor MM, editor. The obesity epidemic: why a social justice perspective matters. Cham: Springer International Publishing; 2018. p. 21–38.

[10] Ghoochani OM, Torabi R, Hojjati M, Ghanian M, Kitterlin M. Factors influencing Iranian consumers' attitudes toward fast-food consumption. Br Food J 2018.

Chapter 43

Deleterious Impact of Smog on the Intestinal Bacteria

L.R. Pace[1], C.M. Wells[1], R. Awais[1], P. Shrestha[2], R.D. Parker[2] and T.Y. Wong[2]

[1]*Department of Biological Engineering, University of Memphis, Memphis, TN, United States;* [2]*Department of Biological Sciences, University of Memphis, Memphis, TN, United States*

INTRODUCTION

On topics of smog, please refer to reviews [1,2]. Emeran Mayer's book *The Mind-Gut connection* [3] has provided an authoritative insight on the microbiota–body connection. The disruptive effects of pollutants on gut microbiota have been reviewed extensively by Jin et al. [4]. *Strict and Facultative Anaerobes: Medical and Environmental Aspect* by Nakano et al. [5] has provided a comprehensive description of the ecology and metabolomes of the gut bacteria. Topics of toxicity and risk of environmental nanoparticles have been reviewed by Fu et al. [6].

WHAT IS SMOG AND HOW POLLUTANTS ENTER INTO THE GUT?

Smog is a mixture of atmospheric pollutants, dust, and smoke combined with fog under the sun, the formation of which is directly related to the weather and landscape of a general area. Under normal conditions, warm air rises to widely disperse pollutants, preventing smog formation. However, at times when a temperature inversion occurs, the upper air mass is warmer than the air below it, therefore preventing the vital air flow, which spreads and dilutes pollutants. The landscape also plays a role in creating a temperature inversion, with cities built on low basins or valleys more prone to smog accumulation.

Smog is a naturally occurring phenomenon. It forms when sunlight reactions of many plants produce volatile organic compounds (VOCs), and when volcanoes and forest fires produce airborne ash, and when gases with high concentrations of VOCs, NO_x, SO_2, and H_2S mix together. Smog is also man-made. Pollutants generated by cars, trucks, trains, airplanes, and ships are called nonpoint sources of pollution. Nonpoint pollution is the major source of smog in most cities.

The burning of coal and fossil fuels by power plants and incinerators produces pollution that is considered as stationary sources. Additionally, gases released from landfills and sewage treatment plants often contain various toxic gases and even pathogens, including *Streptococcus pneumoniae*, *Aspergillus fumigatus*, and human *adenovirus C* [7]. These gases can contribute to smog [1,7].

Particulate matter (PM) is a key pollutant present in the smog. PM is a complex mixture of extremely small particles and liquid droplets suspended in the air. The PMs are often subclassified based on their sizes (PM_{10}, $PM_{2.5}$, and UFPs [ultrafine particles]). PM_{10} is defined as particulates, such as dust, pollen, and mold, with a diameter smaller than 10 microns. $PM_{2.5}$ is defined as particulates with a diameter smaller than 2.5 microns. UFPs are extremely small particulates, with less than 1 micron.

Combustion particles often belong to the $PM_{2.5}$ and UFP classes. $PM_{2.5}$ and UFP particulates pose the greatest health risk because many of them can directly get into the cell. Very often, many VOC and metals are attached to the PMs. Transition metals found in smog include chromium (Cr), copper (Cu), iron (Fe), nickel (Ni), and vanadium (V). These metals have the potential to produce reactive oxygen species (ROS) in biological systems. In addition, heavy metals, such as cadmium (Cd), lead (Pb), and mercury (Hg), are some of the most common air pollutants emitted by industrial activities, combustion, extraction, and processing activities. Many of these metals exhibit oligodynamic effect and can induce ROS to kill bacteria.

Microbiome and Metabolome in Diagnosis, Therapy, and other Strategic Applications. https://doi.org/10.1016/B978-0-12-815249-2.00043-9

Contaminated foods and drinks can bring these pollutants into the gut directly. The lung inhales pollutants from the air. These pollutants accumulate in the alveoli. The mucociliary clearance system transports the mucus containing the foreign particles to the larynx, where it is swallowed and consequently transferred into the gut.

ECOLOGY OF THE GUT MICROBIOTA

Metagenomic studies suggest that half of the bacteria in the gut are metabolic generalists, while others are specializing and feeding on specific substrates, such as carbohydrates, proteins, or lipids [8]. It should be noted that, unlike the aerobic environment where NADP and PADH can readily dispose their electrons to oxygen (the potential difference between NADH to O_2 is about 1.15 V), the concentration of oxygen in the gut is very low. In a reduced environment, microbes must rely on other less efficient acceptors, to dispose of their electrons and protons.

Bacterial Consortia and Metabolic Synergy

Many anaerobes use the hydrogenase, to oxidize their NADH or FADH and produce hydrogen gas. However, hydrogen gas is also a strong feedback inhibitor of the hydrogen-producing process. Therefore, the hydrogen producers need the support of other bacteria, such as methanogens, nitrate reducers, and sulfate reducers, to reduce the hydrogen levels in the gut. In fact, in the gut, consortia of microbes often formed single units, to obtain their nutrients in a step-by-step manner.

Novel methods, such as adding a fumarate or carboxyl group to active hydrocarbons, hydrolysis of ATP to provide low potential electrons to rescue aromatic compounds, and using B_{12}-mediated reactions to generate free radicals for dehydration reactions, are common in the gut environment, to allow the flow of nutrients between species [5]. These interspecies interactions sometimes require individuals to be in close contact. The gut bacterial community is well-organized, and they seem to be preprogrammed to develop in an orderly manner [9].

Spatial Organization and Resource Sharing

An imaging technique, called combinatorial labeling and spectral imaging—fluorescence in situ hybridization (CLASI-FISH), can show the spatial locations of many different species of bacteria on a surface simultaneously. CLASI-FISH analysis of human gut microbiota, in a gnotobiotic mouse model, confirms that at the border between the lumen and mucosa, there is a distinct biogeography of some groups of bacteria, relative to one another [10]. This spatial organization of gut microbiota at the mucosa reflects a higher level of cooperation between the gut bacteria and their host.

Agonism and Antagonism in the Gut Environment

Microbial ecologists often use the term "Red Queen Hypothesis" [11], from Lewis Carroll's book *Alice in Wonderland*, Chapter 2, where The Red Queen said to Alice, "*It takes all the running you can do, to keep in the same place.*" The theory assumes bacteria in a community would interact with each other antagonistically. An advance in one species may displace other species. Such negative interactions include predation, virus–bacteria antagonists, and competition for the same resource. For example, *Bacteroides* stay fit in the human gut, by acquiring the carbohydrate metabolic genes through horizontal gene transfers [12].

Positive interactions, in turn, include the sharing of a preferred environment, commensalism, and mutualism. The interaction between two species may also benefit a third party indirectly (through scavenging free radicals from the area, for example), the so-called "Black Queen Hypothesis" [13]. This theory suggests that individuals in a community, after a long time, could become highly dependent on each other through gene loss.

It theorizes that keeping an infrequently used function is expensive. Losing a dispensable function could conserve a bacterium's resources and increase the fitness. For example, free radicals (ROS) are common by-products of normal metabolism. Most bacteria would produce superoxide dismutase, catalase, and peroxidases to remove ROS. Some bacteria, such as the *Clostridia, Ruminococcus*, and *Bacteroides*, do not produce these detoxifying enzymes. They take advantage of the ROS-free environment created by the other bacteria in the gut to avoid oxidative damage.

Ruminococcus bromii, a minority in the human's gut bacterial community, is considered a keystone species because it initiates the breakdown of dietary fibers and resistant starches in potato, banana, wheat, and maize. Without this partial digestion, many bacteria in the gut cannot utilize the carbohydrates [14]. Actinobacteria, another minority in the gut microbiota, is also considered keystone because of the high degree of ecological connectedness [7].

SUCCESSION OF GUT MICROBIOTA

An independent study of $\approx 500,000$ DNA metagenomic sequences, from fecal samples of infants, revealed that the original lactate (milk sugar) utilization bacteria were soon replaced by bacteria capable to metabolize plant polysaccharides, before the introduction of solid food [8,9]. These observations meant that an infant's gut microbe community could anticipate the future, and prime itself for the eventual arrival of solid food, such as cereal and pea. Preprogramming of bacterial succession sets the foundation for the future gut diversity and resilience against diseases [3,9].

Table 43.1 summarizes the many physiological links to the disruption of gut microbiota.

SMOG-INDUCED OXIDATIVE STRESS IN BACTERIA

Free radicals are involved in a wide range of biochemical processes in isomerization, reduction, and rearrangement reactions. They are unstable and rapidly engaged in other secondary reactions by binding and potentially damaging, DNA, proteins, lipids, and other biomolecules. Common free radicals are hydroxyl radical, superoxide anion radical, hydrogen peroxide, oxygen singlet, hypochlorite, nitric oxide radical, and peroxynitrite radical.

Lipid Peroxidation

When binding to unsaturated lipids, free radicals can induce lipid peroxidation. Lipid peroxidation is a chain reaction that generates more reactive products such as aldehydes and lipid radicals. Malondialdehyde, the end-products of lipidation, is also toxic to many bacterial cells [15]. Most bacteria are equipped with superoxide dismutase, peroxidase, and catalase to stop the free radicals from causing harm to the cell. However, some obligate anaerobic bacteria, including many clostridia and methanogens in the gut, lack these protective enzymes.

Electron Transfer Reactions and Transition Metals

These bacteria are protected from the anaerobic environment of the gut, by other members of the community that can scavenge the free radicals. Energy- or electron-transfer reactions often generate ROS. For example, electron transfer between succinate and fumarate of the TCA cycle by the obligate anaerobes *Bacteroides fragilis* would generate ROS [16].

TABLE 43.1 Some Physiological Effects Resulting From Dysbiosis of the Human Gut

System	Condition	Symptom/Description
Nervous	Posttraumatic stress disorder	Microbiota metabolites mediate changes in brain function and performance; stress-induced changes in gut microbiome may increase potential for future stress-related disorders; probiotics can reduce onset of mental health disorders related to stress
	Alzheimer's	Decreased microbial diversity, distinct composition of microbiome
	Brain	Neural pathway is operational through the enteric nervous system. The postnatal gut microbiota influences the developing brain. Chronic systemic inflammation may also alter brain development
	Autism (ASD)	Probiotics can facilitate restoration of dysbiosis in individuals with ASD, while improving their compliance and adherence
	Circadian rhythm	Microbiota can regulate host circadian rhythms
	Anxiety	Gut bacteria are involved and play role in the programming of the hypothalamic—pituitary—adrenal axis early in life and stress reactivity over the life span
Immune	Inflammation	If dietary 2,3,7,8-tetrachlorodibenzofuran inhibits the bile acid receptor signaling pathway, significant inflammation may be triggered
	Multiple sclerosis (MS)	Modulation of microbiome can either improve or worsen the symptoms of demyelinating diseases. Risk factors for MS include exogenous stimuli such as diet, smoking, alcohol consumption, and vitamin D insufficiency, which are all responsible for changing composition of gut microbiota

The electron transport chain is the most common site for ROS production [17]. In fact, one of the functions of the electron transport chain is to slow down the flow of the electrons to prevent excessive ROS formation [7]. When a cell cannot balance its free radicals with other enzymatic or nonenzymic defenses, it undergoes oxidative stress. Transition metals in smog can penetrate the cell membrane and uncouple the normally regulated electron transfer. The uncontrolled transfer of electrons could cause excessive ROS formation and induce oxidative stress of the bacterium.

Nitrogen Oxides

Nitrogen oxides (NO_x) are a component of smog. NO_x can disrupt the gut microbiota in many ways. For example, peroxyacetyl nitrate, a ubiquitous air pollutant formed of NOx, increases bacterial mutation by base-pair substitution [18]. Bacteria can catalyze NO_x to N-nitrosamines, a highly mutagenic chemical. NO_x are also free radicals that can transform into a highly active molecule (peroxynitrite) when reacting with superoxide.

Sulfur Dioxide

Sulfur dioxide (SO_2) is another common component of smog. The lactic acid bacteria are important symbionts of the gut. They are not only antagonistic to many infectious pathogens but also form a complex molecular cross-talk with the human. Lactic acid bacteria are very sensitive to SO_2. In fact, SO_2 is used commercially as an additive, to inhibit lactic acid spoilage in many wines and fruit juices [7].

Particulate Matter

$PM_{2.5}$ and UFP can adsorb to the carboxyl and phosphate groups of the bacterial cell protein, by a process called as biosorption, and cause protein coagulation. These particulates also disrupt the energy coupling processes. These effects can lead to leakage of cellular contents, resulting in bacterial death [19]. More importantly, transition metals found in $PM_{2.5}$ and UFP can integrate into the lipid membrane, and uncouple the electron transfer. Uncoupling of the electron transport chain reduces ATP formation and causes ROS formation [1]. Many of these metals in $PM_{2.5}$ and UFP exhibit oligodynamic effect against bacteria, by reacting to the thiol and amine groups of the protein.

Volatile Organic Compounds

VOCs in smog can also disrupt the gut community. VOCs, such as acetaldehyde,1,3-butadiene, polycyclic aromatic hydrocarbons, and benzene, are genotoxic and have been linked to birth defects, cancer, and other serious illnesses [20]. These chemicals are also tested positive in the *Salmonella*-based mutagenicity Ames test [20]. Incidentally, bacteria themselves are important producers of a number of VOCs, which can have clinical applications, when measured in breath, fecal, or urinary metabolome (volatolome).

SMOG CAN INDUCE INFLAMMATION

Despite the protective barrier provided by the epithelial layer, smog particulate can penetrate the epithelial cells. When the gut epithelium is irritated or injured, it activates polymorphonuclear neutrophils and macrophages. These cells would cross the blood vessel and migrate to the infected site. Once activated neutrophils arrive at the site, they generate a ROS burst, which can eliminate bacterial pathogens, as well as contribute to inflammation. PMs have been shown to induce acute and chronic inflammatory responses in the intestine [21].

Short-term treatment of mice with PMs altered immune gene expression, enhanced proinflammatory cytokine secretion in the small intestine, increased gut permeability, and reduced the response of T and B lymphocytes, dendritic cells, and macrophages. Long-term treatment with PMs increased proinflammatory cytokine expression in the large intestine and altered short-chain fatty acid concentrations and microbial composition [21]. Given the fact that close contacts are needed for some groups of bacteria [10], the killing of one group will disrupt the spatial organization of gut microbiota (Fig. 43.1).

PERSPECTIVE

Oxidative stress seems to be the common disruptive pathway, leading to the death of bacteria in the gut [1,4,9,20–24]. Obviously, a healthy gut demands clean air and healthy diets, rich in complex carbohydrates and antioxidants, such as

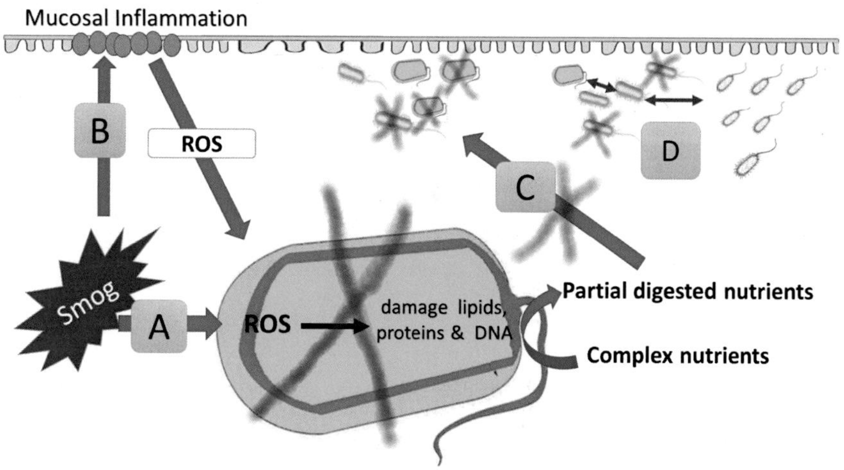

FIGURE 43.1 Proposed bactericidal mechanisms of smog in the intestine. Particulate matters, volatile organic chemicals, oxides of nitrogen and sulfur, and metals in smog get into the intestine by contaminated foods and drinks and by swallowing mucus expelled from the lung. Once arrived, these pollutants can kill the gut bacteria by (A) inducing oligodynamic effect and uncoupling the electron transfers, leading to free radicals (ROS) formation and oxidative stress; (B) causing inflammation of the intestinal mucosal layer. ROS released at the site of inflammation causes oxidative stress to the surrounding bacteria. (C) The decrease in keystone species reduces the food supply/protection of those bacteria that depend on the keystone species. (D) The loss of the native species opens space for other species to colonize. This will initiate new rounds of antagonistic competition.

those found in green plants and mushrooms [25–27]. New methods are needed to evaluate the risk of xenobiotics, including smog, in the gut. It is essential to identify, at the molecular level, the roles of the major factors influencing the gut events. The sequence-based methods for bacterial classification from the gut metagenomics data are ineffective, partially due to the rampage of horizontal gene transfers (HGT) among bacteria.

The average ratio of the stop codons (TAA:TAG:TGA), on the three reading frames of the genes in a bacterial genome, is unaffected by HGT [28,29]. This 9-vector method of phylogenetic clustering can greatly reduce the processing time of the metagenome data. The Pathway Tools–based software has been developed, for nutrient flux analysis in a bacterium [30]. This innovative method can be extended and expanded, to consider a consortium of gut bacteria as a single unit, to learn how nutrients are transferred between the bacteria in the gut. Knowing the nutrient fluxes can help to pinpoint specific keystone species, in a particular metabolic process.

Deep-minding statistical techniques, such as the state-dependent stochastic simulation models [31], has been used to automatically characterize stochastic biochemical rare events (such as cancer), over a long period of time. This method can be combined with the phylogenetics and the fluxes data, to assess the risk of smog in the bacteria–gut–brain connection.

REFERENCES

[1] Wong T-Y. Smog induces oxidative stress and microbiota disruption. J Food Drug Anal 2017;25(2):235–44.

[2] Eckelman MJ, Sherman J. Environmental impacts of the U.S. Health care system and effects on public health. PLoS One 2016;11(6):e0157014.

[3] Mayer E. The Mind-Gut Connection: how the hidden conversation within our bodies impacts our mood, our choices, and our overall health. New York, NY: HarperCollins Publishers; 2016.

[4] Jin Y, Wu S, Zeng Z, Fu Z. Effects of environmental pollutants on gut microbiota. Environ Pollut 2017;222(Suppl. C):1–9.

[5] Nakano MM, Zuber PA. Strict and facultative anaerobes: medical and environmental aspects. Wymondham, Norfolk, England: Horizon Bioscience; 2004.

[6] Ray PC, Yu H, Fu PP. Toxicity and environmental risks of nanomaterials: challenges and future needs. J Environ Sci Health, Part C 2009;27(1):1–35.

[7] Slonczewski J, Foster JW. Microbiology: an evolving science. 4th ed. New York: W. W. Norton & Company; 2016.

[8] Vieira-Silva S, Falony G, Darzi Y, et al. Species–function relationships shape ecological properties of the human gut microbiome. Nat Microbiol 2016;1:16088.

[9] Koenig JE, Spor A, Scalfone N, et al. Succession of microbial consortia in the developing infant gut microbiome. Proc Natl Acad Sci U S A 2011;108(Suppl 1):4578–85.

[10] Mark Welch JL, Hasegawa Y, McNulty NP, Gordon JI, Borisy GG. Spatial organization of a model 15-member human gut microbiota established in gnotobiotic mice. Proc Natl Acad Sci U S A 2017;114(43):E9105–14.

[11] Castrodeza C. Non-progressive evolution, the Red Queen hypothesis, and the balance of nature. Acta Biotheor 1979;28(1):11–8.

[12] Hehemann J-H, Correc G, Barbeyron T, Helbert W, Czjzek M, Michel G. Transfer of carbohydrate-active enzymes from marine bacteria to Japanese gut microbiota. Nature 2010;464(7290). https://doi.org/10.1038/nature08937.

[13] Morris JJ, Papoulis SE, Lenski RE. Coexitence of evolving bacteria stabilized by a shared Black Queen function. Evolution 2014;68(10):2960–71.

[14] Ze X, Duncan SH, Louis P, Flint HJ. Ruminococcus bromii is a keystone species for the degradation of resistant starch in the human colon. ISME J 2012;6(8):1535–43.

[15] Ni YC, Wong TY, Lloyd RV, et al. Mouse liver microsomal metabolism of chloral hydrate, trichloroacetic acid, and trichloroethanol leading to induction of lipid peroxidation via a free radical mechanism. Drug Metabol Dispos 1996;24(1):81–90.

[16] Meehan BM, Malamy MH. Fumarate reductase is a major contributor to the generation of reactive oxygen species in the anaerobe Bacteroides fragilis. Microbiology 2012;158(Pt 2):539–46.

[17] Xiong R, Siegel D, Ross D. Quinone-induced protein handling changes: implications for major protein handling systems in quinone-mediated toxicity. Toxicol Appl Pharmacol 2014;280(2):285–95.

[18] Claxton LD, Matthews PP, Warren SH. The genotoxicity of ambient outdoor air, a review: Salmonella mutagenicity. Mutat Res 2004;567(2–3):347–99.

[19] Wang L, Hu C, Shao L. The antimicrobial activity of nanoparticles: present situation and prospects for the future. Int J Nanomedicine 2017;12:1227–49.

[20] Fu PP, Xia Q, Sun X, Yu H. Phototoxicity and environmental transformation of polycyclic aromatic hydrocarbons (PAHs)-light-induced reactive oxygen species, lipid peroxidation, and DNA damage. J Environ Sci Health C Environ Carcinog Ecotoxicol Rev 2012;30(1):1–41.

[21] Kish L, Hotte N, Kaplan GG, et al. Environmental particulate matter induces murine intestinal inflammatory responses and alters the gut microbiome. PLoS One 2013;8(4):e62220.

[22] Stockmann C, Ampofo K, Pavia AT, et al. Clinical and epidemiological evidence of the red queen hypothesis in pneumococcal serotype dynamics. Clin Infect Dis 2016;63(5):619–26.

[23] Cole SA, Stahl TJ. Persistent and recurrent Clostridium difficile Colitis. Clin Colon Rectal Surg 2015;28(2):65–9.

[24] Fu PP, Xia Q, Hwang H-M, Ray PC, Yu H. Mechanisms of nanotoxicity: generation of reactive oxygen species. J Food Drug Anal 2014;22(1):64–75.

[25] Mayer EA, Savidge T, Shulman RJ. Brain-gut microbiome interactions and functional bowel disorders. Gastroenterology 2014;146(6):1500–12.

[26] Chung KT, Wong TY, Wei CI, Huang YW, Lin Y. Tannins and human health: a review. Crit Rev Food Sci Nutr 1998;38(6):421–64.

[27] Shrestha P, Joshi B, Joshi J, Malla R, Sreerama L. Isolation and physicochemical characterization of laccase from ganoderma lucidum-CDBT1 isolated from its native habitat in Nepal. BioMed Res Int 2016;2016, 3238909.

[28] Chen S, Deng L-Y, Bowman D, et al. Phylogenetic tree construction using trinucleotide usage profile (TUP). BMC Bioinf 2016;17(Suppl. 13):381.

[29] Wong T-Y, Kuo J. A new drug design strategy: killing drug resistant bacteria by deactivating their hypothetical genes. J Environ Sci Health, Part C 2016;34(4):276–92.

[30] Latendresse M, Krummenacker M, Trupp M, Karp PD. Construction and completion of flux balance models from pathway databases. Bioinformatics 2012;28(3):388–96.

[31] Roh MK, Daigle Jr BJ, Gillespie DT, Petzold LR. State-dependent doubly weighted stochastic simulation algorithm for automatic characterization of stochastic biochemical rare events. J Chem Phys 2011;135(23):234108.

FURTHER READING

[1] Dominguez-Bello MG, Costello EK, Contreras M, et al. Delivery mode shapes the acquisition and structure of the initial microbiota across multiple body habitats in newborns. Proc Natl Acad Sci U S A 2010;107(26):11971–5.

[2] Azad MB, Konya T, Maughan H, et al. Gut microbiota of healthy Canadian infants: profiles by mode of delivery and infant diet at 4 months. CMAJ (Can Med Assoc J) 2013;185(5):385–94.

Block VI

Challenges and Promises for the Future

Chapter 44

New-Generation Probiotics: Perspectives and Applications

Dinesh Kumar Dahiya[1,a], Renuka[2,a], Arun Kumar Dangi[3], Umesh K. Shandilya[4], Anil Kumar Puniya[5,6] and Pratyoosh Shukla[3]

[1]Advanced Milk Testing Research Laboratory, Post Graduate Institute of Veterinary Education and Research (Rajasthan University of Veterinary and Animal Sciences, Bikaner), Jaipur, India; [2]Department of Veterinary Physiology & Biochemistry, Post Graduate Institute of Veterinary Education and Research, (Rajasthan University of Veterinary and Animal Sciences, Bikaner), Jaipur, India; [3]Enzyme Technology and Protein Bioinformatics Laboratory, Department of Microbiology, Maharshi Dayanand University, Rohtak, India; [4]Animal Biotechnology Division, National Bureau of Animal Genetic Resources, Karnal, India; [5]College of Dairy Science & Technology, Guru Angad Dev Veterinary & Animal Sciences University, Ludhiana, India; [6]Dairy Microbiology Division, ICAR-National Dairy Research Institute, Karnal, India

INTRODUCTION

The burden of lifestyle diseases such as obesity, cardiovascular disease, and diabetes has increased many folds, and their management faces high cost, adverse effects, and other serious challenges [1—4]. Natural dietary products have been used for disease treatment since the ancient times [5—7]. Fermented foods, and fruit-based/natural products, were often introduced to satisfy hunger and to overcome various diseases. The predominant microorganisms present in fermented dairy foods are the lactic acid bacteria (LAB) [8]. Among LAB, *Lactobacilli* and *Bifidobacteria* are the most encountered, both encompass important probiotics, and they are classified as GRAS (generally recognized as safe) [9,10].

Lactobacillus plantarum, isolated from several sources including the fermented milk beverage "raabadi," is another probiotic bacteria [11]. These probiotics are known to exert many health benefits in experimental studies, such as antiinflammatory, antiobesity, antidiabetic, hypocholesterolemic, and hypolipidemic [12]. However, the word probiotic can only be used for a LAB strain that passes the set criteria of assessment, as probiotic attributes vary from strain to strain [13]. Many fermented food formulations exhibit bioactive compounds [14,15] and will be soon available in the market. Moreover, new probiotics are being selected by means of modern DNA sequencing technologies [16].

Nondigestible fibers known as prebiotics may enhance or replace probiotics, although the proliferative effect varies. Both probiotic and prebiotic supplementation can impose better results than alone. Computational biology tools have been used for determination of catalytic interactions, and molecular docking studies of bile salt hydrolase, which acts as prebiotic toward stimulating the growth of probiotic bacteria [17].

Gut microbiota produces bioactive compounds [18,19] such as some antibiotics and bacteriocins, which inhibit the growth of methicillin-resistant *Staphylococcus aureus* and *Listeria* spp. inadvertently ingested through food [20,21]. A number of mechanisms have been proposed, by which altered gut microbiota lead to inflammatory events in the host [22,23]. Probiotics and prebiotics restore the microbiota by influencing some key microbial communities and production of short-chain fatty acids (SCFAs). Restoration of gut microbiota also helps in improving the disease condition. Physical activity may also be relevant in this context, as revealed by Clarke et al. [24].

---

a. Both authors contributed equally to this work

Microbiome and Metabolome in Diagnosis, Therapy, and other Strategic Applications. https://doi.org/10.1016/B978-0-12-815249-2.00044-0

NEXT-GENERATION PROBIOTICS

Next-generation probiotics (NGPs) have been proposed in recent years, including *Akkermansia muciniphila*, *Faecalibacterium prausnitzii*, *Bacteroides* spp., *Eubacterium halli*, and *Clostridium* cluster IV [25]. As these microbes are of human origin, ascertaining their safety before application is required. Also, newer techniques to formulate functional food with these NGPs are the need of the hour.

EXPERIMENTAL AND CLINICAL STUDIES

Akkermansia muciniphila

This is a gram-negative, obligate anaerobe that comes under the family Verrucomicrobiaceae. It is one of the most abundant microorganisms residing in the human intestine. The assumption that the human body has the ability to synthesize "prebiotics," or microbial substrate, is the main reason behind the discovery of *A. muciniphila* [26]. In 2004, Muriel Derrien for the first time isolated and characterized this mucin-degrading bacteria, from human fecal samples [27,28]. Administration of certain prebiotics has stimulated a several-fold rise in the relative abundance of *A. muciniphila*, whereas it increases up to 4.5% in high-fat diet [29–32].

In humans, *A. muciniphila* was documented to be in less abundance in obesity, type 2 diabetes, hypertension, and inflammatory bowel diseases [33–36]. Administration of *A. muciniphila*, either live or in pasteurized state, has been followed by significant reduction in body weight and insulin resistance, with no inflammatory infiltration in the adipose tissue of mice [37]. The pili-like surface protein Amuc_1100 is reported to be highly stable during pasteurization and could be the main reason behind the effects of pasteurized *A. muciniphila*. Further, *A. muciniphila* improved the intestinal barrier function, decreased mucosal layer thickness and production of antimicrobial peptides such as islet-derived 3-gamma (RegIII-γ), to levels similar to lean individuals. Obesity and metabolic status was also improved [37].

MECHANISM OF ACTION

The host might benefit from *Akkermansia* spp. because of the capability to colonize in mucus, and to produce SCFAs as the result of mucus degradation [27]. The SCFAs can send signals to the host via the Gpr-43/41 receptor and trigger a cascade of signaling pathways, for the activation of immune response [38]. *A. muciniphila* also enhances the production of certain bioactive lipids such as 2-oleoylglycerol (2-OG) that belongs to the endocannabinoid system. These cannabinoids further contribute in controlling the intestinal inflammation and the production of enteroendocrine peptides, such as the glucagon-like proteins (GLP-1 and GLP-2), which regulate the satiety (glucose regulation) and gut barrier integrity [39,40]. *A. muciniphila* also upregulates several genes involved in immune response activation, such as antigen presentation, complement cascade activation, cell adhesion, and maturation of B and T cells [38].

The pili-like protein that is present on the bacterial surface, i.e., Amuc_1100, also participates in the interaction of bacteria with cell surface receptors, such as Toll-like receptors (TLR-2 and TLR-4). This interaction would further activate the signaling cascade, involved in the immune regulation, restoration of tight junctions and mucus thickness in gut mucosa, and improvement of metabolic disorders such as obesity and type 2 diabetes [41]. Both live and pasteurized *A. muciniphila* grown on a synthetic medium have been safely supplied to humans [39].

Faecalibacterium prausnitzii

This nonmotile and nonspore-forming bacterium is a single member of genus *Faecalibacterium*, and 16S rRNA analysis showed it as a close relative of clostridial cluster IV. *F. prausnitzii* is a gram positive and extremely oxygen sensitive (EOS) bacterium, which constitutes about 3%–5% of human fecal microbiota, and can increase up to 15% in some individuals. Modulation of several intestinal metabolites due to altered *F. prausnitzii* revealed it as a highly active member, having the ability to influence diverse host pathways [42]. Altered abundance of this microbe has been reported in inflammatory diseases, such as Crohn's disease (CD), and depressive disorders, which show its crucial role in human health [43].

Oral administration of *F. prausnitzii* in the colitis mice model decreased the severity of the disease, as well as rectified the associated dysbiosis [44]. Very little is known about its safety aspects, but the antiinflammatory attributes of *F. prausnitzii* make it a well-deserving candidate, for inclusion in the list of next-generation prebiotics [45].

MECHANISM OF ACTION

F. prausnitzii has the ability to digest prebiotics to produce several SCFAs, especially butyrate. Butyrate and other bioactive molecules lead to the inhibition of the NF-κB pathway and thereby show protective effects in inflammatory models [46]. Butyrate also provides energy for the epithelial cells and might interact with CD103$^+$ dendritic cells (DCs) present in the lamina propria, thus stimulating their movement to mesenteric lymph nodes (MLNs) and instigating colonic regulatory Treg cells.

The ability of *F. prausnitzii* to induce high levels of IL-10 in antigen-presenting cells may increase the suppressive activity of Foxp3$^+$ Tregs and block Th17 cells induced by proinflammatory stimuli [45]. The decreased production of inflammatory cytokines, such as IFN-γ and IL-12, indicates that *F. prausnitzii* also helps in the maintenance of the gut barrier, along with immune functions in Crohn's disease [46].

Bacteroides spp.

Bacteroides species are gram-negative, anaerobic, bile-resistant bacteria residing in the gut and constitute approximately 25% of the intestinal gut microbiota. These commensal bacteria can affect the intestinal immune system, interacting with the host, or by production of certain molecules that ultimately alter the intestinal immune response. Various *Bacteroides* strains have been studied for their probiotic-like attributes. *Bacteroides fragilis* produces polysaccharide A (PSA), known for its immunomodulatory properties, and activates the T cells. It also activates TLR-2 in PSA-dependent manner, upgrading regulatory T cells for immune tolerance and maintenance of intestinal homeostasis [47,48].

B. fragilis has also been reported to produce fragilysin, which has been implicated as a risk factor for colon cancer development [49]. This feature would not be a desirable trait for its inclusion in the list of NGPs. Some *Bacteroides* spp. are also known for their ability to degrade plant-derived polysaccharides in the gut and produce intermediates for energy. *Bacteroides xylanisolvens,* such as *B. fragilis,* has been shown to increase the production of antibodies (IgM), against Thomsen–Friedenreich (TFα) antigen, in a dose- and time-dependent manner [50]. The increased TFα-specific antibodies would prompt the host immune response, for the immune surveillance against tumorigenic cells present in the system. *B. xylanisolvens* has been considered among the NGPs because of its anticancer attributes, with no evidences of toxicity [51].

Bacteroides uniformis, B. acidifaciens, and *B. dorei* are equally mentioned. Oral administration of *B. uniformis* to high-fat–fed mice improved lipid profile, leptin levels, and TNF-α production by DCs concerning phagocytosis [52]. *B. uniformis* has been reported to ameliorate the immunological dysfunctions and metabolic disorders, related to intestinal dysbiosis in obese mice. Acute administration of this strain did not show any adverse effects.

B. acidifaciens has also been reported to have an antiobesity potential, as suggested by a study carried out on the knockout mouse [53]. The mechanism of diet-induced obesity prevention is by promoting the energy expenditure via activation of TGR5 (bile acid membrane receptor)–PPARγ pathway in the adipose tissue and GLP-1 increase in plasma, which is probably caused by reduction of dipeptidyl peptidase IV [53].

Eubacterium halli

This anaerobic bacterium belongs to the family Lachnospiraceae, important butyrate producers residing in the intestinal microbiota. Butyrate and other SCFAs, through different mechanisms, inhibit intestinal inflammation, maintain the intestinal barrier, and also modulate gut motility [54]. The endocrine function is benefitted via the synthesis of GLP-1 and GLP-2, for maintenance of insulin sensitivity and glucose tolerance. Administration of *Eubacterium halli* to obese and diabetic db/db mice increased energy metabolism and insulin sensitivity, with no effect on body weight and food intake. The increased dosage of *E. halli* did not show any toxicity [55].

Clostridium Clusters

The *Clostridium* clusters XIVa and IV, which form a substantial (10%–40%) part of gut microbiota, possess some beneficial attributes. *Clostridium butyricum* MIYAIRI has been studied for several decades, for its cholesterol-lowering effects [56], in cancer prevention [57] and treatment of *Clostridium difficile* infection [58]. The *Clostridium* clusters XIVa, IV, and XVIII isolated from human fecal samples have also been documented for the ability to accumulate Treg cells in the colon and thereby suppress inflammation. The SCFAs produced by these bacteria on degradation of dietary fibers are proposed to influence the expression of *FoxP3,* which is the main gene for the regulation of Tregs generation [59].

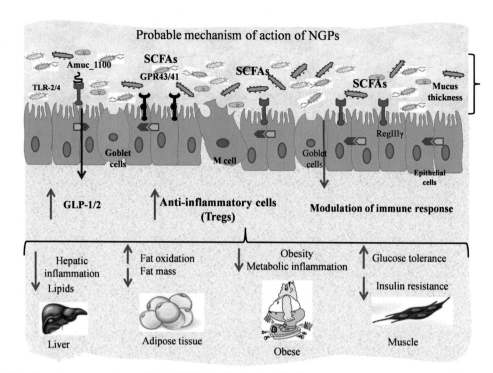

FIGURE 44.1 The effect of next-generation probiotics (NGPs) on the host health. The arrows (↑) shows upregulation and (↓) showdown-regulation. Illustration of NGPs–host interaction might have a beneficial effect on its physiology via different mechanisms. This figure presented here gives the overview that how NGPs exert their effects on the host intestine. The SCFAs derived from the digestion of food components via action through G-protein–coupled receptors (GPR-43/41) trigger the release of glucagon-like peptides (GLP-1/2) that have a diverse metabolic role, including the decreased insulin resistance and increase glucose tolerance. Some bacterial strains contribute to the mucosal defense system against pathogens by instigating mucin production and stimulating the secretion of antimicrobial peptides such as RegIIIγ. This interaction also ensures the stability of the intestinal microenvironment via modulation of immune response and generation of the regulatory T cells, which thereby reduces metabolic endotoxemia and inflammation. The increased fat oxidation and reduced metabolic inflammation also help in the reduction of body weight.

The population of these *Clostridium* clusters was found to be significantly low, in fecal material of inflammatory bowel disease (IBD) patients. Repopulation could significantly reverse the process of dysbiosis [59]. The general mechanism of action of NGPs is illustrated in Fig. 44.1.

Several studies related to in vitro or in vivo effects of NGPs are compiled in Table 44.1.

PRODUCTION CONSTRAINTS

Microbes intended to promote health should reach the gut in required numbers, to poise the desired effects. Candidate probiotics should be resistant to low gastric pH, bile salts, and digestive enzymes. Moreover, success in preclinical evaluation does not preclude failure, during clinical probiotics tests.

Microencapsulation aids in the process of gastrointestinal pass, however, encapsulating materials and technology, along with probiotic cell release, are crucial for therapeutic outcome. The biggest limitation with NGPs is that they are nearly strict anaerobes, so their isolation, propagation, and maintenance are not only tedious but also requires specific expertise. Formulating food products with NGPs is also a challenge, in comparison to conventional probiotics, as foods undergo many stages during industrial processing, which might interfere with the viability of the strain. Again, compatibility between strain and all food components needs to be confirmed.

SAFETY ASPECTS OF NGPS

The demand for probiotics has tremendously increased in the last couple of years, as reflected by market trends. Prebiotics should not be overlooked in this context, for example, fructooligosaccharides [60]. In India, the Food Safety and Standards Authority of India (FASSI) have passed a regulation in the year 2016, concerning the foods containing prebiotics or probiotics, or novel ingredients with health claimable benefits [61].

TABLE 44.1 Experimental Studies and Outcomes Involving Principal NGPs

Strain	Effects	Outcomes	Safety	Refs.
Akkermansia muciniphila	↓ body weight, metabolic endotoxemia, hypertension, insulin resistance, and inflammation ↑ gut peptide and glucose tolerance	Weight control, increase glucose tolerance, reduce insulin sensitivity, and inflammation	Potential NGP, safety under consideration	[28,29,33,34,36]
Faecalibacterium prausnitzii	↓ inflammatory cells (Th17), NF-κB, and IFN-γ ↑ regulatory T cells and IL-10	Antiinflammatory in Crohn's disease	Safety evaluation under process	[39,42–45]
Bacteroides spp.	Immune response, antibodies (IgM), and antiobesity potential ↑ energy expenditure, improve lipid profile, leptin	Decreased immune dysfunction and obesity	Safe only in nonviable form	[45,47–49,51]
Eubacterium halli	Gut motility, improve glucose tolerance and energy metabolism, and reduce insulin sensitivity	Increase insulin sensitivity and manage diabetes	No toxic effects shown	[53,54]
Clostridium cluster	Treatment of Clostridium difficile infection, increase regulatory T cells	Decrease cholesterol, cancer, and Clostridium difficile infection	No toxic effects found out	[56–58]

The type of health claim and nutrient involved should be clearly stated on the food, and prior to mass application, their efficacy should be accurately evaluated in human studies. Also, the food should be approved by the food authority. The use of symbols or pictures to claim the cure of a disease such as cancer is strictly prohibited. To date, no specific guidelines regarding the use of NGPs have been formulated.

In the European Union, microbe-based foods are regulated by the European Food Safety Authority (EFSA). Before the launch, these should pass the Qualified Presumption of Safety (QPS) parameters, established by EFSA in 2007 [62]. A limited numbers of microbes are covered in the QPS list. The panel formulated by EFSA on biological hazards will evaluate the QPS status of a microbial culture and classify it into a risk group [63]. The microbes of ancient use are deemed as traditional food ingredients and legally allowed for human usage without EFSA assessment.

In 2015, the EFSA panel on Dietetic Products, Nutrition and Allergies approved *Bacteroides xylanisolvens* DSM 23964, used as a primary culture for the fermentation of milk products. Only heat-treated inactivated cells will be allowed in the final product [64]. The bacterium was not included in the QPS list, due to lack of enough experimental studies, as judged by an EFSA panel on biological hazards.

A. muciniphila is a potential weight-lowering bacterium, but human trials in this concern are lacking. Also in some cancer patients, their population was found significantly increased [65]. The beneficial properties vary from strain to strain, and one cannot give a generalized statement for whole species. As a recently isolated microorganism, it is not included in the QPS list. Yet, safe administration to humans, of both live and pasteurized bacteria, has already been announced [39].

Similarly, *F. prausnitzii* is not included in the QPS list because of its nontraditional use, resistance to several antibiotics, and the lack of toxicological assessment [66].

In the United States, microbes with a long history of use are considered as GRAS (generally regarded as safe) and do not require premarket approval from the Food and Drug Administration (FDA). However, before claiming the health benefits of a probiotic in the form of supplement or food ingredient, the safety assessment and efficacy need approval and should be in accordance with the Dietary Supplement Health and Education Act (DSHEA).

Recently, the FDA has provided guidelines on newer food ingredients, microbes earlier not in use, which are unlikely to be allowed as food ingredients [67]. That creates many doubts for NGPs introduction.

CONCLUSION AND FUTURE DIRECTIONS

Probiotics may restore the disrupted gut microbiota, in turn modifying the host phenotype. Like any novel food ingredient, the use of NGPs should be in strict adherence to the regulatory authorities. This is one of the major constraints that NGPs might face, as their safety for human consumption has not been fully evaluated. Also, a good regulatory framework specifically for the use of NGPs should be prescribed by the regulatory agencies.

Many questions are still to be elucidated in context to NGPs. Basically the in silico approach should be utilized, to compare the different strains of a particular NGP, to gain insight on the "molecular islands" implicated in positive health effects. In addition, in vitro studies, especially on cell lines, are also required to better anticipate the in vivo effects.

The effect of NGPs on the gut virome is still to be elucidated, as viruses also play an important role in maintaining the health and disease state [68]. Formulation of newer delivery means, for the supply of these NGPs to the customer, is also a challenge.

Progress in the field of NGPs, and of bacterial therapy in general, should be accelerated by the utilization of new scientific tools such as systems biology, evolutionary engineering, metabolic engineering using recombinant DNA technology, and particularly gene editing using CRISPR-Cas [69–73] These approaches can be very helpful, to design and reconstruct the metabolic pathways of bacteria, toward the development of more effective probiotics [74,75].

REFERENCES

[1] Chan M. Obesity and Diabetes: the slow-motion disaster. Milbank Q 2017;95(1):11–4.

[2] Pischon T, Nimptsch K. Obesity and risk of cancer: an introductory overview. Springer Obes Cancer 2016:1–15.

[3] Rodgers RJ, Tschop MH, Wilding JP. Anti-obesity drugs: past, present and future. Dis Model Mech 2012;5(5):621–6.

[4] Madura JA, DiBaise II JK. Quick fix or long-term cure? Pros and cons of bariatric surgery. F1000 Med Rep 2012;4.

[5] Renuka, Agnihotri N, Singh AP, Bhatnagar A. Involvement of regulatory T cells and their cytokines repertoire in chemopreventive action of fish oil in experimental colon cancer. Nutr Canc 2016;68(7):1181–91.

[6] Renuka, Kumar S, Sharma B, Sharma P, Agnihotri N. n-3 PUFAs: an elixir in prevention of colorectal cancer. Curr Colorectal Cancer Rep 2015;11(3):141–9.

[7] Negi AK, Renuka, Bhatnagar A, Agnihotri N. Celecoxib and fish oil: a combination strategy for decreased inflammatory mediators in early stages of experimental mammary cancer. Inflammopharmacology 2016;24(1):11–22.

[8] Goyal S, Raj T, Banerjee C, Imam J, Shukla P. Isolation and characterization of probiotic microorganisms from curd and chili sauce. Int J Probiotics Prebiotics 2013;8(2/3):91.

[9] Linares DM, Gómez C, Renes E, Fresno JM, Tornadijo ME, Ross RP, Stanton C. Lactic acid bacteria and bifidobacteria with potential to design natural biofunctional health-promoting dairy foods. Front Microbiol 2017;8.

[10] Collins J, Thornton G, Sullivan G. Selection of probiotic strains for human applications. Int Dairy J 1998;8(5–6):487–90.

[11] Yadav R, Puniya AK, Shukla P. Probiotic properties of Lactobacillus plantarum RYPR1 from an indigenous fermented beverage Raabadi. Front Microbiol 2016;7.

[12] Nagpal R, Kumar A, Kumar M, Behare PV, Jain S, Yadav H. Probiotics, their health benefits and applications for developing healthier foods: a review. FEMS Microbiol Lett 2012;334(1):1–15.

[13] Dahiya DK, Puniya AK. Evaluation of survival, free radical scavenging and human enterocyte adherence potential of Lactobacilli with anti-obesity and anti-inflammatory CLA isomer-producing attributes. J Food Process Preserv 2015;39(6):2866–77.

[14] Dahiya DK, Puniya AK. Isolation, molecular characterization and screening of indigenous lactobacilli for their abilities to produce bioactive conjugated linoleic acid (CLA). J Food Sci Technol 2017;54(3):792–801.

[15] Dahiya DK, Puniya AK. Optimisation of fermentation variables for conjugated linoleic acid bioconversion by Lactobacillus fermentum DDHI27 in modified skim milk. Int J Dairy Technol 2017;71(1):46–55.

[16] Yadav R, Shukla P. An overview of advanced technologies for selection of probiotics and their expediency: a review. Crit Rev Food Sci Nutr 2017;57(15):3233–42.

[17] Yadav R, Singh PK, Puniya AK, Shukla P. Catalytic interactions and molecular docking of bile salt hydrolase (BSH) from Lactobacillus plantarum RYPR1 and its prebiotic utilization. Front Microbiol 2017;7:2116.

[18] O'Hara AM, Shanahan F. The gut flora as a forgotten organ. EMBO Rep 2006;7(7):688–93.

[19] Ubeda C, Djukovic A, Isaac S. Roles of the intestinal microbiota in pathogen protection. Clin Transl Immunology 2017;6(2):e128.

[20] Sharma S, Sharma V, Dahiya DK, Khan A, Mathur M, Sharma A. Prevalence, virulence potential, and antibiotic susceptibility profile of Listeria monocytogenes isolated from bovine raw milk samples obtained from Rajasthan, India. Foodborne Pathog Dis 2017;14(3):132–40.

[21] Sharma V, Sharma S, Dahiya DK, Khan A, Mathur M, Sharma A. Coagulase gene polymorphism, enterotoxigenecity, biofilm production, and antibiotic resistance in Staphylococcus aureus isolated from bovine raw milk in North West India. Ann Clin Microbiol Antimicrob 2017;16(1):65.

[22] Dahiya DK, Renuka, Puniya AK. Impact of nanosilver on gut microbiota: a vulnerable link. Future Microbiol 2017;13(4):483–92. https://doi.org/10.2217/fmb-2017-0103.

[23] Dahiya DK, Renuka, Puniya M, Shandilya UK, Dhewa T, Kumar N, Kumar S, Puniya AK, Shukla P. Gut microbiota modulation and its relationship with obesity using prebiotic fibers and probiotics: a review. Front Microbiol 2017;8(563).

[24] Clarke SF, Murphy EF, O'sullivan O, Lucey AJ, Humphreys M, Hogan A, Hayes P, O'reilly M, Jeffery IB. Exercise and associated dietary extremes impact on gut microbial diversity. Gut 2014;63(12):1913–20. https://doi.org/10.1136/gutjnl-2013-306541.

[25] O'Toole PW, Marchesi JR, Hill C. Next-generation probiotics: the spectrum from probiotics to live biotherapeutics. Nat Microbiol 2017;2:17057.

[26] Ouwehand AC, Derrien M, de Vos W, Tiihonen K, Rautonen N. Prebiotics and other microbial substrates for gut functionality. Curr Opin Biotechnol 2005;16(2):212−7.

[27] Derrien M, Vaughan EE, Plugge CM, de Vos WM. *Akkermansia muciniphila* gen. nov., sp. nov., a human intestinal mucin-degrading bacterium. Int J Syst Evol Microbiol 2004;54(5):1469−76.

[28] Cani PD, Everard A. Talking microbes: when gut bacteria interact with diet and host organs. Mol Nutr Food Res 2016;60(1):58−66.

[29] Everard A, Lazarevic V, Gaïa N, Johansson M, Ståhlman M, Backhed F, Delzenne NM, Schrenzel J, Francois P, Cani PD. Microbiome of prebiotic-treated mice reveals novel targets involved in host response during obesity. ISME J 2014;8(10):2116.

[30] Zhu L, Qin S, Zhai S, Gao Y, Li L. Inulin with different degrees of polymerization modulates composition of intestinal microbiota in mice. FEMS (Fed Eur Microbiol Soc) Microbiol Lett 2017;364(10).

[31] Liu TW, Cephas KD, Holscher HD, Kerr KR, Mangian HF, Tappenden KA, Swanson KS. Nondigestible fructans alter gastrointestinal barrier function, gene expression, histomorphology, and the microbiota profiles of diet-induced obese C57BL/6J mice. J Nutr 2016;146(5):949−56.

[32] Reid DT, Eller LK, Nettleton JE, Reimer RA. Postnatal prebiotic fibre intake mitigates some detrimental metabolic outcomes of early overnutrition in rats. Eur J Nutr 2016;55(8):2399−409.

[33] Png CW, Lindén SK, Gilshenan KS, Zoetendal EG, McSweeney CS, Sly LI, McGuckin MA, Florin TH. Mucolytic bacteria with increased prevalence in IBD mucosa augment *in vitro* utilization of mucin by other bacteria. Am J Gastroenterol 2010;105(11):2420.

[34] Belzer C, De Vos WM. Microbes inside—from diversity to function: the case of *Akkermansia*. ISME J 2012;6(8):1449.

[35] Zhang X, Shen D, Fang Z, Jie Z, Qiu X, Zhang C, Chen Y, Ji L. Human gut microbiota changes reveal the progression of glucose intolerance. PLoS One 2013;8(8):e71108.

[36] Dao MC, Everard A, Aron-Wisnewsky J, Sokolovska N, Prifti E, Verger EO, Kayser BD, Levenez F, Chilloux J, Hoyles L, Dumas ME. *Akkermansia muciniphila* and improved metabolic health during a dietary intervention in obesity: relationship with gut microbiome richness and ecology. Gut 2015;65(3):426−36. https://doi.org/10.1136/gutjnl-2014-308778.

[37] Everard A, Belzer C, Geurts L, Ouwerkerk JP, Druart C, Bindels LB, Guiot Y, Derrien M, Muccioli GG, Delzenne NM, De Vos WM. Cross-talk between *Akkermansia muciniphila* and intestinal epithelium controls diet-induced obesity. Proc Natl Acad Sci U S A 2013;110(22):9066−71.

[38] Derrien M, Van Baarlen P, Hooiveld G, Norin E, Müller M, de Vos WM. Modulation of mucosal immune response, tolerance, and proliferation in mice colonized by the mucin-degrader *Akkermansia muciniphila*. Front Microbiol 2011;2.

[39] Plovier H, Everard A, Druart C, Depommier C, Van Hul M, Geurts L, Chilloux J, Ottman N, Duparc T, Lichtenstein L, Myridakis A, Delzenne NM, Klievink J, Bhattacharjee A, van der Ark KC, Aalvink S, Martinez LO, Dumas ME, Maiter D, Loumaye A, Hermans MP, Thissen JP, Belzer C, de Vos WM, Cani PD. A purified membrane protein from *Akkermansia muciniphila* or the pasteurized bacterium improves metabolism in obese and diabetic mice. Nat Med 2017;23(1):107−13.

[40] Shin NR, Lee JC, Lee HY, Kim MS, Whon TW, Lee MS, Bae JW. An increase in the *Akkermansia* spp. population induced by metformin treatment improves glucose homeostasis in diet-induced obese mice. Gut 2013;63(5):727−35. https://doi.org/10.1136/gutjnl-2012-303839.

[41] Ottman N, Reunanen J, Meijerink M, Pietilä TE, Kainulainen V, Klievink J, Huuskonen L, Aalvink S, Skurnik M, Boeren S, Satokari R. Pili-like proteins of *Akkermansia muciniphila* modulate host immune responses and gut barrier function. PLoS One 2017;12(3):e0173004.

[42] Li M, Wang B, Zhang M, Rantalainen M, Wang S, Zhou H, Zhang Y, Shen J, Pang X, Zhang M, Wei H. Symbiotic gut microbes modulate human metabolic phenotypes. Proc Natl Acad Sci U S A 2008;105(6):2117−22.

[43] Miquel S, Martin R, Rossi O, Bermudez-Humaran LG, Chatel JM, Sokol H, Thomas M, Wells JM, Langella P. *Faecalibacterium prausnitzii* and human intestinal health. Curr Opin Microbiol 2013;16(3):255−61.

[44] Sokol H, Pigneur B, Watterlot L, Lakhdari O, Bermúdez-Humarán LG, Gratadoux JJ, Blugeon S, Bridonneau C, Furet JP, Corthier G, Grangette C. *Faecalibacterium prausnitzii* is an anti-inflammatory commensal bacterium identified by gut microbiota analysis of Crohn disease patients. Proc Natl Acad Sci U S A 2008;105(43):16731−6.

[45] Breyner NM, Michon C, de Sousa CS, Boas PB, Chain F, Azevedo VA, Langella P, Chatel JM. Microbial anti-inflammatory molecule (MAM) from *Faecalibacterium prausnitzii* shows a protective effect on DNBS and DSS-induced colitis model in mice through inhibition of NF-κB pathway. Front Microbiol 2017;8.

[46] Quévrain E, Maubert MA, Michon C, Chain F, Marquant R, Tailhades J, Miquel S, Carlier L, Bermúdez-Humarán LG, Pigneur B, Lequin O. Identification of an anti-inflammatory protein from *Faecalibacterium prausnitzii*, a commensal bacterium deficient in Crohn's disease. Gut 2015;65(3):415−25. https://doi.org/10.1136/gutjnl-2014-307649.

[47] Deng H, Li Z, Tan Y, Guo Z, Liu Y, Wang Y, Yuan Y, Yang R, Bi Y, Bai Y, Zhi F. A novel strain of *Bacteroides fragilis* enhances phagocytosis and polarises M1 macrophages. Sci Rep 2016;6:29401.

[48] Moncrief JS, Duncan AJ, Wright RL, Barroso LA, Wilkins TD. Molecular characterization of the fragilysin pathogenicity islet of enterotoxigenic *Bacteroides fragilis*. Infect Immun 1998;66(4):1735−9.

[49] Obiso R, Azghani AO, Wilkins TD. The Bacteroides fragilis toxin fragilysin disrupts the paracellular barrier of epithelial cells. Infect Immun 1997;65(4):1431−9.

[50] Ulsemer P, Toutounian K, Kressel G, Goletz C, Schmidt J, Karsten U, Hahn A, Goletz S. Impact of oral consumption of heat-treated *Bacteroides xylanisolvens* DSM 23964 on the level of natural TFα-specific antibodies in human adults. Benef Microbes 2016;7(4):485−500.

[51] Ulsemer P, Toutounian K, Schmidt J, Karsten U, Goletz S. Preliminary safety evaluation of a new *Bacteroides xylanisolvens* isolate. Appl Environ Microbiol 2012;78(2):528−35.

[52] Cano PG, Santacruz A, Moya Á, Sanz Y. *Bacteroides uniformis* CECT 7771 ameliorates metabolic and immunological dysfunction in mice with high-fat-diet induced obesity. PLoS One 2012;7(7):e41079.

[53] Yang JY, Lee YS, Kim Y, Lee SH, Ryu S, Fukuda S, Hase K, Yang CS, Lim HS, Kim MS, Kim HM. Gut commensal *Bacteroides acidifaciens* prevents obesity and improves insulin sensitivity in mice. Mucosal Immunol 2017;10(1):104−16.

[54] Engels C, Ruscheweyh H-J, Beerenwinkel N, Lacroix C, Schwab C. The common gut microbe *Eubacterium hallii* also contributes to intestinal propionate formation. Front Microbiol 2016;7.

[55] Udayappan S, Manneras-Holm L, Chaplin-Scott A, Belzer C, Herrema H, Dallinga-Thie GM, Duncan SH, Stroes ES, Groen AK, Flint HJ, Backhed F. Oral treatment with *Eubacterium hallii* improves insulin sensitivity in db/db mice. NPJ Biofilms Microbiomes 2016;2:16009.

[56] Kobashi K, Takeda Y, Itoh H, Ji H. Cholesterol-lowering effect of *Clostridium butyricum* in cholesterol-fed rats. Digestion 1983;26(4):173−8.

[57] Shinnoh M, Horinaka M, Yasuda T, Yoshikawa S, Morita M, Yamada T, Miki T, Sakai T. *Clostridium butyricum* MIYAIRI 588 shows antitumor effects by enhancing the release of TRAIL from neutrophils through MMP-8. Int J Oncol 2013;42(3):903−11.

[58] Woo TD, Oka K, Takahashi M, Hojo F, Osaki T, Hanawa T, Kurata S, Yonezawa H, Kamiya S. Inhibition of the cytotoxic effect of *Clostridium difficile in vitro* by *Clostridium butyricum* MIYAIRI 588 strain. J Med Microbiol 2011;60(11):1617−25.

[59] Atarashi K, Tanoue T, Oshima K, Suda W, Nagano Y, Nishikawa H, Fukuda S, Saito T, Narushima S, Hase K, Kim S. Treg induction by a rationally selected mixture of *Clostridia strains* from the human microbiota. Nature 2013;500(7461):232.

[60] Yadav R, Singh PK, Shukla P. Production of fructooligosaccharides as ingredients of probiotic applications. Microbial Biotechnology: An Interdisciplinary Approach. 2016. p. 311.

[61] India FSaSAo. Food safety and standards (Health supplements, Nutraceuticals, food for special dietary use, food for special medical purpose, functional food and novel food) Regulations. 2016.

[62] Barlow S, Chesson A, Collins JD, Dybing E, Flynn A, Fruijtier-Pölloth C, Hardy A, Knaap A, Kuiper H, Le Neindre P, Schans J. Introduction of a qualified presumption of safety (QPS) approach for assessment of selected microorganisms referred to EFSA. Opin Sci Comm EFSA J 2007;587:1−16.

[63] Panel EB. Scientific opinion on the maintenance of the list of QPS biological agents intentionally added to food and feed (2013 update). EFSA J 2013;11(11):3449.

[64] Authority E. Scientific opinion on the safety of heat-treated milk products fermented with *Bacteroides xylanisolvens* DSM 23964'as a novel food. EFSA J 2015;13(1).

[65] Weir TL, Manter DK, Sheflin AM, Barnett BA, Heuberger AL, Ryan EP. Stool microbiome and metabolome differences between colorectal cancer patients and healthy adults. PLoS One 2013;8(8):e70803.

[66] Brodmann T, Endo A, Gueimonde M, Vinderola G, Kneifel W, de Vos WM, Salminen S, Gómez-Gallego C. Safety of novel microbes for human consumption: practical examples of assessment in the European Union. Front Microbiol 2017;8:1725.

[67] Green JM, Barratt MJ, Kinch M, Gordon JI. Food and microbiota in the FDA regulatory framework. Science 2017;357(6346):39−40.

[68] Renuka, Dahiya DK. The gut virome: a neglected actor in colon cancer pathogenesis. Future Microbiol 2017;12(15):1345−8. https://doi.org/10.2217/fmb-2017-0159.

[69] Gupta SK, Shukla P. Gene editing for cell engineering: trends and applications. Crit Rev Biotechnol 2017;37(5):672−84.

[70] Kumar Singh P, Shukla P. Systems biology as an approach for deciphering microbial interactions. Briefings Funct Genomics 2014;14(2):166−8.

[71] Dangi AK, Dubey KK, Shukla P. Strategies to improve *Saccharomyces cerevisiae*: technological advancements and evolutionary engineering. Indian J Microbiol 2017:1−9.

[72] Gupta SK, Srivastava SK, Sharma A, Nalage VH, Salvi D, Kushwaha H, Chitnis NB, Shukla P. Metabolic engineering of CHO cells for the development of a robust protein production platform. PLoS One 2017;12(8):e0181455.

[73] Gupta SK, Shukla P. Advanced technologies for improved expression of recombinant proteins in bacteria: perspectives and applications. Crit Rev Biotechnol 2016;36(6):1089−98.

[74] Yadav R, Kumar V, Baweja M, Shukla P. Gene editing and genetic engineering approaches for advanced probiotics: a review. Crit Rev Food Sci Nutr 2016:1−12.

[75] Yadav R, Singh P, Shukla P. Metabolic engineering for probiotics and their genome-wide expression profiling. Curr Protein Pept Sci 2016;19(1):68−74.

Chapter 45

Fecal Microbiota Transfer and Inflammatory Bowel Disease: A Therapy or Risk?

Krista M. Newman, Carlos G. Moscoso, and Byron P. Vaughn

Division of Gastroenterology, Hepatology and Nutrition, University of Minnesota, Minneapolis, MN, United States

INTRODUCTION

Inflammatory bowel diseases (IBDs), Crohn's disease (CD), and ulcerative colitis (UC) are manifested by chronic intestinal inflammation that can have a devastating impact on patients' lives and represent a significant burden to the healthcare system. The incidence of IBD has been on the rise since the 20th century and now affects over 1.5 million individuals in the United States [1]. The etiology IBD is not clear, although it appears to be related to an inappropriate immune response to the intestinal microbiota (or their products), in a genetically susceptible individual [2].

Although numerous genes are associated with IBD, only a small number contribute to development of IBD [2–4]. There has been rising interest in the role of dysbiosis or deviations from a healthy intestinal microbial community in the development and treatment of IBD [2,5,6].

INVESTIGATING THE MICROBIOME: THE OMICS ERA

Nucleic acid–based approaches that target 16S ribosomal RNA have identified 10^{14} bacteria that inhabit the human intestine. However, one strong limitation to 16S rRNA sequencing is that it is limited to broad taxonomic profiles and cannot reliably distinguish species or strain variants to define bacterial function [7]. To overcome this challenge, next-generation shotgun metagenomic, transcriptomic, proteomic, and metabolomic profiles of bacterial function known collectively as "omics studies" have further revolutionized investigations of the microbiome [7,8]. These omics studies allow for expanded resolution to the bacterial strain level, microbial function, as well as evaluation of fungi, archaea, and viruses.

Healthy Intestinal Microbiome

Current applications of the Human Microbiome Project have revealed marked interindividual diversity in Western populations. In addition, Knights et al. observed that longitudinally stable human intestinal microbiomes may not necessarily exist. Rather, there is rising evidence that individual microbial communities are dynamic and vary continuously along a complex multidimensional distribution [9–11], highlighting the need for ongoing evaluation of the intestinal microbiome. Nevertheless, there appear to be hallmark shifts in microbial communities related to general health and disease.

Nucleic acid sequencing techniques have identified that the human intestinal microbiome is composed of 500–2000 different commensal species of bacteria that line the gastrointestinal tract in axial (from mucosa to lumen) and longitudinal (proximal to distal) gradients [12–16]. Aberrations from a healthy core microbiome and related function are associated with IBD. Such states of dysbiosis or imbalance between proinflammatory and antiinflammatory bacterial populations generally include globally reduced alpha diversity (species richness), specific alterations in particular bacterial populations, and temporal instability [6,17].

Alterations of Bacterial Populations in Inflammatory Bowel Disease

In patients with IBD, there are decreased proportions of protective bacteria and increased pathogenic species, as well as an overall decrease in bacterial diversity. Typically, species of the Firmicutes and Bacteroidetes phyla are proportionately decreased, while there is an increase in the class Gammaproteobacteria of the Proteobacteria phylum [18]. There are also unique differences in microbial populations between patients with UC and CD.

In UC, these alterations specifically include decreased quantities of *Clostridium coccoides* group of the Firmicutes phyla and increased sulfate-reducing bacteria. Interestingly, patients with UC who are in remission have increased levels of *Faecalibacterium prausnitzii* of the Firmicutes phyla [19,20], whereas in CD, *F. prausnitzii* is diminished and *Mycobacterium avium*, subspecies *paratuburculosis* (MAP), is found [21].

Diminished Protective Bacteria

Bacterial phyla that reduce or regulate intestinal inflammation, such as Bacteroidetes (e.g., *Bacteroides fragilis*), Firmicutes (e.g., *Lactobacillus*, *F. prausnitzii* and *Clostridium* strains), and Actinobacteria (e.g., *Bifidobacterium*), are reduced in IBD. There are several mechanisms by which these beneficial bacteria modulate immune responses, including expansion of T_{regs}, stimulation of antiinflammatory cytokines (interleukin [IL]-10), regulation of proinflammatory cytokines (IL-17), and related pathways (nuclear factor-κB [NF-κB]).

Within the Bacteroidetes and Firmicutes phyla, *Bacteroides* and *Clostridium* species can promote expansion of regulatory T cells known as T_{regs}. One specific example of this includes the capsular exopolysaccharide, polysaccharide A (PSA), on *B. fragilis*. In murine models, PSA downregulates pathogenic T_H17 cell differentiation, in turn downregulating synthesis and secretion of the proinflammatory cytokine IL-17. Additionally, PSA promotes IL-10 production by Fox $P3^+$ $T_{regs.}$ [19,20]. Similarly, *Clostridium* strains have immune-suppressing activity, through stimulation of $T_{regs.}$

Clostridium species produce short-chain fatty acids, such as butyrate, that stimulate epithelial cells to produce transforming growth factor (TGF)-β, a key cytokine for differentiation and expansion of $T_{regs.}$ [22]. The beneficial effects of the microbial induction and proliferation of T_{regs} have been observed in several murine models [19,20,23]. For example, Atarashi et al. noted reduced intestinal inflammation, mediated by T_{reg} induction and accumulation in germ-free mice, inoculated with fecal samples rich in *Bacteroides* and *Clostridium* species (T_{reg} inducers) [22].

In addition to activation of T_{regs}, other commensal bacteria can downregulate proinflammatory cytokines. A prime example of this is *F. prausnitzii* of the Firmicutes phylum, which is generally diminished in IBD. *F. prausnitzii* regulates inflammation by production of a unique protein, termed microbial antiinflammatory molecule, which mediates antiinflammatory effects, through suppression of the prototypical proinflammatory NF-κB pathway [24,25]. Further, *F. prausnitzii* induces IL-10-secreting $T_{regs.}$ [25,26].

Increased Pathogenic Bacteria

There is mounting evidence that particular members of commensal bacteria, termed pathobionts, expand with dysbiosis and as a result exert pathogenic effects on the host [27–29].Increased populations of these potentially pathogenic bacteria are thought to be deleterious in patients with IBD. Generally, patients with IBD have a proportional increase in Proteobacteria and Fusobacteria [30–35]. These bacterial phyla can be pathogenic in particular circumstances and can increase inflammation through different mechanisms, including epithelial cell invasion with resultant epithelial damage, loss of mucin-secreting goblet cells, and activation of proinflammatory cytokines [36].

Crohn's Disease Precipitating Agents

In CD, pathogenic bacteria involved in granuloma formation are suspected to be involved in the onset of disease, such as MAP (phylum Actinobacteria), and adherent-invasive *Escherichia coli* (AIEC) (phylum Proteobacteria). Both species have been isolated from ileal biopsies of patients with CD [21,33]. Darfeuille-Michaud et al. found an increased percentage of AIEC, from ileal biopsies of CD patients with chronic lesions, new lesions, and stable disease, when compared to controls [37].

Identification of MAP was more specific. Hulten et al. identified MAP in granulomatous CD but not in non-granulomatous disease, or non-IBD patients [38]. This data suggest specific roles for these bacteria in different disease states. Initially, MAP was felt to be the primary pathogen responsible for CD, with additional evidence implicating this species in granuloma formation in animal models. Timms et al. demonstrated the ability of MAP to cause chronic granulomatous enteritis in cattle [39], though clinical data are conflicting [40] (Table 45.1).

TABLE 45.1 Changes in Bacterial Populations in Patients With Inflammatory Bowel Disease

Increased Proinflammatory Bacterial Populations	Reduced Protective Bacterial Populations
Proteobacteria *Escherichia coli,* adherent-invasive *Proteus mirabilis* *Klebsiella pneumoniae* *Bilophyla wadsworthia*	**Bacteroidetes** *Bacteroides fragilis*
Fusobacteria *Fusobacterium varium*	**Firmicutes** *Lactobacillus* *Faecalibacterium prausnitzii* Groups IV and XIV A Clostridium

Selby et al. observed symptomatic improvement with antibacterial treatment of MAP. However, there was no change in inflammatory markers or on endoscopic assessment [41]. The second species, AIEC, has a particularly high prevalence in the ileal mucosa of CD patients [37]. This strain of pathogenic *E. coli* invades epithelial cells, replicates in macrophages, and induces large amounts of the proinflammatory cytokine tumor necrosis factor alfa in vitro [33,34]. Ultimately, the end result of AIEC is tissue damage and inflammation.

Contribution of Proteobacteria Pathobionts

Proteobacteria are associated with experimental colitis in murine models [35,42]. T-bet$^{-/-}$ x RAG2$^{-/-}$ knockout mice (TRUC) are useful models of experimental colitis that develop spontaneous colitis with histologic features similar to UC in humans. Garrett et al. found that the presence of *Klebsiella pneumoniae, Proteus mirabilis,* and *Bilophyla wadsworthia* (phyla Proteobacteria) correlated with colitis in TRUC animals [35].

Further Implications

Although changes in bacterial populations are well described in patients with IBD, the question remains whether this is due to chronic intestinal inflammation, a result of prior therapy for IBD (including antibiotics), or if the bacteria are driving the inflammation. To address this question in part, Gevers et al. examined the mucosal and stool-related microbes in treatment-naïve pediatric patients with CD [43]. Consistent with prior work, these patients show significantly higher Proteobacteria, with reduced Bacteroidetes and Firmicutes phyla, suggesting that dysbiosis is likely an early change in IBD, rather than a result of therapy.

Interestingly this cohort had marked differences between mucosal and stool-related dysbiosis. Examination of stool samples only showed a weak correlation to the marked dysbiosis found in mucosal samples [43]. This observation is noteworthy as the majority of correlative data rely on stool specimens.

FECAL MICROBIOTA TRANSPLANTATION IN INFLAMMATORY BOWEL DISEASE

Fecal microbiota transplantation (FMT) is not only a more robust and complete method of manipulating the gut microbiome, through reconstitution of the entire microbial population, compared to antibiotics or probiotics but also offers more stable engraftment [44], which is associated with maintenance of remission. In contrast to antibiotics, FMT increases, rather than decreases, the diversity of intestinal bacteria. Compared to probiotics, FMT donor material has a substantially higher bacterial count and diversity. One gram of donor fecal material contains approximately 10^{11} bacterial cells, in addition to viruses, fungi, and archaea [45].

Traditionally, there has been great success with FMT in recurrent *Clostridium difficile* infections (CDIs) [46,47] to restore a healthy intestinal microbiome. This has spurred great interest for the potential utility of FMT to restore a healthy gut microbiome in patients with IBD.

Ulcerative Colitis

Three systematic reviews analyzing the data in FMT for UC clearly show significant increases in clinical and endoscopic remission [48−50]. The most recent systematic review and meta-analysis were conducted by Narula et al. with four randomized controlled trials (RCTs) and 277 patients [51—54]. FMT was indeed associated with higher combined clinical and endoscopic remission compared to placebo, with a number needed to treat of five [50].

A 2015 RCT by Moayyedi and colleagues demonstrated significant remission in UC patients, given FMT via weekly retention enema for 6 weeks, when compared to placebo [51]. Several of the patients who received FMT and underwent remission received fecal material from the same donor, raising important concerns regarding the importance of thorough donor selection. A subsequent trial compared patients with mild to moderately active UC who received FMT via the nasoduodenal route from healthy donors versus autologous fecal microbiota as a control [52].

There was no statistically significant difference in endoscopic or clinical remission found between these two groups, suggesting that FMT is not effective in increasing rates of remission. However, the authors noted some shortcomings in this study, namely the administration of only two rounds of FMT (once every 3 weeks) compared to six in the Moayyedi trial, as well as the nasoduodenal route of administration. Additionally, in responders, there were distinct signature changes in the microbiota, specifically a gain in *Clostridium* clusters, again undergirding the importance of donor selection and donor-patient matching (Table 45.2).

A subsequent trial attempted to determine the efficacy of intensive rounds of multidonor FMT, delivered by retention enema, on induction of clinical and endoscopic remission in 85 patients [53]. Significant reduction in remission was achieved using five weekly enemas for 8 weeks (27% vs. 8% in the control group), with distinct microbial changes in responders and nonresponders. The intensive nature of the dosing limited the practical applicability of this study, as different administration schedules were not simultaneously evaluated. Thus far, the evidence is variable, and the myriad dosing schedules and routes of administration make direct comparative assessment challenging.

Recently, a study comparing multidonor, frozen-and-thawed, colonoscopically administered FMT found that FMT was more effective than placebo (autologous FMT), at induction and maintenance of remission at 8 weeks, when delivered three times in 1 week [54]. This study included rigorous clinical endpoints, including a mandatory oral corticosteroid taper. Those unable to discontinue steroids were categorized as nonresponders. This study provided further evidence that FMT is a feasible modality to achieve clinical remission, though it did not address the optimal administration schedule nor did it compare multiple versus single FMT donor composition.

A systematic review and meta-analysis of 29 studies including 514 patients [49], including the aforementioned RCTs, demonstrated a lower rate of IBD symptomatic worsening with FMT for IBD compared to CDI. Additionally, discordance between rates of worsening IBD activity was observed between RCT and observational data, with RCTs demonstrating low IBD activity rates with lower gastrointestinal (GI) delivery (colonoscopy, retention enema) compared to upper GI delivery (nasogastric tube, nasoduodenal tube or upper endoscopy), in contrast to findings in observational data. However, the low rate of IBD activity worsening observed in RCT may be outweighed by the benefits observed with FMT, bolstering its use as a novel therapeutic for IBD treatment.

TABLE 45.2 Fecal Microbiota Administration in Ulcerative Colitis[a]

Preparation (Reference)	Route	Frequency	Duration (Weeks)	Control	N, Total	Outcome	Endoscopic or Clinical Remission
Anaerobically prepared, pooled donor stool (3—4 donors) (56)	Colonoscopy and enema	Colonoscopy at time 0 and two enemas on day 7	8	Autologous stool	73	Positive	Both
Pooled donor stool (3—7 donors) (55)	Colonoscopy and enema	Colonoscopy at time 0, then five enemas per week	8	Saline enema	81	Positive	Clinical
Single donor (53)	Enema	Weekly	6	Water enema	65	Positive	Endoscopic
Single donor (54)	Nasoduodenal	One at time 0 then week 3	6	Autologous stool	48	No difference	—

[a]Pooled rate for combined endoscopic and clinical remission is 27.9% for treatment and 9.5% for control intervention, with Number needed to treat (NNT) of 5 (95% confidence interval: 4—10) [48,50—54].

Similar, although smaller data exist for children, a small pediatric pilot RCT comparing single-donor, frozen-and-thawed, colonoscopically administered FMT to autologous FMT saw a significant benefit to FMT [55]. All children were given mesalamine for the duration of the trial.

Crohn's Disease

Data supporting efficacy of FMT for CD is more limited. There are no RCTs evaluating the efficacy of FMT in this population. An older (2014) systematic review and meta-analysis of FMT in IBD found a 36.2% clinical remission rate for UC and CD, with a 60.5% remission rate in CD alone [56]. A prospective open-label study of nasogastrically administered FMT in nine pediatric patients, with mild-to-moderate CD, demonstrated that 7/9 patients (78%) achieved clinical remission by week two, though only 5/7 maintained remission by week 12 [57].

A cohort study, completed by Cui et al., that included 30 patients with mid-gut refractory CD had favorable results, with 77% clinical remission at 1 month, following a single round of FMT through nasoduodenal delivery [58]. Additionally, a prospective study including 19 subjects treated with colonoscopic FMT found a significant increase in microbial diversity following transplantation. However, only 58% of patients (11/19) demonstrated a clinical response, as measured by a Harvey−Bradshaw Index, decrease by greater than three points [59].

SAFETY

Studies evaluating safety in this cohort have only reported mild, self-limiting side-effects, such as fever or abdominal cramping. However, reports of more potentially serious complications exist. A systematic review of FMT in refractory CDI and other intestinal disorders including IBD only reported a limited number of the most serious events. This review included 10 publications that reported death as the most serious outcome. Of these, only one patient death in 1089 recipients was directly related to FMT (aspiration during sedation for colonoscopy) [60].

The potential and concern for IBD flares following FMT is also present. Khoruts et al. noted a 25% increased risk of flare in patients with IBD, who received FMT for multiple recurrent CDI [47]. However, it is notable that CDI is a likely confounder for this observation. Recently, Qazi and colleagues evaluated the rate of worsening in IBD activity in FMT, in a systematic review in 2017 [49]. Overall, high-quality studies and RCTs related only a marginal risk of worsening in IBD activity (4.6%, 95% confidence interval: 1.8%−11%).

Although initial studies may be promising with at least short-term safety for FMT in IBD, overall long-term longitudinal evidence is still scarce. Further, it should be noted that current safety evaluations have not addressed the presence of potentially deleterious species or products (sulfate reducing and proteolytic bacteria) in donor samples. Further, there is sparse understanding of the fungal and viral components included in FMT [45]. Additional questions remain that seek to identify the ideal mode, duration, and frequency of treatment, along with selection of ideal donors for individual recipients.

TECHNICAL CHALLENGES, FUTURE CONSIDERATIONS, AND NEXT-GENERATION FECAL MICROBIOTA TRANSPLANTATION

FMT continues to emerge as a promising alternative to conventional therapies in IBD. A growing understanding of the role of the gut microbiome, the implications of dysbiosis, and the therapeutic effect of healthy microbiome reconstitution, as well as maintenance of engraftment, are driving much of the research into this treatment modality (Fig. 45.1). Further advances in the field include the development of stool-derived mixtures of defined microbe populations, so-called microbial ecosystem therapeutics that could define the next-generation of FMT [61].

Donor Screening and Selection

So far, protocols for donor screening and preparation of fecal suspension samples remain unstandardized. Donor selection is an important component of FMT from regulatory, patient safety, and disease control perspectives. While in the case of CDI, a single-donor FMT dose appears sufficient, it is clear from the experience in UC that diversity and frequency are essential elements for efficacy in IBD. The current strategy of donor screening places significant emphasis on infectious disease; however, in the case of IBD, it may be important to stratify fecal material on other parameters such as diversity, short-chain fatty acid, bile acid, or H_2S content.

Many donors are asymptomatic carriers of pathogenic species including *C. difficile*, rotavirus, and *Blastocystis hominis* [53,62,63]. Additionally, screening for viral opportunistic infections via polymerase chain reaction has become an

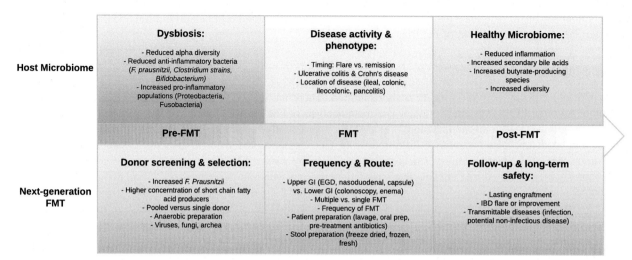

FIGURE 45.1 Schemata of host microbiota alterations and key themes prior to, at administration, and post-FMT for IBD. *FMT*, fecal microbiota transplantation; *IBD*, Inflammatory bowel disease; *EGD*, Esophagogastroduodenoscopy

important consideration. While this may not be relevant for CDI, it may have profound implications in FMT for IBD. Addressing these various components could limit the potential pool of suitable donors, though the lack of long-term safety data undergirds the importance of robust donor selection protocols to prevent adverse events of FMT.

The American Gastroenterological Association has launched the FMT National Registry to assess short- and long-term outcomes of FMT, which will yield important data to guide FDA regulatory efforts. However, this will mostly comprise of FMT for CDI, and it is not clear how generalizable this will be to IBD. Rather than use of donor-directed FMT material, another possibility is to generate a predefined bacterial consortium. This is appealing in that it would limit infectious risk and be reproducible in terms of content and abundance.

However, which bacteria to include in the consortium are unknown. It is likely that larger scale trials, with donor-directed FMT and subsequent network analysis, will be needed to determine which bacteria are associated with mucosal and clinical outcomes.

Formulation of Fecal Samples

Research into the optimal formulation of FMT exposes concerns regarding availability of fresh samples as opposed to the convenience offered by frozen samples with respect to storage and shipping [64]. Lee et al. examined whether frozen FMT was noninferior to fresh, standard FMT, with respect to efficacy in treatment of recurrent or refractory CDI. In this RCT with 232 patients, frozen-and-thawed FMT was deemed to be noninferior to fresh FMT prepared in the standard manner [65].

Using the frozen formulation can reduce the number and frequency of donor screenings, provide immediately available samples, and offer frozen FMT to facilities lacking on-site laboratory capabilities. More recently, novel freeze-dried encapsulated preparations have shown promise as well. Preliminary studies have shown the capability to freeze-dry samples, for further ease of storage and preparation in capsules, with comparable efficacy for treatment of recurrent CDI [66,67].Given that the frequency and intensity of FMT appears to be important in IBD, formulation and storage will be relevant for clinical trial and potential future clinical use [54].

Route of Delivery

Early studies employed a nasogastric or nasoduodenal route of administration. Significant barriers need to be overcome regarding patient acceptance of this modality, given its inherently insalubrious esthetic. Alternative routes have also been studied. For the treatment of CDI, a pooled analysis of studies comparing nasogastric versus colonoscopic routes revealed comparable efficacy despite some procedural differences mostly revolving around concomitant use of lavage with and without antibiotics [68].

Lee et al. looked at a case series of 94 patients, using single and multiple FMTs delivered by retention enema and examined clinical resolution rates as well as a longitudinal safety assessment. The cumulative CDI resolution rate was 86%

following four or more FMTs and increased to 92% when antibiotic therapy was employed between FMTs. No reported adverse events as well as a lack of recurrence at 6–24 months of follow-up suggested that FMT delivered via retention enema is effective and safe for treatment of recurrent or refractory CDI [69]. As described earlier, meta-analyses of FMT in IBD have demonstrated overall increased safety and efficacy of FMT administered via the lower gastrointestinal route.

Frequency of Delivery

In UC, administrations schedules in RCTs have ranged from once weekly for 6 weeks [51] to five times weekly for 8 weeks [53], to once every 3 weeks [52], and to three times a week [54]. In CD, single administration has been employed in the three studies [57–59].At this time it is unknown what the optimal frequency should be for IBD.

Engraftment

The baseline richness of the donor microbiome, as defined by the number of different bacterial operational taxonomic units, has been shown to be higher in healthy donors compared with IBD patients [70], and higher bacterial richness was identified in donors, with stools resulting in a response to FMT. In this study, Vermeire et al. assessed the response to FMT in 14 refractory patients, eight with UC and six with CD, through both nasojejunal and rectal routes of delivery. There was no significant difference between the baseline richness of patient versus donor stool between responders and non-responders, suggesting that it is the absolute richness of the donor stool sample and not the difference in bacterial richness between donor and recipient that is a correlate of successful FMT.

There was also a trend toward maintenance of bacterial richness in patients with maintenance of remission at week eight (p = .05833), suggesting a correlation between successful engraftment of donor microbiota and maintenance of remission in IBD. The relevant transferred phylotypes in this study in UC patients were identified as *Roseburia* and *Oscillibacter* of the Firmicutes phylum.

In CD patients, four transferred phylotypes were identified, including members of both the Firmicutes and Bacteroidetes phyla. However, these findings did not achieve statistical significance. *Roseburia* is a butyrate-producing genus, and its colonization in successful FMT has been previously reported [71], as well as an inverse relationship between *Roseburia* abundance and UC activity has been reported [72].

CONCLUSIONS

Over the past decade, great advances in mechanistic elucidation in IBD have been achieved, providing significant insight into our understanding of intestinal microbiota and into implications in intestinal inflammation. Unique alterations in commensal bacteria in CD and UC and related bacterial function and products have inspired investigations with novel therapies, including FMT. This therapy has proven remarkably effective in treatment of recurrent CDI infections and in its early stages shows promise in IBD as well.

While FMT appears to be safe for the treatment of recurrent CDI, the safety profile in IBD remains to be established. Challenges remain with uniformity in trials to date. There is great heterogeneity in route, dose, delivery, frequency, and donor selection. Future targeted studies to identify specific donor profiles may improve therapeutic outcomes.

LIST OF ACRONYMS AND ABBREVIATIONS

CD Crohn's disease
CDI *Clostridium difficile* infection
FMT Fecal microbiota transplant
HMP Human Microbiome Project
IBD Inflammatory Bowel Disease
RCT Randomized controlled trial
UC Ulcerative colitis

REFERENCES

[1] Molodecky NA, Soon IS, Rabi DM, et al. Increasing incidence and prevalence of the inflammatory bowel diseases with time, based on systematic review. Gastroenterology 2012;142(1):46.
[2] Italia L, Claudio R. Inflammatory bowel disease: genetics, epigenetics, and pathogenesis. Front Immunol 2015;6:551.
[3] Pillai S. Rethinking mechanisms of autoimmune pathogenesis. J Autoimmun 2013;45:97–103.

[4] Halme L, Paavola-Sakki P, Turunen U, Lappalainen M, Farkkila M, Kontula K. Family and twin studies in inflammatory bowel disease. World J Gastroenterol 2006;12(23):3668—72.

[5] de Souza HS, Fiocchi C. Immunopathogenesis of IBD: current state of the art. Nat Rev Gastroenterol Hepatol 2016;13(1):13—27.

[6] Manichanh C, Rigottier-Gois L, Bonnaud E, et al. Reduced diversity of faecal microbiota in Crohn's disease revealed by a metagenomic approach. Gut 2006;55(2):205—11.

[7] Sadowsky MJ, Staley C, Heiner C, et al. Analysis of gut microbiota — an ever changing landscape. Gut Microb 2017;8(3).

[8] Sartor RB, Wu GD. Roles for intestinal bacteria, viruses, and fungi in pathogenesis of inflammatory bowel diseases and therapeutic approaches. Gastroenterology 2017;152(2).

[9] Human Microbiome Project C. Structure, function and diversity of the healthy human microbiome. Nature 2012;486(7402):207—14.

[10] Turnbaugh PJ, Ley RE, Hamady M, Fraser-Liggett CM, Knight R, Gordon JI. The human microbiome project. Nature 2007;449(7164):804—10.

[11] Knights D, Ward TL, McKinlay CE, et al. Rethinking "enterotypes". Cell Host Microbe 2014;16(4):433—7.

[12] Bäckhed F, Fraser CM, Ringel Y, et al. Defining a healthy human gut microbiome: current concepts, future directions, and clinical applications. Cell Host Microbe 2012;12(5):611—22.

[13] Dave M, Higgins PD, Middha S, Rioux KP. The human gut microbiome: current knowledge, challenges, and future directions. Transl Res J Lab Clin Med 2012;160(4):246—57.

[14] Khanna S, Vazquez-Baeza Y, Gonzalez A, et al. Changes in microbial ecology after fecal microbiota transplantation for recurrent *C. difficile* infection affected by underlying inflammatory bowel disease. Microbiome 2017;5(1).

[15] Nature H. Structure, function and diversity of the healthy human microbiome. Nature 2012.

[16] Tap J, Mondot S, Levenez F, et al. Towards the human intestinal microbiota phylogenetic core. Environ Microbiol 2009;11(10):2574—84.

[17] Ott SJ, Musfeldt M, Wenderoth DF, et al. Reduction in diversity of the colonic mucosa associated bacterial microflora in patients with active inflammatory bowel disease. Gut 2004;53(5):685—93.

[18] Sokol H, Seksik P. The intestinal microbiota in inflammatory bowel diseases: time to connect with the host. Curr Opin Gastroenterol 2010;26(4):327—31.

[19] Mazmanian SK, Round JL, Kasper DL. A microbial symbiosis factor prevents intestinal inflammatory disease. Nature 2008;453(7195):620—5.

[20] Round JL, Lee SM, Li J, et al. The Toll-like receptor 2 pathway establishes colonization by a commensal of the human microbiota. Science (NY, NY) 2011;332(6032):974—7.

[21] Sokol H, Lay C, Seksik P, Tannock GW. Analysis of bacterial bowel communities of IBD patients: what has it revealed? Inflamm Bowel Dis 2008;14(6):858—67.

[22] Atarashi K, Tanoue T, Oshima K, et al. Treg induction by a rationally selected mixture of Clostridia strains from the human microbiota. Nature 2013;500(7461):232—6.

[23] Shen Y, Giardino Torchia ML, Lawson GW, Karp CL, Ashwell JD, Mazmanian SK. Outer membrane vesicles of a human commensal mediate immune regulation and disease protection. Cell Host Microbe 2012;12(4):509—20.

[24] Llopis M, Antolin M, Carol M, et al. Lactobacillus casei downregulates commensals' inflammatory signals in Crohn's disease mucosa. Inflamm Bowel Dis 2009;15(2):275—83.

[25] Quévrain E, Maubert MA, Michon C, et al. Identification of an anti-inflammatory protein from Faecalibacterium prausnitzii, a commensal bacterium deficient in Crohn's disease. Gut 2016;65(3):415—25.

[26] Sokol H, Pigneur B, Watterlot L, et al. Faecalibacterium prausnitzii is an anti-inflammatory commensal bacterium identified by gut microbiota analysis of Crohn disease patients. Proc Natl Acad Sci USA 2008;105(43):16731—6.

[27] Chow J, Tang H, Mazmanian SK. Pathobionts of the gastrointestinal microbiota and inflammatory disease. Curr Opin Immunol 2011;23(4):473—80.

[28] Zechner EL. Inflammatory disease caused by intestinal pathobionts. Curr Opin Microbiol 2017;35:64—9.

[29] Hajishengallis G, Lamont RJ. Dancing with the stars: how choreographed bacterial interactions dictate nososymbiocity and give rise to keystone pathogens, accessory pathogens, and pathobionts. Trends Microbiol 2016;24(6):477—89.

[30] Ohkusa T, Okayasu I, Ogihara T, Morita K, Ogawa M, Sato N. Induction of experimental ulcerative colitis by Fusobacterium varium isolated from colonic mucosa of patients with ulcerative colitis. Gut 2003;52(1):79—83.

[31] Ohkusa T, Yoshida T, Sato N, Watanabe S, Tajiri H, Okayasu I. Commensal bacteria can enter colonic epithelial cells and induce proinflammatory cytokine secretion: a possible pathogenic mechanism of ulcerative colitis. J Med Microbiol 2009;58(Pt 5):535—45.

[32] Lupp C, Robertson ML, Wickham ME, et al. Host-mediated inflammation disrupts the intestinal microbiota and promotes the overgrowth of Enterobacteriaceae. Cell Host Microbe 2007;2(2):119—29.

[33] Meconi S, Vercellone A, Levillain F, et al. Adherent-invasive *Escherichia coli* isolated from Crohn's disease patients induce granulomas in vitro. Cell Microbiol 2007;9(5):1252—61.

[34] Glasser AL, Boudeau J, Barnich N, Perruchot MH, Colombel JF, Darfeuille-Michaud A. Adherent invasive *Escherichia coli* strains from patients with Crohn's disease survive and replicate within macrophages without inducing host cell death. Infect Immun 2001;69(9):5529—37.

[35] Garrett WS, Gallini CA, Yatsunenko T, et al. Enterobacteriaceae act in concert with the gut microbiota to induce spontaneous and maternally transmitted colitis. Cell Host Microbe 2010;8(3):292—300.

[36] Ni J, Wu GD, Albenberg L, Tomov VT. Gut microbiota and IBD: causation or correlation? Nat Rev Gastroenterol Hepatol 2017;14(10):573—84.

[37] Darfeuille-Michaud A, Boudeau J, Bulois P, et al. High prevalence of adherent-invasive *Escherichia coli* associated with ileal mucosa in Crohn's disease. Gastroenterology 2004;127(2):412—21.

[38] Hulten K, El-Zimaity HM, Karttunen TJ, et al. Detection of *Mycobacterium avium* subspecies paratuberculosis in Crohn's diseased tissues by in situ hybridization. Am J Gastroenterol 2001;96(5):1529—35.

[39] Timms VJ, Daskalopoulos G, Mitchell HM, Neilan BA. The association of *Mycobacterium avium* subsp. Paratuberculosis with inflammatory bowel disease. PLoS One 2016;11(2).

[40] Sartor RB. Does *Mycobacterium avium* subspecies paratuberculosis cause Crohn's disease? Gut 2005;54(7):896—8.

[41] Warwick S, Paul P, Brendan C, et al. Two-year combination antibiotic therapy with Clarithromycin, Rifabutin, and Clofazimine for Crohn's disease. Gastroenterology 2007;132(7):2313—9.

[42] Devkota S, Chang EB. Interactions between diet, bile acid metabolism, gut microbiota, and inflammatory bowel diseases. Dig Dis 2015;33(3):351—6.

[43] Gevers D, Kugathasan S, Denson LA, et al. The treatment-naive microbiome in new-onset Crohn's disease. Cell Host Microbe 2014;15(3):382—92.

[44] Alexa W, Antonio G, Yoshiki V-B, et al. Dynamic changes in short- and long-term bacterial composition following fecal microbiota transplantation for recurrent *Clostridium difficile* infection. Microbiome 2015;3(1):1—8.

[45] Bojanova DP, Bordenstein SR. Fecal transplants: what is being transferred? PLoS Biol 2016;14(7).

[46] Newman KM, Rank KM, Vaughn BP, Khoruts A. Treatment of recurrent *Clostridium difficile* infection using fecal microbiota transplantation in patients with inflammatory bowel disease. Gut Microb 2017;8(3).

[47] Alexander K, Kevin MR, Krista MN, et al. Inflammatory bowel disease affects the outcome of fecal microbiota transplantation for recurrent *Clostridium difficile* infection. Clin Gastroenterol Hepatol 2016;14(10):1433—8.

[48] Shi Y, Dong Y, Huang W, Zhu D, Mao H, Su P. Fecal microbiota transplantation for ulcerative colitis: a systematic review and meta-analysis. PLoS One 2016;11(6).

[49] Qazi T, Amaratunga T, Barnes EL, Fischer M, Kassam Z, Allegretti JR. The risk of inflammatory bowel disease flares after fecal microbiota transplantation: systematic review and meta-analysis. Gut Microb 2017.

[50] Narula N, Kassam Z, Yuan Y, et al. Systematic review and meta-analysis: fecal microbiota transplantation for treatment of active ulcerative colitis. Inflamm Bowel Dis 2017.

[51] Moayyedi P, Surette MG, Kim PT, et al. Fecal microbiota transplantation induces remission in patients with active ulcerative colitis in a randomized controlled trial. Gastroenterology 2015;149(1).

[52] Rossen NG, Fuentes S, van der Spek MJ, et al. Findings from a randomized controlled trial of fecal transplantation for patients with ulcerative colitis. Gastroenterology 2015;149(1).

[53] Paramsothy S, Paramsothy R, Rubin DT, et al. Faecal microbiota transplantation for inflammatory bowel disease: a systematic review and meta-analysis. J Crohn's Colitis 2017;11(10):1180—99.

[54] Costello SP, Waters O, Bryant RV, et al. Short duration, low intensity, pooled fecal microbiota transplantation induces remission in patients with mild-moderately active ulcerative colitis: a randomised controlled trial. Gastroenterology 2017;152(5):S198—9.

[55] Michail S. Fecal microbial transplant in children with ulcerative colitis: a randomized, double-blinded, placebo-controlled pilot study. J Pediatr Gastroenterol Nutr 2017. Publish Ahead of Print: 1.

[56] Colman RJ, Rubin DT. Fecal microbiota transplantation as therapy for inflammatory bowel disease: a systematic review and meta-analysis. J Crohn's Colitis 2014;8(12).

[57] Suskind DL, Singh N, Nielson H, Wahbeh G. Fecal microbial transplant via nasogastric tube for active pediatric ulcerative colitis. J Pediatr Gastroenterol Nutr 2015;60(1).

[58] Cui B, Feng Q, Wang H, et al. Fecal microbiota transplantation through mid-gut for refractory Crohn's disease: safety, feasibility, and efficacy trial results. J Gastroenterol Hepatol 2015;30(1).

[59] Vaughn BP, Vatanen T, Allegretti JR, et al. Increased intestinal microbial diversity following fecal microbiota transplant for active Crohn's disease. Inflamm Bowel Dis 2016;22(9).

[60] Wang S, Xu M, Wang W, et al. Systematic review: adverse events of fecal microbiota transplantation. PLoS One 2016;11(8).

[61] Petrof EO, Gloor GB, Vanner SJ, et al. Stool substitute transplant therapy for the eradication of *Clostridium difficile* infection: 'RePOOPulating' the gut. Microbiome 2012;1(1):1—12.

[62] Kazerouni A, Burgess J, Burns LJ, Wein LM. Optimal screening and donor management in a public stool bank. Microbiome 2015;3(1):1—8.

[63] van Nood E, Vrieze A, Nieuwdorp M, et al. Duodenal infusion of donor feces for recurrent *Clostridium difficile*. N Engl J Med 2013;368(5):407—15.

[64] Hamilton MJ, Weingarden AR, Sadowsky MJ, Khoruts A. Standardized frozen preparation for transplantation of fecal microbiota for recurrent *Clostridium difficile* infection. Am J Gastroenterol 2012;107(5).

[65] Lee CH, Steiner T, Petrof EO, et al. Frozen vs fresh fecal microbiota transplantation and clinical resolution of diarrhea in patients with recurrent clostridium difficile infection: a randomized clinical trial. J Am Med Assoc 2016;315(2):142—9.

[66] Staley C, Hamilton MJ, Vaughn BP, et al. Successful resolution of recurrent *Clostridium difficile* infection using freeze-dried, encapsulated fecal microbiota; pragmatic cohort study. Am J Gastroenterol 2017;112(6):940—7.

[67] Staley C, Vaughn BP, Graiziger CT, et al. Community dynamics drive punctuated engraftment of the fecal microbiome following transplantation using freeze-dried, encapsulated fecal microbiota. Gut Microbes 2017;8(3):276—88.

[68] Postigo R, Kim JH. Colonoscopic versus nasogastric fecal transplantation for the treatment of *Clostridium difficile* infection: a review and pooled analysis. Infection 2012 (1439—0973 (Electronic)).

[69] Lee CH, Belanger JE, Kassam Z, et al. The outcome and long-term follow-up of 94 patients with recurrent and refractory *Clostridium difficile* infection using single to multiple fecal microbiota transplantation via retention enema. Eur J Clin Microbiol Infect Dis 2014;33(8):1425—8.

[70] Vermeire S, Joossens M, Verbeke K, et al. Donor species richness determines faecal microbiota transplantation success in inflammatory bowel disease. J Crohn's Colitis 2016;10(4):387−94.

[71] Angelberger S, Reinisch W, Makristathis A, et al. Temporal bacterial community dynamics vary among ulcerative colitis patients after fecal microbiota transplantation. Am J Gastroenterol 2013;108(10). https://doi.org/10.1038/ajg2013257.

[72] Machiels K, Joossens M, Sabino J, et al. A decrease of the butyrate-producing species Roseburia hominis and Faecalibacterium prausnitzii defines dysbiosis in patients with ulcerative colitis. Gut 2014;63(8):1275−83.

FURTHER READING

[1] Lloyd-Price J, Abu-Ali G, Huttenhower C. The healthy human microbiome. Genome Med 2016;8(1):51.

[2] Bäckhed F, Roswall J, Peng Y, et al. Dynamics and stabilization of the human gut microbiome during the first year of life. Cell Host Microbe 2015;17(6):852.

Chapter 46

Precision Medicine: the Microbiome and Metabolome

Joel Faintuch[1] and Jacob J. Faintuch[2]

[1]Department of Gastroenterology, Sao Paulo University Medical School, Sao Paulo, Brazil; [2]Department of Internal Medicine, Hospital das Clinicas, Sao Paulo, Brazil

INTRODUCTION

Since the advent of biochemical tests, microbiological assays, diagnostic imaging, endoscopic investigations, electronic heart and brain registration devices, tissue biopsies, and other diagnostic procedures, medicine surpassed its ancestral empirical and philosophical approach. Of course detailed observations and generations-old experience, as accumulated since Hippocrates of Kos (460–370 BC), are relevant and will never be abandoned.

Modern health care relies on numbers, in other words, on quantitative, objective information, to categorize patients into defined strata. Thus the appropriate health care is delivered, in exact amounts, neither too much nor too little. Lord Kelvin (1824–1907) already insisted that "If you can express it with numbers, you know something about it" although Lord Kelvin was not a physician and did not mind it, in a sense that was the starting point of modern precision medicine. Others call it personalized medicine, a related concept.

Every creature is to some extent unique, as such he or she deserves tailor-made assistance for optimal results. One hotly debated example is cancer chemotherapy. Depending on the tumor and the drug protocol, success rate may substantially fluctuate, with an estimated average of just one patient in three responding. However, pharmacogenomic and metabolomic fingerprinting can remarkably improve the outcome. Polymorphism of genes encoding drug-metabolizing enzymes, drug transporters, and targets, along with other "omic" tools, are crucial for optimal results.

Precision as well as personalized medicine are mostly by-products of the Human Genome Project started in the 1990s and still going on [1]. By elucidating the pattern, expression, and function of human genes, both in the general population and in single individuals, this initiative paved the way for personally adjusted drugs, diets, and treatments based on chances of response and risks of adverse effects.

Pharmacogenetics, Pharmacogenomics, Pharmacometabolomics

As known for a long time, inherited genetic variations influence the response of the organism to therapy, be it pharmacological, dietetic, surgical, or bacteriological (pharmacogenetics, nutrigenetics, etc). At the same time, these therapies or interventions may play a role on gene expression (pharmacogenomics, nutrigenomics, etc). Their combination does not mean squaring the circle. On the contrary, they complement each other and tend to illuminate therapeutic efforts not only in cancer but also in nononcological contexts, including cardiovascular, inflammatory, autoimmune, and metabolic diseases. By means of targeted and more effective action and by identifying specific biomarkers such as genes, metabolites, proteins, lipids, and other molecules, health-care costs could be reduced by means of enhanced treatment effectiveness and reduced adverse effects [2].

Pharmacogenetics, pharmacogenomics, and pharmacometabolomics have been studied since the 1960s [3]; however, acceleration coincided with the Human Genome Project [4]. Along with other omics techniques and refined information processing with the help of artificial intelligence and machine learning, great strides were made, concerning the interaction between host genes and response to multiple drugs, nutrients, and procedures in an array of clinical contexts.

Bacterial Genomics

If one accepts that this is the "second genome", as alluded to in the Introduction of the book, then precision medicine without microbiome studies is hardly conceivable. Nevertheless, just spreadsheets or maps with taxonomic clusters and profiles do not confer much advantage to clinical practice. In the same way, piles of numbers alone would not make Lord Kelvin happy. He was after univocal and unambiguous values, for instance, of the absolute zero temperature, which he determined, or the first and second laws of thermodynamics, also among his sharp formulations [5].

In other words, one needs to move beyond simple associations toward comprehensive and targeted genotyping of both host and microbiome. Microbiomic fingerprinting, with distinct signatures, is the current goal to draw the desired benefits in precision medicine. The same applies to the metabolome, lipidome, and proteome as avenues for deep phenotyping of the host—microbiome universe.

Molecular Pathology

A further stride to both Hippocratic medicine and conventional disease recognition is that molecular pathology disentangles clinical and genetic heterogeneity within diseases until molecular alterations in cells and their interaction with the surrounding microenvironment are asserted for a given individual (the unique disease principle). One way or another, the molecular pathology paradigm relies on molecular signatures for all diagnostic and therapeutic interventions [6].

Precision Medicine and Metabolomic Signatures

The theoretical framework of metabolomic fingerprinting does not essentially conflict with molecular pathology. Both aim to underpin medical assessments, procedures, and predictions on a validated molecular basis, which by definition is unique to each subpopulation and even to each patient. General goals do not differ much either: (1) predictive, prognostic, diagnostic, and surrogate markers; (2) molecular mechanisms of diseases; (3) subclassification and stratification based on metabolic pathways; (4) biomarkers for drug response phenotypes (pharmacometabolomics); (5) one metabotype for each genotype (functional readout for genetic variants); and (6) monitor response and recurrence of diseases [7].

MICROBIOME SIGNATURES

Nonalcoholic Fatty Liver Disease

The association between the most common liver abnormality in the world and the gut microbiome has been investigated for decades. One of the principal challenges is differential diagnosis between simple steatosis, a benign modality, and steatohepatitis, which tends to progress to liver fibrosis, cirrhosis, and failure. Boursier et al. were able to predict steatohepatitis, fibrosis, or both on the basis of the abundance of two gut microbiome genera, namely *Bacteroides* and *Ruminococcus*. This could be a step toward early identification of severe cases [8].

Following a different lead, a nominal increase in Proteobacteria and *E coli*, along with reduction of Firmicutes, a Random Forest classifier model was built to predict advanced fibrosis. Not less than 37 bacterial genera were included, achieving robust accuracy [9].

Porphyromonas gingivalis, a ubiquitous mouth pathogen also present in the intestine, is not infrequent in the scientific literature of cardiovascular disease (CVD), arthritis, and other systemic inflammatory conditions. It could play a role in tissue damage during NAFLD as well as antibody titers against *P. gingivalis fimbriae* type 4 correlated with advanced fibrosis. Simultaneous metabolomic deviations included elevated monounsaturated/saturated fatty acid ratio and attenuated expression of the related enzymes stearoyl-CoA desaturase 1 (*Scd1*) and elongation of very long chain fatty acids (*Elovl6*) [10].

Alcoholic Hepatitis and Liver Cirrhosis

Until the recent past, searching for bacteria in the blood, other fluid compartments, or internal organs and tissues of nonseptic, nonbacteremic subjects would seem a waste of time and resources, if not insanity, given the germ theory of disease by Louis Pasteur (1822—95) and the infection postulates by Robert Koch (1843—1910). Both implied that microorganisms were able to trigger disease, and very often did it as soon as they gained entry into the internal milieu in sufficient numbers.

These dogmas were already questioned a long time ago, and microbiome studies contributed to turn them inside out. Relationships between bacteria and internal host tissues are more complex than just health versus disease and sterility versus contamination. The immune system condones many patterns of coexistence beyond classic phlegmons, abscesses, and septicemia.

Lipopolysaccharide (LPS), an endotoxin of gram-negative bacteria, is regularly measured in the plasma of infection-free individuals, exhibiting from obesity to undernutrition and from liver disease to cancer [11]. Therefore translocation of whole bacteria or cellular components across the gut mucosal barrier is not a pathological phenomenon, except when massive and followed by systemic abnormalities.

Population density is obviously a key variable. Even in the days of Pasteur and Koch it was perceived that depending on the size of the experimental inoculum and the virulence of the germ, sickness induction could fail, mirroring micro-organism neutralization or tolerance by the immune system.

Modern metagenomic techniques are able to track bacteria in any number from the exponential ones as in the colon and rectum to those so small that conventional culture yields negative growth. This last assumption was explored in a protocol of alcoholic hepatitis. Bacteroidetes were diminished in the plasma of heavy drinkers without hepatitis compared with nondrinking controls and in patients with both moderate and severe alcoholic hepatitis, contrasting with increase of Fusobacteria in the same groups. Endotoxemia (LPS) was also a hallmark of alcoholic hepatitis, as expected. These statistically confirmed findings pave the way for personalized diagnosis and follow-up in alcohol addicts, both with and without chronic hepatitis [12].

In the protocol of Jie et al. [13], *Veillonella* spp. was strongly discriminant for liver cirrhosis. This result is consistent with that of a metaanalysis, which found important correlations with *Fusobacterium nucleatum* and *Streptococcus* besides *Veillonella* [14]. *F. nucleatum* will be again alluded to in connection with colorectal cancer, suggesting at least partially shared pathological pathways.

A dearth of *Lactobacillus* spp. is also a hallmark of alcohol addiction and alcoholic liver cirrhosis. This is a relevant genus that produces short-chain fatty acids (SCFAs) and includes well-regarded probiotics. Replenishment would be a logical step; however, an even more innovative and precise intervention would be administration of metabolites. Supernatants of *Lactobacillus* cultures have been demonstrated to ameliorate inflammation and hepatic levels of tumor necrosis factor alfa and toll-like receptor 4 (*TLR*-4) in experimental animals. Expression of intestinal tight junction proteins is enhanced, preserving intestinal barrier function [15].

Cancer

Colorectal cancer (CRC) is a natural candidate for microbiome fingerprinting. Its anatomical location coincides with the highest concentration of bacteria and their metabolites. Furthermore, since the 1980s, before the microbiome concept was clearly established, it was already suspected that the advantageous role of dietary fibers against CRC was partly mediated by local bacterial fermentation, including SCFAs and other metabolites [16].

According to one protocol, the tumor microenvironment is endowed with a distinctive microbiome, featuring elevated amounts of *Fusobacteria* and *Providencia*. Prognosis might be inferred by the severity of the profile as virulence-associated genes were also enriched in the circumstances [17].

The oral microbiome is also focused as this is the original home of *Fusobacterium nucleatum* that could play a major role in CRC along with suspects from the genus *Porphyromonas* [18]. *F. nucleatum* is associated with microsatellite instability, CpG island methylator phenotype, and suppression of T cells in the tumor microenvironment [19]. Within this context, chemoresistance to combinatorial chemotherapy of 5-flourouracil (FU) and oxaliplatin in CRC could also occur as a result of the activation of the autophagy pathway by *F. nucleatum* [20].

Peptostreptococcus stomatis and *F. nucleatum* are more abundant in CRC fecal samples, whereas *Streptococcus salivarius* is reduced in the same circumstances [14].

Bacterial colonies often form biofilms in connection with CRC, particularly on the right colon. These are true for *Bacteroides fragilis* as well as for oral pathogens *F. nucleatum*, *Parvimonas micra*, and *P. stomatis*, all involved in CRC tumorigenesis. Although a universal pattern has not been unveiled, symbionts of oral origin are overrepresented in more than 80% of the cases of CRC [21].

Breast cancer should not be overlooked as the most prevalent oncologic pathology in women. Even though breast tissue is technically as sterile as plasma and other internal compartments, a typical microbiome was characterized in cancer patients, which was unique for tumors of this location. *Fusobacterium*, which was alluded to in connection with CRC, as well as *Atopobium, Gluconacetobacter, Hydrogenophaga,* and *Lactobacillus* were all overrepresented [22]. Some of these could have intestinal origin as gut, mouth, and urinary dysbiosis are recognized in this illness as well [23].

Crohn's Disease

The microbiome is so closely linked to this entity that in early times Crohn's disease (CD) was believed to be an infection, particularly an intestinal mycobacteriosis. Viruses, yeasts, fungi, and other bacteria have also been incriminated [24]. Now the immunoinflammatory essence of the illness is crystalized; however, the close association with intestinal bacteria remains true as well.

This has led to fruitful characterizations in a number of occasions. In a protocol based on microbiome and transcriptome, the ileum was identified as the primary site of CD. Proteobacteria were increased during active disease, as had been reported in ulcerative colitis and other modalities of dysbiosis. Firmicutes exhibited several shifts as well. Changes in the host similarly occurred in the form of increased activity of antimicrobial dual oxidase and reduced expression of APOA1 gene. A model based on microbiome findings and APOA1 successfully predicted 6-month remission, an important step toward therapeutic planning in such circumstances [25].

In a metaanalysis delving into inflammatory bowel disease (IBD), *Bifidobacterium bifidum* and Lachnospiraceae were revealed as discriminant for condition [14].

Such results are reinforced by a series of operated patients with CD. During surgery the expected increase in Proteobacteria, along with *Bacillus* spp., materialized. After 6 months, large numbers of Lachnospiraceae and *Enterococcus durans* signaled recurrence, whereas a robust microbial community, especially rich in *Dorea longicatena* and *Bacteroides plebeius*, corroborated remission. This information allows more effective monitoring of surgical outcomes [26].

Obesity and Diabetes

In groundbreaking investigations with germ-free mice and other models, since the 1990s, Gordon's team presented compelling evidence about the role of the gut microbiome in energy harvesting and storage, consequently in weight control [27–29], Firmicutes were promptly recognized as major SCFA synthesizers and thus obesity inducers. In contrast, Bacteroidetes, the other major intestinal phylum, were less efficient in the fermentation and energy extraction from dietary fibers in the form of SCFAs. A Firmicutes/Bacteroidetes ratio (FBR) emerged as a tool for separating obesity-prone from obesity-resistant gut environments.

Many years and clinical protocols later, the importance of the microbiome for obesity and diabetes has never been refuted. Indeed a recent case of massive weight gain after fecal transplantation from an overweight donor for *Clostridium difficile* infection lends further credence to the theory [30], even though confirmation is lacking [31]. Yet the diagnostic usefulness of FBR has been shaken. It is still accepted as an index of energy production and absorption in the intestine [32], but not necessarily as the hallmark of obesity.

Variability in the taxonomic profile of obesity exceeds differences between lean and obese individuals, thus compromising relatively nonspecific signatures such as FBR [33]. Concerning type 2 diabetes that similarly displays links to Firmicutes and Bacteroidetes, experimental studies suggest that reduction of both phyla are simultaneously recommended to elevate incretins such as Glucagon like peptide-1 (GLP-1) [34] and improve glucose control.

In a recent metaanalysis based on strain-specific markers, instead of phylum-level abundance, a statistically useful pattern was reached for obesity as well as diabetes. However, it was less discriminating than for liver cirrhosis, CRC, or IBD [14]. In a Chinese investigation [13], differences were sharper. *B. vulgatus* was robustly discriminating for diabetes and *Dorea longicatena* for obesity.

Occasional obstacles notwithstanding, obesity microbiome fingerprinting has a future in precision medicine, notably within some contexts and in association with the metabolome. In one experience [35], excessive weight gain after dieting was explored. It was demonstrated that 6 months after weight loss the microbiome composition was still unchanged. This "obese" gut environment was associated with exacerbated secondary weight gain.

Antibiotics remodeled the unfavorable microbiome as well as fecal transplantation with "normal" microorganisms, thus abrogating exaggerated weight recovery. The metabolome unveiled a role for flavonoid supplementation as well. The postdieting gut quickly degraded dietary flavonoids, which play a role in fat cycling and obesity prevention. Once additional flavonoids were supplemented, accelerated weight gain failed to materialize [35].

These results are strengthened by microbial exchange studies in mice [36]. Mice are coprophagic animals; thus they are all the time colonizing themselves with other microbiomes. To investigate the role of diets in this setting, animals are ordinarily segregated. By sequential cohousing, an intimate connection between diet, gut microbes, and obesity was recently reaffirmed in certain settings.

As confirmed by transplanting the respective microbiomes into germ-free animals, some bacteria sabotage the dieting plan. Microbiome sequencing and manipulation, by means of selective addition and deletion of microorganisms, could be essential to restore the weight loss program. In other words, the right microbiome could amplify the dieting effort, whereas the wrong one will dampen results [36].

Within the context of bariatric surgery, Firmicutes decreased postoperatively, with a corresponding shift in the FBR, and shaped the entire microbiome into a more "lean" profile, differently from dieting, which was unable to induce such pattern. Such "lean" profile included increase in *Faecalibacterium prausnitzii*, along with decrease of *Clostridium*, *Eubacterium*, *Faecalibacterium*, *Dorea*, *Coprococcus*, *Ruminococcus*, and Lachnospiraceae. After a low calorie diet, none of these were noticed. On the contrary, SCFA producers increased, *Butyrivibrio fibrisolvens*, *Clostridium saccharolyticum*, *Eubacterium limosum,* and *Blautia hydrogenotrophic* among them [32].

Atherosclerotic Cardiovascular Disease

Most cardiovascular investigations target specific conditions such as stroke, myocardial infarction, or peripheral arterial disease. Yet, general markers of the disorder emerge as well, with sufficient accuracy to improve diagnosis and monitor therapy. Enterobacteriaceae are highlighted analogously to what is documented in other conditions such as NAFLD, hepatitis B, liver cirrhosis, bone marrow transplantation, and premature infants.

Main representatives are *E. coli*, *Klebsiella* spp., and *Enterobacter aerogenes*.

Similarly, overexpression of microorganisms typical of the oral cavity materializes in the gut in the form of *Streptococcus* spp., *Lactobacillus salivarius*, *Solobacterium moorei*, and *Atopobium parvulu*. A few other intestinal commensals are abundant, such as *Ruminococcus gnavus*, and *Eggerthella lenta*, whereas butyrate-producing bacteria *Roseburia intestinalis* and *Faecalibacterium prausnitzii*, along with *Bacteroides* spp., *Prevotella copri*, and *Alistipes shahii*, become scarce.

A metagenomic linkage group—based classifier was successful in the selection of atherosclerotic patients on the basis of such findings [13].

Arterial Hypertension

In volunteers a high salt challenge decreased *Lactobacillus* spp. Similar results had already been observed in experimental animals, and the outcome was increase in both TH17 cells and arterial blood pressure. Replenishment of missing *Lactobacillus* in the laboratory model ameliorated experimental hypertension [37].

SCFAs produced by *Lactobacillus* could thus play a role along with other microbial metabolites such as neurotransmitters serotonin, dopamine, norepinephrine, p-cresol sulfate, and indoxyl sulfate much studied in connection with renal disease and trimethylamine N-oxide (TMAO) involved in pathogenesis of atherosclerosis [38]. A specific gut—brain hypertension axis has been postulated as well [39]. Both salt-dependent and salt-independent mechanisms could be active in pathogenesis of hypertension and amenable to precision medicine interventions.

Rheumatological and Autoimmune Conditions

Behçet's disease is not ubiquitous; however, similarities with Crohn's disease have prompted microbiome investigations in the gut as well as oral cavity [40,41]. Fecal samples revealed depletion of *Roseburia* and *Subdoligranulum* along with low butyrate production. Oral metagenomics pointed to a different direction with abundant *Haemophilus parainfluenzae* and depleted *Alloprevotella rava* and *Leptotrichia* spp. Although encouraging, these signatures will require confirmation before they can be translated to clinical practice.

Gout is a condition with solid metabolomic credentials, represented by derangements of purine metabolism and uric acid. However, they stem from the host, not the microbiome. In this sense it is somewhat surprising that *Bacteroides,* Porphyromonadaceae*, Rhodococcus*, *Erysipelatoclostridium,* and Anaerolineaceae were upregulated in fecal samples [42]. Although a mechanistic explanation is lacking, integration between metabolome and microbiome shifts could emerge in future studies.

Rheumatoid arthritis is another challenge. In early disease an expansion of *Prevotella copri* has been registered [43]. In a more heterogeneous cohort an abundance of Actinobacteria, along with reduction in conventional phyla, was the most visible shift. At the genus level, higher proportions of *Collinsella*, *Eggerthella*, and *Faecalibacterium* occurred. Collinsella was associated with proinflammatory cytokine production, as well as gut permeability disruption, and disease severity [44].

Gut Dysbiosis

Despite its enormous importance, dysbiosis is still a poorly defined entity, and an attempt to cast any microbiological features radically conflicts with the different and often specific profiles unearthed for each illness. All these criticisms notwithstanding, a number of species have been pointed out as general markers of ecological stress and unstable microbial communities: *Streptococcus* salivarius, *Streptococcus anginosus*, *Veillonella parvula*, *Roseburia intestinalis*, and *Coprococcus comes*. Particularly the two *Streptococcus* species and *Veillonella*, all proceeding from the oral cavity, could be early signals of an unhealthy microbiome as they are common in multiple conditions [14].

Point-of-care microbiomics: "Gut on a chip"

Miniaturized mechanical/electronic devices emulating the gut, complete with human epithelium, villi, microfluidic systems, and even peristalsis, have been developed. They are not meant exactly for bedside use as preliminary cell growth and epithelization takes days to weeks. Also, microbiome colonization is not an instant phenomenon. Yet, they could expand and simplify access to valuable information, which currently simply does not exist or would require much more time and effort to be collected. One of them is already in the market (Emulate Inc, Boston, MA, USA) [45].

When bacteria are inoculated, a host—microbe ecosystem can be produced. A personalized microgut in which disease models, nutrients, probiotics, drugs, and even fecal transplantation could be simulated is nothing less than revolutionary [46]. Such personalized models have been called "Avatar chips", or "Your gut in a chip", and studies aiming at IBD are progressing [47].

The convenience and safety of the chip, as well as low cost in comparison with clinical trials, could pave the way for precise investigations of the interactions between the microbiome and specific diseases, with age, gender, ethnicity, life style, and comorbidities. Each patient could be individually tested with his own intestinal epithelium and microbiome colonizing the model before therapeutic interventions were conducted [47].

METABOLOMIC FINGERPRINTING

Obesity and NAFLD

Also, within the framework of metabolomics, NAFLD is a preferred subject because of its background of unbalanced diet, obesity, and metabolic syndrome, thus representing an overarching aberration. Indeed, cross talk between the gut microbiome and the liver encompasses the full range of hepatic transcriptome, proteome, and metabolome.

Cytochrome P450, a crucial enzyme for drug detoxification, is downregulated. Gene expression and epigenetics are equally impacted among others by SCFAs such as butyrate and propionate. Moreover, pleiotropic bile acids and GLP-1 are modulated by the gut microbiome, interfering with multiple metabolic responses of the host [48].

About half of NAFLD patients display a distinct metabolomic serum signature, featuring elevated fat molecules (triglycerides, diglycerides, fatty acids, ceramides, and oxidized fatty acids), along with depressed S-adenosyl methione (SAMe). Both subtypes, with and without this profile, could progress no nonalcoholic steatohepatitis, with additional corresponding markers. These included amino acids, bile acids, fatty acids, and other lipids. As a matter of fact, SAMe supplementation in a laboratory model improved both liver biochemistry and histology. In such circumstances, disease progression and strategical interventions can be more reliably planned [49].

Slightly divergent evidence in favor of circulating metabolomic signals was accrued in similar contexts by others, such as phospholipids and amino acids (Feldman) along with amino acids and bile acids [50].

The SCFA acetate has been experimentally demonstrated to promote glucose-stimulated insulin and ghrelin secretion eventually resulting in hyperphagia and obesity. Activation of the parasympathetic system is the putative afferent route [51].

Urinary metabolites should not be neglected in such context, even though those of intestinal origin could be masked or diluted by further downstream processing. In a cohort of obese and/or NAFLD-affected children, characteristic urinary metabolites predictably related more to diet and endogenous metabolism than the gut microbiome. Nevertheless, high urinary p-cresyl sulfate (p-CS) was a marker of obesity without NAFLD. At the same time, diminished p-CS, butyrate, and adipic acid pointed toward small intestinal bacterial overgrowth. These are all gut-related molecules [52].

Diabetes

The identification of SCFA receptors in pancreatic beta cells lends further credence to the important interface between the microbiome and type 2 diabetes. Therapeutic interventions are being envisaged [53].

Longevity

The original claim to fame of probiotics, as advocated by Methnikoff, focused improved longevity [54]. This was a tall claim, and little progress was achieved in the last 100 years, even though commensal microorganisms have been incriminated in inflammation, cognition, and cardiovascular health, all germane to longevity [55].

It cannot be denied either that the intestinal environment is altered by old age with relative scarcity of highly prevalent families Ruminococcaceae, Lachnospiraceae, and Bacteroidaceae, all partly replaced by subdominant genera [55]. Yet, these are somewhat hazy shifts, and few practical interventions have resulted.

Metabolomic resources enabled bigger strides. In the plasma, statistically relevant citric acid intermediates, linked to the mitochondria-based tricarboxylic acid cycle, have been tracked along with certain bile acids and metabolites [56]. Both could at least partially stem from microbiome patterns of old age, as further highlighted in laboratory models.

Indeed, longevity pathways are not a secret anymore. The intestinal ecological network is not far away from mammalian target of rapamycin protein (mTOR), c-Jun N-terminal kinase (JNK), and insulin/IGF mitochondrial signaling, clearly pertinent to longevity. A gut polysaccharide potentially suppressed by certain bacteria and acting on the alluded to pathways, namely colanic acid, enlarged lifespan in an invertebrate model. According to some optimists, clinical trials could be just a few steps away [57].

Cancer

Breast cancer metabolome has also received much attention, even though this is a complex topic. Not only is this a heterogeneous malignancy with multiple subtypes but also is a metabolic reprogramming, described in this context. Namely, substrate utilization and pathways undergo successive changes as the cancer cell faces hypoxia and other challenges to enable its survival [58]. Metabolome fingerprinting in such circumstances becomes quite tedious.

All these factors notwithstanding, shifts in glucose, glutamine, amino acid, and lipid pathways are already established in breast cancer subtypes, potentially leading to targeted interventions. The plasma microbiome expands the knowledge collected in cell cultures with abnormalities detected for glucose, lactate, pyruvate, alanine, leucine, isoleucine, glutamate, glutamine, valine, lysine, glycine, threonine, tyrosine, phenylalanine, acetate, acetoacetate, β-hydroxy-butyrate, urea, creatine, and creatinine. In one series, lactate levels are inversely correlated with the tumor size in early breast cancer. The gut microbiome could be involved with several of these molecules, particularly SCFA (acetate, acetoacetate, and β-hydroxy-butyrate) and also lactate and certain amino acids [59,60],

Cancer Chemotherapy and Immunotherapy

Inhibitory signaling pathways or immune checkpoints can suppress cytotoxic T-cell anticancer activity, thus abrogating immune defense of the host. Immune checkpoint inhibitors targeting cytotoxic T lymphocyte–associated protein 4 (CTLA-4), programmed cell death protein 1 (PD-1), programmed cell death ligand 1 (PD-L1), and additional targets (PD-L2, T-cell immunoglobulin and mucin domain-3 (TIM-3), Lymphocyte activation gene-3 (LAG-3), T-cell immunoreceptor with immunoglobulin (TIGIT), B and T lymphocyte attenuator (BTLA), and V-domain Ig suppressor of T cell activation (VISTA)) prevent blockade of surface receptors of immune cells, thus restoring disrupted signaling and, consequently, the immune response of cytotoxic T cells.

Monoclonal antibodies against CTLA-4 such as ipilimumab and tremelimumab, anti- PD-1 including nivolumab, pembrolizumab, and pidilizumab, or anti-PD-L1 encompassing atezolizumab, durvalumab, and avelumab have been welcomed as a paradigm shift in the management of solid tumors, including selected hematological malignancies.

Ongoing investigations list melanoma, sarcoma, Hodgkin's disease, leukemia, multiple myeloma, and lymphoma, along with cancer of the lung, pancreas, ovary, uterus, kidney, bladder, prostate, breast, head and neck, gastric, colorectal, and squamous cell carcinoma.

In keeping with the immune properties, major side effects of these agents often engage immunoinflammatory routes, particularly enterocolitis akin to IBD. Microbiome imbalances are closely involved with gut dysfunction; therefore delving into such ecology was inevitable.

A French group [61] has emphasized that *Bacteroides* spp. adversely mediated both response and complications during melanoma therapy by ipilimumab. In a cohort of such patients affected by colitis, compared to those with an abundance of *Bacteroides*, patients exhibiting larger proportions of *Faecalibacterium* and other Firmicutes displayed longer progression-free and total survival. In other protocols, *Bacteroides* hegemony was not confirmed, *Bifidobacterium longum*, *Collinsella aerofaciens*, and *Enterococcus faecium* being encountered in responding cases. Fecal transplantation of germ-free mice with material from responding patients leads to greater efficacy of anti−PD-L1 therapy [62]. Ruminococcaceae have also been highlighted as adjuvants to immune checkpoint inhibitors, suggesting that different species, or even strains, might be involved depending on the circumstances [63].

In epithelial tumors analogously managed by immune checkpoint blockade the involved intestinal bacteria was *Akkermansia muciniphila*. Fecal transplantation to germ-free animals reproduced the positive role of the microorganism as nonresponding animals supplemented by *A. muciniphila* oncologically improved in an IL-12−dependent manner [64].

The spectrum of microbiome interference in cancer chemotherapy outcomes extends well beyond immune checkpoint inhibitors. The acronym TIMER was coined to summarize the primary actions: translocation, immunomodulation, metabolism, enzymatic degradation, and reduced diversity and ecological variation. Among widely used agents affected by such response, 5-FU, cyclophosphamide, irinotecan, oxaliplatin, gemcitabine, methotrexate, and others are highlighted. Also, TLR agonists, vaccines, and adoptive T cell transfer therapies undergo modulation by the microbiome [65].

In an experimental model of CRC, tumor-associated Gammaproteobacteria were able to inactivate gemcitabine by means of an isoform of the enzyme cytidine deaminase, resulting in chemotherapy failure. Use of antibiotics restored cancer response, reinforcing the direct role of the microorganisms in the oncological course of the disease [66].

Atherosclerosis

The relevance of TMAO produced in the gut and the potential of inhibiting TMAO-producing bacteria for prevention of atherosclerosis is already being explored [13]. Further targets are indole- and phenyl-derived metabolites, at least partly expressed by gut microorganisms. In one experience, indole (tryptophan), indole-3-propionic acid, and indole-3-aldehyde in plasma were negatively associated with advanced atherosclerosis, whereas the kynurenine/tryptophan ratio was positively associated. Tryptophan and indole-3-propionic acid concentrations are correlated with the ankle−brachial index, a traditional marker of arterial stiffness [67].

A related family of plasma metabolites such as betaine, choline, phosphocholine, and α-glycerophosphocholine stem from the choline pathway, besides the well-known TMAO. In a protocol targeting CVD in general, the betaine/choline ratio was inversely associated with CVD. The choline metabolite score was directly associated both with CVD and stroke [68].

Intestinal Barrier Function

Intestinal permeability physiologically allows the passage of microorganisms and their metabolites into the host as widely demonstrated [11,12]. Nevertheless, limits need to be observed, or an array of deleterious consequences, potentially including endotoxemia and sepsis, will ensue (leaky gut syndrome). Pathological increases in intestinal permeability, consistent with defective barrier function, are detected in IBD, liver cirrhosis, trauma, burns, shock, cancer radiotherapy and chemotherapy, bone marrow transplantation, HIV/AIDS, and other immune and inflammatory abnormalities.

Aromatic amino acids are degraded by the gut microbiome, and resulting metabolites can be measured in serum. One of them is indolepropionic acid (IPA), originated from dietary tryptophan. In a protocol with *Clostridium sporogenes*, genetically engineered to produce IPA, in germ-free specifically colonized mice, intestinal permeability changed in a pregnane X receptor-dependent fashion. This study paves the way for intervention trials in human conditions of leaky gut [69].

Chronic systemic inflammation is one of the consequences of the leaky gut syndrome; however, it can be equally precipitated by metabolic derangements such as obesity or diabetes (metabolic inflammation). Irrespective of increased intestinal permeability, it has been associated with elevated cardiometabolic risk and mortality.

Escherichia coli, a classic gut commensal, synthesizes a protein mimetic of a peptide hormone, α-melanocyte−stimulating hormone (α-MSH). As a hormone this protein molecule is a major player in energy homeostasis. At the same time it is a potent immune modulator, inhibiting proinflammatory cytokine release from macrophage-like cells. Mortality in mice receiving lethal doses of LPS has been averted, demonstrating the interest of this molecule in inflammatory conditions and critical illness [70].

VOLATILE ORGANIC COMPOUNDS AND THE ELECTRONIC NOSE

The volatolome or breathome is a promising subarea of the metabolome, directed toward volatile organic compounds (VOCs), although nonvolatile metabolites are eventually included. These molecules typically do not exceed 10 carbons and were originally sought in the breath. This is the fundamental difference regarding classic metabolome. However, the concept has been enlarged to all modalities of body secretions and excretions, such as plasma, urine, stools, sputum, sweat, saliva, milk, pleural effusions, surgical drainage material, and obviously breath, in which small molecules are analyzed.

Fecal bacteria are responsible for about 300 VOCs [71], many of which are absorbed into the bloodstream, emerging in virtually any body solid, fluid, or air compartment. That is a pathophysiological bridge between the gut microbiome and metabolomics, although regional microbiomes should not be neglected. In the case of respiratory diseases, both benign and malignant, the bronchopulmonary microbiome is meaningful. Similar parallels may be drawn for emanations in oral, urinary, genital, and skin diseases and the respective microbiomes. Of course there is no dearth of nonmicrobiome metabolites originated from the host, food, drugs, breathed air, or environmental pollutants.

Gas chromatography, mass spectrometry, and nuclear magnetic resonance spectroscopy are usually used. Analogous to what occurs with the general metabolome, already discussed in other chapters, such techniques are expensive and labor-intensive and, however, extremely informative.

Bedside devices (electronic nose) have received attention because they are easy to carry, affordable, and provide information in real time. They have been called as biomimetic cross-reactive sensor arrays (B-CRSAs). Few of them contain metabolite-identifying sensors. One gets a nonspecific scattergram or cluster map, depicting the distribution of the most relevant molecules in the circumstances without characterization. This may already represent a signature, corroborated by pattern recognition statistics, thus allowing for bedside screening, a valuable resource in precision medicine. Nevertheless, full analysis of the volatolome by means of mass spectrometry and nuclear magnetic resonance spectroscopy is still the standard approach.

Obesity and NAFLD

Conventional point-of-care diagnosis of obesity has never been an insurmountable challenge; however, metabolic fingerprints can provide precious insights into deranged pathways and potential complications.

In a pediatric series measuring urinary VOCs, 14 molecules exhibited distinct profiles. Statistics confirmed the importance of 5-methyl-3-hexanone and 4-methyl-2-heptanone, both methyl-ketones, potentially synthesized by bacteria and fungi. A number of alcohols were also abundant, and an excess of *Bacteroides* in the gut could be incriminated for their appearance [72].

In another experience with children aiming breath examination, the culprits were isoprene, 1-decene, 1-octene, ammonia, and hydrogen sulfide being significantly higher in the obese group. Although the source of several of these could be the gut, particularly as concerns ammonia and hydrogen sulfide, different bacteria are involved, suggesting nonoverlapping pathways between breath and urinary findings [73].

The objective was somewhat different, namely NAFLD, in a cohort of adult overweight and obese participants, and fecal samples were collected. Fecal ester compounds were the VOC finding consistent with NAFLD. The authors attribute such pattern to an overexpression of *Lactobacillus,* Lachnospiraceae, *Dorea, Robinsoniella,* and *Roseburia,* along with less abundance of *Oscillibacter (*Ruminococcaceae) [74].

Inflammatory Bowel Disease

Mining for biomarkers in this domain has been traditionally rewarding, even though controversies remain. In Crohn's disease, upregulation of fecal heptanal, 1-octen-3-ol, 2-piperidinone, and 6-methyl-2-heptanone called attention at the same time that methanethiol, 3-methyl-phenol, SCFAs, and ester derivatives became scarce [71].

In pediatric IBD, specific molecules were not pinpointed (electronic nose methodology); however, a pattern mimicking the altered microbiome and potentially useful for diagnosis was emphasized as well [75].

Cancer

CRC has been addressed in previous items and deserves further notes. Breath tests disclosed cyclohexane, methylcyclohexane, 1,3-dimethyl- benzene, and decanal as the most prominent features of the disease [76]. Such outcome conflicts with that of another study [77], in which acetone and ethyl acetate were elevated and ethanol and 4-methyl octane depressed, emphasizing that subpopulations of CRC could exist with distinct volatolome profiles.

One center encountered 14 breath metabolites, which could distinguish benign from malignant gastric illnesses: 2,3-butanediol, 1,3-dioxolan-2-one, hexadecane, undecane, 3,8-dimethyl, *N,N*-dimethylacetamide, phosphonic acid, (*p*-hydroxyphenyl), 1,3-dioxolane-2-methanol, 3,5-decadien-7-yne, 6-t-butyl, 2,2,9,9-tetramethyl, 1,6-dioxacyclododecane-

7,12-dione, caprolactam, 5,7-octadien-2-one, 3-acetyl, nonanal, 5-hepten-2-one, 6-methyl, and benzothiazole. A few could be ascribed to *H. pylori* metabolism, whereas the origin of others is still unknown [78].

Urinary VOCs are compellingly defended for pancreatic ductal adenocarcinoma, a challenging and ominous malignancy that was successfully screened, with a field-asymmetric ion mobility spectrometry portable VOC analyzer [79]. In serum, elevated pyroglutamine, gamma-glutamylphenylalanine, phenylpyruvate, N-acetylcitrulline, and stearoylcarnitine were discriminating for prostate cancer [80].

Lung Diseases

Cyclohexane and xylene were increased in lung cancer and correlated both with disease progression (elevation) and successful treatment (reduction) [81]. A distinctive pattern (electronic nose) for bronchopulmonary dysplasia in infants has also been announced [82].

Many other respiratory diseases are currently being investigated, such as bacterial pneumonia, tuberculosis, chronic obstructive pulmonary disease, asthma, acute respiratory distress syndrome, and pulmonary fibrosis [83].

LIPIDOME

Lipids regulate very diverse cellular processes from ATP synthesis and the activation of essential cell-signaling pathways to membrane organization and plasticity.

Multiple Sclerosis

Multiple sclerosis is another focus of concentrated attention as reported in this book. Among other markers, lipidomic studies have suggested depressed plasma lipid 654 expression as a potential diagnostic tool for this condition. This lipid molecule is synthesized by a number of Bacteroidetes present in both the mouth and the gastrointestinal tract [84].

Obesity and Diabetes

As already mentioned with regard to the microbiome and again in the context of breast cancer, which is counted among the obesity-related tumors, SCFA and other metabolomic markers are firmly ingrained in the scientific literature. However, one should be careful not to generalize. A recent study with odd chain fatty acids, also linked to glucose intolerance, dispelled their relationship with the microbiome. Circulating C15:0/C17:0 are probably processed by the host in the presence of dietary and other substrates, independently of the gut microenvironment [85].

PROTEOME

Proteomics is highly promising, but also the most challenging item in the omics universe. The hardware is more complex and expensive, typically liquid chromatography tandem mass spectrometry. Immunohistochemical confirmation may be required as well. Among the >21,000 intestinal taxa, there are more than 63,000,000 proteins [86]. Just for Crohn's disease, as many as 2900 were identified during host-microbiome fingerprinting [87]. Indeed, both sides express proteins and peptides, which need to be screened and classified. As a consequence, large protocols are infrequent, most investigations recruiting a few dozen participants or less. All these barriers notwithstanding, knowledge is being accrued.

Obesity

The involvement of Bacteroidetes gained traction with one publication, and subset classification of obese subjects could be enhanced by unveiled protein signatures [88].

Inflammatory Bowel Disease

In a mouse model, differential expression of 276 proteins was recently recognized. Meaningfully, microbial proteases and hydrolases were overrepresented in the inflamed gut environment [89]. In humans, four protein modules could be characterized on the basis of mucosal lavage samples collected in different topographical areas of the gut. These encompassed a total of 28 proteins and were related to either Crohn's disease or ulcerative colitis [90].

GENETICALLY ENGINEERED MICROORGANISMS

Prophylactic and therapeutic applications for genetically modified bacteria are not a new idea. In the early 1990s such proposals were widely debated, even though CRISP/Cas9 (clustered, regularly interspaced, short palindromic repeats), zinc-finger nucleases (ZFNs), transcription activator—like effector nucleases (TALENs), and other gene engineering technologies were not available [91].

Much earlier, in the 1940s, viruses with affinity for solid tumors [92] and subsequently bacteria with similar properties such as *Salmonella typhimurium* were already receiving attention. The hypothesis of genetic manipulation of such microbes to convert them into vehicles for anticancer agents and other drugs was subsequently envisaged.

Biotechnological drug manufacturing, relying on genetically engineered organisms such as *Escherichia coli* and the yeasts *Saccharomyces cerevisiae* and *Pichia pastoris* is nowadays an industry standard. The gut microbiome could provide the long-term personal platform to manufacture as well as release low doses of the required molecule, especially if ordinary commensals were selected for genetic manipulation. Such tailored treatment would be an elegant complement to precision medicine.

Salmonella typhimurium can colonize the human gut. As a tumor-seeking organism, a manipulated strain expressing flagellin B (FlaB) suppressed tumors and metastasis and assured prolonged survival through a TLR-5—dependent mechanism in a mouse model [93].

A modified strain of the same species has been experimentally endowed with antimelanoma properties by means of expression of interferon gamma [94].

Genetically engineered *E. coli* Nissle 1917 strains for the treatment of urea cycle disorder and phenylketonuria are advancing. Inflammation-sensing bacteria able to both diagnose and orchestrate a therapeutic response in patients with IBD are also being pursued [95]. At least one molecular basis for such antiinflammatory result, namely α-MSH, has already been elucidated, as previously alluded to in the study by Qiang et al. [70].

FINAL CONSIDERATIONS

Precision medicine was officially born in 2015, and the National Institutes of Health promptly endorsed the initiative. The aim was to provide a robust, personalized diagnosis to each patient, consequently leading to safer and more successful prognosis, therapy, and quality of life. Large databases, concerning the human genome and microbiome, are providing the background. Metabolomics, transcriptomics, proteomics, and lipidomics exponentially increase the knowledge. Precision medicine cannot be divorced from molecular and cellular pathology because these are indispensable to underpin disease processes and therapeutic targets. Such tools have been applied to the microbiome with fruitful results and are being further enriched by the deep phenotyping resources of the metabolome.

REFERENCES

[1] Stranger BE, Brigham LE, Hasz R, Hunter M, Johns C, Johnson M, Kopen G, Leinweber WF, Lonsdale JT, McDonald A, Mestichelli B, Myer K, Roe B, Salvatore M, Shad S, Thomas JA, Walters G, Washington M, Wheeler J, Bridge J, Foster BA, Gillard BM, Karasik E, Kumar R, Miklos M, Moser MT, Jewell SD, Montroy RG, Rohrer DC, Valley DR, Davis DA, Mash DC, Gould SE, Guan P, Koester S, Little AR, Martin C, Moore HM, Rao A, Struewing JP, Volpi S, Hansen KD, Hickey PF, Rizzardi LF, Hou L, Liu Y, Molinie B, Park Y, Rinaldi N, Wang L, Van Wittenberghe N, Claussnitzer M, Gelfand ET, Li Q, Linder S, Zhang R, Smith KS, Tsang EK, Chen LS, Demanelis K, Doherty JA, Jasmine F, Kibriya MG, Jiang L, Lin S, Wang M, Jian R, Li X, Chan J, Bates D, Diegel M, Halow J, Haugen E, Johnson A, Kaul R, Lee K, Maurano MT, Nelson J, Neri FJ, Sandstrom R, Fernando MS, Linke C, Oliva M, Skol A, Wu F, Akey JM, Feinberg AP, Li JB, Pierce BL, Stamatoyannopoulos JA, Tang H, Ardlie KG, Kellis M, Snyder MP, Montgomery SB. Enhancing GTEx by bridging the gaps between genotype, gene expression, and disease. Nat Genet 2017;49:1664—70.

[2] Kashyap PC, Chia N, Nelson H, Segal E, Elinav E. Microbiome at the frontier of personalized medicine. Mayo Clin Proc December 2017;92(12):1855—64.

[3] Peters JH. Genetic factors in relation to drugs. Annu Rev Pharmacol 1968;8:427—52.

[4] Kalow W. Pharmacogenetics and pharmacogenomics: origin, status, and the hope for personalized medicine. Pharmacogenomics J 2006;6:162—5.

[5] Thomson W, Tait PG. Mathematical and physical papers. UK: Cambridge University Press; 1882—1911.

[6] Dietel M, Jöhrens K, Laffert MV, Hummel M, Bläker H, Pfitzner BM, Lehmann A, Denkert C, Darb-Esfahani S, Lenze D, Heppner FL, Koch A, Sers C, Klauschen F, Anagnostopoulos IA. 2015 update on predictive molecular pathology and its role in targeted cancer therapy: a review focussing on clinical relevance. Cancer Gene Ther 2015;22:417—30.

[7] Beger RD, Dunn W, Schmidt MA, Gross SS, Kirwan JA, Cascante M, Brennan L, Wishart DS, Oresic M, Hankemeier T, Broadhurst DI, Lane AN, Suhre K, Kastenmüller G, Sumner SJ, Thiele I, Fiehn O, Kaddurah-Daouk R, For "Precision Medicine, Pharmacometabolomics Task Group"-Metabolomics Society Initiative. Metabolomics enables precision medicine: "a white paper, community perspective". Metabolomics 2016;12:149.

[8] Boursier J, Mueller O, Barret M, Machado M, Fizanne L, Araujo-Perez F, Guy CD, Seed PC, Rawls JF, David LA, Hunault G, Oberti F, Calès P, Diehl AM. The severity of nonalcoholic fatty liver disease is associated with gut dysbiosis and shift in the metabolic function of the gut microbiota. Hepatology 2016;63:764–75.

[9] Loomba R, Seguritan V, Li W, Long T, Klitgord N, Bhatt A, Dulai PS, Caussy C, Bettencourt R, Highlander SK, Jones MB, Sirlin CB, Schnabl B, Brinkac L, Schork N, Chen CH, Brenner DA, Biggs W, Yooseph S, Venter JC, Nelson KE. Gut microbiome-based metagenomic signature for non-invasive detection of advanced fibrosis in human nonalcoholic fatty liver disease. Cell Metab 2017;25:1054–62. e5.

[10] Nakahara T, Hyogo H, Ono A, Nagaoki Y, Kawaoka T, Miki D, Tsuge M, Hiraga N, Hayes CN, Hiramatsu A, Imamura M, Kawakami Y, Aikata H, Ochi H, Abe-Chayama H, Furusho H, Shintani T, Kurihara H, Miyauchi M, Takata T, Arihiro K, Chayama K. Involvement of Porphyromonas gingivalis in the progression of non-alcoholic fatty liver disease. J Gastroenterol 2018;53:269–80.

[11] Aleman JO, Eusebi LH, Ricciardiello L, Patidar K, Sanyal AJ, Holt PR. Mechanisms of obesity-induced gastrointestinal neoplasia. Gastroenterology 2014;146:357–73.

[12] Puri P, Liangpunsakul S, Christensen JE, Shah VH, Kamath PS, Gores GJ, Walker S, Comerford M, Katz B, Borst A, Yu Q, Kumar DP, Mirshahi F, Radaeva S, Chalasani NP, Crabb DW, Sanyal AJ, TREAT Consortium. The circulating microbiome signature and inferred functional metagenomics in alcoholic hepatitis. Hepatology October 30, 2018;67:1284–302.

[13] Jie Z, Xia H, Zhong SL, Feng Q, Li S, Liang S, Zhong H, Liu Z, Gao Y, Zhao H, Zhang D, Su Z, Fang Z, Lan Z, Li J, Xiao L, Li J, Li R, Li X, Li F, Ren H, Huang Y, Peng Y, Li G, Wen B, Dong B, Chen JY, Geng QS, Zhang ZW, Yang H, Wang J, Wang J, Zhang X, Madsen L, Brix S, Ning G, Xu X, Liu X, Hou Y, Jia H, He K, Kristiansen K. The gut microbiome in atherosclerotic cardiovascular disease. Nat Commun 2017;8:845.

[14] Pasolli E, Truong DT, Malik F, Waldron L, Segata N. Machine learning meta-analysis of large metagenomic datasets: tools and biological insights. PLoS Comput Biol 2016;12:e1004977.

[15] Bluemel S, Williams B, Knight R, Schnabl B. Precision medicine in alcoholic and nonalcoholic fatty liver disease via modulating the gut microbiota. Am J Physiol Gastrointest Liver Physiol 2016;311:G1018–36.

[16] Jacobs LR. Relationship between dietary fiber and cancer: metabolic, physiologic, and cellular mechanisms. Proc Exp Biol Med 1986;183:299–310.

[17] Burns MB, Lynch J, Starr TK, Knights D, Blekhman R. Virulence genes are a signature of the microbiome in the colorectal tumor microenvironment. Genome Med 2015;7:55.

[18] Flynn KJ, Baxter NT, Schloss PD. Metabolic and community synergy of oral bacteria in colorectal cancer. mSphere 2016;1(3).

[19] Li YY, Ge QX, Cao J, et al. Association of Fusobacterium nucleatum infection with colorectal cancer in Chinese patients. World J Gastroenterol 2016;22:3227–33.

[20] Yu T, Guo F, Yu Y, Sun T, Ma D, Han J, Qian Y, Kryczek I, Sun D, Nagarsheth N, Chen Y, Chen H, Hong J, Zou W, Fang JY. Fusobacterium nucleatum promotes chemoresistance to colorectal cancer by modulating autophagy. Cell 2017;170:548–63. e16.

[21] Drewes JL, White JR, Dejea CM, Fathi P, Iyadorai T, Vadivelu J, Roslani AC, Wick EC, Mongodin EF, Loke MF, Thulasi K, Gan HM, Goh KL, Chong HY, Kumar S, Wanyiri JW, Sears CL. High-resolution bacterial 16S rRNA gene profile meta-analysis and biofilm status reveal common colorectal cancer consortia. NPJ Biofilms Microbiomes 2017;3:34.

[22] Hieken TJ, Chen J, Hoskin TL, Walther-Antonio M, Johnson S, Ramaker S, Xiao J, Radisky DC, Knutson KL, Kalari KR, Yao JZ, Baddour LM, Chia N, Degnim AC. The microbiome of aseptically collected human breast tissue in benign and malignant disease. Sci Rep 2016;6:30751.

[23] Mani S. Microbiota and breast cancer. Prog Mol Biol Transl Sci 2017;151:217–29.

[24] Carrière J, Darfeuille-Michaud A, Nguyen HT. Infectious etiopathogenesis of Crohn's disease. World J Gastroenterol 2014;20:12102–17.

[25] Haberman Y, Tickle TL, Dexheimer PJ, Kim MO, Tang D, Karns R, Baldassano RN, Noe JD, Rosh J, Markowitz J, Heyman MB, Griffiths AM, Crandall WV, Mack DR, Baker SS, Huttenhower C, Keljo DJ, Hyams JS, Kugathasan S, Walters TD, Aronow B, Xavier RJ, Gevers D, Denson LA. Pediatric Crohn disease patients exhibit specific ileal transcriptome and microbiome signature. J Clin Invest 2015;125:1363.

[26] Mondot S, Lepage P, Seksik P, Allez M, Tréton X, Bouhnik Y, Colombel JF, Leclerc M, Pochart P, Doré J, Marteau P, GETAID. Structural robustness of the gut mucosal microbiota is associated with Crohn's disease remission after surgery. Gut June 2016;65(6):954–62.

[27] Bry L, Falk PG, Midtvedt T, Gordon JI. A model of host-microbial interactions in an open mammalian ecosystem. Science 1996;273(5280):1380–3.

[28] Bäckhed F, Ding H, Wang T, Hooper LV, Koh GY, Nagy A, Semenkovich CF, Gordon JI. The gut microbiota as an environmental factor that regulates fat storage. Proc Natl Acad Sci USA 2004;101:15718–23.

[29] Ley RE, Turnbaugh PJ, Klein S, Gordon JI. Microbial ecology: human gut microbes associated with obesity. Nature 2006;444:1022–3.

[30] Alang N, Kelly CR. Acute weight gain after fecal microbiota transplantation. Open Forum Infect Dis 2015;2. ofv004.

[31] Fischer M, Kao D, Kassam Z, Smith J, Louie T, Sipe B, Torbeck M, Xu H, Ouyang F, Mozaffarian D, Allegretti JR. Stool donor body mass index does not affect recipient weight after a single fecal microbiota transplantation for C. difficile infection. Clin Gastroenterol Hepatol December 12, 2018;16:1353–61.

[32] Damms-Machado A, Mitra S, Schollenberger AE, Kramer KM, Meile T, Königsrainer A, Huson DH, Bischoff SC. Effects of surgical and dietary weight loss therapy for obesity on gut microbiota composition and nutrient absorption. BioMed Res Int 2015;2015, 806248.

[33] Finucane MM, Sharpton TJ, Laurent TJ, Pollard KS. A taxonomic signature of obesity in the microbiome? Getting to the guts of the matter. PLoS One 2014;9:e84689.

[34] Hwang I, Park YJ, Kim YR, Kim YN, Ka S, Lee HY, Seong JK, Seok YJ, Kim JB. Alteration of gut microbiota by vancomycin and bacitracin improves insulin resistance via glucagon-like peptide 1 in diet-induced obesity. FASEB J 2015;29:2397–411.

[35] Thaisss CA, Itav S, Rothschild D, Meijer MT, Levy M, Moresi C, Dohnalova L, Braverman S, Rozin S, Malitsky S, Dori-Bachash M, Kuperman Y, Biton I, Gertler A, Harmelin A, Shapiro H, Halpern Z, Aharoni A, Segal E, Elinav E. Persistent microbiome alterations modulate the rate of post-dieting weight regain. Nature 2016;540:544—51.

[36] Griffin NW, Ahern PP, Cheng J, Heath AC, Ilkaveya O, Newgard CB, Fontana L, Gordon JI. Prior dietary practices and connections to a human gut microbial metacommunity alter responses to diet interventions. Cell Host Microbe 2017;21:84—96.

[37] Wilck N, Matus MG, Kearney SM, Olesen SW, Forslund K, Bartolomaeus H, Haase S, Mähler A, Balogh A, Markó L, Vvedenskaya O, Kleiner FH, Tsvetkov D, Klug L, Costea PI, Sunagawa S, Maier L, Rakova N, Schatz V, Neubert P, Frätzer C, Krannich A, Gollasch M, Grohme DA, Côrte-Real BF, Gerlach RG, Basic M, Typas A, Wu C, Titze JM, Jantsch J, Boschmann M, Dechend R, Kleinewietfeld M, Kempa S, Bork P, Linker RA, Alm EJ, Müller DN. Salt-responsive gut commensal modulates TH17 axis and disease. Nature 2017;551:585—9.

[38] Kang Y, Cai Y. Gut microbiota and hypertension: from pathogenesis to new therapeutic strategies. Clin Res Hepatol Gastroenterol November 1, 2018;42:110—7.

[39] Zubcevic J, Baker A, Martyniuk CJ. Transcriptional networks in rodent models support a role for gut-brain communication in neurogenic hypertension: a review of the evidence. Physiol Genom 2017;49:327—38.

[40] Coit P, Mumcu G, Ture-Ozdemir F, Unal AU, Alpar U, Bostanci N, Ergun T, Direskeneli H, Sawalha AH. Sequencing of 16S rRNA reveals a distinct salivary microbiome signature in Behçet's disease. Clin Immunol 2016;169:28—35.

[41] Consolandi C, Turroni S, Emmi G, Severgnini M, Fiori J, Peano C, Biagi E, Grassi A, Rampelli S, Silvestri E, Centanni M, Cianchi F, Gotti R, Emmi L, Brigidi P, Bizzaro N, De Bellis G, Prisco D, Candela M, D'Elios MM. Behçet's syndrome patients exhibit specific microbiome signature. Autoimmun Rev 2015;14:269—76.

[42] Shao T, Shao L, Li H, Xie Z, He Z, Wen C. Combined signature of the fecal microbiome and metabolome in patients with Gout. Front Microbiol February 21, 2017;8:268.

[43] Maeda Y, Kurakawa T, Umemoto E, Motooka D, Ito Y, Gotoh K, Hirota K, Matsushita M, Furuta Y, Narazaki M, Sakaguchi N, Kayama H, Nakamura S, Iida T, Saeki Y, Kumanogoh A, Sakaguchi S, Takeda K. Dysbiosis contributes to arthritis development via activation of autoreactive T cells in the intestine. Arthritis Rheumatol 2016;68:2646—61.

[44] Chen J, Wright K, Davis JM, Jeraldo P, Marietta EV, Murray J, Nelson H, Matteson EL, Taneja V. An expansion of rare lineage intestinal microbes characterizes rheumatoid arthritis. Genome Med 2016;8:43.

[45] Mertz L. Omics tech, gut-on-a-chip, and bacterial engineering: new approaches for treating inflammatory bowel diseases. IEEE Pulse 2016;7:9—12.

[46] Kim HJ, Lee J, Choi JH, Bahinski A, Ingber DE. Co-culture of living microbiome with microengineered human intestinal villi in a gut-on-a-chip microfluidic device. J Vis Exp 2016;(114).

[47] Lee J, Choi JH, Kim HJ. Human gut-on-a-chip technology: will this revolutionize our understanding of IBD and future treatments? Exp Rev Gastroenterol Hepatol 2016;10:883—5.

[48] Fu ZD, Cui JY. Remote sensing between liver and intestine: importance of microbial metabolites. Curr Pharmacol Rep 2017;3:101—13.

[49] Alonso C, Fernández-Ramos D, Varela-Rey M, Martínez-Arranz I, Navasa N, Van Liempd SM, Lavín Trueba JL, Mayo R, Ilisso CP, de Juan VG, Iruarrizaga-Lejarreta M, delaCruz-Villar L, Minchólé I, Robinson A, Crespo J, Martín-Duce A, Romero-Gómez M, Sann H, Platon J, Van Eyk J, Aspichueta P, Noureddin M, Falcón-Pérez JM, Anguita J, Aransay AM, Martínez-Chantar ML, Lu SC, Mato JM. Metabolomic identification of subtypes of nonalcoholic steatohepatitis. Gastroenterology 2017;152:1449—61. e7.

[50] Han J, Dzierlenga AL, Lu Z, Billheimer DD, Torabzadeh E, Lake AD, Li H, Novak P, Shipkova P, Aranibar N, Robertson D, Reily MD, Lehman-McKeeman LD, Cherrington NJ. Metabolomic profiling distinction of human nonalcoholic fatty liver disease progression from a common rat model. Obesity 2017;25:1069—76.

[51] Perry RJ, Peng L, Barry NA, Cline GW, Zhang D, Cardone RL, Petersen KF, Kibbey RG, Goodman AL, Shulman GI. Acetate mediates a microbiome-brain-β-cell axis to promote metabolic syndrome. Nature 2016;534:213—7.

[52] Troisi J, Pierri L, Landolfi A, Marciano F, Bisogno A, Belmonte F, Palladino C, Guercio Nuzio S, Campiglia P, Vajro P. Urinary metabolomics in pediatric obesity and NAFLD identifies metabolic pathways/metabolites related to dietary habits and gut-liver Axis perturbations. Nutrients 2017;9. pii: E485.

[53] Priyadarshini M, Wicksteed B, Schiltz GE, Gilchrist A, Layden BT. SCFA receptors in pancreatic β cells: novel diabetes targets? Trends Endocrinol Metab 2016;27:653—64.

[54] Metchnikoff E. The prolongation of life: optimistic studies. London: William Heineman; 1907.

[55] Biagi E, Franceschi C, Rampelli S, Severgnini M, Ostan R, Turroni S, Consolandi C, Quercia S, Scurti M, Monti D, Capri M, Brigidi P, Candela M. Gut microbiota and extreme longevity. Curr Biol 2016;26:1480—5.

[56] Cheng S, Larson MG, McCabe EL, Murabito JM, Rhee EP, Ho JE, Jacques PF, Ghorbani A, Magnusson M, Souza AL, Deik AA, Pierce KA, Bullock K, O'Donnell CJ, Melander O, Clish CB, Vasan RS, Gerszten RE, Wang TJ. Distinct metabolomic signatures are associated with longevity in humans. Nat Commun 2015;6:6791.

[57] Gruber J, Kennedy BK. Microbiome and longevity: gut microbes send signals to host mitochondria. Cell 2017;169:1168—9.

[58] Ogrodzinski MP, Bernard JJ, Lunt SY. Deciphering metabolic rewiring in breast cancer subtypes. Transl Res 2017;189:105—12.

[59] Lerner A, Aminov R, Matthias T. Transglutaminases in dysbiosis as potential environmental drivers of autoimmunity. Front Microbiol January 24, 2017;8:66.

[60] Richard V, Conotte R, Mayne D, Colet JM. Does the 1H-NMR plasma metabolome reflect the host-tumor interactions in human breastcancer? Oncotarget July 25, 2017;8(30):49915—30.

[61] Chaput N, Lepage P, Coutzac C, Soularue E, Le Roux K, Monot C, Boselli L, Routier E, Cassard L, Collins M, Vaysse T, Marthey L, Eggermont A, Asvatourian V, Lanoy E, Mateus C, Robert C, Carbonnel F. Baseline gut microbiota predicts clinical response and colitis in metastatic melanoma patients treated with ipilimumab. Ann Oncol 2017;28:1368−79.

[62] Matson V, Fessler J, Bao R, Chongsuwat T, Zha Y, Alegre ML, Luke JJ, Gajewski TF. The commensal microbiome is associated with anti−PD-1 efficacy in metastatic melanoma patients. Science 2018;359:104−8.

[63] Gopalakrishnan V, Spencer CN, Nezi L, Reuben A, Andrews MC, Karpinets TV, Prieto PA, Vicente D, Hoffman K, Wei SC, Cogdill AP, Zhao L, Hudgens CW, Hutchinson DS, Manzo T, Petaccia de Macedo M, Cotechini T, Kumar T, Chen WS, Reddy SM, Sloane RS, Galloway-Pena J, Jiang H, Chen PL, Shpall EJ, Rezvani K, Alousi AM, Chemaly RF, Shelburne S, Vence LM, Okhuysen PC, Jensen VB, Swennes AG, McAllister F, Sanchez EMR, Zhang Y, Le Chatelier E, Zitvogel L, Pons N, Austin-Breneman JL, Haydu LE, Burton EM, Gardner JM, Sirmans E, Hu J, Lazar AJ, Tsujikawa T, Diab A, Tawbi H, Glitza IC, Hwu WJ, Patel SP, Woodman SE, Amaria RN, Davies MA, Gershenwald JE, Hwu P, Lee JE, Zhang J, Coussens LM, Cooper ZA, Futreal PA, Daniel CR, Ajami NJ, Petrosino JF, Tetzlaff MT, Sharma P, Allison JP, Jenq RR, Wargo JA. Gut microbiome modulates response to anti-PD-1 immunotherapy in melanoma patients. Science November 2, 2018;359(6371):97−103.

[64] Routy B, Le Chatelier E, Derosa L, Duong CPM, Alou MT, Daillère R, Fluckiger A, Messaoudene M, Rauber C, Roberti MP, Fidelle M, Flament C, Poirier-Colame V, Opolon P, Klein C, Iribarren K, Mondragón L, Jacquelot N, Qu B, Ferrere G, Clémenson C, Mezquita L, Masip JR, Naltet C, Brosseau S, Kaderbhai C, Richard C, Rizvi H, Levenez F, Galleron N, Quinquis B, Pons N, Ryffel B, Minard-Colin V, Gonin P, Soria JC, Deutsch E, Loriot Y, Ghiringhelli F, Zalcman G, Goldwasser F, Escudier B, Hellmann MD, Eggermont A, Raoult D, Albiges L, Kroemer G, Zitvogel L. Gut microbiome influences efficacy of PD-1−based immunotherapy against epithelial tumors. Science November 2, 2018;359(6371):91−7.

[65] Alexander JL, Wilson ID, Teare J, Marchesi JR, Nicholson JK, Kinross JM. Gut microbiota modulation of chemotherapy efficacy and toxicity. Nat Rev Gastroenterol Hepatol 2017;14:356−65.

[66] Geller LT, Barzily-Rokni M, Danino T, Jonas OH, Shental N, Nejman D, Gavert N, Zwang Y, Cooper ZA, Shee K, Thaiss CA, Reuben A, Livny J, Avraham R, Frederick DT, Ligorio M, Chatman K, Johnston SE, Mosher CM, Brandis A, Fuks G, Gurbatri C, Gopalakrishnan V, Kim M, Hurd MW, Katz M, Fleming J, Maitra A, Smith DA, Skalak M, Bu J, Michaud M, Trauger SA, Barshack I, Golan T, Sandbank J, Flaherty KT, Mandinova A, Garrett WS, Thayer SP, Ferrone CR, Huttenhower C, Bhatia SN, Gevers D, Wargo JA, Golub TR, Straussman R. Potential role of intratumor bacteria in mediating tumor resistance to the chemotherapeutic drug gemcitabine. Science 2017;357:1156−60.

[67] Cason CA, Dolan KT, Sharma G, Tao M, Kulkarni R, Helenowski IB, Doane BM, Avram MJ, McDermott MM, Chang EB, Ozaki CK, Ho KJ. Plasma microbiome-modulated indole- and phenyl-derived metabolites associate with advanced atherosclerosis and postoperative outcomes. J Vasc Surg December 13, 2017.

[68] Guasch-Ferré M, Hu FB, Ruiz-Canela M, Bulló M, Toledo E, Wang DD, Corella D, Gómez-Gracia E, Fiol M, Estruch R, Lapetra J, Fitó M, Arós F, Serra-Majem L, Ros E, Dennis C, Liang L, Clish CB, Martínez-González MA, Salas-Salvadó J. Plasma metabolites from choline pathway and risk of cardiovascular disease in the PREDIMED (prevention with Mediterranean diet) study. J Am Heart Assoc 2017;6. pii: e006524.

[69] Dodd D, Spitzer MH, Van Treuren W, Merrill BD, Hryckowian AJ, Higginbottom SK, Le A, Cowan TM, Nolan GP, Fischbach MA, Sonnenburg JL. A gut bacterial pathway metabolizes aromatic amino acids into nine circulating metabolites. Nature 2017;551(7682):648−52.

[70] Qiang X, Liotta AS, Shiloach J, Gutierrez JC, Wang H, Ochani M, Ochani K, Yang H, Rabin A, LeRoith D, Lesniak MA, Böhm M, Maaser C, Kannengiesser K, Donowitz M, Rabizadeh S, Czura CJ, Tracey KJ, Westlake M, Zarfeshani A, Mehdi SF, Danoff A, Ge X, Sanyal S, Schwartz GJ, Roth J. New melanocortin-like peptide of E. coli can suppress inflammation via the mammalian melanocortin-1 receptor (MC1R): possible endocrine-like function for microbes of the gut. NPJ Biofilms Microbiomes 2017;3:31.

[71] Smolinska A, Bodelier AG, Dallinga JW, Masclee AA, Jonkers DM, van Schooten FJ, Pierik MJ. Investigation of faecal volatile organic metabolites as novel diagnostic biomarkers in inflammatory bowel disease. Aliment Pharmacol Ther May 2017;45(9):1244−54.

[72] Cozzolino R, De Giulio B, Marena P, Martignetti A, Günther K, Lauria F, Russo P, Stocchero M, Siani A. Urinary volatile organic compounds in overweight compared to normal-weight children: results from the Italian I.Family cohort. Sci Rep 2017;7:15636.

[73] Alkhouri N, Eng K, Cikach F, Patel N, Yan C, Brindle A, Rome E, Hanouneh I, Grove D, Lopez R, Hazen SL, Dweik RA. Breathprints of childhood obesity: changes in volatile organic compounds in obese children compared with lean controls. Pediatr Obes 2015;10:23−9.

[74] Raman M, Ahmed I, Gillevet PM, Probert CS, Ratcliffe NM, Smith S, Greenwood R, Sikaroodi M, Lam V, Crotty P, Bailey J, Myers RP, Rioux KP. Fecal microbiome and volatile organic compound metabolome in obese humans with nonalcoholic fatty liver disease. Clin Gastroenterol Hepatol 2013;11:868−75. e1-3.

[75] van Gaal N, Lakenman R, Covington J, Savage R, de Groot E, Bomers M, Benninga M, Mulder C, de Boer N, de Meij T. Faecal volatile organic compounds analysis using field asymmetric ion mobility spectrometry: non-invasive diagnostics in paediatric inflammatory bowel disease. J Breath Res April 25, 2017;12:016006.

[76] Bhattacharyya D, Kumar P, Mohanty SK, Smith YR, Misra M. Detection of four distinct volatile indicators of colorectal cancer using functionalized Titania nanotubular arrays. Sensors 2017;17(8).

[77] Amal H, Leja M, Funka K, Lasina I, Skapars R, Sivins A, Ancans G, Kikuste I, Vanags A, Tolmanis I, Kirsners A, Kupcinskas L, Haick H. Breath testing as potential colorectal cancer screening tool. Int J Cancer 2016;138:229−36.

[78] Tong H, Wang Y, Li Y, Liu S, Chi C, Liu D, Guo L, Li E, Wang C. Volatile organic metabolites identify patients with gastric carcinoma, gastric ulcer, or gastritis and control patients. Cancer Cell Int November 21, 2017;17:108.

[79] Arasaradnam R, Wicaksono A, O'Brien H, Kocher HM, Covington JA, Crnogorac-Jurcevic T. Non-invasive diagnosis of pancreatic cancer through detection of volatile organic compounds in urine. Gastroenterology November 9, 2018;154:485−7.

[80] Huang J, Mondul AM, Weinstein SJ, Koutros S, Derkach A, Karoly E, Sampson JN, Moore SC, Berndt SI, Albanes D. Serum metabolomic profiling of prostate cancer risk in the prostate, lung, colorectal, and ovarian cancer screening trial. Br J Cancer 2016;115:1087—95.

[81] Oguma T, Nagaoka T, Kurahashi M, Kobayashi N, Yamamori S, Tsuji C, Takiguchi H, Niimi K, Tomomatsu H, Tomomatsu K, Hayama N, Aoki T, Urano T, Magatani K, Takeda S, Abe T, Asano K. Clinical contributions of exhaled volatile organic compounds in the diagnosis of lung cancer. PLoS One 2017;12:e0174802.

[82] Berkhout DJC, Niemarkt HJ, Benninga MA, Budding AE, van Kaam AH, Kramer BW, Pantophlet CM, van Weissenbruch MM, de Boer NKH, de Meij TGJ. Development of severe bronchopulmonary dysplasia is associated with alterations in fecal volatile organic compounds. Pediatr Res November 22, 2018;83:412—9.

[83] Devillier P, Salvator H, Naline E, Couderc LJ, Grassin-Delyle S. Metabolomics in the diagnosis and pharmacotherapy of lung diseases. Curr Pharm Des 2017;23:2050—9.

[84] Farrokhi V, Nemati R, Nichols FC, Yao X, Anstadt E, Fujiwara M, Grady J, Wakefield D, Castro W, Donaldson J, Clark RB. Bacterial lipodipeptide, Lipid 654, is a microbiome-associated biomarker for multiple sclerosis. Clin Transl Immunol 2013;2:e8.

[85] Jenkins BJ, Seyssel K, Chiu S, Pan PH, Lin SY, Stanley E, Ament Z, West JA, Summerhill K, Griffin JL, Vetter W, Autio KJ, Hiltunen K, Hazebrouck S, Stepankova R, Chen CJ, Alligier M, Laville M, Moore M, Kraft G, Cherrington A, King S, Krauss RM, de Schryver E, Van Veldhoven PP, Ronis M, Koulman A. Odd chain fatty acids; new insights of the relationship between the GutMicrobiota, dietary intake, biosynthesis and glucose intolerance. Sci Rep 2017;7:44845.

[86] Erickson AR, Cantarel BL, Lamendella R, Darzi Y, Mongodin EF, Pan C, Shah M, Halfvarson J, Tysk C, Henrissat B, Raes J, Verberkmoes NC, Fraser CM, Hettich RL, Jansson JK. Integrated metagenomics/metaproteomics reveals human host-microbiota signatures of Crohn's disease. PLoS One 2012;7:e49138. 10.1371.

[87] Wilmes P, Heintz-Buschart A, Bond PL. A decade of metaproteomics: where we stand and what the future holds. Proteomics 2015;15:3409—17.

[88] Kolmeder CA, Ritari J, Verdam FJ, Muth T, Keskitalo S, Varjosalo M, Fuentes S, Greve JW, Buurman WA, Reichl U, Rapp E, Martens L, Palva A, Salonen A, Rensen SS, de Vos WM. Colonic metaproteomic signatures of active bacteria and the host in obesity. Proteomics 2015;15:3544—52.

[89] Mayers MD, Moon C, Stupp GS, Su AI, Wolan DW. Quantitative metaproteomics and activity-based probe enrichment reveals significant alterations in protein expression from a mouse model of inflammatory bowel disease. J Proteome Res 2017;16:1014—26.

[90] Braun J. Microgeographic proteomic networks of the human colonic Mucosa and their association with inflammatory bowel disease. Cell Mol Gastroenterol Hepatol 2016;2:567—83.

[91] Millis NF. Second international symposium on the biosafety results of genetically modified plants and microorganisms. May 1992, Goslar, FRG. Australas Biotechnol 1992;2:237—9.

[92] Syverton JT, Berry GP. The superinfection of the rabbit papilloma (shope) by extraneous viruses. J Exp Med 1947;86:131—44.

[93] Zheng JH, Nguyen VH, Jiang SN, Park SH, Tan W, Hong SH, Shin MG, Chung IJ, Hong Y, Bom HS, Choy HE, Lee SE, Rhee JH, Min JJ. Two-step enhanced cancer immunotherapy with engineered *Salmonella typhimurium* secreting heterologous flagellin. Sci Transl Med 2017;9(376).

[94] Yoon W, Park YC, Kim J, Chae YS, Byeon JH, Min SH, Park S, Yoo Y, Park YK, Kim BM. Application of genetically engineered *Salmonella typhimurium* for interferon-gamma-induced therapy against melanoma. Eur J Cancer 2017;70:48—61.

[95] Bourzac K, Bender E, Dolgin E, Mullard A, Savage N, Gruber K. Therapeutic developments: masters of medicine. Nature 2017;545:S4—9.

Microbiome and Metabolome Glossary

Joel Faintuch[1] and Jacob J. Faintuch[2]

[1]*Department of Gastroenterology, Sao Paulo University Medfical School, Sao Paulo, Brazil;* [2]*Department of Internal Medicine, Hospital das Clinicas, Sao Paulo, Brazil*

16S rRNA genes The most used bacteria and archaea genes for metagenomic analysis. The small ribosomal subunits of these microorganisms (particle size 16 Svedberg) are good markers for taxonomic classification.

16S sRNA sequencing Most metagenomic studies rely on this modality of sequencing. Taxonomic classification usually proceeds up to genus level, although species identification may be possible as well. Fungi, viruses, and other microorganisms that lack 16S rRNA genes are not recognized.

18S rRNA genes Alternative markers for metagenomic classification. Virtually all centers prefer the 16S rRNA genes.

Alpha diversity See diversity.

Amplicon sequencing Metagenomic sequencing based on a single 16S rRNA marker gene.

Axenic Animals born in aseptic conditions (cesarian section) and maintained in sterile laboratories with autoclaved food and water. The same as germ-free animals.

Beta diversity See diversity.

Bioengineered bacteria See Genetic engineering.

Clades A taxonomic group or unit with a common ancestor. Most authors prefer the phylum-based classification, using morphofunctional features.

Commensal Microorganism that benefits from the gut environment without causing any harm.

Conventional (animals) See Holoxenic.

Diversity A quality related to the amount of microbial taxa and their distribution. Alpha diversity describes the multiplicity of taxonomic units within a sample or community (the different trees in a single forest). Beta diversity addresses the diversity structure between a number of samples or communities (the comparison between different forests).

Diversity may also consider richness (how many kinds of microorganisms) as well as evenness (whether all kinds are equally abundant).

Dysbiosis Qualitative or quantitative imbalances in the microbiome, associated with diseases, functional impairments, or health risks. Dysbiosis is often linked to diversity troubles (alpha or beta diversity), as well as to abundance (excessive, insufficient) of bacteria taxa.

Ecosystem. A collection of microorganisms and their habitat which operates as a system.

Epigenetic Alteration in gene expression that is independent of the corresponding DNA or RNA sequence. One example is gene methylation, which may be influenced by the microbiome.

Eukaryotes All organisms, including humans, with membrane-bound intracellular organelles, particularly a nucleus containing chromosomal DNA.

Evenness A measure of biodiversity of the microbiome. It mirrors how similar the different taxa are numerically. In other words, an overwhelming growth of a single taxon can be suspected by this index.

Exposome The sum total of environmental influences (nongenetic), from birth onward, that shape human health and disease, together with the genome. This is a wide-reaching concept, including not only absorbed molecules and epigenetics but also lifestyle, urban environment, climate, and education. With a narrower focus, targeting the microbiome only, some use exposome as a synonym for xenobiotics (See Xenobiotics).

Genetic engineering The synthesis, deletion or insertion in the genome of nucleotide sequences, which may be artificial or exist in natural conditions, aiming to change both the genotype and the phenotype of a microorganism.

Genome The collection of genes and noncoding nucleotides of an organism.

Genotype The assemblage of sequenced genes of the host or microorganisms, which, together with epigenetic and environmental factors, shapes the phenotype.

Germ-free See Axenic; Gnotobiotic.

Gnotobiotic Animals colonized by known microorganisms only. These are originally axenic or germ-free animals subsequently exposed to a known microorganism to investigate its effects.

Holomicrobiome The aggregate or collective microbiome covering bacteria, viruses, phages, and fungi. Some include eukaryotes (protozoa, amebae, flagellates, and algae). See Whole-genome shotgun sequencing.

Holoxenic animals Animals with the standard microbiome of the species.

Metabolite See metabolome.

Metabolome The profile of metabolites (small molecules generated by metabolic activity, with <1500 Da) of a given fluid, tissue, or organism.

Metabonome The larger metabolome, or total metabolite pool, encompassing, for instance, plasma, urine, and fecal samples. It corresponds to the sum total of the metabolome or "meta-metabolome."

Metagenome The genes and genomes of the participants in a microbiome. They are usually evaluated by gene sequencing of total DNA extracted from a sample.

Metaproteome The assemblage of all proteins (or just small proteins), generated by a microbiome.

Metataxonomy The taxonomic tree of the organisms present in a microbiome. Taxonomic classification, up to the level of genus and species, whenever possible, relies on the amplification and sequencing of taxonomic marker genes.

Metatranscriptome The set of transcripts of the genes expressed by microorganisms, potentially related to a phenotype. Both community-wide and single-organism metatranscriptomics may be addressed.

Microbiome (Microbiota) A community of microorganisms that occupy a specific habitat or environment. Some limit microbiome to the genes or genomes only, the microbial collection being designated as microbiota. However, in both circumstances, the taxonomic catalog is the same, typically based on analysis of 16S rRNA gene sequences (metagenomics). As a consequence, many groups use the terms interchangeably.

Mutualism Interaction between different microorganisms that are mutually beneficial.

NGS/Next generation sequencing High-throughput sequencing technique for improved results.

OTU/Operational taxonomic units Similar clusters of 16S rRNA, identified during nucleotide sequencing. For many years their number was used to reflect microbiome diversity; however, they are just proxies of bacterial species. Formal taxonomic identification of each genus and species, and not just OTU findings, is currently recommended.

Pathobiont A commensal or symbiont organism that may eventually become pathogenic, precipitating disease.

PCR/Polymerase chain reaction One of the gene sequencing tools for microbiome studies.

Phenotype Collection of physical, biochemical, and psychological traits or features of an individual. The phenotype is susceptible to both genetic and environmental impacts.

Phylogenetics Taxonomic classification of genome sequences of microorganism populations, generated by microbiome analysis, into existing phyla or major microbial divisions.

Prokaryotes Unicellular organisms that lack membrane-bound organelles such as nucleus or mitochondria. They include bacteria and archea.

Proteome See metaproteome

Resilience The ability of the microbiome to regenerate or restore itself after a period of dysbiosis.

Sequence analysis Procedures for estimation of structure, features, and functions of sequenced DNA or RNA strands. This leads to the identification of genes as well as of nontranscribing (noncoding) nucleotides. Alignment, clustering, chimera detection, phylogenetics, assembly, and annotation are typical steps in this process.

Shotgun sequencing A high-throughput system that randomly breaks down DNA/RNA strands into small pieces and then reassembles them by computer programs that juxtapose overlapping fragments. This is the most modern technique in use.

Symbiont A commensal microorganism that benefits from the human environment, however, also potentially benefitting the host. See also Mutualism.

Taxon (Taxa) Microbial taxonomic units classified as kingdoms, phyla, classes, orders, families, and genera.

Transcriptome see Metatranscriptome.

Volatile organic compounds/VOCsSmall molecules generated by human or bacterial metabolism, as well as contaminants originating from ingested or absorbed foods, drugs, tobacco products, and other environmental sources. Small molecules generated by human or bacterial metabolism, as well as contaminants originating from ingested or absorbed foods, drugs, tobacco products, and other environmental sources.

They may be identified in the breath (breathome or volatolome), as well as in other samples (feces, urine, or secretions).

Volatolome See Volatile organic compounds.

Whole-genome (shotgun) sequencing Does not limit itself to 16S rRNA genes and is thus able to identify all types of microorganisms, up to the species level. See also Shotgun sequencing; Holomicrobiome;

Xenobiotics Foreign molecules that are absorbed by the human organism are detected in metabolome analysis and undergo biotransformation by host cells, the microbiome, or both. They include environmental pollutants, drugs, cosmetics, food additives, insecticides, carcinogenic molecules, etc. Some include food in the list as certain food-derived molecules also behave as xenobiotics and require detoxification.

FURTHER READING

[1] Marchesi JR, Ravel J. The vocabulary of microbiome research: a proposal. Microbiome 2015;3:31.

[2] Ursell LK, Metcalf JL, Parfrey LW, Knight R. Defining the human microbiome. Nutr Rev 2012;70(Suppl. 1):S38−44.

[3] Shetty SA, Hugenholtz F, Lahti L, Smidt H, de Vos WM. Intestinal microbiome landscaping: insight in community assemblage and implications for microbial modulation strategies. FEMS Microbiol Rev 2017;41:182−99.

[4] Luketa S. New views on the megaclassification of life. Protistology 2012;7:218−37.

[5] Faintuch J, editor. Microbiome, dysbiosis, probiotics and bacteriotherapy. Sao Paulo (Brazil): Manole; 2017.

Index

bile acids and FXR, 266
endotoxin and liver inflammation, 266
gut microbiome in NAFLD patients, 266
HFD, 267
intestinal inflammasomes, 266
Intestinal epithelial cells (IECs), 129
Intestinal homeostasis, 271, 349
Intestinal infections, 205–206
Intestinal inflammasomes, 266
Intestinal inflammation-associated diseases, 104
Intestinal microbiome, 85, 139, 149, 236, 273–274
Intestinal microbiota, 62, 127, 155, 195, 219, 272
  BA effects, 141
Intestinal mononuclear phagocytes (iMPs), 130
Intestinal permeability, 254, 442
Intestinal physical barrier, 134
Intestine, 253, 271
Intracellular metabolome, 44
Intrauterine stress, 382
Invariant natural killer T cells, 310
Invasive pulmonary aspergillosis (IPA), 297
Ion collision cross section (CCS), 48
Ion mobility spectrometry (IMS), 48, 69
Ion source, 47
Ion suppression effect, 47
IPA. *See* Indolepropionic acid (IPA); Invasive pulmonary aspergillosis (IPA)
Ipilimumab, 441
IR. *See* Insulin resistance (IR)
Irritable bowel syndrome (IBS), 7, 155–156, 197, 252, 358
  FMT and, 159–162
IS. *See* Indoxyl sulfate (IS)
Isolated lymphoid follicles (ILFs), 130
Isoprene, 72
Isoprostanes, 72–73

J

Jackson Laboratory (JAX mice), 368
JAMA metanalysis, 198
JNK. *See* c-Jun N-terminal kinase (JNK)

K

K/DOQI. *See* National Kidney Foundation Kidney Disease Outcome Quality Initiative (K/DOQI)
α-KB. *See* Alpha-ketobutyrate (α-KB)
Kidney Disease: Improving Global Outcomes (KDIGO), 354
κ kinase, 401
Klebsiella spp., 328, 439
  K. pneumoniae, 427
Knockout mice (KO mice), 236
Knowledge discovery, 53
Kynurenic acid, 229
Kynurenine/tryptophan ratio, 442

L

L-GPC. *See* Lineoleoylglycerophosphocholine (L-GPC)
LAB. *See* Lactic acid bacteria (LAB)
Laboratory procedures, 34
Lachnospiraceae, 326, 328, 438–439, 441
Lactate-producing bacteria, 326
Lactic acid, 110–111
Lactic acid bacteria (LAB), 105, 131, 149, 202, 244–245, 417
Lacto-ovo vegetarians, 394
Lactobacillaceae-enriched probiotic, 327
Lactobacilli, 202, 307
Lactobacillus plantarum 299v (Lp299v), 257, 258f
Lactobacillus sp., 4–6, 38, 202, 223, 300, 326–328, 393, 437
  L. acidophilus, 174, 202
  L. bulgaricus, 386
  L. casei, 174, 202
    BL23, 368
  L. crispatus, 109
  L. helveticus, 328
  L. iners, 109
  L. murinus, 17
  L. plantarum, 202, 368, 417
  L. reuteri, 4–6
  L. rhamnosus, 38, 327
  L. sakei CTC 494, 202
  L. salivarius, 439
  Lactobacillus-dominated vaginal microbiota, 110
Lactococcus lactis, 386
Lamina Propria (LP), 130, 221
Lantibiotics, 122
Late-onset sepsis (LOS), 86, 279
  microbiome role in, 281–282
LBP. *See* LPS binding protein (LBP)
LC. *See* Liquid chromatography (LC); Lung cancer (LC)
LDA. *See* Linear discriminant analysis (LDA)
LDL. *See* Low-density lipoprotein (LDL)
Ldlr. *See* Low-density lipoprotein receptor (*Ldlr*)
Leaky gut, 254, 273
  syndrome, 442
Lectin regenerating islet-derived 3 α (Reg3α), 376
Leptin, 235
Leptotrichia spp., 439
Leucine, 217
Leukotrienes, 73
LFF571 (Bacteriocin), 179
LHMP. *See* Lung HIV Microbiome Project (LHMP)
Lifestyle repercussions, 153, 222–223
  diet, nutrition, and exercise as, 152–153
Linear discriminant analysis (LDA), 52
Lineoleoylglycerophosphocholine (L-GPC), 217
Lingual region, 308–309
Lipid(s), 444
  metabolism, 217

peroxidation, 411
  and oxidative stress, 77
  signaling, 217
Lipidome
  multiple sclerosis, 444
  obesity and diabetes, 444
α-Lipoic acid, 229
Lipopolysaccharides (LPS), 17, 134, 201, 216, 244–245, 253, 266, 273, 309, 350, 369, 393–394, 437
  LPS-CD14 complex inducing NF-κB, 218
Lipoprotein lipase (LPL), 266
Liquid chromatography (LC), 35
  LC-MS, 35
Liver, 271
  cirrhosis, 271, 273, 436–437
  disease, 273
  inflammation, 266
  injury, 273
Liver X receptor α (LXRα), 143
Longevity, 441
LOS. *See* Late-onset sepsis (LOS)
Low-density lipoprotein (LDL), 18
Low-density lipoprotein receptor (*Ldlr*), 18
Low-grade immune response, 254
Low-protein diet, 351–354
LP. *See* Lamina Propria (LP)
Lp299v. *See* Lactobacillus plantarum 299v (Lp299v)
LPL. *See* Lipoprotein lipase (LPL)
LPS. *See* Lipopolysaccharides (LPS)
LPS binding protein (LBP), 218, 369
Luminal bacteria, 14
Lung
  bacterial environment, 297
  diseases, 444
  microbiota, 297
    interaction between immune system and, 298
Lung cancer (LC), 75
  applications in, 75–76
  breath as sample for LC biomarkers searching, 75
  EBC as sample for LC biomarkers searching, 75
  marker interpretation, and overlapping with tumors, 75–76
Lung HIV Microbiome Project (LHMP), 308
LXRα. *See* Liver X receptor α (LXRα)
Lymphoid tissues, adaption of, 14
Lyophilization, 187
Lysine, 87
Lysogenic phage cocktails, 230

M

m/z ratio. *See* Mass-to-charge ratio (*m/z* ratio)
mAb. *See* Monoclonal antibodies (mAb)
Macrophage scavenger receptor (MSR), 309
Malignant diseases, 203–204
Malondialdehyde, 411
MAMPS. *See* Microbial-associated molecular patterns (MAMPS)
Man Rogosa Sharpe (MRS), 202

Printed in the United States
By Bookmasters